BIRDS IN SCOTLAND

Birds in Scotland

by VALERIE M. THOM

Published for the Scottish Ornithologists' Club

T & A D POYSER

Calton

First published 1986 by T & AD Poyser Ltd
Print-on-demand and digital editions published 2011 by T & AD Poyser,
an imprint of A&C Black Publishers Ltd, 36 Soho Square, London W1D 3QY

www.acblack.com

ISBN (print) 978-1-4081-3837-3
ISBN (epub) 978-1-4081-3836-6
ISBN (e-pdf) 978-1-4081-3835-9

A CIP catalogue record for this book is available from the British Library

This is a print-on-demand edition produced from an original copy.

It is produced using paper that is made from wood grown in managed sustainable
forests. It is natural, renewable and recyclable. The logging and manufacturing processes
conform to the environmental regulations of the country of origin.

Printed in Great Britain by Martins the Printers, Berwick upon Tweed

Contents

List of Plates

Foreword

The publication of this book would have pleased beyond words both Dr Baxter and Miss Rintoul. Their *The Birds of Scotland* (1953) contained the fruits of almost two lifetimes spent observing and collecting birds throughout Scotland, examining meticulously old records, much correspondence, and the unearthing of new facts. All this at a time when the watching of birds was in its infancy, was almost entirely the occupation of amateurs, and was not at all a suitable pursuit for ladies! Professional study had only just begun.

It was always the hope of these two 'good ladies' of Scottish ornithology that their book would be continually up-dated, and they would have been glad indeed that this has been so competently done by Miss Valerie Thom to cover the past 30 years. Valerie has taken as her base-line for measuring change the position in the early 1950s as described by Dr Baxter and Miss Rintoul, whose investigations of the past had left so very few stones unturned.

It must be remembered that it is only quite recently that sight records, properly vouched and documented, have become acceptable. Before that only the dead bird provided an unassailable record – what was hit was history, what missed mystery. As a result of the improved accuracy of field observations 'hoodwinks' are perhaps less common now than once they were.

Looking through the pages of *Birds in Scotland* one might get the impression that many of the rarities and vagrants turn up more frequently now than once they did. I do not think that this is true: the increase in records of such strays nowadays is more a measure of the increased number of birdwatchers and the now general acceptance of sight records than anything else. There were probably as many occurrences in the past but fewer were observed. A map of the records of such birds would, as Valerie points out, be as much an indication of the distribution of birdwatchers as of the occurrences of bird rarities.

The biggest changes in status, particularly local status, in the past 30 years, are due to changes in habitat, as for instance the disappearance of so many wet places, including farm ponds, the pushing back of moorlands, the continuing increase of 'improved' pasture, the creation of huge expanses of coniferous plantation, the vanishing of the last remnants of the ancient pine and hardwood forests, and the reclamation of mudflats. Some of this is no doubt inevitable but not all. Change there must be and birds are to some extent adaptable – but only if the change is slow.

We have now got two datum lines – the situation as described in *The Birds of Scotland* and the picture, 30 years later, as described for us in the present book by Valerie Thom. An assessment again, possibly early in the next century, will undoubtedly reveal further change. Some existing habitats will to all intents have gone but new ones will have been created. Some of our present avifauna will have disappeared – or become very rare. Other species, unexpectedly, may have appeared. Only continued monitoring, such as has so successfully been summarised and brought up to date by Valerie, can demonstrate this clearly. She has produced something invaluable and very much to be praised. I commend her book without reservation and wish it the wide and enthusiastic reception that it deserves.

W. J. EGGELING

Preface

Since *The Birds of Scotland* was published in 1953 there has been a quite remarkable expansion of interest in birds. Professional ornithologists, once rarities, are now numbered in hundreds, while 'hobby' birdwatchers have multiplied even faster. There has been a corresponding growth in the intensity with which birds are studied; each year now sees a veritable army of enthusiasts counting, ringing, mapping distribution, carrying out behavioural, ecological and biological studies, and, of course, 'twitching' the rare birds that occasionally reach Britain by accident. Within the last decade or two there has also been growing concern for habitat conservation, and appreciation of the need to know which factors present the greatest potential threats to the well-being of bird populations.

One outcome of all this activity has been a tremendous flow of publications about birds. These range from 'picture books' and field guides, through atlases of breeding distribution, tomes covering the whole of the western Palearctic, monographs on individual species and endless papers in scientific journals, to local bird reports and checklists. To find out about every aspect of a species occurring regularly in Scotland the birdwatcher may have to consult a whole series of publications, each of which provides only part of the picture. What I have attempted to do here is to summarise within one book what is currently known about the numbers, distribution and movements of those birds that are resident in, regular visitors to, or occasional vagrants in Scotland, the ways in which their populations have been changing in recent years, and the possible causes of these changes.

Inevitably, compromise has been necessary in order to keep the book to a manageable size. Identification and breeding biology – both very adequately covered in many general works – are not dealt with at all here, while references to migratory movements and discussion of ecological and behavioural topics are largely confined to Scottish examples. Information on local distribution and numbers is now so abundant – for some species and some parts of Scotland – that it has not been possible to include all the available detail; comparatively little space is devoted, for example, to species status on individual islands. To try to compensate for such omissions and limitations, however, I have included both a fairly extensive bibliography (mainly of relatively recent works, many of which themselves give comprehensive references to earlier studies) and appendices listing local reports and island checklists; these will, I hope, assist readers interested in a particular species or area to obtain the additional information they seek. The appendices on local reports and island checklists will also, hopefully, draw attention to those parts of Scotland currently least well documented – as evidenced by the absence of a local report or the out-datedness of an island checklist – and so help to stimulate recording in such areas. As will be apparent from the species accounts, however, it is not only the remoter Highlands and less accessible islands that offer much scope for further study; so, too, do many of our commoner species, throughout Scotland.

A site list has been included as an appendix in order to reduce the need for repetition of county names in the text, to help those not familiar with Scotland to locate the places mentioned, and to provide a convenient cross-reference between old county names, post-1974 regions and the national grid. The last seems likely to become the basis for all bird recording in the near future.

The Birds of Scotland was published soon after I became 'seriously' interested in birdwatching, and I was fortunate at about the same time to make contact with Dr Baxter, who encouraged me to start regular wildfowl counting on the Ythan Estuary; both book and personal contact did much to stimulate my interest in and love of birds. It is my sincere hope that this successor to 'B&R' will also, in some small measure, lead others towards greater understanding and enjoyment of Scotland's birds – a priceless heritage now recognised as of European significance but all too vulnerable to modern man's materialistic attitudes. I believe that we who derive so much pleasure from birds have a particular responsibility to regard ourselves as 'stewards' of their welfare, and consequently to take an active role in their conservation.

An eagle soaring high on a thermal,
A peregrine stooping swift to the kill,
Divers wailing on lonely lochans –
Such sights and sounds have the power to thrill.

Seabirds aswirl in an avian blizzard,
Filling the air with their raucous cries.
Frost-pink dawn and the grey geese flighting –
We strive to count them, with straining eyes!

The tinkling song of the Willow Warbler
Heralding spring as the birches green,
Or night-time calls from migrating waders –
It matters not that the birds are unseen.

A sudden glimpse of an unknown stranger –
Perhaps its a 'first' for the Scottish list? –
Or the feel of a bird in the hand for ringing.
Each new experience adds its own twist.

The 'everyday' birds such as Wren and Robin
And the rarities blown here from distant lands
All have their place in God's great pattern –
But the future for many will lie in man's hands.

Some of us count and some of us capture,
Some simply listen and watch and enjoy;
Lord, grant that we none of us fail to remember
How much, by our thoughtlessness, we may destroy.

V.M.T

Acknowledgements

In Scotland we are fortunate in having the Scottish Ornithologists' Club as a focal point for birdwatchers. The Club's journal *Scottish Birds* carries many important papers, its local recorders collate the records submitted annually for the *Scottish Bird Report*, and its members play a major part in national surveys such as the wildfowl counts; much of the information presented in this book has, naturally, been gathered by SOC members. Additionally, the SOC Council's decision to sponsor the preparation of *Birds in Scotland* has enabled financial assistance to be obtained from charitable bodies; I thank Council for this support.

On the Club's behalf, I acknowledge with gratitude the financial backing received. The bulk of this appropriately came – through the good offices of Dr John Berry and Dr W. J. Eggeling – from Miss E. V. Baxter's Ornithological Trust and Miss L. J. Rintoul's Publication Trust. These Trusts were established by the 'good ladies' for the purpose of assisting with publications relating to Scottish birds, and the Trustees generously decided to make over the total monies held in them to the SOC; the unused balance will continue to be held as a fund available for the support of future publications. A substantial grant from The George Lodge Trust, given specifically to allow more lavish illustration than would otherwise have been possible, is also gratefully acknowledged, as are the contributions received from BP Petroleum Development Ltd, Shell UK Ltd, The Scottish International Education Trust, D. G. Hutchison's Charitable Trust and the Late Lord Rootes' Charity Trust.

This book could not have been written without an enormous amount of help from a great many people – I am most grateful to all who contributed to it, in whatever way. My requests for assistance met with so generous a response that in many cases the data received were much more than I could incorporate within the limited space available. If I have inadvertently omitted the names of any who offered assistance I tender my sincere apologies.

Much help has been received from three national ornithological bodies: the British Trust for Ornithology (BTO), the Wildfowl Trust and the Royal Society for the Protection of Birds (RSPB). The BTO undertook to abstract summaries of all Scottish ringing recovery data for me; I thank the Trust's Ringing & Migration Committee for authorising this work and Raymond O'Connor, Bob Spencer and Chris Mead for organising it. I am grateful also to: John Marchant – for the abstraction of Scottish CBC data, Peter Lack – for pre-views of Winter Atlas maps and text drafts, Mike Moser – for pre-publication results of the 1984/85 Winter Shorebird Survey and other wader counts, and Kenny Taylor – for the unpublished results of the 1983 Buzzard Survey. The Wildfowl Trust kindly prepared summaries of recent Scottish wildfowl count data (the full count records are forwarded annually to the SOC) and also separate indices of relative abundance, to permit comparison of Scottish population trends with those in England and Wales. I thank Myrfyn Owen for preparing the necessary computer programmes and for allowing me to read drafts of the species accounts for the forthcoming second edition of *Wildfowl in Great Britain*, David Salmon for responding rapidly to a variety of queries and requests, and Malcolm Ogilvie for supplying unpublished goose count figures. RSPB staff both in Scotland and at Sandy have provided assorted information; most are mentioned individually later but I wish to record special thanks to Frank Hamilton, John Hunt, David Minns, Ian Bainbridge, Lennox Campbell and Mike Everett for their ready help.

I am grateful to Bill Harper, Alwyn Easterbee and Alan Stewart, librarians with the SOC, Nature Conservancy Council (NCC) and Countryside Commission for Scotland respectively, for assistance in locating references. Permission to quote from studies commissioned by Britoil and the Shetland Oil Terminal Environmental Advisory Group (SOTEAG) is gratefully acknowledged, as is permission to quote from unpublished NCC and RSPB reports.

Turning now to individuals, I am deeply indebted to Dougal Andrew and Andrew Macmillan for the use of the detailed summaries of rarer species prepared by them with a view to producing a Scottish checklist. These covered the period up to 1972 and consequently obviated the need for much laborious and time-consuming search of the literature prior to the establishment of *Scottish Birds*. The following (most of whom are – or at the time were – local recorders) kindly completed current breeding status schedules for me: Dr E. S. Alexander, N. K. Atkinson, Drs M. V. Bell and S. Buckland, A. W. & L. M. Brown, E. D. Cameron, Mrs P. M. Collett, R. F. Coomber, I. G. Cumming, W. A. J. Cunningham, A. Currie, R. H. Dennis, N. Elkins, Dr E. Fellowes, I. P. Gibson, Major J. J. Gordon, M. J. P. Gregory, M. I. Harvey, Dr C. J. Henty, J. A. Love, K. S. Macgregor, J. Mitchell, R. D. Murray, Dr C. J. Spray, R. J. Tulloch and A. D. Watson. I thank them all, and also John Kirk (Snr), who undertook both the tedious task of transferring information from schedules to species sheets and the preparation of the island checklist appendix.

Unpublished results of species studies were generously provided by: Alan Allison, Norman Atkinson, Mark Brazil, Roger Broad, David Bryant, Steve Buckland (NE Scotland Atlas material), Stan da Prato, Pete Ellis, Pete Ewins, Bob Furness, Iain Gibson, Paul Green, Jeremy Greenwood, Alan Heavisides, Martin Heubeck, Gareth Jones, John Kirk (Jnr), Derek Langslow, Bob McMillan, Mick Marquiss, John Massie, John Mitchell, Mike Nicoll, Tony Prater (1984 Ringed Plover breeding survey), Graham Rebecca, Henry Robb, Martin Robinson, Alistair Smith, Chris Spray, Patrick Stirling-Aird, Ian Strachan, David Stroud, Bob Swann, Mark Tasker, Iain Taylor, Lance Vick, Alan Walker, Donald Watson, Jeff Watson and Alan Wood. For permission to quote other unpublished results I am indebted to NCC / Tim Reed (moorland wader densities), NCC / Phil Shaw (farmland birds), ITE / Don French (woodland songbird densities), RSPB / Roy Dennis (monitoring data for rare breeding species) and RSPB / Lennox Campbell (moorland wader densities). Stewart Angus, Colin Bibby, Morton Boyd, Colin Campbell, David Cant, David Counsell, Andrew Currie, Tom Dougall, Chris Eatough, Jack Gibson, Cliff Henty, David Lea, Ray Murray, Dick Roxburgh, Bill Sinclair, M. F. Stevenson, Doug Weir, Gordon Wright and John Young all made available unpublished notes or other data. Dr G. A. Best (Clyde River Purification Board) provided information on pollution levels in the Clyde and its tributaries, and M. Rutherford (Scottish Development Department) on estuarine reclamation. I am grateful to them all.

Many people commented helpfully on drafts at various stages. I am especially grateful to: Mike Rogers (Sec., BBRC) for checking all the rarity accounts; Roy Dennis – all semi- and Scottish rarities, all rare breeding species, and the chapter summarising changes in status; Roger Broad – a large and motley collection of species for which there was no obvious 'expert' referee; Mike Pienkowski – all waders; Bill Bourne, Mike Harris and Sarah Wanless – between them all the seabirds; Myrfyn Owen, Malcolm Ogilvie and David Salmon – all the commoner wildfowl; Desmond Neth-ersole-Thompson – assorted highland birds; Adam Watson – various hill birds and the chapter on uplands; Stan da Prato – warblers, thrushes and some buntings; Ian Newton – finches; Jim McCarthy (NCC) – all the habitat chapters and the one on conservation and protection; and George

Dunnet – most corvids, Fulmar, Starling and the chapter on coastal habitats.

Others who read and commented upon the introductory chapters were: Mike Locke, Gunnar Godwin, Steve Petty and Dorian Moss (woodlands), Brian Morrison (fresh-water), Ken Runcie and Phil Shaw (farmland), Bob Spencer and John Davies (developments in bird study) and Frank Hamilton and his colleagues in the RSPB's Scottish office (conservation & protection). Each of the following commented on one or more of the species accounts: Alan Allison, Sandy Anderson, John Berry, Colin Bibby, David Bryant, Nigel Buxton, James Cadbury, Lennox Campbell, Martin Cook, Brian Etheridge, Pete Ewins, Bob Furness, Colin Galbraith, Iain Gibson, Paul Green, Frank Gribble, Tony Hardy, David Houston, David Jenkins, John Love, Mick Marquiss, Chris Mead, Richard Mearns, Eric Meek, Harry Milne, John Mitchell, Pat Monaghan, Bob Moss, Bryan Nelson, Sandy Payne, Steve Petty, J. Phillips, Nick Picozzi, G. R. Potts, Andrew Ramsay, Derek Ratcliffe, Mike Richardson, Gordon Riddle, Alistair Smith, Bob Smith, Bob Swann, Chris Spray, David Stroud, Judy Stroud (Warnes), Ron Summers, Iain Taylor, Kenny Taylor, Gareth Thomas, Bobby Tulloch, Andrew Village, Alan Walker, Donald Watson, Jeff Watson, Peter Wormell and John Young.

I am most grateful to all these people for doing their best to ensure that the information presented in this book is as accurate as possible. If – as is almost inevitable – errors have crept in the responsibility for them is entirely mine.

My warmest thanks go to my friend Roberta Seath, who spent long hours 'polishing' my English and eliminating ambiguities and many more helping with the correction of proofs, and to Bill Thompson who gave invaluable help with the maps and figures. I greatly appreciate the care with which Donald Watson, in his capacity as Art Editor, chose the team of artists: Keith Brockie, John Busby, Andrew Dowell, John B. Fleming, Birgitte Hendil, Alan F. Johnston, C. E. Talbot Kelly, John A. Love, Rodger McPhail, William Neill, Ian Willis and Bernard Zonfrillo. Their work – together with Donald's own – does much to enhance the appearance of the book. I am grateful also to all who helped me to obtain the photographs I wanted; individual photo credits are given beside the plates. And I could not have wished for a more helpful and accommodating publisher than T. and A. D. Poyser.

Finally, I wish to record my gratitude to Sally Morris, for sowing the idea of 'up-dating B&R' in my mind, and to Joe Eggeling, for the unfailing encouragement and support he has given me throughout what has at times seemed an almost Herculean task. Joe was the first person with whom I discussed the idea; his comments (along with those of Andrew Macmillan and Ian Newton) on my preliminary out-line and trial drafts were immensely helpful in determining the overall pattern of the book; he conscientiously read and constructively criticised virtually every section as it was drafted; and he has written most kindly in his Foreword. I am deeply grateful for all his efforts on my behalf.

1: Scotland – its avifauna and geography

Scotland's geographical location (Map 1), projecting north-west from the main land-mass of Europe, and its wide variety of habitats are the principal factors responsible for the diversity of the 449 species (excluding those square-bracketed – see Chapter 11) currently on the Scottish list. The 300–325 species recorded annually include some 200 which are either residents or regular summer, passage or winter visitors. The others are largely disorientated wanderers, wind-drifted across the North Sea from as far east as Siberia or carried by gales across the Atlantic. For these vagrants – which add so much to the excitement of birdwatching – our scattered islands frequently offer a much-needed landfall. And when tired strangers touch down on one of the barer inhabited islands, and especially on Fair Isle or the Isle of May, there is a good chance that they will be found and identified. In the last 30 years the Bird Observatories on these two islands have recorded 35 species new to Scotland, 19 of which had not previously been reported in Britain (Eggeling 1960 & 1974; Williamson 1965; Annual Reports).

Although such casual visitors have their place, it is with the status and distribution of the birds occurring regularly in Scotland that this book is mainly concerned. These include species at the northern limit of their breeding range, and others which breed no further south than the Highlands. They include seabird and raptor populations of significance in a European context, while the 19 species which in the British Isles nest regularly only in Scotland are of special national interest. Scotland's wintering concentrations of sea-ducks and divers are important too at a national level, and our wintering geese internationally. The presence and distribution of all these groups, and indeed of all but the most adaptable of species, is closely linked to habitat availability. In the following six chapters I describe the principal habitats found in Scotland and indicate the main bird communities associated with them. I also draw attention to major changes that have taken place in the last few decades and to threats currently facing the various habitats.

I have grouped habitats into six categories: coastal, woodland, upland, freshwater, farmland and urban/suburban, including man-made. Within each category appropriate subdivisions are considered separately and in some detail. Although not identical, many of these divisions are similar to the categories used by Fuller (1982) in *Bird Habitats in Britain* but there is one major difference in treatment, in that I have focused attention on those species which are largely dependent upon a given habitat, whereas Fuller assessed the frequency with which the species occurred (irrespective of numbers) in different habitats. The following brief general description is intended to set the scene for readers not fully familiar with the country's geography.* Map 2 shows the principal physical features.

* Non-ornithological references used in compiling these chapters are listed separately at the beginning of the Bibliography.

SPITSBERGEN

GREENLAND

BAFFIN IS.

Arctic Circle

ICELAND

FAEROES

SCOTLAND

FENNO-SCANDIA

SIBERIA

CANADA

USA

IBERIA

MIDDLE EAST

SAHARA

BRAZIL

SOUTH AFRICA

Mercator Projection

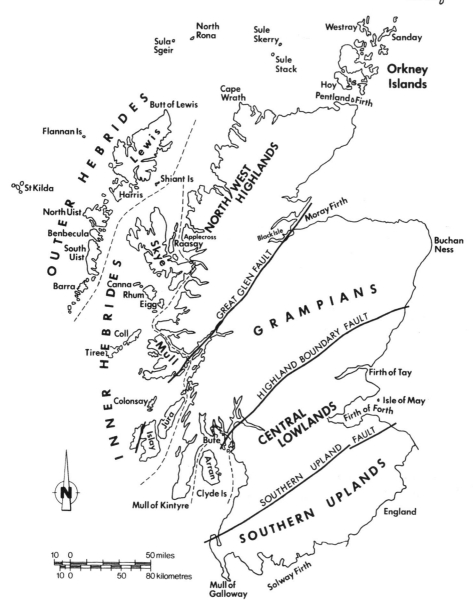

Map 2. Scotland's main physical features

Scotland sits on the continental shelf and is surrounded on three sides by relatively shallow seas supporting rich populations of fish. Off the west coast the islands of the Outer Hebrides provide a sheltering barrier against the Atlantic for many of those closer inshore. A handful of isolated rocky islets is scattered still further to the west, while the archipelagoes of Orkney and Shetland to the north reach latitudes similar to those of Bergen in Norway and the southern tip of Greenland. The east coast has many estuaries of importance for birds, but islands only in the Firth of Forth.

Mountains dominate much of the northern two-thirds of the mainland, in the west often rising steeply from near the coast but in the east separated from it by foothills and fertile coastal plains. In the Southern Uplands lower hill-ranges separate more fertile areas, although here too there is some rugged mountainous country towards the west. Good farmland is restricted to the eastern coastal fringe, the Central Lowlands and some of the broader valleys. A large proportion of the population (c80%) and of the country's industrial developments is concentrated in the Central Lowlands.

Although only 442 km long – from Cape Wrath to the Mull of Galloway – and less than 250 km across at its widest point – from Buchan Ness to Applecross – Scotland has almost 4,000 km of mainland coast, largely due to the many fiord-like sea-lochs indenting the coastline between the Clyde and Cape Wrath. The 126 inhabited and 664 uninhabited islands add a further 6,300 km of coastline. Only 10% of the total of approximately 10,000 km is occupied and used by man on a regular basis – a fact of considerable importance to many coastal birds. Of the total land area of 7,720,000 ha, some 63% is rough grazing and ungrazed uplands, 23% is used for agriculture (including permanent pasture), and 11% is productive forest. Around 12% of the land area is more than 500 m above sea-level, and nearly 50% is over 200 m asl. Large tracts of the uplands are used for sheep grazing, deer forest (often virtually treeless) and grouse moor, but the proportion so used is currently declining as a result of increasing afforestation.

Scotland is abundantly endowed with freshwater, both static and flowing, little of which is significantly affected by pollution. In addition to natural lochs – which include rich shallow lowland waters, deep and nutrient-poor lochs, and *dubh lochans* (small, shallow peaty pools) – there are many dams and reservoirs. The northeast is the only part of the country in which open waters are comparatively scarce, though many of those in the Borders are man-made. Apart from the Clyde, all the major rivers flow east; several are very long and most are relatively slow-flowing in their lower reaches. In the west those north of the Clyde are in general short and steep, tumbling rapidly down rocky hillsides to the sea. Few of the islands have streams of any size, but most are well supplied with lochs, indeed in parts of the Outer Hebrides the area of freshwater actually exceeds that of land. The island groups vary widely in character and ornithological interest, reflecting the diversity of their geology and land-use. Perhaps the most striking contrasts are between Orkney's rolling green pastures and neighbouring Shetland's bleak, peaty moorlands, between the barren rocky hills of Harris and the flower-rich *machair* (see Chapter 2) of the Uists, and between Arran's rugged mountains and Bute's rich farmlands.

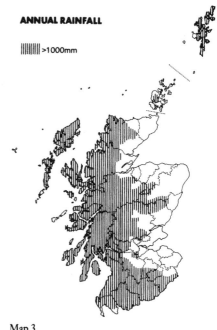

ANNUAL RAINFALL

||||||| >1000mm

Map 3

Among the more important factors affecting the distribution of plants – and consequently of many habitat types – are climate, geology, soil type, landform, and intervention by man. Climatically, Scotland is wet (Map 3), windy and temperate to the west of the main north–south mountain barriers (and in the Northern Isles) and markedly drier but with wider seasonal temperature fluctuations further east. In most seasons snow lies locally throughout the summer in the Cairngorms, where on average only July and August are entirely frost-free. Although perhaps of most significance through its influence on habitat, climate also affects birds directly; for example when winters of exceptional severity may cause high mortality, or persistent cold northerly winds during April and early May result in the late arrival of many migrants.

It is neither appropriate nor practicable to try to cover in detail Scotland's extremely complex geology, but some comment is needed on a few aspects of importance either to the birds themselves or to the distribution of major habitat types. (For further detail see Whittow 1977; Bunce & Last 1981.)

Rocks of different composition and structure vary in the way they are affected by erosion, and such variations determine the extent to which exposed surfaces, whether on the coast or inland, may develop cliff-faces with ledges or niches suitable for nesting sites. Horizontally-bedded rock which includes layers of varying hardness produces ideal seabird colony sites with tiers of shelves crammed with birds, whereas hard rock without definite cleavage planes often weathers to form relatively rounded surfaces with few potential nesting niches. Very brittle rock which readily frost-shatters to form scree is also unlikely to provide many good ledges. Sedimentary rocks such as sandstones, and some of the metamorphic rocks derived from them, more often produce good bird cliffs than do the harder rocks of volcanic origin – though St Kilda, the Bass Rock and Ailsa Craig are notable exceptions to this generalisation.

Soil type and landform are often affected by the underlying rock but are not necessarily determined by it, for much of Scotland's land surface is covered by deposits of sand, gravel, silt and boulder clay left behind by glaciers and rivers. To take two examples: in the northwest a landform due to a platform of very hard, impervious Lewisian gneiss is responsible for the existence of vast areas of wet wilderness where ice-scraped exposures of bare rock are interspersed with lochans and peaty ground. In other parts of the country – for example around Carstairs in the Central Lowlands – the interrupted terrain is often due to the presence of glacial mounds or ridges, many of which are currently vanishing through sand and gravel extraction. Most Scottish soils tend to be acid in character, calcareous soils occurring only very locally where lime-rich rocks outcrop or shell sand has accumulated on the land surface – as on the Hebridean *machairs* – but there is fertile land in the Carboniferous Limestone and Old Red Sandstone areas of the east coast, and also on the flood plains of some of the larger rivers. Good quality agricultural land is largely restricted to the eastern and central lowlands, parts of Ayrshire and the southwest, Caithness and Orkney. In the west and north, however, high rainfall limits the extent to which crops other than grass can be grown. Upland vegetation is also affected by rainfall and drainage, with grasses – particularly purple moor-grass and deergrass – and sedges dominating many western hills, while heather and blaeberry flourish on the drier moors further east. The distribution of natural woodland types is determined by a combination of factors, including soil quality, depth and drainage, altitude and exposure.

In virtually every part of the country, man's influence is superimposed upon the various natural factors determining vegetation types and bird distribution. Only the smallest and most inaccessible islands remain unaffected – and even these are increasingly threatened by contamination with oil and other marine pollutants. Parts of the lowlands have been cultivated for literally thousands of years – but it is only recently that agricultural practices have had a profound effect on wildlife. Most of our native woodlands have long since been cleared, and planted woods have been a part of the Scottish scene for several hundred years – but it is only in the last few decades that very large areas have been covered with a single tree species. People have enjoyed recreational activities in the Scottish countryside for many generations – but it is only in the last ten to twenty years that the pressure of people (as distinct from deliberate destruction by people) has presented any threat to wildlife. The pace of change is still accelerating, making it increasingly important that every effort be made to monitor the effects of environmental changes and, wherever possible, to mitigate those that adversely affect the wildlife of Scotland.

2: Coastal habitats

Scotland's coastal habitats – rocky coasts, soft shores, and inshore waters – are of international significance for seabirds, and of national importance for waders and wintering geese, seaducks and divers. They have been the subject of intensive study in recent years and a great deal is now known about the composition and distribution of the bird communities they support. Further detail on many of the aspects touched upon here can be found in Cramp *et al* 1974; RSPB 1979; Prater 1981; Dunnet 1982; Fuller 1982; Barrett 1983; Blake *et al* 1984; Evans 1984; Green 1984; and Lloyd 1984.

In the last three or four decades, coastal habitats have been affected by a variety of changes. Although generally local in scale, some of these have had a major impact on the status of at least one species (the Little Tern) and its distribution at a national level. Other changes have as yet had no major effects but present serious potential threats, while others have been proposed and may be implemented within the near future.

ROCKY COASTS

Some 90% of the mainland coast is rocky, as are the shores of most of the islands. For the seabird and wader species dependent upon them, two aspects of rocky coasts are of prime importance: the presence of cliffs and the extent of the intertidal zone respectively. Freedom from disturbance and proximity to ample food supplies are major factors influencing seabird breeding distribution. For the most ubiquitous species of rocky shores, the Rock Pipit, coastal formation is unimportant but for the Rock Dove and Chough caves or similar sheltered nesting sites are essential. The ducks most often seen off rocky coasts – Eiders and Red-breasted Mergansers – are not confined to this type of coastline, while for many other species, among them Peregrine and Raven, coastal cliffs and rocky shores are just one of a range of habitats used.

The west mainland has few high cliffs, the shorelines of the great sea-lochs generally sloping gently to give a wide intertidal zone much favoured by fishing Grey Herons. Most of the western islands are formed of hard rock; in much of the Inner Hebrides this is of volcanic origin while in the Outer Hebrides it is Lewisian gneiss. The main islands of the Outer Hebrides are of little importance for seabirds, but the Shiant Islands, the Flannan Isles, Berneray and Mingulay hold large colonies, as do Sula Sgeir and North Rona, while St Kilda is the premier seabird site in western Europe. Although several of these islands are mountainous few have cliffs of any scale – with the notable exception of the St Kilda group, which has the highest cliff in Britain in Conachair's 430 m precipice. Of the inshore islands only Ailsa Craig has extensive cliffs of major seabird importance, but Islay's relatively low cliffs are the stronghold of the very local Chough population, while Rhum's mountains hold a massive colony of Manx Shearwaters. A multiplicity of low rocky islets, often heather or grass-crowned, fringes the mainland and larger islands, providing nest sites for terns and Eiders; some of these small inshore islands have Storm Petrel colonies, and some are grazed by wintering Barnacle and White-fronted Geese.

Many of the cliffs in the north – for example at Noss in Shetland, Marwick Head and the Noup of Westray in Orkney, and Handa in Sutherland – are composed of horizontally-bedded or gently-dipping sedimentary rocks which have eroded to form tiers of ledges; such cliffs hold some of our most spectacular seabird colonies. From the Moray Basin southwards the cliffs vary in rock type and in suitability as nesting sites. Some sections, for example Troup Head, the Buchan cliffs, Fowlsheugh and St Abb's Head, hold large numbers of birds but these are seldom as closely packed as in the northern colonies. Apart from the Bass Rock – a text-book example of a volcanic plug – the Forth islands are low, but together support a good variety of seabirds. The Puffin colony on the Isle of May is by far the largest on the east coast and still growing, as are the populations of several other species, presumably because there is currently abundant food available in local waters. (Map 4 and Table).

Although cliffs often hold very high densities of nesting seabirds only a few species are entirely dependent upon such sites. Kittiwakes, Shags and Guillemots nest mainly on cliff ledges, though the last two occasionally occupy the flat tops of stacks or (rarely) boulder beaches. Gannets were at one time confined to cliff faces but now also use relatively flat areas, eg on the Bass Rock. And Fulmars – the fastest-spreading of all the seabirds – regularly nest on flat ground on some islands and are even colonising sand dunes, old buildings and cliffs quite far inland. Among the three most markedly gregarious cliff-nesters, the densest colonies of Guillemots and Kittiwakes occur on the shelved cliffs of the north, and of Gannets on St Kilda, the Bass Rock and Ailsa Craig. Razorbills find suitable niches not only on cliff faces but also among boulders at the cliff foot, the site most favoured by Black Guillemots; both these species nest solitarily. Shags generally use ledges not far above high tide level, while Cormorants may occupy similar sites but also nest on low islands, on stacks, and occasionally at inland freshwater sites.

The three large gulls all nest on cliffs but vary in the extent to which they do so. Herring Gulls are the most catholic in their tastes and readily nest on low islands, on the ground far inland (as at Flanders Moss) and even on buildings. Lesser Black-backed Gulls occasionally use the first two of these site types, but Great Black-backed Gulls show a distinct preference for inaccessible and flat-topped rock stacks, presumably because of the safety these afford. If farm-.d extends close to the cliff edge few gulls nest there.

The burrow-nesters (the small petrels, Manx Shearwater and Puffin) require an adequate depth of soil, loose scree or some similarly open stony formation; they occur in a variety of situations, including cliff gullies – and in the case of the Storm Petrel old buildings and dry stone walls – and reach their highest densities on little-disturbed island sites. Apart from Rhum's shearwaters, which nest at an altitude of over 650 m, and the Isle of May's Puffins, all the large colonies are relatively inaccessible. Many terns nest on small rocky islets but there are very large concentrations on some Orkney moorlands (and on sandy shores). Island moorlands hold most of the breeding skuas, of which Scotland supports the entire British population; both Great and Arctic Skuas are most abundant in Shetland.

Although a variety of waders make some use of them, rocky shores are of major importance to only two species, the Turnstone and the Purple Sandpiper. Unlike those favouring estuarine feeding grounds, which often congregate in large flocks, these rocky shore waders are usually thinly scattered along suitable stretches of coast. Numbers are highest where rock shelves form part of the intertidal zone or many weed-covered rocks are exposed at low tide, and where there are also adequate areas of rock above high water mark to provide safe roosts. The Outer Hebrides, the Northern Isles, and parts of the north and east coasts are known to hold important numbers of passage and wintering

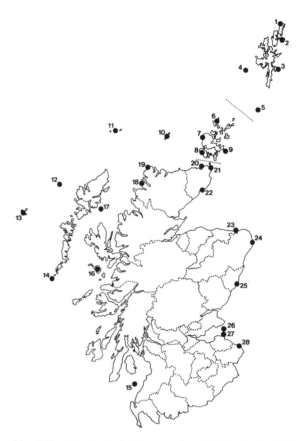

Map 4. Seabird colonies holding more than 10,000 pairs (see table opposite for key)

Turnstones and Purple Sandpipers, but comparatively little is yet known about the populations of the northwest coast and Inner Hebrides.

Oil-related developments, discussed below under 'Inshore waters', present the most obvious danger to our seabird colonies, while quarrying and disturbance by rock-climbers are problems locally on some stretches of coast. However, current and possible future changes in commercial fishing practices represent the most serious and widespread potential threat to seabird populations in general.

SOFT SHORES

These comprise estuarine flats and marshes, and coastlines with sand and shingle beaches, dunes and *machair*. The former provide rich feeding grounds for passage and wintering waders and wildfowl, but hold few breeding species restricted to this habitat type; several are used as safe moulting areas. Sandy shores, in contrast, are most important for their breeding birds, which include tern colonies in the east and notably high densities of breeding waders on the *machair* of the Outer Hebrides.

The ability of estuarine flats to support birds is determined largely by the nature of the substratum – whether predominantly of mud / silt or sand – and by the amount of organic matter it contains. These are the principal factors controlling

The relative size and diversity of Scottish seabird colonies holding more than 10,000 breeding pairs (Based on Cramp et al 1974, with up-dating of species numbers)

Map code	Site	No. of species breeding	Groups present*												Est. total breeding pairs
			F	P	MS	G	C	S	Sk	G/R	Pu	BG	Gulls	Terns	
1	Hermaness	14	+	−	−	+	+	+	+	+	+	+	+	?	10,000–100,000
2	Fetlar	18	+	+	+	−	−	+	+	+	+	+	+	+	10,000–100,000
3	Noss	16	+	?	−	+	?	+	+	+	+	+	+	?	10,000–100,000
4	Foula	17	+	+	+	+	−	+	+	+	+	+	+	+	100,000+
5	Fair Isle	16	+	+	−	+	−	+	+	+	+	+	+	+	10,000–100,000
6	Westray/Papa Westray	18	+	−	−	−	?	+	+	+	+	+	+	+	100,000+
7	Marwick Head	10	+	−	−	−	+	+	−	+	+	+	+	−	10,000–100,000
8	Hoy	16	+	?	+	−	?	+	+	+	+	+	+	+	10,000–100,000
9	Copinsay	14	+	−	−	−	+	+	+	+	+	+	+	+	10,000–100,000
10	Sule Stack/Sule Skerry	12	+	+	−	+	−	+	−	+	+	+	+	+	10,000–100,000
11	Sula Sgeir/North Rona	14	+	+	−	+	−	+	+	+	+	+	+	+	10,000–100,000
12	Flannan Isles	10	+	+	−	+	−	+	?	+	+	?	+	+	10,000–100,000
13	St Kilda (group)	16	+	+	+	+	−	+	+	+	+	+	+	+	300,000+
14	Berneray & Mingulay	10	+	−	−	−	−	+	−	+	+	+	+	+	10,000–100,000
15	Ailsa Craig	12	+	?	−	+	+	+	−	+	+	+	+	−	10,000–100,000
16	Rhum	12	+	?	+	−	−	+	+	+	+	+	+	+	10,000–100,000
17	Shiant Islands	11	+	?	−	+	−	+	+	+	+	+	+	+	10,000–100,000
18	Handa	13	+	−	−	−	−	+	+	+	+	+	+	+	10,000–100,000
19	Clo Mor	8	+	−	−	−	+	+	+	+	+	+	+	−	10,000–100,000
20	Dunnet Head	11	+	−	−	−	+	+	−	+	+	+	+	−	10,000–100,000
21	Duncansby Head	12	+	−	−	−	+	+	+	+	+	+	+	−	10,000–100,000
22	Caithness Cliffs	11	+	−	−	−	+	+	−	+	+	+	+	−	100,000+
23	Troup & Pennan Heads	9	+	−	−	−	−	+	−	+	+	?	+	−	10,000–100,000
24	Buchan Cliffs	9	+	−	−	−	−	+	−	+	+	−	+	−	10,000–100,000
25	Fowlsheugh	7	+	−	−	−	−	+	−	+	+	−	+	−	10,000–100,000
26	Isle of May	10	+	−	−	−	−	+	−	+	+	−	+	+	10,000–100,000
27	Bass Rock	9	+	−	−	+	?	+	−	+	+	−	+	−	10,000–100,000
28	St Abb's Head	9	+	−	−	−	+	+	−	+	+	−	+	−	10,000–100,000

Key: F – Fulmar; P – Storm &/or Leach's Petrel; MS – Manx Shearwater; G – Gannet; C – Cormorant; S – Shag; Sk – Arctic &/or Great Skua; G/R – Guillemot & Razorbill; Pu – Puffin; BG – Black Guillemot; Gulls – at least one of the following: Lesser Black-back, Herring, Great Black-back, Kittiwake; Terns – at least one of the following: Sandwich, Common, Arctic, Little.

the intertidal invertebrate life upon which most of the birds feed. Slow-flowing rivers deposit much fine silt at their mouths, and those receiving generous outfalls of sewage or other organic effluent have much larger, but possibly less diverse, invertebrate populations than rivers with few towns or industries on their banks. The extent of tidal scouring also has an effect, with relatively enclosed estuaries tending to be richer than those with funnel-shaped mouths. Studies have shown that wader numbers per square kilometre of mudflat vary widely: the Tay (reputedly one of the cleanest major estuaries in Europe) holds fewer than $300/km^2$, as do the Beauly and Dornoch Firths. The Ythan Estuary, Cromarty Firth, Loch Fleet and North Solway support between 300 and $600/km^2$; the Forth and Clyde around $1200/km^2$; and the almost enclosed Montrose Basin and Eden Estuary more than $1600/km^2$ (Bryant & McClusky 1977; Standring 1978).

Apart from those on the Solway and around Skinflats on the Forth, Scotland's saltmarshes are rather limited in extent, widely scattered and usually grazed. Neither Clyde nor Tay has any significant area of saltmarsh, but the Tay does possess the greatest expanse of tidal reedbed in Britain. This supports a fairly high density of breeding Sedge Warblers and Reed Buntings, is an important late summer roost site for hirundines, Pied Wagtails, Starlings and Sedge Warblers, and holds good numbers of Snipe, especially in spring (McMillan 1979).

In terms of total numbers of passage and wintering waders and wildfowl, the Solway and the Forth are outstanding; each holds well over 50,000 birds at times of peak population, and is used regularly by some 19 species of wader and 18 of wildfowl. The Solway supports Scotland's only sizeable wintering flock of Pintail, and the entire Spitsbergen population of Barnacle Geese, which feed mainly on the adjoining saltmarsh and farmland. It also holds the largest concentrations of Oystercatcher, Lapwing and Ringed Plover and is the only estuary on which really large numbers of Sanderling have been recorded; Sanderlings are typical of sandy beaches but in May 1982 more than 9,000 were found resting on the North Solway shore during spring migration. The Forth has the biggest numbers of Knot and Dunlin, and is of special significance for Shelduck – as a safe moulting site, a wintering area and a breeding ground (Table).

Although the other estuaries shown on Map 5 hold relatively much smaller total numbers, most have their own particular importance. On the Clyde, which is the only significant west-coast site other than the Solway, Redshank are present in higher numbers than elsewhere, although they have decreased in the last few years. In the east, the Eden is notable for Black-tailed Godwit and Grey Plover, Montrose Basin for moulting Mute Swans and good numbers of Wigeon, Knot and Dunlin, the Ythan for its Eider nursery, the Beauly for moulting Canada Geese, and the Cromarty Firth for its wintering Wigeon.

Apart from those already mentioned, the only abundant waders largely dependent upon estuaries in winter are the Curlew and Bar-tailed Godwit, both of which are well scattered. Golden Plovers occasionally occur in sizeable flocks on mudflats but, like Lapwings, are as likely to be found feeding on farmland. Similarly, while Mallard, Teal and grey geese are present on many estuaries in winter, often using them as roosts, they occur at least as regularly in other habitats.

The reclamation of estuarine flats, which is of course irreversible, has already had a considerable effect on wildfowl and wader distribution – and is still continuing. Large areas in Nigg Bay, Longman Bay, Invergowrie Bay and on the Forth around Kinneil and Longannet (some 700+ ha in all, of which more than 300 ha are in the Forth Estuary) have been infilled since 1950 or are currently being reclaimed, and there have been smaller developments at other sites. The resultant loss of wader and wildfowl feeding grounds has had some evident effects on status: for example, tipping followed by industrialisation at Longman Bay has affected wintering Pintail, while Invergowrie Bay is a much less important wader site now than in B&R's time. More recently the effects of reclamation at Kinneil have been monitored by Bryant (1980), who found the resultant distributional changes more marked than any decline in numbers. Symonds *et al* (1984) have suggested that since the use made of estuarine flats by different species varies in both temporal and spatial patterns, the loss of significant areas of mudflat might have serious consequences for a substantial proportion of the wintering population of certain species. Proposals for a major barrage on the Solway Firth, which would have a serious impact on the tidal flats and saltmarshes there, fortunately seem to have been greatly reduced in scale and may never be implemented.

Sandy beaches, which make up only c8% of Scotland's coastline (Ritchie & Mather 1984), are often comparatively bird-less. Only two species are more dependent upon them than upon other habitats, one a summer and the other a winter visitor. They are the Little Tern, which nests only on sand and shingle and often chooses a site not far above the high tide line, and the Sanderling, which feeds at the

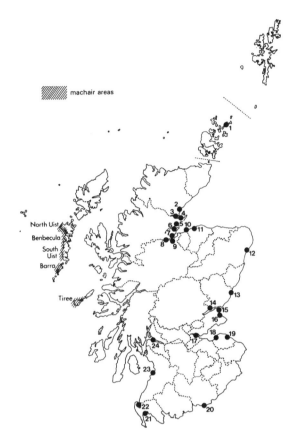

Map 5. Soft shore sites of ornithological importance

1	Sanday	13	Montrose Basin
2	Loch Fleet	14	Invergowrie Bay
3	Dornoch Firth	15	Tentsmuir
4	Morrich More	16	Eden Estuary
5	Nigg Bay	17	Forth Estuary
6	Udale Bay	18	Aberlady Bay
7	Munlochy Bay	19	Tyninghame Estuary
8	Beauly Firth	20	North Solway
9	Longman Bay	21	Luce Bay
10	Culbin Bar	22	Loch Ryan
11	Findhorn Bay	23	Ayr coast
12	Ythan Estuary	24	Inner Clyde

water's edge. Ringed Plovers, Oystercatchers and Common Terns are among other beach-nesting birds, while gulls, terns and Oystercatchers often rest in considerable numbers on open beaches. Many more species breed among sand dunes, the variety increasing with the stability of the dunes and the consequent increase in structural diversity of the vegetation, but all also breed in other habitats. Those for which dunes are probably the most important breeding habitat include the Shelduck and the Sandwich Tern – the latter often nesting in close association with Black-headed Gulls. The colony of Eiders on the dunes and heaths at the Sands of Forvie is the densest breeding concentration in Britain.

Many mainland beaches and dunes are heavily used for recreation, with a consequent high level of disturbance, while those on the west coast, although beautiful to look at, are

Wildfowl and wader numbers at the principal Scottish estuaries (Data for 1969–75 from Prater (1981) and for 1980–84 from WWC)

	Waders				Wildfowl			
	1969–75			1980–84		1969–75		1980–84
No. of species regularly recorded	No. of spp. in nat./int. imp. numbers	Av. peak count *	Av. peak count (n) *	No. of species regularly recorded	No. of spp. in nat./int. imp. numbers	Av. peak count *	Av. peak count (n) *
Firth of Forth	19	9/6	45,000	33,000	20	10/4	29,000	18,540 (4)
North Solway	19	12/6	67,000	39,400 (2)	18	7/4	20,200	21,860 (4)
Inner Clyde	11	2/1	20,500	13,600 (4)	12	4/0	5,600	4,220 (4)
Eden Estuary	18	5/2	14,000	12,600 (3)	12	2/1	6,100	4,400 (4)
Montrose Basin	12	2/1	12,000	8,600 (3)	15	2/0	7,000	13,900 (3)
Dornoch Firth	17	0/0	2,400	6,400 (3)	12	6/1	6,400	10,640 (4)
Ythan Estuary	14	0/0	5,100	9,000 (2)	17	4/2	6,000	4,400 (4)
Cromarty Firth	14	5/2	12,900	7,700 (1)	12	6/2	10,000	19,920 (4)
Beauly Firth	14	0/0	3,000	2,500 (1)	12	5/2	5,700	10,940 (2)
Firth of Tay	13	4/3	11,500	10,300 (3)	18	3/3	14,500	11,430 (4)
Moray Basin	12	4/2	16,700	6,200 (1)†	16	7/4	12,600	9,120 (4)†

*based on sum of highest average monthly peaks for each species   †Inner Moray Firth only

relatively sterile. The areas most important for breeding birds are the Sands of Forvie and Morrich More; both are well-protected from disturbance but in rather different ways – the former is a National Nature Reserve and the latter a firing range! Two sandy areas once of great ornithological importance for ground-nesting species – Tentsmuir and Culbin – are now largely forest-covered or heavily grazed, but the offshore sand-bars at both continue to provide roosting sites for large numbers of wildfowl and waders, and resting places for terns and gulls. In the Northern Isles most of the terns nest on moorland and there are few beaches of major importance for birds, though moulting Eiders walk ashore to rest on some that are cliff-bound and so safe from disturbance.

The increasing recreational use of beaches has been sufficiently extensive to endanger the future of one species, the Little Tern. This bird, which has always been local in Scotland, shows a marked preference for the type of east coast beach that is also attractive to people; frequent disturbance results in poor, if any, breeding success and it is now only within the protection of nature reserves that the Little Tern has much prospect of raising young. Sand extraction at one time presented a potential threat to dunes in several areas, but the introduction of stricter planning control, based on the findings of a comprehensive survey of Scotland's beaches (Ritchie & Mather 1984), has helped to reduce this threat to tern and Shelduck breeding sites.

The *machair* of the Outer Hebrides is in a class by itself. An intricate mosaic of shallow lochs, damp and dry grassland, crops and fallow land covers this coastal plain of windblown shell sand, which lies behind the beaches of the western seaboard. The machair is most widespread in the Uists and Benbecula, but also occurs more locally in Harris, on low-lying Tiree in the Inner Hebrides and on some of the smaller islands. Much of it sheltered from Atlantic gales by dunes, and backed by marshy rough grazing and peat-covered moorland, this habitat holds an exceptionally large breeding population of waders. Studies covering more than 130 km² in the southern Outer Hebrides have shown that the overall density of waders exceeds 90 pairs / km², far above the numbers found elsewhere other than in very small areas of particularly favourable habitat. The 12,000+ pairs found included more than 2,000 pairs of Ringed Plovers – mostly on the drier ground – and similar numbers of Dunlins and Redshanks, together with c500 pairs of Snipe in the wetter areas (Green 1984; Galbraith *et al* 1984). These populations are of national significance, while the breeding densities of Dunlin and Ringed Plover are probably exceptional at an international level.

There is currently much concern regarding the potential effects on the machair of implementation of the European Economic Community-funded Integrated Development Programme (IDP) for the Western Isles (ie the Outer Hebrides). This offers financial incentives for crofters to abandon traditional methods of machair management and intensify their farming practices. The drainage and fertiliser / pesticide usage which might be involved in such intensification could present a very real threat to the internationally important breeding wader populations of this unique habitat. It is at present too early to assess the extent to which the IDP may affect crofting in the Outer Hebrides, but the changing agricultural situation is being monitored.

INSHORE WATERS
Tidal waters, on the open coast and within estuaries, are important for divers, grebes, wildfowl and seabirds, and are used not only as feeding grounds but also as roosting and resting areas. Some species are present throughout the year, others only in winter or summer, and some move freely between the open sea and more sheltered estuarine waters, while others use only one of these habitats. In summer the areas of most importance are those close to the large seabird colonies, where vast numbers of auks bathe and rest before flying further out to sea to feed, and Manx Shearwaters

gather on the water at dusk before flighting in to their nesting burrows. In winter the larger east coast firths hold the most important concentrations of seaducks and grebes, and the inter-island sounds the biggest gatherings of divers. Several estuaries are used by grey geese for roosting, while ducks normally found on freshwater resort to tidal areas during periods of hard frost.

Apart from those species using the water simply as a safe roost, bird distribution on inshore waters is determined by the availability of the relevant foods. For auks and Kittiwakes the most important foods are sprats and sandeels, the former being taken also by the sawbill ducks and the latter by terns. Sprats are widely dispersed in inshore waters in summer, with larger concentrations further offshore during the spawning period; they move nearer the coast in winter, distribution patterns varying between years. In some seasons very large shoals occur in the Moray Basin and the Firth of Forth, but numbers there (assessed from commercial catches) declined markedly during the early 1980s when the highest winter concentrations were found in the southern North Sea (Blake *et al* 1984). The nationally important gatherings of Goosander and Red-breasted Merganser in the Beauly Firth and – less regularly – in the Forth Estuary are probably associated with sprat movements. Sandeels of the size most frequently taken by birds are abundant in inshore areas throughout the summer (and also occur right across the North Sea) but become less available in winter, possibly because they bury themselves in the sea-floor sediments. The diet of the remaining seabirds is less dependent upon only one or two prey species. Gannets, Shags and Cormorants take a variety of fish and the petrels feed largely upon plankton, although Fulmar distribution in spring and autumn is perhaps linked also with commercial fishing activity, the birds taking offal thrown overboard.

The seaducks feed largely on molluscs and crustaceans, and for them the location of mussel beds is often an important factor in determining distribution; this is especially so in the case of the Eider, for which mussels are the principal food. The Moray Basin holds the largest British concentrations of Common and Velvet Scoters and Long-tailed Ducks, and the Tay of wintering Eiders; important flocks of Long-tailed Ducks also occur in Scapa Flow. Other diving ducks, such as Goldeneye, Scaup and Pochard, tend to gather closer inshore where the water is enriched by discharges from sewers or distilleries. Changes in the quantity or nature of such discharges can cause major changes in duck numbers and distribution. In the Forth, for example, the reduction since 1978 in raw sewage discharge has resulted in the loss of the nationally important Scaup flock which formerly wintered there, and also a significant decrease in Goldeneye and Pochard numbers.

Divers and grebes are in general rather thinly scattered around the coast, but a few important concentrations do occur. The Outer Firth of Forth has the only sizeable winter gathering of Great Crested Grebes in Scotland, and fair numbers of Slavonian Grebes. Shetland and the Moray Basin are the most important areas for divers. Many of our wintering Great Northern Divers are in Shetland waters from October to April, while numbers of Red-throated Divers are high there during much of the year and very high in the Moray Basin during the autumn. Coastal waters

around the Hebrides, and between the mainland and the islands, hold good numbers of Red-throated Divers too.

The most publicised threat to coastal habitats and their bird communities, especially in the north and east, comes from the oil industry. Developments in the North Sea have been so rapid and widespread that it is easy to forget just how recent they are. Exploratory drilling started only in the early 1960s and the first oil came ashore in 1975. By 1984 there were 25 offshore oilfields (in four main areas) in production in the UK sector of the northern North Sea (Map 6). Terminals capable of handling large tankers had been established at Sullom Voe in Shetland, Flotta in Orkney, Nigg in Easter Ross, and Hound Point in the Firth of Forth, with undersea pipelines laid to all but the last. The Beatrice field is the closest inshore – less than 20 km from the Sutherland coast and within 60 km of six major auk colonies. Development of wells to the west of Shetland is possible within the next ten years, licences for test drilling in the Firth of Clyde were issued early in 1984, and exploration in Outer Hebridean waters is expected in the near future.

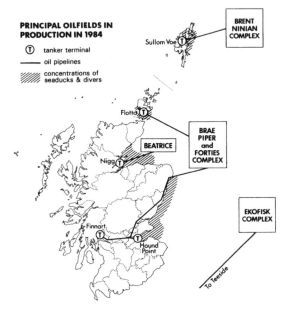

Map 6

The increase in tanker movements resulting from these developments (especially in Shetland and the Forth, where navigation can be tricky), the transfer of oil between pipelines, tanks and tankers at the terminals, and the exploitation of wells at sites increasingly close to the coast, all present risks for local populations. To date the *Esso Bernicia* spill at Sullom Voe in December/January 1978/79, has been the most serious incident directly attributable to North Sea oil operations. That spillage, involving some 1,700 tons of fuel oil, killed at least 3,700 birds. Among them were nearly 150 Great Northern Divers and more than 600 Black Guillemots, representing substantial proportions of Shetland's wintering and resident populations respectively. More than 50 oiling incidents have since occurred at Sullom; most have involved only small numbers of birds, but the possibility of a major disaster clearly exists, both there and at the other terminals. The northern seabird colonies, the seaducks of Scapa Flow and the Moray Basin, and the ducks and grebes that winter in the Forth are the populations at greatest risk.

Although attention is often focused on the threat presented by North Sea oil activities, the bulk of seabird mortality from oiling is not at present attributable to this source. Industrial spillages onshore, the discharge of tank washings, and leakage of fuel oil at sea are believed to account for nearly 80% of bird deaths from oiling.

Another environmental factor thought to be affecting some coastal birds, although comparatively little is yet known about its extent or origin, is the presence of marine pollutants such as polychlorinated biphenyls (PCBs) and toxic heavy metals. Varying levels of one or both of these substances have been found in many seabirds, although there is little evidence that they are significantly affecting seabird breeding success.

However, relatively high levels of PCBs and heavy metals have also been found in the eggs of coastal-nesting Peregrines, which have markedly poorer breeding success than those inland, while the seabird wreck in the Irish Sea in 1969 was believed to be at least partly due to the effect of PCBs in reducing the birds' ability to withstand severe gales.

Little is yet known about the relationships between seabird numbers and breeding success and the commercial fishing of small prey species, such as sandeels, which are particularly important during the breeding season. Sandeel populations expanded while the larger predatory cod and haddock, together with herring and mackerel, were being over-exploited; now the efforts of commercial fisheries are being directed increasingly towards sandeels. Seabird monitoring over the next few decades should help to demonstrate whether this change in fishing practice has any significant effect on populations as a whole.

3: Woodlands

Little now remains of Scotland's native woodlands, which were formerly dominated by Scots pine and birch north of the Central Lowlands, with oak and ash on the better soils at low altitudes and to the south. These natural woodlands included a variety of tree and shrub species, and of age groups; in consequence they possessed the structural diversity so important to birds and also offered a varied food resource. Today only small patches of semi-natural woodland survive, and even the planted broadleaf woodlands characteristic of the big lowland estates are dwindling. In their place, and also over stretches of country which had been bare of woodland for hundreds of years, spreads an ever-expanding forest of conifers. The area of woodland in Scotland increased by c75% between 1947–49 and 1979–80 (Map 7). There are now 10 ha of conifers for every hectare of broadleaf woodland – and the ratio is still widening.

Although only a few bird species are entirely restricted to one particular woodland type, the structure of the bird community in broadleaf woodland is different from that in coniferous forest. Changes in woodland type and distribution consequently have an influence on the status and distribution of woodland birds, but the afforestation of land previously carrying rough grass or heather moor has a more pronounced effect.

In considering woodland bird communities the ornithologist is primarily concerned with three main aspects: species diversity, total breeding numbers, and the presence of species with a restricted distribution. All three are valid criteria in assessing the value of a wood to birds, but conservationists usually regard the first and last as of most importance. There are exceptions to this generalisation, of course – for example, the presence of a sizeable heronry in an otherwise species-poor conifer plantation would result in a higher placing on the value-judgement scale than if there were no herons nesting there. On the whole, however, it is species-rich woodland types, and especially those supporting birds not found nest-

ing in other habitats, which merit the highest rating – and unfortunately it is this group which is vanishing at an alarming rate.

More detailed consideration of many of the aspects mentioned below can be found in the Forestry Commission's *Census of Woodlands and Trees – Scotland, 1979–82*; Fuller 1982; Elton 1966; Yapp 1962; Simms 1971; and various papers by Kenneth Williamson and Dorian Moss.

WOODLAND BIRD COMMUNITIES

The diversity and density of woodland bird communities are determined by a combination of factors, many of which are related to forest structure and composition. The age and management of a woodland, as well as its size and the plant species it contains, affect its suitability for a particular bird at any given time. In general, mixed conifer/broadleaved woods hold the highest number of songbird species, at the greatest densities, and pure stands of well-grown conifers both fewest species and lowest densities (Newton 1982). Large woods of any given type support more species but at lower densities than do small woods of the same type. And even-aged woods hold fewer species than those containing trees of mixed ages. Suitable nesting sites are clearly of prime importance, to birds feeding largely or entirely outwith the wood as well as to those entirely dependent upon it; and their availability may determine whether or not a particular species is present. Among hole-nesters, those which excavate their own holes – for example Willow Tits and woodpeckers – must have access to dead or dying trees; others, such as the Pied Flycatcher and Redstart, will use nest-boxes to compensate for a lack of natural holes. Scrub-nesters and ground-nesters are often scarce in, or absent from, heavily-grazed woodlands, which usually lack ground cover and have little or no shrub layer. Conifer plantations, which grow very rapidly, offer dense ground cover when very young, dense scrub cover just before the thicket stage, and neither ground nor scrub cover as they approach maturity, unless heavily thinned. In any type of woodland open glades or rides add valuable habitat diversity and increase the extent of 'edge', along which some species prefer to feed. The availability of suitable song-posts and/or clearings is important to some birds. The Tree Pipit, for example, occurs only in relatively open woods or along the forest edge, where it can launch into its song-flight from the top of a prominent tree. For the Chiffchaff a combination of tall trees and dense ground cover is necessary. Species such as these are absent from woods with continuous closed canopy.

For strictly woodland species, the food supply available on, under or between the trees is equally important. In general, the more diverse the woodland the greater the food resource it represents. Mixed woodlands with nut, cone and berry-bearing trees and a good understorey of shrubs hold a greater variety of birds than one-species woods, while broadleaved trees, especially oaks, support a much wider range of insects than do conifers – and a plentiful supply of caterpillars is essential to the breeding success of many small passerines. Predator numbers are greatly influenced by the abundance of passerine or small mammal prey. Many of the birds breeding in woodland are not, however, dependent upon it for food, and for these easy access to surrounding feeding grounds is important.

The geographical location of a wood also influences its bird population. Several of the scarcer migrant passerines

vary markedly in their northern breeding limit between years, possibly as a result of the weather conditions prevailing during spring passage as well as of variations in winter mortality. Species diversity decreases towards the north and west and is generally lower in island woods than on the neighbouring mainland. Species diversity and density also decrease with altitude, and – to a more limited extent – from the east to the wetter west. These gradients are closely associated with land productivity, which is largely determined by soil quality, altitude and exposure.

Attempts to assess and compare the bird communities of different woodland types have involved a variety of approaches and methods (see Chapter 8). None of these is entirely reliable but together they have provided a good deal of information on the *relative*, as distinct from absolute, situation in woods of different composition and structure. It is, however, advisable to interpret the results of such studies with caution. Assessments of breeding numbers are influenced not only by the study method adopted and factors such as observer variability, weather conditions and time of day, but also by the size of the study plot; 'edge-effect' can greatly inflate the numbers breeding in a relatively small census plot. And since many species show large annual fluctuations in breeding numbers, results obtained in different years cannot validly be compared. Table (a) overleaf presents figures derived from a series of studies. They are not directly comparable, nor may they all even be typical for a particular woodland type, but they will serve to indicate the wide differences in both species diversity and overall numbers that occur among woodland bird communities. Figures for breeding densities given both here and in the species accounts are best regarded as indicative of relative diversity and abundance, rather than as accurate measures of population size.

SCOTTISH WOODLANDS TODAY

Much of the 76,000 ha of what the Forestry Commission classes as Broadleaved High Forest currently in existence is in the south and east. Planted woodlands with a good diversity of species are most widespread in the lowlands, particularly in the Borders, where the many large estates have traditionally planted for amenity and sport as well as shelter and commercial timber production. Most of the surviving semi-natural oakwood is in the west, especially in Argyll, Dunbarton and Stirling, while birch is predominant in the north.

Although variations in survey methods and definitions make direct comparisons difficult, it appears that the total area of broadleaved woodlands decreased by some 20,000 ha between 1947 and 1980. While areas of sycamore, ash and mixed broadleaves have increased, there have been major decreases in birch, oak and beech. Much of the remaining oak and beech was planted prior to 1900; these species are commercially mature at about 90–120 years old, and start to degenerate after a similar time-span, so many of the pre-1900 trees – if they escape clearance – will be 'going back' by the year 2000. Unless planting policies change radically the area of oak and beech will be much reduced by the middle of next century.

Birch woodland, currently the most widespread and typical broadleaf woodland over much of Scotland, is also

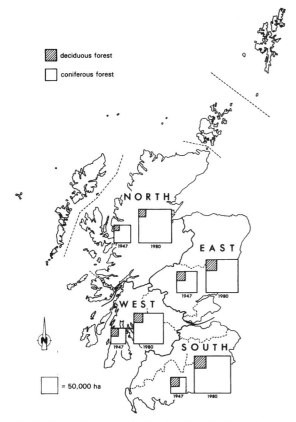

Map 7. Changes in area and distribution of woodland types between 1947 and 1980 (based on FC Woodland Census data)

(a) *Some examples of the variation in numbers and species diversity of birds breeding in different woodland types (Data from sources listed).*
(Note – see text for explanation of the limited comparability of these figures)

Location	Woodland type	No. of species	No. of terri-tories/km²	Source
Wester Ross	S Pine – semi-natural	11	217	Williamson 1969
	S Pine with deciduous regeneration	12	444	,,
	Birch	20	1,100	,,
	Oak	20	1,320	,,
Galloway	Mixed broadleaf	22–35	650–840	Williamson 1976
Loch Lomond	Mixed/oak – mainland	30	1,150–1,160	Williamson 1974
	– islands	19/31	880–1,600	,,
Rhum	Policy woodland	23	c1,000	Williamson 1975
Speyside	S Pine – planted (sparse undergrowth)	13	155–215	Newton & Moss 1977
	S Pine – semi-natural (heavy undergrowth)	14	385–471	,,
	Birch – natural (heavy undergrowth)	16	418–455	,,
Deeside	S Pine – semi-natural (mature)	7–12	250–530	Jenkins *et al* 1984
	S Pine – young plantation	5–7	270–500	
	Broadleaf – birch domin.	9–13	230–450	D. French/ITE
	Broadleaf – oak dominant	11–16	490–730	,,
	Mixed broadleaf/conifer	12–19	330–1,100	,,
Dumfries	Spruce plantation	13	351–598	Moss 1978a
	Regenerating birch/pine	24	880	,,

vanishing fast. Much shorter-lived than oak and beech, the birch reaches its prime at about 50 years old. Nearly 75% of the birch 'High Forest' recorded in the 1979–82 census was more than 30 years old, and only a little over 2,000 ha was less than 20 years old. Birch seeds readily in open ground but heavy grazing, by domestic stock or deer, prevents regeneration on most moorlands and in many older woods, where glades formed as trees die and fall provide the light conditions necessary for seedling establishment. Unless action is taken to permit regeneration in suitable areas, the extent of birch woodland is likely to decrease progressively.

There has also been a substantial reduction (nearly 50%), especially in the last two decades, in the area classified as 'scrub' – predominantly birch woodland of little commercial value but of considerable importance as bird habitat. Although part of this apparent reduction is attributable to reclassification as High Forest and part to clearance for replanting, a substantial proportion has been cleared and reclaimed for agricultural use. There has also been extensive loss, in recent years, of hedgerow and other elms, as a result of Dutch elm disease.

Scots pine now comprises only about 19% of the total conifer forest, and native pinewoods extend over c 11,000 ha, of which only 1,600 ha is in relatively dense stands (Goodier & Bunce 1977); since 1970 only 4% of the coniferous area afforested has been planted with this species. Conservationists have long been concerned for the surviving remnants of the native pinewoods, not only because they have been dwindling in extent, but also because few of those remaining are actively regenerating – largely due to heavy browsing and grazing by deer and sheep. Climatic change, as well as management practices, may be contributing to this situation. Whatever the cause, it seems likely that the overall area of semi-natural pinewood will continue to decline, despite the fact that a substantial proportion is now protected by nature reserve status or by designation as a Site of Special Scientific Interest (see Chapter 9). From an ornithological

viewpoint it is fortunate that the Scottish Crossbill (whose status as a full species is debatable) is the only bird strictly confined to such woodland; other primarily pinewood species, such as the Crested Tit and the Capercaillie, are successfully colonising pine plantations, albeit at a lower density than in native pinewoods.

Introductions from America account for more than 60% of the conifer area, 88% of which has been established since 1940; about half the post-1940 afforestation has been with Sitka spruce. The planting of Lodgepole pine, which can tolerate conditions too wet and poor for satisfactory growth of Sitka spruce, has increased rapidly since the 1950s. This species has been widely used, often with Sitka spruce, in the north and on some of the islands, both areas in which large-scale afforestation has recently taken place.

Although the Forestry Commission still has the largest holdings, there has been a progressive increase since the 1950s both in the area of privately-owned conifer forest and in the proportion 'dedicated' under schemes which offer financial assistance to owners undertaking to manage their woodlands in accordance with a plan designed to secure sound forestry practice. In many cases the management of private woodlands is now undertaken by one of the commercial forestry firms, rather than by the owner himself. Whereas the Forestry Commission is obliged under the Countryside (Scotland) Act 1967 to 'have regard for conservation (and amenity)', no such constraints are imposed upon 'dedicated woodlands'. For these the only requirement is that they are managed on the basis of 'sound forestry practice' – which may fall short of what is desirable from the conservation point of view, as demonstrated by the clear-felling in the early 1980s of part of Abernethy native pinewood on Speyside. Afforestation based on purely commercial interests involves the planting of conifers on virtually all usable ground within the forest boundary. In such situations there is little prospect that sufficiently large areas of heather moorland to maintain Merlin, Golden Plover and Red Grouse populations will remain unplanted, or that a buffer zone of open ground will be left around a hill loch.

Much public antagonism against 'blanket afforestation', involving the planting of large areas with only one or two species, is based on the visual effects of such treatment, but there is now ample evidence that such homogeneous woodlands are also the least valuable in conservation terms. One recent study led to the conclusion that, so far as species diversity is concerned, the presence of a large number of fairly small discrete patches of different woodland habitat types is most advantageous (Jenkins *et al* 1984). The mosaic of conifer blocks interspersed with mixed – predominantly broadleaf – woodland and farms, typical of many areas on the Continent, is both visually more attractive and better for wildlife. Unfortunately, as Ian Newton has commented (1982), 'the patchy afforestation which the conservationist would favour is likely to come only with planning control'.

AFFORESTATION IN THE UPLANDS

The last three or four decades have seen a striking extension of planting on ground which had not carried significant tree cover for hundreds of years. Galloway and the Borders were among the first areas affected, with planting largely on grass and heather moorland. These new forests have already had a marked effect on bird distribution. Conifers now occupy much of the eastern end of the grassy Ochils and many Perthshire moorlands which once supported Red Grouse and Golden Plover. In Sutherland and Caithness, where a large proportion of the planting has taken place on wet moorland and involved Lodgepole pine, more than 30,000 ha of conifers were established between 1950 and 1980, and a further 8,000 ha in 1980–83. There has also been quite extensive afforestation on some of the islands. Arran currently has some 6,000+ ha of conifers, Mull more than 9,000 ha, and Skye more than 7,500 ha; about 90% of these totals has been planted since 1950. Elsewhere on the islands plantation areas are much smaller but are gradually expanding. Lewis and Harris together now have c700 ha and in 1983–84 some 1,400 ha of hill land on Islay were sold for afforestation. Such plantings in situations which had long been virtually treeless are bound to have an influence on the distribution of woodland birds, allowing viable populations to become established in new areas.

In its Census report the Forestry Commission presents figures for woodland area as a percentage of the total land and water area in each of its Conservancies; these figures tend to obscure the extent of moorland loss – an aspect of land-use change currently causing concern. The Commission has kindly provided the corresponding figures for selected smaller areas, for which the area of land lying between the 200 m and 500 m contours (on which the bulk of the recent afforestation has taken place) has also been abstracted (Table b). Although this gives only a very rough indication of the true situation, since some afforested land lies below or (less often) above the chosen levels, it does give a better idea than can be obtained from the FC figures of the proportion of these formerly moorland areas now under trees.

The extent to which afforestation of the uplands worries ornithologists depends largely upon the type of vegetation previously present. The loss of heather moorland, which often supports species which are both scarcer numerically and of much more restricted distribution than those found on grassy uplands, is of growing concern at the present time. This topic is dealt with more fully in the next chapter.

THE EFFECTS OF WOODLAND CHANGES ON BIRDS

Fuller (1982) and others have documented the most frequently occurring species in woodlands of various types: in Scotland these are generally the Willow Warbler, Wren, Chaffinch and Robin. Much has also been written about the succession of bird populations in conifer plantations – an initial increase in both species and overall numbers gradually gives way to relatively high densities of a small number of species, notably Goldcrest and Coal Tit. Rather than repeat this documentation, I want to direct attention first to those woodland species most at risk as a result of the changes currently taking place, and then to species which are benefiting from these changes – albeit only temporarily in some cases.

The first group are essentially birds with specific habitat requirements which they are able to satisfy only in woods containing a particular tree species, or in semi-natural or relatively unmanaged woodlands which have one or more of the following attributes: open clearings, a diversity of vegetation height and density (which generally implies the presence of trees, shrubs and herbs), and dead or dying trees. Woodland management solely concerned with the economical production of timber aims at full utilisation of the available ground, minimal competition from other plants, and

(b) The extent of conifer afforestation in selected sample moorland areas. (Data – other than land areas between 200 m and 500 m asl – from Forestry Commission, figures rounded)

Sample location	Total land & water area (ha)	Area between 200 & 500 m asl (ha)	Area of mainly conif. high forest in 1980 (ha)	Planted since 1951	since 1971	Main species
Glentrool	123,000	77,300	38,000	c90%	25%	Sitka (c70%)
Moffat/Langholm	187,000	162,700	53,000	c90%	45%	Sitka (c75%)
Ochils	55,000	28,500	3,400	c77%	47%	Sitka (c58%)
Pitlochry	129,000	72,700	14,100*	77%	16%	mixed

* It is estimated that a further 4,000 ha or thereby was planted in this area between the census date (1980) and 1984.

the removal of all dead and dying wood (which represents a potential source of pests and diseases). Coppicing, a form of management which perpetuates the ability of a tree to produce usable material and at the same time maintains a sequence of vegetation heights, virtually ceased in Scotland many years ago. The species likely to be adversely affected by woodland change and having the most restricted current distribution are the Scottish Crossbill and the Willow and Marsh Tits. The first appears to be dependent upon native Scots pine and the second upon wet semi-natural woodland with abundant dead trees; both these habitats are diminishing in extent. The Marsh Tit, on the northern fringe of its British breeding distribution, is so scarce and local that it would be irrelevant to consider its future in the context of woodland change.

The species likely to be most adversely affected by the loss of oakwoods is the Wood Warbler, which is virtually confined to oakwoods with relatively little understorey. The Pied Flycatcher also shows a preference for oak, but its other major requirement – suitable nesting holes – can be met by the provision of nest-boxes. Most of the other species suffering local decreases as a result of conversion from broadleaf to conifer woodland are either sufficiently common and widespread, or sufficiently catholic in their habitat usage, for the population as a whole to be at no real risk.

Turning now to species which increase with the spread of afforestation, the most lasting benefit is clearly to those for which coniferous woodland is the preferred habitat. At the top of the list must be the Goldcrest and Coal Tit, both now well established and abundant in areas which previously held little or no suitable woodland. The Crested Tit is largely dependent upon Scots pine and nests in dead stumps, but

its recent expansion of range and use of nest-boxes suggest that it may be able to maintain its population – despite some further loss of native pinewood – provided that these requirements are met in pine plantations adjoining its present distribution. Like the Crested Tit, the Capercaillie has expanded its range as the new forests matured and was until very recently more widespread than at any time since its reintroduction last century. Its future spread may, however, be limited to woods containing a high proportion of Scots pine or larch. Other species breeding mainly or solely in conifers include the Common Crossbill, Siskin and Goshawk, while the Sparrowhawk is now more abundant in some conifer forests than in many mixed or broadleaf woodlands. Undoubtedly the least popular inhabitant of conifer plantations is the Woodpigeon, which benefits from the safe nesting sites they afford.

The Hen Harrier is only a temporary resident in plantations, leaving before the trees reach the thicket stage – although it may continue to nest in forest clearings if there is open moorland within a short distance. Short-eared Owls occur both following initial afforestation and at appropriate stages in subsequent rotations, when they occupy large felled and replanted areas until the canopy closes. Most of the small passerine species that increase temporarily in the early stages – such as Willow Warbler, Meadow Pipit and Whinchat – are common and widely distributed, but the much scarcer Grasshopper Warbler is also often present. Bullfinches take advantage of plantations from the thicket stage onwards. Most of the other species that make some use of coniferous woodlands, such as Woodcock, Jay, Blackbird and Song Thrush, breed at greater densities in other habitats.

4: Uplands

The term 'upland' is used here to cover a variety of open moorland, bog and mountain habitats ranging in altitude from sea-level to the high tops of the Cairngorms. These habitats carry many different types of vegetation but they have one important factor in common: they are exposed to relatively little direct impact by man, and such impact as occurs is usually only local and is often only seasonal. Although they may largely be the result of such actions as woodland clearance, burning and over-grazing in the past, many upland areas have until recently been considered of little commercial value, due to one or more of the following adverse factors: poor soil, waterlogging, deep peat cover, high altitude, steepness of slope, and exposure to wind or salt spray. Some such areas have been used as extensive hill sheep grazings, deer forest or grouse moor, while others have remained virtually unused except by the occasional climber or hill walker. With modern technology it is now possible to strip off peat and reclaim the land for agriculture, to drain bogs sufficiently to allow afforestation to take place, and to convey thousands of people to mountain tops. Changes of these kinds have considerable implications for the bird communities occupying our upland habitats. Fuller accounts of many of the aspects touched upon in this chapter can be found in Watson *et al* 1970; Ratcliffe 1977a; Nethersole-Thompson & Watson 1981; and the RSPB's recent report *Hill Farming and Birds* 1984.

In general, upland habitats have little structural diversity and their bird communities consequently include a high proportion of ground-nesting species. Crags, wooded gullies, streams and lochans offer nesting sites for species which are absent from areas where such features are lacking, and some of the raptors and scavengers that forage over the uplands nest in adjoining woodlands. Food resources are often limited in both quantity and variety, and may also be seasonal in availability; in consequence many species that breed in upland habitats desert them in winter, either migrating or simply dispersing to less harsh environments. Comparatively few species breed only in the uplands – but these few include some of Scotland's rarest and most interesting birds.

As different upland types often merge gradually into one another, and some have a wide altitudinal range, categorisation from an ornithological point of view is not easy. Sites which carry very different vegetation may support similar bird communities, so there seems little point in sub-dividing on the usual botanical basis. I therefore consider uplands under only three main headings: moorlands, bogs and mountains. It should be understood, however, that these categories both overlap and intermingle in many areas. The upland birds dependent upon open water are discussed in the chapter on freshwaters.

MOORLANDS

Included in this category are all heather and grass dominated uplands that are well-drained; they may be virtually flat – as in many low-level coastal heaths, gently undulating – as on many grouse moors, or relatively steeply-sloping – as on the grassy hills of the Southern Uplands and the lower slopes of the Cairngorms. These examples typify the main contrasts in moorland types, and will be used to illustrate their respective importance for birds.

Because they are so exposed, low altitude coastlands often share some ecological characteristics with what are more appropriately described as uplands. Such coastal – or maritime – heaths, often carrying a stunted cover of heather, crowberry and sedges, are present on many of the islands and locally on the mainland. They are particularly widespread in the Northern Isles, where they form the primary nesting habitat of Great and Arctic Skuas, and of Whimbrel; Scotland holds the entire British breeding population of these three species, all of which are very local in their distribution. Other local or scarce species associated with this habitat type include Arctic Tern, the largest colonies of which are on coastal heaths, Snowy Owl and Dunlin.

Many of these maritime heaths were formerly part of the crofters' common grazings, providing a sparse living for small numbers of sheep and remaining unimproved because the land was held by a community rather than an individual. With the apportionment of sections of common grazing to specific crofts, and the advent of grant-aid for fencing, fertilising and reseeding, there has been a progressive increase in land enclosure and 'improvement', resulting in the conversion of heath to permanent pasture, with a consequent increase in grazing. In Shetland, where oil revenue funds are available to help finance agricultural development, there is a very real possibility of a major decrease in this type of habitat – with serious implications for the Whimbrel population especially.

Moorland of the grouse moor type, dominated by heather but with a good range of low berry-bearing shrubs also present, lies mainly between 200 and 600 m asl and is predominantly undulating, though often broken by steeper slopes or rocky outcrops. Management by rotational burning, to ensure a fairly constant proportion of young shoots for the benefit of both grouse and sheep, results in a mosaic of long old heather and younger growth, with grassy patches in the most recently burned and wetter areas. Even where moorland management has ceased or has never existed, variations in soil depth, drainage and exposure to wind often produce a somewhat similar mosaic. Moors of this type are the primary breeding habitat for Red Grouse, Hen Harrier and Merlin, and are widely used by Golden Plover, Dunlin and Short-eared Owl. Where they have scattered shrubs, such as gorse or bracken, they may support Whinchat and Stonechat, and if rocky streams and gullies are present so too are likely to be Ring Ouzels.

Long-continued over-grazing and over-burning have been responsible in many areas, especially in the wetter west, for the replacement of heather by poor quality moorland grasses and deer sedge. More recently heather moorlands have been extensively afforested, in many cases because they were no longer commercially viable as grouse moors or hill sheep farms. And in Orkney, where the underlying soil is good, the moorland edge is gradually creeping up the hillsides as land is reclaimed for agriculture. The best moors for grouse lie between the Tay and the Moray Basin, on the eastern fringes of the Grampians; these still hold fair populations of Merlin, but in some other areas there has been a marked decline in this species. Afforestation renders heather moors unsuitable for Golden Plover immediately, and for Red Grouse within a few years. As a result these species are now scarce in areas where they were formerly abundant. Hen Harriers and Short-eared Owls benefit in the early stages of afforestation, and Black Grouse occupy the forest fringe – as long as adjacent open moorland remains

available. All the other species (eg Skylark, Meadow Pipit, Curlew and Lapwing) breeding on moorlands of this type also occur widely in other habitats. With increasingly intensive management of farmland, however, these species may become progressively more dependent upon moorland breeders for maintenance of their overall numbers.

Grassy moorlands include both the grass-covered sheepwalks of the Border hills and the Ochils, and the grasslands on the lower slopes of the mountains further north and west. These uplands support no species that does not also occur in other types of habitat, and they generally have a low species diversity. Among the most typical birds are Wheatear, Meadow Pipit, Skylark and Curlew. Some examples of wader breeding densities on different moorland types are given in the Table.

Relative abundance of four wader species breeding on moorland in different areas of Scotland. (Data from Reed et al 1983 (1), NCC/T. Reed (2), NCC/Pitkin & Easterbee (3) and RSPB/L. H. Campbell*)

Study area	Average number of territories/km²			
	Golden Plover	Dunlin	Curlew	Lapwing
Caithness (1)	2.0	+	0.76	0.47
Sutherland (2)	1.9	0.75	0.21	0.10
Sutherland	0.7	0.6	0.06	0.02
Moray	1.6	0.13	0.34	0.39
E Perth	0.59	0	4.9	1.15
Lanark	1.07	0.08	2.35	2.26
S. Scotland (3) (several counties)	2.1	0	3.1	0.73

* RSPB study areas were selected on the basis of anticipated high total wader populations.

Provided that suitable crags or trees are available for nesting, raptors such as Golden Eagle, Peregrine, Buzzard and Kestrel, and scavengers such as Raven and Carrion/Hooded Crow, will hunt or forage over both heather and grassy moorlands. For most of these species the available food supply exceeds demand, but in some areas a decrease in sheep carrion, resulting from afforestation and improved sheep management, has already caused a serious decline in Raven numbers.

BOGS

Large areas of upland Scotland are boggy in character. In some places, such as Rannoch Moor and parts of Sutherland, deep blanket bog covers many hectares of flat or gently undulating ground up to around 1,000 m asl, often forming a mosaic of small lochans and peaty pools fringed with cotton grass interspersed with sodden patches of *Sphagnum* and slightly drier humps with heather, cross-leaved heath, purple moor grass and deer sedge. In other areas, for example Wester Ross and many of the Hebridean islands, similar boglands lie in basins scattered among the hills. Although more limited in individual extent, they too cover a considerable area in total. And in the southwest there are extensive boggy flows among the Kirkcudbright and Wigtown hills.

The bird populations of areas such as these are seldom large but are sometimes quite varied and include several species of very limited distribution, for example Greenshank and Wood Sandpiper, while the Arctic Skua nests on some Caithness flows. These upland bogs are the traditional breeding-ground of the native Greylag Goose; they also support Teal and Wigeon, and considerable numbers of Golden Plover, Dunlin, Curlew and Redshank. Where there are abundant *dubh lochans* Red-throated Divers are often present, while the very local Common Scoter favours a somewhat similar type of habitat (see also Chapter 5). Other species often breeding on upland bogs include Common and Black-headed Gulls, but these, like many of the waders, are widely distributed in other habitats too. Some boglands at low altitudes are important as feeding and/or roosting areas for geese; this is especially so in the case of Greenland White-fronted Geese.

Until quite recently physical difficulties, associated with inaccessibility and low potential productivity, have protected most upland bogs from development for commercial purposes. Peat-cutting for domestic fuel has altered the surface form and vegetation of some areas, especially on the islands, but such influences have generally been only local in their impact and the great expanses of blanket bog have remained virtually undisturbed. This situation is now changing. Powerful modern equipment has made it possible to drain many such areas well enough to permit afforestation with wet-tolerant species, while in other areas peat extraction on a commercial scale is already taking place or has been proposed – and is likely to be followed by reclamation for agricultural use. Since 1950 about 30,000 ha of conifers have been planted in Sutherland and Caithness, and afforestation is continuing rapidly, with Lodgepole pine often the dominant species. Much of the more recent planting has been on former peat bog, as for example between Lairg and Altnaharra; new techniques, including deep-ploughing and aerial fertiliser application, have only recently made practicable the planting of such areas.

Changes in the status and distribution of bird species inevitably follow such major habitat change, and will continue to occur as the new forests mature. Most of the species benefiting from the newly-established woodlands are already widely distributed in the country as a whole, whereas several of those being displaced from the boglands occur in few other parts of Scotland, are present in only small numbers, and require relatively large areas of undisturbed habitat for successful breeding. The identification of those flows and bogs of major importance for such species is clearly a matter of some urgency in order that adequate habitat protection can be ensured before change has progressed too far.

MOUNTAINS

Some 14% of Scotland's land surface is more than 500 m asl; summits rising above 1,000 m are scattered throughout the central and northwest Highlands, with a few outliers on the islands, most notably on Skye (Map 8). By far the largest continuous stretch of montane country is that including the Cairngorms, the Monadhliaths and the Grampians. This area is, in effect, a deeply dissected plateau, much of which lies between the 900 m and 1,000 m contours. The relatively gentle upper slopes of this plateau support the majority of our breeding Dotterels, while the high corries of the Cairngorms, with their jumbled rocks and late-lying snow, hold most of the very small population of breeding Snow Buntings. The Ptarmigan, the only other regularly-breeding species strongly associated with arctic-alpine

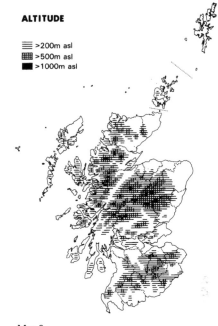

ALTITUDE

≡ >200m asl
⊞ >500m asl
■ >1000m asl

Map 8

habitat, is more widely distributed throughout the Highlands, occurring in exposed sites in the far north at altitudes of down to c200 m asl and also breeding, though in only small numbers, on Skye and Mull.

The more westerly mountains are in general more sheer and less rounded than those of the central Highlands, with many steep scree slopes and peaked or narrowly ridged summits. They consequently have little habitat suitable for species such as Dunlin and Golden Plover which occur at altitudes of up to about 1,000 m asl further east. The peaks of Rhum have a special claim to fame, however, in the vast numbers of Manx Shearwaters that breed high on their slopes. Golden Eagles and Peregrines are widely distributed in most mountainous areas, although in some districts their numbers may be limited by a shortage of suitable nesting sites. Neither species is, however, confined to the high tops; a few Golden Eagles nest on sea-cliffs or in trees at relatively low altitudes, while Peregrines regularly occupy coastal as well as inland cliffs.

In the last two or three decades access to the mountains has become progressively easier. New estate roads give deer stalking parties vehicular access to areas which could formerly be reached only on foot or pony, while chairlifts, initially provided for the use of skiers, are used in summer to transport thousands of people to the summits of Cairn Gorm and the Cairnwell. Over the years increasing concern has been expressed by conservationists regarding the impact of such developments on mountain birds and other wildlife, especially in respect of the Cairngorms (eg Watson *et al* 1970; Curry-Lindahl *et al* 1982), and each new proposal for additional skiing facilities now gives rise to controversy. The most visible result of heavy visitor pressure on the mountains is damage to the vegetation, followed by progressively increasing erosion; the impact on birds is less obvious but is nevertheless considerable. With many bird species disturbance at the nest readily leads to breeding failure, and for mountain-top breeders the risk of loss through the chilling of eggs or young chicks is especially high. Perhaps even more serious a threat comes from the 'new' bird inhabitants of the summits – crows and gulls attracted to the more popular areas by the easy pickings available in the form of picnic remains. Both these species will quickly take unattended eggs. Even the ski-lifts themselves present a hazard – many Ptarmigan are known to have been killed through flying into cables. Indeed the Ptarmigan population has declined drastically in the vicinity of the Coire Cas ski slopes on Cairn Gorm (Watson 1981).

The increased ease of access resulting from new roads and tracks into the glens and up mountainsides has naturally resulted in more people penetrating further into what were once virtually 'wilderness' areas. Although few hill-walkers and climbers would wish deliberately to disturb an incubating Golden Eagle, such disturbance inevitably occurs since eagles are wary enough to leave the nest while intruders are still a considerable distance away. Even today large areas of the Highlands are accessible only on foot – and it is to be hoped that most of them will remain that way. In the past there have been proposals to drive roads through Glen Tilt and across Rannoch Moor – both fortunately abandoned largely due to the prohibitive costs involved – and more recently plans for a rack-railway up Ben Wyvis and a chairlift from the A9 to the summit plateau at Drumochter. Such developments present a threat not only to the breeding birds of the high tops but also to the scenic and wilderness quality of our mountains – features which are of economic significance in terms of tourism as well as of conservation importance.

5: Freshwater habitats

The freshwater habitats considered here fall into three main groups: standing open water, running water, and fens. Scotland's 157,000 ha of inland water include a wide diversity of static freshwaters, ranging from small peaty hill lochans to shallow nutrient-rich lowland lochs. The former are particularly important for upland breeders such as Red-throated Diver and Greenshank, and the latter for nesting grebes and ducks and for wintering wildfowl. Few standing waters in Scotland are entirely man-made but a surprisingly high proportion of lochs have been deepened and enlarged by damming. Gravel pits are scarce.

Running waters include only a comparatively small stretch of canal but an abundance of rivers and small streams or burns. Most of the principal rivers are relatively fast-flowing, and of the 50,000 km graded according to water quality only 200 km is classed as grossly polluted and a further 300 km as 'poor quality'. Although many of the surviving fens (freshwater marshes) are associated with shallow standing waters, the largest continuous expanse of fenland, the Insh Marshes, lies on the flood-plain of the River Spey. Such areas are important for breeding waders and also for breeding and wintering wildfowl; they also often support rails and good numbers of small breeding passerines. Changing agricultural practices in the catchment areas and increasing recreational use have affected many lochs and rivers in recent years. Drainage to permit cultivation has decreased the area of fen, especially in the lowlands.

More detailed consideration of waters important to wildfowl (Map 9) can be found in Owen *et al* 1986; Fuller (1982) deals with a wider range of freshwaters and gives further references; Allison *et al* (1974) gives a full account of Loch Leven.

STANDING OPEN WATER

Few districts of Scotland are entirely devoid of standing water, although lochs are relatively scarce in the northeast and the Borders. In terms of the bird populations they support, there are major differences between the highland lochs and most of those in the lowlands. Highland lochs are in general either shallow and peaty or deep, cold and clear, according to whether they are situated in the uplands or lie along one of the many steep glaciated valleys. Both types are poor in nutrients and have little submerged vegetation, but all support invertebrates and most contain fish. The very large, deep lochs, eg Loch Tay and most of the hydro-electric dams, are of little ornithological importance, often holding only small numbers of saw-billed ducks and Mallard, but the remoter upland lochs provide breeding habitat for divers, and their shores are valuable feeding grounds for waders such as Greenshank and Dunlin. The two divers occupy contrasting types of loch, the Red-throated nesting by small lochans and flying quite long distances to feed, whereas the Black-throated both breeds and feeds on larger waters, often choosing to nest on an islet. Moorland-breeding wildfowl, such as Wigeon and the native Greylag Geese, take their young to upland lochs soon after hatching and the very small breeding population of Common Scoter is also associated with similar habitat. Common Gulls often nest by lochs with stony shores, while Black-headed Gull colonies on upland lochs may produce marked changes in the surrounding vegetation, as the gulls transfer nutrients from low-ground feeding areas to the impoverished bogs by the hill lochs.

Lochs in areas where relatively calcareous rock occurs are visibly richer in both plant and bird life than those on acid moorland. Patches of reeds (*Phragmites*) replace the

Map 9. Freshwater sites of national or regional importance

1	Loch Spiggie	14	Kilconquhar Loch
2	Loch Harray	15	Duddingston Loch
3	Lochs Heilen & St John	16	Gladhouse Reservoir
4	Loch Eye	17	West Water Reservoir
5	Loch of Strathbeg	18	Loch Ken
6	Loch of Skene	19	Castle & Mochrum Lochs
7	Loch of Lintrathen	20	Lochwinnoch
8	Lochs Rescobie & Balgavies	21	Loch Lomond
9	Dunkeld/Blairgowrie Lochs	22	Insh Marshes & Loch Insh
10	Drummond Pond	23	Loch Bee
11	Dupplin Loch	24	Loch Druidibeg
12	Carsebreck	25	Loch Hallan
13	Loch Leven		

(Wetlands of international importance are listed in Appendix 4)

sparse stands of sedges and rafts of white water-lily typical of the latter, and these provide nesting cover for Slavonian and Little Grebes. Both the variety and the numbers of breeding ducks are greater; in addition to Teal, Wigeon and Mallard, Tufted Duck and Pintail may be present, and on Speyside Goldeneye now regularly breed in nest-boxes provided for them – though they apparently prefer sites near rivers rather than those by lochs.

The shallow and nutrient-rich lowland lochs support by far the greatest species diversity and the largest wildfowl and grebe populations (Table). These lochs hold abundant submerged vegetation and aquatic invertebrates as well as fish, and often have adjoining fen or willow carr which provides valuable nesting cover; islands and densely-vegetated banks are added assets in the breeding season. Up to seven duck species may breed on such lochs, as they do on Loch Leven (Shelduck, Wigeon, Gadwall, Teal, Mallard, Shoveler and Tufted Duck). The very small and local population of Black-necked Grebes and most of the Great Crested Grebes are associated with this type of water. These rich lowland lochs are also the principal freshwater resorts for wintering wildfowl; many of them are surrounded by farmland, which provides feeding grounds for grey geese roosting on the lochs and offers additional feeding opportunities for Whooper Swans and some duck species. Most such waters lie in the central and northeastern lowlands, but there are outliers locally in the Borders and southwest, Easter Ross, Orkney and the Outer Hebrides.

Loch Leven is outstanding in terms of wildfowl species diversity and numbers, with the highest density of breeding Tufted Ducks in Britain and notable wintering populations of geese and ducks. It provides a vivid illustration of the potential effects of nutrient enrichment – currently one of the more serious threats to lochs in farming areas. During the 1960s, run-off of nitrogenous fertilisers from the surrounding farmland, together with phosphates from detergents discharged in industrial and domestic effluents, stimulated particularly heavy algal growth. This was sufficiently dense to reduce both oxygen and light penetration in the water, leading to fish deaths and a loss of submerged plants. These changes seriously affected Loch Leven's famous trout fishery for a few years, and also caused a marked drop in the Mute and Whooper Swan populations. Better effluent treatment and the development of slow-release nitrogenous fertilisers have since improved the condition of the loch, but swan numbers have not returned to their former levels. Similar nutrient-enrichment is occurring in most lowland lochs set amidst farmland, though generally on a less dramatic scale. For some species, notably Shoveler, enrichment may be beneficial, since it increases the crop of small invertebrates feeding on the algae in the surface water, but in the long-term it may render a water unsuitable for predominantly vegetarian species.

Another recent development adversely affecting the ornithological value of some lochs has been their increasing use for recreation. Of the water-based sports, water-skiing and the speed-boating associated with it are the most likely to cause serious disturbance, not only driving birds from a loch but also in some cases producing sufficient wash to dislodge or flood floating nests. Canoeing, sailing and windsurfing present fewer problems, though the first two may result in nest disturbance if vessels approach reedbeds too closely, or people land from them on areas not normally used by humans. Angling, too, can cause problems. Stocking with trout is often accompanied by pike control, which frequently results in birds such as Cormorants and Tufted Ducks being drowned in nets; heavy use by boats in relation to the water area available may drive ducks from a loch; and the careless disposal of damaged fishing-gear can lead to injury and death through birds swallowing hooks or becoming entangled in lines. Recreational pressures such as these can be overcome relatively simply, by co-operation and care. A row of floats to demarcate a no-go zone adjoining

Freshwaters holding nationally (*) or regionally (+) important numbers of more than two wildfowl species at least once (between September and March) during the period 1980/81 to 1983/84 (Data from Wildfowl Counts and SBR; see footnote for qualifying levels)

Site	Mute Swan	Whooper Swan	Greyl. Goose	Pink-f. Goose	Wigeon	Gadwall	Teal	Mallard	Pintail	Shov-eler	Pochard	Tufted Duck	G/E	Goos-ander
L. Leven	*	*	*	*	+	*	*	+		*	*	*	*	*
L. of Strathbeg	*	*	*	*	+		+	+		*	*	*	*	*
L. Eye		*	*		*		*	+	+					
L. Harray	*	*			+						*	*	*	
L. Ken		*	*		+				+	+				
Slains Lochs	+		*	*	+		*							
L. Scarmclate		*	*		+		+	+						
Hoselaw Loch		*	*					+		*				*
L. of Skene		*										+	*	*
L. of Lintrathen		*			+			+						
Drummond Pond	+	*						+						
Carsebreck			*		+			+			+			
Kilconquhar L.										*		+	*	
Castle Loch		*		*										*
L. Lomond					+		*			*				
Carron V. Res.							*	+			+			
L. Spynie	+		*					+						
R. Tweed/Teviot	+	*						+					*	*
Boardhouse L.		+			+			+		*				
Cameron Res.			*	*								*		
Bute Lochs			*		+		+							
L. Watten					+			+				+	*	
L. Ore								+			*	*		

Notes: (1) Qualifying levels for national importance (from WWC 1983–84): Mute Swan – 180; Whooper Swan – 50; Pink-footed Goose – 900; Greylag Goose – 900; Wigeon – 2,000; Gadwall – 50; Teal – 1,000; Mallard – 4,000; Pintail – 250; Shoveler – 90; Pochard – 500; Tufted Duck – 600; Goldeneye – 150; Goosander – 50.

(2) Arbitrarily selected levels for 'regional', ie Scottish, significance *on freshwater sites*: Mute Swan – 100; Wigeon – 500; Teal – 500; Mallard – 1,000; Pintail – 100; Shoveler – 50; Pochard – 300; Tufted Duck – 300

(3) Internationally important numbers of Whooper Swans (>100), Pink-footed Geese (>900) and Greylag Geese (>2,000) occur at many sites (roosts in the case of the geese) and of Pochard (>2,000) on the Loch of Harray – see Appendix 4.

reedbeds important to nesting birds or as a feeding area for wintering birds, care in the timing and carrying-out of pike netting operations, and a more responsible attitude by the minority of anglers who behave inconsiderately – all these steps would benefit the birds yet impose relatively few constraints upon recreational users. Many commercial fish-farms have been established in the last decade and these attract Grey Herons and Kingfishers – and even the occasional Osprey. The first is the most serious potential predator, and Grey Herons are quite often illegally shot although inexpensive deterrent measures have been shown to be effective in minimising predation.

Probably the most important measure to safeguard the waterbird populations of lowland lochs must, however, be the establishment of reserve status over an adequate number of them. This need not prevent other uses being made of a loch (angling and regular, though carefully controlled, wildfowling both take place on Loch Leven) but it does ensure that the interests of the wildlife are given a high priority.

A more recent threat, and one which affects both upland and lowland waters and may prove prohibitively expensive to overcome, is that presented by the imminent implementation, under a directive from the EEC, of the Reservoirs Act 1975. This requires that all water bodies retained by a dam and holding more than 25,000 cubic m (5 million gallons) must be subjected to annual inspection; those failing to satisfy the safety standards must be repaired or reduced in capacity by breaching or lowering of the dam. In the past many hill lochs were created or enlarged by landowners for fishing purposes and the dams retaining these are unlikely to satisfy the requirements of the Act. Inspection costs alone would be prohibitive in the case of isolated hill lochs – often important diver sites – and reconstruction to present-day standards quite out of the question. It remains to be seen whether approaches aimed at obtaining exemption for waters which by virtue of their location and size present no significant threat to human populations will be successful.

There is also currently concern about the possible effects of airborne oxides of sulphur and nitrogen, derived from industrial and domestic combustion processes and deposited either dry or in the form of acid rain. In Sweden acid rain has killed aquatic vegetation, fish and trees – with obvious implications for the associated birdlife. There is evidence that some lochs in the southwest of Scotland are affected, especially in areas where the underlying rock is granite (itself acid in character), and attempts are now being made to assess the scale and extent of the problem. A reduction in polluted emissions by British industry should not only minimise the risk of serious damage in Scotland but also help the situation in Scandinavia, the unfortunate recipient of much of Europe's aerial pollution.

RUNNING WATER
Five of Scotland's six major rivers – the Tweed, Forth, Tay, Dee and Spey – flow east, the only west-flowing river of comparable length being the Clyde. All six pass through relatively flat country in their lower reaches, but most have their headwaters high in the hills and consequently flow rapidly over rocky ground for considerable stretches. The Forth is an exception, meandering through flat land for most of its length. These major rivers and many smaller ones are subject to seasonal flooding in some areas, and the temporary wetlands so formed are locally important for wildfowl and waders. Virtually all support trout in some of their reaches, and the Tweed, Tay, Dee and Spey are famous for their salmon. These larger fish-stocked rivers are the main breeding habitat of the Goosander and also hold fair numbers of Red-breasted Mergansers. The faster-flowing rivers of the hill ground are the chief haunt of Grey Wagtail and Dipper, though both occur also on stony lowland stretches. The Kingfisher is most abundant as a regular breeder in the southwest and on the Clyde and some of its tributaries, occurring only rarely north of the Highland Boundary Fault. Many rivers have shingle banks or islands with nesting Ringed Plovers and Common Terns, even well inland; those with steep sandy banks frequently have nesting Sand Martins. Common Sandpipers and Oystercatchers are widespread, the former especially along upland rivers, and Moorhens breed on lowland backwaters and oxbow lakes.

The fact that only a small proportion of the land area is either industrialised or densely populated means that most of Scotland's rivers and streams are free from serious pollution. Only c200 km, mostly of the Clyde and its tributaries, are now classed as 'grossly polluted' – a big improvement on the situation some 10–15 years ago, when many rivers were subject to much heavier pollution from both industrial and domestic sources. Within the last three decades big improvements have been made in the treatment of domestic sewage and of distillery and other organic industrial wastes, and much more stringent measures to control the discharge of toxic wastes have been introduced. The efforts made to clean up the Clyde were fully rewarded in 1983 when salmon came up the river for the first time for 120 years. Local pollution occasionally results from inadequate sewage treatment, industrial discharges, water seeping from flooded mineworkings, silage effluent, or accidental spillages of oil or agricultural chemicals. A more recent matter for concern in the drier east has been the increasing use of irrigation on farm crops, with consequent lowering of the level and rate of flow in the rivers from which water is drawn. While such local incidents and activities may cause serious mortality among fish they appear seldom to have any widespread effect on bird populations. River and stream straightening and bank clearance, with consequent loss of vegetative cover, are fortunately carried out on a limited scale and present problems only very locally.

Recreational use of rivers, especially for canoeing, causes disturbance in a few areas, but a much more potentially serious problem is that created by the presence of feral mink. Escaped mink have established themselves in many districts, and are presumed to have considerable impact on the waterbird population, although positive evidence of this is difficult to obtain. Angling interests have long presented a threat to species believed to take significant numbers of young trout and salmon, notably the Goosander, Red-breasted Merganser and Cormorant, large numbers of which were formerly shot on the principal salmon rivers. Although these species are now given general protection under the Wildlife and Countryside Act, they can still be killed under licence if such action can be shown to be necessary for the purpose of 'preventing serious damage ... to fisheries'. Despite the fact that research has shown that these birds do not take commercially significant numbers of fish (Mills 1962, 1965) many estates and fishing syndicates continue to have them shot – often without the required licence.

FENS
Freshwater marshes and swamps, here collectively referred to as fens, occur either in association with standing water or on the floodplains of rivers. In both situations the water table is near or above ground level for much of the year and the ground seldom if ever dries out. A sequence of vegetation is characteristic of such sites. Reeds, reed sweet-grass and bogbean are typical of the wettest areas; rushes, sedges, willows and alder predominate on slightly drier ground; and as the plant litter gradually rises clear of the water table colonisation by woodland species starts. In time, if left to itself, a fen will eventually become woodland. Over the years many areas of this type have been drained for agriculture, with varying degrees of success. Most of those surviving would be difficult or prohibitively expensive (even with modern techniques) to bring into cultivation – which is a good thing from the conservationist's point of view.

There are few large freshwater reedbeds in Scotland, and none of the species dependent upon reedbeds in other parts of Britain – Reed Warbler, Bearded Tit, Marsh Harrier and Bittern – breeds here regularly. Nevertheless many lowland lochs have considerable adjoining areas of reed-dominated fen and these are the main habitat of Reed Bunting and Sedge Warbler and are also valuable for Teal, Snipe and Water Rail, and as autumn roosts for species such as Sand Martin, Sedge Warbler, Pied Wagtail and Starling. Among the largest fens of this type are those at the Loch of Strathbeg, Lochs Davan and Kinord, and Lochwinnoch, but there are also many smaller areas which are locally important. The much more varied and greater expanse of fenland at the Insh Marshes is of particular significance for its breeding waders and wintering wildfowl: 11 species of wader (including the very local Wood Sandpiper) are recorded there regularly, up to 8 species of duck visit, and a sizeable herd of Whooper Swans winters on the marsh and Loch Insh itself. Somewhat similar habitat also occurs in association with some shallow reservoirs.

In the past proposals have been put forward for draining both the Insh Marshes and the Barr Loch at Lochwinnoch. In the case of the Insh Marshes they were not implemented owing to the costs involved and the Spey's reputation for frequent and uncontrollable flooding, and at the Barr Loch the system of sluices and lades installed functioned effectively for only a short time, after which the area reverted to marsh. Both sites are now RSPB reserves. Management work at the Insh Marshes is creating more open pools and shallow muddy areas to increase the site's attraction for wetland birds, and at Lochwinnoch the water is maintained at a level considered to be optimum for the waterbirds inhabiting the reserve. Similar works carried out at Morton Lochs NNR involved the formation of new islands.

Apart from the possibility of drainage, the principal factor resulting in vegetational change in fenland is the natural one of progressive drying-out and colonisation by woodland. This is normally a very slow process but it may be accelerated where farmland run-off enriches the water, as at Stormont Loch, or where large colonies of Black-headed Gulls become established, as at Kinnordy Loch.

6: Farmland

The period since the end of World War II has seen the most rapid and widespread changes ever experienced in British agriculture. These have involved not only the much-publicised use of pesticides and removal of hedges but also a big increase in fertiliser use and the abandonment of traditional practices which had been little altered for generations or even centuries. Productivity per acre, in terms of both crops and livestock, and capital investment in buildings and machinery have increased greatly, while the labour force has declined by c60%. Financial incentives, in the form of government grants, and powerful equipment have made it feasible to carry out drainage, scrub clearance and other land reclamation work which would earlier have been impracticable, while the application of technology has greatly reduced the diversity of plants and insects present among both crops and grassland (Table a). These changes – and other less obvious ones such as farm building design and livestock management techniques – have inevitably affected the farmland habitats available to birds. Although much has been written about relationships between farming and wildlife in general, there is little dealing primarily with birds or with Scotland, and farmland is not covered in Fuller's *Bird Habitats in Britain* (1982). I have therefore considered this important habitat in rather more detail than the others.

Scotland currently has c1,200,000 ha of arable land and c563,000 ha of 'improved' permanent pasture, together amounting to some 23% of the total land area. Nearly 30% of the arable and permanent pasture area is in categories A+, A, and B+ of the Department of Agriculture and Fisheries, Scotland, land quality classification, ie it is inherently capable of growing a wide range of crops (though climatic conditions may effectively limit the options). The 174,000 ha of highest quality land (A+ and A) are largely on the east coast (Map 10) and it is here that many of the intensive farming practices introduced since 1950 have

had their most obvious impact. Change came more slowly on the poorer land in the straths and glens and on the islands, but here too there have been changes with far-reaching effects. Not all agricultural developments have been detrimental to birds but some have resulted in drastic population declines in species largely, if not wholly, dependent upon

(a) *Some habitat changes in lowland agricultural areas between the 1940s and 1970s, expressed as a percentage of the total area (1) or length (2) of each habitat type present in the 1940s (Data from Langdale-Brown et al 1980)*

| | Habitat type (see footnote for definitions) | | | | |
	Wetland (1)	Unimproved grassland (1)	Dwarf shrub heath (1)	Treeline (2)	Hedgerow (2)
Borders	–	−27.5	−76.5	−26.0	−20.5
Lothian	–	–	–	−11.5	−17.5
Fife	–	–	−96.2	−23.8	−32.5
Central	–	–	–	−21.0	−33.0
Tayside	–	–	–	−10.4	−28.7
Dumfries & Galloway	−12.7	–	−89.9	−17.5	−23.1
Strathclyde	−20.0	–	−17.0	–	−21.1

Notes: (a) Results showing changes of less than 10% are omitted.
(b) Habitat types are defined as follows:
Wetland – seasonally or permanently waterlogged areas eg fens & marshes.
Unimproved grassland – no signs of drainage or other surface treatments.
Dwarf shrub heath – *Calluna, Empetrum, Erica & Vaccinium* dominated areas.
Treeline – lines of trees along field boundaries without hedges.
Hedgerow – shrub or tree species forming low barrier between fields, sometimes with occasional tall trees.
(c) The actual areas of a habitat type lost will vary – in some cases widely – between regions showing a similar percentage decrease, as the 1940s totals differed considerably.
(d) Substantial further decreases have doubtless occurred during the early 1980s, probably most notably in unimproved grassland and hedgerows.

farmland; those most seriously affected to date have been the Corncrake and Corn Bunting, but there is increasing concern regarding those waders which breed mainly on damp rough grazing.

Many of these changes are discussed in *Farming and Wildlife* (Mellanby 1981) and in reports of various farming and wildlife conferences, but it should be noted that these publications are mainly concerned with the situation in England, which differs in many ways from that in Scotland. The proportion of Scotland's land area under intensive arable cultivation is comparatively small, hedges have never existed in many districts, and managed water-meadows are unknown; in consequence several of the agricultural developments that have caused great concern elsewhere in Britain are of only very local significance here. The Common Birds Census (CBC) farmland data, which are representative only of conditions in lowland England, do not reflect population changes on British farmland as a whole (Fuller *et al* 1985).

Although they are to a large extent inter-related, it seems best, in the interests of clarity, to consider recent developments in crop production and in grassland management and livestock handling separately. The final section of this chapter discusses conflicts – both real and imagined – between birds and agricultural interests. The question of the reduction in rough grazing is dealt with in the chapter on uplands, since most of the land of this type is in moorland areas. Virtually all of the c450,000 ha lost from agriculture to forestry in 1960–82 were in the uplands.

The considerable pressures on farmers to maximise the productivity of every hectare of land, and the attraction of the elevated prices resulting from farm price support, eventually led to surpluses over market demand, and thence to the recent introduction of quotas (which limit an individual farmer's production) by the European Economic Community. It may be that the next decade will see a reversal towards lower-cost farming, with reduced applications of pesticides and fertilisers, and less effort directed towards bringing additional land into cultivation. If this does, in fact, happen it is likely to be of general benefit to both birds and other wildlife, although more intensive management might result locally.

CROP PRODUCTION

Since 1950 there have been major changes in the relative proportions of the crops grown, in the cultivation and harvesting techniques used, and in the control of crop pests and diseases. The resultant changes in the quantity and quality of the food supplies available on arable land have affected a wide variety of bird species. For some, notably the geese, the changes have been largely beneficial; for some, such as the Lapwing, their effects have been insidious rather than obvious; for a few they have apparently been at least temporarily disastrous. It is the last group, which includes the Corncrake, that usually receives the widest publicity.

The cropping change of most significance from the ornithological point of view has been the vast expansion in the area of barley grown; this increased by more than 525% between 1950 and 1980, when it totalled c450,000 ha. The oats area dropped by c90% in the same period while wheat remained fairly constant. Barley must be fully ripe when harvested, and at this stage bad weather or rough handling can cause many heads to break off; each head carries about 25 seed grains, so unfavourable conditions at harvesting may result in large amounts of grain being left on the stubbles. Many granivorous birds take advantage of this abundant food supply, among them the grey geese. During the 1960s the autumn distribution of Pink-footed and Greylag Geese showed a progressive concentration in arable east-central Scotland, where the barley area was expanding particularly rapidly, a trend which has since reversed as barley growing has spread to other areas. Storm-laid crops also occasionally attract very large numbers to individual farms or fields, as happened in Easter Ross in November 1981 when some 38,000 Greylags gathered on a large field which had remained unharvested owing to wet weather. It was during the period when great concentrations of geese spent weeks or even months within a relatively small area, gleaning the stubbles and potato fields then moving onto grass, that farmers became most concerned about the extent to which they might be competing with domestic stock – a slightly ironic situation since it was the farmers' change in cropping practice which had led to the problem.

Now, in the mid 1980s, there is a swing towards autumn sowing of barley (in the past, wheat was the only cereal regularly sown in autumn), which not only means that the crop is likely to be harvested earlier and with less grain loss but also that the stubble is available to birds for only a brief period before ploughing. Recent years have also seen the introduction of oilseed rape – whose brilliant yellow flowers now brighten the countryside in early summer; it too is autumn-sown, necessitating rapid cultivation after harvesting of the previous crop, with consequent reduction in stubble-feeding opportunities (though Whooper Swans have already been observed feeding on rape in Perth – J. Kirk). Concurrent with these developments there has been a change in the autumn distribution of the grey geese, which appear

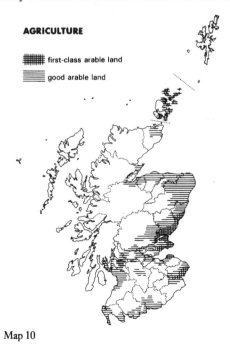

AGRICULTURE

▦ first-class arable land

≡ good arable land

Map 10

to be reverting to their former practice of dispersing and moving southwards quite soon after reaching their traditional arrival points in east-central and northeast Scotland.

While the increase in barley area has made more grain available in stubble fields for at least a short period in autumn, changes in harvesting methods have greatly reduced the availability of cereal seeds later in the winter. Until the use of the combine harvester became universal in the 1960s cereal production involved a whole series of operations: cutting and binding into sheaves, stooking in the field, carting and stacking, and finally threshing – which was usually carried out at intervals throughout the winter. At each stage grain was shaken off or spilled and became accessible to birds – and to small mammals. Red Grouse used to feed on oat stooks and I have even seen a cock Capercaillie feeding on a field stack of oats; Rooks, House Sparrows and Corn Buntings were among the species foraging in and around the stackyard; and the mice and rats living in the stacks provided a ready winter food supply for Barn Owls. Now these traditional harvesting methods survive only in the remoter glens and on the islands, and the bulk of the cereal crop is cut, threshed and transferred to indoor storage in a single operation. I know of no studies that have attempted to assess the overall effects of these changes on farmland birds, but they must surely have been considerable, and possibly contributed in no small measure to the decline of the Barn Owl.

The coming of the combine harvester was also a major factor in initiating hedge-removal. For maximum economy large, expensive machines must be as fully utilised as possible – and time spent turning or moving between fields is unproductive. So fields were enlarged and gates widened to accommodate the growing number of large implements. In Scotland, where there was no tradition of hedge-laying and many field boundaries were already drystone walls or wire fences, the loss of habitat through field enlargement has not been nearly as great as in England. In arable areas considerable stretches of hedgerow have certainly been removed, but these were often of stunted or virtually pollarded hawthorns, offering little more than limited nesting cover and a sparse berry crop. In the more pastoral western half of the country, and along roadsides in the east, many hedges still remain. Some are grazed so bare at the base as to be of little value for either wildlife or stock shelter but others have the wide, densely-vegetated ground layer that provides valuable cover not only for nesting birds but also for important prey items such as insects and small mammals.

Much has been written about the value of hedges for wildlife in general and birds in particular (eg Pollard, Hooper & Moore 1974; Arnold 1983; Osborne 1984), but virtually all the work described has been carried out in conditions very different from those existing in Scotland. However, recent work in lowland Scotland (NCC/P. Shaw; da Prato 1985) suggests that the presence of hedgerow trees more than doubles the bird numbers present per km of hedge, irrespective of hedge condition, while roughly twice as many birds are likely to be seen along an 'unmanaged' hedge as in one which is well-trimmed (Table b). CBC results indicate that farmland is a second-best habitat for many species; when the population is high, hedge nesting sites are occupied but following a hard winter the decline in numbers nesting in hedges is greater than in adjoining woodland. Where any

(b) The effect of hedgerow management on bird numbers in lowland Scotland (Data from NCC/P. Shaw)

| Hedgerow trees | Numbers of birds sighted per 1 km of hedge | |
	Hedge clipped/pollarded	Hedge overgrown/remnant
present	125.4	188.2
absent	43.0	87.6

sort of woodland is scarce – and there are still relatively few farmland areas in Scotland where this is the case – then good hedges provide useful habitat, but even quite small areas of scrub woodland are likely to hold greater numbers and diversity of birds than an equivalent area of hedge. The presence of a shrub layer is more important than woodland area in encouraging species diversity (Table c).

(c) The effect of shrub layer density on species diversity in small woods on farmland in lowland Scotland (Data from NCC/P. Shaw)

| Size of wood | Average number of species holding territory | |
	Shrub layer dense	Shrub layer sparse/absent
less than 0.5 ha	6.9	4.0
0.5 to 1.5 ha	9.6	6.8

One advantage to the farmer in replacing hedges by fences is the greater ease of cultivation right to the edge of the field. Not even content with the small additional area of land so gained, some farmers go on to spray the ground along the fenceline with a herbicide, destroying not only any 'noxious weeds' that may be present but also the rough vegetation which provides overwintering cover for insect species which are actually beneficial to agriculture.

The progressively expanding use of herbicides and other pesticides – of which insecticides and fungicides have the most serious implications for birds – has been one of the most emotive and widely debated recent developments in agriculture. The effects, most notably on the breeding performance of birds at the top of the food chain, of persistent organochlorine insecticides accumulated in the body fat are well documented. The whole subject of pesticide use was thoroughly covered by Mellanby (1967) and subsequently brought up-to-date by the same author in 1981. It suffices here to say that much spraying with insecticides is, regrettably, carried out as a preventive measure rather than when there is a proven need. The careless use of organophosphorus seed-dressings still occasionally causes mortality among birds; the routine spraying of growing cereals with fungicides (as well as insecticides) is widespread; and herbicides are commonly used not only to ensure weed-free crops but also to kill off all vegetation prior to the surface cultivation which has to some extent replaced ploughing.

Herbicides present a less obvious threat to birds than do insecticides, but recent work by the Game Conservancy has indicated that cornfield weeds such as knotgrass and the mayweeds are important for Partridges, not only because they provide food for the adult birds but also because they support insects essential for the survival of the chicks. Leaving untreated even as little as one spray-width around the edge of a field may make the difference between survival and death for young chicks in wet weather, simply by enabling them to find sufficient food with minimal expenditure of energy. This work has also shown that the spraying of cereal crops with fungicides may markedly affect insect populations, so reducing the food available to birds. Corn Buntings, which raise their young on insect foods, have probably suffered in the same way as Partridges. Pest-free cereal monocultures may appear desirable to the farmer, but they are clearly far from ideal for birds.

GRASSLAND AND LIVESTOCK MANAGEMENT

Both grassland management and livestock handling have also undergone many changes over the last three decades, most of them involving intensification of management. Grazing is more closely controlled to reduce wastage and more fertiliser is used to increase yields. Weed control and the use of selected seed mixtures have decreased the diversity of plant species – and consequently insects – present. The area of grassland (mainly short-term leys, ie sown grassland intended primarily for harvesting, as opposed to grazing) cut for preservation as winter feed has increased by 60% to over 350,000 ha and harvesting takes place earlier and is more efficient, while preservation methods now frequently involve storage in a form which provides little scope for foraging birds. Higher densities of cattle and sheep per unit area have been achieved by these improvements in grassland management, together with better housing, handling methods and disease control, and the mechanisation of many tasks which were formerly labour intensive. Cattle numbers increased by c43% between 1951 and 1981, but – apart from some expansion in marginal and upland districts – the land area devoted to cattle has not increased. Sheep numbers in 1981 were c14% higher than in 1951, yet the area of sheepwalk has decreased with the afforestation of upland grazings. Several of these changes have implications for birds, though probably few have had a major impact other than locally.

The change most significantly affecting birds has been the progressively earlier cutting of grass for preservation as silage or hay. In the past hay was seldom cut before mid July, by which time the young of ground-nesting birds were fledged and mobile. Today the first cut for silage is usually made in mid-late May and hay is often mown in June. Mowing not only exposes nests, causing desertion, but also sometimes kills incubating birds and newly-fledged young which are unable to escape. The species affected most seriously by this change is the Corncrake, which was widespread even in lowland Scotland prior to World War II but is now confined to the islands and a few glens where late-cut hay meadows still provide the cover necessary for breeding success. Although difficult to prove in retrospect, the Corncrake's decline has been attributed largely to these changes in grassland management. Other species must also have been affected; for example Iain Gibson has found in Ayrshire that Yellow Wagtails have lower breeding success in fields cut for silage than in hayfields.

Preservation as silage involves storage in covered pits or tower silos, and even when fed to stock on open ground the amorphous wads of 'pickled' grass are not attractive to birds. Hay, which is a potential source of winter food for seed-eaters, is now often protected by plastic wrappings or stored in big tightly-packed bales which make access by birds

difficult. Even grazed fields have in many cases become less favourable to breeding birds, with closer grazing reducing the incidence of cover-providing tussocks, while more thorough weed control reduces the incidence of plants such as thistles, which are valuable for seed-eaters in winter. Damp pastures with scattered tufts of rushes, which formerly provided good breeding habitat for several wader species, have become progressively scarcer as new techniques have made effective drainage practicable. This last type of habitat loss is currently causing particular concern, as recent surveys (Galbraith & Furness 1983, 1984) have shown that wader populations on farmland have decreased in many areas, most markedly so where extensive drainage has been carried out.

The increasing use of fertilisers throughout the summer has probably had little direct effect on birds, although spraying with slurry (semi-liquid manure) may perhaps cause some chick mortality. It is the side-effects of this development – the nutrient-enrichment of streams and lochs resulting from the run-off of nitrogen-rich drainage water – that are most likely to affect birds through habitat change (see Chapter 5).

Improvements in upland sheep management have included the introduction of more convenient – and hence more widely used – supplementary winter feeding in the form of cubes or pellets, advances in disease prevention and control, and, to a more limited extent, indoor winter housing of stock. As a result mortality among ewes has declined and the survival rate of lambs has improved. Sheep carrion is consequently scarcer and in some hill sheep areas this has reduced the food resources available to species such as the Golden Eagle, Raven and Carrion Crow. The conversion of upland sheepwalk to forest has, of course, had a similar effect.

Most of the other marked changes have taken place around the farm steading. Traditional stone and slate buildings, which offered nesting sites for a variety of species, are progressively being replaced by wide-span concrete and asbestos sheds, which have comparatively few holes and sheltered ledges suitable for nest-building. Swallows, House Martins and Barn Owls have probably suffered most from this loss, but carefully-sited nesting shelves have proved an alternative acceptable to Barn Owls. The equally traditional adjacent midden or dung-heap, with its seething summer insect population, has virtually vanished from the farming scene. On most stock farms today manure either accumulates in cattle courts under cover until it is taken out to the fields, or else is swilled through slatted floors and pumped as slurry into storage tanks and onto grassland. These changes, and the very much more hygienic conditions under which livestock are now kept, must have greatly reduced the supply of insects available around farm steadings – and have perhaps contributed to the apparent decline in both Swallows and House Martins in some areas.

There have been losses too in the availability of water. Concreted yards lack the muddy puddles needed by hirundines for nest construction. The piping of drains and the provision of water troughs instead of streamside drinking places for cattle have reduced potential sources of mud and feeding sites for waders. And the silting-up or infilling of redundant mill-dams (and curling ponds) has further reduced the open water areas available on or near farmland.

Some recent developments have certainly benefited a few

species, however. Houses and steadings abandoned due to farm amalgamations are rapidly colonised by Jackdaws and Starlings – and on Islay are now being occupied by Choughs – while House Sparrows sometimes take up permanent residence in grain stores. On the whole, however, life has been made more difficult for most farmland birds as a result of the changes which have occurred in the last 30 years.

CONFLICT OR COMPROMISE?

From an agricultural viewpoint, birds inhabiting farmland fall into two main categories: those that are injurious to farmers' interests and those that are not. From an ornithological viewpoint they may be grouped according to whether or not they are adversely affected by current farming practices. All too often claims of actual or anticipated damage or loss are exaggerated – by both sides. The still accelerating rate of agricultural change makes the need for better understanding between farmers and conservationists a matter of some urgency. The recent establishment of Farming (Forestry) and Wildlife Advisory groups in many parts of the country is an encouraging sign that there is a growing awareness of the value of seeking possibilities for compromise, rather than pursuing a pattern of confrontation.

The species a farmer regards as in conflict with his interests include the obvious crop-damagers, of which by far the most important economically is the Woodpigeon. Pigeons consume large amounts of clover and grain, and in severe weather take crops exposed above snow level, such as kale and swedes. The recently-introduced oilseed rape crop is also liable to suffer winter damage, while peas and beans are often taken at an early stage of growth. Control of Woodpigeons is not easy – but nobody is likely to object to a farmer doing all that the law allows to decrease the numbers on his farm. The situation is not always so clear cut however. Many farmers consider that Rooks do significant damage and regularly shoot out rookeries on their land, yet Rooks are known to eat large numbers of invertebrates injurious to crops – and the shot-out rookery is likely to be rapidly recolonised by birds from the surrounding area. With both these species food shortage at certain times of year (late winter for the Woodpigeon and early summer for the Rook) is probably as effective in limiting numbers as any artificial form of control.

There are also divergent views about geese. An annual outcry about goose damage has become almost traditional in areas where large numbers winter, and there is little doubt that some farmers suffer financial loss, especially as a result of the puddling caused by geese feeding on young cereals in wet weather and the reduction or delaying of the early bite of spring grass. However, few farmers take active steps to scare off the geese when they first arrive in a field, and once use has been established the birds are difficult to discourage. A potentially more serious goose problem seems imminent in some areas as a result of the rapid recent spread of breeding Canada Geese; being largely sedentary these geese feed throughout much of the year in the same restricted area. Concentrations of moulting birds are particularly likely to cause problems, as a flock of moulting Mute Swans has already done at one site.

Several of the other 'pest' species (which may be killed by authorised persons at any time), such as the Carrion Crow and Great Black-backed Gull, take advantage of the opportunities afforded by sickly lambs or sheep unable to move for one reason or another, and attack them while they are still alive. Although improving standards of shepherding on hill farms should help to limit this problem, the fact that crow numbers are increasing with the decline in gamekeeping means that it is unlikely to vanish altogether. Claims are sometimes made that Golden Eagles are taking live lambs – and there is evidence that the occasional 'rogue' bird does so, though most such cases – as with the crows – are likely to involve sick or weakly animals. With a Schedule 1 specially-protected species like an eagle very good proof that the charge is well-founded must clearly be a prerequisite of any authorisation for retaliatory action.

Birds, especially gulls and Starlings, are sometimes held responsible for the spread of disease among domestic livestock, possibly the greatest potential risk being the spread of *Salmonella* (a bacterium which can cause enteritis in calves and food-poisoning in man) by gulls feeding on garbage tips and then resting on grassland used by cattle. Contamination of domestic water supplies has certainly occurred where reservoirs have been used by roosting gulls.

Turning to the opposite side of the picture, a continuation of the present policies of maximising crop production by routine pesticide spraying and the bringing into cultivation of all potentially suitable (and some barely suitable) land may be expected to produce a continuing decline in farmland bird populations. A more selective and careful use of pesticides, treating crops only when necessary and avoiding field boundary verges, would benefit many species, since the young of most granivorous birds as well as insectivores are reared on insects. Retaining hedges and any wasteground carrying a semi-natural mixed flora, and refraining from draining damp low-lying land unless it is certain that drainage will be sufficiently effective in the long-term to justify this action, would also make a valuable contribution towards conservation. More positive steps to improve the situation could include rehabilitation of mill-dams and similar ponds and the planting-up of areas which are awkward to cultivate. All these suggestions involve financial considerations, however, and the prospects for the kind of farming changes that ornithologists would wish to see are likely to depend to a large extent upon the Common Agricultural Policy and the availability of subsidies and grants for different types of agricultural development – a matter currently undergoing review.

7: Urban-suburban and man-made habitats

In addition to the categories already dealt with, there are various habitats strongly dominated by man yet widely used by birds. These include industrial buildings – and even oil installations – as well as the more obviously hospitable city parks and gardens. Motorways, bridges, rubbish tips and pipelines are among the other man-made habitats utilised by the more adaptable and opportunist species.

Suburban gardens often hold much higher breeding densities of species such as Blackbird, Blue Tit and Robin than do more natural habitats. As the British Trust for Ornithology's Garden Birds Survey has shown, a very wide range of species visits gardens in winter and benefits from the food and water provided by man. Among the most recent additions to the list of regular garden visitors is the Siskin, which now appears at peanut feeders in many areas, especially in severe weather. In Scotland, Magpie distribution is concentrated around the larger conurbations in the Central Lowlands and northeast, while reports of the Hawfinch are also most frequent in suburban and park-like situations. A less welcome feature of recent years has been the increasing numbers of Carrion Crows scavenging in towns; these, with Jackdaws, are probably responsible for many of the losses of garden bird nests not attributable to domestic cats. Both Black-headed and Herring Gulls are also more frequent visitors to inland towns than formerly, and appear to have a marked preference for sitting on the highest roofs available in a particular area.

Feral pigeons often reach pest proportions in town centres, while winter roosts of Starlings create noise and dirt problems locally, for example on some of Glasgow's principal streets, where they are presumably taking advantage of the warmth provided by the street lighting. Starlings also roost in large numbers on other man-made structures such as bridges, while Pied Wagtails have been recorded roosting

in hundreds in greenhouses, factories and among beer bottle crates, in each case benefiting from the shelter provided by these unlikely refuges.

Motorways attract not only carrion-eating crows and gulls, which scavenge on the rabbits and birds killed by fast-travelling vehicles, but also Kestrels hunting over the verges. Oystercatchers too seem to be increasingly making use of these stretches of mown grassland – where the chance of chicks surviving to fledging is unfortunately very poor. Pied Wagtails often feed on insects disturbed by the slipstream of the traffic, and seed-eaters glean spilt grain at the kerbside. Inadequately protected layby litter bins attract Black-headed Gulls and Jackdaws, Chaffinches are frequent visitors to picnic places, and in winter migrant Snow Buntings now regularly gather at ski-resort car parks in search of crumbs.

The formation of open rubbish tips, where untreated waste (which nowadays often includes much discarded food) may lie uncovered for several weeks, has been responsible in many areas for attracting large concentrations of gulls – of all ages outwith the breeding season and non-breeders during the summer. Dumps near shellfish processing plants act in a similar way. The discharge, into rivers and the sea, of waste from distilleries and food processing plants is now strictly controlled, so this is no longer a major artificial food source, as it was in the 1960s. Sewage disposal is also more strictly regulated than in the past – and Scotland has never had many sewage farms of the type that act as magnets for both birds and birdwatchers further south.

Many birds use man-made structures as nesting sites, and species such as Swifts, Swallows and House Martins are almost entirely dependent upon them. Among those fairly recently taken over, since they have not long been available, are old war-time 'Mulberry harbours' – used by nesting Cormorants in Loch Ryan – and the sloping stonework faces of dams and the jointing sections of large-diameter hydro-electric pipelines – both used by nesting Common Gulls. Other species have adapted to man-made sites, often as natural ones became fully occupied: Fulmars, Kestrels, Oystercatchers, Herring Gulls and Kittiwakes now nest on buildings, Common Terns have colonised a disused area of Leith Docks, and many species – from Blue Tit to Goldeneye – readily occupy nest-boxes erected for their benefit. North Sea oil installations attract large numbers of tired migrants. The North Sea Bird Club's Bulletin 34, reporting on a 'bird blizzard' on 7 November 1984, quotes observers' reports of '200 to 300 thousand mixed birds massing around the platform … owls kept coming aboard with a captured bird in each foot' and '16 owls, mostly Short-eared, rose silently, hovering above us and below at varying heights. We counted 44 in the air at one time but many more were hidden from sight by the ironwork of the platform. They hunted throughout the day, roosting occasionally'. Oil installations also attract semi-permanent populations of sea-birds – and there is even a record of Starlings breeding on a barge well out in the North Sea.

The Musselburgh lagoons, developed at Cockenzie Power Station for the disposal of waste ash, have provided a valuable, little-disturbed, roost for many waders feeding in that section of the Forth. Artificial water-bodies, both static and flowing, are important bird habitats in many areas. So too are abandoned railways, derelict industrial sites and spoil heaps which have been colonised by vegetation – all of which may provide oases of undisturbed cover, especially in intensively farmed areas. Quarries offer nesting ledges for Peregrines and Kestrels, sandpits hold the biggest colonies of Sand Martins – and where would the Swifts nest if no roofs and towers were available?

The extent to which a species is able to adapt to, and utilise, unnatural food resources and nesting sites may be an increasingly important factor in population maintenance. Those with the most specific range of requirements clearly have poorer prospects than do more adaptable species in a world which is changing as rapidly as ours is today. This is especially true of species dependent upon what are, and may long have been, habitats of limited extent and very local distribution.

8: Developments in bird study

When *The Birds of Scotland* was being written, organised bird-watching had barely begun. Ringing had been carried out since the early 1900s certainly – but only of fledglings or birds caught at observatories or in small wire traps. There had been a few national censuses of selected species – with most of Scotland inadequately covered – and regular counting of wildfowl had just started. Field notes on rarities were beginning to be accepted as a satisfactory alternative to a corpse, though often subject to rejection. Organisations such as the Scottish Ornithologists' Club and the British Trust for Ornithology (both established in the mid 1930s but in suspended animation during World War II) had not yet been able to make much progress in organising the collection and analysis of data. Evelyn Baxter and Leonora Rintoul were consequently dependent largely upon their own observations and the notes of those with whom they corresponded.

Our knowledge and understanding of bird populations, habitat needs and behaviour has since expanded greatly, as a result both of the increasing numbers of birdwatchers and of the introduction of new study approaches. Advances in field techniques and equipment, the application of modern technology in the recording and analysis of results, and the growing opportunities for international, as well as national, co-operation have all contributed towards this progress. Even the writing of this book has been done with the aid of a computer – which has greatly facilitated my task and has enabled me to incorporate data received within only weeks of the publisher's deadline.

It is not possible to deal with these matters fully here, but it seems desirable to include a brief summary of the developments that have taken place, and to provide references for those who wish to explore any aspect more fully. Rather than giving a strictly chronological account, I have chosen to deal with material under five main headings: distribution, numbers and population fluctuations, ecological and biological studies, movements and migration, and co-operative projects. Several of these topics overlap, as will be apparent, but I hope that by grouping them in this way I shall present a reasonably comprehensive overall picture, not just a catalogue of projects and dates.

DISTRIBUTION

'Where can I find it?' is an obvious question for anyone keen to see a particular species, while information on the extent of a species' breeding range, the sites holding large migratory or wintering flocks, and the habitats upon which the various species depend, is important to those interested in conservation. Systematic approaches to obtaining such information started – not surprisingly – with relatively easy species: large and generally colonial nesters, such as the Heron (first censused in 1928), Gannet (1939) and Fulmar (1941), or those which occur only on a limited range of sites, such as the Great Crested Grebe. Wildfowl counts – initiated in 1947 by the British Section of the International Wildfowl Inquiry Committee and run since 1954 by the Wildfowl Trust – were more ambitious in scope, covering many species and involving regular counts at monthly intervals. From small beginnings this scheme gradually grew, building up a bank of data on the whereabouts of many large concentrations of open-water wildfowl during the autumn and winter. The picture presented (Atkinson-Willes 1963; Thom 1969) was by no means complete, however, as many waters remained uncounted and little attention was paid to seaducks, which are notoriously difficult to count even when tide and weather conditions are ideal – let alone on a set date. The introduction in 1967 of a January International Wildfowl Census provided the impetus for observers to visit waters not previously counted, and so filled in some of the blanks in the jigsaw, while since the mid 1970s aerial surveys have added greatly to our knowledge of seaduck distribution. Not all wildfowl spend the daylight hours on open waters, however, and other techniques were needed to assess goose and Whooper Swan populations; these are considered later.

Operation Seafarer, inspired by Bill Bourne, organised by the Seabird Group and carried out in 1969–70, was more ambitious still. It aimed 'to find out precisely where all our seabirds were nesting and to estimate as accurately as possible their present numbers'. One has only to think of the inaccessibility of the St Kilda islands, and the scale of their seabird colonies, to appreciate the enormity of this task, which involved not only visiting many miles of precipitous cliff and countless islands but also attempting to assess the breeding numbers of burrow-nesting species such as the Puffin and of densely-packed colonially-nesting Guillemots and terns. By the end of the project (Cramp *et al* 1974) more was known about the distribution and population sizes of Britain's breeding seabird species than about all but a very few land birds. Work was already in hand, however, to make good the paucity of information on the breeding distribution of even the commonest species.

Planning for the *Atlas of Breeding Birds in Britain and Ireland* had started at the BTO in the mid 1960s and survey work covered five seasons, 1968–72. Recording was on the basis of 10 km grid squares, with status assessed as definitely, probably or possibly breeding. This massive cooperative effort succeeded in achieving far better coverage, even in the remoter parts of Scotland, than the organisers had dared to hope, and resulted in the production of a set of maps giving a clear picture of the breeding distribution of some 220 species (Sharrock 1976). Because records were cumulative over the five years, and a square holding a single isolated pair in only one year scored the same as a square regularly holding many pairs of the same species, the *Atlas* does not, however, give a reliable indication of population size. It nevertheless represented a major step forward, and has served as a model for many other European countries; fieldwork for a *European Atlas*, based on the 50 km grid, is scheduled for 1985–88.

With systematic recording of breeding distribution fairly satisfactorily achieved, and a baseline established for future reference, attention turned again to wintering populations. Birds dependent upon estuarine habitats, which are not only limited in distribution but also often threatened by reclamation, were covered first – by the Birds of Estuaries Enquiry, which initially ran from 1969–75 but is now an on-going project. This pinpointed the sites of major importance, during migration and in winter, for the waders and wildfowl occurring regularly on estuaries, and also accumulated much data on their numbers (Prater 1981). But even among waders there are species which do not feed on mudflats, and others which are as likely to be found inland as on an estuary, so once again the picture remained far from complete. By the early 1980s the concept of a *Winter Atlas*, to complement the breeding season one, was being explored by the BTO and a trial survey was mounted in 1980/81. The aim this time was to determine where different species are between November and March and also to obtain some indication of relative numbers. Survey work was carried out between 1981 and 1984 – fortuitously covering both unusually severe and unusually mild winters – and the publication of the results is due shortly.

NUMBERS AND POPULATION FLUCTUATIONS

Early experience with Heron censuses had highlighted the way in which bird populations fluctuate, and over the years increasing attention has been paid to the problems of both assessing and explaining such fluctuations. With species which can be counted with a fair degree of accuracy, because their overall numbers are small and their nesting requirements rather specialised, repeat full censuses are clearly the most satisfactory way of monitoring population changes. For many species the time and effort required to carry out a full census make such an exercise impracticable, while for some the obtaining of accurate figures presents insuperable difficulties. In such cases – and they include most of the commoner species – regular counts at sample sites, at approximately the same dates and times in consecutive years, provide the best available means of assessing fluctuations.

Repeat surveys of several species have demonstrated progressive declines, though the cause is by no means always clear. There is no very obvious reason why the Roseate Tern, for example, should have decreased almost to the point of extinction, nor why the Red-necked Phalarope should have vanished from areas which apparently still offer suitable breeding habitat. What has become evident, however, as a result of continuing work on monitoring much commoner species, is that many bird populations fluctuate markedly. But before this conclusion can be reached in respect of any species it is necessary to be confident that the apparent fluctuations are in fact real, and not attributable to inaccuracies in the counting techniques used or to invalid comparisons.

The difficulties experienced during Operation Seafarer in estimating the breeding numbers of some species, together with the widely varying estimates made by different observers at the same site, focused attention on the high level of error likely to be involved in assessing seabird numbers. Mike Harris and others (see Chapter 11) have since put much effort into identifying the main sources of error, which have been found to vary according to species. Intensive observation at selected sites, for example on the Isle of May, has shown that the time of day, weather conditions, and stage of the breeding cycle are among the factors affecting the proportion of a population present at the colony at any one time. Such knowledge can help to ensure that counts are made under optimum conditions, thus minimising observer error and making comparisons between years a more reliable indication of population fluctuations.

Scattered populations of small birds present even greater problems of assessment than do seabird colonies. To measure trends among such populations the Common Birds Census technique, developed by Ken Williamson, is now generally used. This involves observers in making a series of breeding season visits each year to the same area and recording on maps all signs of territorial behaviour seen or heard and any nests found. At the end of the season the maps are returned to the BTO and the records analysed to determine the number of territories held by each species present in the census area that year. An index for each species is then calculated by summing the totals for all plots counted in two consecutive years (excluding invalid comparisons) and calculating the percentage change between years; a datum level of 100 is set for an arbitrarily selected baseline year, in most cases 1966 (Batten & Marchant 1976; Marchant 1983). Changes in the CBC index from one year to the next give an indication of overall population increase or decrease – though the extent to which they reflect population fluctuations in a particular region or habitat is dependent upon the adequacy of coverage. Such indices have confirmed that wide variations exist between resident species in their ability to withstand severe winters, and have also drawn attention to disasters which have hit migrants outwith this country – most notably to the drought conditions in the Sahel area

PLATE 1. *Shetland and Orkney differ widely but both are important for breeding seabirds and wintering seaducks. In the 1970s oil developments aroused fears of major disasters but commercial fishing of sandeels is at least as serious a potential threat to seabird populations.*

1a. *(above)* In this satellite photo of the Northern Isles pale-coloured farmland contrasts with darker moorland. *Image processed by NRSC Space Department, RAE, Farnborough.*

1b. *(top right)* Fair Isle in 1962. Since then seabird numbers have increased and Gannets have colonised the island. *Aerofilms Ltd.*

1c. *(centre)* High cliffs and rich fishing grounds enable the Noup, Westray, to support a huge seabird colony (1980). *Valerie M. Thom.*

1d. *(right)* A catch of sandeels off Shetland, a vital food for many young seabirds. *HIBD Picture Library.*

1e. *(below)* Sullom Voe oil terminal (April 1982). Many divers, seaducks and Black Guillemots winter around the Yell Sound islands. *British Petroleum.*

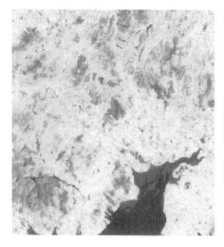

PLATES 2 & 3. *The most striking and extensive land-use change since 1950 has been the afforestation of moorlands. As a result, the extent of breeding habitat for species such as Golden Plover, Dunlin and Greenshank, and of hunting ground for raptors such as Merlin and Golden Eagle, is progressively decreasing. In the southwest large areas of what were once heather moors and grassy hills are now conifer-covered and in central Scotland many former grouse moors have been planted up.*

2a. *(top left)* The forests of Eskdalemuir and Ae, and those between Loch Ken and Newton Stewart, stand out clearly in this 1982 satellite photo. More recent plantings are less easy to distinguish. *Image processed by NRSC Space Department, RAE, Farnborough.*

2b. *(top right)* This area near Loch Kinnardochy, on the eastern slopes of Schiehallion, was one of the first Perthshire grouse moors to be afforested (photo mid-1970s). *Valerie M. Thom.*

2c & d. *(below)* 'Blanket' forest is now such a familiar sight in Galloway that it can be difficult to remember what the landscape formerly looked like. The appearance of Loch Trool changed greatly between 1950 – *The Scotsman;* and 1985 – *M. J. Bannister.*

3a. *(above)*   Only very high or rocky ground seems likely to remain entirely unplanted, but even in such areas eagles' winter food supplies may be reduced when lower slopes become tree-covered, as in Glen Etive (1981). *D. A. Ratcliffe.*

3b. *(bottom left)*   During the early and mid 1980s there has been a rapid spread of afforestation into the remote 'wildernesses' of the far north, hitherto regarded as largely unsuitable for economic timber production. The flowlands now under threat, or already ploughed and planted, are the breeding ground of several species that are both numerically scarce and of very local distribution. *Image processed by NRSC Space Department, RAE, Farnborough.*

3c. *(centre)*   Moorland in west Sutherland (Suilven in the distance) ploughed ready for planting (1984). *S. Angus.*

3d. *(bottom right)*   A typical flowland pool system, with recently planted ground beyond; Strathy Bog, east Sutherland (1984). *S. Angus.*

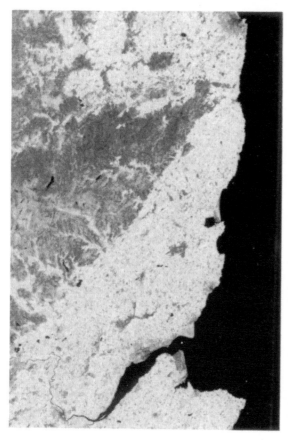

PLATES 4 & 5. *Land-use changes and habitat losses have also been considerable during the last two or three decades in both the lowlands and the glens. Some have been local in impact but affected species of restricted distribution, while others have been widespread and have occurred so insidiously that at the time they passed almost unremarked.*

4a. *(top left)* In many parts of Scotland the boundary between farmland and rough ground is well-defined, nowhere more so than along the line of the Highland Boundary Fault. *Image processed by NRSC Space Department, RAE, Farnborough.*

4b. *(top right)* The dwindling stands of native pinewood, habitat of the Scottish Crossbill, continue to suffer loss. Clear felling on Speyside in 1978. *B. M. S. Dunlop.*

4c. *(centre)* Many natural birchwoods, already under pressure from over-grazing, have been cleared or 'converted' to conifer forest. Under-planted birches in Strathtummel, mid 1970s. *Valerie M. Thom.*

4d. *(below)* Reclamation has reduced the area of intertidal mudflats available to waders and wildfowl on all the larger east coast estuaries. Nigg Bay NNR viewed across the Britoil terminal. *Marcus Taylor/Britoil.*

5a & b. *(above)* Although much has been written about the loss of hedges and hedgerow trees in arable districts, there was surprisingly little change in this intensively-farmed part of East Lothian between 1965 and 1985. 1965 (left) *East Lothian District Council*; 1985 *Ian Fullerton.*

5c & d. *(below)* A much more widespread change, affecting virtually every crop-growing farm in Scotland, has been the loss of the stackyard, in the past a valuable winter food source for many species. Steadings, too, have become less suitable for birds, though few have changed as completely as the one pictured here. Balinbreich Farm, Fife: 1953 – *The Scotsman*; 1985 – *John Watt.*

PLATE 6.   *Over the last 20 years increasing recreational pressure has had an adverse impact locally from the high tops to the lowland lochs, affecting the breeding success of species as diverse as the Ptarmigan and the Mute Swan.*

6a &b. *(above)*   Many changes have taken place around Loch Morlich and on Cairn Gorm since 1956, though even then visitor numbers were increasing steadily. The network of erosion scars on the ski slopes is clearly visible in the 1985 picture, but the car park and buildings at the lower station of the ski-lift are largely hidden by the trees. 1956 – *The Scotsman*; 1985 – *D. Gowans.*

6c & d. *(below)*   Some lochs are sufficiently large, or human pressures are sufficiently limited or controlled, to allow waterbirds and recreational uses to co-exist, but on some waters in the central lowlands disturbance from sailing and fishing has resulted in changes in the breeding distribution of species such as Mute Swan and Great Crested Grebe. Left, Loch Ken – *B. S. Turner*; right, Castle Loch LNR – *R. T. Smith.*

PLATE 7.   In the Hebrides a mosaic of open water, marsh, damp meadow and strip cultivation, backed by peaty moorland, provides conditions attractive to species now absent from, or becoming scarce in, other parts of the country. These machair habitats are very vulnerable to modern agricultural practices, which have only recently become a potential threat in the islands.

7a. *(top left)*   The sand-sweetened machair shows up clearly as the light-coloured strip, dotted with dark areas of standing water, down the west coast of the southern Outer Hebrides. *Image processed by NRSC Space Department, RAE, Farnborough.*

7b. *(top right)*   A 1965 view of Balmartin, North Uist, showing the crofts, with their narrow strips of cultivation, lying between the sand dunes and the moorlands. *Aerofilms Ltd.*

7c & d. *(below)*   The close proximity of rich, shallow lochs with marshy margins, mixed-species meadows and arable crops (only rarely treated with chemicals) makes the machair ideal breeding habitat for Corncrakes and for waders such as Dunlin and Ringed Plover. Left – Loch Hallan; right – Rudha Ardvule, both South Uist. *P. Wakely/Nature Conservancy Council.*

PLATE 8. *With the rate of habitat change continuing to increase, the ability of birds to adapt to environmental change, and the extent to which man endeavours to provide substitutes for losses of natural habitats, are likely to be of increasing significance in the future.*

8a. *(above)*   Estuarine reclamation usually ends in complete loss of habitat but the affected area may attract gulls and other birds during the intermediate stages. The ash-settling lagoons at Musselburgh, associated with the nearby power station, have provided safe roosting areas for large numbers of waders – and become a mecca for birdwatchers. *Eric Thorburn/SSEB.*

8b. *(below)*   Habitat improvement is now a regular practice on many nature reserves. The provision of islands – as at Morton Lochs NNR – and the excavation of shallow scrapes are particularly valuable for wildfowl and waders. *Valerie M. Thom.*

8c. *(below)*   There is also a growing interest in habitat creation, among both conservationists and landowners/occupiers. The establishment of undisturbed wetlands and woodlands is an urgent need in the lowlands, where pressures on land and water are greatest; semi-derelict sites often present good opportunities for this type of development. Shewalton SWT Reserve, Ayrshire, shown during excavation in 1982, was created in an abandoned sandpit originally destined to be used as a tip. *Garth Foster.*

south of the Sahara, the most dramatic result of which to date has been a population crash in the Whitethroat between 1968 and 1969 (Winstanley *et al* 1974).

The CBC results can also be used to assess breeding densities, but the level of error involved is higher, since each individual sample or plot is small and also because species vary greatly in their detectability. Whereas birds which sing loudly from prominent song-posts, or indulge in frequent territorial battles, are likely to be fully recorded, those that skulk and remain silent may be overlooked. Such variations in detectability present particular problems in woodland, where visibility is often restricted and even some of the common and abundant species, such as the Treecreeper, are very unobtrusive. A further complication in interpreting figures for woodland bird populations is the edge effect. Many birds nest along a woodland edge but feed largely over adjoining ground; a small wood with a proportionately long edge is consequently likely to hold a higher density of breeding birds than is a larger wood of the same type. It is advisable therefore to regard CBC figures as giving an indication of relative densities rather than as a reliable basis for estimating total populations. So far as Scotland is concerned even more caution is desirable when basing estimates on CBC densities, as the number of Scottish plots covered in both the CBC and the corresponding Waterways Bird Survey (20 and 12 respectively in 1983, compared with 259 and 91 in England and Wales) is too small to give representative figures for the country as a whole. Nevertheless there is sufficient evidence to suggest that the densities of many species are appreciably lower here than further south. A recent analysis has shown that on farmland Wrens, Dunnocks and Blackbirds occur at lower densities in Scotland than in any other region of Britain, Skylarks are more abundant than in Wales but less so than in England, and Willow Warblers reach their highest densities in Scotland and Wales (Fuller *et al* 1985). CBC results also confirm that in some species winter mortality is appreciably higher in Scotland than in England (Table).

The winter Wildfowl Count scheme, in which many waters are counted more or less regularly, but by no means all are covered every month of every year, was the first to introduce the use of indices of relative abundance. Having satisfied themselves that the overall level of counting error was acceptably low (±10% for most species), the organisers devised this method of making maximum use of the available data to examine population trends. This technique is most accurate for strictly aquatic feeding ducks, of which the bulk of the population is concentrated at relatively few sites. The marsh and grassland feeding ducks are inevitably more difficult to assess, while seaducks present special problems (Owen *et al* 1986).

Because they spend much of their time feeding on land, geese and Whooper Swans are not adequately covered by the routine wildfowl counts and different methods must be used to assess their populations, a large proportion of which winters in Scotland. Most effort has been directed towards an annual November census of Pink-footed Geese and Greylags, which involves observers in all areas known to be used by these species counting on the same day. Counts are generally made when the birds are flighting from or to the roost, since finding them on the ground can be surpris-

The effects of the severe winter of 1981/82 on breeding populations of Wrens and Robins in Scotland and England, as shown by the Common Birds Census (Data from BTO)

	Farmland		Woodland	
	Scotland	England	Scotland	England
Wren territories				
1981	30	1,193	136	1,817
1982	13	692	64	1,235
decrease	−57%	−42%	−53%	−32%
Robin territories				
1981	34	1,003	184	1,909
1982	17	738	134	1,444
decrease	−50%	−26%	−27%	−24%
Number of plots	3	71	10	85

(*Note: see text for explanation of the limited reliability of the Scottish figures*)

ingly difficult; dawn flights are preferable to those at dusk, as some geese do not come in to roost until after dark. Although the results of these counts give a fair indication of population trends, the scope for error is both considerable and variable, even when experienced counters are involved. Early morning mist can reduce visibility to such an extent that the birds can only be heard and not seen, while a bright moon may encourage them to remain on the fields all night. An assessment of counting accuracy was made in the early days of these censuses, but they have not been subjected to the rigorous investigation that has been given to seabird counting – and perhaps such an investigation is now due. The concentration of effort in areas known to have held significant numbers of geese in previous years is another potential source of error. Although strongly traditional by nature, geese undoubtedly adapt their habits in response to changes in farming practices – and in fact provide one of the few convincing examples of birds benefiting from habitat changes.

ECOLOGICAL AND BIOLOGICAL STUDIES

As knowledge of distribution and numbers expanded, the need for greater understanding of the factors affecting breeding numbers and success and of habitat requirements was also receiving attention. Research studies, most often by professional biologists, investigated many individual species, eg work on the Red Grouse, started by David Jenkins, and intensive studies of the Gannet (Bryan Nelson), Eider (Harry Milne) and Puffin (Mike Harris). Comparative work on several members of a group, eg waders and wildfowl, has shed light on differences in habitat usage and so assisted prediction of the likely effects of habitat change or loss. Aberdeen University's Culterty Field Station took the lead in much of this work, with the other Universities, the NCC and the RSPB becoming increasingly involved in the 1970s and 1980s. With the advent of computers it became possible to analyse more fully the vast amount of data on breeding biology accumulated through the Nest Record Scheme (started in 1939) and to explore the relationship between breeding success and the CBC indices. The effect of severe weather on mortality rates in different species and on garden bird feeding behaviour was studied, and there was growing awareness of the implications of habitat loss and other environmental changes.

Two species stand out as examples of the way in which population trends can direct attention towards environmental changes: the Peregrine and the Corncrake. The census of Peregrines in 1961–62, carried out in response to concern about losses of homing pigeons, revealed that the population was not increasing but instead was declining rapidly. The cause of the decline was eventually traced to the effects of persistent organochlorine pesticides, a discovery which alerted conservationists to the potential threat to many species presented by the increasing use of such chemicals. With the Corncrake the evidence was more circumstantial, but the progressive shrinking of its breeding range towards the areas where traditional farming practices have persisted longest supports the view that changing agricultural methods have been an important contributory factor in this case.

Concern over habitat loss focused first on farmland, eventually leading in 1969 to the Silsoe conference on Farming and Wildlife (Barber 1970), forerunner to the Farming (Forestry) and Wildlife Advisory Groups now active in many parts of Britain. As pressure on habitats of various types increased there was need to identify those sites of greatest conservation value. This led to yet another massive co-operative exercise by birdwatchers. The BTO's Register of Ornithological Sites, compiled in the mid 70s, lists and documents those sites throughout Britain considered to be of high ornithological value, ranking their bird communities according to three main criteria – population size, species diversity and rarity. The information it contains has not been published, since this would give undesirable publicity to individual sites, but complete copies of the Register are held by the NCC and the RSPB. Analyses of data in the Register form the basis for the descriptions of habitats and their bird communities given by Fuller (1982); these contain a wealth of information on the species likely to be found in various types of habitat at different seasons. Farmland is unfortunately not covered as a specific habitat.

The extent to which a species or population is dependent upon a particular habitat or site obviously varies according to whether the species in question is a specialised feeder or has very specific nesting requirements, and whether it is sedentary, dispersive or migratory. An understanding of movements and migration has therefore an important role in conservation planning.

MOVEMENTS AND MIGRATION

Until comparatively recently the re-catching or recovery dead of ringed individuals provided virtually all the positive evidence of bird movements, but the last two decades have seen many innovations in this field. Colour-marking – by dyeing the plumage or fitting collars, patagial flags or colour rings – enables local or dispersive movements of birds such as waders, gulls and wildfowl to be followed without the need to keep re-catching them. By using these techniques much has been learned about the inter- and intra-estuary movements of different wader species, and the gradual southward movement of Whooper Swans following their autumn arrival. Radio-telemetry is also now regularly used to track local movements, for example in work on Sparrowhawks (Marquiss & Newton 1982), and is especially valuable in woodland.

Little of this work would have been practicable without the advances in bird-catching methods made in the last 30 years. Prior to 1950 most ringing of fully fledged birds was carried out at the observatories on the Isle of May and Fair Isle, where fixed Heligoland traps had been constructed, and in gardens. In the late 1950s the Wildfowl Trust started to catch feeding geese with rocket nets. Cannon-netting of wader flocks followed, while dazzling at night is used with ground-roosting species, and Canada Geese and Mute Swans are rounded up for ringing while flightless during moult. But the most significant development was undoubtedly the introduction of mist nets in 1956. These revolutionised the bird-catching scene, enabling ringers to operate in a wide variety of new situations. Many more small passerines could now be caught in their breeding habitat, large numbers could be netted at communal roosts, and nets could be set up at Storm Petrel colonies (where they are often used in conjunction with tape-lures) or where passage migrants were passing through. Nor were the opportunities limited to this country: mist nets could easily be taken abroad and used to catch passage and wintering birds in countries where no bird-ringing had previously taken place. As a result of these developments the numbers of birds ringed, and also the control/recovery rates, increased enormously, shedding new light on the movements and migration of many species and confirming what had been suspected about others. With many more species and individuals being handled the opportunities for examining birds in the hand naturally increased greatly, facilitating the study of such matters as moult and weight changes during migration, while the higher rates of recovery helped in building up data on life expectancy and causes of death. The study of migration remains, however, the primary interest for many ringers. (See *Ringing and Migration* Vol. 5, June 1984, for a general review of past progress and future possibilities.)

Much migration work had already been done in Scotland by the time *The Birds of Scotland* was written and the last three decades have seen a continuation of this work. The observations of B&R themselves on the Isle of May and of Eagle Clark on Fair Isle had established the value of these islands as migration sites long before observatories were founded on them in 1934 and 1948 respectively (Durman 1976). B&R's comments on the influence of weather on the scale of visible migration were amplified in Kenneth Williamson's drift theory, developed from his study of the relationship between large falls of migrants on Fair Isle and the weather systems prevailing over Scandinavia, the North Sea and Iceland. This and other weather-related migratory movements are discussed in detail by Norman Elkins in *Weather and Bird Behaviour*. Radar has been widely used to investigate migration, while observers on North Sea oil installations have contributed useful evidence to support the view that grounded birds represent only a very small proportion of those moving at the time, and that in all but the most adverse weather conditions the majority of migrants pass high – and often unseen – overhead (Bourne 1980).

So much has now been written about migration that it would be unrealistic to attempt even a summary here. Instead I will simply cite a few examples of discoveries which strike me as having particular interest. Not all of these examples refer to birds ringed or recovered in Scotland.

The fact that Swallows return to breed at their natal site has long been known, but there is now evidence that individuals tend to return to the same wintering sites in Africa. Several warbler species, wagtails and Redstarts ringed in Africa have been re-trapped in subsequent winters at exactly the same site – on the identical bush in some cases (Moreau 1972). The fastest journey on record was made by a female Blackcap, which flew against a window in Shetland 1000 km away from the Dutch site where it had been ringed the previous day. Other species known to have travelled at speeds of 500 km/day or more include such diverse types as Manx Shearwater, Teal, Knot, Wheatear, Ring Ouzel, Sedge Warbler and Starling. The furthest-travelled recovery so far has been an Arctic Tern in New South Wales, more than 18,000 km from its ringing site, with Manx Shearwater, Storm Petrel, Common Tern, Swallow and Spotted Flycatcher all recovered more than 10,000 km away. These, and many other, intriguing details are included in the appendices of *Enjoying Ornithology*, the BTO's golden jubilee publication.

It is not only migratory movements that are of interest and importance, however. Bad-weather movements are important to the winter survival of many species, and reference has already been made to the fact that some waders apparently need to move between estuaries during the course of the winter, and to the possible implications of such movements for the conservation of these species. A knowledge of seabird movements is similarly of potential value in predicting the likely effects of oil-spills on breeding populations, and much has been learned about these movements in the last decade. Intensive sea-watching from headlands has resulted in many useful observations of inshore movements, including the discovery of the regular large-scale passage of Pomarine Skuas off the Outer Hebrides in May. Offshore study presents much greater problems, but concern over the risks associated with offshore oil extraction led to the three year study *Seabird distribution in the North Sea* (Blake *et al* 1984). Systematic observations from ships, oil installations and aircraft were used to prepare maps showing the distribution of each species in summer and winter. These highlighted the areas and periods at which the various seabirds were most at risk, and showed that the danger offshore was low in comparison to that in coastal waters.

In the case of many species it is not sufficient, however, to study their distribution (and the potential threats facing them) only in this country. Information is needed too on conditions and numbers along migratory routes and on the wintering grounds.

NATIONAL AND INTERNATIONAL COOPERATION

There is now a wide variety of organisations and groups whose primary aim is to facilitate contact between ornithologists with common interests. Some are active only in this country but several have a European – or even wider – remit.

At the national level the British Trust for Ornithology is at the centre of most co-operative work carried out in Britain by amateur ornithologists – other than wildfowl projects, which are often under the aegis of the Wildfowl Trust. The BTO co-ordinates many of the schemes described earlier and also provides the forum for exchange of information among ringers and observatory staffs, as well as its wider membership. The Royal Society for the Protection of Birds, founded in 1889, has always led the field in non-statutory bird conservation efforts. Other important ornithological bodies include the long-established British Ornithologists' Union, whose Records Committee is responsible for maintaining an up-to-date list of the species and subspecies of birds reliably recorded in Britain and Ireland. This Committee is concerned only with first records, however, and it is the British Birds Rarities Committee and Rare Breeding Birds Panel that check and record all reports of birds in these latter categories. In Scotland we have, of course, the Scottish Ornithologists' Club, which celebrates its golden jubilee in 1986 and of which Miss Baxter and Miss Rintoul were the first (joint) Presidents.

There are also a number of British or Scottish organisations concerned with particular groups or species, for example the Game Conservancy Trust, the British Association for Shooting and Conservation (formerly the Wildfowlers' Association of Great Britain and Ireland), the Seabird Group (formed in 1966), the Greenland White-fronted Goose Study Group (1978), the Gull Study Group (1979), and the more local East Scotland Mute Swan Study Group (1982). The British-based Wader Study Group, founded in 1970, provides a contact point for those carrying out wader research on both sides of the Atlantic. EURING, the European Union for Bird Ringing, was founded in 1963 to promote co-operation between European ringing schemes; as a result of its efforts a common code, suitable for computer use, has been introduced to facilitate the exchange of recovery data between countries. A European Ornithological Atlas Committee, set up in 1971, is responsible for the organisation of the planned *European Atlas of Breeding Birds*, fieldwork for which should be completed by 1988.

The International Council for Bird Preservation (ICBP), the British Section of which operates from the Institute of Biology in London, and the International Waterfowl Research Bureau (IWRB), now based at Slimbridge, were among the first world-wide ornithological bodies to be established. Their work is discussed in the next chapter.

This list is by no means comprehensive but will serve to give some idea of the number and variety of ornithological organisations now in existence, many of which have been established in the last two or three decades. A more complete inventory can be found in the *Birdwatcher's Yearbook* (Pemberton 1984), which also lists enquiries and projects currently under way.

It would be wrong to conclude this brief review of developments in bird study without some reference to those ornithologists popularly referred to as 'twitchers'. These are the people whose aim in life is to see and tick on their 'life list' as many different species as possible. Armed with powerful binoculars, telescopes and cameras they race up and down the country in response to rumours of rarity sightings, some even flying to Shetland from the Scillies in order to add another 'lifer' to their list. At times these twitchers undoubtedly cause much harassment to tired vagrants, and nuisance to the people upon whose land such birds have been found, but their tireless efforts have added to the field descriptions available of birds seldom if ever recorded previously in this country.

9: Protection and conservation – progress and problems

The last three decades have seen considerable progress in bird protection and conservation, through legislation, the establishment of reserves and education. By no means all problems of wilful damage have been overcome, however, while recent legislation has in some respects created more problems than it has solved, and bird mortality due to accidental causes is undoubtedly higher than it was before 1950. Although factors such as persecution and accidental deaths from pollution may have significant effects on certain species or in particular areas, there is no doubt that loss of habitat is a much more important and widespread cause of population decrease.

LEGISLATION RELATING TO BIRD PROTECTION

The earliest laws designed to protect wild birds (as distinct from gamebirds) from exploitation established close seasons for all species, but allowed such periods to be varied by Special Orders applicable to different counties. As a result there was, by 1950, such a plethora of lists and dates that the situation was very confused and the law virtually unenforceable. The passing of the Protection of Birds Act 1954 greatly simplified matters, extending protection to all but a limited number of 'pest' or 'sporting' species and reducing the possibilities for creating Special Orders. This Act also made provision for the establishment of sanctuaries, for controlling the methods used to kill or take wild birds and the conditions under which captive wild birds could be kept, and for restricting the sale of eggs and dead birds.

The Protection of Birds Act 1967, which amended the 1954 Act, introduced the concept of licences to permit individuals to carry out actions prohibited by the law, eg the taking of birds or eggs for scientific study or falconry, the catching of birds for ringing and release, and the disturbance for scientific or photographic purposes of birds on the specially protected list – usually referred to as Schedule 1 species.

The 1967 Act also prohibited the sale of dead wild geese at any time (formerly permissible outwith the close season) – a restriction which caused some controversy, coming as it did at a time when goose numbers were increasing steadily. In addition this Act made provision for special protection for quarry species (mostly wildfowl and waders) in severe weather.

Some interesting changes took place over the years in the species listed as 'pests' and in the exceptions authorised by Special Orders. Goosanders and Red-breasted Mergansers were classed in both Acts as pest species in Scotland; an Order permitted the (traditional) taking of Gannets outwith the close season on Sula Sgeir; and the Collared Dove's population explosion resulted in its transfer from the specially protected to the pest category in 1967, only ten years after the first breeding record. In Skye and Argyll, Ravens were classed as pests, and the shooting of Barnacle Geese was permitted only on islands off the mainland of the counties of Argyll, Inverness, Ross and Cromarty, west of longitude 5° west, from 1 December until 31 January.

The Wildlife and Countryside Act 1981 incorporated most of the provisions of the Protection of Birds Acts (now repealed) but also covered a much wider field, including the protection of all forms of wildlife and of habitats. Some of its habitat protection provisions have had considerable repercussions, mainly by either creating or exacerbating situations where conservation and other land-use interests come into conflict.

The 1981 Act further reduced the list of so-called pest species (Table), which may be killed or taken by authorised persons (ie landowner or occupier or person with landowner's permission) at all times, removing from it the Cormorant, Goosander, Red-breasted Merganser, Rock Dove and Stock Dove, and repealing the Orders relating to Ravens and Barnacle Geese in specified areas. (Sula Sgeir Gannets may still – 1985 – be taken, as may gulls' eggs and (before 15 April) Lapwing's eggs – but only for food and if a licence to do so has been granted.) It also restricted the range of quarry species other than gamebirds which may be killed or taken outwith the close season, giving full protection to more wildfowl and waders. All the strictly marine ducks are now fully protected, as are all waders other than Golden Plover, Snipe and Woodcock. In Scotland the only geese that may be shot (outside the close season) without a licence are Canada, Greylag and Pink-footed.

Birds which may be killed or taken by authorised persons at all times (Wildlife and Countryside Act 1981, Schedule 2 Part II)

Crow	Magpie
Dove, Collared	Pigeon, Feral
Gull, Great Black-backed	Rook
Gull, Lesser Black-backed	Sparrow, House
Gull, Herring	Starling
Jackdaw	Woodpigeon
Jay	

As in the previous Act provision has been made for the granting of licences to allow otherwise prohibited actions to be taken for specified purposes. The most controversial of these relate to the taking of birds for falconry and the killing of birds 'for the purposes of preventing serious damage to livestock, foodstuffs for livestock, crops, vegetables, growing timber or fisheries'. The issue of licences for falconry purposes is controlled by an Advisory Committee upon which both conservation and falconry interests are represented. Views differ as to the desirability of permitting the taking of any raptor eggs and young from the wild but if such action were confined to those holding the necessary licences there would be little real cause for concern. However the protection law is not easy to enforce, and much illegal taking of raptors occurs. Licences for killing birds in order to prevent damage to crops or fisheries are issued by the Department of Agriculture and Fisheries for Scotland, apparently in at least some instances with little evidence that the threat of damage is serious or that adequate attempts have been made to scare away the offending birds. The birds most frequently involved in such cases are geese – especially grey geese in east-central Scotland and Barnacle Geese on Islay, Cormorants and the saw-billed ducks. According to the RSPB (*Birds* 10:4) no fewer than 907 Cormorants, 347 Goosanders and 523 Red-breasted Mergansers were killed in Scotland under licence in 1983; the full toll was doubtless much higher. Other fish-eating birds killed without licences having been granted include Herons – which can be discouraged from taking advantage of fish farms much less drastically at little cost. Events on Islay since the 1981 Act came into force support the view that the law is being abused, licences issued on the grounds of damage to grassland being used to enable sport-shooting of Barnacle Geese to take place. These are matters which will require to be reviewed, though they are unlikely to be easy to resolve to the satisfaction of all parties. Concern over the seemingly casual manner in which licences were being issued led the RSPB to make a formal complaint to the EEC, on the grounds that current

practice is in contravention of the Commission's Directive on Bird Conservation (see below).

The other aspect of the 1981 Act at present arousing controversy relates to habitat protection, covered previously by the National Parks and Access to the Countryside Act 1949. Under the 1949 Act the Nature Conservancy (now Nature Conservancy Council) was required to establish and manage nature reserves and to provide scientific advice on conservation. The latter requirement led to the notification of Sites of Special Scientific Interest, which remained in private ownership and received only minimal protection against potentially damaging developments – which excluded agricultural and forestry operations. In response to growing concern regarding habitat loss, the 1981 Act introduced new measures to safeguard SSSIs, although their conservation is still largely dependent upon the voluntary co-operation of the owner and occupier. The NCC is now required to inform SSSI owners / occupiers of all operations considered likely to damage the scientific interest of the site; the owner/ occupier in turn must inform NCC of any plans for such operations. If the parties cannot reach agreement over management of the site, NCC may formally request that grant be refused on conservation grounds; if the case for conservation is upheld and grant refused, the Council is required to compensate the owner / occupier for not carrying out the 'damaging operation' – generally one that will increase the income derived from the land in question.

The financial implications of this legislation are potentially enormous. Examples of sites over which agreement has not been reached and compensation has been sought include wetland areas (the Somerset Levels is probably the most publicised of these), ancient forests such as Abernethy (where clear-felling and replanting, or even underplanting, with introduced conifers would increase the commercial value but seriously reduce the scientific interest of the woodland), and moorland, eg Creag Meagaidh (on which afforestation could produce increased income). Such cases may involve hundreds of hectares, and the income lost through refraining from development is on-going. Compensation, in theory at any rate, can be claimed over many years – but the financial resources currently available to NCC are quite inadequate to cope with such situations. There is the additional concern that some owners / occupiers may put forward development proposals, knowing them to be unacceptable, simply in order to obtain compensation. As yet it is impossible to anticipate how the situation may develop over the next few decades. The effectiveness of the Act in preventing further loss of scarce and scientifically important habitats will be largely determined by the speed with which the necessary formalities are completed and the extent to which financial resources are made available by Government.

For some species even the habitat protection measures now available are inadequate, since site designation is either unlikely to ensure the conservation of a sufficiently large area, for example the hunting range of a pair of Golden Eagles, or is inappropriate for a particular type of habitat, as in the case of the crofting ground which holds the relict Corncrake population. Problems can arise too where comparatively small areas support a relatively high proportion of the total breeding or wintering population of a species. Whimbrels in certain parts of Shetland and Barnacle Geese on Islay are examples of this type of situation, in which the land-use practices of only a few landowners or tenants may have a disproportionately serious effect on the population as a whole. Similar, though less obviously acute, problems arise in connection with upland afforestation and the agricultural improvement of rough grazing land, since some of the areas most attractive to developers hold much higher than average breeding densities of species with a limited distribution, for example waders, grouse and Merlins. The loss of such areas may result not only in loss of the birds previously breeding on them, but also in a decrease in the level of recruitment into less productive areas of apparently similar habitat. Possible solutions to such problems are among the topics discussed in *Nature Conservation in Great Britain* (1984), which proposes objectives and a strategy for conservation in the future.

This brief summary outlines only those aspects of the 1981 Act of most relevance to wild birds in Scotland. The Act also covers the registration of captive birds, caging requirements for captive birds, the importation of endangered species, and the introduction of new species. This last seems a case of shutting the stable door when the horse has gone as it makes it an offence deliberately to release into the wild several species which have already established feral populations, eg Canada Goose.

PROBLEMS IN ENFORCING THE PROTECTION LAWS

The fact that a practice is made illegal does not, of course, mean that it ceases to occur. Every year eggs of Schedule 1 species are taken by collectors, and young raptors (especially Peregrines) are removed from the nest for falconry purposes by unauthorised people. Some offenders are brought to court, and the fines imposed on them are now more realistic as a deterrent than in the past, but many either escape detection or remain unconvicted – despite the increasing effort put in by the RSPB and others to protect the rarer species. One of the problems is that the suspect must be caught with the eggs or chicks actually in his possession, and to achieve this involves anticipating where the thieves will strike, having watchers there at the right time, and ensuring that adequate police backup is at hand. The Table below, which summarises suspected cases of nest robbing of some Schedule 1 birds in Scotland in 1979–84, indicates the scale of the problem.

The threat to raptors from illegal activities by gamekeepers is still serious, although there are many fewer keepers now than in the past and the attitudes of younger keepers are generally more enlightened. Nevertheless persecution of Golden Eagles and Hen Harriers still occurs, particularly on grouse moors, where eagles are held responsible for preying on grouse and harriers for scaring them and so upsetting grouse drives. The methods used include not only shooting

Numbers of nests of some Schedule 1 species believed to have been robbed of eggs or young in Scotland 1980–84 (Data from RSPB)

Species	1980	1981	1982	1983	1984
Golden Eagle	2	4	6	2	2
Peregrine	12	14	37	43	33
Osprey	0	1	5	3	5

and the destruction of nests (by fire in some cases), but also trapping and poison baits. Neither of these last methods is specific in its victims, and many birds other than the target species may suffer. Short-eared Owls often fall victims to pole-traps, the use of which has mercifully decreased in recent years, though not yet entirely ceased.

Poison baits are even less selective: the RSPB's report *Silent Death* (Cadbury 1980a) details the alarming record of deaths due to the misuse of poisons in 1966–78. Among species that died through eating poisoned eggs intended to kill crows and foxes were Hen Harrier, Raven, Rook, Magpie – and dogs; from meat baits intended for eagles and Buzzards – Sparrowhawk, Goshawk, Magpie – and dogs; and from poisoned grain intended for Woodpigeons and Collared Doves – Rook, Jackdaw, gulls, gamebirds and Moorhen. Despite considerable publicity regarding the risk to those involved in preparing poison baits – and one gamekeeper died in central Scotland from this cause in 1982 – their use continues, apparently in at least some cases with the knowledge and approval of the landowner concerned. The chemicals responsible for most of the confirmed poison-bait bird deaths in Scotland are alphachloralose (readily available for mouse control), Mevinphos (an organophosphorus insecticide), and strychnine (used to kill moles underground). In theory their use is restricted and control is exercised over their sale; in practice this is not the case.

BIRD DEATHS FROM ACCIDENTAL CAUSES

Poisoning of birds occurs accidentally as well as through the deliberate use of baits and it may sometimes be difficult to distinguish between the two. It has been suspected that mass deaths of grey geese have at times been due to the deliberate spreading, or spilling, on the surface of the ground of grain treated with organophosphorus seed-dressing. An instance of Whooper Swans in Perthshire becoming paralysed and dying after feeding on growing winter wheat was also due to the ingestion of seed-dressing, in this case rendered unusually accessible by weather conditions loosening the soil so that as the birds grazed they pulled the plants up, seed and all.

The whole picture of pesticides – and other pollutants – and their effect on birds is still by no means either clear or resolved. Bird deaths and poor breeding success definitely attributable to such causes have decreased greatly since the late 1960s, and the most persistent of the pesticides have been withdrawn, but there is still little room for complacency. Recent work by Mick Marquiss (1983) has shown that Herons in the intensively-farmed northeast lowlands have a higher rate of egg breakage than those feeding mainly on trout and salmon parr at altitudes above 200 m. In another study (Cooke *et al* 1982) the levels of DDE (a metabolite of DDT) in the livers of Merlins were three times higher than those in any of the other predators examined – and low breeding success in some Merlin populations is currently causing concern.

One of the big difficulties in the field of pesticide studies is the widely varying response by different species to comparable levels of the same pollutant. Grey geese are particularly susceptible to organophosphorus poisoning whereas some other species are apparently unaffected by similar doses. Polychlorinated biphenyls (PCBs) are suspected as being the cause of poor breeding success among coastal Peregrines, but experiments have shown that Puffins can tolerate quite high levels without any adverse effect on breeding (Harris & Osborn 1982). What is clear is the fact that a continuing programme of monitoring will be necessary so long as new chemicals, or new uses for existing compounds, are being introduced into the environment, whether intentionally or by accident. A recent proposal to introduce legislation aimed at controlling the use of pesticides, if carried through, should improve the situation, though enforcement is unlikely to be easy.

Another continuing cause of mortality is oiling of birds at sea. Despite international laws designed to reduce its incidence, and increased surveillance aimed at detecting offenders in the act so that court action can be taken, the deliberate discharge of waste oil by ships currently causes some 20% of deaths due to oiling. Headline-hitting leakages of crude oil from tankers, whether accidental or deliberate, are responsible for only a further 20%, though they may have marked local effects. The remaining 60% is attributable to oil reaching the sea from land-based sources (RSPB 1979). When North Sea oil operations started conservationists expressed grave concern regarding the probability of a marked increase in oiling incidents, and there was indeed an increase in the number of spills reported from the mid 1970s until 1980. Since then there has been a progressive decrease (Richardson *et al* 1982), probably at least in part due to the firm action taken in respect of sub-standard tankers and seamanship in Shetland waters.

The RSPB's Beached Bird Survey results provide a general picture of the regional occurrence of oiled birds – although weather conditions influence the numbers actually driven ashore – and also demonstrate which species are at greatest risk. The proportion of birds oiled is higher on the east coast than on the west, and in southern Britain than in the north. Divers and auks are the groups most at risk, with on average 60+% of beached birds found oiled; seaducks follow closely at 40+%. All these species gather in large concentrations and spend much time sitting on the surface of the sea. There is evidence to suggest, however, that even massive bad-weather wrecks of auks have little effect on breeding populations, probably because there are enough non-breeders available to make good the losses among established pairs. Although more than 31,000 auks came ashore between Orkney and Kent in a large-scale wreck early in 1983 (Underwood & Stowe 1984), there was no significant reduction in breeding numbers or success that season on the Isle of May (Harris & Wanless 1984).

A number of other causes of accidental deaths have become commoner over the years, although probably none is responsible for any significant level of mortality other than at a local level. Traffic deaths have escalated as vehicle speeds increased. Collisions between bird flocks and aircraft, most often involving gulls, Lapwings or Starlings, now present a risk to human life as well as to birds on some airfields. Low level power lines kill Mute Swans in the Outer Hebrides and locally on the mainland, often resulting in breakdowns in the electricity supply – which could be avoided if such lines were not erected across swans' regular flight lines. Many waterbirds die through anglers' carelessness when disposing of damaged tackle, though the problem of swan poisoning through ingestion of lead shot originating in fishing tackle is not so serious here as south of the Border, largely because there is little coarse fishing. A good many divers, grebes, seabirds and diving ducks are drowned in fixed nets set in lochs or inshore waters, and plastic waste – bags, netting, can-rings and twine – causes numerous deaths.

Direct protection measures are not the only way, nor even perhaps the most important way, of assisting bird conservation. Habitat protection and management are also vital and now receive at least as much attention.

POSITIVE STEPS TOWARDS CONSERVATION
The first action to protect sites of special value to birds (as opposed to the more general nature conservation provisions of the 1949 Act) came through the establishment of bird sanctuaries, under the 1954 Protection of Birds Act. A Sanctuary Order protected all wild birds within the site concerned and often also imposed restrictions on entry; such Orders were the means initially used to protect many areas that are now reserves, for example Loch Garten, Inchmickery and Horse Island. Maintenance of bird populations often involves more than just protecting a species from disturbance and destruction by humans, however. Management of the habitats upon which a species is dependent for nesting and feeding is frequently necessary, and action to control competing and more aggressive species may also be required. As mentioned earlier, the 1949 National Parks and Access to the Countryside Act had made provision for the establishment and management of nature reserves, leading in due course to the network of National Nature Reserves with which we are now familiar. Most NNRs were selected primarily as representative of particular habitat types, and many are valuable for their bird populations. This is especially true of the big upland reserves, which may be able to support several pairs of large raptors, as well as major wetland sites and those holding seabird colonies.

As part of its responsibility for providing scientific advice on conservation, the Nature Conservancy Council also undertook a major investigation of natural and semi-natural sites which resulted in the publication of *A Nature Conservation Review* (Ratcliffe 1977b). This identified many additional sites of national importance, some of which have since become NNRs while others have been notified as SSSIs, but again concentrated mainly on plant communities as indicators of habitat rather than on birds. The identification of sites important to birds involves assessment of migrant and wintering numbers as well as breeding populations, and these are not comprehensively covered by the NNR network. The BTO's 'Register of Ornithological Sites' attempts to do for birds what *A Nature Conservation Review* had done for ecosystems in general.

While all this was going on the RSPB and other organisations had also been busy establishing reserves. By 1984 the RSPB's Scottish holdings numbered 36 and covered 17,630 hectares, of which 10,250 were owned outright (Map 11). They range in size from the mere one hectare of Inchmickery to nearly 4,000 ha of North Hoy in Orkney and include sites representative of all the major bird habitat types – commendable progress indeed, when one remembers that the first Scottish reserve, Horse Island, was established only in 1961.

Some of the Scottish Wildlife Trust's reserves are impor-

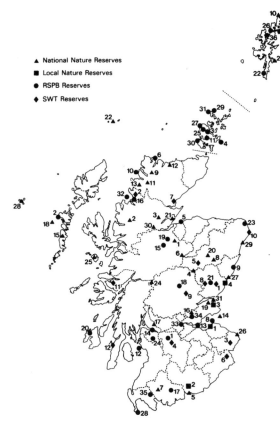

Map 11. Nature reserves of ornithological importance (note: all RSPB reserves and LNRs are included but the NNR and SWT reserve lists are restricted to those of particular ornithological significance)

● Royal Soc. for the Protection of Birds reserves
1 Barons Haugh, Lanark
2 Balranald, North Uist
3 Birsay Moors & Cottasgarth, Orkney
4 Copinsay, Orkney
5 Culbin Sands, Nairn
6 Eilean Hoan, Sutherland
7 Fetlar, Shetland
8 Forth Islands, East Lothian
9 Fowlsheugh, Kincardine
10 Handa, Sutherland
11 Hobbister, Orkney
12 Horse Island, Ayr
13 Inchmickery, Midlothian
14 Inner Clyde Estuary, Renfrew
15 Insh Marshes, Inverness
16 Isle Martin, Wester Ross
17 Ken/Dee Marshes, Kirkcudbright
18 Killiecrankie, Perth
19 Loch Garten, Inverness
20 Loch Gruinart, Islay
21 Loch of Kinnordy, Angus
22 Loch of Spiggie, Shetland
23 Loch of Strathbeg, Aberdeen
24 Lochwinnoch, Renfrew
25 The Loons, Orkney
26 Lumbister, Shetland

27 Marwick Head, Orkney
28 Mull of Galloway, Wigtown
29 North Hill, Orkney
30 North Hoy, Orkney
31 Noup Cliffs, Orkney
32 Priest Island, Wester Ross
33 Skinflats, Stirling
34 Vane Farm, Kinross
35 Wood of Cree, Wigtown
36 Yell Sound Islands, Shetland

◆ Scottish Wildlife Trust reserves:
1 Balgavies Loch, Angus
2 Benmore Coigach, Ross & Cromarty
3 Duns Castle, Berwick
4 Falls of Clyde, Lanark
5 Glenmuick & Lochnagar, Aberdeen
6 Hoselaw Loch, Roxburgh
7 Loch Fleet, Sutherland
8 Loch of Lintrathen, Angus
9 Loch of the Lowes, Perth
10 Longhaven Cliffs, Aberdeen
11 Rahoy, Argyll
12 Rhunahaorine, Argyll

▲ National Nature Reserves:
1 Abernethy Forest, Inverness
2 Beinn Eighe, Wester Ross
3 Ben Wyvis, Ross & Cromarty
4 Caenlochan, Angus/Aberdeen
5 Caerlaverock, Dumfries
6 Cairngorms, Inverness/Aberdeen
7 Cairnsmore of Fleet, Kirkcudbright
8 Glen Tanar, Aberdeen
9 Gualin, Sutherland
10 Hermaness, Shetland
11 Inchnadamph, Sutherland
12 Invernaver, Sutherland
13 Inverpolly, Wester Ross
14 Isle of May, Fife
15 Loch Druidibeg, Outer Hebrides
16 Loch Leven, Kinross
17 Loch Lomond, Stirling/Dunbarton
18 Monach Isles, Outer Hebrides
19 Morton Lochs, Fife
20 Muir of Dinnet, Aberdeen
21 Nigg & Udale Bays, Ross & Cromarty
22 N Rona & Sula Sgeir, O. Hebrides
23 Noss, Shetland
24 Rannoch Moor, Perth/Argyll
25 Rhum, Inner Hebrides
26 St Abb's Head, Berwick
27 St Cyrus, Kincardine
28 St Kilda, Outer Hebrides
29 Sands of Forvie & Ythan Estuary, Aberdeen
30 Strathfarrar, Inverness
31 Tentsmuir, Fife

■ Local Nature Reserves:
1 Aberlady Bay, East Lothian
2 Castle & Hightae Lochs, Dumfries
3 Eden Estuary, Fife
4 Montrose Basin, Angus

tant for birds, especially those on wetlands or extensive uplands, and so too are several of the mountain areas and islands belonging to the National Trust for Scotland. Local authorities have powers to establish nature reserves but few have made use of them, and those that have done so have sometimes found that demands for recreational use conflict with conservation needs. The most valuable of the Local Nature Reserves so far established are Aberlady Bay, the Eden Estuary and Montrose Basin (a joint venture between the local authority and the Scottish Wildlife Trust). The benefit of LNR status for these sites lies in the fact that the local authorities, like the NCC, can make bye-laws and so introduce and enforce control over wildfowling. (In the absence of bye-laws, wildfowling on the foreshore is a public right in Scotland.) That wildfowling need not be incompatible with conservation has been amply demonstrated at Caerlaverock (Harrison 1974) and Loch Leven, but both a permit system and a sanctuary zone are necessary adjuncts to a satisfactory situation. Caerlaverock represents a pioneering example of the way in which the interests of conservation, agriculture and shooting can be catered for in a single area, through careful planning and control.

Many of the quarry species hunted by wildfowlers are migratory and depend for maintenance of their populations on the availability of an adequate supply of suitable feeding grounds in widely scattered areas. It was in connection with the needs of such long-distance migrants, which may cross several frontiers in the course of their travels, that the importance of international co-operation in conservation was first appreciated.

THE INTERNATIONAL CONSERVATION SCENE

Talk about international co-operation in nature conservation had gone on for many years before much real action took place (Nicholson 1970). Of the major conservation organisations whose work has an impact on bird protection – the International Council for Bird Preservation (ICBP), the International Union for the Conservation of Nature and Natural Resources (IUCN), the International Waterfowl Research Bureau (IWRB), and the Inter-Governmental Maritime Consultancy Organisation (IMCO) – only ICBP was in existence at the beginning of this century. The others were not established until after World War II – when advantage was taken of the contacts and communication channels available through the United Nations – although efforts to set up an international conservation organisation had started many years earlier. More recently the European Economic Community (EEC), which Britain joined in 1973, has played an increasing role in this field, its directives having a powerful influence on the conservation activities of member nations.

Three aspects of international co-operation as it affects birdlife will serve to illustrate the types of benefit it can offer. The IMCO, a United Nations agency, was responsible for formulating the Convention for the Prevention of Pollution of the Sea by Oil (implemented in 1958 but difficult to enforce) and the Convention for the Prevention of Pollution by Ships. Although oil pollution is still far from being eliminated, and some nations have refused to become signatories to the conventions, these measures have resulted in considerable improvement in some aspects of marine pollution.

The IWRB was instrumental in initiating the first European Meeting on Wildfowl Conservation, held at St Andrews in 1963, and has since played a major part in three International Conferences on the Conservation of Wetlands and Waterfowl: at Ramsar, Iran, in 1971; Heiligenhafen, W. Germany, in 1974; and Cagliari, Italy, in 1980 (IWRB 1980). These conferences resulted in the adoption of a 'Convention on Wetlands of International Importance Especially as Waterfowl Habitat' (known as the Ramsar Convention), which aims at encouraging contracting states to plan for wise use of wetlands and requires them to designate at least one site of international significance. Sites currently assessed as of international importance include those which:

(a) regularly support 10,000 ducks, geese and swans *or* 10,000 Coots *or* 20,000 waders; or
(b) regularly support 1% of the individuals in a population of one species or subspecies of waterfowl; or
(c) regularly support 1% of the breeding pairs in a population of one species or subspecies of waterfowl.

By 1984 ten Scottish wetlands had been designated, though only one of these, Loch Leven, is of international importance for waterfowl; the rest are outstanding for other aspects of biology. Possible future additions include the Upper Solway flats and marshes and the Insh Marshes on Speyside (see Appendix 4). While there is still a long way to go before an adequate series of wetland refuges is available throughout the range of all migratory waterfowl, at least some progress towards this goal has been made possible by international co-operation. An increase in site designation can be expected to result from the fourth conference, held in the Netherlands in May 1984. However, since these conventions have no legal standing in any country, designation does not guarantee protection.

A further step was taken with the adoption in 1979 of the European Community Directive on Bird Conservation, which covers a much wider remit. It imposes obligations on EEC member states to take special measures to protect the habitat for *all* migratory species and certain rare species, and forms a legal basis for a Community policy on bird conservation and nature reserves for birds. (Annex I, which lists the species to be the subject of special conservation measures concerning habitat, includes 17 species that breed regularly in Scotland.) Although not dealing specifically with hunting, the directive prohibits both the taking of birds on spring migration and the use of 'all means of hunting, killing and capture used for large-scale or non-selective capture or killing'. These measures, if enforced, should eventually eliminate the traditional large-scale destruction of passerines as they move north through Spain and France. Other aspects covered include protection laws and research.

The accumulated knowledge and experience of ICBP and IWRB were heavily drawn upon in formulating the directive, and will continue to be of benefit in the listing and assessment of areas to be protected as nature reserves. The basis for forming a European network of reserves for migratory birds has been created for the first time by this international action. What progress is made in the next few decades will depend upon how the directive is implemented by EEC member states. This will in turn doubtless be influenced by the relative strengths of national conservation and opposing interests (although member states violating the directive may be brought before the European Court of Justice) and also by the availability of funds. As John Temple Lang (1982) has said 'The most important thing now lacking is a European Environment Fund with power to finance the purchase of nature reserves for birds' – to which one might add 'and adequate funding to permit effective management'.

10: Recent changes in status and distribution

Before summarising the changes that have taken place between 1950 and 1984 I feel it desirable to make some cautionary general comments regarding the comparability of data and the interpretation of apparent changes. These comments should be borne in mind when reading the species accounts, especially those which indicate that a slight change in status may have taken place.

In Chapter 8 I outlined the developments in bird study by means of which, over the last three decades, we have built up a more comprehensive and accurate knowledge of numbers and distribution than was available in the past. Although we are still far from having a complete picture of every species, and equally far from understanding all the reasons for population fluctuations, we do now have a reasonably realistic idea as to whether any particular species is present in tens, hundreds, thousands or millions, and also of the limits of its distribution. In 1950 there was very little firm information of this kind. Any assessment of change over the period as a whole – at least in the case of the more widely distributed and abundant species – consequently involves comparison between fairly objective estimates and almost entirely subjective ones. Such comparisons cannot be regarded as providing reliable evidence of status stability or change.

Where rarities and semi-rarities are concerned, the great increase in the numbers, competence and mobility of bird-watchers means that the chances of a vagrant being detected are very much higher than they used to be. For many of the scarcer species an increase in the numbers recorded may consequently be due as much to more intensive watching as to a real increase in frequency of occurrence. In the case of those species which can be reliably identified only in the hand, the use of mist-nets and more intensive trapping at the observatories must have resulted in many reports of birds for which sight records alone would be unacceptable. And improvements in optical instruments now enable birds to

be identified, and often photographed, at much greater distances than in the past – a point of particular relevance to offshore sightings of uncommon seabirds. Even now, however, birdwatcher coverage of Scotland is so uneven that the distribution of rarity records must be much influenced by the distribution of observers.

Where status and distributional changes have undoubtedly taken place one is always tempted to try to explain them. This is relatively easy where cessation of persecution or an obvious increase or decrease in the availability of breeding habitat or food supply is the underlying cause, but such straightforward explanations are by no means applicable in every case. Climatic change – sometimes described as amelioration and sometimes in terms of a returning Ice Age – is often held responsible for colonisation by new species (from both north and south) and for contractions of range, but there is little positive evidence to support this claim. As Norman Elkins (who is a meteorologist and should know!) says in *Weather and Bird Behaviour* 'many aspects of climatic change are little understood and often difficult to correlate with variations in bird populations'. Climatic variability, involving marked differences from the long-term average, is undoubtedly responsible for short-term population fluctuations, however, and may also be part of the explanation for some of the small-scale and short-lived attempts at colonisation that occur.

In the following summary of recent changes I have grouped the material in three main sections: species recorded for the first time and / or reported breeding for the first time, significant increases or decreases in numbers and / or breeding range, and other apparent or suspected changes in status.

RECENT ADDITIONS TO THE SCOTTISH LISTS
From 1950 to 1984 inclusive 84 species (excluding those classed as Category C or D – see Chapter 11) were recorded in Scotland for the first time, 24 of these also being first

records for Britain (Table a). Interestingly, they include nine species mentioned by B&R but either classed as escapes or square-bracketed because the identification was open to doubt: Black-browed Albatross, Lesser White-fronted Goose, Bufflehead, Buff-breasted Sandpiper, White-winged Black Tern, Lesser Spotted Woodpecker, Dusky Thrush, Parrot Crossbill and Firecrest. There have also been occurrences of two species recorded by B&R but subsequently rejected: Isabelline Shrike and Needle-tailed Swift. Six other species (Great White Egret, Blue-winged Teal, Killdeer, Bonaparte's Gull, Nutcracker and Hawk Owl) had not been recorded since the 19th century.

Of these 'first records' 35 were of American origin, 18 from Asia/Siberia, 30 from Europe (including England), the Middle East, and northern Africa, and one, the Black-browed Albatross, from the southern hemisphere. The albatross is fast qualifying for a 'long service' medal; what is presumed to be the same bird was seen first in 1967 on

(*a*) *Species first recorded in Scotland between 1950 and 1984 inclusive (Category C & D species are not included)*

* indicates first British records

Pied-billed Grebe	Citrine Wagtail *
Black-browed Albatross	Siberian Rubythroat *
Little Shearwater	Isabelline Wheatear
Magnificent Frigatebird *	Siberian Thrush *
Cattle Egret	Hermit Thrush *
Little Egret	Swainson's Thrush
Lesser White-fronted Goose	Grey-cheeked Thrush*
Baikal Teal *	Eye-browed Thrush
American Black Duck	Dusky/Naumann's Thrush
Ring-necked Duck	American Robin
Bufflehead	River Warbler *
American Kestrel *	Great Reed Warbler
Sandhill Crane * (previous Ireland)	Thick-billed Warbler *
Western Sandpiper *	Olivaceous Warbler
Least Sandpiper	Dartford Warbler
White-rumped Sandpiper	Spectacled Warbler
Sharp-tailed Sandpiper	Sardinian Warbler
Stilt Sandpiper	Rüpell's Warbler *
Buff-breasted Sandpiper	Orphean Warbler
Marsh Sandpiper	Pallas's Warbler
Greater Yellowlegs	Radde's Warbler
Terek Sandpiper	Bonelli's Warbler
Spotted Sandpiper	Firecrest
Wilson's Phalarope *	Bearded Tit
Mediterranean Gull	Isabelline Shrike *
Laughing Gull	Trumpeter Finch
Franklin's Gull	Pine Grosbeak
Ring-billed Gull	Evening Grosbeak *
Caspian Tern	Tennessee Warbler *
Bridled Tern	Cape May Warbler *
White-winged Black Tern	Yellow-rumped Warbler
Collared Dove	American Redstart
Rufous Turtle Dove	Ovenbird *
Great Spotted Cuckoo	Common Yellowthroat
Black-billed Cuckoo	Song Sparrow *
Common Nighthawk	White-crowned Sparrow *
Needle-tailed Swift	Dark-eyed Junco
Lesser Spotted Woodpecker	Cretzschmar's Bunting *
Calandra Lark	Yellow-browed Bunting *
Bimaculated Lark	Pallas's Reed Bunting *
Crested Lark	Bobolink
Olive-backed Pipit	Northern Oriole

(*b*) *Species recorded breeding for the first time since 1950 (the year given is the date of first proved breeding – not necessarily successful)*

1951	Green Woodpecker
1952	Montagu's Harrier
1957	Collared Dove
1958	Little Owl
1959	Wood Sandpiper
1967	Snowy Owl
	Fieldfare
	[Red-legged Partridge – late 1960s]
1968	Little Ringed Plover
	Bluethroat (nesting – no male seen)
1969	Wryneck
1970	Great Northern Diver
	Goldeneye
1973	Reed Warbler
1974	Golden Oriole
1975	Spotted Sandpiper
1977	Shore Lark
	Lapland Bunting (may have bred earlier)
	Red-backed Shrike (may have bred in 1930s)
1978	Purple Sandpiper
1979	[Ruddy Duck]
1980	Ruff
1982	Scarlet Rosefinch

the Bass Rock, but later moved to Shetland, where it has appeared annually at the Hermaness gannetry since 1972.

During the period under review breeding by 21 'new' species has been confirmed (ie eggs or young have been found for the first time) (Table b); two of these, Lapland Bunting and Red-backed Shrike, had earlier been suspected of breeding but definite proof had not previously been obtained. Two further species (Marsh Harrier and Long-tailed Skua) have attempted to breed, and the Turnstone is also believed to have done so. Successful recolonisation by the Osprey has taken place, and the Goshawk is also well on the way to re-establishing itself as a regular breeder. Reintroduction of the White-tailed Eagle resulted in successful breeding for the first time in 1985.

Those species which have bred only once or twice and then vanished again (Montagu's Harrier, Little Ringed Plover, Bluethroat, Great Northern Diver, Reed Warbler, Golden Oriole, Spotted Sandpiper and Shore Lark) are clearly of less significance than those showing signs of colonisation. The Spotted Sandpiper merits special mention, however, in view of the chances against a pair of American waders arriving in Scotland in the same place, at the same time and at the start of the breeding season. Also noteworthy is the single breeding record of the Reed Warbler – in Shetland, where its customary habitat is scarce, to say the least. The Snowy Owl proved to be only a temporary addition to the breeding list, the Wryneck is by no means firmly established, the Little Owl is making little progress in expanding its range, and it is too soon to guess what will happen with the Purple Sandpiper and Scarlet Rosefinch – or the Brambling, which was recorded breeding successfully for the first time in 1982. The Wood Sandpiper and Fieldfare maintain only a somewhat tenuous breeding presence and their future status remains uncertain – as is that of the Redwing, which first bred in 1925 and reached a peak of well over 60 pairs in 1983. The only species which have established themselves firmly during this period (other than the recolonists) are the Green Woodpecker, Collared Dove and Goldeneye; all three look as though they are here to stay.

Although not strictly eligible as 'firsts', breeding records of several other species are worthy of comment. The Ruddy Duck, now well-established as a feral species in England, bred for the first time in 1979. Glaucous and Mediterranean Gulls have been recorded breeding in mixed pairs with Herring and Black-headed Gull respectively, in the former case successfully. And Blue-winged Teal and Black Duck have attempted to pair with Shoveler and Mallard respectively.

SIGNIFICANT CHANGES IN NUMBERS AND/OR
BREEDING RANGE

By far the most impressive expansion of breeding range has been that of the Collared Dove, which only arrived in Scotland in 1957 yet was breeding in every county and all the island groups by 1965. Presumably its success has been largely due to the fact that it was able to exploit a previously unoccupied niche. The Green Woodpecker has also done well; from the Borders, where breeding was first proved in 1951, it has spread throughout the lowlands and north as far as Aberdeen and Inverness. But it is still by no means abundant, and its future may be much influenced by woodland changes. In 30 years Osprey breeding numbers have risen from one pair to more than 30, and nesting has occurred in widely separated areas. The Hen Harrier has also extended its range, spreading south from its pre-war Orkney stronghold and moving into many moorland areas where gamekeeping has ceased; it now breeds regularly in most counties with suitable habitat, though the exact area occupied varies with changing land-use.

Several seabird species have also markedly expanded their numbers and/or range since 1950 (Table c). Among those showing the fastest growth is the Great Skua, which has increased its population more than five-fold and also expanded its range, though only to a limited extent. The Arctic Skua has undergone a similar expansion of range, but not nearly such a marked population increase. Many other seabirds have continued to increase: the Gannet has established three new colonies, the Fulmar has colonised virtually all coastal cliffs and is now nesting inland, the large gulls have all increased – in some areas to pest proportions, and auk populations in the Firth of Forth have expanded rapidly in recent years. The reasons for these increases are not always clear and may often involve several factors. The skuas are presumably benefiting from reduced persecution and perhaps an increase in food supply as colonies of their principal 'prey' species expand. The gulls have learned to exploit the increasing amount of waste food made available to them by man. And the auk expansion is possibly related to changes in the local availability of sprats and sandeels.

The situation in respect of passerines is less clear, as there is little information available upon which to base an assessment of population changes. It seems safe to claim, however, that the Carrion Crow has increased as a consequence of reduced persecution and to the disadvantage of other species; it is now a common bird in towns and readily visits gardens. Woodpigeon numbers have also almost certainly increased. Although breeding in rapidly increasing numbers since the mid 1970s, the Goldeneye has not yet spread far from its initial nesting area, but if the population continues to expand at the present rate an extension of range must surely occur in the near future. The only other breeding species which has undoubtedly undergone a marked increase is an undesir-

(c)   Species for which there is good evidence of status change in the period 1950–1985

Increases

Breeding species showing marked population expansion:

Fulmar	Herring Gull
Gannet	Kittiwake
Shag	Guillemot
Canada Goose	Razorbill
Osprey	Puffin
Goldeneye	Collared Dove
Arctic Skua	Green Woodpecker
Great Skua	Carrion Crow

Breeding species showing local increase or expansion of range:

Slavonian Grebe	Goldcrest
Greylag Goose	Crested Tit
Hen Harrier	Coal Tit
Buzzard	Jay
Golden Eagle *	Siskin
Peregrine *	Redpoll
Dotterel	Crossbill
Lesser Black-backed Gull	Reed Bunting

Passage/winter visitors showing substantial, and generally continuous, increase:

Pink-footed Goose	Barnacle Goose
White-fronted Goose (Greenland)	Grey Plover
Greylag Goose	

Decreases

Breeding species showing substantial overall decrease:

Black-throated Diver	Nightjar
Corncrake	Sand Martin
Red-necked Phalarope	Chough
Roseate Tern	Tree Sparrow
Little Tern	Twite
Barn Owl	Corn Bunting

Breeding species which have decreased locally:

Mute Swan	Redshank
Merlin	Snipe
Red Grouse	Ring Ouzel
Grey Partridge	Willow Tit
Golden Plover	Raven
Lapwing	Linnet

Passage/wintering species showing substantial, and generally progressive, decrease:

Bean Goose	Pochard
Mallard	Scaup
Pintail	Ortolan Bunting
Knot	

* Recent increase following marked declines in the 1960s.

able introduction, the Canada Goose. Originally confined to a small number of sites where feral flocks had become established, it has now spread to many more waters. Unless steps are taken to prevent further expansion this species is likely to become a nuisance to farmers in some areas.

Assessment of change is even more difficult in respect of winter visitors, and only in the case of wildfowl are there sufficient reliable data to justify an evaluation. There is good evidence of substantial increase over the last 20 years in the numbers of Pink-footed, Greylag and Barnacle Geese wintering in Scotland; much of this increase is directly attributable to the introduction of protection measures, especially in the case of the Spitsbergen Barnacle Goose population wintering on the Solway. Wintering numbers of Scaup and Pochard have recently shown a marked decrease, however, largely due to a reduction in sewage discharge into the Forth, which formerly held major concentrations of these species.

Turning now to the gloomier side of the picture, although no species which formerly bred regularly has ceased to do so in the last 30–40 years several have shown a marked decline in numbers or contraction of range (Table c). In the case of the Corncrake and Corn Bunting, the decline is attributable largely, if not entirely, to agricultural changes; these are probably also implicated in the decrease in Grey Partridges and Barn Owls. Several factors have probably contributed to the Red Grouse decline, among them less efficient moor management, disease and moorland afforestation. The reduction in Little Terns has been largely due to disturbance on the nesting grounds, in Ravens to local decreases in the availability of sheep carrion and to poisoning, and in Red-necked Phalaropes and Black-throated Divers (currently having very poor breeding success) probably at least partly to habitat loss, though in these last two cases other factors may also be involved (including the problematic 'climatic change'). The Merlin situation is also complex; there has been substantial loss of habitat in some areas but there is also evidence of poor breeding success due to the effect of pesticides. The reasons for the almost complete loss of the Roseate Tern are obscure, and probably involve factors outwith Britain.

OTHER APPARENT OR SUSPECTED CHANGES IN
NUMBERS OR DISTRIBUTION

Had this book been written twenty years ago the preceding section would undoubtedly have started with reference to declining Peregrines and concern over poor breeding success among Golden Eagles – both due to the effects of pesticides. Fortunately, the populations of both species have now recovered and are in a healthy state. There has probably been some increase in Buzzards and Sparrowhawks too over the last two decades, but this is likely to have been due as much to decreased gamekeeping as to reduction in pesticide levels.

Other relatively minor changes are less well documented and assessment of change is in consequence largely subjective. Among the wildfowl, Eiders and Tufted Ducks have

maintained a slow expansion of range, the distribution of Pintail has changed as a result of habitat loss, and Shelduck have started moulting on the Forth Estuary. Bean Geese have decreased, there have been some changes in the wintering pattern of White-fronted Geese and Whooper Swans, and the breeding Mute Swan population in central Scotland has declined.

There have undoubtedly been local declines in, and contractions in the breeding distribution of, several wader species as a result of habitat change, though the extent to which these are affecting populations as a whole is not yet clear. Among the species thought to be most affected are Snipe and Redshank (through drainage of wet grasslands), Golden Plover and Dunlin (through afforestation of moorland), and possibly Lapwing (through decreasing availability of insects on farmland). Monitoring of wader breeding populations over the next few years should show just how serious the situation may be.

The spread of afforestation has resulted in an expansion in the breeding range of species able to find adequate food supplies among conifers – most notably the Coal Tit and Goldcrest and, though to a lesser extent, the Blue Tit, Crested Tit, Siskin, Crossbill and Capercaillie. Jays, Magpies and Sparrowhawks have increased in numbers in some extensively forested areas, and Black Grouse have probably also extended their range as new forests have been established. At the same time the breeding distribution of those species dependent upon old woodland, such as the woodpeckers and the Willow Tit, has contracted, at least on a local scale, as this habitat has become scarcer. Nightjar numbers have decreased and Twite too seem to be declining; in neither case is there an obvious reason.

Conditions outwith Scotland have had a major impact on the breeding numbers of some summer visitors in the last two decades. The classic example is the Whitethroat crash due to drought conditions in the Sahel region where the species winters; numbers have not again reached their pre-1968/69 level. The prolonged drought in that area has affected other species too – for example the Sand Martin, whose population in 1984 was estimated to be about 30% of the previous year's and less than 10% of that in the mid 1960s.

Finally, there have been a few apparent changes in the frequency with which some rarities and passage visitors occur. Three species seen during the first half of this century have not been reported since 1950: Squacco Heron, Baillon's Crake and Blyth's Reed Warbler. A further eight have not been recorded since last century. A number of American waders apparently turn up much more frequently than they did in the past, as do some of the scarcer warblers – but this may be largely if not entirely due to more intensive observation. There is no doubt, however, that more Grey Plovers and Blackcaps now remain to winter in Scotland, and that the numbers of Ortolan Buntings reaching our shores during migration have decreased.

11: Background to the species accounts

PERIOD COVERED

The main emphasis is on the period 1950–83, but records of rarities and details of counts up to spring 1985 have been included wherever possible; the brief summaries of earlier data are based on the accounts in B&R. Assessments of migration dates, and numbers of the scarcer migrants and winter visitors, are based on the SBR and consequently cover only the period from 1968 onwards.

ARRANGEMENT OF MATERIAL

The species accounts follow the sequence and nomenclature of Voous's *List of recent holarctic bird species* (1977), as given in *The 'British Birds' List of Birds of the Western Palearctic* (1984). Species not included therein are placed according to the listing given in *A Complete Checklist of the Birds of the World* (Howard & Moore 1984).

All species included in B&R, and those for which records have been published in SB, SBR or local checklists, appear in the appropriate position; those not on the British list, or for which the only Scottish records are believed to relate to escapes, are square-bracketed, and species in the BOU classes C (originally introduced, now a self-sustaining feral population) and D (reasonable doubt as to whether they have ever occurred in a wild state) are indicated. It is noteworthy that several of the species listed as escapes in B & R's Appendix 1 have since been accepted for inclusion in the British list. Unless otherwise stated, 'first records' refer to Scotland only. Where a species is given 'special penalty' protection under the Wildlife and Countryside Act 1981 this is indicated by the entry [1] indicating Schedule 1, in the text heading. The text for each regularly-occurring species starts with a brief summary of status and distribution.

Old county names (in the abbreviated form, eg Perth = Perthshire, unless otherwise stated) are used throughout. Map 12 shows both the old county and the new regional

boundaries. All the places referred to (apart from a few that are mentioned only once) are listed, together with their four figure national grid reference, county and region, in Appendix 1; only the place-name is usually given in the text. The limits of the areas referred to as Outer Hebrides, Inner Hebrides and Clyde Islands are shown on Map 2, Chapter 1. The name Forth Estuary is used for the area between the Forth and Kincardine Bridges; the Firth of Forth includes both this area and the outer Firth as far east as Largo and Gullane. The Moray Firth is used for the area between Inverness and Fort George; the Moray Basin includes the Dornoch, Cromarty and Moray Firths and east to a line from Brora to Spey Bay.

Unless otherwise indicated, all statements refer solely to the situation as it exists in Scotland. 'Britain' implies Scotland, England and Wales; 'Ireland' implies Northern Ireland and Eire combined. The use of an oblique in a date, eg 1981/82 or May/July, implies a continuous period; a hyphen, eg 1981–83, is used where data refer only to particular seasons in several years, ie where the recording period is not a continuous one.

The abbreviations used are listed at the beginning of the General Index; those relating to publications are also referred to later in this section. Information derived from personal communications is indicated by references including either initials or first-names ('pers. comm.' is used only where reference is made to both published and unpublished material by the same author).

THE SPECIES MAPS

These are intended to give a generalised picture of distribution, and do not purport to represent accurate boundaries; within the areas outlined, breeding will always be dependent upon the presence of suitable habitat. Where appropriate an attempt has been made to distinguish between areas in

Map 12. Local authority divisions: old county boundaries (dashed line) and post-1974 regions (bold type and solid line)

which a species breeds regularly and is relatively abundant, and those in which it is a scarce or sporadic breeder. The distribution shown in *The Atlas of Breeding Birds of Britain and Ireland* (Sharrock 1976) has in many cases been modified to take account of recent changes reported by local recorders. Details of wildfowl and wader concentrations are drawn from the sources listed below.

SOURCES

For many species the principal sources are *Scottish Birds* (SB), the *Scottish Bird Reports* (SBR), Baxter & Rintoul's *The Birds of Scotland* (B&R), *The Atlas of Breeding Birds of Britain and Ireland* (Atlas or 1968–72), *Breeding Birds of Britain and Ireland* (Parslow) and *The Status of Birds in Britain and Ireland* (BOU); references to these are abbreviated as indicated. The 'Winter Atlas', in preparation at the time of writing, is referred to as (Winter Atlas). Abbreviations are used in references to *British Birds* (BB) and local reports, eg *Grampian Ringing Group Report 1980* is referred to as (Grampian RG Rep. 1980) and *The Borders Bird Report 1983* as (Borders BR 1983). Local reports and checklists (listed in Appendix 2) have been widely used, and brief assessments of current breeding status have been obtained from the local recorders in all areas for which recent checklists were not available. Individual references are given in the text only where an important paper is involved; where a series of related studies exists only the most recent reference is usually given (provided that this contains a reasonably comprehensive bibliography). Details of breeding ranges, migrations and wintering areas outwith Britain are drawn from Voous (1960), Harrison (1982), *Birds of the Western Palearctic* Vols. 1–3 (BWP) (1977 *et seq.*) and, for American species, Peterson (1959). Important additional sources of data for particular groups are detailed below.

VAGRANTS

Most species occurring less than annually, or irregularly and in only small numbers, are included in this category; an indication of the probable area of origin is given in the introductory summaries for these species. *Scottish Birds* (first published in 1958), *Scottish Bird Reports* (from 1968), the annual reports on rarities published in *British Birds* (BB), and *Rare Birds in Britain & Ireland* (Sharrock & Sharrock 1976) are the main sources of data on vagrants. Descriptions of recent additions to the British List can be found in *Birds New to Britain* (Sharrock & Grant 1982) and BB. The lists of species considered by the British Birds Rarities Committee and species classed as rare in Scotland are subject to periodic review; reference should be made to BB and SBR respectively for information on the current situation. Details of the records upon which accounts of the rarest species (fewer than 20 records to the end of 1984) are based have been deposited in the Waterston Library of the Scottish Ornithologists' Club.

RINGING RECOVERIES

Full computer print-outs of Scottish ringing recovery data have been prepared annually by the BTO for the SOC since 1979 and are held in the Waterston Library. Summaries of earlier data, or full listings in the case of species for which there have been few recoveries, together with details of foreign-ringed birds recovered in Scotland, were kindly provided by the BTO for use in the preparation of this book. These data have also been deposited in the Waterston Library, but may not be used for other publications without the permission of the BTO.

SOURCES SPECIFIC TO PARTICULAR GROUPS

Wildfowl. Wildfowl have been systematically and regularly counted for longer than any other group; monthly winter wildfowl counts were started by the Wildfowl Trust in the late 1940s and annual censusing of grey geese in the late 1950s (see Chapter 8). The unpublished results of these studies are important sources of data on numbers. (A full print-out of Scottish wildfowl counts, covering the period since 1960, is available for reference in the Waterston Library.) Other important sources for this group include *Wildfowl in Great Britain* (Atkinson-Willes 1965; Owen *et al* 1986), which gives much information on individual sites, *Wild Geese* (Ogilvie 1978), and the annual summaries of *Wildfowl and Wader Counts* (WWC), published jointly by the Wildfowl Trust and the BTO. The current qualifying levels for nationally and internationally important concentrations of wildfowl and waders are given in WWC (note that these levels change as new data on population sizes become available).

Waders. Much published material relating to this group has resulted from the many wader studies that have taken place since the late 1960s. These studies include the Birds of Estuaries Enquiry (BoEE), cannon-netting and dye-marking of passage and wintering birds and, more recently, assessments of breeding populations. The formation of the Wader Study Group (WSG) in 1970 helped to stimulate such work and to encourage close co-operation between workers in different areas, both within Britain and further afield. Sources used here include *Estuary Birds* (Prater 1981), the raw data collected during the BoEE and preliminary 'Tables of average wader counts' prepared for NCC by Rowe (1978), Wader Study Group Bulletins, and local ringing group reports. Although the movements and winter populations of some species, notably those which show a strong preference for habitats of limited extent and/or very localised distribution, are now fairly well documented, there are still big gaps in our understanding of many common species, especially those which are widely dispersed in winter (Summers *et al* 1984) or breed at low densities over large areas. Recent work on the breeding waders of agricultural land (Galbraith *et al* 1984) and of moorlands (Reed *et al* 1983a, and RSPB/L. H. Campbell unpub.) has provided base-line information of value for future monitoring. The BTO/WSG Winter Shorebird Count in 1984/85 added to knowledge of the birdlife of rocky coasts, especially in the northwest.

Seabirds. Many major sources are common to the Fulmar-to-Shag group and the skuas to auks; these groups are therefore dealt with together here. Although the populations of some individual species, such as the Fulmar and the Gannet, were studied earlier, it was not until the Seabird Group (founded in 1965) initiated 'Operation Seafarer' that any co-ordinated attempt was made to census all breeding seabirds. *The Seabirds of Britain and Ireland* (Cramp *et al* 1974), which summarises the results of this massive exercise (referred to as '1969–70'), is the principal single source used for this group; the detailed survey counts (deposited in the NCC Library in Edinburgh) have also been consulted. Operation Seafarer drew attention to the difficulties of counting seabirds, and there have since been many studies designed to assess the extent to which variations in timing and techniques may affect the comparability of counts (eg Lloyd 1975; Harris 1976; Hope Jones 1977; Evans 1980; Richardson *et al* 1981; Furness 1982; Wanless *et al* 1982; Harris *et al* 1983; Wanless & Harris 1984). Intensive studies have also been made of the seabird populations of particular areas, including the Hebrides (Bourne & Harris 1979), Berwickshire (da Prato & da Prato 1980), Caithness (Mudge 1979), St Kilda (Harris & Murray 1978), Foula (Furness 1983), the Isle of May (Harris & Galbraith 1983) and Canna (Swann & Ramsay 1984). The planned repetition in 1985–86 of the Operation Seafarer Survey will permit early up-dating of many of the figures quoted here. Regular monitoring of sample populations, particularly of cliff or open-ground nesters, has shown that natural population fluctuations from one year to the next may be quite marked, and has consequently cast doubt on some of the earlier claims of major changes in status.

Raptors and relatively rare breeding species. Populations of the larger raptors have been regularly monitored in some areas for many years, and more comprehensively censused at least once. Study of the smaller species has been less co-ordinated, but the recent establishment of raptor study groups covering most of Scotland (SB 13:162-166) should help to ensure that the overall national situation can be assessed in future. Unpublished records relating to several of the raptors and to other relatively rare breeding species have kindly been made available by the RSPB and NCC; these have been used to illustrate annual and regional variations in densities and breeding success, and confidentiality has been maintained for all but the most widely-known 'public' sites.

Passerines. Much of the published information on passerine breeding densities is based upon Common Birds Census (CBC) and Waterways Bird Survey (WBS) results; as rather few CBCs and WBSs are carried out in Scotland, the results are heavily biased towards the situation prevailing south of the Border. To enable me to present figures reflecting a more realistic assessment of the Scottish situation, the BTO kindly abstracted and summarised the available data. For Scotland, however, these turned out to be too limited to do more than indicate very roughly the range of breeding densities that may occur here and, for a few species, the greater effect of severe winters on bird populations in the north of Britain. Work on farmland bird populations in lowland Scotland, by Stan da Prato (1985) and by Phil Shaw (commissioned by NCC and as yet unpublished), has added greatly to the available information. Unpublished density data for a number of species have been kindly provided by people who have studied these species locally in Scotland.

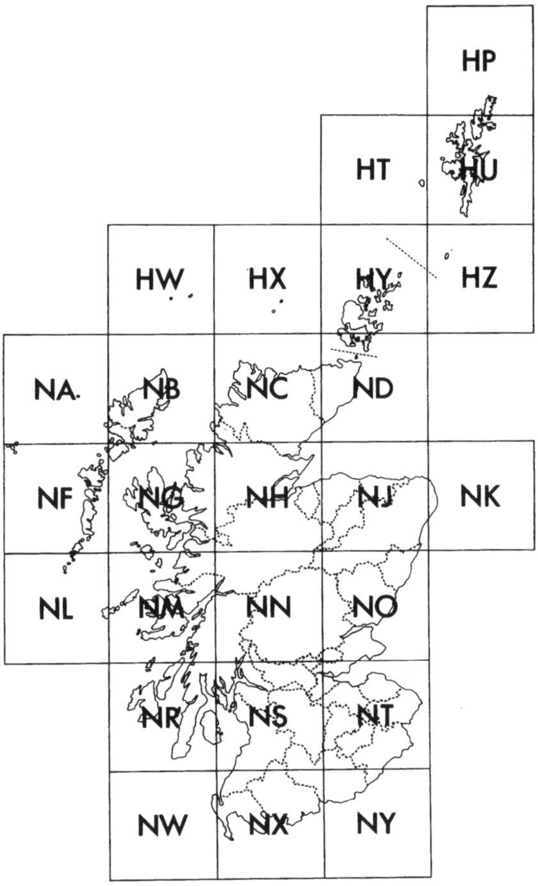

Map. 13. The grid on this map is the National Grid taken from the Ordnance Survey map with the permission of the Controller of Her Majesty's Stationery Office

The species accounts

Red-throated Diver *Gavia stellata* [1]

Breeds on hill lochs in the north and west, from Arran to Shetland, with the largest numbers in the Northern Isles. In winter most abundant in east coast waters, where immigrants are probably also present.

The Red-throated Diver nests beside quite small and shallow moorland waters, often flying to larger lochs and the sea to feed. A considerable extension in breeding range took place in the first half of the 20th century (Atlas) and this is apparently continuing. Numbers too have been increasing, and the current population is probably in the region of 1,000–1,200 pairs. Shetland, Orkney and the Outer Hebrides hold the bulk of the population, but substantial numbers also breed in Sutherland, Caithness, Wester Ross and the Inner Hebrides and there are smaller populations in Inverness, Perth, Argyll (including Kintyre) and Arran. Surveys in the early 1980s located 700 pairs in Shetland (Gomersall *et al* 1984), 90–95 pairs in Orkney (Booth *et al* 1984) and 11 pairs in Arran (Arran BR 1983); no comprehensive figures are available for other areas. Breeding has also taken place, or been attempted, at least sporadically in Nairn (1978), west Stirling (1974), Renfrew (one or two pairs 1972–1981 but never successful), Ayr (1957–58 but not since), Bute (1980), Kirkcudbright (1955), and Galloway (1973 – successful). In 1980 a juvenile was seen in Moray, though breeding was not proved there, and birds have recently summered on an inland loch in Aberdeen.

Breeding densities are very high in some parts of Shetland, and especially on Foula (Merrie 1978). Detailed observations of breeding, feeding behaviour and other aspects of the species' ecology have been carried out on Foula over many years, and trapping of breeding adults has shown that they are extremely faithful to their chosen nesting waters, which there range in size from 15 m² to 24,000 m² (Furness 1983). The breeding biology of the Red-throated Diver has also been studied in Unst and Yell (Bundy 1976, 1978) and in Orkney (Booth 1982). These studies have shown that breeding success is much affected by disturbance, 'wilful' disturbance by birdwatchers and photographers being a more frequent cause of failure than incidental disturbance by fishermen or peat-cutters. Fluctuations in water-level may also affect breeding success, but the suggestion by Bundy (1978) that high numbers of skuas and Great Black-backed Gulls may also influence success rate seems likely to be correct only if high predator numbers are associated with frequent human disturbance. On Foula, where Great Skua densities are higher than anywhere else in Britain but disturbance is minimal, breeding success is good, ranging in 1973–79 from 0.55 to 1.0 and averaging 0.63 chicks fledged per pair (Furness 1983); the average for Fetlar is about 0.47 (Gomersall *et al* 1984). Taken in conjunction with site-fidelity, it is clear that this species' vulnerability to disturbance is likely to be a major factor in determining where, and to what extent, it will be able to maintain its numbers in the future. The

RED-THROATED DIVER

▨ main breeding area

▨ scarce or sporadic

Breeding success of monitored pairs of Red-throated Divers in the north of Scotland and the Northern Isles (Data from SBR/R. H. Dennis et al)

	1972	1973	1974	1975	1976
No. of pairs checked	62	101–111	130	106	96
No. of young reared	42	55	75	53	48/49
Young/territorial pair	0.68	0.52	0.58	0.50	0.50

very great fluctuations in productivity between years are well brought out by the results of the RSPB's regular monitoring of selected sites (Table).

Red-throated Divers are at their breeding lochans from late March until August or early September, thereafter moving to the coast. Quite marked coasting movements are sometimes observed between the end of August and early October, for example 72/hour travelling south off Buchan Ness and over 50/hour off Rattray Head. Numbers vary considerably from year to year (Elkins & Williams 1974), but are always high on the east coast, especially off the Aberdeen coast between the Don and Collieston. There was an exceptionally high count of 1,470 there on 15 September 1979; totals of 500–800 are more usual. Large numbers also occur in the Moray Basin, where c1,500 were counted in October 1982 (Barrett & Barrett 1985); nearly 500 flew east past Tarbat Ness in early December 1978. South of Aberdeen numbers tend to be lower and counts of over 100 are relatively infrequent, while on the west coast and around the Northern Isles winter gatherings of 50 or more are seldom recorded. Occasional birds occur inland in winter.

Comparatively few Red-throated Divers remain in Shetland waters during the winter and only seven were found oiled after the mid winter *Esso Bernicia* spill of 1978, in an area where nearly 150 were present in April (Heubeck & Richardson 1980). However, more than 50 were beached between Berwick and Aberdeen in the oiling incident of January/February 1970 (SB 6:235–250), and the big concentrations off the Aberdeen coast must be increasingly at risk as oiled-related activity in that area expands. A threat which may be serious on a more local scale has been described from the Outer Hebrides, where breeding divers are sometimes caught in gill-nets set close inshore (Buxton 1983). Net drownings have also been recorded in Shetland and in the Moray Basin; one ringed bird was found in a skate-net at a depth of 14 m.

Ringing recoveries indicate that Scottish Red-throated Divers disperse widely in winter, some moving south as far as the French coast in mid winter, while others remain in Scottish waters. First-summer recoveries include one caught in a fishing-net off Norway, and Shetland birds in the Hebrides and Northern Ireland. Most winter immigrants are probably from Norway and Sweden, but at least a few come from Greenland; there is no evidence to date that Icelandic Red-throats winter in British waters (BWP). Native birds start moving north again in February and most have settled into their breeding quarters by late March.

Black-throated Diver *Gavia arctica* [1]

Scarce breeder on the larger and remoter lochs of the north and west Highlands and the Hebrides. Winters in coastal waters.

The Black-throated Diver nests close to the shore, often on small islets, and is very vulnerable to fluctuations in waterlevel, and to disturbance – which can facilitate predation of the exposed eggs by crows or gulls. Breeding success is seldom high and in some years few young are reared. With a total population of not more than 100 pairs, and numbers declining in some parts of the range, the future of the Black-throated Diver must give cause for concern. A new potential threat in some districts is the afforestation of moorlands around the breeding lochs; as yet these plantations are so young that the open moorland atmosphere has not been lost, but as the trees mature and reduce the birds' view of the surrounding area such lochs may well be deserted.

In 1977 Bundy (1979) studied the breeding and feeding habits of the species in three areas in northwest Scotland. Although 40–42 breeding pairs were located, breeding was proved for only 13 pairs, which reared a total of seven young. Black-throated Divers were seen on less than half the apparently suitable waters, ie deep, island-studded lochs at least 10 ha in extent, and with a shallow-water feeding area less than 5 m deep. None bred successfully on regularly-fished lochs smaller than 45 ha, although several young were reared on larger lochs which were fished almost daily.

Only minor distributional changes have taken place this century, but there has been some spread southwards. B&R give 19th century records for Moray and Coll and old records for Orkney; breeding has not been proved in any of these places this century. The Black-throated Diver colonised Skye, Jura and Mull in the first half of the century (Reed *et al* 1983) but there has been no recent breeding record

Breeding success of monitored pairs of Black-throated Divers in the Highlands (Data from RSPB/R. H. Dennis et al)

	1972	1973	1974	1975	1977	1981	1983	1984
No. of pairs checked	58–59	28	45–47	45	31	47–56	28	55
No. of successful pairs	10	6	11	12	11	11–12	6	12
No. of young reared	13	6	16	15	14	12–13	8	15
Young/territorial pair	0.22	0.21	0.35	0.33	0.45	0.24	0.28	0.27

concentrations are known, but fair numbers apparently winter on the west coast, where more than 20 have been seen together near Gairloch. Coastal numbers are smaller on the east, seldom exceeding 10–15 in any one area, but during the autumn there is occasionally a considerable build-up on some freshwaters, for example Linlithgow Loch, where a peak of 66 was recorded in October 1980.

Although Scotland is near the northern limit of the Little Grebe's European breeding range, and the population is largely sedentary, some immigration undoubtedly occurs. The birds involved are possibly from among the small numbers breeding in the extreme south of Norway and Sweden, or the rather larger Danish population. Virtually nothing is known about the scale, regularity or origins of any such movements across the North Sea, but one might suppose that they are influenced by icing conditions around the Skaggerak and Kattegat. The most obvious recent evidence of immigration was in mid October 1979, however, too early in the season for icing to be a causative factor. On that occasion there was a marked influx into Shetland, with 18 records during November and the first-ever Little Grebe on the Out Skerries.

Natural hazards, such as flooding of nests and predation of chicks, appear to be the main threats to successful breeding in this species. Other factors may at times affect local numbers, for example disturbance and wave action caused by water sports, loss of food supply when a loch is poisoned with Rotenone to kill off pike, and perhaps disturbance and predation by mink, but there seems no likelihood of these having any significant impact on the population as a whole.

Great Crested Grebe *Podiceps cristatus*

Breeds on most suitable lowland waters but distribution is currently changing, probably due to increasing recreational pressures in the Central Lowlands. In winter most move to coastal waters; the Firth of Forth holds the largest concentration, thought to include many immigrants.

For breeding the Great Crested Grebe prefers shallow, standing waters with sufficient emergent vegetation for attachment and concealment of its floating nest, and an adequate supply of eels and small fish. Most lochs of this type are at low altitudes, but nesting has been recorded up to about 250 m asl. Breeding success is much affected by water-level fluctuations; a sudden rise due to heavy rain, or a progressive fall due to prolonged drought, can be equally disastrous and may at times result in complete failure on some waters. This species has an extended breeding season, however, and frequently re-lays after losing a clutch.

The gradual colonisation of lowland Scotland since 1877 has been documented in detail by B&R. By 1931, when a census was first attempted, the overall pattern of distribution was much as it is today, and most subsequent changes have been in local numbers rather than on a wider geographical scale. Following reports of relatively high levels of organo-chlorine residues in Great Crested Grebes in some parts of England, a second census was carried out in 1965; this showed that Fife, Perth and Angus held about 50% of the

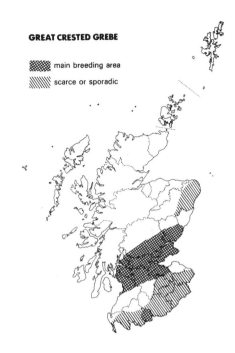

GREAT CRESTED GREBE

▓ main breeding area

⦚ scarce or sporadic

Scottish population and that the most rapid increases were occurring in these counties (Prestt & Mills 1966). A partial survey in Scotland in 1973 (Smith 1974) was later followed by a further national census in 1975 (Hughes *et al* 1979). The summarised results (Table) indicate little change in overall numbers since 1965; it is important to note, however, that the results of the three counts are not directly comparable, since coverage was not consistent throughout, and also because grebes, especially non-breeding birds, may move between waters within a few days. One point that the 1975 census did make quite clear was that the Scottish situation was stable in comparison to that in England, where numbers had increased by around 50% since 1965 and more than 400 new waters had been occupied, many of them man-made (eg old gravel pits and disused canals).

Perth still holds the largest numbers and, although the total now seems relatively static, distribution is changing, with four new sites occupied in at least one season since 1975, and several old-established ones now of much reduced importance (Thom & Cameron 1980). Declines have, however, taken place in Fife and, to a lesser extent, in Angus; in both these counties there has been much recent development of water sports and angling, and several waters which previously held grebes no longer do so. Equally marked increases have occurred in Ayr, Renfrew and the southwest, especially Kirkcudbright. Continuation of this trend could result in an east to west swing in the main breeding concentration of this species.

B&R recorded breeding in Moray and Berwick in the first half of this century but there have been no recent reports from these counties. The reported breeding on Loch Maree, which they describe as 'curious', is disregarded in *The Status of Birds in Britain and Ireland* (BOU). There appears to be no record of breeding in East Lothian, Kincardine, Banff, Nairn, Inverness, Ross and Cromarty, Sutherland or Caithness, although there are very occasional summer reports from as far north as Orkney and Shetland, and from the Outer Hebrides. A pair with young on Ardnave Loch, Islay, in June 1969 (SBR) is the only island breeding record to date.

In their report on the 1975 census Hughes *et al* stress the need for further investigation into the local factors responsible for decreases in the breeding population. No formal investigations have yet been mounted, but examination of the history of selected lochs suggests that the factors (other than fluctuating water-levels) most likely to be influencing Great Crested Grebe populations in Scotland are intensive mangement for angling, and disturbance from water sports. In Fife and Perth several of the lochs holding most birds in 1965 were subsequently developed for angling; this can involve treatment with Rotenone and/or netting to eliminate pike, stocking with well-grown fish (thus altering the natural proportions of different sizes), and frequent disturbance of nesting areas by boats. The lochs most affected (Lake of Menteith, Butterstone and Lindores) showed substantial reductions in population between 1965 and 1973–75.

Many other lochs at which the level of disturbance has increased in the last 20 years (eg Lochs Fitty and Gelly in Fife and most of the Lanark lochs) have also shown a decline in breeding numbers (Smith 1974). That recreational use need not have such an adverse effect has, however, been demonstrated at Linlithgow Loch, where a reedbed and adjoining water area are marked off as out-of-bounds to small boats; there the grebes have continued to thrive and to breed successfully. It seems desirable that wherever major recreational use is introduced on waters used by breeding Great Crested Grebes action should be taken to ensure that suitable areas are protected from disturbance.

At Loch Leven colonial nesting has been recorded, with up to 15 nests in an area of c15 m × 5 m (A. Allison). Numbers there were low (less than ten pairs) in the late 1960s and early 1970s but have been increasing fairly steadily since 1975 to reach an all-time peak of 35 pairs in 1984 (G. A. Wright).

Great Crested Grebes are not easily caught for ringing so little is known about their movements. Many of the breed-

Breeding season distribution of Great Crested Grebes (Data from Smith 1974 and Hughes et al 1979)

	Estimated total number of adult grebes			Number of waters holding grebes		
	1965	1973	1975	1965	1973	1975
E. Central						
Perth	76	c79	84	18	17	16
Angus	28	(12)	16	7	(3)	5
Fife	66	52+	21	12	15	7
Kinross	8	4–8	18	1	1	1
Clackmannan	9	?	3	2	?	1
W. Lothian	6	9	?	1	1	?
Midlothian	16	13	28	5	6	6
W. Central						
Stirling	10	17+	14	5	7	5
Dunbarton	2	0	?	1	0	?
Renfrew	20	34	38	2	9	6
Lanark	26	18	19	9	8	8
Ayr	4	10	23	1	4	7
South-west						
Dumfries	10	19	18	3	5	4
Kirkcudbright	7	22–24	30	3	6	6
Wigtown	0	14	2	0	3	1
South-east						
Selkirk	8	8	8	4	3	4
Roxburgh	5	6+	4	2	3	2
Peebles	0	5	2	0	2	1
Berw. & E. Loth.	0	0	?	0	0	?
North-east						
Aberdeen	6	16	9	2	5	2
Banff, Moray & Kincardine	0	0	?	0	0	?
North						
Nairn & Invern.	0	0	?	0	0	?
R & C, Sutherl. & Caithness	0	0	?	0	0	?
Argyll	2	?	?	1	?	?
Total	309	370–380	337	79	98+	82

Notes – () indicates incomplete cover of formerly occupied sites.
? indicates no records

ing lochs are certainly deserted in mid winter, while others may hold large numbers; Loch Leven for example has held over 100 in December. Return dates are determined largely by the severity of the winter, with birds reappearing as early as February in open weather. Gatherings of 20–50 grebes occur from August onwards off Lunan Bay and in the Tay and Clyde Estuaries, and many more in Loch Ryan (peak to date 255 in November 1978), but by far the largest concentration is in the Firth of Forth. In January 1980 a record 1,200 were off Edinburgh (SBR), and in February that year there were 650 further up-river, between Bo'ness and Kinneil (Bryant 1980).

In February 1978 a relatively small oil-spill at Leith Docks killed at least 200 Great Crested Grebes, some 75% of them first-winter birds (Campbell 1980). As it was feared that this high mortality might have serious consequences for the Scottish breeding population a partial census was organised that summer; this showed that there had, in fact, been no significant change in breeding numbers. This incident demonstrated clearly the vulnerability of such major concentrations of birds to even relatively minor oiling incidents.

Ringing recoveries provide some evidence of immigration into England by grebes from Denmark and the Netherlands, but as yet none to indicate whether any, most, or all of those wintering in the Forth originate from the other side of the North Sea. Campbell (1980) has suggested that the Forth mid winter population consists mainly of immigrants; if his supposition is correct, birds from the northwestern part of the species' extensive Eurasian range, ie southern Scandinavia and Denmark, are perhaps the most likely to be involved.

some seasons reports are widely scattered as far west as the Outer Hebrides while in others few are seen away from the Forth. Little is known about the movements of this species, which breeds locally – and often sporadically – from Denmark, West Germany, southern Sweden and Finland east through Russia, in the Balkans and in North America. The north and northwest European breeders are thought to winter in the Baltic, and off North Sea and Atlantic coasts (BWP). Presumably the size of the Scottish wintering population is influenced by conditions in the Baltic and the North Sea at the time of the post-breeding dispersal. The only record of the American race *P.g.holbollii* is of one shot in Wester Ross in September 1920.

Since the early 1970s the number of Red-necked Grebes summering here has also increased, although it is still very small indeed. The first bird in breeding plumage at an inland site was in Perth from late April to 8 June 1974, and in 1975 one was seen displaying in Lanark in late May. In 1976 and 1977 birds were again present during the summer. 1978 was a blank season but in 1979, following a marked influx of grebes early in the year (Chandler 1981), there were at least three summer records – though of single birds only – in east Scotland and in Sutherland. In 1980 a pair was present for the first time. They built a nest and were seen copulating, but it is not known whether eggs were laid, so there is as yet no proved breeding record of this species in Scotland. Two summered separately in 1981 and in 1983 there were two at an inland loch in June. It is interesting that the Red-necked Grebe apparently requires a considerably smaller but more densely vegetated water area for successful breeding than does the Great Crested Grebe (BWP).

Red-necked Grebe *Podiceps grisegena*

Winter visitor, usually present in the outer Firth of Forth from July to April; sporadic elsewhere on the coast and inland. A few occasionally summer and in 1980 a pair nested for the first time.

B&R described the Red-necked Grebe as 'an occasional winter visitor' but it is now recorded annually, though it is uncertain whether this indicates a real increase in wintering numbers or simply reflects more effort and skill directed towards locating the birds. By far the largest numbers are seen on the East Lothian coast, in Gosford Bay and from Aberlady to Gullane; counts there have increased fairly steadily since 1968, reaching a record total of 58 at the end of July 1981. Red-necked Grebes are also seen fairly regularly, although in much smaller numbers, elsewhere on the east coast north to Shetland – the highest count so far outwith the Forth has been 19 off the Dornoch Firth in November 1982. On the west coast occurrences are more sporadic, and inland they are rare. Since 1968 there have been reports from all coastal counties and island groups, and from all inland counties except Selkirk and Clackmannan.

Arrival has tended to be earlier in recent years, with several July records, but in 1982 there was a return to the former situation, with few reported until well into August. Numbers appear to be lower in mid winter than in autumn and spring, presumably because the birds move further south along the coast. There are also clearly variations between years. In

Slavonian Grebe *Podiceps auritus* [1]

Scotland holds the entire British breeding population, which is confined to the Highlands. Widely distributed around the coast in autumn and winter, generally in fairly small numbers but with a major concentration – probably including immigrants – in the Forth.

*Numbers and distribution of Slavonian Grebes (pairs) on Scottish breeding sites * (Data from Atlas & RSPB/R. H. Dennis)*

	1974	1975	1976	1977	1978	1979	1980	1981	1982	1983
Inverness N. of Gt. Glen	19–20	23	21–22	24–25	24	19–20	24–26	22	16–19	21
S. of Gt. Glen	29–31	31–33	41–43	35–36	48–50	45–50	46	38–43	38–40	43
Strathspey	1	1	3	?	2	2	2	2	4	6
Moray	5–6	4	4–6	4	4	5–6	3–5	5	4	3–4
Other (Caithness/Perth/Aberdeen)	5	3	1	1	2	1	1	1	1	0
Total	59–63	62–64	70–75	64–66	80–82	75–81	76–80	68–73	63–68	73–74

* not all pairs actually nested.

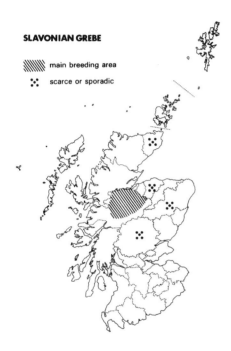

SLAVONIAN GREBE

main breeding area

scarce or sporadic

This is the most northerly of the grebes, with a circumpolar distribution; elsewhere in Europe it breeds only in Iceland, Faeroe (sporadically) and Fenno-Scandia. It usually nests on less productive, and much less heavily vegetated, fresh-water lochs than the other species, and also occurs at higher altitudes. Breeding was first recorded in Inverness in 1908, and the main area of expansion has been based around the original site, although colonisation has also occurred in two other parts of the county. Nesting was recorded in Suther-land in 1929, when four pairs were present, and took place irregularly there until the early 1960s (SB 2:176); an adult was seen again at this site in May 1980. In 1929 breeding also occurred for the first time in Caithness, where it con-tinued fairly regularly, with a maximum of ten pairs, until 1975, since when there has been no further record of nesting. The next county to be colonised was Moray; breeding has taken place regularly there since the 1950s, from three to seven pairs generally being present at one or two sites. In Aberdeen breeding has been sporadic, with single pairs nest-ing in 1960–62 and in 1974 but not since. The first breeding south of the Grampians was in 1973, when a pair nested in Perth, at the Loch of the Lowes; at least one pair has

since bred annually in the county, and in 1982 four sites were occupied although breeding was confirmed at only one. There is an unconfirmed and undated report of breeding in Orkney (SB 5:103) and a single adult was seen there in July 1977. Summering birds have also been recorded in Argyll and Dumfries.

Although numbers fluctuate from year to year, there has been a fairly steady increase in the last 15 years, involving expansion of the range and a growing number of sites as well as an increase in the total population (Table). Breeding success, as with all grebes, is very variable and probably not more than 50 young have fledged in any one season. Losses of young are sometimes heavy; the 1973 Loch of the Lowes pair lost all four of their chicks – one taken by a Coot (in full view of the hide!) and the rest by pike. In addition to the natural hazards of changing water-levels and predation, Slavonian Grebes are subjected to considerable pressure from disturbance – often, regrettably, by bird-watchers – and from egg collectors. The fact that slow popu-lation growth continues in spite of such heavy losses has led to the suggestion that some immigrants are still joining the breeding stock (Atlas).

Birds often stay on the breeding lochs well into October, and in 1978/79 two remained at an Inverness site throughout the winter. During September and October Slavonian Grebes are quite often seen on inland waters – occasionally rivers – where they do not breed, but by November most have moved to the coast. Wintering birds are thought to include immigrants from Iceland and northern Norway, and perhaps also, on the east coast, birds from the Baltic (BWP), but nothing is known about the relative proportions of native and immigrant birds, or even whether Scottish birds move south in winter. The first passage birds often reach Fair Isle in August and numbers in the country as a whole appear to peak in mid winter and remain fairly high until March, by which time Scottish breeding sites are being reoccupied.

In winter this is the most widely distributed of the grebes, recorded from all coastal counties and all island groups. The biggest gatherings occur in relatively sheltered waters, but even there numbers are seldom large. The main concen-tration is in the Firth of Forth, where peak numbers in the Gosford to Gullane area increased from around 30–35 in the early 1970s to an all-time record of 154 in January 1979. Counts have since been slightly lower with maxima of 83 in 1979/80, 120 in 1980/81 and 64 in 1982/83. It may be that conditions in the western Baltic affect the numbers wintering in the Forth. Similar numbers probably winter in the Northern Isles; the Shetland wintering population has been estimated at 50–100 (Kinnear 1978) and that of Orkney may be nearly as large. Sullom Voe and Scapa Flow have held 43 and 33 respectively in February. In the west peak counts are lower, for example 22 in the Outer Hebrides (Broad Bay), 23 in the Inner Hebrides (Loch Sligachan, Skye), 25 in Argyll (Sound of Gigha), and 13 in Wigtown (Loch Ryan).

Because wintering Slavonian Grebes are often close inshore they are potentially at risk from oiling. To date the only mortality of any significance from this cause has been in Shetland, where 8–16% of the estimated wintering population died following the *Esso Bernicia* spill in the winter of 1978/79 (Heubeck & Richardson 1980). A major spill in the outer Firth of Forth could have much more serious consequences.

Protection from both disturbance and egg collectors seems likely to be an important factor in enabling the Slavonian Grebe to continue its slow expansion. Regrettably, despite the special penalty protection afforded under the 1981 Act, the controlling of over-enthusiastic birdwatchers and of egg collectors remains a difficult problem to solve.

Black-necked Grebe *Podiceps nigricollis*   [1]

The scarcest of the breeding grebes, confined to a few sites in central Scotland. In autumn and winter the very small numbers at widely scattered localities on the coast probably include immigrants.

For successful breeding the Black-necked Grebe requires shallow and highly productive waters, with abundant submerged and emergent vegetation but not necessarily a large area of open water. In parts of its range such waters are liable to dry out, and it was when particularly arid conditions were experienced in the Caspian area, especially during the 1920s and 1930s, that the species expanded into northwest Europe, where it now breeds very locally and often sporadically from southern Sweden to Spain (BWP). It first bred in Wales in 1904, Ireland in 1915, England in 1918 and Scotland in 1930. Although a single Irish colony held some 250 pairs in 1929–30, breeding has not been confirmed in that country since 1966, and has occurred only sporadically in England. The small Scottish population, which has probably never exceeded 25 pairs, is consequently of national importance.

Since breeding was first confirmed, in Midlothian in 1930 (B&R), it has occurred in 'Tay' (1931, B&R), Fife (1937, B&R), Angus (1945, B&R), Perth (1949, B&R), Renfrew (1956, SB 1:5) and Selkirk (1965, SB 6:26) – the dates given

are those of the first reported breeding. B&R's statement that a few pairs bred in Sutherland in 1949–50 must surely be discounted; neither details nor references are given and the birds involved were in all probability Slavonian Grebes. More recently, growing appreciation of the need to protect these very rare birds from disturbance has meant that not even counties are referred to in published reports of breeding. Such confidentiality, by minimising disturbance, has probably been a major factor in enabling Black-necked Grebes to continue breeding at the two principal sites.

Both numbers and breeding success vary greatly from year to year even at these established sites, neither of which suffers to any significant extent from fluctuations in the water-level. At the more easterly, from two to six pairs have attempted to breed each year since 1970, raising between one and six young per season. At the more westerly site the corresponding figures have been 5–14 pairs and 5–16 young. Losses of chicks are often heavy, possibly due to predation by other birds (eg Coots or gulls) or pike; in 1978 of 24 chicks known to have hatched only about 12 were successfully reared. During the 1970s and early 1980s breeding has also taken place on two other Central Lowland waters and on one slightly further north; none of these can yet be considered firmly established sites, and on most occasions only a single pair has been involved. There was an interesting report in 1972 of possible interbreeding between Black-necked and Slavonian Grebes. At one of the regular Black-necked Grebe sites, a Slavonian Grebe was seen feeding a chick on the back of a Black-necked Grebe, and *vice versa* (Dennis 1973).

Black-necked Grebes start to arrive on the breeding lochs in March but the main build-up usually occurs during April; early in the season many more may be present than eventually remain to breed. At this time birds also frequently make brief visits to waters on which they do not nest. By mid October the breeding lochs are deserted. Most casual sightings inland are during spring and autumn, and most winter records are from the coast. Although there have been reports since 1968 from most mainland coastal counties and island groups (there are no 20th century records for Orkney and the first for the Outer Hebrides was as recent as 1982), the only regular wintering area appears to be Loch Ryan. Even there numbers are very small: a peak of 23 in 1965–74 (Dickson 1975), rather fewer in the late 1970s, and a recent maximum of 18 in February 1980. Elsewhere records generally refer to single birds, but up to nine have been present at one time in the Aberlady-Gullane area. The movements of this species are little known, but it is believed to winter mainly within the breeding range, presumably dispersing more widely, and resorting to coastal waters, where freezing of inland waters is likely to occur.

With such a very small breeding population, apparently dependent upon so few sites, the future of the Black-necked Grebe in Scotland must be uncertain. Little in the way of direct action can be taken to try to ensure its survival, but restraint on the part of birdwatchers would undoubtedly help in giving it every possible chance. Because these birds favour dense cover they can be difficult to see; attempts to observe them are liable to cause considerable destruction to vegetation, and perhaps also to nests and young, quite apart from the possibility that disturbance might cause desertion of a site.

Black-browed Albatross *Diomedea melanophris*

Vagrant (S Atlantic) – annual in Shetland (one individual) since 1974.

This oceanic bird of the southern hemisphere has been reported off British and Irish shores with increasing frequency since the 1950s (Sharrock & Sharrock 1976). Unidentified albatrosses were seen off Orkney in July 1894 and Fair Isle in May 1949 but the first definite record of the Black-browed Albatross in Scotland was in 1967, when one took up residence among the Gannets on the Bass Rock (BB 61:22–27). It was present from May to 28 September, attracting widespread attention, and returned the following year, arriving on 13 April and leaving on 20 July. What was presumably the same bird was seen at sea off Eyemouth in early February 1968, off St Abbs in February 1969, and off Elie in August the same year, but there were only four sightings on the Bass in 1969, all between 10 April and 3 May. An albatross off Hoy in mid August 1969 might also have been the same bird.

An unconfirmed report from Hermaness in 1971 was the next indication that an albatross was still – or again? – frequenting Scottish waters but it was not until July 1972 that the presence of an adult Black-browed Albatross in the gannetry there was confirmed. This bird was not seen at all in 1973 but has since appeared annually and in 1976 constructed a nest for the first time. In most seasons it arrives in March but in 1981 did so on 21 February and in 1984 on 29 February; the last sighting is generally in August but in 1977 it was still present on 20 September. Away from Unst there have been reports, probably of the same bird, off Fife in August 1972 (four days after the latest Hermaness sighting), in Hoy Sound in August 1975 and off Lerwick in July 1976 (both within the period of residence at Unst), and near Rockall on 16 March 1980 (the day after the first Shetland sighting that season).

Fulmar *Fulmarus glacialis*

Scotland holds about 90% of Britain's breeding Fulmars; the largest concentrations are in the north and west but breeding occurs in almost every coastal county and inland nesting has recently started. Non-breeders wander widely over the North Atlantic and

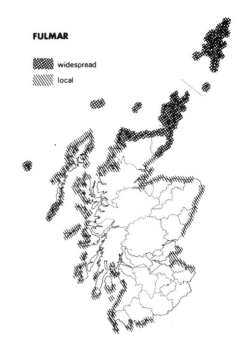

FULMAR

▓▓ widespread

░░ local

the North and Arctic Seas; breeding birds are at the colonies virtually throughout the year.

Since the end of the 19th century the British Fulmar population has shown a spectacular increase, the earlier stage of which was fully documented by James Fisher (1966); it is currently estimated as well over 300,000 pairs and still growing. St Kilda, the first British site, is thought to have been occupied for 800–900 years, but it was possibly birds from Iceland and Faeroe that colonised Foula in 1878 and started the subsequent remarkable expansion in range and numbers. Virtually every suitable part of the Scottish coast has now been colonised and also much of the rest of the British and Irish coastline and as far south as Brittany. The cause of this vast expansion remains uncertain. It has been attributed variously to increased availability of food resulting from developments in commercial fishing, to genetic changes which enabled birds to breed successfully in small straggling colonies as well as large dense ones, and to oceanographic factors associated with the gradual warming of the North Atlantic over the last 100 years (Cramp *et al* 1974). Perhaps the fortunes of the Fulmar over the next 100 years will shed light on this question.

Fulmars are often difficult to count from the land; 'prospecting' non-breeders may occupy ledges during much of the year, estimates of breeding numbers made at different times may not be directly comparable, and the coverage of sites prior to Operation Seafarer was incomplete. The figures in the Table do not, therefore, represent a valid comparison of numbers but they do indicate the scale of the increase in the period 1959 to 1969–70. On Fair Isle the population expanded from c5,000 occupied sites in 1959 to 17,000+ in 1969–70 and 25,000+ in 1975, while monitoring studies elsewhere (Harris 1976, Mudge 1979, Furness 1983) suggest that the population as a whole is continuing to increase,

Increase in sites occupied by Fulmars between 1959 and 1969–70. (Data from Cramp et al 1974 – figures rounded)

| | Number of occupied sites | |
	1959	1969–70
Shetland	36,000	117,000
Orkney	14,000	47,000
Caithness to W. Ross	13,500	43,500
E. Ross to Berwick	4,000	7,000
Outer Hebrides (excluding St Kilda)	20,000	25,000
St Kilda *	c37,500	c36,600

* estimated from partial counts

though probably less rapidly than in the second quarter of this century.

A recent feature of the population expansion has been the growing tendency towards inland and level-ground nesting. In the early 1960s Fulmars in Unst were nesting on fields more than 1 km from the sea; in 1972 they bred on Vane Crags above Loch Leven, about 15 km from the Firth of Forth; during the 1970s they established a colony among the dunes at Sands of Forvie (Anderson 1982); and in 1982 they bred successfully for the first time on Salisbury Crags, Edinburgh (SB 12:228).

Of the many thousands of Fulmars ringed as nestlings only a few hundred have been recovered away from breeding colonies. Analysis of the recoveries (Macdonald 1977) has shown that after fledging birds disperse to the western North Atlantic, the Norwegian Sea and the European Arctic. When four years old their distribution contracts, possibly owing to developing association with prospective nesting areas. Adults spend all their time in the eastern half of the oceanic range, remaining within several hundred kilometres of their colonies at all seasons. The most distant recovery, of a Fair Isle chick in its first November, was from Nova Scotia, nearly 4,700 km away. There have been several from the Murmansk-Archangel area and the Barents Sea, mostly of birds at least two years old, and to the south from as far as northern Spain. Chicks ringed in Iceland, Faeroe and Norway have been recovered in Scotland.

Large-scale coasting movements of Fulmars occur regularly, often involving as many as 1,000+ birds/hour in the Northern Isles in autumn (Jones & Tasker 1982). Similar movements, though on a smaller scale, are also seen off the east coast, and occasionally on the west, where the majority are heading south. Gales in September sometimes result in hundreds of dead and dying juveniles being cast up on northern beaches and birds occasionally turning up far inland. Few Fulmars appear to suffer from oiling – but many other birds are oiled by Fulmars! The species' habit of spitting stomach-oil at intruders puts at risk both potential predators and migrants unfortunate enough to seek shelter on a Fulmar cliff (Broad 1974). Among those recorded as suffering in this way have been Honey Buzzards and a young White-tailed Eagle released on Fair Isle.

Fulmars belonging to the so-called 'blue' phase occur in small numbers and have occasionally bred in Scotland.

These darker-coloured birds are predominant in most, but not all, of the high-arctic colonies.

Aberdeen University's long-term study of the Eynhallow colony, started in 1950, has added greatly to knowledge of this species. Fulmars do not breed until they are, on average, about ten years old, and only about 6% of young birds surviving to this age return to their natal colony. Adults have a mean annual survival rate of 97% and when breeding may forage up to 600 km from the colony. Breeding success, adult survival, recruitment of new breeders and population size have been shown to vary substantially between years (Dunnet & Ollason 1978, 1982; Dunnet et al 1979; Ollason & Dunnet 1978).

Cory's Shearwater *Calonectris diomedea*

Vagrant (Mediterranean / southern N Atlantic) – annual in very small numbers, mainly in early autumn.

Cory's Shearwater breeds well to the south of Britain and winters in the South Atlantic. Although large numbers occasionally wander into British waters during the post-breeding dispersal, the species remains a scarce vagrant off the Scottish coast. A marked increase in the numbers seen off Ireland in the last few years may be partly due to greater sea-watching effort, but there is also some evidence of a true increase; in 1980 an exceptional influx occurred, with over 2,800 in Britain and some 14,400 in Ireland (BB 11:485) – but only three were seen in Scotland.

The first Scottish record was one off Aberdeen on 10 September 1947 (BB 41:88). From then until 1970 there were reports in only four years; one of these involved the only real influx to date, in 1965, when 83 were seen in the vicinity of Fair Isle between 18 and 23 September (Dennis 1966). Since 1973 up to nine, but usually less than five, have been reported annually. Most sightings are in August/September but there are a few June, July and November records and one in April. Cory's Shearwater has been reported from Shetland, Fair Isle, Orkney, Caithness, Sutherland (east and west), Easter Ross, Aberdeen, Angus, Fife, the Isle of May, East Lothian, Ayr, Argyll mainland, Islay, North Uist and Lewis, and also in the Forties Oilfield. More than half the records are from east coast sites popular with sea-watchers.

Great Shearwater *Puffinus gravis*

Vagrant (S Atlantic) – occurs almost annually, mainly in autumn.

This oceanic species breeds on South Atlantic islands and moves into the northern hemisphere in the non-breeding season, when it is widely dispersed from Greenland and Newfoundland east to Rockall. Adverse weather conditions and strong westerly winds sometimes drive birds further east than usual, occasionally in considerable numbers.

B&R give the impression that the Great Shearwater occurred regularly and was relatively common in Outer

Hebridean waters around the turn of the century and it has recently been described as 'not uncommon' at sea around St Kilda (Harris & Murray 1978). It only rarely comes close inshore, however, and although recorded in most years since 1962 the annual total has only twice exceeded 20. In September 1965 (when over 5,000 passed Cape Clear, Cork, in two days) there were nearly 100 around Fair Isle (Dennis 1966) and in 1976 about 70 were counted in August/September, mostly on the east coast from Easter Ross to Kincardine. Nearly 90% of records are for August/September, with a few in June/July and October. The occasional occurrences in other months include three in February 1981, while the latest date is 10 November (off Rockall). The pattern of reports probably reflects the distribution of sea-watchers as much as that of birds; most are from Shetland, Aberdeen and Easter Ross, but there are records for all coastal mainland counties except Angus and Kirkcudbright, and for Orkney and the Outer and Inner Hebrides but not the Clyde Islands.

Sooty Shearwater *Puffinus griseus*

Passage visitor, occurring annually, sometimes in large numbers and most often in August / September.

Although breeding only in the Southern Ocean, the Sooty Shearwater disperses widely outwith the breeding season and regularly reaches the North Atlantic (Phillips 1963). Since 1970 annual totals have ranged from a few hundreds to several thousands. Spectacular movements were recorded in September 1969, when up to 800 were visible at one time off North Ronaldsay, and on 30 August 1983, when 1,360 flew past Papa Westray in three hours. Peak counts are generally in the low hundreds, however, and there are many records of single birds and small parties.

Sooty Shearwaters are seen most frequently around the Northern Isles and Hebrides but have been recorded off all counties with exposed sea coasts and occasionally wander quite far into the larger firths, eg a group of eight off Hound Point, West Lothian, in October 1970. On the east coast movement is generally northwards, in Orkney and Shetland waters the birds are often heading west, and off Islay virtually all are south-bound. In late summer individuals are widespread at sea and it is at this period, especially between mid

August and the end of September, that concentrations visible from the coast are likely to occur. There are a good many July and October records and scattered reports for every other month except February.

B&R classed this species as an uncommon visitor; the apparent increase in occurrences over the last few decades is almost certainly attributable to the great expansion in sea-watching.

Manx Shearwater *Puffinus puffinus*

An offshore-feeding species; breeds on widely scattered islands in the north and west and winters off eastern South America.

The only large Manx Shearwater colony in Scotland is on Rhum. The others, scattered from Shetland to Arran, are mostly very small; those on Horse of Burravoe (Yell) and Sanda (Kintyre) were only discovered as recently as 1978 and 1979 respectively (Maguire 1978, Fowler 1979). In 1983 birds were heard ashore on Mull and in 1984 breeding was proved on Muck (Dobson 1985). Manx Shearwaters come to the colonies at night, from February to September, and often gather in large rafts offshore during the evening. Their weird crows and cackles as they flight in after dark help to make visits to nesting sites a truly memorable experience. On Rhum, where the greatest concentration is on Hallival, the colony lies largely above 650 m asl and up to 3 km from the sea, and on Foula the main colony is on the 248 m high Noup. Although there are colonies on Hirta, Soay and Dun, the total St Kilda population is probably not very large (Harris & Murray 1978).

As with other nocturnal, burrow-nesting species, estimation of breeding populations is extremely difficult and there is little information on many of the known sites. Only two

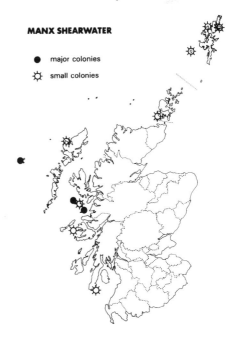

MANX SHEARWATER

● major colonies

☼ small colonies

colonies have been studied in any detail. On Rhum, Wormell (1976) assessed numbers by measuring the extent of green, bird-fertilised vegetation at the colony and counting burrows in sample areas; this gave an estimated 100,000+ occupied burrows. Subsequent counts on sample areas suggested that a 25% increase had taken place over a ten-year period, which would give a total of about 150,000 pairs (Thompson & Thompson 1980). The more accessible Canna colony has been monitored since 1970 by Swann & Ramsay (1984). They estimated the population as around 1,500 pairs and found that breeding success varied widely between years, with numbers of young fledged ranging from 25% to 83% of eggs laid. Rats sometimes caused almost complete failure. At their breeding colonies shearwaters are predated by gulls, Peregrine, Buzzard and Golden Eagle, forming 10% of the latter's diet on Rhum (Corkhill 1980). More surprisingly they are also occasionally eaten by red deer, which take juveniles at the mouths of their burrows – a situation in which they are very vulnerable (Wormell 1976).

Some thousands of Manx Shearwaters have been ringed but comparatively few recovered. Analysis of the available records suggests that during the summer birds feed within 200 km of their colony, probably on young herring, and that three-year olds returning to the breeding area from the South Atlantic do so via the eastern Atlantic (Swann & Ramsay 1976). There are several winter (October / March) recoveries from South American waters and one from the Gulf of Mexico; the most distant is from Argentina and one juvenile was 9,830 km from the ringing site only eight weeks after marking. Birds from Skokholm and Bardsey (Wales) have been recovered on the west coast as far north as the Clyde and Kintyre, and one ringed on Copeland (Eire) was caught on Canna the following year.

Apparently perfectly healthy birds are occasionally recorded far inland, for example the one I saw skimming over Loch Tay in July 1961. The regular (two or three per season) finding of Manx Shearwater remains in Peregrine nests along the Great Glen and in Glen Moriston, and of young birds grounded during September fogs in the same area, suggests that at least some overland passage to the North Sea takes place (R. L. Swann). Away from the breeding colonies the largest numbers are seen in August / September, when big gatherings are occasionally recorded on the east coast as well as the west. Among recent reports are an exceptional 1,350 in Gullane Bay in mid August, and 4,000 and 6,000+ passing Corsewall Point and Turnberry respectively in early September. Most Manx Shearwaters have left Scottish waters by the end of October but there are a few later records, including one at Ailsa Craig in mid December and one freshly-dead in Orkney in January.

The British population, estimated at 175,000–300,000 pairs and confined to Atlantic and Irish Sea coasts, belongs to the nominate race *P.p.puffinus* which breeds only in the eastern north Atlantic, from Iceland to Madeira and the Azores. The Balearic or West Mediterranean race *P.p.mauretanicus* was first recorded in 1874, when one was shot in the Firth of Forth in mid August; there were no further reports until one was seen west of Lewis in September 1966. After the next report, of four birds together off Fife in 1969, there was again a gap of several years, but since 1976 there have been sightings almost annually. Totals in any one year

have ranged from one to six. More than half the records are from the east coast and all are between 18 August and 14 October. Birds of this race are slightly larger and generally browner than *P.p.puffinus* and lack the sharp contrast between upper and underparts (BWP). After breeding this population disperses from the Mediterranean into the Atlantic and moves north off Europe. Balearic Shearwaters occur regularly off southwest Britain in July / October and may well be more frequent in Scottish waters than the records suggest.

Little Shearwater *Puffinus assimilis*

Vagrant (Atlantic islands, from Madeira south) – one record, 1974.

A Little Shearwater was seen off Islay, Inner Hebrides, on 30 June 1974 (SB 9:380). During a period of intensive sea-watching in the 1960s this species was reported fairly regularly in autumn off southwest Ireland but there have been fewer records in recent years.

Wilson's Petrel *Oceanites oceanicus*

Vagrant (Antarctic / sub-Antarctic) – one old record, 1891.

A bird caught alive on Jura, Inner Hebrides, on 1 October 1891 (ASNH 1892:18) is the only Scottish record. Elsewhere in Britain Wilson's Petrel has been reported less than ten times and only from Cornwall and Ireland.

White-faced (Frigate) Petrel *Pelagodroma marina*

Vagrant (sub-tropical N Atlantic) – one old record, 1897.

The only British record is of a young female caught alive on Colonsay, Inner Hebrides, on 1 January 1897 (ASNH 1897:88).

Storm Petrel *Hydrobates pelagicus*

Breeds on islands in the north and west and is often seen on crossings from the mainland to Orkney, Shetland and the Hebrides. Winters at sea off southern Africa.

The Storm Petrel is our smallest seabird. A colonial hole-nester, it breeds at many locations from Shetland to Kintyre and there may well be other colonies still undiscovered. Since 1970 eggs or chicks have been found for the first time at nine sites (Table) and there are many more at which breeding has been suspected but not proved. Because it nests among boulders, down burrows or in drystone walls, and flights to and from the nest only at night, this species is extremely difficult to census. The fact that colonies are often on small isolated islands, or steep and inaccessible slopes, adds to the problems. Wandering non-breeders visit breeding sites in considerable numbers, so population estimates based on mist-net catches are far from accurate. Most colonies are apparently small but a few are believed to hold 100–1,000

pairs and three to be very much larger. The St Kilda population is possibly more than 10,000 pairs (Bourne & Harris 1979), that of Priest Island about 10,000 pairs (SBR), and that of Foula 1,000–10,000 (Furness 1983).

The breeding season is protracted; the first birds come ashore in late April, laying usually occurs in June and chicks are sometimes still on land in late November. The 'purring' of Storm Petrels on the nest and their musky smell are useful guides to the location of occupied holes or burrows. One of the most accessible, and certainly one of the most picturesque, places where these may be experienced is the broch of Mousa, which holds a sizeable colony. It is interesting that this species is attracted not only by recordings of its own calls but also by those of Leach's and Wilson's Petrels (Zonfrillo 1982). The use of tape-lures has recently resulted in catches in the Firth of Forth and on the Ayr coast, both areas without any known Storm Petrel colony. Birds ringed in Shetland, Caithness and the Summer Isles have been caught on the Isle of May within 22 days (in one case only eight days later) and one ringed there was caught on Sanda 21 days later (Zonfrillo 1983).

Storm Petrel colonies at which breeding has been confirmed for the first time since 1970. (Data from SBR, Maguire 1978)

	Year found
Caithness	
Stroma	1971
Orkney	
Switha Holm	1973
Green Holms	1973
Muckle Skerry	1973
Holm of Papa Westray	1973
Little Linga	1973
Holm of Huip	1973
Eynhallow	1975
Argyll	
Sanda	1978

The Storm Petrel breeds only in the eastern North Atlantic, from Iceland and Norway south to Iberia and possibly the Canaries, and in the Mediterranean. Large numbers nest in Ireland, especially in the southwest, and on islands off Wales and southwest England. Well over 500 of the many thousands ringed in Scotland have subsequently been re-caught during the breeding season; the records indicate that much of the apparent interchange between islands involves birds, originally ringed on casual visits, which later settle to breed in another colony (Mainwood 1976; Fowler *et al* 1982). There is considerable interchange with Irish, Welsh and English colonies. Scottish-ringed birds have also been trapped in summer in the Lofoten Islands, Faeroe and Iceland (Fowler & Swinfen 1984), and recovered in November/February off western and southern Africa, from Liberia round Cape Province to Natal. In the autumn and early winter Storm Petrels are occasionally recorded at lighthouses and North Sea oil installations; under normal weather conditions they seldom come close to the coast in daylight.

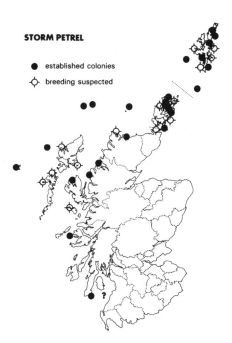

STORM PETREL

● established colonies
◇ breeding suspected

Leach's Petrel *Oceanodroma leucorhoa*   [1]

Nests on remote islands and winters at sea, mainly in the Tropics. Scotland holds the entire British breeding population.

Although much has been learned about this oceanic species since the pioneering studies of the 1930s (Ainslie & Atkinson 1937), Leach's Petrel remains the least-known of our seabirds. Only six colonies have been found. Those on St Kilda, the Flannan Isles, North Rona and Sula Sgeir have been known for many years but two others have been located only recently – on Foula in 1974 (SB 8:321–323) and Ramna Stacks in 1980 (SB 12:86–87). Birds with brood patches have been caught on Sule Skerry and breeding is also suspected on several other islands. Most colonies are in relatively inaccessible situations. Its nocturnal and burrow-nesting habits make this species impossible to census accurately and there is little information available on numbers. Estimates based on calling birds or numbers caught in mist-nets vary greatly; for example the North Rona population was assessed as c5,000 by Bagenal & Baird (1959) and as a few hundred pairs by Love (1978). However, the St Kilda colony is probably by far the largest (Bourne & Harris 1979).

The picture is further complicated by the fact that non-breeders wander widely and may at times comprise a substantial proportion of the birds flighting at a site. Once established as breeders, Leach's Petrels show a strong tendency to return to the same colony (Love 1978). The use of tape-lures has increased the numbers caught in recent years but, although many have been ringed, there have not yet been any recoveries away from breeding areas.

Like the Storm Petrel, this species has a protracted breeding season; the main arrival is in early May but chicks are sometimes still in burrows in October. Because the birds are in North Atlantic waters so late in autumn they are occasionally caught up in severe gales which result in spectacular wrecks. The last such occasion was in 1952, when over 500 dead and dying Leach's Petrels were picked up during October/November, in 24 of the 33 Scottish counties. Surprisingly, none were reported from Orkney (Wynne-Edwards 1953). Under more normal conditions Leach's Petrel is seldom seen close to land and is more likely to be sighted at sea in Shetland, Orkney and Outer Hebridean waters than in the North Sea, where it is recorded only rarely. Elsewhere in the eastern North Atlantic the only known colonies are on the Faeroe, Westmann and Lofoten Islands, but far larger numbers breed in the western Atlantic and there are several Pacific populations.

Gannet *Sula bassana*

Colonial breeder, nesting at a small number of colonies – all but one of which are in the north and west – and dispersing in winter. Scotland holds roughly half the expanding world population of this species, which breeds only in the North Atlantic.

Five old-established sites hold nearly 90% of the breeding total of c100,000 pairs (Table); all are on steep cliffs from which the birds can readily become airborne. The population as a whole apparently increased by at least 3% per annum between 1939 and 1969 (Nelson 1978) but growth has not been uniform and continuous throughout this period or at all colonies, and the vast St Kilda colony – the species' largest – seems currently to be stable (Murray 1981, Wanless & Wood 1982). The Bass Rock colony is still expanding – as are those in Shetland – and nesting birds now spread from the cliff edge well up onto the summit slopes.

Of the three most recently-founded colonies, that on Roareim in the Flannan Isles was established prior to 1969–70, while those on Fair Isle and Foula have been documented from the 'prospecting' stage. Cliff-patrolling Gannets were first noted at Fair Isle and Foula in the early 1960s, birds were ashore in 1969 and 1970 respectively, the first nests were constructed in 1974 and 1975, and the first chicks hatched in 1975 and 1980 (FIBOR, Furness 1983). An apparently occupied nest was found on the Shiant Islands in 1984 (Buxton 1985), while breeding could soon take place on Bearasay (off Great Bernera, west of Lewis), where adults

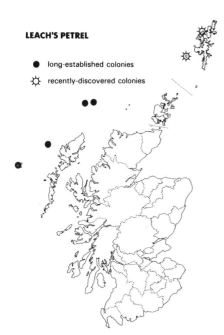

LEACH'S PETREL

● long-established colonies

☼ recently-discovered colonies

Numbers and distribution of breeding Gannets. (*Data from Cramp* et al *1974, FIBO Reports, SBR, Murray 1981, Murray & Wanless 1983, Furness 1983, Blake* et al *1984, S. Wanless & M. P. Harris and as listed – figures rounded*)

	Date of first known breeding	Number of pairs occupying nests		
		1949	1969–70	Latest count**
St Kilda	Pre-9th cent.	17,000	52,000	40,000 (1979)
Bass Rock	Pre-1521	4,800	9,000	18,200 (1984) J. B. Nelson
Ailsa Craig	Pre-1526	5,000	13,000	21,500 (1983)
Sula Sgeir	Pre-1549	6,000	9,000	
Sule Stack	Pre-1710	2,000	4,000	
Scar Rocks	Pre-1883 – deserted & recol. 1939	100	450	770 (1984) North Solway Ringing Group
Noss	1914	3,000 (1946)	4,300	6,900 (1984) NCC
Hermaness	1917	3,800	6,000	14,400 (1984) SOTEAG
Flannan Is.	Pre-1969*	0	16	40 (Oct/Nov 1983) D. Emerson
Fair Isle	1974	0	0	130 (1984) FIBO
Foula	1976	0	0	210 (1984) R. W. Furness
Shiant Islands	1984	0	0	1 (1984) N. E. Buxton

* established when found during Operation Seafarer
** 1984 counts expressed as occupied sites

have been coming ashore since the mid 1970s, and perhaps in the longer-term at St Abbs, where they first did so in 1979.

Large numbers of Gannets have been ringed, the majority on the Bass Rock, and there have been some 2,000 recoveries. Analysis of those for first-year birds (Thomson 1975) has shown that after a wide dispersal through northern European waters in August/October, there is a gradual southward movement; some travel as far as tropical west Africa, some wander into the Mediterranean, and some remain to winter in northern waters. Few adults move further than the Bay of Biscay and birds start returning to their colonies in January, after only a brief absence in early winter, though numbers fluctuate markedly until laying begins in April. Many non-breeding immatures and prospecting adults are present during much of the breeding season, often occupying 'club' areas in which there are no nests. Some young birds breed away from their natal colony. Such movements must be largely responsible for the very rapid increase that takes place at new colonies once 20–35 pairs are present; this is apparently the threshold level at which social stimulation is sufficient to achieve successful breeding (Nelson 1978).

Gannets fish mainly within the limits of the continental shelf and can be seen inshore most of the year, especially in the vicinity of colonies, although numbers in the North Sea are small from December to February. When they spot shoaling fish they provide spectacular diving displays, sometimes actually inside a harbour mouth. Large-scale movements take place, with several thousand birds passing per hour; in autumn such movements involve mainly immature birds, often north-bound up the east coast and west-bound through the Pentland Firth. The strongest passage yet recorded was in spring, when some 4,500 Gannets headed north past Fraserburgh in 40 minutes on 12 April 1973. Birds are occasionally reported far inland and small numbers of first-year birds may cross overland from the Forth to the Clyde (SB 9:298).

The present population expansion follows a period of decrease last century, perhaps due partly to persecution when the Gannet was an important food source for many remote communities. (Sula Sgeir is the only colony at which harvesting of young *gugas* is still (1985) permitted, by special licence issued only to the men of Ness in Lewis.) Changing climatic or oceanographic conditions affecting the availability and

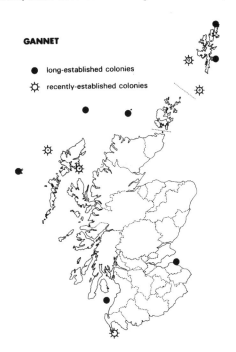

GANNET

● long-established colonies

☼ recently-established colonies

distribution of the surface-shoaling fish on which the Gannet depends may also be involved. Provided that such food supplies – including mackerel, herring and sandeels, all of which are now being commercially fished – continue to be abundant, expansion should continue as there appear to be plenty of suitable breeding areas. Many Gannets found dead on beaches are oiled or entangled in nylon, and toxic chemicals have been found in some birds in the Clyde; as the Gannet is a long-lived species such mortality is unlikely to affect populations.

Many aspects of the Gannet's behaviour, population dynamics and ecology have been studied by Bryan Nelson on the Bass Rock and recorded in a monograph on the species (Nelson 1978). The accessibility of this colony gives less dedicated Gannet-watchers a fine opportunity to see – and smell – this handsome bird at close quarters and to observe for themselves its fascinating behaviour. The Ailsa Craig colony has also been studied over a long period (Gibson 1951, Murray & Wanless 1982, Wanless 1983).

[Double-crested Cormorant *Phalacrocorax auritus*]

On 22 December 1963 a live bird of this American species was found in the hold of a cargo ship which had arrived in Glasgow from Newfoundland (SB 3:98).

Cormorant *Phalacrocorax carbo*

Breeds at widely scattered colonies around the coast and at a very few inland sites. In autumn birds disperse in a southerly direction, though few leave British waters; some winter inland. Scotland holds about a third of the British and Irish breeding population.

Most Cormorants nest on rocks in coastal situations but they are dependent upon relatively sheltered shallow waters in which to fish, hence the large numbers found among the Northern Isles and Outer Hebrides (Table). More than 75% of the population breeds north of the Great Glen. Although many sites are continuously occupied, some populations move between two or more locations (Smith 1969) and local populations sometimes fluctuate markedly. On the east coast of Caithness, for example, numbers were estimated at more than 800 nests in 1969–70 but less than 300 in 1977 (Mudge 1979). On the Forth islands, however, the total of c220 nests has remained virtually static for about 15 years, although the number of islands occupied has varied from one to four (R. W. J. Smith). Since 1969–70 breeding has occurred sporadically in Aberdeen, Kincardine and Angus, and there has been an expansion of the Berwick population, with 52 pairs nesting in 1984.

A detailed account of the history up to 1969 of all known colonies is given by Smith (1969). At the old-established freshwater colony in Wigtown, which has been in existence since the 17th century, numbers have ranged from 100 to 370 pairs, and the birds have moved between Mochrum and Castle Lochs. Elsewhere in Wigtown Cormorants colonised

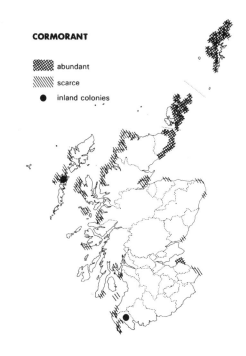

CORMORANT

▓▓ abundant

▨▨ scarce

● inland colonies

war-time 'Mulberry Harbour' constructions, later moving to nearby cliffs, and have been progressively displaced from the Scar Rocks by the expanding Gannet colony (Young 1968). The freshwater site at Loch an Tomain on North Uist, probably established in the 1940s, held an estimated 200 birds in 1978 (Cunningham 1983). Inland nesting has also occurred sporadically in Inverness.

From the several thousand young Cormorants ringed in Scotland there have been more than 1,500 recoveries. The data were analysed in the late 1960s by Coulson & Brazendale (1968) and Balfour *et al* (1967), but there has been no subsequent analysis. Less than 10% of recoveries were outside British and Irish waters; most of these were from France and Spain but there were also a few from Norway, Denmark and the Netherlands. The evidence suggests that each colony has its own dispersal pattern, and that the Cormorant's reluctance to undertake long open-sea crossings may be a contributory factor in keeping populations genetically discrete. Young birds from Orkney move south, reaching Caithness in August and the Forth in September; a substantial part of the population winters away from the islands. Birds from the Farne Islands also winter in the Forth and

Numbers and distribution of coastal breeding Cormorants in 1969–70. (Data from Cramp et al 1974)

County	No. of pairs	County	No. of pairs
Kirkcudbright	140	Orkney	590
Wigtown	360	Shetland	460
Ayr	100	Caithness	850
Bute	14	E. Ross	200
Argyll	60	Moray to Fife	1
W. Inverness	180	E. Lothian	150
W. Ross	80	Berwick	1
W. Sutherland	105		
O. Hebrides	380	*Total*	c.3,700

as far north as the Ythan, while many from west coast colonies cross to the east. The wide pattern of dispersal is well demonstrated by the fact that in August/November of their first year Mochrum birds have been recovered in Aberdeen, Kent, Portugal, Wales and Ireland. The recovery in Unst of a first-winter Norwegian bird shows that some immigration occurs.

In autumn and winter Cormorants are frequently seen on freshwaters well inland, sometimes in fair numbers. Up to 80 have been present on Loch Lomond and Loch Leven and a few occur annually on Loch Rannoch. At inland sites they often roost on trees – a regular nesting site in other parts of the species' Eurasian range – which may be killed by their droppings. Among the well-established tree roosts are those at Munlochy Bay, Earnmouth, and near Kinclaven on the Tay. Coastal feeding movements occasionally produce very large concentrations, for example c650 near Inchgarvie in the Forth in March and 470 off Caerlaverock in December.

Cormorants are much persecuted in areas where salmon and trout fishing is of economic importance and in the past bounty schemes resulted in large numbers being shot. In 1965 nearly 800 were destroyed on the Tweed, but in 1966 the bounty scheme there was dropped, largely as a result of work by Derek Mills (1965) which demonstrated that predation by Cormorants was unlikely to have a significant effect on fish stocks. Birds feeding in coastal waters are equally unlikely to be in real conflict with commercial interests (Mills 1969). Persecution is now probably appreciably less than it used to be and there are no other major threats to the population, although many are drowned in nets. As this is a relatively easy species to census it should be possible to monitor future population changes with a fair degree of accuracy.

Shag *Phalacrocorax aristotelis*

Breeds in all coastal counties with exposed, rocky coasts. Numbers, currently increasing, are largest in the north and west. In winter some dispersal occurs but many adults remain at or within 100 km of their colonies.

Shags nest in caves and under cliff-foot boulders as well as on ledges and are consequently less easy to census than Cormorants. They also nest at greater densities, sometimes forming huge colonies, as on Foula where c1,000 pairs occupy the Wick of Mucklebreck (Furness 1983). Shetland holds about 25% of the British breeding population, with more than half the county total on Foula and Fair Isle, both of which have all-round access to good fishing grounds. South of the Great Glen numbers are small except in the Firth of Forth (Table). There the population has increased, slowly at first but faster in the last 20–30 years, since breeding first occurred on the Isle of May in 1918. Numbers of apparently occupied nests on the May rose to over 1,000 in 1973 but dropped again to 364 in 1974 (Galbraith 1981), since when there has been a further big increase, with 1,855 nests built in 1983 (SB 13:69). Craigleith and the Lamb – both colonised since 1950 – had 344 and 230+ nests respectively

Numbers and distribution of breeding Shags in 1969–70. (Data from Cramp et al 1974)

County	No. of pairs	County	No. of pairs
Kirkcudbright	8	Caithness	1,560
Wigtown	130	E. Ross	25
Ayr	180	Moray	0
Bute	26	Banff	50
Argyll	2,080	Aberdeen	250
W. Inverness	1,840	Kincardine	40
W. Ross	540	Angus	15
W. Sutherland	2,050	Fife	880
O. Hebrides	2,790	E. Lothian	440
Orkney	3,580	Berwick	125
Shetland	8,600	*Total*	c.25,200

in 1982. Shags first nested on Fidra in 1971, on Carr Craig in 1973, and on Inchkeith in 1974.

Studies in Shetland (Harris 1976), on the Isle of May (Galbraith 1981, Harris & Galbraith 1983) and on Canna (Swann & Ramsay 1984) have shown that local populations are liable to fluctuate widely, sometimes dropping suddenly by a third or even more. Shellfish poisoning caused an 80% mortality among Shags on the Farne Islands in 1968 but there has been no evidence of similar catastrophes in Scotland and the reason for these fluctuating numbers remains obscure. Despite these periodic declines, and occasional bad weather wrecks, numbers on both Canna and the Isle of May have increased at an average rate of more than 8% per annum in the last decade.

There have been some 3,000 recoveries of Shags, c70% of them within 100 km of the ringing site. Analyses of the data for northwest Scotland (Swann & Ramsay 1979), the Isle of May (Galbraith *et al* 1981), and Foula (Furness 1983) show that first-winter birds disperse further than adults in all three areas but that dispersal patterns in the east differ

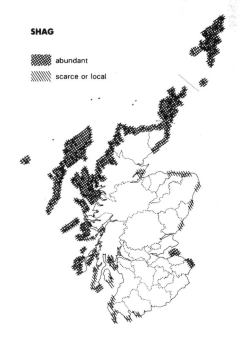

SHAG

▨ abundant

▧ scarce or local

from those in the west. Whereas birds from the Isle of May appear to make eruptive autumn movements, involving quite a number of inland occurrences, those from the northwest colonies, for whom the sheltered inter-island waters around the Uists and Barra are the most important wintering grounds, do not. Only 1% of all recoveries are from foreign waters, most of these being from North Sea coasts between Norway and France. Some interchange between the Forth and the Farne Islands takes place.

The numbers seen offshore vary greatly, even within a single season and at one locality, as a result of feeding and dispersive movements. The possibility that these variations may be related to the varying prevalence of onshore winds, and consequent food availability, has been suggested by Elkins & Williams (1974). Shags feed in deeper water than Cormorants and largely on free-swimming, as opposed to bottom-feeding, fish. Although they do not present a serious threat to commercial fishing interests (Mills 1969a) they are shot – often for food – and accidentally caught in nets. About half the ringed birds recovered have suffered one or other of these fates, but in recent years oiling has been responsible for an increasing proportion of recoveries. Inshore oil-spillages could cause high mortality, especially in Shetland or the Forth.

[White Pelican *Pelecanus onocrotalus*]

Escape / vagrant (Category D)

In May 1973 several White Pelicans, presumed to be escapes, were reported: four were on Islay between 1st and 10th, and one or two on Gadloch, Lanark, between 3rd and 6th (SBR 1973). A pelican, not identified as to species, reported from the Isle of May on 8 August 1960 was also considered to be an escape.

Magnificent Frigatebird *Fregata magnificens*

Vagrant (tropical Atlantic) – one record, 1953.

A female Magnificent Frigatebird was caught alive on Tiree, Inner Hebrides, on 9 July 1953 (BB 47:58); its skin is now in the Royal Scottish Museum. This is the only definite record but an unidentified frigatebird was seen off Forvie, Aberdeen, on 20 August 1960.

Bittern *Botaurus stellaris*   [1]

Irregular visitor, occurring almost annually, most often between October and February and in the south and east. Formerly bred.

In Britain the Bittern is at the extreme northwest limit of its Eurasian range. Its past history in Scotland is somewhat unclear, although it is reported to have been present and probably nesting in 'a good many parts' up to about the end of the 18th century (B&R). It is said to have bred in Berwick until 1830, since when there has been no definite breeding record. During the first half of the 19th century the species became virtually extinct in Britain, due to persecution and to drainage of its wetland habitat, but recolonisation took place in the early 1900s and there has since been a gradual spread. The nearest colony, which is currently expanding steadily, is at Leighton Moss in Lancashire.

Many of the Bitterns recorded in Scotland are dead or dying when found and have probably travelled here in the course of dispersive movements during spells of hard weather, to which this species is very vulnerable. About a third of all reports are in January and over 80% between August and February; May is the summer month for which there are most records. Although the Bittern has occurred as far north and west as Shetland and Lewis most records are from south of the Forth/Clyde. In some years reports come from widely scattered areas, as in 1983 when Bitterns were seen in Caithness, Perth, Mull, Renfrew, Kirkcudbright and East Lothian, but numbers recorded never exceed single figures, although more may, of course, remain unnoticed. 'Booming' birds in summer are more likely to be detected; the most recent such record was in the Borders in June 1980, but there was no indication that breeding was attempted.

American Bittern *Botaurus lentiginosus*

Vagrant (N America) – two records this century (1931 and 1981/82) and several older ones.

An adult male American Bittern was on Benbecula, Outer Hebrides, on 27 December 1932, and from 4 November 1981 to 9 January 1982 there was one at Kilmacolm, Renfrew (BB 75:485). Like all bitterns, this species is difficult to see well and identify with certainty. The greater frequency of records in the past, which date back to the middle 1800s and were mostly in autumn or early winter, is probably attributable to the fact that bitterns were formerly shot.

Little Bittern *Ixobrychus minutus*   [1]

Vagrant (continental Europe, excluding Scandinavia) – ten occurrences 1950–83.

Although B&R refer to 'a good many old records' of the Little Bittern, they give only one dated record, a bird on Fair Isle in April 1940. From 1950 to 1983 inclusive there were ten reports: six in May, three in June and one in July. They came from Angus (1958), Ayr and Shetland (1965), Fife (1970), Orkney (1971), Ayr and Inverness (1973), Dunbarton (1978), and East Lothian and Fife (1979). Of the six birds that were sexed, four were males. The Little Bittern occurs more frequently in southern England than elsewhere in Britain, but even there numbers have been low in the early 1980s. It is a skulking bird which could easily escape detection in the dense reedbeds that it favours.

Night Heron *Nycticorax nycticorax*

Escape/vagrant (Europe) – probably less than ten true vagrants since 1950.

In Europe the Night Heron, a nearly cosmopolitan species, occurs locally from the Netherlands and Czechoslovakia south to the Mediterranean and the Black Sea. The young disperse widely prior to the autumn migration to Africa. There has been a free-flying colony of the American race *N.n. hoactli* at Edinburgh Zoo since 1951; its existence greatly increases the probability that birds seen in central and southern Scotland are escapes. This colony held about 30 birds in mid 1984, having peaked at 60+ a few years earlier (M. F. Stevenson). A colour-ringing programme has recently been started to help in tracing the movements of young birds leaving the colony.

There are several pre-1900 records of the Night Heron but only one in the period 1900 to 1950, an immature near Glasgow in November/December 1926. Since 1950 there have been a number of reports from Midlothian, all of which are assumed to refer to escapes. The status of an adult on the Isle of May in May 1960 and an immature at Neidpath, Peebles, on 27 July 1980 is less certain, while birds seen in Orkney (November 1961 and May 1982), Aberdeen (May 1975), Shetland (May/June 1981 and April 1983) and Dumfries (December 1982/April 1983) are likely to have been true vagrants.

Squacco Heron *Ardeola ralloides*

Vagrant (S Europe) – last record 1913.

There has been no record of the Squacco Heron since one was seen on Lewis, Outer Hebrides, in June 1913 (SN 1913:211). B&R also give three earlier records, from Orkney and central Scotland. The European population decreased considerably as a result of persecution in the 19th and early 20th centuries but has since shown some signs of recovery, with suspected breeding in France in 1981. This species occurs more often in southern Britain than in Scotland.

Cattle Egret *Bubulcus ibis*

Vagrant/escape (S Europe) – one record, 1979.

The only accepted record of the Cattle Egret is of one at the Loch of Kinnordy, Angus, from 10 to 19 May 1979 (BB 73:494). An earlier report from Dumfries in July 1964 was considered to refer to an escape (BB 58:356), and a record of one long-dead in East Lothian in February 1980

was not accepted by the Rarities Committee. The species is notable for periodic major extensions of range from its principal bases in Africa and southeast Asia; in the present century it has colonised the Americas and been introduced into Australia. In Europe it is very locally distributed; the nearest old-established colony is in the Camargue but a few pairs bred in northwest France in 1981, so there is an increasing possibility of nesting in Britain.

Little Egret *Egretta garzetta*

Vagrant (S Europe) – occurred in 14 years, 1954–84.

The Little Egret, readily identified by its yellow feet, was first recorded in 1954, when there were four reports, involving at least two birds, in Perth/Lanark and Shetland/Sutherland (BB 47:127–128). From then to the end of 1984 there were reports in 13 years, seven of them between 1968 and 1975, after which there was a gap until 1981. The Little Egret has occurred in Shetland, Orkney, Sutherland, Wester Ross, Inverness, Moray, Aberdeen, Angus/Kincardine, Perth, East Lothian, Lanark, Dumfries, Wigtown, Mull, Islay and both the Uists. Three-quarters of the records are in spring, mostly April/June, and the rest in autumn or early winter. Most reports are of single birds but there were four together on Mull in October/November 1969; individuals often remain in one area for several weeks. The Little Egret breeds near shallow lowland waters, ranging from fresh to saline, from Iberia eastwards through Eurasia and in Africa and Australia. In the late 1970s it nested for the first time in northern France and the Netherlands; as might be expected, in Britain it occurs most frequently in the south.

Great White Egret *Egretta alba*

Vagrant/escape (Europe, very locally) – three recent records, 1978–83, and two old ones.

Great White Egrets were shot in East Lothian and Perth last century but none were recorded from 1881 until the 1970s. In 1978 there were reports from North Ronaldsay (28 April) and the Loch of Strathbeg (23–28 June), and in 1980 one was at Caerlaverock from 8 to 11 April. One found dead at North Roe in Shetland in March 1971 was identified as belonging to the eastern subspecies, *E.a. modesta*, and was therefore considered to be an escape. The Great White Egret is almost cosmopolitan but very local in Europe; in the 1970s it nested for the first time in the Netherlands, very much further west than the Hungarian colonies which were previously the closest sites to Britain.

Grey Heron *Ardea cinerea*

Widely distributed, absent as a breeding bird only from Shetland. Numbers drop markedly following severe winters, but increased productivity in subsequent years enables the population to recover rapidly. Few native birds disperse more than 50–100 km from their natal area. Some immigration occurs in winter.

The Grey Heron was the first species for which a national census was attempted, in 1928; repeat censuses were carried out in 1954, 1964 and 1985. Despite the fact that this is a large and conspicuous bird, which often nests on tree-tops and in sizeable groups, it can be very difficult to locate all the heronries in an area. Many are small, consisting of only one or two nests, and some are in dense conifer plantations or on remote stretches of coast. The assessment of population changes is made even more difficult by the fact that heronries are abandoned when trees are felled or wind-blown, and are sometimes deserted for no apparent reason. Five of the

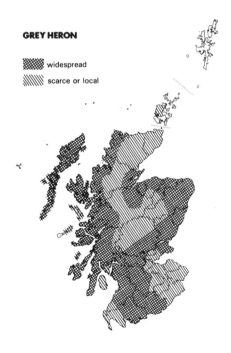

GREY HERON

▓ widespread

▨ scarce or local

seven colonies counted in Perth in 1954 were extinct ten years later (Thom 1966). In Scotland none of the earlier censuses achieved full coverage, and only the results of the first two have been published; those for the 1954 census are given in Table a. In that year most heronries (67%) held two to ten nests and 18% were single nests. Only one (in Lanark) had more than 30 nests. Tree-nesting birds (84%) favoured Scots pine and spruce the most. Cliff-nesting (7%) was recorded in Orkney, Caithness, Sutherland, Wester Ross and the Outer Hebrides. The remaining 9% were breeding in scrub on islands in lochs or on coastal islets, in reedbeds (on Colonsay and Oronsay), or on ruins (in Argyll and on the Monach Isles). Nesting has been recorded from c400 m asl (B&R) to near sea-level.

Atlas survey work located heronries in 286 10 km squares in Scotland, and subsequent fieldwork by Mick Marquiss and others has added to the available information. The data for five study areas (Table b) illustrate the problems that

(a) Numbers and distribution of breeding Grey Herons in 1954. (Data from Garden 1958)

	No. of occupied heronries	No. of breeding pairs
Aberdeen	10	48–49
Angus	4	18–21
Banff	4	33
Kincardine	3	11
Moray	4	35–42
Nairn	1	1
Argyll (mainland)	c20	c148
Mull & Ulva	9	c89
Islay	1	6
Colonsay/Oronsay	2	6
Jura	2	20
Arran	2	15
Bute	5	14
Caithness	2	12
Inverness (mainland)	5	20
Skye	14	74
Uists/Benbecula	8	84
Lewis/Harris	4	28
Ross & Cromarty (mainland)	10	30
Sutherland	7	33–34
Orkney	1	6
Dunbarton	5	12
E. Lothian	1	11
Fife	3	17
Lanark	1	33
Midlothian	0	0
Perth	7	c28
Renfrew	4	13
Stirling	1	19
Ayr	9	71–72
Berwick	3	10–11
Dumfries	14	56–59
Kirkcudbright	8	52–57
Peebles	1	c5
Roxburgh	2	c7
Selkirk	0	0
Wigtown	?	?
Grand Total	177	1,065–1,087
Average colony size	6	

? indicates no record

(b) *Numbers of breeding Grey Herons in selected study areas (Data from M. Marquiss). Numbers of 10 km squares*

	In study area	Containing breeding Herons in				
		1928–29	1932–49	1954	1968–72	1981–82
Dumfries	16	8 (93/10)*	?	9 (60/10)	9	9 (99/11)
Fife & Kinross	26	6 (33/6)	?	5 (26/5)	9	15 (160/21)
Perth & Angus	40	5 (60/5)	?	9 (43/10)	9	11 (146/13)
Aberdeen	29	5 (16/6)	?	9 (53/9)	14	17 (130/20)
Uists & Benbecula	18	?	11 (328/20)	?	9	11 (183/16)

* bracketed figures are number of nests counted/number of colonies
? indicates no record

arise when trying to assess population changes; it is not known whether variations in the totals recorded are due to differences in population size or in the proportion of all heronries found. By comparing the figures for his thoroughly-searched study areas with Atlas results for the same areas, and extrapolating on the basis of this comparison, Mick Marquiss has estimated the 1982 population as about 3,000 pairs, distributed in 450 colonies. This calculation assumes, however, that the study areas are representative of the population as a whole, and this is probably not a valid assumption (M. Marquiss).

Annual sample censuses in England and Wales since 1928 have shown that Grey Heron numbers may range from 45% below to 32% above the 'normal' average according to the severity of preceding winters (Reynolds 1979). At regularly monitored colonies in Scotland there was an overall population decline of 11% following the severe winter of 1981/ 82, a significantly smaller decrease than in 1978/79, when the hard weather was much more prolonged (Marquiss *et al* 1983). More young per pair were reared in 1982 than in 1981, so that despite the reduced population the total production of young was only 4% down in 1982. National ringing analyses indicate that first-winter birds are most seriously affected by hard winters (North 1979) but there was no significant increase in overall mortality among young birds in Scotland in 1981/82, although a higher proportion of deaths than usual occurred in January. There was evidence, too, of much wider dispersal than normal; whereas the average recovery distance in December/February for previous years was only 36 km from the nesting site, in 1981/82 it was 86 km.

The information on mortality and movements of Scottish birds derives largely from extensive ringing and wing-tagging of nestlings carried out since 1976 by Mick Marquiss and the Tay Ringing Group. Fewer than 30% of native birds move 100 km or more from their natal colony, young birds dispersing more widely than adults. Of 215 recoveries (from 2,700 birds ringed), 89% were in Scotland, 5% in England and 4% in Ireland. Only 2% were abroad – single recoveries, or sightings of wing-tagged birds, in Denmark, Belgium, France, Spain and Morocco; all were between October and March and involved first-winter birds.

Herons are very opportunist in their feeding habits and take a wide variety of prey, including fish, eels, small mammals, insects, earthworms, shrimps and crabs. In Scotland many feed around estuaries and along the seashore. Being at the end of a food chain they are vulnerable to pesticide pollution, and in the 1960s some English heronries were badly affected, with many thin-shelled eggs being laid and adult birds behaving in an aberrant manner, resulting in either egg breakage or chick death. The work of the Institute of Terrestrial Ecology on the Grey Heron as an 'indicator' species to monitor pollutants in the environment has shown that Scottish Grey Herons breeding at over 150 m asl, and most of those on the west coast, are relatively 'clean'. The most polluted birds are those feeding around east coast estuaries, where substantial egg breakage occurs due to thin shells; at the worst affected colonies up to half the clutches have at least one egg broken. A high incidence of egg breakage has also been recorded amid intensively farmed land in the northeast (Marquiss 1982 & pers. comm.). Fortunately Herons readily re-lay, and second or third clutches contain progressively lower levels of pollutant; there is consequently a good chance of at least some young being reared.

The Grey Heron has long suffered from persecution – and still does so. In the past the birds were shot because they were believed to have a serious effect on game-fish stocks. First-year birds were the most frequent victims but the introduction of protection in 1954 resulted in a reduction in the mortality rate for this age group from 70% to 60% (Mead *et al* 1979). More recently the main conflict has centred on fish-farms, where much financial loss may be caused by birds scarring fish as well as eating them. Under the Wildlife and Countryside Act 1981 a licence can be issued for Grey Heron control at a fish-farm if evidence is provided that alternative means of control have failed, but most of the shooting that takes place is illegal. Shooting is in any event neither the most effective nor the cheapest method of solving the problem (Meyer 1981). Herons walk into the water, so either steeply-shelving banks or wires at a suitable height and distance from the edge are considerable deterrents. With small fish-ponds netting is the most effective protection.

In addition to pollution and persecution, land management changes may pose a threat to Grey Heron numbers. Drainage of boggy fields and the canalisation of water-courses decrease potential feeding grounds, excavation and ditching which results in silting and loss of fish will reduce available food supplies, and afforestation of uplands may also have a detrimental effect through reducing water-flow. Despite the problems involved in counting heronries it is clearly desirable that periodic censuses should be carried out in order to monitor the changing situation.

Little is known about the scale of winter immigration but there is obvious passage through the Northern Isles, where the resident population is very small. Most of the immigrants are from Norway, where the population is partly migratory; smaller numbers of Swedish and Danish birds have also been recovered here. More than 75% of foreign-ringed recoveries are first-winter birds.

Purple Heron *Ardea purpurea* [1]

Vagrant (Europe) – recorded in eight years, 1950–83.

Although the Purple Heron has become an annual visitor in some numbers to southeast England, it continues to appear only as an irregular vagrant in Scotland. It was first recorded in 1872, when specimens were shot in East Lothian and Aberdeen, but only two were reported during the first half of this century: in Caithness in 1907 and Berwick in 1917. Between 1950 and 1983 the Purple Heron occurred in only eight years: on Fair Isle (1965, 1969 and 1970), in Dumfries (1973), Shetland (1977 and 1981), Orkney (1980 and 1982), South Uist (1980) and East Lothian (1982). Most of the post-1950 records have been in spring, between 6 April and 17 June, and the remainder between 2 August and 4 October. The 1969 Fair Isle bird had been ringed as a chick in Holland in June 1967; most of those occurring in Scotland presumably originate from that area, which is at the northern limit of the species' south-central Eurasian range.

Black Stork *Ciconia nigra*

Vagrant (Europe) – recorded in six years, 1946–83.

The Black Stork was first recorded in May 1946, when one was seen flying high over Longniddry, East Lothian (BB 39:344). It has since occurred in Orkney (June 1972), North Uist (August 1974), Shetland, Sutherland, Nairn and Inverness – these four reports possibly related to the same bird (May/July 1977), Perth (August 1980) and Peebles/East Lothian (June 1983). Although Black Storks are kept in captivity in this country, and some of the English records are believed to be of escapes, there has been no suggestion that those occurring in Scotland are not wild birds. The species now breeds only very locally in western Europe, having ceased to do so in Belgium and southwest Germany last century, and in Sweden and Denmark during the 1950s. The colonies nearest to Britain are in northern Germany, eastern France and Iberia; the main breeding range stretches from Poland and the Balkans eastward through Asia.

White Stork *Ciconia ciconia*

Vagrant (Europe) – annual in very small numbers since 1975. Has bred.

Although the north-European sector of its breeding range has contracted in the last 200 years, the White Stork has been recorded here with increasing frequency since 1965 and annually since 1975. It breeds locally in Germany, Denmark, the Netherlands and northern France but has ceased to do so in Belgium and Sweden; it winters in Africa. As some individuals apparently move around rapidly, while others remain in one place for several days or even weeks, it is difficult to assess the actual number of birds appearing in Scotland. It seems likely that not more than ten birds have been involved in any one year, even when there have been many more reports (as in 1977 and 1979), and in most years numbers are much smaller. With such a conspicuous species it is unlikely that many pass unnoticed.

The White Stork has been recorded in all mainland counties except Wigtown, Selkirk, Peebles, Kinross, Clackmannan, and Dunbarton (the Berwick and Midlothian records are old ones), and in Shetland, Orkney, Fair Isle, Skye and Islay. Most reports are from the eastern half of the country and between April and mid June, but there are single records for January (Glasgow 1976 – possibly an escape?), February (Aberdeen 1977) and August to October, the latest date being one at Dalry, Ayr, on 30 October 1980. In May 1982 one was watched following a plough, presumably in search of worms.

The sole British breeding record of this species is of a pair nesting on the top of St Giles' Cathedral, Edinburgh, in 1416.

Glossy Ibis *Plegadis falcinellus*

Vagrant/escape (E Europe) – two records since 1950, formerly more frequent.

The only recent records of the Glossy Ibis are of one on Tiree in October 1958 and one at Dornoch in December 1962. That it was formerly a much more frequent visitor is evident from the list of earlier records given by B&R, most of which are for the second half of the 19th century and the first decade of the 20th. Occurrences during this period were widely scattered, from Dumfries and Roxburgh to Shetland and the Outer Hebrides; most were in autumn but a few in spring and summer. A particularly interesting record is that of 19–20 in Orkney in September 1907; the ten birds shot were all immatures. The breeding distribution of the Glossy Ibis is discontinuous, very local and irregular. There has been a marked contraction at the western end of the range, perhaps largely due to drainage and disturbance; the former colonies in France, Spain and Austria have gone and those in the Balkans are now the most westerly.

Spoonbill *Platalea leucorodia*   [1]

Vagrant (Europe) – less than annual, in very small numbers.

B&R regarded the Spoonbill as a rare visitor and knew of only two mainland records, but it has occurred with greater frequency since 1960. In the 1950s it was recorded in one season, in the 1960s in four seasons, and in the 1970s annually, but there were no reports in 1980 or 1982, only two (probably referring to the same individual) in 1981, and one in 1983. Numbers seldom exceed five but occasionally larger influxes occur, as in 1975 when at least 11 birds were present in late May/early June. The maximum number seen together in recent years is three but B&R note an old report of a party of nine in Orkney. Though most arrive in May/July or September/October there are records for every month except February and March; individuals sometimes remain in the same area for several weeks. There is no regular pattern to the occurrences, which are scattered from Shetland to Kirkcudbright and from the Isle of May to Lewis, but there are more records from Aberdeen than from any other county. This is one of the few species on the Scottish list not yet recorded on Fair Isle.

There have been several recoveries or sightings of ringed birds. Two colour-ringed as juveniles in Holland in 1974 were seen at Strathbeg in May 1975 and July 1977, while a bird shot near Perth in November 1964 (probably the one seen a week earlier at Blairgowrie) had been ringed as a nestling in Yugoslavia in June the same year. Spoonbills feed in slow-moving shallow water and those arriving in early winter can have only poor prospects of survival if they do not rapidly move south, as the first hard frosts will make it impossible for them to feed. Several of the late-season records are of birds already dead; one at Loch Fleet in December 1975/January 1976 became so weak that it was caught and taken to a zoo in an unsuccessful attempt to keep it alive.

[Flamingo *Phoenicopterus spp.*]

Escape/vagrant (Category D)

Although it is not impossible that the Greater Flamingo, *P. ruber,* which breeds in Iberia and southern France, might occur as a vagrant, it is probable that all flamingos recorded in Scotland are escapes from captivity. There have been many reports since the mid 1960s, scattered from Berwick to Shetland and from the Solway to the Outer Hebrides. Most birds showed the characters (grey legs and pink 'knees') of the Chilean Flamingo *P. chilensis* but a few were sufficiently brightly-coloured to suggest the Caribbean race *P.r. ruber* and a few were considered to belong to the race breeding in Europe *P.r. roseus.* In some instances the consecutive dates of reports from neighbouring areas suggest that the same bird was variously identified as Chilean and Greater. While some individuals have remained for months, or even years, in the same area, others clearly move over considerable distances in a relatively short time. Despite the fact that these birds are almost certainly escapes, there is undoubtedly a

thrill to be had from spotting a flamingo 'in the wild'; my own first sighting was when I discovered that two of the Mute Swans I was counting on Loch Bee, South Uist, in 1969, were pink!

Mute Swan *Cygnus olor*

Resident, widely though often thinly distributed on the mainland but scarce or absent on many of the islands. Large moulting flocks gather in late summer and a few sites hold important wintering concentrations.

The Mute Swan's breeding distribution is closely related to the availability of still or slow-running waters of sufficient size to permit take-off and with abundant submerged vegetation at a suitable depth. Although breeding in every mainland county, it is consequently absent from most upland areas and from those with predominantly deep or acid lochs. In Shetland it occurs only as a straggler, the most northerly breeding site being on North Ronaldsay, Orkney. Nesting has not been recorded on Lewis, Harris, Skye or the Small Isles, and there is usually only a single pair on Arran. B&R had no records for Banff, Moray, Nairn or Caithness but birds were nesting in these counties by 1968–72.

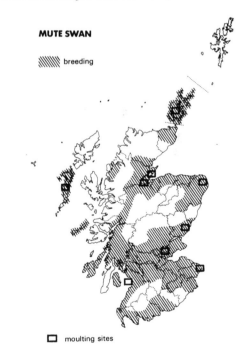

MUTE SWAN

▨ breeding

☐ moulting sites

Attempts to assess the breeding population have been made in 1955–56, 1978 and 1983 (Table a) but none has achieved complete coverage and population trends are consequently not always clear. The 1978 census (Ogilvie 1981) showed that the highest breeding densities per 100 km square in Britain are in the Outer Hebridean islands south of the Sound of Harris (20+ pairs/100 km²) and in the extreme northeast of Aberdeen (19+ pairs/100 km²). The islands held an average of 122 breeding pairs in 1979–82 (C. Spray) and the Loch of Strathbeg 31 pairs in 1978. Although the estimated total for 1978 showed little change from the 1955–56 census figures there were marked changes in distribution (Table a), while the ratio of breeding:non–breeding birds had increased from 1:2.5 to 1:1.8; in 1983 the ratio, 1:1.6, was very similar to that in 1978. The breeding population in the north of Scotland expanded greatly between the mid 1950s and 1978 but has apparently since declined again slightly; 1983 was, however, a poor year with 46–61 pairs in the north, northeast and Orkney holding territory but failing to breed. There has been a considerable recent decline in the Central Lowlands, with the total population in the Lothians dropping by c50%, from 249 birds in 1961 to c115 in 1982, and the number of territorial pairs decreasing from 59 to 30 over the same period (Brown & Brown 1984a). The 1983 figures suggest that elsewhere in central and western Scotland there has probably been a similar decrease (Brown & Brown 1985).

In July moulting birds, largely non-breeders, gather on lochs or tidal water, often in the vicinity of distillery or other organic waste discharges. The most important sites are Loch Bee, South Uist, where about 400 full-grown birds are usually present in July/August (C. Spray), the Loch of Strathbeg (c370 birds in 1975), and the Montrose Basin (c350 in 1980–81). Ringing and colour-marking has shown that the Strathbeg moult flock consists largely of birds from the northeast; that at Montrose includes swans from as far afield as Glasgow, Northumberland and Cumbria (C. Spray). Several hundred moult at the mouth of the Tweed, about 200 each on the Beauly at Inverness, in Udale Bay and on Lochs Harray and Stenness, and up to 100 on the Ayr coast between Doonfoot and the town of Ayr. Some 500 formerly moulted on Loch Leven but when water plants decreased in abundance during the 1960s the swan numbers declined to under 50; ringing recoveries showed that they came largely from central Scotland (Allison *et al* 1974). About 200 currently moult there. Where large moult flocks start walking onto fields to feed they may become a problem for farmers, as has happened in one area.

The Mute Swan is generally regarded as a very sedentary species but there are scattered records from Fair Isle, the Isle of May and St Kilda. However, although long-established in the Uists, the species was not recorded in Lewis until 1974. Recent work in the Outer Hebrides has shown that immatures occasionally move considerable distances: for example one South Uist bird reached Kintyre in its first summer and County Derry the following year (Spray 1982). Evidence of a much longer movement is provided by the recovery in Dumfries of a bird ringed as a juvenile in Lithuanian SSR four years earlier. The discrete Hebridean population, introduced late last century, was studied intensively in 1978–82 by Chris Spray, who found that, although brackish lochs held a large proportion of the population, those nesting by machair freshwater lochs had much the highest rate of breeding success (Spray 1980 & pers. comm.).

The maximum number of Mute Swans recorded in Scottish wildfowl counts is usually 1,700–1,800, but this figure does not include the Uists/Benbecula population of 800–900 (Spray 1981), nor many of the scattered small parties else-

(a) *Numbers and regional distribution of breeding Mute Swans in 1955–56 (incomplete), 1978 and 1983. (Data from Ogilvie 1981, Brown & Brown 1985 and C. Spray)*

Region	Numbers of breeding pairs		
	1955–56	1978	1983
South-east: Lothians/Borders	73+	54	57
East-central: Fife/Angus/Perth/Kinross	56+	83	50
North-east: Abdn/Kinc/Moray/Banff/Nairn	71+	85+	75–76
North: Inv/Ross & Crom/Suth/Caith.	30	72	40–41
West-central: Arg/Stir/Clack/Dunb/Renf/Lan.	101+	52++	83–85
South-west: Wigs/Kirk/Dumf/Ayr	84+	76+	75
Outer Hebrides	22++	116	112
Orkney	26	75	67–68
Estimated total of breeding birds	c1,000	c1,300	1,118–1,128
Estimated total of non-breeding birds	c2,500	c2,350	1,792–1,814

(b) *Peak counts of Mute Swans at sites regularly holding wintering flocks of national importance – qualifying level 180. (Data from Wildfowl Trust & SBR)*

	1976/77–1980/81		1981/82–1983/84	
	Average	Range	Average	Range
Loch of Strathbeg	330	280–430	300	280–314
Loch Eye	200	65–270	22	10–30
Montrose Basin	200	180–230	238	230–245
Lochs Harray & Stenness	190	150–250	188	140–240

where. The figure of 3,750 suggested by Owen *et al* (1986), and based on the count data for the last ten years, is likely to be a fair estimate of the wintering population. Total numbers change comparatively little through the season, although there may be substantial local fluctuations as birds disperse from the moulting areas. No Scottish sites hold gatherings of international importance (more than 1,200) but several support nationally important numbers (Table b), while flocks of over 100 have also been seen recently on Lochs Watten, Spynie and Ken, the Cromarty Firth at Udale Bay, Drummond Pond near Crieff, and the River Teviot at Nisbet.

There have been several serious oiling incidents affecting swans; in 1974 many of the 126 oiled on the Cromarty Firth at Dalmore died or had to be destroyed, and 209 of the 211 swans on the Montrose Basin in February 1980 were badly affected when a faulty fuel tank leaked several hundred gallons of oil. At least 37 died in the latter incident and it is believed that the final death toll was probably about 60 (Atkinson 1981). A major, though less spectacular, cause of mortality is collision with power lines, especially the low-voltage lines supplying farms and villages, which often pass at no great height close to regularly-used waters. In Lewis one of three immatures collar-marked in Uist died in this way in 1980 (Crummy 1981), while in parts of Fife swans are responsible for frequent power breakdowns. Lead poisoning (from fishing-line weights) is a serious problem in England (Ogilvie 1981) but is unlikely to be as significant in Scotland, where coarse fishing is not so widespread; that it can also result from ingestion of shot is evident from the fact that a Mute Swan found dead near Elgin contained 944 gunshot pellets (Spray 1983). Increasing disturbance and changes in aquatic vegetation, such as took place at Loch Leven, are probably the main reasons for the declining breeding population in the lowlands.

[Black Swan *Cygnus atratus*]

This Australian species has long been kept in captivity in Europe. As free-flying birds are obviously escapes the few recent reports probably do not adequately reflect the frequency of their occurrence. In 1979 two immatures were on Montrose Basin from 12 July, in 1980 there were reports from Angus, Aberdeen, East Lothian and Ayr, and in 1983 one was at Tyninghame from April to October.

Bewick's Swan *Cygnus columbianus* [1]

Winter visitor, occurring annually in small but increasing numbers, especially at Caerlaverock, Dumfries.

Bewick's Swans were at one time quite abundant, particularly in Tiree and the Uists, through which flocks passed *en route* to Ireland from the breeding grounds in northern Russia, but during the first half of this century they decreased until no more than occasional stragglers (B&R). Numbers remained low until the end of the 1960s but have since increased progressively (Table). There has been a marked increase at the species' principal British haunts, Slimbridge and the Ouse Washes, during the last 20 years (Owen *et al* 1986); a record 4,500 were on the Ouse Washes in December 1984.

By far the largest numbers in Scotland occur at the Wildfowl Trust's Eastpark Refuge, Caerlaverock, where the peak count increased from 30 in 1977 to 77 (including 13 first-year birds) in November 1980; in 1980–84 annual peaks were between 60 and 70. This large flock feeds partly on grain put out at Eastpark and partly on farmland, including flooded potato fields. This is the only site at which nationally important numbers (more than 50) have been recorded. Elsewhere flocks of more than ten birds have been reported from Islay, Ayr, Kirkcudbright, Wigtown, East Lothian and Kinross, but the bulk of the records relate to parties of less than five. Since 1956 there have been reports from all mainland counties and island groups other than Peebles, Kincardine, Banff, Moray, Inverness, Sutherland and the Clyde Islands.

Autumn arrival is usually from mid October, but sometimes not until November; in some years there are no reports before January. The earliest recent record is of one with

Numbers of Bewick's Swans recorded in Scotland 1960–84. (Data largely from SBR)

	Annual total	
	Average	Range
1960/61–1968/69	17	11–23
1969/70–1973/74	35	28–43
1974/75–1979/80	79	46–100
1980/81–1983/84	91	71–131

Whooper Swans on 18 September; this bird had probably come from Iceland, where the species has recently occurred in summer – those from USSR do not arrive until at least four weeks later. Peak numbers are present in January/February and departure starts in March, with most birds gone by early April; the latest date is 7 May. A Bewick's Swan colour-ringed at Slimbridge was at Summerston, near Glasgow, in January 1975 and there were three marked adults in the Caerlaverock flock in 1979/80; a ringed and yellow-dyed bird seen in Caithness in February/March 1978 had also been marked at Slimbridge. Four colour-marked at Caerlaverock in February 1982 were in the Netherlands in October/December that year.

The old records of the closely-related American Whistling Swan, square-bracketed by B&R, remain unaccepted.

Whooper Swan *Cygnus cygnus* [1]

Widely distributed winter and passage visitor, often in small parties but occasionally in flocks of several hundreds. Although many still feed in the traditional manner in shallow water, an increasing number forage over farmland. A few summer each year and breeding occurs sporadically.

Scotland holds a large proportion of the British and Irish wintering population of Whooper Swans, estimated at c7,000 when numbers are at their November peak (Owen *et al* 1986). The birds are widely dispersed, with many family parties and small groups occupying remote lochs and marshes; this pattern of distribution makes the assessment of numbers difficult and the regular monthly wildfowl counts inadequate

WHOOPER SWAN

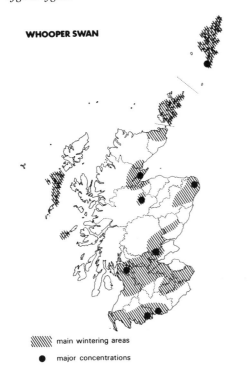

▨▨▨ main wintering areas

● major concentrations

(a) *Numbers and regional distribution of Whooper Swans in November. (Data from Boyd & Eltringham 1962, M. Brazil, J. Kirk (Jun.) and as listed)*

Region	1960 census	1979 census	Other counts
Shetland	180	(275) (a)	486 (Nov. 1982, Shetland BR)
Orkney		450	1,813 (Nov. 1982, SBR)
Hebrides – Outer	50	160	359 – Uists & Benbecula only (mid Oct. 1981, C. Spray)
– Inner		23 (b)	
North	250	2,237	
Northeast	290 (c)	473+	
Central	630	240 (d)	
West	160	210	
Southeast	90	185	
Southwest	180	167 (e)	
Total – Scotland	1,830	4,145+ (f)	
England & Wales	389	558	
Ireland	NA	2,014	

(a) Counted c2 weeks before main census and not included in total
(b) Excluding Tiree
(c) Excluding Loch of Strathbeg, where 631 in November 1961
(d) Excluding Stirling and Dunbarton
(e) Excluding Wigtown
(f) An additional 64 were recorded in Shetland wildfowl counts in mid Nov.

for this purpose. Two attempts at complete censuses have been made: in 1960/61, when an enquiry was organised by the Wildfowl Trust and the British Trust for Ornithology, and in 1979 by Kirk & Brazil (Table a). Neither census achieved complete cover but the Scottish count in 1979 represented 62% of the British and Irish total. Most, and probably all, of the Whooper Swans visiting Scotland come from Iceland, where the population was recently estimated at 10,000–11,000 (Gardarsson & Skarphedinsson 1984). Although some birds do overwinter in Iceland, the difference between this estimate and the totals counted in Britain and Ireland suggests that many are missed in census attempts here.

The first Whoopers often arrive in early September and the main influx takes place from late September to November. The use of neck-collars has shown that some move around extensively within Britain: one marked in Iceland on 25 July was at Loch Eye on 15 October, the Loch of Strathbeg on 1 November, Caerlaverock from 30 November to 2 December, and the Ouse Washes from 7 December to 12 March. It was in Caithness by 25 March and was back in Iceland on 19 May. Flocks do not necessarily move as a unit and considerable interchange in membership apparently takes place (Brazil 1983) but many birds do show quite a high return rate to a favoured area, such as Caerlaverock (Black & Rees 1984). Some movement to Ireland takes place in late autumn and there are fewer Whooper Swans in Scotland in mid winter. Numbers increase again in March as the birds start returning to their breeding grounds. Most have gone by late April though a few sometimes linger into May.

The largest concentrations in recent years have been around the Cromarty Firth, where numbers exceeded 500 each year from 1977 to 1981 and reached a remarkable 1,300+ in November 1979, when Loch Eye alone held 1,200. This unusually large gathering was due to a profuse growth of the submerged herb *Ruppia*, which provided an exceptionally abundant food supply. In November 1977 there were 683 on the Invergordon/Barbaraville stretch of the firth and only 65 on Loch Eye, and in 1980/81 405 on the loch and 314 on the firth at Dalmore. The other major haunts are widely scattered and include a variety of habitat types (Table b). Numbers have declined since the 1960s at Loch Leven, where there were c430 in autumn 1970, and the upper Forth valley, and also in the Blairgowrie area, where flocks of more than 100 formerly occurred fairly regularly. Between 1976

(b) *Peak numbers of Whooper Swans at selected sites – no. of counts in brackets. (Data from Wildfowl Trust, SBR, Bell 1981)*

	1976/77–1980/81		1981/82–1983/84	
	Average	Range	Average	Range
Loch Spiggie	114 (3)	76–174	190 (2)	44–336
Loch Eye/Cromarty	705 (4)	405–1,200	157 (3)	50–310
Insh Marshes	110 (5)	78–176	135 (2)	128–141
Loch of Strathbeg	383 (5)	248–502	519 (3)	382–633
Tweed/Teviot	141 (3)	71–192	154 (3)	135–183
Islesteps	104 (4)	83–139	91 (3)	54–142
Loch Ken	106 (4)	85–118	71 (3)	18–101

and 1984 flocks of national importance (more than 50 birds) were reported from many sites, scattered from Shetland (Loch Spiggie) to Wigtown (Moss of Cree).

Considering the 1960/61 survey results in relation to earlier counts, Boyd and Eltringham (1962) concluded that numbers had remained fairly static from 1952 to 1960, after increasing considerably between 1948 and 1952. They also recorded a change in the pattern of distribution, with many more Whooper Swans wintering in central Scotland and relatively fewer in the Outer Hebrides, described by earlier writers as the chief wintering place. The 1979 survey indicated a further increase in the population and a change in November distribution, with much larger numbers in the Northern Isles and the Tweed valley, while central Scotland was relatively less important. The explanation for any such increase is not obvious, as the Icelandic population has not enjoyed high breeding success more than occasionally in recent years. From 1974 to 1980 the proportion of juveniles present in autumn averaged 15% and ranged from 4.8% to 22%; in 1981 there were 14.5% young birds, in 1982 18.2% and in 1983 14%. It is possible that the apparent increase is largely due to variations in census coverage.

Attention was first drawn to the growing habit of feeding over farmland when there were complaints of damage to crops in central Scotland in the late 1950s. This habit has since become more widespread. Whooper Swans studied in Orkney (Reynolds 1982) and the Forth valley (Brazil 1980) fed almost exclusively over stubble up to the end of the year, partly on stubble and partly on loch vegetation, grass or floodwater in January/February, and increasingly on grassland from March until they left in April. The current swing towards increasingly early ploughing of stubble will doubtless affect this feeding pattern. In some areas Whooper Swans also feed on waste potatoes and young winter cereals, sometimes causing localised damage to the latter as a result of 'puddling' the soil in a wet season.

This change in feeding habits is not without risk to the swans. In 1968 nearly 40 birds, about a third of the local population, were found dead or dying in the Blairgowrie area after feeding on young winter wheat. In this case the ground was loose after frost and the swans uprooted the seedlings and gained access to the grain, which still carried considerable quantities of seed-dressing (SB 5:111–112). Large-scale mortality also occurred in Wigtown in 1969 and in Roxburgh in 1979, but in the latter case was thought due to lead poisoning (Badenoch 1980); poisoning through the ingestion of gunshot lead has been recorded at Possil Marsh and on the Ythan (C.Spray). It seems possible that such incidents may have lasting effects on distribution, perhaps because the Whooper Swans of a regularly wintering flock in Scotland represent the bulk of the breeding population from a particular locality in Iceland; the Blairgowrie group has certainly never recovered to its former size.

Birds suffering from the after effects of sub-lethal poisoning, as well as injured swans, are among those remaining to summer in Scotland. Although some of these are on waters apparently suitable as breeding sites few actually attempt to breed. The last recorded breeding prior to the late 1970s was in 1947, when a pair nested successfully on Benbecula. Then in 1978 a pair reared three cygnets on Tiree, followed by two in 1979; although the pair was still present in 1980

there was no sign of breeding. Also in 1979 a feral pair, believed to be a wild male mated to an escaped female, reared one young on Loch Lomond; two young were reared there in 1980 and again in 1982, when two feral pairs nested.

Bean Goose *Anser fabalis*

Scarce and very local winter visitor, occurring mainly on arable land and rough pasture and in marshy areas. Most sightings are between October and March.

Bean Geese were at one time the most abundant of the grey geese visiting Scotland but around the turn of the century their numbers declined greatly and they are now uncommon; the species' earlier status changes are documented by B&R. Flocks of 200–400 were recorded on the main wintering ground in Solway until the mid 1940s, since when there has been a further decrease (Table). The principal haunts

Peak counts of Bean Geese at main Galloway haunts, 1952–1984. (Data from A. D. Watson)

	Average	Range
1952/53–1958/59	193	150–240
1959/60–1968/69	78	17–150
1969/70–1978/79	47	14–70
1979/80–1983/84	23	3–40

in that area are around Threave, Gelston and Mid Kelton, near Castle Douglas, though small groups are occasionally seen at Loch Ken, Loch Ryan and the Moss of Cree. In recent years the flock has been more elusive than formerly, often appearing for only a few days if at all. Peak numbers usually occur between late December and early February and often coincide with a westward spread of cold weather from the Continent; two of the largest groups recorded recently, 38 in January 1979 and the same number in December 1981, arrived shortly after the onset of severe weather (Watson, in press). The advent of milder weather generally results in the disappearance of the flock.

A smaller group used to winter around the southeast corner of Loch Lomond but no longer does so regularly. In the early 1950s this flock numbered up to 30 but during the 1960s it never exceeded 17. There have been few records from there since 1970. The discovery in February 1981 of a flock of 73 at Carron Valley Reservoir (only 20 km further east) suggests that the birds may simply have changed their feeding area – and that it may be worth searching other areas of rough pasture for this species. The Carron Valley held flocks of 52 in October 1982, 46 in October 1983 and 160 in late September 1984. As there is no overlap in dates, it is possible that the Bean Geese seen in the Carron Valley and those in Galloway are the same birds.

Away from the regular sites, Bean Geese have been recorded at least occasionally in most mainland counties and all the island groups. They occur fairly regularly in Aberdeen, especially around the Slains lochs, where there was an unusually large influx in October/December 1968 with a peak of 107 on 23 November. Ogilvie (1978) mentions Deeside as a regular haunt but the only recent record from that area appears to be of two shot near Aboyne on 13 January 1979. These birds were considered to be of the Russian race *A.f. rossicus*, whereas most of those wintering in Britain are believed to belong to the western race *A.f. fabalis*, which breeds in Scandinavia (Ogilvie 1978). The frequency of sightings in the Northern Isles, Caithness and Easter Ross has increased during the early 1980s, and in 1980/81 about 25 overwintered around Munlochy Bay. Reports were unusually numerous and widely scattered that season; many birds arrived during the first half of October, presumably having been driven off-course while migrating from Scandinavia to the main wintering grounds in the Netherlands. The earliest arrival date noted is 16 September and most have left by mid April. The occasional summer sightings, some of which may involve escapes, include June/July records from Foula (1956), St Kilda (1974), Inverness (1971) and Orkney (1982). Few reports relate to flocks of more than 15 birds.

The Bean Goose is fully protected under the Wildlife and Countryside Act 1981 but is quite likely to be shot in error, as it can easily be mistaken for a Pink-footed Goose. A slight increase at the only regular English wintering site, in Norfolk, has meant that overall numbers in Britain have changed little in the last 20 years, but in the Netherlands there has been a marked increase (Owen *et al* 1986). Some 50,000 Bean Geese now winter there and others in Denmark, Germany, France and southern Sweden but the exact origins of these flocks are not known (Ogilvie 1978). A Bean Goose Research Group was established in 1981, under the aegis of the IWRB, to study the populations and movements of the four recognised races of the species, whose breeding range stretches from Fenno-Scandia to Kamchatka.

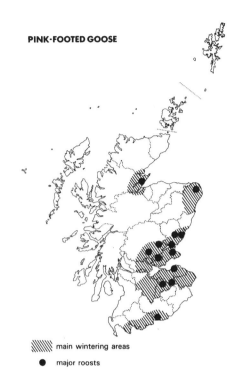

PINK·FOOTED GOOSE

▨▨▨ main wintering areas
● major roosts

B

Pink-footed Goose *Anser brachyrhynchus*

Winter visitor, feeding over rich arable farmland in lowland areas and using traditional and well-established roosts. Distribution varies through the winter and between years, but in autumn east-central and southeast Scotland hold a large proportion of the Iceland/Greenland population, all of which winters in Britain.

The first Pink-footed Geese generally reach Scotland during the first half of September, and the main influx between the middle of the month and the end of the first week in October. Arriving birds follow a roughly north–

south route over mainland Scotland, often reaching their initial destination during the night or soon after dawn. When strong westerly winds occur during migration the geese are drifted down-wind; under these conditions they can be seen struggling head-on into the wind as they move inland from the coast, for example up the Tay Estuary. The Loch of Strathbeg, Strathearn, Loch Leven and Fala Flow are among the most important arrival points; on reaching these areas the geese may rest for 24 hours before starting to feed. Numbers usually reach their peak before the annual November census.

On the basis of the sparse information available at the time, B&R described the Pink-footed Goose as having increased markedly between 1880 and 1950, attributing this to extension of the breeding range in Greenland and Iceland; during that period the Pink-foot supplanted the Bean as the predominant goose in Scotland. The overall trend has since been one of continuing slow increase. Between 1960 and 1983 census totals for Britain ranged from 47,000 in 1960 to 101,000 in 1983, increasing steadily from 1960 to 1966 and fluctuating thereafter. During much of this period there was a progressively increasing concentration of the population north of the Border, but this trend has been reversed in the early 1980s, with more birds moving into England. The regional distribution of the birds has varied greatly from one autumn to another (Fig a) with east-central Scotland most often holding the highest proportion.

While in Scotland, Pink-footed Geese feed almost exclusively on farmland and generally roost on water – either fresh or tidal – although in some areas flocks rest on moorland (Newton *et al* 1973). In autumn their preferred feeding ground is barley stubble, and regional variation in the amount of grain shaken off before or during harvest is a

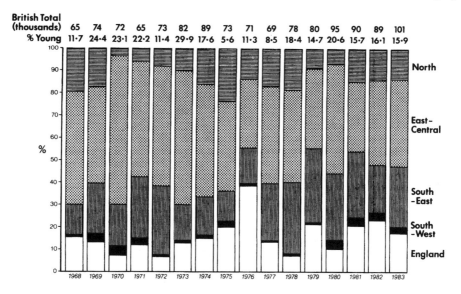

(a) Distribution of Pink-footed Geese in autumn (from November census data)

major factor in determining their distribution then and in early winter. Current agricultural changes, involving very early autumn cultivation and a consequent reduction in stubble feeding opportunities, are already being reflected in distribution patterns. The availability of safe roosts is also important. Where disturbance is minimal and there are good feeding grounds within 10–15 km very large numbers may build up. The highest count yet recorded at a single roost is 27,500, at Dupplin Loch near Perth in November 1973; West Water Reservoir (first used a a roost in 1966) held 19,200 in mid October 1983, and the Loch of Strathbeg 20,000 in October 1984. This species is more wary than the Greylag, preferring to feed in large open fields and generally roosting on bigger expanses of water.

Most roosts have been in use for many years, although the numbers of birds using them may fluctuate considerably (Table). Several coastal roosts of importance early in the

(a) Numbers of Pink-footed Geese counted at the roosts most consistently important in 1976–83. (Date from Owen et al 1986, M. A. Ogilvie & SBR – figures rounded)

	1976–81		1981–83
	November census counts		
Roost	Average	Range (peak year)	Peak count
Dupplin Loch	9,400	3,340–19,550 (1980)	10,000+ (Sept)
Loch Leven	6,300	4,000–8,750 (1980)	12,460 (Oct)
Aberlady Bay	6,000	3,200–11,930 (1980)	9,300 (Oct)
Gladhouse Reservoir	5,600	2,000–9,600 (1978)	13,700 (Oct)
Ythan Est./Slains			
Lochs	5,400	2,900–7,800 (1978)	17,500 (Oct)
Loch of Strathbeg	4,900	1,400–7,500 (1977)	20,000 (Oct)
West Water Reservoir	4,800	0–12,340 (1981)	19,200 (Oct)
Cameron Reservoir	4,700	3,000–6,150 (1981)	8,000 (Nov)
Arbroath/Tay Estuary	4,300	645–7,330 (1979)	3,680 (Nov)
Carsebreck	3,600	630–5,680 (1980)	4,920 (Nov)

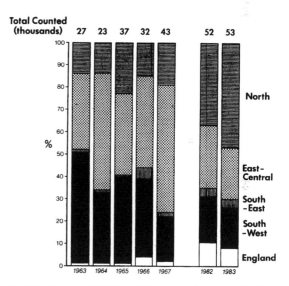

(b) Distribution of Pink-footed Geese in spring (from March census data)

century – Montrose Basin, the Eden Estuary and Aberlady Bay – became less attractive to geese during the war and in the early 1950s, due to disturbance first from aircraft and target practice and later as public access and wildfowling increased. Since Aberlady Bay became a nature reserve, numbers there have built up again, and the same pattern has been repeated at Montrose Basin within three years of its establishment as a reserve. More than forty of the principal roosts of this species were of international importance during the period 1976–81, with many of them holding numbers well above the qualifying level of 900 birds.

As winter progresses the distribution of the geese changes and the population becomes more fragmented. From January to April many more Pink-footed Geese are in southwest Scotland – which may hold as much as 25% of the population at that time – and in England. Although some goose movements are clearly due to food shortages or weather conditions – for example an influx of 10,000 at Caerlaverock during severe weather in January 1979 and the 36,500 in Lancashire at the end of January 1982 – others take place for no obvious reason. Nor is it yet apparent why the geese do not take advantage of all the seemingly suitable feeding areas in the country.

In late autumn and winter the geese eat potatoes remaining in the fields after harvesting, and growing cereals, but before they leave in April grass becomes their principal food. It is at this time of year, when they compete with livestock for the early bite of spring grass, that the only real conflict between Pink-footed Geese and agriculture occurs (Newton & Campbell 1973). In March there are still substantial numbers in the southwest (Fig b), especially at the end of a severe winter like 1981/82, but northwards movement starts before the end of the month. The main emigration takes place in successive waves during April and early May; last sightings are generally around mid May but occasionally later, while a few (probably injured) birds often remain to summer in this country. Flocks often set off at dawn in anti-cyclonic weather and take advantage of passes such as Drumochter to avoid the highest ground as they fly north. Some birds head out over the Atlantic from the Butt of Lewis and a flightline is known to cross Scotland from the Kyle of Sutherland northwards over Lochs Shin and More. The fact that there are few spring records from Orkney, Shetland and Fair Isle suggests that there is no regular spring passage corresponding to the autumn one over these islands.

Very large numbers of Pink-footed Geese were caught for ringing in the 1950s and 1960s, both while flightless on the Icelandic breeding grounds and by rocket-netting in Scotland. An early analysis was carried out by Hugh Boyd (1955), but no full examination has yet been made of the 7,000 or so ringing recoveries now available. The following examples have been selected to illustrate various aspects of distribution and movement. Scottish recoveries include birds from east and northeast Greenland and one ringed in Spitsbergen. Scottish-ringed birds have been found in Iceland between April and early October, in Greenland between May and August, and in Norway in September/October. Within-season movements are demonstrated by the appearance in England (mainly near the Wash) in January/February of birds caught in east Scotland the previous autumn. The most unlikely fate on record is surely that of a bird killed by a train near Crianlarich.

The increase in Pink-footed Geese in Scotland during the last 30 years is probably attributable to several factors. Increasing use of inland roosts, including reservoirs protected from disturbance, has brought the birds within reach of new feeding grounds, while farming changes have increased food supplies. The banning – in 1968 – of the sale of dead wild geese, with consequent lessening of shooting pressure, may have improved winter survival rates, but breeding success (although still exceeding the mortality rate) has fallen in recent years, probably due to shortage of suitable nesting and rearing areas (Owen *et al* 1986). It remains to be seen whether inclusion in Schedule 2 of the Wildlife and Countryside Act 1981, which allows Pink-footed Geese to be shot under licence during the close season if causing damage to crops, will have any significant effect on numbers.

In the long-term the size of the population may largely be determined by what happens on the breeding grounds. Following bad weather in the breeding season the proportion of first-winter birds present at the autumn census may be less than 10%, compared with 30% or more in a good year (Fig a). A sequence of poor seasons could therefore produce a substantial drop in numbers. A more drastic reduction could, however, result from hydro-electric development in the Thjorsarver area of central Iceland (now a National Park), which holds more than half the breeding pairs in the Iceland–Greenland population (Kerbes *et al* 1971). Until recently this was the only large breeding concentration known, but in 1979 a second densely-populated colony, holding over 800 nests, was located (Lok & Vink 1979). The importance of the Greenland breeding grounds may well have been underestimated in the past (Owen *et al* 1986).

Other threats to Pink-footed Geese in Scotland include both poisoning and oiling. Most poisoning incidents have involved the consumption of grain treated with an organophosphorus seed-dressing, carbophenothion, to which grey geese are particularly susceptible (Hamilton & Stanley 1975). Although this was banned as a seed-dressing in Scotland during the 1970s, occasional poisonings still occur; one of the most recent was near Montrose in the winter of 1982/83, when both Pink-footed and Greylag Geese were affected (C. Eatough). Geese have also died from oiling at coastal roosts – 1,000 suffered in the Cromarty oil-spill in 1972.

Although the 1980 census count was well above the 90,000 predicted for that year by Ogilvie & Boyd (1976), the rapid increase in the population experienced during the 1960s and early 1970s is now showing signs of levelling out. At the moment it seems unlikely that environmental changes either in Iceland or here will ever seriously threaten the population, but there can be no certainty that the situation will remain unchanged. It is very much to be hoped that the spectacular flights of really large flocks of this species, which are one of the most thrilling features of Scotland's birdlife today, never become a thing of the past.

White-fronted Goose *Anser albifrons*

Winter and passage visitor, occurring mainly in the west, where Islay is by far the most important resort. Seldom seen elsewhere in Scotland outwith passage periods.

About half of the Greenland White-fronted Goose population, which belongs to the race *A.a. flavirostris* and is estimated at c15,000 birds (Ruttledge & Ogilvie 1979), winters in Scotland. The European White-fronted Goose, *A.a. albifrons*, which breeds in northern Russia and winters mainly in the Netherlands but has a minor wintering population in the south and southwest of Britain, occurs only as a vagrant.

Considerable changes have taken place in the distribution of this species over the last 50 years. B&R recorded thousands in the Outer Hebrides in the 1890s and described this as the commonest goose in Orkney; not more than 100 now winter in either of these areas. In the 1930s John Berry reported large flocks around the Tay Estuary, especially during spring migration; only odd stragglers now appear there. These former flocks were almost certainly of the European race (Ruttledge & Ogilvie 1979). The current trend is for an increasingly large proportion to winter on Islay, which held about 56% of the British population in 1982/83 (Stroud 1984). Such a concentration increases vulnerability and is giving some cause for concern. Smaller numbers occur on other Inner Hebridean islands, on the mainland and inshore islands of Argyll, in Bute, Galloway and Caithness, and around Loch Lomond.

Fluctuations in the numbers counted on Islay (Table) reflect variations in both breeding success and counting efficiency (Stroud 1984). The birds are widely scattered in smallish flocks, usually of less than 100 birds but sometimes up to 400, and feed over stubble and improved grassland as well as in the wetter types of pasture and bog with which they are traditionally associated. Complete coverage is consequently difficult to achieve, but it has improved since the initiation of monthly counts in 1982. Sightings of birds colour-ringed on the breeding grounds have shown that Greenland White-fronts are exceptionally faithful to their wintering sites; in consequence they are likely to be very vulnerable to changes which may drive them from traditional areas. On Islay afforestation and commercial peat-working

Peak counts of wintering Greenland White-fronted Geese. (Data from Ruttledge & Ogilvie 1979, Owen et al 1986, Stroud 1983, 1984, N. E. Buxton & A. D. Watson)

		1950s	1970s	1979/80	1980/81	Nov. 1982	Nov. 1983
Northeast:	Caithness	10–100	500	400	270+	393	261
	Orkney	NA	50–80	42	25	34	49
	Loch Eye	30–40	c50	70	70	c30	5
Northwest:	O. Hebrides	c370	c200	120–140	120–140	89	99*
	Skye & Sm. Is.	NA	NA	115	60	96	98*
N. Argyll:	Tiree	} 50–100	} 200–600	NA	329	372	357
	Coll				NA	343	435
	Other sites					158	193
S. Argyll:	Islay	2,500–3,000+	3,700	2,920	4,300	3,250	4,592
	Kintyre	550	940	790	1,160	1,276	940
	L. Lomond	10	110	115	100	118	134
	Other sites					329	268
Galloway:	Loch Ken	400–500	250	260	260	c300	290
	Stranraer	0	280	270	390	280	350
	Other sites	NA	c90	NA	NA	15	43
Total in Scotland		4,150–5,150	6,460–7,230			7,083	8,114

* includes estimates based on November 1982 counts
+ probably an underestimate
NA no count available

GREENLAND WHITE-FRONTED GOOSE

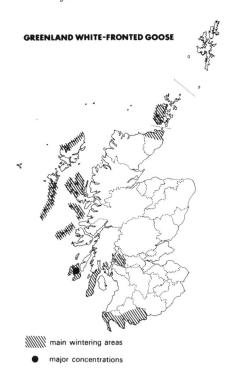

▨▨▨ main wintering areas

● major concentrations

in moorland areas holding important roosting lochs present a threat to the stability of the population (Stroud 1984). Current proposals include a plan to drain and cut for peat one of the most important wintering sites in Britain, which regularly holds at least 600 birds; planning permission for this development was granted in 1984.

Important flocks of Greenland White-fronted Geese also winter around Rhunahaorine and Machrihanish in Kintyre, where there has been some recent increase (Table). In Galloway total numbers have remained fairly constant at around 500–700, but Loch Ken has become relatively less and the Luce Bay/Stranraer area relatively more important. The numbers wintering in Caithness, where the flocks are usually scattered, and in Easter Ross have probably also declined. A detailed account of numbers at the less important but fairly regularly used sites is given by Ruttledge & Ogilvie (1979) and Fox & Stroud (in press).

In other parts of Scotland the Greenland White-fronted Goose is chiefly a passage migrant, occurring sometimes in substantial numbers away from its regular haunts, in both autumn and spring. Four out of 30 ringing recoveries have been away from the regular wintering areas: in Moray, Angus and Lanark. Research has shown that most birds stop-off in Iceland during autumn migration (Fox *et al* 1983); although departure from the breeding grounds starts in late August, the earliest arrival recorded in Scotland is 12 September (70 passing south over Canna in 1980) and the main influx takes place in October and early November. Birds start to leave in mid April and most have gone before the end of the month, with occasional stragglers remaining into May; the latest sighting date is 19 May. During spring migration, skeins of 80–200 have been recorded over central and northern Scotland, while flocks of 300 and 420 have

been seen heading north from Lewis and St Kilda respectively in late April/early May. Colour-ringing has confirmed that some movement takes place between Islay and Ireland, where numbers have declined in the last 20 years, and birds may also move between Islay and Kintyre at times (Stroud 1985). Bi-monthly counts on Islay in the winter of 1983/84 suggested that numbers on the island dropped from a peak of 4,590 on 23 November to 3,330 on 20 January and had risen again to 4,200 by 27 March, but these apparent changes may have been due to the fact that flocks are smaller and more dispersed – and consequently more difficult to locate – in mid winter than in autumn and spring (D. A. Stroud).

With an overall average production of young well under 20%, the breeding success of the Greenland race of White-fronted Goose is consistently poor in comparison with other races, for which the figure is 30–40%. Only a low proportion of adults breed successfully but those that do rear relatively large broods. The proportion of young birds in the Scottish wintering population is sometimes less than 10% and seldom more than 15%. This is clearly a matter of concern since Britain may hold up to 50% of the world population of this race. Studies on the breeding grounds have not yet satisfactorily explained this low productivity rate, but have suggested that the condition of the adults on arrival is likely to be a major contributory factor (Fox *et al* 1983). Research into this aspect of the species' ecology is continuing under the aegis of the Greenland White-fronted Goose Study Group. Marking with prominent Darvic rings has shown that birds from the same breeding area may be widely dispersed on the wintering grounds (Fox *et al* 1983). This scatter seems to be superimposed on a broader pattern of 'leap-frog' migration; birds from the northern part of the range winter in Ireland whilst the more southerly-breeding birds winter mainly in Scotland. The flocks wintering in Ireland have consistently higher breeding success than those in Scotland (Ruttledge & Ogilvie 1979).

Since the 1950s the Scottish Greenland White-fronted Goose population has increased by about 20% (Owen *et al* 1986), and the Irish population has declined by 50%. The species is fully protected in Scotland under the 1981 Wildlife and Countryside Act, a measure aimed at protecting the very small Greenland population, and in 1982–85 it was also given temporary protection in Ireland. At present mortality probably exceeds annual recruitment for the population as a whole and is believed to be lower in Scotland than in Ireland, where there has been considerable habitat loss. Although there are no complaints of serious damage by White-fronted Geese, there had been a recent increase (until protection was introduced under the 1981 Act) in the annual shooting bag on Islay – from 200 to 300 (Ogilvie 1983). This, together with the disturbance resulting from increased shooting of Barnacle Geese, is probably responsible for a slight increase in the mortality rate there. On Islay peatland development may adversely affect the population in the future, and Tiree is likely to be affected by agricultural developments, as drainage of the area most favoured by the geese, the Reef, may reduce its attraction at least for a period (D. A. Stroud).

Records of the European White-fronted Goose, which occurs only as a vagrant, are widely scattered but seldom relate to more than a few birds, parties of 20–30 in Aberdeen in 1969 and 1977 and in Wigtown in 1980 being exceptional. A report of c40 on Islay in March/April 1962, quoted in Booth (1981), should probably be regarded as suspect. There have been two recoveries, in Lanark and Angus, of birds ringed in previous winters in the Netherlands.

Lesser White-fronted Goose *Anser erythropus*

Vagrant/escape (N Scandinavia/Siberia) – occurs irregularly and in very small numbers.

The Lesser White-fronted Goose was first recorded in Scotland in 1953, when its presence was suspected among Bean Geese near Castle Douglas. Thereafter up to three were seen regularly with the Bean Goose flock until February 1959, since when there have been no further records from that area. Elsewhere the Lesser White-fronted Goose has been reported from Cambus, Clackmannan, where an adult was shot on 20 January 1960, and from Wigtown (March 1960), Lanark (February 1974) and Islay (March 1980). A bird near Kinclaven, Perth, in February 1976 was considered likely to be an escape. This species, which breeds from northern Scandinavia to northeast Siberia, occurs more regularly in southern England, most often with European White-fronted Geese at Slimbridge.

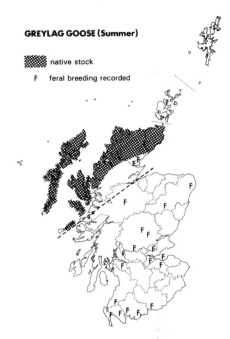

GREYLAG GOOSE (Summer)

native stock

F feral breeding recorded

Sutherland, Wester Ross and the islands that censusing is virtually impossible, even from the air; the available information is consequently patchy and by no means comprehensive. The total native stock is currently estimated as 2,500–3,000 birds (Owen *et al* 1986), and recent figures suggest there are between 500 and 700 breeding pairs. In a survey of the Uists and Benbecula in 1982 some 230–243 pairs were found

Greylag Goose *Anser anser*   *[1]

Resident breeder and winter visitor. Native birds breed in the Hebrides and Highlands, and feral flocks are scattered throughout the country. The bulk of the Icelandic breeding population winters in Scotland, the largest concentrations being on farmland in the eastern lowlands.

The Greylag Geese breeding in northwest Scotland represent all that now remains of a formerly widespread indigenous population, whose gradual decline is documented by B&R. They are so widely and thinly scattered over Caithness,

* native breeding stock in close season

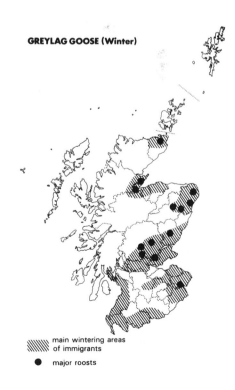

GREYLAG GOOSE (Winter)

main wintering areas of immigrants

major roosts

breeding, with the total post-breeding population estimated as c1,680 birds (Pickup 1982). Numbers on the Loch Druidibeg NNR, formerly the principal haunt, have declined in the last decade, while the North Uist population has increased. Lewis holds c10–15 pairs, Harris two or three pairs and the northern Sound of Harris islands up to 30 pairs (N.E. Buxton). A few breed on Skye and two or three pairs on Coll; successful breeding was recorded on Canna in 1980 and Rhum (two pairs) in 1982.

In Wester Ross there are c15 pairs on the Summer Isles and probably a few around Loch Maree, where four broods were seen in 1973. The Sutherland population is very thinly scattered and no figure for breeding pairs is available; the situation is further complicated by the presence of feral flocks at Loch Brora (about 100 birds) and Spinningdale, the former originating from introduced Hebridean stock which subsequently inter-bred with wild birds (Angus 1983). Some 250–300 moulting birds, with up to 50 young, gather in summer on Badanloch near Kinbrace. Little detailed information is available for Caithness but up to 75 non-breeders have been noted in June, and flocks of up to 300 in August; at least 11 pairs bred successfully in 1977 and 48 juveniles were on one loch in July 1980. Ringing recoveries indicate that the Sutherland birds winter in Caithness, while the Outer Hebridean birds are largely sedentary. The native breeding stock in the Outer Hebrides, Caithness, Sutherland and Wester Ross is protected by special penalties during the close season.

The largest feral population (129 successful pairs in 1971) is in the southwest, where introductions of Hebridean birds were made at Lochinch (Wigtown) in the early 1930s. Despite the subsequent spread, described by Young (1972), this continues to be the main site, with 40 breeding pairs in 1971, when a total of 600 failed and non-breeding birds was found in the area. Breeding has been recorded at several other Wigtown sites, on Loch Ken and elsewhere in Kirkcud-bright, in Dumfries and in Ayr. Other long-established feral flocks are those at Duddingston Loch and at Lochs Tummel and Faskally (Perth). Known introductions have taken place at Loch Achray (Perth) – a BASC (formerly WAGBI) Reserve, Loch Laggan (Inverness) and Loch Muick (Aberdeen) and there have doubtless been others elsewhere. Escapes from a collection have bred on Loch Lomondside and a pair, presumably originating from the Duddingston flock, reared young on Inchkeith in 1980. Elsewhere sporadic breeding probably occurs when one or both birds of an established pair is unable to migrate owing to injury. Small flocks of injured geese, often of mixed species, summer in many areas and hybridisation with the Canada Goose occurs quite frequently. As observers tend to ignore feral flocks, this is probably not a full account of the present situation.

Icelandic Greylag Geese often start to arrive in Scotland towards the end of September but the first major influx is generally during the first half of October and arrivals continue until about mid November. Between 1968 and 1983 the number of Greylag Geese counted in Scotland at the autumn census averaged 68,500 and ranged between 52,000 and 91,000; the corresponding figures for the British population as a whole are 71,000 and 56,000–96,000. (At 96,000 in 1981 the Greylag was for the first time the most abundant goose but it was overtaken again by the Pink-footed Goose in 1983.) There has been an overall increase of about 50% during this period, rather less than that recorded for 1960–70 (Boyd & Ogilvie 1972), with the proportion of young birds present in autumn ranging from 5.9% to 30.2%. Breeding success appears to be inversely correlated with population size and the recent increase in the population is probably wholly attributable to a decrease in winter mortality (Owen *et al* 1986).

At the time of the November census over 75% of the geese are in northern and east-central Scotland, occupying

Numbers of Greylag Geese counted in 1976–83 at sites regularly holding more than 2,000.
(Data from Owen et al 1986, M. A. Ogilvie & SBR – figures rounded)

| Site | 1976–81 November census counts | | 1981–83 |
	Average	Range (peak year)	Peak count (month)
Loch Eye	11,500	3,000–20,800 (1981)	38,000 (November)
Drummond Pond complex*	8,800	3,400–15,000 (1979)	4,500 (November)
Blairgowrie Lochs/ R. Tay/R. Isla complex	8,200	4,100–12,700 (1977)	6,200 (November)
Loch of Strathbeg	5,700	4,000–8,300 (1976)	9,600 (November)
Beauly Firth/Black Isle	4,900	400–11,000 (1981)	9,800 (November)
Lintrathen/Kinnordy	3,900	1,100–7,000 (1978)	5,000 (November)
Caithness Lochs	3,700	300–6,800 (1978)	7,400 (November)
Davan/Kinord	2,900	500–4,300 (1980)	11,200 (November)
Loch Leven	2,900	1,100–5,600 (1980)	2,500 (November)
Loch of Skene	2,800	700–5,700 (1981)	4,100 (November)
Carsebreck	2,500	1,100–5,100 (1980)	4,300 (November)
Hoselaw Reservoir	2,500	1,700–3,200 (1979)	4,100 (October)
Haddo	NA	NA	11,900 (December)

* believed to include some overestimates

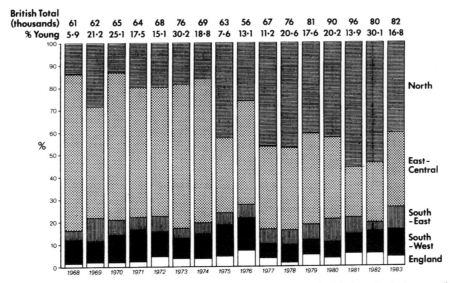

(a) Distribution of Greylag Geese in autumn (from November census data – feral flocks omitted where possible)

more than 50 major roosts, many of which have held internationally important numbers (more than 900 birds). Those currently of most importance are listed in the Table. A marked change has recently taken place in the pattern of distribution (Fig a); since 1975 many more Greylag Geese have remained in the north while the east-central lowlands have become relatively less important. A decrease has also occurred on Bute, which held up to 7,000 in the early 1960s; damage to swede turnip crops led to active disturbance of the geese and numbers there are now down to 2,000–3,000. Elsewhere there has been little overall change although the relative importance of different roosts has varied. In the last few years there has been a tendency for flocks to remain longer in the Northern Isles.

Autumn distribution is largely determined by harvest conditions, with the geese concentrating in areas where there has been most spillage of grain on the stubbles. In 1981, for example, the harvest was in general early and clean but a small area of barley in Easter Ross remained unharvested; there were over 20,000 Greylag Geese there at the time of the November census and almost 40,000 two weeks later. As winter progresses the geese feed on waste potatoes, turnips, grass and, to a limited extent in late spring, young corn. They come into conflict with farming interests particularly when on young grass in early spring, during snowy weather when they feed extensively on turnips, and in wet conditions when they puddle the ground in fields of young corn (Newton & Campbell 1970, 1973).

The other major factor determining distribution is the availability of suitable roosts. Greylag Geese are less wary than Pink-footed Geese and will roost on comparatively small pools or on riverside banks. They use tidal waters less often and generally flight less than 5 km from roost to feeding grounds (Newton *et al* 1973). These characteristics are reflected in the more scattered distribution and smaller average flock size of this species in comparison with the Pink-footed Goose.

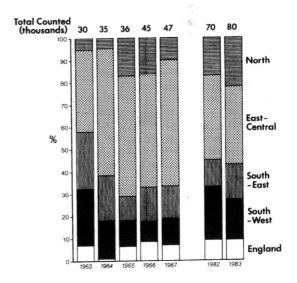

(b) Distribution of Greylag Geese in spring (from March census data – feral flocks omitted where possible)

During the winter the geese disperse in smaller flocks over a wider area. They also tend to move towards the south and west, appearing in small numbers in districts such as Ayr and Renfrew where they are scarce in autumn, and in much larger concentrations in the southwest. Although attempts at censusing in March have been of limited success owing to the difficulty of locating all the birds, the available counts do give some idea of the spring distribution (Fig b); the southwest holds many more geese at that season than in autumn. By the end of March the birds are beginning to move north and most leave during the second and third weeks of April, with a few remaining into May. During spring passage Greylag Geese are regularly seen over Handa, where more than 1,000 have been counted passing in one hour. No full analysis has yet been made of the ringing recovery data.

In the last 15 years Greylag Geese have been exposed to a variety of threats. In the early 1970s the increasing concentration of geese in east-central Scotland led to growing antagonism from the farming community and, in 1974, to a proposal to place both Greylag and Pink-footed Geese on the 2nd Schedule of the 1954 Protection of Birds Act in the six east-central counties – in effect, classifying them as pest species which could be shot at any time. The proposal was dropped the following year, after objections by conservation bodies, and subsequent changes in distribution have to some extent eased the conflict between Greylag Geese and agriculture in the Central Lowlands. Current farming changes seem likely to reduce the scale of the problem locally, as earlier ploughing of stubbles leads to more rapid dispersal of the geese.

During the 1970s Greylag Geese also suffered from poisoning and oiling. In 1971 some 500 died after eating newly-sown grain dressed with carbophenothion, an organophosphorus insecticide, and in 1974 four more cases occurred in Perth and Angus (Hamilton & Stanley 1975). It was suspected that in some cases treated grain had deliberately been made available to the geese; under normal farming conditions there should be little risk to geese from this cause. Nor would Greylags often be at risk from oiling, since they seldom roost on open estuarine waters. However more than 500 were affected, and many died, in an oiling incident on the Cromarty Firth in November 1975.

Despite these misfortunes the Greylag Goose population has continued to increase and has now far exceeded the total of c74,000 predicted for 1980 by Ogilvie & Boyd (1976), who anticipated no substantial increase between 1976 and 1980 'unless there is some important change from the circumstances prevailing in the last seven years'. No major change appears to have taken place in this country, and the production of young during this period was close to the long-term average; possibly grassland improvement in Iceland is resulting in reduced mortality on the breeding grounds. Ogilvie & Boyd's other prediction – that Greylags would increase in the Moray Basin and decline in Angus and Perth – has proved correct.

Although much has been learned about grey geese over the last 20 years, we still do not fully understand all the changing patterns of distribution and numbers. This became obvious in 1982 when, despite a very successful breeding season (over 30% young in a sample of 3,500), only 80,000 Greylag Geese were counted at the November census, compared with 96,000 the previous year. The total expected, on the basis of average mortality and the unusually high proportion of young birds, was 105,000–110,000, so where were the missing 25,000+? (M. A. Ogilvie unpub.). There must inevitably be a substantial margin of error in counts of large numbers of birds, especially when they feed and flight in dense flocks; with the goose census, poor weather on the appointed day can make counting virtually impossible. The inexplicability of the 1982 census results should perhaps be taken as a salutary reminder that population estimates are not always as reliable as published figures may suggest.

[Bar-headed Goose *Anser indicus*]

This species breeds in Mongolia and Tibet and winters in India. It is quite widely kept in captivity and there was at one time a feral flock in southern Sweden (Ogilvie 1978). Free-flying birds were first recorded in Scotland in January 1965, when I found two feeding in a turnip field with Greylag Geese near Coupar Angus. By the end of 1983 Bar-headed Geese had been reported in 11 years and from as far north as Shetland.

Snow Goose *Anser caerulescens*   [1]

Vagrant/escape (N America) – annual in winter, in very small numbers. Feral breeder in the Inner Hebrides.

Reports of Snow Geese usually refer to the Lesser race *A.c. caerulescens* (which occurs in two colour phases, blue and white) and often involve escapes; any genuine vagrants come from arctic America. Blue phase Lesser Snow Geese are seen fairly regularly with Greenland Whitefronts in the west of Scotland and may be considered wild (Ogilvie 1978). Most other records, including those of the Greater Snow Goose *A.c. atlanticus* and Ross's Goose *A. rossii*, probably relate to escapes; in recent years many may have originated from the feral flock established on Mull. The problems of subspecific identification and escapes are discussed in SB 2:306 and SB 3:138.

There are published records of Snow Geese for every mainland county except Selkirk, Clackmannan, Kincardine, Banff and Nairn and for many of the islands. Most reports are between September and May. Birds are often in company with Greylag or Pink-footed Geese, while in 1958 a blue phase Lesser Snow summered with Canada Geese on the Beauly. The number of reports has increased since the mid 1970s, with over 40 records in 1979 and 1980, but it is difficult to assess the number of birds involved, as individuals appear to move independently of the grey geese they accompany.

During the 1960s a pair of Ross's Geese, both ringed, wintered regularly in central Scotland and summered in Iceland, where they attempted to nest in 1963, among Greylags. The nest was flooded out. The last report of these birds was of one found dead at Loch Leven in January 1972.

In autumn 1983 the entire feral flock formerly at Calgary, Mull – at that time numbering 43 birds – moved to Coll; they were still there in autumn 1984.

Canada Goose *Branta canadensis*

The small and very local introduced resident population is expanding rapidly. In summer birds from England moult on the Beauly Firth. There are sporadic records of what appear to be genuine transatlantic vagrants.

Canada Geese were introduced into Scotland more than 200 years ago. B&R noted that they bred in a wild state 'in many parts' but gave little information on numbers or breeding sites and it was not until the 1953 census that data on this species were systematically recorded (Blurton-Jones 1956). The resident Scottish population was then about 200, centred mainly in Dumfries, where 17–18 pairs bred at Kinmount, but with small groups also at Loch Leven, Perth and Colonsay. A survey in 1967–69 located only about 100 birds on breeding waters; Loch Leven was no longer occupied but a new colony had become established in Renfrew. A third census was carried out in July 1976, when the Scottish total was reported as 140 (Ogilvie 1977). Counts from breeding areas in the early 1980s suggest a population of at least 100 breeding pairs and a total of well over 500 birds. This represents a more rapid increase than the 8% per annum estimated for Britain as a whole, so it seems probable that some breeding groups were overlooked in 1976.

In Dumfries, breeding has been recorded at Locharbriggs, Thornhill, Duncow and Dalswinton as well as Kinmount. Twelve pairs nested on lower Loch Ken in 1979, about eight pairs on the Stranraer lochs in 1981, five to eight pairs at Mellerstain and a few pairs at other sites in Berwick in 1981, c25 pairs at Rowbank Reservoir (Renfrew) and several lochs in Ayr in 1980, one pair in East Lothian in 1979, probably at least 20 pairs on four lochs in Perth in 1982 (resident population estimated at 200–300 in 1983), and several pairs at a feral colony established at Inverlochy (Inverness) in 1976. On Colonsay the flock decreased from 80+ in the early 1960s to only three pairs in 1983 (R. Coomber). Where Canada and Greylag Geese occur in the same area hybridisation often takes place.

The moult migration of Canada Geese from Yorkshire to the Beauly Firth has been documented by Walker (1970), and the capture of the flightless geese for ringing by Dennis (1964). The growth of this moult flock has been spectacular and was particularly rapid during the early 1970s. From around 18–20 birds in 1950–55 it doubled to 40+ by 1961, and then increased four-fold in the next five years, passing the 200 mark by 1970. Over the next five years there was again a four-fold increase but thereafter expansion slowed, reaching a peak of 1,100 in 1981 but dropping again to 900–1,000 in 1982–83. In the early years only Yorkshire birds were believed to be present but the flock has recently included many from other areas; in 1980 30% of the ringed birds caught were from the west Midlands and in 1978 there were two from Sussex. Beauly-ringed geese occur on the Tay at Kinclaven (Perth) in winter; while most of the flock there is probably locally bred it contains some birds which were at one time Yorkshire-based (J. Kirk). About half the Beauly flock are one or two-year olds, the rest being older non-breeders (Owen *et al* 1986).

Movement north to the Beauly starts in May, with the first birds generally arriving towards the end of the month. Numbers reach their peak in July and return passage takes place from mid August to mid September. During passage periods flocks of Canada Geese are seen in many parts of eastern and central Scotland, and can often be identified by their colour rings as Beauly/Yorkshire birds. Most records come from Aberdeen and East Lothian but there have also been reports from Tomintoul, Strathspey and Glen Clova, suggesting that some birds follow a largely coastal route while others travel inland. The occasional occurrence of this species in the Northern Isles is probably due to birds overflying their intended destination – which may also have been responsible for the establishment of a small moulting flock on Loch Rangag in Caithness, first found in 1973 and holding up to 25 birds. This flock is believed to be of English origin, as are the 40–45 birds moulting on Loch Leven. There are also moulting flocks, thought to represent local stock, at Mellerstain (up to 50) and Kinmount; the latter move to the Solway shore in August/September, when more than 200 are sometimes present.

There are many races of Canada Goose and subspecific determination is very difficult because they tend to grade into one another (Ogilvie 1978). The original introductions to Britain were almost certainly large pale birds of the Southern Group and most of the records refer to this type. The sporadic reports of small birds, often with Greenland Barnacle Geese or Greenland White-fronts, probably involve transatlantic vagrants. Individuals seen in Islay in April 1958 and Orkney in March 1981 were considered to be Cackling Geese *B.c. minima*, and possible sightings of Richardson's Goose *B.c. hutchinsii* have been reported from Loch Leven (October 1977) and the Mire Loch at St Abbs (October 1982).

Because Canada Geese are in general both tame and very sedentary they can become a local problem, sometimes causing considerable damage to crops and grassland in the neighbourhood of breeding waters (Ogilvie 1977). The technique of attempting to control numbers by pricking eggs is relatively ineffective, and the taking of eggs is illegal, but this species is recognised as a potential pest and can be shot under licence during the close season. Its introduction or deliberate release is now (belatedly) prohibited under the Wildlife and Countryside Act 1981. With an estimated 33,000 in 1983/84 (19,000 in 1976) Canada Geese are now present in such numbers in Britain that, as Owen *et al* (1986) comment, only co-ordinated action can prevent the species assuming pest proportions.

Barnacle Goose *Branta leucopsis*

Birds from the Spitsbergen and the east Greenland populations winter on the Solway, and on Islay and other western islands, respectively. During migration both sizeable flocks and stragglers occur outwith the regular wintering areas.

Virtually all our knowledge of the origins and habits of the Barnacle Geese that visit Scotland has been gained since 1950. B&R make no mention of the breeding grounds of this species and were clearly unaware that two quite separate populations are involved. They do, however, mention the Solway and Islay as important haunts and comment on an apparent decrease over the previous 100 years, attributing this largely to disturbance.

The Spitsbergen and Greenland populations are discrete (as is a third group, which breeds in western Siberia and winters in the Netherlands). As a result of ringing on the main Spitsbergen breeding ground, it was realised in 1963 that that entire population winters on the Solway. With numbers there down to a few hundreds in the early 1950s, concern over the population's future prospects led to it being given full protection – in Britain in 1954 and throughout its range by 1961/62 – and to the establishment of the NNR at Caerlaverock, one of its favourite haunts, in 1957 (Harrison 1974). Over the last 30 years wintering numbers there have increased progressively, reaching a record 10,500 in 1984 (Table a). The rate of increase has accelerated since 1970, when the Wildfowl Trust established a refuge at East-

park, adjoining the Reserve. The refuge includes 96 ha of farmland managed primarily for the benefit of the geese.

The first birds arrive on the Solway in the last week of September, having stopped-off on Bear Island on their way south, and the bulk of the population is present by mid October (Owen *et al* 1986). Most spend the early part of the winter at Caerlaverock, but some thousands move to Rockcliffe Marsh, Cumbria, in mid winter; 1,000–2,000 winter on farmland behind Southerness Point. Departure is in mid to late April and on their way north the birds spend some time on islands off the coast of Norway.

While at Caerlaverock the geese – the most strictly coastal of the species occurring abundantly in Scotland – feed either on the saltmarsh 'merse' or farmland. In the past they spent the first few months on the merse, feeding mainly on the

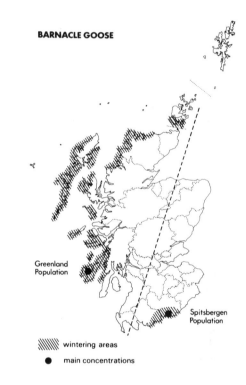

BARNACLE GOOSE

Greenland Population

Spitsbergen Population

///// wintering areas

● main concentrations

(a) *Peak counts of Barnacle Geese at Caerlaverock (Spitsbergen population) and Islay (Greenland population) from 1955 to 1984. (Data from Ogilvie 1978, WWC)*

	Caerlaverock		Islay	
	Average	*Range*	*Average*	*Range*
1955/56–1959/60 (1)	1,190	810–1,650	5,240	3,000–7,100
1960/61–1964/65	3,330	2,800–4,250	8,440	6,800–10,400
1965/66–1969/70	3,860	3,700–4,200	12,720	9,000–16,500
1970/71–1974/75 (2)	4,320	3,200–5,200	17,420	15,100–19,400
1975/76–1979/80 (3)	7,300	6,050–8,800	20,620	17,300–24,000
1980/81–1983/84	8,930	8,300–10,500	17,030	14,000–20,500

(1) Caerlaverock NNR established 1957
(2) Eastpark Refuge established 1970
(3) Shooting season on Islay extended 1978

stolons of clover (Owen & Campbell 1974), and only moved onto the fields in mid winter, but in recent years this sequence has been reversed, probably due to a reduction in disturbance. On farmland they feed largely on grass, but also on barley spillage on the stubbles. Their reluctance to feed among long stubble has been overcome at Eastpark by cutting the straw very short at harvest. Other management techniques used to increase the goose-carrying capacity of the Refuge include grassland improvement by re-seeding and fertilising, heavier summer grazing of the merse, and removal of stock from both merse and fields during the winter to avoid disturbance to the feeding birds.

The proportion of young birds in the autumn flocks averaged 23% from 1959 to 1978 but ranged widely, from 3.5% to 49.2% (Owen 1982). These fluctuations may reflect variations not only in breeding success but also in the mortality of adults during the moult (Owen & Campbell 1974). Since the establishment of Eastpark Refuge, and consequent improvement in wintering conditions, the survival rate of adult birds has not fluctuated so widely.

The Greenland population of the Barnacle Goose is estimated at 25,000–30,000, of which more than 50% winter on Islay. Numbers there increased steadily from the 1950s to a peak of 24,000 in November 1976, but have since declined again (Table a) (Ogilvie 1983, 1983a). In 1978 pressure from the farming community, concerned over the growing numbers of geese grazing the pastures, resulted in extension of the shooting season from the previous restricted period of December and January to the full season, 1 September to 31 January, and to 20 February below high tide mark. Until this change was made the autumn distribution of the geese was very localised, with a few large flocks remaining close to the principal roosts, Loch Indaal and Loch Gruinart. Later in the winter these flocks break up and smaller parties become more widely scattered. Disturbance from shooting, which in the early 1980s was often conducted illegally as a commercial enterprise, is liable to make the flocks disperse earlier.

The geese graze largely on improved grassland and trials have shown that they can have a significant effect on yields (Patton & Frame 1981). Both a reduction in numbers and wider dispersal of the flocks are likely to be of benefit to the tenant farmers most affected. Although this species has been given complete protection under the 1981 Wildlife and Countryside Act, licences for shooting may be granted where damage is being caused to crops; in 1984 the use, without success, of alternative means of dispersing the geese was introduced as a pre-requisite of licence granting. During the 1960s and early 1970s about 500–700 Barnacle Geese were shot annually on Islay; by the early 1980s the bag had risen to 1,200–1,400 and since 1976 losses have on average exceeded recruitment (Ogilvie 1983). In 1983 the RSPB established a 1,200 ha reserve beside Loch Gruinart; by providing good and undisturbed grazing there the Society hopes to reduce the conflict between geese and farmers elsewhere on the island.

The remainder of the Greenland population wintering in Scotland is scattered over a large number of islands, from the Sound of Jura to the Pentland Firth, with some occasionally visiting mainland headlands, as at Durness and Kintyre. Most of the islands used are small and uninhabited and

(*b*) *Numbers of Greenland Barnacle Geese counted in aerial surveys of their main haunts (Data from Ogilvie 1983a – figures rounded)*

	March 1973	March/April 1978	March/April 1983
Islay	15,000	21,500	14,000
Treshnish Is.	420	610	620
Tiree/Coll	145	390	620
Sound of Barra	335	455	375
Isay (Skye)	295	290	250
Monach Is.	640	760	640
Sound of Harris	980	1,330	1,555
Shiant Is.	450	420	580
Scottish total	c19,740	c28,060	c20,820
Irish total	4,400	5,760	4,430
Total	c24,140	c33,820	c25,250

any attempt at a full census is feasible only from the air (Boyd 1968). To date eight censuses have been made, the most recent in 1983, when c4,640 birds were found on the Inner and Outer Hebrides (Table b) (Ogilvie 1983). On other occasions flocks of more than 300 have been recorded on islands off the north coast of Sutherland and on Hoy and some of the Scapa Flow islands in Orkney. On the Sound of Harris islands numbers have increased since the 1960s, when they held c600; at 1,550 this was the second most important site in March 1983. There has also been some increase on Tiree and Coll, but the Monach Isles have declined in importance since the 1960s.

The influx of Barnacle Geese from Greenland does not take place until two to three weeks after that of the Spitsbergen birds, with the main arrival usually in the third or fourth week of October. The population on Islay may peak in early November, soon after the birds arrive, or not until March. Up to 2,000–3,000 additional birds are believed to arrive on Islay in mid February, perhaps attracted from other sites by the better feeding there (Ogilvie 1983). Departure takes place from mid April and there are often stragglers into May. Large numbers regularly pass north over North Uist in spring, the maximum count recorded being 5,000 on 1 May 1972.

Single Barnacle Geese or small groups occur from time to time in most mainland counties, and in Shetland and the Clyde Islands; occasionally much larger numbers land in the Northern Isles or near the east coast, as in 1983, when 715 were between Barnsness and Aberlady on 1 October. It is seldom possible, unless they are ringed, to be certain to which population such birds belong, but recoveries and sightings of colour-ringed birds indicate that passage birds seen in autumn on Fair Isle, Shetland, the Beauly Firth and moving south down the Angus coast are from Spitsbergen. Further south and west stragglers from either population may occur, although few Greenland birds have been identified with the Solway flocks and only three away from the west coast (in Caithness and Perth). Several Solway birds have been seen in Islay. A Barnacle Goose ringed in the Netherlands – presumably a straggler from the Siberian population – has been recovered in Fife, and a Spitsbergen bird in Spain.

Brent Goose *Branta bernicla*

Passage and irregular winter visitor, in variable but generally small numbers and at widely scattered locations. There is no regular Scottish wintering ground.

Both the European races of the Brent Goose occur in Britain. The Dark-bellied Brent *B.b. bernicla* breeds in northern Siberia and winters in western Europe, its most important British haunts being in southeast England. More than 85% of the Scottish records of this form are from the eastern half of the country, between Shetland and East Lothian. Two populations of the Light-bellied Brent *B.b. hrota* occur: one breeds in Spitsbergen and winters at Lindisfarne and in northern Denmark, and the other breeds in northern Greenland and the Queen Elizabeth Islands (north of Canada's Northwest Territories) and winters in Ireland (Ogilvie 1978). Scottish birds include individuals from both these populations but most probably belong to the Irish wintering flock. A bird on Islay in May 1979 was wearing a plastic collar fitted on Bathurst Island (one of the Queen Elizabeth group) in July 1975 and a Spitsbergen-ringed bird was recovered in Nigg Bay (Easter Ross) the following winter.

The Brent Goose was formerly much more abundant than it is today but its numbers have apparently always fluctuated (Atkinson-Willes & Matthews 1960). B&R knew it as a regular visitor to Aberlady Bay and noted that up to 4,000 occurred in the Cromarty Firth, but give no date for this record. Large flocks were generally of the Dark-bellied form. Marked changes in numbers and distribution, and probably in feeding habits, resulted from a widespread die-off of eelgrass (*Zostera spp.*) in the 1920s and 1930s (Owen *et al* 1986).

Annual totals of Brent Geese in Scotland currently range from under 30 to around 300. In most years about half the birds are subspecifically identified; between 1968 and 1980 83% of 728 birds were Light-bellied. The species is recorded fairly frequently in most coastal counties and as an occasional straggler inland. It occurs regularly in Shetland, Orkney and both Inner and Outer Hebrides, and occasionally on Arran and Bute; there are several records from Fair Isle, the Isle of May and St Kilda. No Scottish sites regularly hold populations of international importance (qualifying levels for the Canadian/Greenland and Spitsbergen populations are 150 and 100 respectively). Many of the bigger groups occur on the western islands, from Lewis south to Islay, where they probably pause only briefly on passage to Ireland, and the only recent record of a large flock was 3,000–4,000 on Islay in April 1971 (Booth 1981). Flocks of up to 150 Brent have been reported from Harris, Skye and Moray; although the birds in the last two areas were not subspecifically identified it is probable that they, like the Harris ones, were Light-bellied, as were smaller flocks recorded in the same areas. Parties of 50–100 have also occurred recently in Islay, Barra and South Uist; elsewhere numbers are generally very much smaller. Apart from one surprising record of 35 far inland at Loch Mallachie (Inverness) in late September 1977, no recent report of Dark-bellied birds has related to a group of more than ten.

Arrival generally takes place in September or October, although in some years none appear until much later; for example in 1976/77 there were no reports until January.

More than half the records relate to September/November, and the earliest date is 2 September. A few overwinter with grey geese or Barnacles, but many appear to move on, presumably to Ireland. Considerable numbers pass through on spring passage, when flocks of 100 and 60 have occurred in South Uist and Islay during April, and a few stragglers linger on into May. Occasional individuals summer – one was with Canada Geese on the Beauly from July to September 1974 and there is a June record from St Kilda. Although Brent Geese normally prefer to feed in tidal areas, in Scotland they are as likely to be found on pasture as on saltings. On some of their English wintering grounds they have recently taken to feeding largely on arable land and have consequently come into conflict with farming interests.

Red-breasted Goose *Branta ruficollis*

Escape/vagrant – two occurrences since 1950.

The Red-breasted Goose, which breeds in Siberia and winters mainly in Rumania and Bulgaria, is commonly kept in wildfowl collections and it is likely that most if not all, Scottish records refer to escapes. The only reports since 1950 are of one with Greylag Geese on the Beauly Firth in the winter of 1956/57 and five near Greenlaw, Berwick, in March 1966. The Red-breasted Goose occurs more often in southern England, where it is generally in company with White-fronted or Brent Geese and may be a genuine vagrant. Occurrences in central and western Europe have increased in recent years, so any free-flying birds seen in Scotland in future may possibly be wild ones (J. Berry).

[Egyptian Goose *Alopochen aegyptiacus*]

There is a well established feral population of this introduced species in Norfolk and the occasional Scottish reports presumably refer to birds from there or from other wildfowl collections. There have been three records in recent years – from Kinross, the Black Isle and West Lothian, the first two in 1977 and the other in 1979. It is unlikely that these published reports represent a true picture of the frequency of occurrence of this bird.

Ruddy Shelduck *Tadorna ferruginea*

Escape/vagrant (Iberia/Morocco)

The Ruddy Shelduck is so commonly kept in collections that observers probably do not always submit records of sightings, assuming them all to involve escapes. Its nearest breeding ground is in Morocco, and birds from there winter in southern Iberia, but the chances of a genuine vagrant occurring here are small. There were about seven records prior to 1950, including what may have been a true irruption of some 30 birds at Durness in June 1892, and there have been two reports since: a male near Montrose in May 1971 and another on Loch Lomond in April 1979. Both are assumed to have been escapes, as are all other British records in the last 50 years (Rogers 1982).

Shelduck *Tadorna tadorna*

Widespread breeder on sandy coasts and at a few inland freshwater sites. Many make a moult migration to the Helgoland Bight but increasing numbers now moult on the Forth Estuary. The biggest concentrations occur in mid to late winter on the estuaries of the Forth, Eden, Solway and Clyde.

An estimated 10,000 pairs of Shelducks breed in Britain (Owen *et al* 1986), and the Scottish breeding population is possibly in the region of 1,500 pairs. The species feeds largely on intertidal invertebrates and nests mainly in rabbit holes; distribution is consequently governed by the availability of mud- or sand-flats and banks or dunes suitable for burrows. The major east coast estuaries, and the many sandy stretches between them, accommodate large numbers of breeding Shelducks but on the west, where much of the coastline is rocky, distribution is more local. Assessment of breeding densities is complicated by the presence in summer of many non-breeding birds, and comparatively few data are available. Relatively high densities occur at Tentsmuir, where there were 130 pairs in 1972, and the Ythan Estuary (45–60 pairs). The Forth Estuary holds about 115 pairs, the coastline from the Ythan north to Rattray Head more than 24 pairs, and the Inner Clyde more than 20 pairs. Few figures are available for other mainland coasts but counts of 180 unfledged young at Findhorn Bay and 284 on Kirkconnell Merse (Dumfries) give some indication of breeding numbers. Berwick is the only coastal county in which breeding has not been proved.

Shelducks breed on all the island groups, with the largest numbers (15–30 pairs) on the machair islands of the Outer Hebrides, and Orkney's northern islands. In Shetland they are scarce, with only about ten pairs in south Mainland and single pairs on Burra Isle and near Lerwick. There are sporadic breeding records for the Isle of May.

The most important inland sites are Loch Leven and Loch Lomond. Numbers have been increasing on Loch Leven, which now holds some 40–55 pairs, but breeding success has declined as the population expanded (G. A. Wright). Breeding success is also low in the Loch Lomond population, which includes 25–35 pairs plus 50–70 non-breeding birds and is relatively stable (Bignal 1980). Single pairs nest inland in other areas, such as the Sidlaw Hills above the Tay, and take their broods on long treks to the coast, sometimes covering several miles of rough country and crossing main roads. In 1976 a pair bred at 240 m asl on Cairnsmore of Fleet (Kirkcudbright) but it is not known whether they reached the Solway safely. The ducklings' survival prospects are not good even in areas capable of supporting large numbers of Shelducks, as the very aggressive territorial behaviour of the adults means that they frequently leave their broods unattended and vulnerable to predation by gulls (Pienkowski & Evans 1982). At Dun's Dish in Angus all nests failed in 1979 due to predation by mink, which also affects breeding success at Loch Lomond.

Departure for the moulting grounds starts in mid July, with non-breeding birds leaving first, and by August only a few adults remain in charge of the creches of young. It has been known for more than 30 years that a very large proportion of Europe's Shelducks moult at the Helgoland Bight, the British birds gathering mainly at the Knechtsand area of the Wadden Sea. In the late 1950s a moulting flock was found in Bridgwater Bay in Somerset but it was not until 1975 that a gathering of flightless adults was discovered in Kinneil Bay on the south side of the Forth Estuary (Bryant 1978a). This flock has since expanded greatly, from about 100 birds in full wing-moult in August 1975 to 800 in 1978

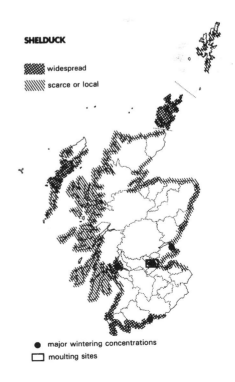

SHELDUCK

⬛ widespread

\\\\\ scarce or local

● major wintering concentrations
☐ moulting sites

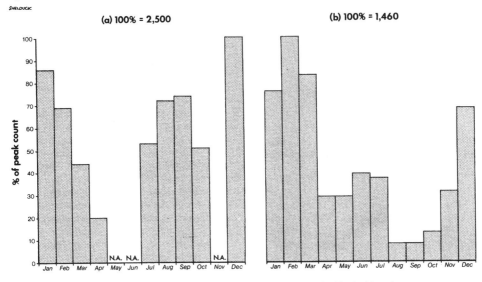

Seasonal fluctuations in Shelduck numbers on the Forth (a) and Eden (b) Estuaries (based on average monthly counts during period 1974–81; NA = no complete count available)

and about 3,000 in 1980. The proportion of flightless birds is highest during the second week in August and the flock spends much of its time well offshore, feeding by up-ending in a close pack. Ringing recoveries have shown that many of the birds are one- and two-year olds, and that birds from Aberlady Bay, Loch Lomond and the Ythan are among those involved (Pienkowski & Evans 1979).

The moulting flocks start to disperse in September, and during the early part of the winter the birds gradually move towards their breeding areas, not reaching Orkney, Shetland and the Hebrides until January/March. The numbers recorded in wildfowl counts are highest in January/March, when the total Scottish population may be 10,000–15,000; the average total in Britain at that time is estimated as 75,000–80,000. Although some immigration is known to occur further south, it is thought unlikely to have much effect on the numbers in Scotland (Owen *et al* 1986). On the Forth Estuary, which regularly holds internationally important flocks (more than 1,250 birds), numbers peak in December/January; maxima there in 1976–81 averaged 2,650 and ranged from 2,360–2,785 (D. M. Bryant). These figures are appreciably higher than the highest average monthly count of 1,500+ given by Prater (1981) for the whole Firth of Forth in 1969–75, presumably due to inclusion of incomplete counts in the latter case rather than an increase in the population. From the limited data available for the whole Firth of Forth it seems likely that the area holds about 3,000 Shelducks most winters (D. M. Bryant).

The Eden Estuary also holds a flock of international importance, with an average peak in 1976–81 of 1,500+. Data for the Forth Estuary are not complete for the same period but counts since 1974 have been used (Fig) to illustrate the different timing of the movements into and out of these two areas. Prater (1981) gave a highest average monthly count of 1,460 for north Solway, which ranks this area also as internationally important, but counts in 1982–84

produced mid-winter peaks of only around 400 (Moser 1984).

On the Clyde between Woodhall and Erskine the average peak in 1971–77 was 1,600 but there has since been a decrease there, with maxima of only 525 in 1979/80 and 495 in 1983/84. Several hundred Shelducks are also regularly present from December at the Montrose Basin (1982–84 average peak 380), and from January on the Cromarty Firth, and rather smaller numbers (100–250) on the Beauly Firth, the Dornoch Firth and Loch Fleet, and Loch Gruinart in Islay. Elsewhere counts seldom exceed 100.

During migration periods Shelducks are frequently recorded far inland, where they might be either prospecting for nest sites or making an overland journey to or from the moulting grounds. Examples of such movements include records from Glencoe in May, the Lake of Menteith in late July and November, and Loch Insh in February.

Much detailed and important research on this species has been carried out on the Ythan, the Forth and the Clyde. The topics studied include: feeding behaviour in relation to food supply (eg Bryant & Leng 1975, Buxton & Young 1981, Thompson 1981, 1982), breeding behaviour and population regulation (eg Jenkins *et al* 1975, Evans & Pienkowski 1982, Pienkowski & Evans 1982, Patterson *et al* 1983, Patterson 1983), and the moulting flock on the Forth (Bryant 1978a, Pienkowski & Evans 1979).

[Wood Duck *Aix sponsa***]**

Escape/vagrant (Category D)

This North American species breeds at various collections in England, and occasionally in Scotland, and many of the birds are free-flying. The five Scottish records, which probably relate to birds from this source rather than to transatlantic vagrants, are of a female in Caithness in December 1972, and males in Shetland in October 1977, Fair Isle in November 1979, and Renfrew and Stirling in 1980.

Mandarin *Aix galericulata*

Feral breeder (Category C)

A native of eastern Asia, the Mandarin has long been popular in wildfowl collections. It has bred ferally in Britain for many years and in 1971 was admitted to the official British and Irish list. There have been Mandarins on the River Tay in the city of Perth since the early 1960s; from 1968 to 1983 numbers fluctuated between 22 and 65. The birds still return for winter feeding to the nearby collection from which the flock originated and several pairs have bred successfully in owl nest-boxes. Outwith Perthshire there have been records during the 1970s and early 1980s from as far north as Orkney; they include a report of possible feral breeding in Berwick in 1980. Birds seen at Loch Lomond and Inverness, in 1974–77 and 1979–80 respectively, are believed to have come from local collections.

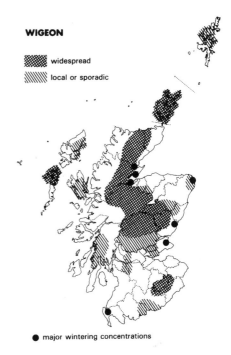

WIGEON

▨ widespread

◤ local or sporadic

● major wintering concentrations

Wigeon *Anas penelope*

Widespread as a breeder north of the Highland Boundary Fault but more local in southern Scotland and the islands. Passage and winter visitor in large numbers, with the biggest concentrations on coastal sites, especially around the Moray Basin.

B&R documented the gradual expansion of the Wigeon's breeding range, from the first nesting in 1834 up to 1950, and considered it still to be continuing. They described the species as widely distributed north of the Clyde / Forth Valley and as the commonest breeding duck in Caithness, central Ross and parts of Sutherland but not common on the islands. No significant change in status has since taken place, although there have been minor expansions or contractions in some areas. Over most of its range the Wigeon breeds

near pools and lochs in moorland, up to 600 m asl; and its principal breeding grounds lie along a broad band up the centre of Scotland, from Perth to the Pentland Firth. Further south there are well-established colonies at Loch Leven and in the Border hills. On the islands breeding takes place regularly in Orkney and the Outer Hebrides. In other parts of the country numbers are small and breeding often sporadic.

At Loch Leven, where Wigeon nest closer together than any other duck, numbers increased from about eight pairs in 1959 to 40 pairs in 1980. The colony in the Borders, centred around the headwaters of Ettrick, Yarrow and Teviot, has been estimated as 25– 35 pairs (R. D. Murray), while up to 50 pairs probably breed in Badenoch and Strathspey (Dennis 1984). In Perth in the late 1960s I found Wigeon most abundant on hill lochs on either side of the Tay, Tummel and Garry valleys; few were seen west of Kinloch Rannoch, where the species was formerly said to be common. Little change appears to have taken place this century in Caithness and Sutherland where Wigeon nest thinly in moorland areas. The limited information available on breeding densities suggests that the population estimate of up to 500 pairs in Britain (Atlas), at least 75% of which are likely to be in Scotland, is a realistic one.

Parslow (1973) recorded breeding in Stirling since 1954 and there are Atlas records for Dumfries, Fife, Dunbarton, Argyll, Kincardine, Moray, Nairn and Ross and Cromarty but no more recent data from these areas. Nor are there recent records for Wigtown – where B&R noted breeding – or Renfrew, where Wigeon nested at Castle Semple Loch in 1971. Breeding has not been recorded in Berwick, East Lothian, West Lothian, Clackmannan, Ayr, Lanark or Banff. The position in the islands is less clear. In 1955 the Venables

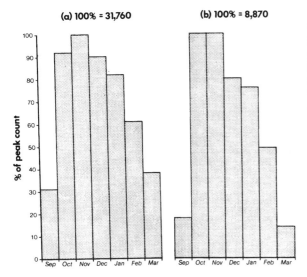

Seasonal fluctuations in the numbers of Wigeon counted (a) in Scotland and (b) on the Cromarty Firth (based on average monthly counts (a) 1976–81 (b) 1978–81)

said of Shetland 'a few breed', but had no first-hand evidence of breeding; a brood on Sandwater in 1976 was later claimed as the first proved breeding record. Wigeon bred in Shetland again in 1977 and 1978 and probably did so in 1983. Parslow (1973) reported a decrease in Orkney since the 1940s and said Wigeon no longer nested, but in 1975 breeding was reported from several islands, and Lea & Bourne (1975) estimated the population as between ten and 99 pairs; at least six pairs bred at four sites in 1983. B&R noted breeding in the Outer Hebrides and mentioned North and South Uist and Benbecula; there are recent records for North Uist – where Wigeon have nested at Balranald since 1969 – and Benbecula, but numbers appear to be very small. On the Inner Hebrides breeding has occurred since 1950 on Skye and possibly on Islay and Tiree. Wigeon have not been proved to breed on the Clyde Islands since the 1920s, when several pairs nested on Bute, but nesting has been suspected on Arran.

Immigration starts in September and the bulk of the wintering population has arrived by mid October. Numbers on the Cromarty Firth, where the largest concentration occurs, are highest in October/November but may not drop appreciably until well through January (Fig.). Ringing recoveries suggest that most Wigeon arriving in the north and northwest are of Icelandic origin and that birds from Norway, Finland and Russia probably form a larger proportion of the flocks in the east and south. There is a southward movement during the winter (Owen & Williams 1976), and there have been first-winter recoveries in Ireland, Denmark and France of Wigeon ringed as chicks in the Highlands.

The size of the British wintering population is much influenced by weather conditions on the Continent; in severe winters larger influxes occur. Peak numbers in Britain as a whole are usually recorded in January, when the average population is currently estimated to be c200,000 (Owen *et al* 1986). In Scotland, where numbers are higher in November than in January, the maximum winter count averaged 33,400 between 1976 and 1980 and ranged from just under 30,000 to over 40,500 (in November 1980); in 1983/84 the peak was 38,500, in October.

The estuaries around the Moray Basin are by far the most important Wigeon sites, with several holding populations of international importance (more than 5,000 birds). On the Cromarty, Dornoch and Inner Moray Firths numbers have increased in the last decade (Table), but there are considerable annual variations in Wigeon distribution both between the estuaries and at localities within them. For example Tain Bay, on the Dornoch Firth, peaked at under 1,000 in 1977–79, after holding 5,500 in October 1976 and having an average peak of over 3,000 from 1965 to 1976. Longman Bay, in contrast, showed an increase from an average peak of 1,770 in 1969–76 to 2,880 in 1977–81. Montrose basin and the Eden Estuary also support large flocks (Table) but the only site in the west to have held more than 2,000 since 1976 is Loch Ryan/Soulseat. Loch Eye is the only freshwater site supporting internationally important flocks, with counts of 6,000 in October 1979 and 8,000 in November 1980 – when the birds were feeding in association with Whooper Swans (SBR). The Loch of Strathbeg and Loch Ussie occasionally hold flocks of over 2,000 but no other freshwater did so between 1976 and 1984.

Elsewhere in the country Wigeon are widely distributed but in much smaller numbers, with few waters regularly

Peak counts of Wigeon at major passage/wintering sites (Data from Owen et al 1986, WWC & SBR – figures rounded)

	1977/78	1978/79	1979/80	1980/81	1981/82	1982/83	1983/84
Cromarty Firth a.	NA	6,970	7,200	10,810	15,000	9,380	10,000
Dornoch Firth b.	6,720	4,640	5,420	3,520	4,030	8,280	12,000+
Loch Eye	500	3,000	6,000	6,000	80	60	300
Montrose Basin	1,500	1,100	2,500	3,200	NA	3,860	3,400
Eden Estuary	1,350	960	880	1,480	1,740	1,300	1,270
Caerlaverock	500	550	550	890	1,000	1,120	810
Loch Ryan/Soulseat	620	1,460	2,300	2,010	1,000	2,500	1,070

Notes: NA – not available

a. Nigg Bay + Udale Bay + Alness ⎫ These figures represent minima for the areas as they do not include any allowance for birds
b. Edderton + Dornoch Sands ⎬ present at nearby sites not counted on the same day.

holding more than 1,000 birds. In Shetland, the Hebrides and the western Highlands flocks seldom hold more than a few hundred and are often much smaller; in Orkney and Bute they sometimes exceed 1,000. Passage birds can be seen moving south along the coast in autumn, and in September 170 an hour have been counted passing St Abbs. Wigeon leave for the breeding grounds from February onwards; the return movement through Shetland takes place in April/May, with occasional stragglers into June.

This species' partiality for eelgrass (*Zostera spp.*) has long been known and its decline earlier in the century was attributed to the loss of this plant – due to disease – from most of the larger estuaries. It is now recognised that inland pastures are of considerable and increasing importance to Wigeon (Owen & Williams 1976). In Scotland the proportion of the population feeding on grassland and freshwater sites increases progressively during the winter, apart from periods of very severe weather; whereas 70–80% are at coastal sites in September/October only about 40% are on such sites in March. Little is known about the winter distribution of native birds; it may be that these spend the entire winter on inland sites.

The increasing use of inland feeding grounds could be important to the species' future prospects. Although the industrial developments around the Moray Basin seem to have had comparatively little detrimental effect on numbers as yet, this may not continue to be the case. Further reclamation and possibly pollution may make these estuaries less suitable as feeding grounds in the future and so force the birds to move to alternative sites. Recreational activities also present a potential threat in some areas as, given suitable roosts and feeding grounds, disturbance is the main factor in controlling the use of a site by Wigeon (Owen & Williams 1976). The establishment of reserves at Montrose Basin and the Eden Estuary should be of considerable benefit to the species. Although upland afforestation may reduce the amount of apparently suitable breeding habitat in some areas, it seem unlikely that it presents a serious potential threat to the small but widely scattered breeding population.

American Wigeon *Anas americana*

Vagrant/escape (N America) – almost annual, in very small numbers.

First recorded in Scotland in 1907 (B&R), the American Wigeon has been noted more frequently since 1950 than during the first half of the century. Up to 1950 there had been fewer than ten records; from 1950 to the end of 1984 there were reports in 22 years, involving a total of some 26+ birds. Long stays have been a feature in several cases, for example a drake seen on several occasions between 17 January 1965 and 30 November 1967 near Inverness. Most sightings are in winter and spring, usually between January and the end of May. Males (which are more easily identifiable) are reported roughly four times as often as females, and the records are widely scattered over the country, from Shetland to Kirkcudbright and Roxburgh to the Outer Hebrides. Pairs have been seen on Loch Leven (May 1977,

A. Allison) and on Unst (June 1983). Although it is possible that some reports relate to escapes, the record of a juvenile female ringed in August 1966 in New Brunswick, Canada, and shot eight weeks later in Shetland proves that genuine vagrants do occur.

Gadwall *Anas strepera*

Scarce migratory breeder, regular in east-central Scotland and the Outer Hebrides and sporadic elsewhere. Outwith the breeding season there are widely scattered autumn records but few winter ones.

The nucleus of the breeding population is at Loch Leven, where Gadwall have bred since 1909 and c25–40 pairs now nest annually, mainly on St Serf's Island (Allison *et al* 1974 & SBR); B&R do not suggest that the original birds were introduced. This species is the most sensitive to disturbance of all the ducks nesting there (Newton & Campbell 1975). Regular breeding has also taken place for some years in Fife, Perth and Angus, involving a total of around 15–20 pairs. More recently nesting has been recorded in Orkney, where Gadwall first bred on Sanday in 1969 and about four pairs now nest, and the Uists, where breeding was first recorded at Balranald and Loch Hallan in the 1970s. At Balranald numbers have increased slowly and up to ten pairs bred in 1980. Sporadic breeding has occurred in Caithness and Sutherland, the most recent records being in 1977. A female Gadwall with two well-grown young at Breakish, near Kyleakin, Skye, in September 1968, and a report of possible nesting on Tiree in 1983 are the only recent records suggesting breeding on the Inner Hebrides; there are several older records for Tiree. A Gadwall mated successfully with a Mallard in Ayr in 1974.

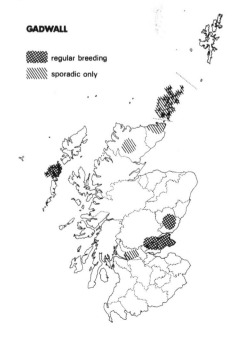

GADWALL

▨ regular breeding

▩ sporadic only

B&R, noting that old accounts suggested that the Gadwall was formerly more abundant, concluded that a marked decrease had taken place since the end of the 19th century, especially in the Hebrides – where the number shot on one estate declined from 27 over a five-year period early this century to nil in the 1970s (per C. Spray). Shooting records for Loch Leven also indicate that wintering numbers decreased sometime in the 1940s. Since wildfowl counting started there have been few records of more than 20 birds together from any site other than Loch Leven and a handful of waters in Fife, notably Kilconquhar Loch, Loch Ore (prior to the creation of the Country Park), and Stenhouse and Cameron Reservoirs. Between 1968 and 1983 there were scattered records, generally at times of passage and involving not more than two or three birds, from every mainland county other than Banff, Nairn, Kincardine, Peebles and Selkirk, and from the Inner Hebrides but not the Clyde Islands. Most records are for August/November and March/May but there are a few winter reports, including one of a pair in Unst on 10 January 1977.

The total Scottish population probably seldom exceeds 200–300 birds and is at its largest in September/October; thereafter it decreases rapidly and few Gadwall are present between mid November and mid March. Autumn numbers on Kilconquhar Loch were relatively high in the early 1970s, with an average peak of c160, but have since reverted to about the former level, peaking at c20–30. At Loch Leven peak numbers have recently shown a progressive increase, from 43 in 1976/77 to 80 in 1978/79, 208 in 1980/81 and 220 in 1983/84. It is not known whether the autumn flock is composed entirely of native birds; with good hatching and survival rates the 60 or so pairs in east-central Scotland could give an autumn total of this order. The occasional birds arriving with Wigeon later in the winter may be of Icelandic origin. Ringing has shown that Gadwall move south-westwards in autumn, to Ireland and sometimes Spain. It is presumed that the scattered passage records, especially those in the north and west, refer mainly to birds from Iceland, where breeding numbers have increased this century (BOU). The northwest European population of the species is estimated as only 6,000–7,000, of which Britain holds more than half (mainly introduced), and has also been slowly increasing (Owen *et al* 1986).

Baikal Teal *Anas formosa*

Vagrant/escape (Siberia) – three records, 1954–83.

The first British record of the Baikal Teal was of a female on Fair Isle on 30 September 1954. Originally considered likely to refer to an escape, the record was not admitted to the British list until 1980 (Wallace 1981). By that time there had been two further reports, of a female shot at Loch Spynie, Moray, on 5 February 1958 and a male at Caerlaverock, Dumfries, from 19 February to 12 April 1973.

Teal *Anas crecca*

Widely distributed both during the breeding season and in winter, when numbers are increased by immigration. Few large flocks occur and a substantial proportion of the population is made up of small parties.

The Teal's secretive habits and preference for dense cover make proof of breeding difficult to obtain, while its catholic taste in habitats – ranging from eutrophic lowland lochs to marshy moorland pools and peaty dubh lochans at up to 700 m asl (B&R) – makes any attempt at assessing the breeding population extremely speculative. The British breeding population is currently estimated at 3,000–4,500 pairs (Owen *et al* 1986) – but this is really little more than a guess.

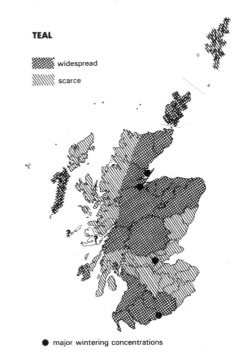

TEAL

▨ widespread
▨ scarce

● major wintering concentrations

Although breeding was recorded in every county, Atlas surveys indicated that Teal are scarcer in the west and northwest Highlands than elsewhere in Scotland. Sharrock (Atlas) has suggested that the quality of the vegetation may be a limiting factor in much of this area, as numbers are highest in the vicinity of limestone outcrops, where the vegetation is much richer. This seems a more probable explanation than his alternative, that wetter summers adversely affect brood survival. The greatest density of proved breeding records was in the southwest, where there are numerous small wetlands set among both agricultural land and moorland. The data on numbers are scanty but the following figures give some indication of regional variations: Kirkcudbright – 20 pairs at Loch Ken; Ayr – 11 pairs at eight sites; Kinross – two or three pairs at Loch Leven; Perth – present on 49 of 124 waters, proved breeding on six; Inverness – c40 pairs at Insh Marshes.

Teal breed on all the main islands in Orkney but are

more local in Shetland, where they are apparently increasing. Breeding was reported from eight Shetland sites in 1977 and proved on Fetlar for the first time in 1980. Although proved breeding is not shown for Lewis in the Atlas, Teal nest there in small numbers (Hopkins & Coxon 1979) but are most abundant in the Outer Hebrides on South Uist and Benbecula; a pair bred on St Kilda in 1974. There are recent breeding records for Islay, Jura, Skye and Rhum, but not for the rest of the Small Isles, nor for Coll, Tiree or Mull, on all of which nesting was reported in the past. Teal breed on all the larger Clyde Islands.

Immigration starts in August. At Loch Leven the first wave of arrivals consists entirely of adults – largely males in full eclipse – and probably involves movement from moulting grounds in Denmark/Scandinavia (A. Allison). In September/October a second wave reaches Loch Leven; this varies in scale from year to year and consists largely of juveniles and adult females. Ringing recoveries show that many of these birds are *en route* from Fenno-Scandia to southwest Scotland and Ireland. The Scottish population as a whole builds up to an October/December plateau and then gradually declines until the birds leave in March and April (Fig.). However, this is probably an over-simplification of the situation, as a marked mid-winter exodus has been noted at Caerlaverock (Wright 1980) and a spring increase at Loch Leven (Allison *et al* 1974). Spring passage is generally evident on Fair Isle until mid June. Ringing recoveries show that in addition to those from Fenno-Scandia, Teal from Russia and Iceland also winter in Scotland. It may be that flocks from these widely separated areas arrive at different times and follow distinct patterns of movement within this country. British breeders are believed to be largely sedentary, moving south and west only in severe weather (Prater 1981).

From 1976 to 1981 the maximum number of Teal counted in Scotland averaged 7,500 and ranged between 5,800 (November 1977) and 9,300 (November 1978); the 1982/83 and 1983/84 peaks were 10,500 and 10,800 respectively, both in October. Because flocks are so widely dispersed and small groups make a substantial contribution to the total, the actual wintering population is likely to be considerably higher, probably at least 15,000. The current estimate of the total British population is 100,000–200,000 (Owen *et al* 1986), so the Scottish element is relatively small. The wide variation in Scottish peak counts suggests that marked fluctuations in numbers occur between years; this is borne out by the figures for seasonal indices of abundance, which show an irregular pattern of peaks and troughs, the timing of which is often not synchronous with changes south of the Border. One possible explanation for this may lie in the origins of the wintering flocks; many Icelandic birds winter in Scotland, whereas those wintering further south are largely of Scandinavian and Russian origin. Variations in breeding success between these populations, or in the extent to which the Scandinavian birds move west rather than southwest, could result in fluctuations of this type. Much more evidence from ringing recoveries would be required to confirm or disprove this hypothesis.

No Scottish site regularly holds flocks of international importance (more than 2,000 birds) and comparatively few are of national importance (more than 1,000 birds). At Loch Eye numbers have shown a marked recent increase and a count of 3,000 there in December 1980 is the largest recorded in Scotland for many years – although still well below the 4,330 present on the Forth Estuary in November 1962. Flocks of more than 1,000 have been recorded since 1976 at several places around the Dornoch, Cromarty and Beauly Firths, the Slains Lochs, Loch Leven, Morton Lochs, Carron Valley Reservoir, the Endrick Mouth, Grangemouth, a small pool near Gifford (East Lothian), Caerlaverock and Loch Connell (Wigtown). Counts of more than 500 are more widespread but neither regular nor common.

The progressive decrease in the numbers of Teal on the Tullibody Island/Kennet Pans stretch of the Forth noted earlier (Thom 1969) has continued. Average peak counts there dropped from 2,430 in 1961–64 to 630 in 1964–72 and 290 in 1972–81; over the same period the timing of the peak changed from November to February/March. The decrease at this site has been associated with a reduction in the discharge of distillery waste in the area, but a more widespread environmental change of significance to both breeding and wintering Teal must be the very extensive drainage schemes being carried out in many parts of the country. In total these must represent a considerable reduction in available habitat, the effects of which may become apparent in the future.

The North American race of the Teal *A.c.carolinensis*, known as the Green-winged Teal, was first recorded in Scotland in 1938, when a male was shot on North Uist. Since then there have been more than 40 records, with annual occurrences since 1972. More than 60% of the records are from the Highlands and islands – mainly the Northern Isles and Outer Hebrides – and most of the remainder from Aberdeen and the Central Lowlands. Green-winged Teal have been reported in all months from October to June and most often between December and February. All records relate to drakes (females, which are not distinguishable in the field from the nominate race, are likely to be identified only if shot).

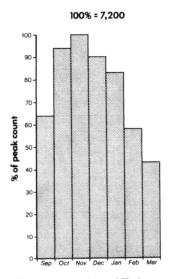

100% = 7,200

Seasonal fluctuations in the numbers of Teal counted in Scotland (based on average monthly counts 1976–81)

Mallard *Anas platyrhynchos*

Breeds in every county and island group and occurs regularly in winter throughout the country. Immigration increases the winter population but flocks are widely dispersed and seldom of any great size. The population is supplemented locally by hand-reared birds.

The Mallard is the most abundant and widely distributed of the ducks; it uses a very wide range of habitats and nest sites and is among the most adaptable and opportunist species in its feeding habits. It is equally at home on a hill loch or in a city park and sometimes becomes largely dependent upon man-provided food supplies. The Atlas shows Mallard to be present in all 10 km squares away from the mainly mountainous areas of the Highlands and north Harris, and breeding has been recorded on many of the smaller islands, such as Lady Isle, Craigleith and Inchkeith. With such a dispersed population the assessment of numbers is extremely difficult and little information is available on breeding densities. The few published figures relate to areas where there is a particularly high concentration eg 400–450 pairs at Loch Leven (Allison *et al* 1974) and 53 pairs at Kinnordy

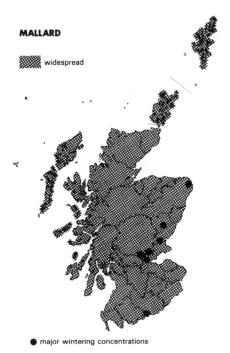

MALLARD

▨ widespread

● major wintering concentrations

Loch (SBR). Sharrock (Atlas) suggested an average density of 20 pairs / 10 km^2 as a minimum figure, which gives an estimated 19,000 pairs for Scotland. The current estimate of the British breeding population is c100,000 pairs (Owen *et al* 1986), but this is very approximate.

Between 1976 and 1981 peak Mallard numbers in the wildfowl counts generally occurred in January, averaged 28,400 and ranged from 23,000 in 1976 to 36,000 in 1981; the 1982/83 and 1983/84 peaks were 28,000 and 25,700 respectively, both in December. Birds are present at many sites not covered by wildfowl counts and, although the number at any one of them is unlikely to be large, together these probably hold nearly as many as the counted sites. The Scottish wintering population may therefore be in the region of 45,000 to 75,000 birds. An estimated 400,000–500,000 regularly winter in Britain as a whole (Owen *et al* 1986).

No British site holds flocks of international importance (more than 10,000 birds) nor are gatherings of national importance (more than 4,000) recorded in Scotland with any regularity. Salmon (1981) gives an average peak count of 3,700+ for the Firth of Forth in 1976–81, but this is a composite of over 40 areas (stretching from Tyninghame and Largo Bay up to Alloa), none of which regularly holds even 500 birds. The peak on the Forth Estuary, between Tullibody Island at Alloa and the bridges, is probably in the region of 1,500–2,000 but the data are not complete enough to be certain of the current situation. A substantial decrease has recently occurred at many places which formerly held 3,000 or more (Table). At only two sites, Loch Leven and the Loch of Strathbeg, does the average peak currently exceed 2,000. Flocks of 1,000+ occur quite often at Gartmorn Dam, Lintrathen Loch, Invergowrie Bay and the North Solway, and less regularly on the Aberdeen and Ayr coasts, several waters in the Borders, Fife and Stirling, and the lochs of Skene, Spynie and Haddo in the northeast. Elsewhere flocks of this size are recorded only very occasionally and the bulk of the population is made up of quite small parties (Thom 1969).

Native birds form a large proportion of the autumn population, but the pattern of seasonal change in numbers is not consistent for all sites. At Loch Leven large numbers of native-bred birds gather in August and numbers peak in September, whereas at the Loch of Strathbeg the influx is much later and numbers do not peak until December (Fig.). By mid winter, when shooting has reduced the native population, immigrants account for about one-third of the Mallard present in Britain (Owen *et al* 1986). Wintering numbers

Peak counts of Mallard at selected major sites (Data from Thom 1969, Wildfowl Trust unpub. & SBR – figures rounded)

	1962/63–1967/68 Av. peak	Max.	1976/77–1980/81 Av. peak	Max.	1981/82 Max.	1982/83 Max.	1983/84 Max.
Loch Leven	NA	c3,500	2,170	2,730	3,690	2,200	1,220
Loch of Strathbeg	c3,000	6,500	2,070	2,800	2,600	2,450	2,100
Lintrathen Loch	c2,700	4,000	1,150	1,200	1,500	700	850
Almond Estuary	c1,100	1,500	340	410	290	430	230
Eden Estuary	c2,390	NA	c300	600	460	300	550

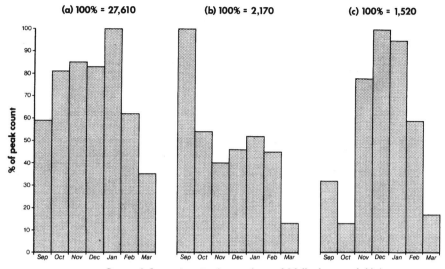

(a) 100% = 27,610   **(b) 100% = 2,170**   **(c) 100% = 1,520**

Seasonal fluctuations in the numbers of Mallard counted (a) in Scotland, (b) on Loch Leven and (c) on the Loch of Strathbeg (based on average monthly counts 1976–81)

as a whole are influenced by the scale of immigration, which in turn is dependent upon weather conditions on the Continent. Ringing recoveries show that birds from a very wide breeding area – stretching from Iceland and the Faeroes through Fenno-Scandia to Russia – visit Scotland in winter. Variations in the proportion of the wintering population deriving from these scattered origins may account for differences in the timing of arrivals as well as the annual fluctuations recorded. No full analysis of the large number of ringing recoveries has been carried out.

In Britain as a whole Mallard numbers declined during the 1970s but increased again following the hard weather of January 1979. The Scottish indices during this period showed a very marked decline in March numbers from 1962 to 1977, since when there has been a small increase, whereas the indices for September and January remained fairly stable. It is not possible to interpret these contrasting patterns with any degree of certainty but the decline in March numbers might be due to a reduction in the breeding stock (which presumably forms a large proportion of the total present at that time), or an earlier departure of immigrants. Although B&R considered Mallard 'less abundant than of yore' due to drainage of breeding haunts, there is no evidence of any recent decline in the breeding population. Indeed the survival rate of native birds that reach fledging may well have improved as a result of the better food supplies consequent upon changed agricultural methods, which leave more waste grain and potatoes on the fields. The more likely explanation seems therefore to be that the decrease is due to alterations in the behaviour of the immigrant flocks, possibly associated with variations in the proportions coming from different parts of the breeding range.

The Mallard's ability to utilise a wide variety of habitats and its dispersed distribution make it much less vulnerable to environmental changes than are species such as Pintail and Wigeon, which are more specialised in their feeding habits and more localised in their distribution. While drainage and industrialisation will doubtless continue to affect local numbers it seems unlikely that either the breeding or the wintering population could ever be seriously endangered by man's activities.

___

American Black Duck *Anas rubripes*

Vagrant / escape (N America) – two records, 1979 and 1981.

This close relative of the Mallard is accepted as occurring in Britain as a wild vagrant, although there is also the possibility of escapes. The first Scottish record was an adult male on Stanley Dam, Paisley, from 22 to 28 December 1979 (SB 11:223) and there was another male on the Beauly Firth at North Kessock from 11 October 1981 to at least March 1982 – when it was paired with a Mallard (BB 75:491). A hybrid was reported in Lanark in November 1982.

___

Pintail *Anas acuta*   z   *[1]

Breeds in very small numbers, often irregularly, and at widely scattered localities. Passage and winter visitor in very small numbers, with concentrations at coastal sites, especially the Solway.

The Pintail's irregularities in distribution and frequency of breeding, which occur throughout Europe and also in America, are difficult to explain as the species seems to be catholic in its tastes, nesting in habitats ranging from fairly open moorland lochs to low-ground marshes. Its main requirement in the breeding season appears to be shallow water in which to feed. The British breeding population is estimated as up to 50 pairs (Atlas); the available data for Scotland suggest that in a good year about 20 pairs may breed but in most seasons numbers are probably much lower. There has been no evidence of any change in breeding status during the last 30 years.

Since 1965 fairly regular breeding has been reported only in Orkney and Caithness. In Orkney pairs were present on four islands in 1977 and nesting took place between 1978 and 1983 on North Ronaldsay, West Mainland and Stronsay;

nine pairs were at one Mainland site in 1982. In Caithness there were four pairs at three sites in 1977 and breeding was proved at four different sites between 1972 and 1982. Elsewhere there have been sporadic records, with breeding reported only once or twice in any one county. The most recent records are: Wigtown (probable) 1982, Kirkcudbright 1980, Dumfries 1970, Berwick c1970 (Atlas), Fife 1972, Kinross (Loch Leven) 1962, Perth 1974, Angus 1970, Moray 1970, Inverness 1983, Ross and Cromarty c1970 (Atlas), Sutherland 1970 (and probably 1980), Shetland 1977, South Uist 1947 (Hopkins & Coxon 1979), North Uist c1970 (Atlas), Tiree 1980 (first record 1951) and Inner Hebrides (probable) 1983. There have been no recent reports of breeding in Aberdeen (classed by Parslow as 'regular'), nor in Midlothian or Selkirk, where breeding was recorded by B&R. The only other areas in which breeding has been recorded are Skye (1889), and Renfrew, Dunbarton and Bute, where records in the 1930s and 1940s possibly involved escaped birds (Gibson *et al* 1980).

Immigrants from Iceland, Fenno-Scandia and possibly Russia start to arrive in September, and gather on estuarine mudflats, building up to a mid-winter peak at the most important wintering ground, the North Solway, which holds a population of international importance (more than 750). Numbers inland are generally small but there have been records in recent years from all mainland counties and island groups. Between 1976 and 1981 the maxima counted in Scotland ranged from 3,380 in November 1976 to about 1,500 in 1980/81; a very large proportion of the population is concentrated in a small number of localities so these figures are likely to give a fairly accurate indication of total wintering numbers. The estimated average British wintering population is 20,000–25,000 (Owen *et al* 1986); the Scottish birds therefore represent only a small proportion of the total. The January indices for Scotland show a substantial and progressive decrease from the early 1960s to 1969, followed by a slight increase during the 1970s and early 1980s. The November indices show a contrasting trend, being appreciably higher in the 1970s than in the preceding decade. The big increase recorded on the Mersey Estuary since 1967/68 (Allen 1974) is not reflected in the figures for the Solway, but it is possible that more birds are pausing in Scotland in autumn on their way to join that wintering flock, and that the slight March increase evident on the Solway may be due to northward movement of birds from the Mersey.

The North Solway is the only area in Scotland that regularly holds internationally important numbers of Pintail. The flocks there move to and fro in the estuary between Airds Point and Annan, the largest numbers generally occurring at Caerlaverock NNR and on the Carse and Mersehead Sands, on either side of Southerness Point. Peak counts at Caerlaverock have varied widely in the last two decades (Table), reaching a maximum of 2,400 in October 1979, when the birds were feeding over stubble.

Away from the North Solway, gatherings of national importance (more than 250) occur regularly at Nigg Bay (where the average mid-winter peak is c350 and 600 were present in January 1982), less often at Longman Bay, and very occasionally in the Dornoch Firth; 1,000 were on the Dornoch Sands in October 1983. The flocks of several hundred recorded sporadically at Wigtown and Luce Bays

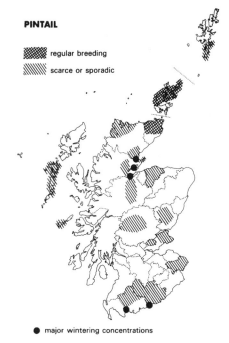

PINTAIL

▓ regular breeding

░ scarce or sporadic

● major wintering concentrations

*in close season

Peak counts of Pintail at Caerlaverock (Data from Thom 1969, Owen et al 1986, Wildfowl Trust & SBR – figures rounded)

	Av.	Range
1970/71–1973/74	580	350–1,200
1974/75–1979/80	1,750	800–2,400
1980/81–1983/84	650	190–1,200

on the Outer Solway, and inland at Hightae Loch, are probably North Solway birds. Flocks of 100+ are fairly frequent on Loch Ken and occasional on Loch Eye, Loch Ryan and the Inner Clyde between Woodhall and Erskine. The Grangemouth/Tullibody Island stretch of the Forth Estuary, Montrose Basin and the Eden Estuary, formerly important resorts (Thom 1969), no longer hold flocks of any size. Elsewhere Pintails occur more often in southwest and east-central Scotland than in the north and west, but seldom in groups of more than 20. Immigrants start leaving in February and virtually all have gone by the end of March.

Its very localised distribution, and the great dependence of the bulk of the population upon estuarine feeding grounds, make this species particularly vulnerable to environmental changes. During the 1970s extensive reclamation of tidal mudflats was carried out at Nigg Bay, Longman Bay and on the Forth Estuary; although Pintail numbers have been maintained so far at Nigg they have declined markedly on the Forth and to a lesser extent at Longman Bay. Earlier reclamation at Invergowrie Bay may be one reason why that site, mentioned by B&R as important, no longer holds more than an occasional few birds. The Solway has not yet suffered any major physical change but plans for a barrage are still extant; such a development, which would create freshwater conditions on the Inner Firth, would be likely to have a serious effect on the Scottish wintering Pintail population.

Garganey *Anas querquedula* [1]

Summer visitor, annual in small numbers since 1961. Has bred.

The Garganey has been known as an irregular summer visitor since the mid 19th century and was first recorded breeding in 1928, when a nest was found at Aberlady (SN 1928:77). B&R noted that it was occurring with increasing frequency during the second quarter of this century and predicted that colonisation might take place. Since 1961 birds have summered on several occasions, but proof of breeding was not obtained again until 1979, when a brood was reared in Ayrshire. Breeding may also have taken place at Aberlady in 1950 and 1954 (EBB 4:72), North Ronaldsay in 1943 (SB 5:103), and Morton Lochs in 1952 (B&R).

Britain is on the extreme western limit of the Garganey's breeding range and numbers in this country are largely determined by the size of the spring arrival (Parslow). From 1961 to 1983 inclusive the numbers recorded in Scotland fluctuated irregularly, ranging from only one in 1962 to 27 in 1976. About 65% of reports are in April/May, 20+%

in June/August, and the remainder in February/March and September/October; the earliest date is 3 February (a male in Lewis in 1984) and the latest 31 October. The Garganey was recorded most frequently during this period in Shetland and around the Clyde. In the Outer Hebrides it occurred on Lewis and the Uists (a pair was at Balranald in May 1982), but the only Inner Hebridean records were from Islay. The most seen together (apart from the breeding record) was seven. A total of 44 pairs was reported but the majority of records relate to males, which are much more easily identified than the females. An interesting record is that of a carcass in a Peregrine eyrie on Islay.

Blue-winged Teal *Anas discors*

Vagrant (N America) – about twelve occurrences since 1970.

First recorded in Scotland in 1858, the Blue-winged Teal has been reported with increasing frequency since 1970. Prior to 1950 there had been only three records, between 1950 and 1970 there were three reports (involving four birds), and from 1970 to 1984 more than 12 reports involving at least 17–18 birds. This species has occurred most frequently in the Outer Hebrides, where there have been several summer records, including one of a female mating with a Shoveler, on North Uist in 1979. Other records are from Shetland, Orkney, Easter Ross, Aberdeen and Dumfries. Most sightings have been in September/November, and the maximum number of birds recorded in any one year has been four.

Shoveler *Anas clypeata*

Small numbers breed regularly in the southwest, the Central Lowlands, Orkney and the Outer Hebrides, and sporadically elsewhere. Passage numbers are small and most leave the country during the severest weather.

The Shoveler's distribution, both during the breeding season and in winter, is limited by its dependence upon shallow, eutrophic waters. It increased as a breeding bird in Britain and elsewhere in Europe during the early 20th century, the main expansion in Scotland taking place between 1900 and 1920 (Berry in Parslow 1973). Most of the districts with suitable waters now hold breeding pairs at least sporadically and considerable numbers nest at the most favoured sites, which are shallow, muddy lochs and marshes with abundant cover. Since 1966 there have been breeding records from all the central and southern counties except West Lothian, Clackmannan and Peebles. The lochs set among the farmlands of Strathmore, east Fife and Kinross are among those occupied most regularly and by the largest numbers; Kinnordy Loch holds 25–35 pairs, Loch Leven about 10–15 pairs, and a group of four Perthshire lochs a total of up to ten pairs. Further north Balranald (North Uist) has a regular summer population of up to 17 pairs, and Orkney has held up to eight pairs in recent years, with

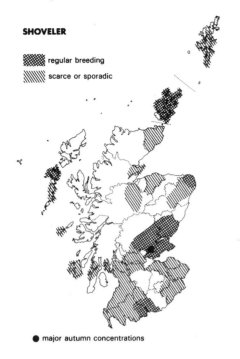

SHOVELER

▨ regular breeding

▧ scarce or sporadic

● major autumn concentrations

Loch Leven is the only site regularly holding several hundred Shovelers. The build-up there starts in late August and most birds usually leave by mid October; some 85% of those present in autumn are adult males moulting out of eclipse (A. Allison). Only a few other waters support 100 or more in most years and all of these are in the Central Lowlands, the Borders and the southwest. On the North Solway, Kilconquhar and the Forfar and Blairgowrie lochs much higher numbers were recorded during the late 1960s and early 1970s than in the last few years; flocks of over 100 also occurred on the Loch of Strathbeg during that period (Thom 1969).

Elsewhere in Scotland Shovelers are comparatively scarce, especially in the northwest and on many of the Hebrides. In Shetland they are most often recorded during the summer, from April to September, but numbers seldom reach ten. Orkney, in contrast, has a small wintering population based on the north islands, where birds are present throughout the year. In the Outer Hebrides a few occur on passage; breeding birds arrive in March and leave by November (Hopkins & Coxon 1979).

There are no obvious threats to Scotland's small breeding population of Shoveler, in fact rather the reverse. The most important breeding waters already have reserve status, while, in the long term, the progressive enrichment of shallow, lowland lochs may be expected to result in an increase in the supply of suitable waters. The wintering population is so small and so irregular in its movements that it is not possible to assess the factors affecting it.

breeding recorded on Sanday, Stronsay, Papa Westray, North Ronaldsay and West Mainland. Nesting occurs sporadically at the Loch of Strathbeg, on Speyside and in Caithness, and more rarely in Moray and Easter Ross. There are also sporadic breeding records for Shetland, South Uist, Tiree, Islay and Kintyre. Bute, which was colonised in the mid 1960s, now holds several pairs, and a pair nested on Inchmarnock in 1974 (Gibson 1979). There is no recent record from Sutherland, where Shoveler bred in 1895.

It has been suggested (Atlas) that the main centre of the Shoveler's breeding distribution in Scotland is the southeast lowlands. The additional data now available indicate that the area between the Forth and the Grampians holds the bulk of the population, which may total around 100 pairs in a good year. This represents not more than 10% of the estimated total of 1,000–1,500 pairs in Britain and Ireland (Owen *et al* 1986).

Most British birds move south to winter in France and Spain, leaving in October before the main influx of continental birds takes place (Ogilvie 1962); Scottish-ringed juveniles have been recovered in Ireland in November/December and in Kent and Spain later in the winter. The autumn peak occurs earlier here than in England, where maximum numbers are present in November, and it may be that in some years immigrants arrive before all the native birds have left. The timing of the first hard frost largely determines the date of departure. Fluctuations in the numbers of immigrants, which originate largely from USSR, result in widely varying peak counts in different years; from 1976 to 1984 the September/October peak count ranged from 378 to 890. These fluctuations are much influenced by numbers at Loch Leven, where the average autumn peak during the same period was under 300, and 700 were present in September 1981 but only 480 in September 1983.

Red-crested Pochard   *Netta rufina*

Vagrant/escape (Europe) – recorded in twelve years, 1950–83.

As the Red-crested Pochard is often kept in wildfowl collections it is uncertain whether Scottish records relate to wild or escaped birds, but the possibility of genuine vagrants occurring has increased recently with the westward spread of the species across Europe. The nearest breeding area is the Netherlands; nesting records in England probably refer to escaped birds (Parslow).

The records from 1950 to 1983 involved a total of about 20 birds. Most reports were for September/December and only nine of the birds were males. There are records this century from Moray, Aberdeen, Perth, Fife, Midlothian, Selkirk/Roxburgh, Berwick, Ayr, Kirkcudbright, Lanark, and Dunbarton, and two older records from Argyll. An account of the earlier status of this species is given by Pyman (1959).

[Rosybill   *Netta peposaca*]

A bird of this South American species seen in Orkney in November 1984 was regarded as an escape (Orkney BR 1984).

[**Canvasback** *Aythya valisineria*]

The record of a male in Peebles on 8 October 1979 (SBR 1979) was not accepted as relating to a genuine vagrant from America.

Pochard *Aythya ferina*

Breeds regularly in a few lowland counties, in small numbers. Widely but in general thinly distributed in winter, with a few large concentrations in the east-central lowlands, Orkney and the northeast. Numbers have fluctuated widely during the last 25 years.

For breeding, the Pochard has a strong preference for shallow waters with abundant plant-life and dense emergent vegetation. Since the first recorded nesting in Scotland in 1871, the species' distribution has shown an expansion followed by a decline. In 1938 it was said to be breeding in seventeen counties but by 1964 this figure had dropped to only six (Atlas). Regular breeding currently (1980s) appears to take place only in Perth, Midlothian, Angus, Renfrew, Wigtown, Kirkcudbright and Aberdeen. Occasional breeding has been recorded since 1950 in Sutherland, Caithness, Inverness, Argyll, Dunbarton, Fife, Ayr, Dumfries, Roxburgh, Berwick and possibly North Uist, but there is no recent record for Loch Leven. A total of 26 pairs was present during the 1976 breeding season on the two main Perthshire sites; elsewhere there are few records of more than four pairs. It seems likely that the breeding population never exceeds 50 pairs and is often substantially smaller. The total British and Irish population is currently estimated as 200–400 pairs (Owen *et al* 1986).

In winter Pochard occur in every mainland county and island group, though in only small numbers in many areas.

Immigrants start to arrive as early as July, with sizeable influxes sometimes occurring towards the end of the month. In 1966 about 1,500 birds arrived at Loch Leven in late July, nearly all of them drakes; in other years the summer influx there has numbered no more than 200. A similarly irregular pattern has also been recorded at other sites. The main arrival takes place between the end of August and November, with many birds not reaching Scotland until October/November. Ringing recoveries suggest that most of the immigrants come from the Baltic countries and USSR. There is considerable regional variation in the timing of the winter peak, with numbers building up faster in Orkney and the northeast than in east-central Scotland. In the Outer Hebrides and the Borders, Pochard are most numerous in autumn and comparatively few remain to winter, but numbers in Scotland as a whole reach a peak in mid winter and drop rapidly from February onwards.

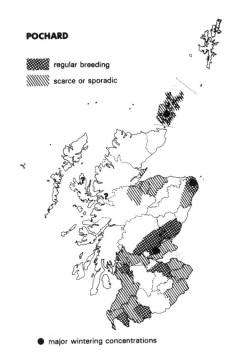

POCHARD

▨ regular breeding

▧ scarce or sporadic

● major wintering concentrations

Wintering Pochard numbers fluctuate markedly between years. In 1976–81 wildfowl count peak totals ranged from 6,000 to over 14,000 (the overall British and Irish wintering total is currently estimated as c50,000); in 1981–84 the range was 4,640 to 10,000. The most striking change has occurred at Duddingston Loch, which was until recently by far the most important resort. Average peak numbers there rose from

Peak numbers of Pochard at selected sites (Data from Wildfowl Trust & SBR – figures rounded)

	Duddingston		Strathbeg		Harray	
	Av.	Range	Av.	Range	Av.	Range
1962/63–1965/66	2,930	2,500–3,500	1,460	500–2,760	3,120	230–1,280
1966/67–1978/79	5,900	1,250–8,200	1,250	650–2,400	1,600	70–3,400
1979/80–1983/84	200	35–350	860	530–1,250	2,980	1,750–4,500

2,500 between 1960 and 1966 to nearly 6,000 between 1966 and 1978 and then dropped to only a few hundred. Similar changes on a smaller scale have taken place at Loch Harray (now the principal site) and the Loch of Strathbeg (Table), though not synchronously with those at Duddingston, which were in line with the general trend for Britain as a whole.

The history of the Duddingston Pochard flock is a fascinating one, since it has both provided answers to questions and posed new ones. Writing in 1968 I commented that this flock deserved further study, as the birds did not feed on the loch but were known to flight towards the Forth at dusk every evening. At that time Pochard were generally described as resorting to brackish or salt water only when driven off freshwater by ice. Subsequent research showed that the Duddingston birds fed regularly in the vicinity of the Edinburgh sewage outfalls, taking vegetable matter and waste grain spilled at the docks nearby. It also became apparent that there is considerable interchange between Duddingston, the various sites on the Forth and many of the Fife resorts.

The regular feeding on tidal water by the birds in this area seems to be very unusual in Europe; only in Scandinavia are Pochard recorded regularly at similar sites, although there they take natural plant food (Hockey 1983). In view of the fact that Edinburgh's new sewage disposal plant came into operation early in 1978 and the Duddingston Pochard population slumped the following winter, it is tempting to suggest that this is a demonstration of cause and effect. The pattern of previous fluctuations in Pochard numbers both at Duddingston and elsewhere suggests, however, that other factors may also be involved (Hockey 1983). The northwest European breeding population is known to have increased during the 1960s and early 1970s and this may have been partly responsible for the build-up in Duddingston numbers at that time; as yet there has been no suggestion of a decline sufficient to explain the recent drop. It will be interesting to see what changes take place in Scotland's Pochard population during the next 25 years.

Apart from Duddingston and Harray no Scottish waters held flocks of international importance (more than 2,500 birds) in 1976–81. Nationally important numbers (more than 500) were recorded at least once during these years on waters scattered from Orkney and the northeast through the Central Lowlands to the southwest. These included both freshwater sites and stretches of the Firth of Forth, most of the latter probably being linked with Duddingston. Elsewhere in Scotland Pochard numbers are in general much lower. A peak wintering total of around 200 has been suggested for Shetland, several Caithness sites hold over 100, and in the Borders maximum counts seldom exceed 40.

Despite the apparent effect on Pochard numbers of reducing organic pollution in the Forth, it is probable that factors other than local environmental changes are largely responsible for fluctuations in the overall Scottish wintering population. The future of the very small breeding population may depend to a considerable extent upon maintenance of the present undisturbed conditions at the most important sites.

Ring-necked Duck *Aythya collaris*

Vagrant (N America) – annual since 1977, in very small numbers.

The Ring-necked Duck was first recorded in Scotland on 2 January 1963, when an immature male was found on Loch Morar, Inverness (SB 2:476). A second male was seen in February 1969, in Aberdeen, after which there were no further reports until early 1977. Since then there has been a big increase in the number of records, with two in 1977, five in 1978, two in 1979, nine (involving up to seven birds) in 1980, seven each in 1981 and 1982, but only two in 1983 and four in 1984. By the end of 1984 there were records for Wigtown, Kirkcudbright, Ayr, Lanark, Renfrew, Argyll, Midlothian, East Lothian, Aberdeen, Inverness, Caithness, Sutherland, Orkney, Shetland, Islay, Mull and Lewis. Most reports are of single birds but two males were on Tingwall Loch (Shetland) in April 1980 and a pair on Loch Soulseat (Wigtown) in late February 1980. Birds have sometimes remained at the same site – usually on freshwater – for two or three months, and it seems possible that some individuals return in successive winters, for example a single male which appeared annually in the Insh Marshes area from 1980 to 1983 inclusive. Most first sightings have been between late September and February, but there have also been March/June reports. The majority of records relate to males (28 males:four females up to the end of 1983), but there is a high possibility of females, which are very similar to Tufted Ducks, being overlooked. A Ring-necked Duck ringed at Slimbridge in March 1979 was recovered in Greenland – on its way home – later the same year, suggesting the possibility of transatlantic colonisation in the future (Owen *et al* 1986).

Ferruginous Duck *Aythya nyroca*

Vagrant (Europe) – seven records since 1972.

B&R quoted five records of the Ferruginous Duck, dating back to the 1850s, but some of these must be considered of doubtful reliability. Between 1950 and 1972 there were no reports, but there have since been records in seven years: 1973, 1974, 1976, 1977, 1978, 1981 and 1982. The records are from Midlothian (two successive years), Roxburgh, Aberdeen, Rhum, Orkney and Possil Marsh in Glasgow. All refer to single birds and most to males. All but one (a May record from Orkney) have been between October and March.

Skye and Tiree. Bute is the only one of the Clyde Islands on which breeding has been confirmed.

Comparatively little information is currently available on breeding numbers, apart from counts at Loch Leven, which is the most important breeding site in Britain. On St Serf's Island Tufted Ducks nest at a remarkably high density, up to 215 nests/ha; 450+ pairs breed in an area of about 16 ha (Allison *et al* 1974). Elsewhere numbers and densities are very much lower even in districts with many suitable lowland waters, while in the northwest Highlands breeding is largely

Tufted Duck *Aythya fuligula*

Partial migrant, widely distributed as a breeder throughout the country but most abundant in the eastern lowlands. Also a passage and winter visitor. Although typical of still freshwater, resorts to estuaries in large numbers during hard weather.

The Tufted Duck was first recorded breeding in Scotland as recently as 1872; since then it has spread out from an initial nucleus in the Tay and Forth Basins to colonise most suitable waters – those at least 1 ha in extent and not more than 5 m deep, preferably with islands and below 400 m asl. Tufted Ducks breed later than most other species and ducklings are seldom seen before July. In 1968–72 breeding was proved in the only counties for which B&R had no nesting records, ie Banff, Nairn and Shetland. Numbers in Shetland, where Tufted Ducks have bred since the 1950s, are small but said to be increasing; 25 pairs were reported in 1977 and 13 pairs at ten sites in 1983. Breeding occurs regularly on the machair lochs of the Uists and Benbecula and was recorded for the first time on Lewis in 1982. In the Inner Hebrides only Islay has a regular, though small, nesting population but sporadic breeding has been recorded on

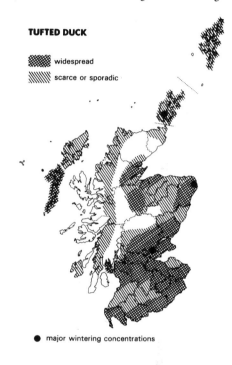

TUFTED DUCK

widespread

scarce or sporadic

● major wintering concentrations

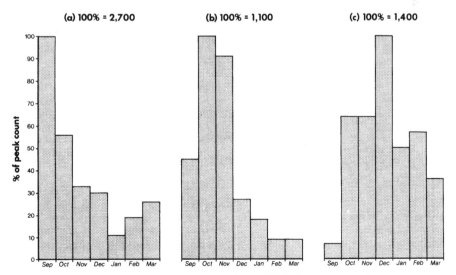

Seasonal fluctuations in the numbers of Tufted Ducks counted on (a) Loch Leven, (b) the Loch of Strathbeg and (c) the Loch of Harray (based on average monthly counts 1976–81)

restricted to the few limestone lochs. From the limited data available it seems possible that the Scottish breeding population may be of the order of 1,000 to 1,200 pairs – but much more detailed surveys would be needed to corroborate this. The presence of Tufted Ducks on a water during the breeding season cannot be taken as a reliable indication of nesting, because not all birds breed every year. The total British breeding population is currently estimated as 7,000–8,000 pairs (Owen *et al* 1986).

Autumn passage is evident in the Northern Isles during September, the main arrival of immigrants generally takes place between mid September and mid October, and there is often a further influx in December, especially in the north (Fig). At Loch Leven numbers peak in September but the birds disperse in October/November and the mid winter population is very small. Ringing recoveries show that many of the Loch Leven birds are in Ireland in December/February and some in France and Spain. Most of the Icelandic Tufted Duck population is believed to winter in Britain and the flocks arriving in Orkney are likely to come from there. Birds reared in Holland, Denmark, Finland and Latvian SSR have also been recovered in Scotland, and many of those wintering on the Loch of Strathbeg and in the Central Lowlands may originate from the more easterly parts of the breeding range. An adult female ringed at the nest on St Serf's and shot near Archangel in October two years later had presumably emigrated with birds from USSR. There have also been summer recoveries in Fenno-Scandia and Germany of Tufted Ducks (mainly males) ringed at Loch Leven in previous breeding seasons.

Seasonal fluctuations in numbers are less pronounced in the Tufted Duck than in some other species but the pattern varies widely between different areas (Fig). Annual maxima in wildfowl counts during the seasons 1976/77 to 1983/84 ranged from about 8,000 to nearly 12,000. A large proportion of the birds are found on waters which are regularly counted, so the peak wintering population is probably in the range 10,000 to 15,000 (at least 62,000 are thought to be in Britain during January (Owen *et al* 1986)). None of the Scottish sites carries a Tufted Duck flock large enough to qualify as of international importance (more than 5,000 birds) but several regularly hold gatherings of national importance (Table) and a further 10–12 occasionally do so. During hard

Peak counts of Tufted Ducks at the most important sites (Data from Owen et al 1986 & Wildfowl Trust – figures rounded)

	Leven		Strathbeg		Harray	
	Av.	Max.	Av.	Max.	Av.	Max.
1970/71–1974/75	2,180	4,000	1,070	1,790	1,520	2,600
1975/76–1979/80	2,050	4,500	1,260	1,500	1,480	1,960
1980/81–1983/84	4,530	4,830	1,400	1,950	1,590	2,280

weather, when many of the freshwaters are frozen, flocks of 800–1,000 birds, and sometimes more, gather on the tidal waters of the Forth Estuary above Kincardine Bridge, on the Tay near Dundee, on the Clyde between Cardross and Dumbarton, and on Orkney's brackish Loch of Stenness. Elsewhere numbers on any one water seldom exceed 500 and on most are considerably smaller.

Whereas the September and January indices of relative abundance have shown a fairly steady increase in England and Wales, no regular trend is apparent in the Scottish data. This is probably because the English breeding population is still expanding, largely due to the increasing availability of suitable habitat on gravel pits and reservoirs, whereas that in Scotland (which was colonised earlier) is now levelling off. An alternative explanation might be different origins for the bulk of the wintering population. On the whole the former seems the more likely. There is no evidence of any significant threat to this very widely dispersed species, which is more tolerant of disturbance and the proximity of man than are most other ducks.

Scaup *Aythya marila*   [1]

Winter visitor to tidal waters. Numbers have decreased since the late 1960s, and the largest gatherings are currently on the north side of the Forth, off Islay and in the Solway. Breeds sporadically.

Scotland holds most of the estimated 5,000–6,000 Scaup wintering off the British coast; although the Scottish population is consequently of national importance it is comparatively insignificant in the context of a northwest European population estimated at 150,000 (BWP). Since the flock which formerly wintered off Leith declined, the only site holding internationally important numbers (more than 1,500 birds) is Largo Bay, with peak counts of 2,000–2,680 in the early 1980s. Loch Indaal, Islay, and the North Solway around Southerness and Carsethorn, which each held up to 1,500 in the mid 1970s, are the next most important sites, but numbers decreased there too in the early 1980s; maxima in 1980–83 were 975 and 1,250 respectively. Flocks of 100 to 500+ occur fairly regularly at Loch Ryan, the Dornoch Firth and Lochs Harray and Stenness, and less often on the Clyde and the Ayr coast, in St Andrews Bay, on the Tay at Dundee and in the Beauly, Moray and Cromarty Firths. Elsewhere numbers are generally small. The suggestion that flocks of 225+ occur regularly in Mull (Prater 1981) appears to be incorrect; there have been few recent records of more than 100 there. Inland records are not uncommon in winter but usually involve only very small numbers.

The spectacular fluctuations in the wintering flock off Edinburgh since 1962 (Table) are thought to have been associated with the discharge of waste grain and changes

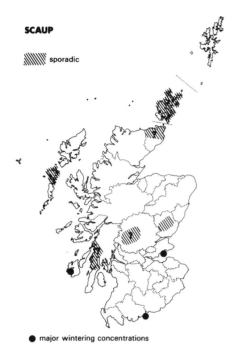

SCAUP

\\\\\\\ sporadic

● major wintering concentrations

in the sewage system. Grain discharge increased progressively up to 1968 and subsequently declined. Not only did the numbers of Scaup decrease as a result of this reduction in the food supply but the pattern of distribution within the Edinburgh area also changed, the main concentration moving from Seafield to Musselburgh as the sewage treatment programme was implemented from 1975 onwards (Campbell 1979). In 1976–79 the average peak on the Firth of Forth as a whole (ie including Largo Bay) was about 4,600 (BoEE) and in 1982/83 it was 2,700 for Largo Bay alone. The comment by B&R that in some years there are thousands on the Firth of Forth suggests that similar fluctuations may have occurred in the past, as they have done on a smaller scale on the Tay.

Some Scaup are in Scottish waters by mid September and there is a gradual build-up during October/November but the main influx does not take place until December. The population is at its peak in January and drops fairly rapidly thereafter, though in some areas numbers may remain stable for several months. Numbers in Scotland as a whole have obviously been much influenced by the size of the Forth flock, which at one time held 95% of the wintering total. In 1976–81 the maximum count in Scotland was 6,220 and there has since been a further decrease; in 1983/84 the peak count was under 2,000.

Peak counts of Scaup wintering on the Forth between Leith and Levenhall (Data from Milne & Campbell 1973, Campbell 1979, 1984 & WWC)

1968/69	1972/73	1975/76	1976/77	1977/78*	1978/79
30,000+	c16,000	10,280	1,528	3,640	2,270

1979/80	1980/81	1981/82	1982/83	1983/84
1,100	44	114	30	6

* data for 1977/78 precede the introduction of sewage treatment.

The tendency of this species to gather in large numbers where sewage and waste distillery grain are discharged makes it very vulnerable to localised disasters such as oil-spills. Writing in 1968, when the Forth flock was approaching its all-time peak, I commented on the potential threat such a disaster might present to the Icelandic breeding stock (Thom 1969). Within four years 200–300 birds were affected by an oil-spill at Musselburgh; mercifully the main flock was unharmed. At that time it was suspected that the entire Icelandic population wintered in Scotland and it seems probable that the regular flocks in Orkney and Islay are from this source; Scaup in these areas arrive and peak earlier than the population as a whole. Birds ringed in Aberdeen have, however, been recovered in the Baltic and USSR – as well as Iceland – suggesting that the large numbers formerly visiting the east coast were probably of continental origin.

As a breeder the Scaup has only a tenuous foothold in this country. Scattered birds are present most summers but there are few records even of attempted nesting. The only recent records of successful breeding are from Orkney (1965, 1969, 1973 & 1978), North Uist (1969), Angus (1971) and a possibly suspect report of three pairs at one Perthshire site in 1970. A female accompanied by a juvenile in Caithness in September 1981 and a pair in Argyll in late May 1979 represent other possible breeding records. There are records from the first half of this century for Ross and South Uist and for Sutherland in 1899 (B&R).

Eider *Somateria mollissima*

Breeds on all open-sea coasts and islands from Berwick to Wigtown, most abundantly in the east and north. Large flocks gather during the moult in July/August, moving later to different areas for the winter, when the largest wintering flock in Britain is at the mouth of the Tay.

The gradual spread of the Eider during the 19th and early 20th centuries is documented in detail by B&R. This spread is still continuing. B&R knew of no nests on the Clyde Islands, although broods had been seen there; at least 250 pairs now breed on Bute and Inchmarnock alone (Gibson *et al* 1980). Numbers on the Ayr coast are still small and, although breeding regularly around the Rhinns of Galloway, Eiders remain relatively scarce on the south-facing coast of Wigtown and along the Solway shore. B&R had

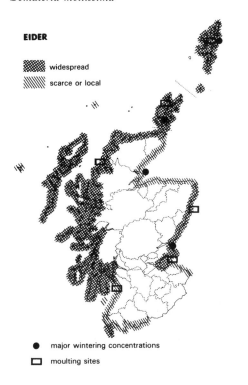

EIDER

▨ widespread

▨ scarce or local

● major wintering concentrations

▢ moulting sites

no breeding record for the coast between Fraserburgh and Berriedale, but colonisation of east Sutherland was well-established by the late 1950s and in 1968–72 breeding was proved in Ross and Cromarty, Nairn, Moray and Banff.

Breeding density on the mainland is greatest at the Ythan estuary, where 1,500–2,000 pairs nested in 1978–80 (NE Scot. BR), and Tentsmuir, with an estimated 1,500–1,800 pairs in 1971–72 (Pounder 1974). Pounder found a further 1,200 or so pairs along the coast from the Tay to Gourdon, and in a land-based survey of the coast from Brora to East Lothian in 1971 L. H. Campbell found 10,500 males. Few data are available for other mainland areas. There has been a recent increase on the Forth Islands with 545 nests found on the Isle of May in 1983 (SB 13:70) and estimates of 180 pairs for Fidra and c100 for Craigleith (R. J. W. Smith). Eiders are abundant and widespread in Shetland and Orkney, but no estimates of the breeding population are available. Numbers are lower in the Outer and Inner Hebrides, although in both areas the species is widespread (Hopkins & Coxon 1979, Ogilvie 1983). The British breeding population is currently estimated as 20,000 pairs (Owen *et al* 1986), of which probably about half are on the east coast of Scotland.

Eiders nest in a wide variety of situations: on marram-covered dunes, vegetated cliff-slopes and heather moorlands, in woodlands at Tentsmuir and East Lothian, and round the edges of arable fields at Montrose Basin. Ducklings may face hazardous journeys, sometimes involving descent from cliffs 100 m high, on their way to the sea, and even those hatched close to suitable feeding grounds are exposed to heavy predation from gulls and crows. Fledging rates are seldom high and sometimes very low indeed; at the Ythan the number of ducklings fledged ranged from 1,200 in 1976 to only 20 the following year.

In July, when the moulting birds are flightless, they gather in large flocks at sites which appear to be selected for their freedom from disturbance rather than for their food supply (Milne 1977). Several of the Shetland moulting areas are off inaccessible cliffs (but within easy reach of sheltered water) and most of those on the mainland are well away from popular beaches. The largest concentrations of moulting birds are in Shetland and on the east coast between the Ythan and North Berwick. Hope Jones & Kinnear (1979) found 15,500 Eiders in Shetland in August 1977, including five flocks of over 1,000 birds, but surveys in 1980–82 located only c8,000–9,000 birds in areas which had held nearly 14,000 in 1977 (SOTEAG per M. Heubeck). In the 1977 survey only 5,000 were found in Orkney, mainly in small and scattered groups, leading Hope Jones & Kinnear to suggest that part of Orkney's breeding population may move north to moult off Sumburgh. Such a movement would explain the relatively high Shetland and low Orkney numbers (in relation to the breeding population) in late summer and also the increased numbers recorded in autumn around Fair Isle, but further study is needed to clarify the situation. Possibly the largest Orkney moulting flock is off Eynhallow, where 650 birds were present in September 1981.

Further south some 12,000–16,000 Eiders moult off the coast between the Ythan and Dunbar and possibly 14,000–20,000 on the entire east coast (Campbell & Milne 1983). The main concentrations are at Murcar near Aberdeen (4,500–10,500) and between Port Seton and the Bass Rock (3,500–7,000). Flocks of up to 3,000 moult off Gourdon and 1,500 off Tantallon, and there are several gatherings of 500–900 elsewhere along the coast. It is noteworthy that there are few birds at the normal wintering areas at this time. The only other large moulting flock known is off the Ayr coast, where numbers have increased steadily since the 1960s; most of the 1,500 birds there in 1980 were males. Little is known about the moulting grounds of the birds breeding on the Hebrides and around the sea-lochs of the west coast, but recent wing-tagging studies suggest that most remain close to their home area, forming only small flocks (C. A. Galbraith). The only recent records of sizeable flocks are 2,000 moulting off Gairloch in 1980 (H. Milne) and up to 450 in the Sound of Taransay (N. E. Buxton).

By mid October movement to the wintering areas is virtually complete and the vast flock at the mouth of the Tay is near or at its peak towards the end of the month. Counts there between late October and March, when the birds start

Numbers and distribution of Eiders in winter (Data from Winter Atlas and BTO/WSG Winter Shorebird Survey 1984/85)

	Total counted
East coast – Border to Grampian	29,000
East coast – north of Grampian	3,400
North mainland	1,000
Orkney	5,000
Shetland	8,800
West coast	3,200
Southwest (Strathclyde south)	7,700

(Coverage of the Outer & Inner Hebrides was incomplete)

to return to their breeding grounds, generally exceed 10,000 and have reached 20,000 – making this the only site of international importance in Britain (Table). Most of the birds breeding between the Tay and the Moray Basin are thought to join this gathering in winter. Movement across the North Sea occurs, but is possibly sporadic. An Eider ringed as a juvenile in the Friesian Islands has been found on the Tay, and some years ago Ythan-ringed birds were recovered at the entrance to and in the Baltic. Despite continued ringing at the Ythan, there have been no recent recoveries from the Baltic, a situation which Owen *et al* (1986) take to indicate that the Tay wintering flock no longer includes immigrants. Birds from southern Scotland, together with some from the Farne Islands, winter in the Firth of Forth. The largest gatherings there were until recently off Leith/Seafield and Leven, where numbers increased from a peak of c2,000–3,000 in the 1960s and early 1970s to c4,000 in 1976/77, since when they have declined again to about 1,000–1,500 (Campbell 1978, 1984).

Elsewhere on the east coast there are wintering flocks of 1,500–2,000 at Montrose Basin, near Rattray Head, and in the Dornoch Firth, off Loch Fleet and between Golspie and Littleferry. At the Ythan mid-winter numbers are fairly low but about 1,800 are present in autumn and up to 5,500 in May. Shetland's wintering population has declined markedly since the mid 1970s, especially in the area around Bluemull and Colgrave Sounds. Monitoring surveys there gave late autumn/early winter peaks of 5,500–6,300 from 1975–78 and 3,300 in autumn 1979. Later that winter unusually large numbers of dead Eiders were found on adjacent beaches; the birds were not oiled and the cause of death was not established. The 1980, 1981 and 1982 survey counts were only 1,300, 930 and 570 respectively, and aerial searches failed to locate any sizeable flocks at new sites in northeast Shetland. It is assumed, therefore, that the decline in numbers was due to abnormally high mortality (SOTEAG per M. Heubeck). Although the *Esso Bernicia* spill in 1978/79 resulted in the deaths of at least 570 Eiders in the Sullom Voe/Yell Sound area, Bluemull and Colgrave Sounds were not affected by floating oil.

In Orkney the total wintering population is around 6,000, with some 2,000 in Scapa Flow in February; a fairly large flock of non-breeding birds arrives later, in May, to summer there (Lea 1980). In the west, where there are few large winter flocks, the most important gathering is on the Clyde; numbers there have occasionally exceeded 3,000 but have been somewhat lower since 1977. Flocks of c500 have been recorded off the Strome Islands in Loch Carron, in the Gareloch, and off the Ayr coast. The Hebridean wintering population is probably in the region of 4,500–6,000 birds (Hopkins & Coxon 1979; Ogilvie 1983).

Because Eiders move from one area to another during the course of the wildfowl count season, and substantial numbers are present in areas which it is impracticable to count regularly or accurately, estimates of the total population are liable to even more error than is the case with many other resident species. Owen *et al* (1986) estimate the British total as c50,000, and suggest that this represents an increase of 30–40% in the last two decades. It seems likely, however, that better coverage is responsible for at least part of the apparent increase, as the estimate of 30,000–40,000 given in Atkinson-Willes (1963) was based on very incomplete data for many parts of Scotland.

The large gatherings of moulting birds and the vast wintering flock at the mouth of the Tay are potentially at great risk from oiling (Hope Jones & Kinnear 1979) and several thousand Eiders have already died from this cause (Pounder 1971). There can be little doubt that large-scale oil-spills in Shetland or east coast waters present the greatest threat to the population as a whole. At the local level, variations in the availability of preferred foods can have a marked influence on breeding numbers and reproductive rates (Milne 1974). This and other aspects of Eider population dynamics have been the subject of long-term studies at the Ythan estuary by Milne and others (eg Milne 1974; Baillie & Milne 1982), while on the Forth feeding behaviour has been studied by Campbell (1978a & 1978b). The recent increase in commercial mussel farming in west coast sealochs has led to concern regarding the potential threat presented by Eiders, which find the cultivated mussels a convenient source of food. Protective nets can be put around the mussel ropes, but these ensnare and drown birds trying to get past them (Dunthorn 1971), while a combination of surface floating nets and scaring techniques is effective but costly (C. A. Galbraith). The current swing towards raft, as opposed to rope, culture may simplify protection, and intending mussel farmers are now being advised to site their operations as far as possible from migration routes regularly used by Eiders.

King Eider *Somateria spectabilis*

Vagrant (Arctic) – annual in very small numbers.

B&R, quoting John Berry, suggested that the King Eider was commoner than the records showed, since only the adult males can be readily identified at a distance. They noted reports of 'large numbers' in the Tay Estuary towards the end of the 19th century, but gave details of only single-bird sightings more recently. Between 1950 and 1965 there were seven reports of this species, whose nearest breeding station is Spitsbergen. Since 1965 it has been recorded annually, with a maximum of 13 birds in any one year, and it seems likely that many reports relate to birds returning to the same area in successive years. The largest group recorded has been a party of three males off Loch Fleet, where they were seen on many occasions in 1975–78 inclusive. Other long-stay locations have been the outer Clyde, Loch Ryan, and the Forth off Culross; the latter site and the Clyde at Langbank were the furthest inshore records.

Since 1950 the King Eider has been most frequently reported in Shetland, where there were up to eight individuals present in 1973 and 1974. The other records have been widely scattered around the coast, as far south as Wigtown on the west and Kincardine on the east; they include reports in every month. Records of adult males greatly outnumber those of females and immatures. In 1979 and 1980 a male mated with a female Eider on Arran but it is not known whether they bred successfully. A bird thought to be a hybrid was seen off Loch Fleet in 1981.

Steller's Eider *Polysticta stelleri*

Vagrant (Alaska / E Siberia) – recorded in 16 years since 1947 (the same birds in several years).

This small Eider was first recorded in Scotland on 5 January 1947, when two males were seen off Gairsay, Orkney (BB 40:253). By the end of 1983 there had been further reports from Orkney (1949, 1974, 1976 and 1978–82), South Uist (1972–83), Fair Isle (1971), Sutherland (1959) and Aberdeen (1970). Two birds, both drakes, remained in the area in which they were first sighted for notably long periods: from October 1974 through 1982 around Westray / Papa Westray, Orkney, and from July 1972 through 1983 off Vorran Island, South Uist. The Vorran Island bird was accompanied by two females on 13 April 1974.

Harlequin Duck *Histrionicus histrionicus*

Vagrant (Iceland / Greenland / N America) – four accepted records, 1931–83.

This normally sedentary species was first recorded on 13 February 1931, when a male was in the Sound of Harris, Outer Hebrides (BB 24:370). A first-winter male was shot in Roxburgh in January 1954, and there have been two reports of pairs (probably the same birds), at Fair Isle in January / February 1965 and at Wick in April / May the same year. The 1933 Shetland record quoted by B&R was subsequently rejected.

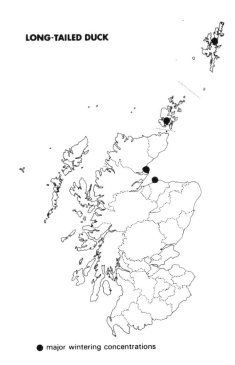

LONG-TAILED DUCK

● major wintering concentrations

Long-tailed Duck *Clangula hyemalis*   [1]

Winter visitor, mainly to the east coast, especially from the Moray Basin to Shetland; elsewhere numbers are generally small. Has bred.

Scotland probably holds the bulk of the British wintering population, about which little was known prior to 1970. Although B&R referred to 'great numbers' at various points around the coast, few parties of more than 200 birds were recorded during the regular wildfowl counts in the 1960s (Thom 1969). Then in 1970 / 71 a flock of 800–1,000 was located off Lossiemouth, about 8 km offshore and invisible from the land (Milne & Campbell 1973). Subsequent studies by Mudge & Allen (1980) and Barrett (1983) have revealed that large numbers of Long-tailed Ducks winter in this area, often feeding quite far out and flighting to inshore, but deep-water, roosts at dusk. The daily pattern of movements is very variable and the birds are often in scattered small groups, making accurate counting extremely difficult. Experience has shown that overall numbers can be best assessed as the birds flight to roost in late afternoon. Counts made by this method in the Moray Basin produced much higher totals than were obtained by aerial or ship-based surveys, and it seems likely that not all the areas used for daytime feeding have yet been discovered.

Three major roost sites have been located in the Moray Basin: at Burghead Bay and Spey Bay on the south coast and off Brora to the north. In 1978 / 79 up to 6,500 of the estimated peak population of 10,000 fed in the western part of this area and roosted 3–4 km offshore in Burghead Bay, while in 1979 / 80 about 4,000 were present (Mudge & Allen 1980). In 1981–83 the three main roosting sites together held peak totals of c18,000 each winter, but the only regular flock of 1,000+ birds was that between Burghead and Hopeman (Barrett 1983). More than 15,000 Long-tailed Ducks were counted at the Burghead roost in mid February 1982 – the largest concentration yet recorded in British waters. The Brora roost, in an area where some 2,000 wintered regularly in the 1970s, held c4,000 in February 1982 and is apparently composed largely of birds which feed to the north and northeast of the roost. More than 1,000, believed to be drawn from the immediate vicinity, used the Spey Bay roost regularly in 1982 / 83. Small numbers start arriving at the roosts up to two hours before dusk, gradually increasing to a peak as dusk falls. Little is known about the night-time behaviour of the roosting flocks, but dawn flighting has been recorded and it is presumed that the flocks disperse before it is light. Day to day variability makes assessment of seasonal trends impracticable but it is thought that after the early winter build-up numbers remain relatively stable until mid March, when they decline rapidly.

Although the Moray Basin is the only location at which a flock of international importance (more than 5,000) has been recorded, sizeable numbers also winter in Orkney and Shetland; Lennox Campbell has suggested a total of 7,000–8,000 (Winter Atlas). The Orkney population, estimated as c6,000 (Hope Jones 1979), is scattered over a number of sites, with the largest numbers in Scapa Flow, where the winter maximum is up to 2,400 (Lea 1980). There too the birds flight from their shallow-water feeding grounds to roost in deep water. This brings them into the central part of the Flow, where they are close to the moorings for tankers loading oil from the Flotta terminal. Although fairly widely scattered, these birds would clearly be at risk in the event of a spill, especially one during the night (Hope Jones 1979). Elsewhere in Orkney there are small flocks at many coastal localities and 150–200 regularly use the brackish Loch of Stenness.

In Shetland the winter maximum has been put at 1,500–2,000, with the largest groups in the north isles (Berry & Johnston 1980) and peak numbers present in the early part of the year; further study may well reveal other concentrations. Between 500 and 750 Long-tailed Ducks have been counted on several occasions in Bluemull Sound, off Hascosay and between Sullom Voe and Yell – all areas where the risk of oiling is considerable.

The situation on other stretches of the east coast is not clear. Only 100–200 Long-tailed Ducks winter off Caithness and numbers along the north coast of Sutherland are also small. Land-based counts suggest that 100–200 winter off the Aberdeen coast, and aerial surveys from Aberdeen to the Bass Rock located only 250–350 in 1980 and 1981, but a flock of at least 800 was off Collieston in February 1976. Movement northwards of several hundred off Peterhead in mid October suggests passage to the Moray Basin – but from where? Up to 300 are regularly recorded off Methil Docks, one of the most accessible places for watching Long-tailed Ducks.

In the west the Outer Hebrides hold at least 500–700 (Hopkins & Coxon 1979), the largest concentrations recorded recently being in Broad Bay and off Luskentyre. The Sound of Harris held 500 in the 1930s but there are no recent data from that area. Probably not more than a few hundred winter in the Inner Hebrides (Ogilvie & Atkinson-Willes 1983) and only small numbers occur further south along the west coast. Inland records, generally of single birds, are not uncommon.

In the absence of ringing recoveries the origin of the Long-tailed Ducks wintering in Scotland is uncertain. Some may be from the comparatively small Icelandic population but the large flocks probably come from Fenno-Scandia or northwest Russia – birds from both these areas are known to winter in the Baltic. The main arrival takes place in early to mid October and maximum numbers are usually present

by December. The proportion of adult males present in wintering flocks varies widely, reaching 90% locally at times (Barrett 1983). Birds start to leave in March but many remain until the end of April and a few stragglers often stay throughout the summer. Before leaving, Long-tailed Ducks form pre-migration parties which indulge in much displaying and calling; at this time they tend to occur more frequently on freshwater (Elkins 1965). During the 19th and the first half of the 20th century the species bred sporadically in both Orkney and Shetland (B&R) but there has been no recent indication of nesting.

There is clearly still much to be learned about the wintering areas and numbers of this species, which in Scotland is at the southwest limit of its wintering range. The data currently available suggest that not all the sites now known are occupied to the same extent every winter. In all three known areas of greatest concentration – the Moray Basin, Orkney and Shetland – flocks are potentially at risk from oiling.

Common Scoter *Melanitta nigra*   [1]

Most abundant in winter, when the main concentration is currently in the Moray Basin. Large moulting flocks summer off the east coast, but the breeding population is small and confined mainly to the north and west.

As is the case with all seaducks, the counts of Common Scoters – which winter at sea and often well offshore – represent far from complete coverage and consequently provide little more than a generalised picture of the current situation. Common and Velvet Scoters often occur in mixed flocks, which increases counting difficulties, and special efforts are necessary to obtain reasonably accurate counts; if these are not sustained results may give a misleading impression of fluctuations. It seem probable that the numbers visiting Scottish waters do, in fact, undergo both long-term and short-term fluctuations. Recent estimates of the British wintering population range from 25,000 to 50,000 (Owen *et al* 1986), of which about half are probably in Scotland.

The Moray Basin is the only area in which gatherings of international importance (more than 10,000) have been reliably recorded and it is not at present clear whether this is a regular, or only an occasional, occurrence. The main concentrations found have been in Burghead and Spey Bays, with smaller flocks at Culbin/Nairn Bars, and off the Dornoch Firth. In the winter of 1973/74 numbers exceeded 10,000 in Burghead Bay alone from November to January, when they peaked at 14,000. Between 1974 and 1977 coverage was incomplete and the highest count recorded was only 5,650, while a special seaduck survey in 1977–79 recorded peaks of 6,000 each season in Spey Bay, which held 65% of the Moray Basin Common Scoters in 1977/78. Aerial surveys in the winters of 1979/80 and 1980/81 found only 2,000–3,000 scoters, mostly Common, in Burghead Bay and few elsewhere in the Basin, but intensive studies in 1981–83 again located very much larger numbers (Barrett 1983).

In 1981–83 totals of more than 10,000 scoters were recorded on five occasions, all between January and March, the peak monthly count being nearly 14,000, in February 1983. Summing the peak counts for individual sites – 5,600 at Spey Bay, 4,800+ at Culbin, 4,500+ at Dornoch and 3,500+ in Burghead Bay – gave even higher totals, with a maximum of 18,000+ in January 1982. However, since these peaks often occurred on different dates, this total may be inflated due to flock movements. Overall numbers were highest during January/February in both seasons, with a particularly marked influx in January 1982. Common Scoters accounted for 60–100% of the total in the first half of the winter and in 1981/82 remained at this level until March; in 1982/83, however, the proportion of this species was lower overall and started to decline rapidly in February.

Further south, nationally important numbers (more than 350) occur regularly in winter at several points between Tentsmuir and Gullane Bay. An average peak of 2,000+ and a maximum of 6,000 were recorded in St Andrews Bay in 1950–68 but numbers there have recently been lower, with few counts of over 2,000. Several hundred winter around the Hebrides (Hopkins & Coxon 1979; Owen *et al* 1986) but the formerly important Solway and Loch Ryan wintering areas are now virtually deserted, with numbers since 1970 seldom exceeding 200. Elsewhere Common Scoters occur in winter in all coastal counties but seldom in flocks of more than 50–100. There are occasional inland winter records, usually of single birds. Although this species is generally classed as a passage migrant, as well as a winter visitor, it is difficult – as B&R noted – to identify the exact pattern of movements and there is little recent evidence of passage.

It has long been known that moulting flocks occur in Scottish waters during the summer and there are old records of very big gatherings on the Solway, and of large numbers off Tentsmuir and the Dornoch Firth (B&R). Up to 2,000 were on the Solway in July during the 1960s (Atkinson-Willes 1963) but this area has held few in recent years and all the large flocks are on the east coast. There has been little attempt to study the moulting flocks and, in the absence of co-ordinated counts, it is not clear whether reports from neighbouring localities on different dates refer to the same or separate flocks. Counts made in July/August between 1975 and 1982 suggest that 2,000–3,000+ currently moult along the Ythan/Aberdeen stretch of coast, 1,000+ between Montrose and Buddon, possibly a further 1,500 off Tentsmuir (although these may be the same birds as those seen in Angus), 1,500 in the outer Firth of Forth off Gullane, and 1,300+ off Dornoch. Some 100–200 moult off Harris. It is not known when these birds arrive in the moulting

areas, nor whether the 1,000+ Common Scoters recorded regularly in May off Gullane/Aberlady remain in the area to moult. Nor is it possible to be certain to what extent the summering birds contribute to the wintering flocks, which total perhaps twice as many.

Little is known about the origin of the birds moulting and wintering in Scottish waters. Iceland-ringed birds have been recovered in Britain but it seems likely that the very big flocks originate from further east, in Fenno-Scandia and Siberia. Large numbers from these areas moult off Denmark and winter off the French and Iberian coasts

The Common Scoter has bred in Scotland since 1885, possibly longer (B&R), but numbers have never been large. Most breeding sites are in remote moorlands, where the birds nest in long heather at least 10 m from the water's edge, but at Loch Lomond and on Islay they are on wooded islands. Breeding has occurred fairly regularly since the late 1960s in Caithness, Perth and Shetland, and probably also in Inverness (where there were at least 15 pairs in 1974), Sutherland and Islay; and has been fairly regular on Loch Lomond since 1971 (Mitchell 1977). There are also recent records for Ross and Cromarty and Argyll, but none since 1958 for Orkney (Booth *et al* 1984). The number of pairs varies from one year to the next at most sites, as does breeding success. Since 1968 up to five pairs/annum have bred in Shetland, 17+ pairs in Caithness (but 45 females 'on territory' in 1976 when 65 juveniles were seen), five pairs in

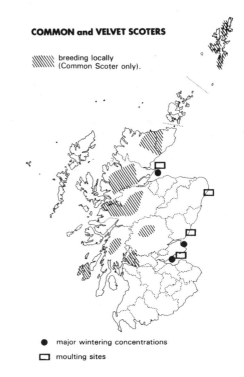

COMMON and VELVET SCOTERS

\\\\\\ breeding locally
(Common Scoter only).

● major wintering concentrations

▭ moulting sites

Perth, five pairs in Ross and Cromarty, 30+ pairs in Inverness, nine pairs on Loch Lomond, seven pairs on Islay and a single pair in Argyll. There are no recent figures for Sutherland. At many sites numbers were highest in the late 1970s, when the Scottish population may have neared 100 pairs;

it seems likely that it is still above the figure of 30–50 pairs suggested by Sharrock (Atlas). Common Scoters also breed in Ireland, where there are 80–100 pairs.

Surf Scoter *Melanitta perspicillata*

Vagrant (N America) – annual in very small numbers.

B&R gave few detailed records of the Surf Scoter but suggested that it was not uncommon in Scottish waters. In the period 1961–84 inclusive the species was recorded in sixteen years, annually from 1974 onwards. Most reports are of single birds, generally adult males, though parties of seven and nine were seen in Spey Bay in 1978 and 1979 respectively. First sightings are usually between September and March but individuals often remain in an area for several weeks, and there have been reports in every month. The distribution of Surf Scoter occurrences is somewhat surprising, in view of the bird's western origin, with records from East Lothian, Fife, Aberdeen, Moray, the Black Isle and east Sutherland as well as from Shetland, Orkney, the Outer Hebrides, Rhum, Islay, west Inverness, Kintyre, Wigtown and Kirkcudbright. There is a single inland record, of a male on Loch Insh, Inverness, on 14 October 1978.

Velvet Scoter *Melanitta fusca* [1]

Locally common in several east coast areas both in winter and in summer, when moulting flocks are present. Numbers fluctuate markedly and there have been recent changes in the areas most favoured.

Most Velvet Scoters are seen in company with Common Scoters and are often well offshore, making accurate counting of the two species difficult and sometimes impossible. Although Common Scoters predominate in most mixed flocks, considerable numbers of Velvet Scoters may at times pass unrecognised, making the count figures liable to be underestimates. This species breeds in Scandinavia and Siberia and winters in large numbers off Denmark. The only evidence of the origin of Scottish wintering birds is from the recovery of one ringed as a juvenile in southern Norway, and drowned in a Solway flounder-net in October the same year.

Winter totals in the last 25 years have ranged from a few hundreds to nearly 9,000. Numbers appear to peak in March/April, when this species is more abundant than the Common Scoter in some areas. The highest count yet recorded in one area has been 8,000+ in the Moray Basin in March 1983, when the principal concentrations were at the mouth of the Dornoch Firth and in Spey Bay; c80% of the Velvet Scoters in these areas at that time were males (Barrett 1983). A flock of 5,000 was in Spey Bay in mid April 1978, but the figures for other years suggest that these high totals do not occur regularly, although coverage is too incomplete and inconsistent for this to be certain. The 2,000–2,500 recorded in the Basin on several occasions dur-

ing the winters of 1973/74, 1974/75, 1977/78 and 1978/79 (Mudge & Allen 1980) may represent a more normal situation.

The Moray Basin – where the main concentrations are scattered along the south shore between Spey Bay and Nairn, and around the Dornoch Firth – is the only site at which internationally important numbers (more than 2,000) have been recorded in recent years. Up to 2,600 were seen in St Andrews Bay during the 1960s but since the early 1970s numbers there have seldom exceeded 100 and have often been much smaller. On the Firth of Forth, too, numbers have been declining; during the early 1970s the annual winter peak at Gullane Bay was 200–300 but since 1976 it has generally been below 100. Elsewhere up the east coast as far as Rattray Head the story is the same; areas described by B&R as frequently holding large numbers of Velvet Scoters now seldom hold any significant numbers and there have been few recent winter records of more than 100 off the Fife, Angus, Kincardine and Aberdeen coasts.

Several hundreds winter in Orkney and smaller numbers along the Caithness coast. About 20 regularly wintered in Sullom Voe prior to the *Esso Bernicia* oil-spill, but few have been seen recently in Shetland waters; in 1980 there were 11 records, none involving more than two birds. In the Outer Hebrides up to 100 may winter (Hopkins & Coxon 1979), although here, too, few have been seen recently. The only other area from which there have been recent reports of nationally important numbers (more than 50) is Argyll, where 60–80 have been seen off Tayinloan in the Sound of Gigha, suggesting that there may be a small flock wintering regularly in this general area. Along the southwest coast Velvet Scoters occur only irregularly and in very small numbers. There are occasional inland records, usually of single birds.

Moulting flocks of Velvet Scoters occur off the Aberdeen, Kincardine and East Lothian coasts. The highest recent counts from these areas are 550 between Blackdog and Murcar in August 1980, 225 off St Cyrus in August 1977, and 300 off Gullane in September 1979. The count coverage at this season is too incomplete to permit any conclusions as to whether these records refer to the same or discrete summering populations, or as to the regularity with which flocks moult in Scottish waters. Records of 150 birds off Brora and Dornoch in May 1975 probably refer to passage birds; at this season passage also takes place along the north Sutherland coast (Angus 1983). It seems possible that both moulting flocks and passage movements occur more often than these few records suggest, and there is clearly still much to be learned about the movements of this species.

The Velvet Scoter has been suspected of nesting on a number of occasions, the most recent in Shetland in 1945, but breeding has never been proved.

Bufflehead *Bucephala albeola*

Vagrant (N America) – one accepted record, 1980.

The first reliable record of the Bufflehead in Scotland was on 14 March 1980, when a male was found on West Loch Bee, South Uist, Outer Hebrides (BB 74:464). Three old records, from Aberdeen, Kincardine and Orkney, were not accepted as sufficiently authenticated to permit admission of the species to the Scottish list and are square-bracketed in B&R.

[**Barrow's Goldeneye** *Bucephala islandica*]

Escape/vagrant (Category D).

Barrow's Goldeneye, which breeds in Iceland, Greenland and North America, is normally very sedentary. The record of a male at Irvine, Ayr, in November/December 1979 (SBR 1979) was the first British report accepted by the Rarities Committee (BB 76:528) – but was admitted only to the Category D list (implying some doubt as to whether it was a genuine vagrant). An older record from Shetland, in March 1913, was not considered acceptable, and a number of recent occurrences in England have been regarded as undoubted escapes.

Goldeneye *Bucephala clangula*   *[1]

Scarce and local breeder. Passage and winter visitor, widely distributed on lochs, rivers, coastal waters and estuaries throughout the country; the largest gatherings are on the east coast and are usually associated with sewage outfalls.

The establishment of the Goldeneye as a breeding species was one of the major ornithological events of the 1970s. B&R predicted that this would eventually happen and drew attention to the fact that, although normally nesting in holes, this species had been known to make use of nest-boxes. During the early 1960s boxes were erected by George Waterston and Pat Sandeman near several lochs on which Goldeneye regularly summered, but it was not until 1970 that proof of breeding was obtained. In July that year a female with four young was found in east Inverness, on a lochan situated in mature woodland and fringed with emergent vegetation. A programme of nest-box erection was subsequently initiated by the RSPB and has proved highly successful (Dennis & Dow 1984). In 1979 14 broods, totalling 110 young, hatched

* in close season

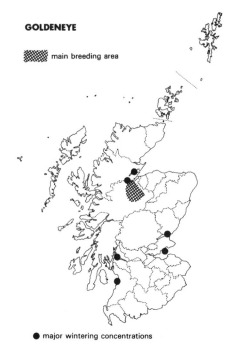

GOLDENEYE

▨▨▨ main breeding area

● major wintering concentrations

from 21 nests, all but two of which were in boxes, but by 1981 more use was being made of natural holes, with only about half of the 29 broods that year hatched in boxes. First broods are on the lochs by mid May and the survival rate appears to be good; 97 young were located in mid July 1981 from 286 known to have left the nest. In 1984 breeding success was very good, with 33 broods, totalling 311+ young. Females ringed during incubation have been recovered the following year in the same boxes or at others up to 3 km away. In 1979 a female with one flightless young was seen on Loch Lomond, but there has been no further evidence of breeding there. Birds regularly summer at many places in Scotland and in 1984 two pairs nested outwith the original breeding area (R. H. Dennis).

Apart from this small – presumably resident – flock, the wintering population is of largely Scandinavian origin; ringing recoveries indicate that many come from northern Sweden. The main arrival does not take place until after mid October and numbers reach their peak in mid winter. Goldeneye leave for their breeding grounds later than most other ducks and numbers remain high until well into April (Fig). Annual maxima in the wildfowl counts during the winters 1976/77 to 1983/84 averaged 5,500 and ranged from 4,800 to 6,400. Because Goldeneye occur on many waters that are not regularly counted, albeit in only small flocks, the total wintering population is probably considerably above this figure, perhaps reaching 7,000 in January, when the British total is thought to peak at c15,000 (Owen *et al* 1986).

Only one site, the Firth of Forth, has held Goldeneye flocks of international importance (more than 2,000 birds). Peak numbers there in 1976–81 averaged over 3,000 and in December 1980 the Forth held nearly a third of the British total (WWC). Flocks are present on both sides of the firth, those off Edinburgh being until recently much the largest. Since 1975 there has been a marked decrease in peak numbers there and also a change in distribution, both apparently resulting from changes in sewage discharge (Campbell 1979, 1983). Peak counts in the area dropped rapidly, following the introduction of sewage treatment in 1978, from 2,000+ to c600 in 1980/81 and less than 200 in 1983/84. Along the north shore between Largo Bay and Kirkcaldy numbers have remained more constant, with an annual peak of 1,000–1,500. There are also Goldeneye resorts associated with sewage outfalls in the Cromarty Firth near Invergordon, and at Inverness, Dundee, Langbank/Erskine and Ayr-/Prestwick; each of these areas regularly holds several hundred wintering birds (Pounder 1976). The only freshwaters regularly holding more than 100 Goldeneye are the Lochs of Strathbeg, Stenness and Kilconquhar. Elsewhere the species is widely distributed but most of the waters used normally hold fewer than 50 birds (Thom 1969).

Wintering Goldeneye flocks show a particularly wide variation in the proportion of adult males present at different sites. This ranges from under 15% at Invergordon to over 70% at Dundee and may be due to differences in feeding habits between the sexes – females being apparently more attracted than males by distillery waste grain (Pounder 1976) – or to differences in tolerance towards such factors as disturbance and exposure (Campbell 1977).

The Wildfowl Trust data show no regular trends in Scottish Goldeneye numbers but much more marked fluctuations than occur in the smaller English population. At present the home-bred birds are too few to affect the trend data, but if the current steady expansion continues they may eventually do so. As yet there have been no winter recoveries of Scottish-reared birds so nothing is known of their wintering habits. Although some of the main coastal concentrations, such as those in the Moray Basin, Scapa Flow and the Forth, are at risk from oiling, the wintering population as a whole is so dispersed that it is unlikely to be seriously threatened. A continuing reduction in the quantities of organic matter discharged at sewage outfalls is probably the factor most likely to affect winter distribution in the future.

100% = 5,370

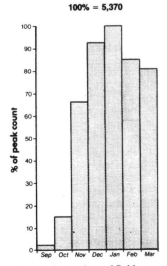

Seasonal fluctuations in the numbers of Goldeneye counted in Scotland (based on average monthly counts 1976–81)

[Hooded Merganser *Mergus cucullatus*]

The only Scottish report of this American sawbill, of an adult male shot in Shetland in July 1884, is not considered reliable. There have, however, been several accepted records in Ireland and Wales, the most recent in Co. Armagh in 1957.

Smew *Mergus albellus*

Irregular visitor in very small numbers, mainly in winter.

Since it was first recorded in the late 18th century the Smew seems always to have been regarded as an irregular visitor. B&R considered that it was becoming commoner and quoted a report of 'an exceptionally fine company' of about 50, including 20 adult drakes, near Glasgow in 1929. There have been no subsequent records of such large parties; the most seen together in recent years has been only seven (in 1965) and about 60% of all reports relate to single birds. Records of 'red-heads' (females or immatures) outnumber those of adult drakes by more than two to one.

Smews occur annually in varying, but usually very small, numbers and most often between November and March. Although most have left by the end of April, occasional birds stay well into the summer and there are records for every month. As individuals vary in behaviour, some remaining on a single water for only a day and others for weeks or even months, it is difficult to assess the total involved at any one time. Some indication of annual fluctuations can, however, be obtained from the numbers of records submitted, which in the period 1961 to 1983 ranged from 12 (in seven counties) to c50 (in 20 counties). The largest recent influx was in the hard winter of 1981/82. Occurrences are most frequent in the Central Lowlands and northeast, while in Shetland the Smew now seems to appear more frequently than in the past. There have been records since 1960 from all mainland counties except Banff and Nairn, but few from the northwest or the west coast islands. Although most are seen on still freshwaters, ranging in size from pools to large lochs, Smews also occur on rivers, estuaries and on the sea.

The Smew breeds in northern Fenno-Scandia and Siberia and winters in Europe west to the North Sea, with a major – and growing – concentration in the Netherlands. In Britain it occurs most regularly in severe winters and in southeast England – but even there numbers have declined markedly in the last two decades.

Red-breasted Merganser *Mergus serrator*

Widely distributed on both fresh and tidal water, breeding most abundantly in the north and west. In late summer moulting flocks gather on the coast; the largest wintering concentrations, which include immigrants, are in the Moray Basin and the Firths of Tay and Forth.

The gradual spread of this species in Scotland is documented by B&R, and its distribution in the late 1950s by Mills (1962). By 1960 it was breeding in all counties north of the Clyde and Tay, and in Renfrew, Ayr and the three southwest counties. Atlas survey work produced the first breeding record for Kinross, at Loch Leven in 1971, and suggested a continuing increase in breeding numbers in the southwest and also in the Hebrides and Northern Isles. Confirmation of breeding in Peebles in 1979 is the only further

extension of range recorded. Red-breasted Mergansers nest near clear water – fresh or salt – which is well-stocked with small fish. The absence of breeding birds from many areas in south and central Scotland probably reflects a shortage of suitable habitat, but may also be due to persecution. If the latter has been a significant limiting factor in the past, some increase in numbers may be expected to follow implementation of the 1981 Wildlife and Countryside Act, in which this species is no longer classed as a 'pest'. Mergansers can still be killed under licence, however, and many are shot annually on the major salmon rivers.

Little information is available on breeding densities and most reports simply use descriptions such as 'common', 'thinly distributed' or 'small numbers'. The Red-breasted Merganser is widespread in Shetland, Orkney, the Hebrides and all the Highland counties, where it is often the commonest duck around the sea-lochs and islands. It is fairly plentiful in Moray and Nairn but is much scarcer in the northeast, where there are few recent proved breeding records. Knox (1977) has suggested that a decline is possibly now taking place in this area, which was colonised only during the first half of this century. In the Clyde Islands an increase in the late 1940s was followed by a decline in the 1960s (Gibson *et al* 1980). In Ayr only small numbers breed, but in Wigtown, Kirkcudbright and Dumfries there appears to have been a considerable expansion of the population since Mills' study in the late 1950s. The current estimate of the British breeding population is 1,500–2,000 pairs (Owen *et al* 1986); Scotland holds the bulk of this population, possibly in the region of 1,200–1,700 pairs.

From May to September birds are present at moulting areas around the coast, numbers generally being at their peak in late August. Since 1976 flocks of international importance (more than 400) have been recorded in the Sound of Gigha (max. 1,700), off the Kincardine / Angus coast (1,000 in 1980) and off Tentsmuir (560 in 1983), and flocks of more than 50 in Wester Ross (Gairloch / Little Gruinard Bay), off the Ythan Estuary and the Ayr coast, and in Orkney (Echnaloch Bay). In Shetland, where flocks seldom exceed 20, moulting parties of 30+ are present during August in Catfirth and Dales Voe on the east side of Mainland. From mid September the moulting flocks disperse and move to the wintering grounds.

In autumn immigrants from Iceland arrive to join the native birds, which are believed to form the bulk of the wintering population. Ringing recoveries show that Icelandic birds are present from October to the end of March and it is likely that those seen on passage at Fair Isle from early September are from this source. Iceland-ringed birds have been recovered in the Northern Isles and Hebrides and on the mainland as far south as the Tay. Recoveries show that there is also some emigration across the North Sea, with Scottish-ringed birds subsequently found in Norway and Denmark; regular or periodic mixing of the widespread northwest European population may occur in this way.

In winter more than 90% of Red-breasted Mergansers are on coastal waters, the majority concentrated in estuaries where food is plentiful. The maximum recorded in wildfowl counts in 1976–84 averaged c1,800 and ranged from 1,200 to 2,500 but the actual population must at times be very much larger, as many areas holding this species are not regu-

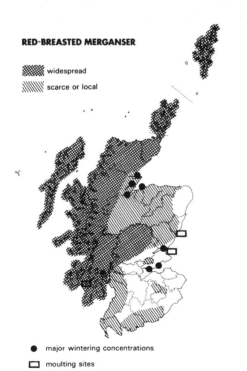

RED-BREASTED MERGANSER

▨ widespread

▧ scarce or local

● major wintering concentrations

▢ moulting sites

larly counted and even the major sites are not always included in the monthly wildfowl count returns. The current estimate for the British wintering population is 6,000–10,000 (Owen *et al* 1986); Scotland probably holds between half and two-thirds of these birds. Trend data for Britain as a whole show a marked recent increase, but in Scotland – where coverage is not complete enough to allow valid comparisons between years – the figures suggest that considerable fluctuations occur, probably due to variations in the size of the influx from the Continent, which does not take place until well into November. An unusually big build-up occurred in the Beauly Firth in late 1980, when large shoals of sprats and herring were present; the Merganser flock there increased from 350 in early November to peak at 2,250 on 25 December (SBR). A similarly abnormal influx was recorded on the Forth in February 1970, when 3,000 were counted between Musselburgh and Portobello. The very marked increase in mid-winter numbers in the late 1970s was not paralleled by the September figures (when virtually all the birds present can be assumed to be native), suggesting that the composition of the population, in terms of origin, may change as winter progresses.

Even without such spectacular influxes, the Moray Basin as a whole supports wintering flocks of international importance; the Beauly, Cromarty and Dornoch Firths and the outer Moray Basin all contribute significantly, but not necessarily at the same time. Maximum counts for these four areas in 1976–81 were: Beauly 2,250, Cromarty 500, Dornoch 860, and outer Moray Basin 800, at Culbin Bar. The Firth of Forth also holds internationally important numbers in winter, the main concentrations generally being in the Kinneil/Culross and Dalgety/Kinghorn areas (Campbell 1978). Prater (1981) gives a highest average monthly count of 420 in October for the Forth; since 1976 few complete counts have been made but Salmon (WWC) gives an average peak of 870 for 1978–83. Flocks of 500+ have also been reported from the Eden/Tentsmuir coast and the Tay, while the Montrose Basin sometimes holds 200–400. Several hundred (each) winter in Shetland, Orkney and the Inner Hebrides; numbers in the Outer Hebrides are said to be highest in the autumn (Hopkins & Coxon 1979), so presumably some birds leave to winter elsewhere. There are flocks of 200+ around Loch Ryan and 100+ on the Ayr coast. Few figures are available for other parts of the west coast, where totals present in the sea-lochs and around the small islands may be considerable, although few flocks of any size are present. Elsewhere flocks of more than 50 are comparatively uncommon.

Goosander *Mergus merganser*

Breeds abundantly over much of northern Scotland and more locally in the south, but not in the Outer Hebrides, Orkney or Shetland. More widely distributed in winter, when the largest concentration is on the Beauly Firth. Moulting birds are present from May to September on both fresh and tidal water.

The Goosander, which nests in holes – either in trees or on the ground – near large, clear lochs or fast rivers, first bred in Scotland in 1871, in Perth. Its subsequent spread up to 1977 is documented by B&R and by Meek & Little (1977a), who disputed Sharrock's (Atlas) conclusion that southern Scotland and northern England are now the species' stronghold in Britain. The data supplied to Meek & Little by local recorders indicated that the original areas of colonisation in Perth, Argyll and Wester Ross still hold large numbers of breeding birds. By the mid 1970s breeding had been recorded in all mainland counties north of the

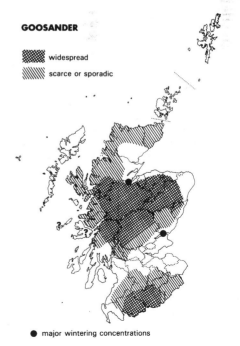

GOOSANDER
widespread
scarce or sporadic
● major wintering concentrations

rivers Tay and Clyde except Caithness, where breeding was suspected in 1977 and proved in 1978. South of the Clyde there is no proof of breeding in Renfrew, and numbers remain small in Lanark, Ayr and Wigtown. The species is well-established in Kirkcudbright and Dumfries, and is increasing steadily in the Borders, but there are no definite breeding records from the Lothians, Fife or Kinross. A reference by Mills (1962) to breeding at Loch Maddy, North Uist, is the only record from the Outer Hebrides. A few pairs nest on Skye and there are old records from Mull and Tiree but no proved breeding on Islay. Shortage of the preferred habitat is presumably a limiting factor in the distribution of the Goosander.

The most recent estimate of the Scottish breeding population was 740–950 pairs in 1976 (Meek & Little 1977a), representing 75–80% of the British total – which forms only a small proportion of the northwest European population. A density figure of one or two pairs per 16 km of river has been quoted for Sutherland, Ross and Aberdeen (Atlas), while the number of pairs per county or group of counties gives some indication of the relative abundance of the Goosander in different areas in the mid 1970s (Table). There is evidence of recent increase in the Borders – where the current estimate is 100+ pairs (R. D. Murray) – and at Loch Lomond and possibly also in the southwest. It seems that the total may now be approaching 1,000 pairs.

During the summer both broods and non-breeding birds move down the rivers and gather to moult on undisturbed fresh or tidal waters. No systematic search for moulting grounds has been carried out but flocks of 50–150 have been seen between May and September on Loch Ken, Loch Doon (Ayr), several Border reservoirs, Tyninghame Estuary, Castle Forbes Loch (Aberdeen), Speymouth and the Beauly

Firth. In 1984 evidence of long-distance moult migration by Goosander drakes was obtained when birds caught and dye-marked in Roxburgh during May were sighted in Finnmark several weeks later (Little & Furness 1985).

In winter the Goosander is widely dispersed but much more abundant in the south and east of the country than in the northwest. Although the bulk of the wintering population is probably made up of native birds, ringing recoveries have shown that there is a marked movement from northern England into the south and southwest of Scotland, with some birds travelling north as far as Sutherland (Meek & Little 1977b), and also that some immigrants from Scandinavia winter here (one ringed in Sweden in July was shot in Perth in December the same year). The wide fluctuations that occur in winter peak counts are presumably due to variations in the number of continental immigrants reaching this country, possibly reflecting the severity of the winter around the Baltic. In most winters Scotland holds more than two-thirds of the British total, estimated as c5,000 native birds plus variable but small numbers of immigrants (Owen *et al* 1986).

The Beauly Firth is the only site in Britain currently of international importance for this species and numbers there do not exceed the qualifying level of 750 in every season. From 1976–84 the average peak was c1,370 and the range 355–2,400 (February 1983). Numbers on the Tay were high (800–1,000+) between 1972 and 1974 but have since been much lower. Gatherings of 50–100 and occasionally more occur regularly in winter on the Loch of Skene, Castle Loch (Dumfries), Loch Leven, the Loch of Strathbeg and several sites in the Borders; flocks of more than 50 have also been recorded in Renfrew (Black Loch and Walton Dam), Ayr (Glenbuck Loch), and on the Forth at Tullibody Island. Some of the more important freshwater sites are used only as roosts, the birds dispersing to nearby rivers during the day. There has been a marked overall increase in September numbers since the early 1970s, but a less consistent increase in mid winter (Wildfowl Trust unpub.); this may be a reflection of the larger numbers of home-bred birds now present in autumn.

The Goosander has been much persecuted on account of its diet, which includes a substantial proportion of salmon and trout parr (Mills 1962). For some years a bounty system was in operation on several important salmon rivers and on many this species continues to be shot – both legally under licence and illegally. In Scotland it was classified as a pest species, and consequently was not afforded any protection under the 1954 Protection of Birds Act, but despite persecution it has been able to extend its range. With breeding season protection now granted to it under the Wildlife and Countryside Act 1981 we may expect to see a continuing increase in the population, although the fact that in the Borders nearly 100 were shot under licence between February and September 1983 demonstrates that Goosanders still suffer persecution in some areas.

Goosander breeding numbers and distribution in 1976. (Data from Meek & Little 1977a)

	Est. no. of pairs
Shetland, Orkney, Caithness, O. Hebrides	no breeding records
Sutherland & Ross	55–75
Inverness – mainland	100–120
– Skye	6–10
Nairn, Moray, Banff	50–100
Aberdeen & N. Kincardine	60–100
S. Kincardine & Angus	15–25
Perth	200+
Argyll & I. Hebrides (mostly on mainland)	50
Dunbarton, W. Stirling, Renfrew, Lanark, Ayr, Arran, Bute (mostly around L. Lomond and in S. Ayr)	10–20
Clackmannan & E. Stirling	only 1 recent record
Kinross, Fife, W. Mid- & E. Lothian	no proved breeding
Berwick	c12
Peebles, Roxburgh, Selkirk	c35
Dumfries	c100
Kirkcudbright & Wigtown (few in Wigtown)	50–100
Total	736–952 pairs

Ruddy Duck *Oxyura jamaicensis*

Feral breeder (Category C)

The British population of this North American species originated from free-flying young reared at the Wildfowl Trust's Slimbridge collection. Feral breeding in England dates from 1960 and the species was admitted to Category C of the official British and Irish list in 1971 (Hudson 1976). The first Scottish record to be accepted as referring to a feral bird rather than an escape was of a male in Unst, Shetland, on 16 May 1974. There were no further records until 1979, when a pair was on the Loch of Kinnordy, Angus, from late May and two ducklings were seen on 14 July. Ruddy Ducks have bred there annually ever since but most ducklings have disappeared at an early age, probably due to predation by pike. By the end of 1983 there had also been records from Orkney, Aberdeen, Perth, Lanark, Berwick and Dumfries, and in 1984 three broods were reared on the Loch of Strathbeg. The release of Ruddy Ducks into the wild is now prohibited under the Wildlife and Countryside Act 1981.

Honey Buzzard *Pernis apivorus* [1]

Scarce passage visitor. Has bred.

The Honey Buzzard has been reported with increasing frequency in the last few years. Prior to 1968 it was recorded irregularly and annual totals never exceeded three; from 1968 to 1983 it occurred every year, the annual total usually being between two and nine, but reaching 27 in 1981. Most records since 1968 have been from the Northern Isles and the remainder mainly from the east coast, though there have been occurrences in Inverness, the Angus glens, Midlothian, Peebles, Roxburgh, Ayr, mainland Argyll, and Mull. The first record for the Isle of May was in June 1983. The earliest spring sighting is 22 April, more than half the records are for May/June, and the rest are fairly evenly spread from July to late September, with stragglers into the first week of October.

A bird of fairly open mature deciduous woodland, the Honey Buzzard breeds over much of Europe and eastern Scandinavia. It feeds largely on insects and winters in equatorial Africa. Very small numbers, possibly not more than 12 pairs, nest in the New Forest and elsewhere in southern England. B&R give 19th century breeding records for Aberdeen and Easter Ross, and a pair possibly nested in northeast Fife in 1949 (SB 2:142).

Black Kite *Milvus migrans*

Vagrant (Eurasia) – recorded in six years since 1950, once earlier.

The first record of the Black Kite was in 1901, when a male was shot in Aberdeen (ASNH 1901:133), and there were no further reports until 1966. Since then there have been records in 1966 (Orkney & Shetland – probably the same bird), 1968 (Orkney), 1970 (Orkney), 1975 (Orkney & Shetland – possibly the same bird), 1976 (North Rona), and 1979 (Aberdeen). Most occurrences have been during the spring migration period, April/June, the exception being an Orkney record in late September 1970. The Black Kite has recently shown a slight north-westerly extension of its breeding range and there has been a corresponding increase in British occurrences, from none in 1958–65 to an average of nearly seven a year in 1979–82.

Red Kite *Milvus milvus* [1]

Vagrant (Europe, including Wales) – recorded in seven years, 1950–83. Formerly bred.

The speed and completeness of this once-abundant bird's extirpation by persecution has been documented in detail by B&R. In most areas the Red Kite was extinct well before the end of the 19th century, but a pair is said to have bred in Glen Garry in 1917; this is by far the latest Scottish breeding record. The 40 or so pairs in Wales now comprise the entire British breeding population.

The post-1950 occurrences have been in 1958 (Aberdeen), 1969 (Sutherland, Angus & Kinross), 1972 (Inverness), 1974 (East Lothian, Aberdeen & Angus), 1975 (Aberdeen, Dumfries & Kirkcudbright), 1980 (Inverness & Banff) and 1983 (at an undisclosed site in southern Scotland from late March to early May). There have been reports in all months except July, September and November. Two were seen together in Angus in April 1969 but all other records are of single birds; it is possible that successive reports from different localities refer to the same individuals. The Kirkcudbright bird, which regrettably was shot, had been ringed in Wales earlier the same year, but some Scottish vagrants may originate from elsewhere in the European breeding range, which extends, discontinuously, from Iberia to south Sweden.

White-tailed Eagle *Haliaeetus albicilla*   [1]

Recent reintroduction; formerly resident breeder

The persecution and gradual extermination of the White-tailed eagle, which was once quite widespread in Scotland, has been documented in detail by Love (1983). The last recorded breeding was in 1916 on Skye; thereafter only occasional vagrants were reported until the reintroduction project started in 1968.

The first reintroductions were made on Fair Isle, which appeared to offer an abundance of suitable prey – seabirds, fish and rabbits – while being sufficiently isolated to discourage this largely sedentary bird from wandering. Four young eagles were flown in from Norway and released after acclimatisation; two vanished within a year. The remaining pair started to kill adult Fulmars, a prey species which was not in the past available in most areas occupied by breeding White-tailed Eagles, and the second female also disappeared. In the August following its release the last bird, a male, was found heavily-soiled with Fulmar oil and is thought to have died a few weeks later (SBR 1969).

In 1975 the project was resumed on Rhum, where there are few Fulmars but abundant other seabirds plus deer and goat carrion. Four eaglets were imported from Norway in 1975; the only male died before release and one female was killed in Argyll soon after, apparently by flying into power cables. The other two females wintered on Rhum and then moved away. From four to ten eaglets have since been imported annually – a total of 65 by the end of 1984. Many of the birds have been located later (c40 in 1984) living wild in the western Highlands, indicating that the survival rate is quite good. Individuals have travelled as far north as Shetland and Fair Isle, immatures have been found dead in Caithness (poisoned) and Skye, and others have been seen in Mull, Canna, North Uist, Lewis, Wester Ross, Sutherland and Orkney, while some have remained on Rhum.

White-tailed Eagles do not breed until they are five or six years old, and the first of the Rhum birds attempted to breed in 1983; two clutches were laid both that year and in 1984. In 1985 four nests with eggs were located and a

single chick was reared. Should the species again become firmly established its presence is likely to affect local populations of potential competitors, such as Golden Eagles, in the areas colonised and perhaps also those of prey species. On Canna Fulmar numbers are declining on a cliff used for roosting by a White-tailed Eagle (R. L. Swann).

This species has been subjected to persecution in other parts of Europe too, and declined in many countries during the late 18th and 19th centuries. More recently reduced breeding success, due to pollution with organochlorines and mercury, has led to further decline. Small numbers still breed in Iceland and around the Baltic, but Norway is the only west European country in which some local increase has been recorded (BWP).

Marsh Harrier *Circus aeruginosus*   [1]

Scarce passage visitor, most frequent in May and June; has attempted to breed. Occasional in winter.

The records suggest that the Marsh Harrier has occurred with increasing frequency during the last 40–50 years. B&R regarded it as an uncommon visitor and gave only seven records for the first half of this century, all of them between 1937 and 1949. There were reports in seven years in the 1950s, with a maximum of four in any one year; in nine years in the 1960s, annual maximum about nine; and annually in the 1970s, maximum 13–14. In 1980–83 between nine and 35 have been reported each year; the record total of 35 in 1982 comprised 23 spring records and 12 in autumn.

On migration Marsh Harriers hunt over damp fields, peat bogs and river margins, as well as over the extensive marshes upon which they depend for breeding. Since 1950 passage birds have been recorded in almost every county and island group; well over half the records are from the east and north, but a fair number of birds reach the west coast and the Hebrides. Nearly 40% of sightings are in May and most of the rest between April and September, but there are a few records for October/March, some of them involving birds which spent the winter in one area. Most reports are of single birds, but up to three are sometimes seen together; females and immatures greatly outnumber adult males.

There is a small breeding population of Marsh Harriers in southeast England, where numbers have been increasing as a result of protection from disturbance and conservation of the important reedbed habitat – a stretch of aquatic vegetation covering at least 100 ha is required for successful breeding. On the Continent, as in England, drainage and persecution caused massive declines in the late 18th and early 19th centuries. Recently, however, the development of new polders has led to an increase in the Netherlands population, the breeding range has extended northwards in Finland, and Marsh Harriers bred in Norway for the first time in 1975 (BWP). Further south and east in Europe numbers are continuing to decline. Birds from the northern part of the range are largely migratory, moving south to winter around the Mediterranean and in Africa.

The Marsh Harrier has not yet been confirmed as breeding in Scotland but has been suspected of doing so on several occasions, for example at Earlshall, Fife, in 1937 (SB 2:142), and at undisclosed sites in 1966 and 1969 (SB 5:25,466). The most positive evidence of attempted breeding so far was obtained in 1980, when a pair in Aberdeen were seen carrying nesting material. Unfortunately one of the birds was shortly afterwards found dead from poisoning; it had been ringed as a nestling in the Netherlands three years previously (SBR 1980). (A Danish-reared three-year old was recovered in Orkney in April 1944.) This species is extremely sensitive to disturbance; anyone finding a pair apparently nesting should avoid publicising their whereabouts.

Hen Harrier *Circus cyaneus*   [1]

Has recolonised much of the mainland after almost complete extermination as a breeder earlier this century and now breeds in nearly every county containing suitable habitat, but is still persecuted in some areas. Many birds move south in winter, when some of those remaining roost communally. Also a scarce passage visitor.

The progressive extermination of the Hen Harrier from mainland Britain, with the possible exception of Kintyre, has been documented by Donald Watson (1977) in his comprehensive monograph on the species; the present account draws heavily upon his work. By 1900 the only districts with viable populations were Orkney and the southern Outer Hebrides, both areas where gamekeeping was minimal. Egg collectors had been responsible for some of the decline, while on grouse moors the Hen Harrier had been persecuted to the point of eradication on the grounds that it not only took grouse chicks but also frightened birds during drives and so reduced shooting bags. With the decrease in gamekeeping during the 1939–45 war the fortunes of the Hen Harrier improved and by 1950 breeding had again taken place in several mainland areas. The rapid spread of afforestation during the early 1950s provided safe breeding grounds when persecution started to increase again on the grouse moors, and over the next two decades there was a substantial expansion in both numbers and range.

Most Scottish Hen Harriers breed on moorland or in young conifer plantations at 200 m to 300 m asl. Local populations fluctuate with the availability of young plantations and the amount of prey-rich open country accessible for hunting. Harriers often abandon conifer plantations when they reach the thicket stage, but Donald Watson has found that in southwest Scotland they may continue to nest in forests well over 20 years old, in which most trees are over 6 m high. Such continuity of nesting apparently occurs only where hunting-grounds on open moorland are available nearby. Although afforestation is currently increasing the habitat available for colonisation in some areas, such as Sutherland, the Ochils, and parts of Perthshire, there is little doubt that, despite its illegality, persistent persecution by shooting, trapping and poisoning on many grouse moors is still limiting the Scottish Hen Harrier population. There has nevertheless been an impressive increase in numbers and expansion of breeding range since B&R said in 1950 'Of late the Hen Harrier is attempting to recolonise the country and we greatly hope it will succeed'. The progress of recolonisation and the current situation in different areas are summarised in the Table.

Detailed studies of the Hen Harrier have been carried out in several well-separated areas. In Orkney, work on breeding biology, food and feeding behaviour was undertaken for many years by the late Eddie Balfour (Balfour & Cadbury 1974, 1979) and was continued to 1981 by Nick Picozzi (Picozzi 1984a,b; Picozzi & Cuthbert 1982). In Kincardine, Picozzi also studied harrier populations on a grouse moor without predator control and concluded that harrier predation may have reduced the numbers of grouse available for shooting in August by not more than 7.4% (Picozzi 1978). Among the results of Picozzi's studies are data on the polygynous habits of the species; up to six females associated with a single male in Orkney, but birds in Kincardine were mostly monogamous (Picozzi 1978, 1984b). In winter communal roosts are formed in some areas; these are usually

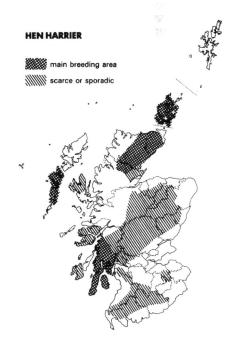

HEN HARRIER

▨ main breeding area

▧ scarce or sporadic

*Recolonisation dates and breeding numbers of Hen Harriers *(Data from A. D. Watson, SBR & local recorders)*

	Approx. date of recolonisation	Population est. in early 1980s
Orkney	(continuous breeding)	c110 breeding birds
Outer Hebrides	(continuous breeding)	c10–12 pairs
Inner Hebrides	?	c50–55 pairs
Clyde Islands	1953	c25–30 pairs
North (Sutherland, Caithness, Ross & Crom)	Sutherland 1946	} 30–35 pairs
Central Highlands (Inverness, Perth)	*Inverness – by 1962 *Perth – by 1946	} 45–60 pairs
Eastern Highlands (Moray, Nairn, Banff, Aberdeen, Kincardine, Angus)	*Moray & Nairn 1944 *Angus 1951, *Kincardine 1953 *Aberdeen 1957	} 35–40 pairs
SW Highlands (Argyll, Dunbarton, W. Stirling)	Stirling 1941 Dunbarton 1947 Argyll 1949	} 40–60 pairs
Southwest (Renfrew, Lanark, Ayr, Kirkcud., Wigtown. Dumf.)	Kirkcud. by 1959, *Renfrew? *Ayr 1960, *Dumfries 1961 *Lanark by 1964, Wigs. by 1965	} 30–45 pairs
Southeast (Peebles, Lothians, Berwick)	*Peebles & *Lothians c1968–72	} 2–4 pairs

* Persecution continues in the counties marked with an asterisk.

small but in the west, southwest and Orkney may involve up to 20–30 individuals (A. D. Watson). Southwest Scotland is the only area where adult males often equal or outnumber 'ringtails' (females and immature males) in winter.

Donald Watson's analysis of ringing recoveries showed that many young birds do not return to their natal area in the first summer after fledging, as demonstrated by some 56% of May / August recoveries of this age group being more than 80 km from the place of ringing (the comparable figure for older birds was only 14%). First-winter birds have been recovered in widely scattered areas: on the Continent from Norway to the Pyrenees, in England south to Norfolk, in Eire, and in many parts of Scotland, but not all disperse so far from their natal area. Young males are apparently likely to disperse further than females (Picozzi 1978). Distant recoveries of older birds have been in Norway, Germany, the Netherlands, England and Eire. Most birds breeding in Orkney are now known from their rings to have been reared on the islands, but in 1984 a female reared in Orkney, which had nested there annually from 1975–81, was found breeding in Argyll (Picozzi & Watson 1985). Balfour & Cadbury (1979) considered that some immigration, especially of males, was necessary for maintenance of the Orkney population; recruitment from elsewhere must be particularly important after a season such as that of 1983, when many young died due to food shortage (SBR).

Although no foreign-ringed Hen Harriers have been recovered in Scotland, it is possible that some winter immigrants from across the North Sea do remain here to breed. The continental breeding range is discontinuous, with scattered populations in southern Norway, Germany, Denmark and the Low Countries, and more extensive breeding areas

in northeast Fenno-Scandia and Russia, and in France and northern Spain. Birds from the more northerly and easterly parts of the range winter largely from the Baltic south to the Mediterranean. Because native harriers disperse so widely, it is not possible to be certain if those occurring in Shetland (where breeding has not been recorded this century) are of continental or British origin. Most such birds are 'ringtails' and appear in September / October, although there are occasional records for most months.

***

Pallid Harrier *Circus macrourus* [1]

Vagrant (E Europe) – one record, 1931.

The only accepted record of the Pallid Harrier is of a second-year male shot on Fair Isle on 8 May 1931 (SN 1932:1). Published reports of birds on Fair Isle in May 1942 and October 1949, and on Shetland Mainland in August 1964, were subsequently rejected.

***

Montagu's Harrier *Circus pygargus* [1]

Vagrant (Europe, including England) – recorded in nine years since 1950. Has bred.

This is probably Britain's rarest diurnal bird of prey, with fewer than five pairs now breeding annually. The old records suggest that sporadic nesting occurred in Scotland in the past but the only proved breeding this century has been in Perth in 1952, 1953 and 1955, and in Kirkcudbright in 1953 (SB 2:349). The female of the Kirkcudbright pair had been ringed as a nestling in Anglesey two years earlier. The breeding habitat of this species is more akin to that of the Hen Harrier than the Marsh Harrier, and the Scottish pairs have nested on heather moorland. Although the Montagu's Harrier is afforded full protection its numbers have declined progressively since the 1950s (Parslow). It has been suggested that climatic variations may be responsible for fluctuations in the northwest limits of the species' breeding range (Harrison 1982).

All the records away from the breeding sites are for late April/May, at which time the birds are moving north from their wintering grounds in Africa and the Mediterranean basin. Since 1950 there have been reports at this season from Renfrew (1950), Shetland (1954 & 1982), Aberdeen (1963), Sutherland (Handa 1979) and the Isle of May (1980). Four of these birds were males, three of them in adult plumage.

Goshawk *Accipiter gentilis* [1]

Scarce resident, breeding at widely scattered localities. Scarce and irregular passage visitor.

The past history of the Goshawk is unclear, owing to problems of misidentification, but the species was still breeding in Perth in the 1880s (B&R). Persecution eventually led to its extermination, and during the first half of this century it was seen only occasionally, such sightings probably relating to falconers' birds at least as often as to migrants. From the early 1950s small numbers were regularly seen in summer in two areas, and from 1964 display and territorial behaviour were observed (SBR 1969). Breeding probably occurred in the central Highlands in 1971 and 1972 and was reported to have done so in 1973; flying young were seen in 1974, by which time nesting was suspected in several areas. Since then numbers have continued to increase slowly: there were reports from 17 mainland counties in 1979, and in 1982 five pairs reared 17 young and at least two further pairs were known to be present. A recent assessment puts the total British population at around 70 pairs and increasing (Newton 1984).

Marquiss & Newton (1982a), reporting on the current status of the Goshawk in Britain, pointed out that this species is normally sedentary; only those in the more northerly part of the range, which covers much of Eurasia, move southwards during the winter. Immigration is consequently unlikely to augment the existing small, and very localised, breeding groups, which are believed to derive from escapes and releases. There have been several recoveries recently of Scottish-ringed birds; few had travelled far from the ringing site but one bird marked in the southwest during its second year was shot in Angus two months later. In many areas Goshawks, despite being protected by special penalties under the 1981 Act, are still heavily persecuted by gamekeepers, by both shooting and pole-trapping. Marquiss & Newton found, however, that although they certainly take some gamebirds, especially partridges and Red Grouse, they also prey extensively on pigeons and rabbits. Substantial numbers of Goshawk eggs and young are taken by collectors and falconers. These pressures are sufficiently great to curtail severely this species' potential for population expansion, and even in some areas to put its continued survival at risk.

Presumed migrants are occasionally recorded in the Northern Isles, Inner Hebrides and Clyde Islands, but the only record from the Outer Hebrides is an old one. One such bird, on North Ronaldsay in the summer of 1981, died as a result of Fulmar oiling (SBR 1981).

The old record of a bird of the American race *A.g. atricapillus* is not among those listed in the BOU's *Status of Birds in Britain and Ireland*.

Sparrowhawk *Accipiter nisus*

Resident breeder, widely distributed on the mainland and regular on the Clyde Islands and Inner Hebrides. The population has now recovered from the pesticide-related decline of the early 1960s, which was not so marked in Scotland as further south. A passage and winter visitor in small numbers.

Sparrowhawks nest in woodlands thick enough to provide good cover yet sufficiently open to permit easy flight between the trees, and often hunt over more open adjacent country (Newton *et al* 1977). They can find suitable conditions in Highland birch and pinewoods as well as in conifer plantations and mixed woodlands on low ground. Although a new nest is built each season, the birds occupy traditional territories, abandoning them only when the habitat becomes unsuitable. Not all territories are occupied annually, however, occupancy rates ranging from 32% to 78%. In low-ground areas with game interests Sparrowhawks were formerly much persecuted, and the wartime reduction in keepering resulted in a marked increase in the population.

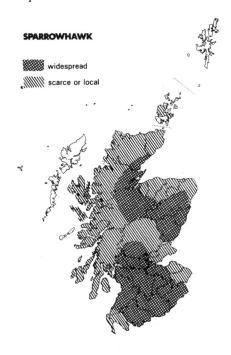

SPARROWHAWK

▓ widespread

░ scarce or local

Numbers were falling again by the late 1940s and in Britain as a whole reached an all-time low in the 1960s, following the widespread use of organochlorine pesticides on arable farms.

In Scotland this decline was comparatively slight, and was most apparent in the Central Lowlands (Prestt 1965), but in parts of England it was so great that in 1963 the Sparrowhawk was given legal protection for the first time. Since pesticide restrictions were introduced there has been a progressive increase in the British population, although DDE continues to cause eggshell thinning in some areas (Newton & Haas 1984). As the amount of suitable habitat in Scotland is now probably greater than at any time in the past, the population is also likely to be at its maximum to date; continuing afforestation should enable yet further expansion to take place in the next few decades.

In many districts Sparrowhawks are thinly scattered, but recent subjective assessments by local recorders suggest that there has been an increase since 1968–72. The words 'widespread' or 'common' were used even in relation to areas, such as East Lothian, where there were few reports of proved breeding in the Atlas period; in the Lothians the increase has occurred largely during the early 1980s. In 1984 more than 12 pairs were known within the Edinburgh city limits (G. Carse). Despite continued persecution, numbers have increased since the early 1970s on Bute (Gibson *et al* 1980) and at least 12 pairs were known on Arran in 1982. On the Inner Hebrides breeding occurs regularly on Islay, Jura, Mull and Skye, and sporadically on Rhum, Raasay, and Canna. Nesting has not been recorded since 1950 in the Outer Hebrides, though the presence of a pair in Stornoway Woods in 1979 and 1980 suggests possible breeding. Successful breeding in Orkney in 1983 was the first reported for many years. There is no record of nesting in Shetland.

Studies by Ian Newton and others have shown that in continuously suitable nesting habitat Sparrowhawk territories are regularly spaced, with nearest neighbour distances ranging from 0.6 km in Annandale and Eskdale to slightly more than 2.0 km in Upper Speyside and Mar Forest (Newton *et al* 1977). Very high densities (equivalent to a maximum of 278 pairs/100 km^2) were found in some districts, but since habitats holding such high concentrations were limited in extent, densities over the whole of each study area (all of which included some farmland) were much lower, ranging from 14 pairs/100 km^2 in upper Speyside to 89 pairs/100 km^2 on lower Deeside. Densities were greatest on productive land at low altitudes, probably because prey is more abundant in such situations. From a survey of Dumfries woodlands, Newton (1972a) concluded that, in a mixed landscape situation, Sparrowhawks preferred to nest in woods not less than 20 ha in size and composed either entirely of conifers or of mixed conifers and hardwoods. However, where there is little choice, Sparrowhawks will nest in tiny scraps of woodland, thick hedges, parkland and thickets of willow, hazel or hawthorn.

Female Sparrowhawks hunt over open ground more often than do males, and take larger prey, including occasional items heavier than themselves, eg rabbit, Woodpigeon, Red Grouse and Pheasant. The principal prey species (in terms of weight consumed) are Woodpigeon, Blackbird and Song Thrush, with Fieldfare and Redwing also important in winter. Laying dates are closely related to the availability of fledgling songbird prey (Newton & Marquiss 1982).

After fledging Sparrowhawks disperse in all directions, making the longest movement of their life within days or weeks of leaving their parents' territory (Newton & Marquiss 1983). Longer movements are made from upland than from lowland areas; among the most distant recoveries is that of a Sutherland-reared chick found in Northern Ireland three months after ringing. Young males stay closer to their birthplace than females, and most Sparrowhawks settle to breed in the general area in which they were reared. Few adults travel far from their nest sites.

On passage through the Northern Isles (where there is virtually no risk of confusion with native birds) Sparrowhawks occur most regularly in late April/early May and between mid September and mid October, but there are also occasional records throughout the winter. Numbers are generally small, with a maximum of only three on Fair Isle at one time. Most immigrants are presumed to come from the northern part of the continental breeding range, which extends well north in Fenno-Scandia; much of the Scandinavian part of the range is deserted in winter. Recoveries of foreign-ringed birds confirm that some come from Norway, but the only record involving a bird ringed as a chick is from Denmark. Sparrowhawks marked on autumn migration on Fair Isle have been found later the same winter in places as widely scattered as Nairn, Angus, the Netherlands and France.

[Red-shouldered Hawk *Buteo lineatus*]

The old report of this American species, square-bracketed by B&R, remains unaccepted.

[Red-tailed Hawk *Buteo jamaicensis*]

An immature Red-tailed Hawk, imported from America for falconry, escaped in Midlothian in January 1968. The following spring it paired with a Buzzard; four eggs were laid but the nest was robbed, probably by crows, just as hatching started (SB 6:34–37).

Buzzard *Buteo buteo*

Resident breeder, widely distributed on the mainland and larger Hebridean islands but absent from predominantly arable areas and scarce where persecuted. A scarce and irregular passage and winter visitor.

Buzzards occupy a variety of habitats, their main requirements being open ground for hunting over and trees or cliffs for nesting on. They feed on varied live prey and carrion. Numbers were reduced nationally by the initial myxomatosis epidemic, which killed most rabbits (Moore 1957), and probably fluctuate locally with winter food supply (Weir & Picozzi 1983).

In the past persecution was the principal factor governing Buzzard distribution (Moore 1957). The progressive dec-

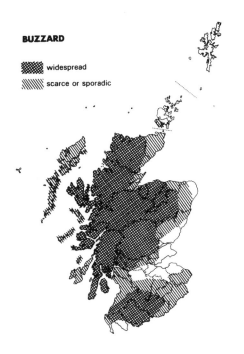

BUZZARD

▓ widespread

▨ scarce or sporadic

rease in gamekeeping this century has allowed an increase in some areas, but in others persecution continues and many Buzzards are illegally killed by poison baits often intended for crows or foxes. In 1978–82 one-third of the ringed Buzzards reported dead (from known causes) had been poisoned, shot or trapped (Taylor 1983). Of 52 Buzzards found dead in Speyside in 1964–72, 54% had been poisoned and 15% shot, while in one year gamekeepers on four estates killed 223 as they moved into vacant ranges (Picozzi & Weir 1976). In 1966–78 there were 25 known incidents of Buzzard poisoning in Scotland, involving 32 birds (Cadbury 1980a); many others have been reported since. One Moray estate had no occupied territories in 1976–80; in 1981 keepering ceased and by 1983 four pairs of Buzzards were breeding (B. Etheridge per K. Taylor).

Despite persecution, numbers have increased and distribution has spread during the past 40–50 years. B&R said that Buzzards no longer bred in Dumfries, 'all the counties in Tweed' or Caithness, and were scarce or absent in Wigtown, Kirkcudbright and Ayr. Much of the low ground in the southwest has been recolonised and breeding has been proved in all the Border counties since 1968. Some afforested uplands have been abandoned, presumably because continuous tree-cover reduces the availability of prey, especially rabbits. Few Buzzards now breed above 100 m asl in Kirkcudbright and Wigtown, where nests were recorded up to 500 m asl in the Atlas survey (G. Shaw per K. Taylor). In the Central Lowlands the Buzzard is a scarce or sporadic breeder. In the east Highlands, Badenoch and Strathspey were recolonised in the late 1940s and ground below 380 m asl was fully occupied by 1969–72 (Picozzi & Weir 1974); this process began later in Deeside and some other straths. The situation in the west Highlands is probably less changed than elsewhere, but there has been a marked increase in Caithness and parts of Sutherland.

Buzzards now breed regularly throughout the Inner Hebrides, except on Coll and Rhum, where nesting is sporadic (Reed *et al* 1983). In the Outer Hebrides nesting was first recorded early this century. By 1945 the Buzzard was described as 'firmly established', and was breeding from Lewis to Barra (B&R); it is now the most abundant raptor on the islands (Cunningham 1983). In Orkney, breeding first took place in 1961 (Balfour 1972) but still occurs only sporadically. There is no breeding record for Shetland.

Where mixed farmland is interspersed with scattered, relatively small, woodlands, Buzzards occur at high densities. In the BTO's 1983 survey densities of 25 occupied territories / 10 km^2 were recorded around Loch Lomond (Mitchell 1984), 22 / 10 km^2 in Moray and Nairn (B. Etheridge) and 14 / 10 km^2 in the Black Isle, where the nearest neighbour distance ranged from 400 m to 850 m (R. H. Dennis). Over much of Scotland Buzzard densities are considerably lower. In Glen Trool and the Cree Valley, where the woodland cover is more continuous, nests were spaced about 3,000 m apart (G. Shaw). In Speyside, there were three pairs / 10 km^2 between Boat of Garten and Kingussie in 1971 (Weir & Picozzi 1984), but at higher altitudes west of Kingussie density was about 1.5 pairs / 10 km^2 in 1975–82 (D. Weir). Some 1,000 km^2 in Lochaber held only 36–40 pairs in 1982 (J. Watson).

Native birds are largely sedentary, and less than 10% of

Scottish recoveries have been over 100 km from the nesting site. Young birds disperse in all directions after fledging, most recoveries being within 20 km of the nest and towards low-lying land (Mead 1973; Picozzi & Weir 1976). In the Northern Isles the Buzzard is scarce and very irregular on passage. There were no records for Fair Isle before 1949, but birds have since occurred almost annually, most often in August/October and May/June. Small numbers are recorded at North Sea oil installations, on which they occasionally land (NSBC). On the mainland it is not possible to distinguish immigrants with certainty, but occasional reports of several Buzzards together on the east coast (where the species is usually scarce or absent) may refer to birds from the Continent, where the breeding range of this race (*B.b.buteo*) includes southern Fenno-Scandia. Birds from the northern part of the range are migratory, wintering mainly between the Baltic and the Mediterranean; there have been no recoveries of foreign-ringed birds and the extent of immigration into Britain is thought to be small (BOU).

Rough-legged Buzzard *Buteo lagopus*

Scarce winter visitor, annual in very small numbers.

The Rough-legged Buzzard is most likely to be seen hunting over open farmland, moorland, coastal wetlands or dunes, and about 70% of reports come from the Northern Isles and east coast. The numbers reaching Scotland are usually very small, and the 35+ recorded in 1982 was more than double the figure for any other recent year; in many seasons only two or three are seen. The main arrival is generally in October or November, but there are a few September records and one in mid August (1982). Numbers often appear to decrease between December and April and then increase again slightly in May. Some birds undoubtedly remain in one area throughout the winter, but the records suggest that mid-winter arrivals also take place. Individuals are occasionally present in June, and in 1980 one was seen displaying in Inverness; July/August reports are infrequent. There has been no recent immigration of the scale noted in 1840–42, when it was said that 'at Dunbar, for example, twenty or thirty specimens were obtained by different collectors' (B&R), but the 1982 influx produced records scattered from Shetland to Berwick and west to Ayr, Argyll and Mull. There have been reports in other years from the southwest and the Outer Hebrides.

This is the most northerly of the Eurasian buzzards, breeding mainly in the tundra zone of Fenno-Scandia and northern Russia. It is migratory, wintering from the Baltic south to the Black Sea and seldom southwest of the Netherlands, Austria and Greece; Britain is consequently well to the west of its main migration route and wintering area. As with many other predators, its movements are influenced by the availability of prey, in this case lemmings and voles. A Rough-legged Buzzard ringed as a chick in central Sweden was recovered in Inverness during its first winter, suggesting Scandinavia as the probable origin of our visitors.

[Spotted Eagle *Aquila clanga*]

The record of a bird 'possibly of this species' on the Isle of May in September 1969 (SB 6:81) was not accepted by the Rarities Committee. B&R also square-bracket a report of one said to have been shot near Aberdeen in 1861; this is described as a 'probable' record in the BOU's *Status of Birds in Britain and Ireland*, which gives two accepted records (in England) early this century.

Golden Eagle *Aquila chrysaetos*   [1]

Resident breeder, most abundant in the western, central, and northwest Highlands and the Hebrides. Scotland holds almost all Britain's breeding Golden Eagles; the population has recovered well after intense persecution last century and a period of low breeding success in the 1960s.

Golden Eagles hunt over extensive, usually upland, 'home-ranges', varying from heather moors with abundant mountain hares and grouse to relatively barren mountainous areas in which sheep and deer carrion are major sources of food. A home-range may contain several alternative eyries, generally on cliffs but quite often in trees; the nest selected for use in the current year is repaired and fully lined early in the season. In any one year 10–25% of pairs may fail to breed, and there is some evidence that the non-breeding percentage is highest in areas where the food supply is poorest (Dennis *et al* 1984). Young eagles continue to be dependent upon their parents for about three months after fledging, and may remain with them until the following January or February. Thereafter they disperse, some moving to lower ground, often on the periphery of the main breeding areas; ringed birds from central Inverness had moved 50–80 km east or southeast to the grouse moors on which they were recovered (R. H. Dennis).

Golden Eagle breeding densities vary widely in different areas. In the east, where wild prey is abundant but the level of persecution relatively high, eagles breed at low density; in the west, where they feed largely on carrion, especially in winter, breeding densities are very much higher. These variations may reflect differences in levels of persecution as much as differences in the type and abundance of the food available. A ten-year study in the early 1960s, covering four widely scattered areas in the northwest, west and east Highlands but not including an eastern grouse moor, indicated that available food was always in excess of requirements (Brown & Watson 1964). In areas where carrion forms a large part of the diet it is possible, however, that seasonal shortages occur. Lockie (1964) suggested that better management of sheep and deer or increased afforestation were likely to affect food supply, through reduction in the availability of carrion in the former case and of live prey in the latter. Improved stock management, involving lower grazing densities, might in the long-term result in an increase in wild prey such as hares, but there can be little doubt that extensive afforestation in an eagle's hunting range will have an adverse effect on food availability.

Breeding success also varies widely, both regionally and between years. A survey of selected areas in 1964–68 showed marked regional variations over the 5-year period: the average numbers of young / breeding pair reared annually were: Deeside 0.82, Wester Ross 0.75, Perth / Strathspey / Monadhliaths 0.53–0.55, south Argyll 0.41, and Galloway 0.19 (Everett 1971a). Such variations are likely to be attributable to the quality of the food supply – live prey being superior to carrion – and to differences in the level of persecution. Regular monitoring of selected sites has confirmed the extent to which bad weather, including late snowstorms and cold, wet springs, can affect breeding success in this species, which is the earliest of the large raptors to lay. For example the 25–27 pairs of Golden Eagles monitored in the northeast reared 20–21 young in the mild spring of 1982 but only 14–17 in the cold, wet spring of 1983 (Payne & Watson 1983, 1984).

In 1982–83 the most comprehensive census yet carried out revealed some 424 pairs of Golden Eagles occupying home-ranges, with single birds present in a further 87 areas (Dennis *et al* 1984). The regional breakdown of the 1982 census results – in terms of numbers of pairs, occupancy of home-ranges, and productivity per pair – is shown on the map. At least 182 pairs bred successfully in 1982, rearing 210 young – an average of 0.52 / occupied home-range; this compares favourably with the theoretical figure of 0.5 believed necessary to maintain a stable population.

The threat posed in the 1960s by pesticide contamination of carrion, due to the use of organochlorine compounds, notably dieldrin, in sheep dip, is fortunately now past. Following voluntary withdrawal of this persistent chemical, breeding success improved markedly in western Scotland. From only 31% rearing young in 1963–65 the success rate rose to 69% in 1966–68, concurrent with a drop in the dieldrin level in eggs from 0.86 ppm to 0.34 ppm (Lockie *et al* 1969). Although dieldrin has been found in eagles in the eastern Highlands, which live largely on wild prey such as Red Grouse, Ptarmigan and mountain hare, levels were not high enough to affect breeding success (Watson & Morgan 1964).

Seeing a Golden Eagle has long been a prime objective for many visitors to Scotland. As access to the remoter parts of the country has become progressively easier so too has the achievement of this ambition, bringing increasing risk of disturbance, and consequent breeding failure, to nesting birds. Despite legal protection, both egg collectors and

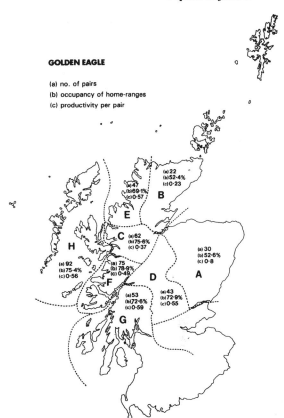

GOLDEN EAGLE

(a) no. of pairs
(b) occupancy of home-ranges
(c) productivity per pair

(a) 22
(b) 52·4%
(c) 0·23

(a) 47
(b) 69·1%
(c) 0·57

(a) 62
(b) 75·6%
(c) 0·37

(a) 30
(b) 52·6%
(c) 0·8

(a) 92
(b) 75·4%
(c) 0·56

(a) 75
(b) 78·9%
(c) 0·49

(a) 53
(b) 72·6%
(c) 0·59

(a) 43
(b) 72·9%
(c) 0·55

Based on data given by Dennis *et al* (1984) in *Brit. Birds* 77 (12)

gamekeepers continue to take a considerable annual toll. In 1980–84 about 25 clutches of Golden Eagle eggs were known to have been stolen, and theft was suspected in a higher number of cases; only in five was there sufficient evidence available to enable court action to be taken. In the same period the destruction of nests, eggs, young, or adult birds, including the use of poisoned baits, was recorded on many occasions. Although seldom provable, it is generally the local gamekeeper or shepherd who is responsible, with or without the knowledge and approval of his employer. Despite evidence to the contrary (see Newton 1972 for refs.), there is a persistent belief in some areas that eagles prey regularly on healthy lambs, and a few farmers and crofters take what they consider to be appropriate action. Persecution of eagles involves the burning of nests, the shooting and trapping of adults and young, and the use of poison baits. Perhaps the most horrific recent case is that in which a newly fledged bird was found with one wing badly broken and the four largest primary feathers torn out of the other (K. Brockie). So long as there are individuals prepared to descend to this level of cruelty, our eagles will always be at risk.

Although two of the factors affecting Golden Eagle distribution and numbers are obvious, namely food supply and persecution, much remains to be learned about the complex relationships between food quality and breeding performance. The extent to which bird species other than those most widely taken as prey can influence Golden Eagles locally also merits comment. On Rhum competition with the growing numbers of White-tailed Eagles has affected Golden Eagles (Love 1983), as has oiling by Fulmars, which regularly form part of the eagles' diet there (Corkhill 1980).

The present situation is encouraging but continuing care will be needed to ensure that Scotland's Golden Eagles – about 25% of the European population of the nominate race – maintain their current status. The problems of ensuring that sufficiently large areas of suitable land remain unafforested, and that nesting sites are adequately protected from disturbance, will not be easy ones to resolve.

Osprey *Pandion haliaetus*   [1]

Migratory breeder; has recolonised the Highlands since 1950 after an absence of more than 40 years. Also occurs as a passage visitor.

Philip Brown and George Waterston (1962) documented the early history of the Osprey's re-establishment as a breeding bird – probably the best-known conservation success story in Scotland – and Waterston (1971) traced progress up to 1970. The present account summarises and updates the situation which has changed so satisfactorily since B&R wrote 'It does seem sad that, owing to the misdeeds of those who lived about a century ago, we should be deprived of the pleasure of seeing Ospreys at their eyries in Scotland. Should any return to breed with us, we trust that public opinion and the forces of the law will prevent their destruction'. Public opinion, influenced by the efforts of the RSPB at Loch Garten and the SWT at the Loch of the Lowes, has certainly helped towards ensuring adequate protection for the recolonising birds, though the 'forces of the law' have been less successful in preventing illegal egg collection.

Once quite common in the Highlands, the Osprey ceased to breed about 1916, after many years of persecution by shooting and stealing of eggs. Although still seen passing through on migration, none attempted to nest again until the 1950s, when a pair settled on Speyside and reared two young in 1954. In 1958, despite an all-night guard, the Loch Garten nest was robbed; this event received wide publicity, and the Ospreys hit the headlines again the following year, when forest fire threatened the eyrie and the three young in it. George Waterston's inspired decision to establish a

Numbers of Ospreys breeding in Scotland (Data from RSPB/R. H. Dennis)

	1960–64	65–69	70–74	75–79	80–
Max. no. of pairs known	2	7	14	25	3
Max. no. of young fledged/year	3	7+	21	30	4
Total young fledged in period	7	21	56	101	22

public hide at Loch Garten ensured that the Osprey soon became a familiar sight to many who had probably never seriously looked at a bird before.

After a slow start the population has gradually expanded (Table) but with breeding success varying widely between years. Egg collectors, bad weather, disturbance during incubation and infertility, in some cases associated with high levels of pesticides, have all been responsible for losses or failures. There have also been cases where the presence of other Ospreys has resulted in a pair failing to rear young. At the Loch of the Lowes two females incubated, unsuccessfully, side-by-side on the same nest in 1976, while in 1983 the male there was so distracted by a visiting female that he failed to feed his mate, resulting in the deaths of the two chicks at about a week old.

Scottish Ospreys all nest in trees, but elsewhere this species, which has an almost world-wide distribution, regularly uses pylons and similar structures. During the 1940s attempts to induce migrants to stop and nest on cartwheels erected beside water proved unsuccessful, but since 1980 the construction of 'foundations' in suitable trees has yielded encouraging results. By providing such 'pre-fabricated' nests, and rebuilding nests destroyed by storms, Roy Dennis has enabled birds to breed more successfully, and in some cases a year earlier, than they would otherwise have done. The first season's occupation of a new site is usually taken up with the building of the basic structure; this task can take so long that by the time it is complete it is too late in the season for the birds to breed. By 1984 twelve artificial sites had been erected by the RSPB, and eight of them occupied (R. H. Dennis).

Most breeding birds return during April, though a few sometimes appear at the end of March; 19th is the earliest record. The majority of nests are in Scots pines, alive or dead, but there are a few in exotic conifers and deciduous trees. Diet varies according to what is most readily available locally; in most areas trout (brown and rainbow) predominate but perch, pike and other species are also taken. The first young are on the wing by mid July, and most of the breeding population has left by mid September. Ospreys are now liable to be seen anywhere in Scotland, especially during passage periods (mainly May/June and September), and there are recent records from all parts of the mainland and every island group. Occasional individuals, probably late-reared young, appear as late as November, and there is one December record, of an emaciated and dying juvenile. Recoveries of foreign-ringed birds show that many passage migrants are from Sweden. Winter recoveries of Scottish birds are mainly from West Africa (Senegal/Gambia/Mauritania), but many are shot or hit overhead wires on their way to the wintering grounds.

The records of breeding and feeding behaviour made over the years at Loch Garten have been analysed and presented in several publications, notably Brown & Waterston (1962) and Green (1976), while Roy Dennis (1983) has monitored and reported on the breeding biology and conservation of the Scottish population as a whole. Analysis of unhatched eggs and dead chicks has shown that the level of chemical contamination in Scottish birds is generally low, though relatively high amounts of DDE and PCB were found in an addled egg from one nest with a history of poor hatching success; very much higher levels of DDE, PCB and mercury were found in Swedish birds which died in this country. Since the early 1970s productivity has averaged 1.2 young/occupied nest and 1.5 young/productive nest; in America production of 1.2–1.3 young/active nest is apparently sufficient to maintain a stable population (Dennis 1983). Regrettably, the principal threat to the rate of expansion of the Scottish population is still the egg collector; between 12 and 14 nests are known to have been robbed in 1982–84.

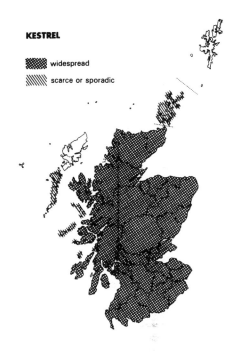

KESTREL

▓ widespread

▨ scarce or sporadic

Lesser Kestrel *Falco naumanni*

Vagrant (S Europe) – one record, 1897.

The only record of this species is of a female shot in Aberdeen in October 1897 (ASNH 1898:51). The few recent reports elsewhere in Britain have mostly been from southern England and Wales. This is not an easy bird to identify with certainty unless good close views are obtained (BB 75:496).

Kestrel *Falco tinnunculus*

The most widespread and abundant of the raptors, absent as a breeder only from Shetland. Dispersal occurs in late summer and some birds winter as far south as Spain. A passage and winter visitor in small numbers.

The hovering Kestrel is now a familiar sight along motorway verges and in cities, a fact which demonstrates the species' ability to exploit a wide range of environments. This is perhaps due partly to its very catholic choice of nest sites and prey species, and partly to its high reproductive rate. Numbers fluctuate locally, especially in uplands, according to the availability of the principal prey species, the short-tailed vole, but CBC results indicate that the population as a whole is fairly stable, and that even very severe winters such as that of 1981/82 do not produce a serious decline.

Kestrels are most abundant in districts which offer many potential nest sites (and there are few areas in Scotland which do not) and a plentiful supply of small mammals. Their relative scarcity over much of the western Highlands may, as Sharrock (1976) suggests, be due to high rainfall affecting hunting efficiency, but shortage of prey or of nest sites seems a more likely explanation. The fact that in parts of Sutherland virtually all the Kestrels move away in winter (Angus 1983) tends to support the latter view. Relatively little is known about the overall size of the population, and Village (1984) has suggested that many published density figures are of questionable validity. The average figure of 75 pairs/100 km² quoted in the Atlas is certainly very much higher than the densities recorded in Scotland. Gordon Riddle (1979), on the basis of a long-term study, estimated the breeding density in Ayr to average 25–33 pairs/100 km² over the county as a whole, but in many areas densities are much lower.

The only evidence of change in distribution this century comes from the islands, where there appears to have been some decrease. There are no recent records for Coll or Tiree, where B&R said the Kestrel still bred, nor has there been evidence of breeding on Lewis (Cunningham 1983, *contra* Atlas 'probable'). A pair seen mobbing a Golden Eagle in south Harris were probably breeding, and nesting occurs fairly regularly on the Uists and Benbecula. Although at one time more abundant, the Kestrel was still fairly common in Orkney in the 1940s; in 1968–72 it definitely bred in 18 10 km squares and probably did so in another two; there have since been signs of a decline, with noticeably fewer ground-nesting pairs reported. The 1945 breeding record for Shetland quoted by B&R is not generally accepted, the last proved breeding being in 1905 (Tulloch 1979).

Studies by Andrew Village (1982a & b, 1983) have shown that at Eskdalemuir Kestrel numbers are correlated with short-tailed vole numbers most closely outwith the breeding season, and at any given vole density are higher in summer than in winter. In the breeding season pairs defend only the area around the nest, and hunting ranges of neighbouring pairs frequently overlap; in winter, when males outnumber females, each bird defends its own range and there is little overlap. In Village's study area territory size varied from 1–2 km² in autumn and 2–5 km² in winter, to 3–6 km² in summer; size is apparently determined by competition, rather than being directly related to food supply. In that part of the Southern Uplands voles formed a large proportion of the diet, other prey being taken more frequently on sheep-walk than in young conifer plantations.

Gordon Riddle carried out a longer-term study, in Ayr, of breeding density and performance in four main habitat types: coastline, lowland farmland with woodland, upland sheepwalk with conifer plantations and heather moor, and urban/suburban situations (Riddle 1979). Densities in young conifer plantations were high in a good vole year, with nine pairs in $9\,km^2$ at one site; in an area of Sitka spruce the population declined from a peak of 15 pairs soon after planting to only five pairs as tree growth made hunting difficult. In this study area Kestrel nests were regularly spaced where suitable sites were readily available, and irregularly spaced where they were scarce; nearest neighbour distance was usually in the range 1–2.5 km, though some nests were less than 100 m apart.

In Riddle's study area the breeding success of Kestrels is high, with an average of 3.6 young per successful nest. More young are produced in warm dry springs, when egg laying is earlier, than in wet spells, when clutch size, hatching rates and chick survival are all lower. Human interference and competition with other raptors also affect breeding performance; 45% of nesting failures in the study area were due to predation by man, while the recent recovery of the Peregrine has resulted in Kestrels losing prime breeding territories which they had used since the 1960s (G. Riddle).

Ringing recoveries show that after fledging young Kestrels initially disperse in all directions (Riddle 1979). Later many move south; a few reach France, northern Spain and, exceptionally, Morocco, but the majority probably do not cross to the Continent (Mead 1973). Some older birds also emigrate, this tendency being more marked in the Scottish population (and those breeding elsewhere in the northern part of the range) than among those from further south; few overwinter in the Hebrides and northwest Highlands. Most Kestrels ringed abroad as chicks and recovered on autumn passage have been from Sweden, Denmark and the Netherlands; one had travelled northwest from the Netherlands, reaching East Lothian less than eight weeks after ringing. There is some ringing evidence of interchange of breeding stocks across the North Sea (Mead 1973), but this has not so far involved Scottish birds.

Kestrels occur regularly on passage in both spring and autumn, but only in the Northern Isles, where the species is scarce or absent as a breeding bird, is it possible to assess the extent of the movement, which generally seems to be small. On Fair Isle most migrants appear in May and late September/early October; there are seldom more than two or three present at a time, though occasionally in autumn there may be up to 15. The maximum recorded at one time on the Isle of May is 31, in mid September 1969, when an obvious influx was also noticeable in east Fife after strong easterly winds.

American Kestrel *Falco sparverius*

Vagrant (N America) – one record, 1976.

Often known as a Sparrow Hawk in its native country, this small falcon was first recorded in Britain in 1976, when a male was on Fair Isle from 25 to 27 May (SB 10:90; BB 74:199–203). The only other British occurrence so far, in Cornwall, was about three weeks later. As well as being noticeably smaller, this species has much more distinct facial markings, in both sexes, than the British Kestrel.

Red-footed Falcon *Falco vespertinus*

Vagrant (E Europe/Siberia) – less than annual, in very small numbers.

The Red-footed Falcon has been reported with increasing frequency since the early 1960s. B&R noted four late 19th century records but only three in the first half of this century. Following a single occurrence in the 1950s, there were eight in the 1960s and 20 in the 1970s; 1973, a year in which a big influx occurred in England, produced six widely scattered Scottish reports. None was seen in 1980 or 1982 and only one in 1983, but in 1981 at least six or seven were present. All records relate to single individuals, which sometimes remain in the same area for several days or even weeks. About half the reports are from Shetland, where four Red-footed Falcons were present in May/June 1981. There are also records from Orkney, Sutherland, Caithness, Inverness, Aberdeen, the Isle of May, East Lothian, Berwick, Dunbarton/Stirling, Ayr, Benbecula and Fair Isle. The species is unusual in that on spring passage from southern Africa to its Eurasian breeding grounds it travels much further west than when on autumn migration. More than two-thirds of occurrences have been in May/June, and the rest in July and August/October, the latest recorded date being 10 October.

The Red-footed Falcon is an insect-eater, often dropping onto its prey from a perch. This habit was turned to advantage in the case of the 1955 Fair Isle bird, which was successfully caught in a clap-net set near a favourite fence-post and baited with mealworms. Plumage differences make it possible to distinguish immature as well as adult males from females; on the breeding grounds males outnumber females but the Scottish records include roughly equal numbers of adult and immature males, together amounting to well under half the total.

Merlin *Falco columbarius* [1]

Partial migrant, breeding on moorlands on the mainland and larger islands; usually thinly scattered and probably often under-recorded. Numbers have been decreasing over a long period and are continuing to do so in some areas. A passage and winter visitor in small numbers.

Merlins are difficult to find on their breeding grounds and casual sightings are infrequent. The location of territories and nests requires much time and effort in searching suitable habitat, and realistic assessments of the species' present status can be made only for areas where intensive survey and monitoring has been carried out. There is evidence that the population has declined throughout its British range over the last three or four decades (Parslow) and the RSPB has expressed increasing concern about the status of the species. This concern has stimulated recent work, the results of which suggest that the situation varies greatly in different parts of Scotland (Table).

Much of Sutherland, Wester Ross, Inverness, Perth and

Argyll has not yet been as adequately searched as have suitable areas of the northeast, though it appears that Merlins are scarce or absent in parts of Kintyre (S. Petty). There are few recent breeding records for West Lothian (where a pair last bred in 1973), Renfrew, Dunbarton or Stirling, and none for Fife, Kinross and Clackmannan. Merlins still nest in the Lothians and the Border counties, where intensive study has recently started (Table). In the southwest there has been a marked decrease in Kirkcudbright: from c15 pairs in the early 1970s, to four or five pairs by 1978, and not more than one pair in the same area in 1981–82 (A. D. Watson). The Merlin is scarce also in Wigtown, while

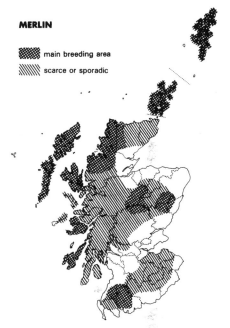

MERLIN

main breeding area

scarce or sporadic

*Numbers and breeding success of Merlins in monitored study areas *(Data from R. H. Dennis, A. Heavisides, E. R. Meek, A. Payne & G. Rebecca)*

	Year	No. of occupied sites checked	No. of pairs breeding successfully	No. of young fledged	Young/ territorial pair
Orkney	1981	13	11	17	1.3
	1982	9	8+	2	0.2
	1983	8	7+	9	1.1
	1984	4	2+	0	0.0
Northeast	1980	17	12	29+	1.7
	1981	15	11	30+	2.0
	1982	21	16+	26	1.2
	1983	48	21	36+	1.7
	1984	42	24	44+	1.0
Central Highlands	1983	27	23	36+	1.3
	1984	41	32+	68+	1.7
Northern Highlands	1983	54	44–45	93+	1.7
	1984	NA	50	129	2.6
Lothians & N. Borders	1984	41	26	39+	1.5+

* Data relate only to territories occupied by pairs, ie they exclude unoccupied sites and those in which only a single bird was present.

NA – not available

in the Kyle and Carrick district of Ayr at least six pairs were found breeding in 1980 (Hogg 1983).

Breeding occurs regularly on many of the Inner Hebrides but little is known about the numbers involved. In the Outer Hebrides Merlins can be seen hunting in the streets of Stornoway (Cunningham 1983) but in the southern islands, despite thorough searching, breeding was not confirmed at any of the 13 sites checked in 1983, ten of which were known to have held nesting pairs in previous years (G. Rebecca). Numbers have declined since about 1980 in Shetland and Orkney, which formerly held important populations. Evidence of egg-shell thinning was detected in both island groups in the early 1980s, and breeding success has since been poor. The Shetland population was estimated at 25–30 pairs in the 1970s (Berry & Johnston 1980), in 1982 only 13 nests were found and from these only 14 chicks fledged (Shetland BR 1982), but in 1984 the 20 pairs located produced 28+ young (P. M. Ellis). In Orkney the situation is more serious; in 1968–72 breeding was proved in nine 10 km squares and all known sites are now checked annually – only six pairs nested in 1984 and four of these failed (E. R. Meek).

Various theories have been put forward to explain the decline, including loss of habitat through afforestation or moorland reclamation, increasing disturbance on the breeding grounds, and exposure to the risk of pesticide accumulation from contaminated prey at lowland wintering grounds. All are probably contributory causes, with some locally more significant than others. In Galloway, for example, Merlins nesting in young conifer plantations continued to hunt on adjacent moorland, upon which they were presumed to be dependent (Watson 1979), whereas in Moray – where all the known sites are threatened by encroaching afforestation – one newly ploughed and planted territory was abandoned between 1983 and 1984 (B. Etheridge). Additional factors not yet identified may also be involved. Because past information on the species' status was so incomplete it is not possible to assess the gravity of the situation in Scotland as a whole, but it is likely that continuing afforestation will reduce the population in some areas in the near future.

Two races of the Merlin are recognised: *F. c. aesalon*, which breeds from the Faeroes and Britain through Fenno-Scandia and Estonia to Siberia, and the larger but otherwise poorly differentiated *F. c. subaesalon*; the latter was formerly thought to be confined to Iceland but equally large birds are now known to breed in Scotland and the north of England (Picozzi 1983). The British population is more sedentary than most, with birds often moving only to low ground or the coast to winter and many returning to within 50 km of their natal site in their first summer. First-winter Scottish birds show a slightly greater tendency to move far south than do adults, but analysis of all British-reared recoveries suggests that more than 60% of both groups remain within 100 km of their natal area (A. Heavisides). Distant recoveries of young birds from the Northern Isles have been in Ireland, England and France, with the most southerly at Bordeaux. Iceland-ringed birds have been recovered in various parts of Scotland, usually in autumn or spring. Many of the large Merlins trapped on Fair Isle and formerly considered from their size to be of Icelandic origin are now known to be dispersing Shetland birds (Robertson 1982). Merlins ringed on autumn passage on Fair Isle have been recovered in Belgium, France, and Germany, as well as in Britain, later the same winter.

In addition to the monitoring work referred to above, studies have recently investigated the growth and development of nestling Merlins as a possible means of sexing the young (Picozzi 1983), and diet during the breeding season (Watson 1979, Newton *et al* 1984).

Hobby *Falco subbuteo*   [1]

Scarce and irregular summer visitor; has bred

Although the Hobby is currently recorded annually, numbers are so small as to suggest that it might justifiably be described as a vagrant. From 1968 to 1983 inclusive not more than 12 were reported in any one year, and in some seasons only one or two. More than half the records are between mid May (6th is the earliest) and the end of June, with smaller numbers reported in July/September, and stragglers to 5 November. Many of the records are from Shetland and Fair Isle; the rest are widely scattered throughout the country, including a few from the Outer Hebrides (St Kilda, Lewis & North Uist) and the Inner Hebrides (Mull & Islay), but there is no recent report from the Clyde Islands. Nearly all involve single birds.

The only accepted breeding record is an old one, at Ballinluig, Perth, in 1887 (B&R). In 1977 a pair was seen displaying in central Scotland but there has not yet been any definite evidence of breeding; these birds were possibly immatures. A small but increasing population of this species, recently estimated at 500+ pairs (Newton 1984), now breeds in Britain, mainly in areas of dry heath and downland which support abundant supplies of large insect and small bird prey. Over much of Europe the Hobby is widespread but it is absent from Norway and northern Sweden; numbers have declined recently in some areas. It winters in tropical Africa.

[Lanner *Falco biarmicus*]

Escape. There have been several reports of this species, which is kept both for falconry and to chase gulls off airfields. Among the most recent records are sightings in Inverness and Shetland.

[Saker *Falco cherrug*]

Escape/vagrant (Category D)

A bird on Out Skerries from 1 to 5 October 1976 was the first British record of this falcon, which breeds from east-central Europe south to Iran and east to Siberia (BB 73:530). A second was on Fetlar from 27 to 29 May 1978. Sakers are quite often kept for falconry so the probability of these individuals being escapes is high – although other eastern species were also present in May 1978 (BB 72:544).

Gyrfalcon *Falco rusticolus* [1]

Very scarce winter visitor; annual in very small numbers and mainly in the north.

Gyrfalcons are recorded most often in Shetland, occasionally in the Outer and Inner Hebrides and Orkney, and very sporadically in other areas. Several have been caught on trawlers, oil installations or weatherships and released ashore. Most arrive in November/December or April; extreme dates since 1970 are 9 September and 27 May. In most years only one or two sightings are reported but occasionally there are several records from widely scattered areas, presumably due to an eruption in the Greenland population.

The Gyrfalcon has a circumpolar distribution, breeding in the Arctic/sub-Arctic south to southern Norway. Most nest on cliffs or crags and, unless located near seabird colonies, hunt over wide areas of sparsely vegetated country. The population fluctuates cyclically on the breeding grounds, in response to the availability of prey, especially Ptarmigan, but there has also been a substantial long-term decline in many parts of the range. Much of this is attributable to persecution – both egg collecting and removal of young for falconry – but there is also recent evidence to show that some birds are being affected by toxic chemicals. A female wintering on Islay in 1978/79 died of alphachlorate poisoning and a first-winter female which died on Fetlar in May 1979 contained very high levels of toxic chemicals. The skin of the Fetlar bird is in the Royal Scottish Museum (BB 73:502–503).

Plumage colour – varying from predominantly white in the northern part of the range to predominantly dark grey in the southern part (including Iceland), with intermediates quite common throughout the range – was formerly regarded as indicative of area of origin; dark birds occurring in Britain were classed as Iceland Falcons and white ones as Greenland Falcons. B&R treated these as separate races, *F.r.islandicus* and *F.r.candicans* respectively, and square-bracketed the Gyrfalcon *F.r.rusticolus*. It is now known that plumage is not a reliable indicator of origin, and also that both Icelandic and Scandinavian populations are largely non-migratory, merely dispersing coastwards during the winter. Birds breeding in high-arctic Greenland are markedly migratory, however, moving south along the coast and sometimes crossing to Iceland to winter. Migration starts in late August and continues into December; return is in April/early May. In years when Ptarmigan are scarce eruptive movements also occur in autumn in southeast Greenland (BWP). Most Gyrfalcons reaching Scotland are white-plumaged and all are probably from Greenland.

B

Peregrine *Falco peregrinus* [1]

Resident breeder, widespread in mountains and on seacliffs but currently absent from much of the east coast. Scotland holds the bulk of the British population, which has recovered well after a serious decline in the 1960s, though breeding success remains poorer on coastal cliffs than inland. In winter those breeding at high elevations or in unproductive areas move to low ground.

Long-term studies of the Peregrine have been carried out by Derek Ratcliffe and others. This account is based largely upon Ratcliffe's monograph on the species (1980), which includes a comprehensive bibliography, and on the results of the most recent census of the breeding population (Ratcliffe 1984). The Peregrine is considered an endangered – or at least threatened – species worldwide, as populations in North America and over much of Europe have been greatly reduced by persecution and the effects of pesticide pollution. The British and Irish population is now one of the largest in Europe and we are fortunate that the Scottish Highlands

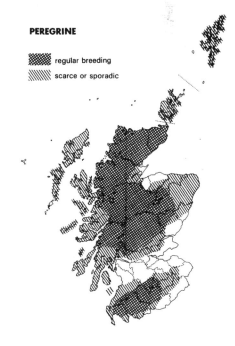

PEREGRINE

▓▓▓ regular breeding

░░░ scarce or sporadic

provide a sanctuary area sufficiently large, remote and productive to have maintained a viable population throughout the critical period in the 1960s, and to provide a source of fresh stock to reoccupy areas where breeding had ceased altogether. Although the risk from pesticides is currently fairly small, persecution, egg collecting and the illegal taking of young continue to present a threat, despite the protection afforded by the Wildlife and Countryside Act 1981.

Peregrines usually return to the same territory in successive years, though females occasionally change territories. At the end of the breeding season the birds disperse, with those in their first winter making the longest movements (Mearns & Newton 1984). The species' fidelity to traditional nesting sites makes the breeding population relatively easy to census, although inaccuracies may occur through failure to locate all the alternative nest sites in a territory, or through early loss of eggs. It is more difficult to assess the numbers of non-breeding birds, some of which also hold territories at least intermittently. Prior to 1961–62 no systematic census had been attempted, but it was estimated that up to the 1939–45 war the British population was reasonably stable at c800 breeding pairs and seemed little affected by natural factors such as climatic fluctuation, all known changes being either directly or indirectly attributable to man. During the war, to protect carrier pigeons used by the services, authorised persons were allowed to take or destroy Peregrines and their eggs in specified areas. The latter included many of the Scottish islands and parts of the mainland, but the numbers killed were apparently insufficient to cause more than local and temporary declines – and were probably to a large extent offset by the reduction in gamekeeping at that time.

By the late 1950s a major decline had become apparent in England and Wales, and by the early 1960s southern Scotland was also affected. A census in 1961–62 revealed that numbers there were down to half the pre-war level, and that the decline was most pronounced on the coast, with Ayr, Galloway and Berwick holding only 20% of their former coastal populations. Large-scale mortality among seed-eating birds, including species preyed upon by Peregrines, had by then been attributed to the increasingly widespread use of organochlorine seed-dressings, and in 1961 the presence of organochlorine residues in Peregrine eggs was recorded for the first time. Sudden deaths, failure to lay, laying of thin-shelled eggs with a high rate of loss through breakage, and smaller broods were all involved in the decline, which was most marked in the cereal-growing districts of the east coast and lowlands. A voluntary ban on the dressing of spring-sown cereals was introduced in 1962 but, although deaths in grain-eaters decreased, the Peregrine population continued to decline.

Recovery started in 1967 following further restrictions on the use of persistent organochlorine insecticides; by 1971 the Scottish inland population was back to about 90% of its pre-war total but coastal numbers were still far below their previous level. The recovery has since continued in most areas (Table a). The slow rate, or absence, of recovery among coastal birds, where breeding success is poorest in areas with seabird colonies, is thought to be due to marine pollutants. Eggs from coastal Peregrine eyries have consistently higher levels of both PCBs and mercury than those

(a) *Peregrine population levels in 1961 and 1981 (Based on Ratcliffe 1980 & 1984)*

	Est. no. of occupied territories 1961	No. of occupied territories 1981	1981 occupation as % of 1930–39 est. level
S Scotland – coastal	7	22	85
– inland	29	45	150
S & E Highland fringe – inland	66	76	107
S & E Highland – W coast	33	38	106
– E coast	7	5	36
S & E Highland centre – inland	77	108	124
N & W Highland – W coast	78	52	59
– E coast	67	54	78
– inland	76	74	78
Scottish Total	440	474	94

(b) *Breeding success of Peregrines in different parts of Scotland* (Data from Ratcliffe 1984, Watson 1982, Payne & Watson 1983 & 1984, Dick 1985, J. Mitchell, P. Stirling-Aird, RSPB/R. H. Dennis)*

	Year	No. of sites checked	occupied†	Broods reared	Young reared/ territ. pair
North	1981	271	144–146	84	1.13
	1982	145	75–76	38	0.91
	1983	136	78+	40	0.99
	1984	101	80	58	1.56
Northeast	1981	NA	41	22–26	1.12
	1982	NA	52	23–28	1.17
	1983	NA	64	21–23	0.63
	1984	NA	86	37	1.00
Central	1981	60	58	36	1.26
	1982	50	50	21	0.98
	1983	64	60	22	0.92
	1984	76	71	39	1.60
Southeast	1981	13	12	8	1.25
	1982	13	13	6	1.31
	1983	20	14	6	1.00
	1984	NA	16	11	0.68
Southwest	1981	55	47	10	0.45
	1982	60	54	29	1.19
	1983	51	44	10	0.55
	1984	NA	74	40–42	0.57

* Regions are those covered by the Raptor Study Groups – see SB13:163

† occupied by an adult pair

from inland. Oiling by Fulmars is thought by several Scottish workers to be a reason for the low populations in Shetland, Orkney and Caithness. In southern Scotland, however, Richard Mearns (1983) found that seabirds formed only a small proportion of the diet of coastal-nesting pairs. Recovery is also limited in the west Highlands, where lack of wild prey due to environmental degradation is probably a contributory factor.

In recent years the breeding success of Peregrines in various parts of Scotland has been regularly monitored and it should consequently be possible in future to detect any changing trends in the population at an early stage. In most areas productivity is currently high (Table b).

The Peregrine's status in the mid 1980s is more encouraging than it has been for many years. New sites are being occupied (several ground nests have been found recently), in some grouse moor areas of eastern Scotland more Peregrines are breeding than were ever recorded previously, and in the Southern Uplands some territories which formerly held only one pair are now occupied by two (Table c). The

(c) Breeding density of Peregrines in inland Galloway and Carrick (From Ratcliffe 1984)

	1930–60	1982
Av. nesting territory size (sq km)	63.3	47.0
Av. nearest neighbour distance (km)	5.4	4.9

recent increase in southern Scotland is reflected in much more frequent winter sightings on low ground around the Solway (Mearns 1984). Some of the most successful eyries are on the outskirts of towns or villages, demonstrating that if the nest itself is not subject to disturbance the birds can flourish in close proximity to man. Egg collecting has probably decreased since pre-war days but there is an apparently increasing demand for young birds for falconry, and it remains to be seen whether the introduction of legislation to enforce the licensing and registration of all falconers' birds will have any effect in reducing the illegal taking of young; on the present evidence it seems unlikely to do so.

Red Grouse *Lagopus lagopus*

Resident; the most widespread of the gamebirds, breeding on moorland in every mainland county and all island groups. Population fluctuations occur over both the short- and long-term and have been the subject of intensive study for many years. Declining numbers are currently causing concern.

Although formerly considered a full species, occurring only in the British Isles, the Red Grouse *L. l. scoticus* is now regarded as a race of the Willow Grouse, which is resident in Scandinavia and across much of northern Eurasia, Canada and Alaska. It occurs most often on heather moors between 300 m and 600 m asl, but occasionally in areas where grass is dominant and heather scarce, and at altitudes ranging from sea-level to c900 m. A few breed on small patches of heather surrounded by arable land and woodland but these are unlikely to be self-maintaining groups.

In 1968–72 Red Grouse bred in all mainland counties, on the Clyde Islands and on all the main Inner Hebridean islands other than Coll and Tiree. In the Outer Hebrides

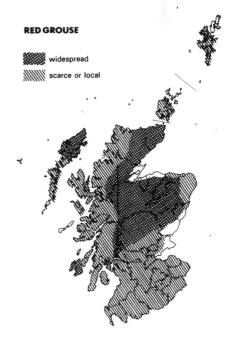

RED GROUSE

▓ widespread

▨ scarce or local

they are most abundant in Lewis, fairly common in Harris and North Uist, and scarce or absent from Benbecula southwards. In Orkney breeding occurs on most of the islands from Rousay south, but not on the northern isles, while in Shetland the small population is confined to the Mainland.

As the Red Grouse is an economically important game-species, especially in eastern Scotland, major fluctuations in numbers give rise to concern, and it was following a decline in the 1940s and early 1950s that research was started in 1956. Since then many aspects of grouse ecology have been studied by David Jenkins, Adam Watson, Robert Moss and others, and much has been learnt about the factors governing population densities and fluctuations (see Watson & Moss 1980 for bibliography). Red Grouse densities are highest, sometimes reaching nearly 100 pairs/km², on heather moors which lie over base-rich rock and are managed – by regular and carefully controlled burning of small areas – to provide a mosaic of long, woody stems for shelter and nesting cover, and young growth of high nutritional value. On western moors, which tend to be wetter, more acid, and less well managed, numbers are very much lower.

The cocks establish territories in autumn, varying in the aggressiveness with which they defend 'their patch'; on good moors a territory may be less than 0.5 ha and on poor ones more than 40 ha. Cocks that fail to claim an adequate area of their own are forced to move away to less suitable ground and nearly all die during the winter. In snowy weather the birds flock to feed on wind-cleared ground, moving to lower moors or even farmland if snow lies deep. Breeding success is determined largely by the food supply available to the hens in spring. There is no evidence that either disturbance or winter predation has a significant effect on breeding numbers in Scotland. Heavy summer predation on adults, eggs and chicks can reduce both August numbers and the breeding population the following spring. In the past this has seldom been a problem, but increasing predation is now occurring in areas where afforestation provides cover for many foxes and crows (Watson 1983).

Red Grouse numbers have long been known to fluctuate cyclically, with a period of six or seven years between population peaks, but during the 1970s a much greater decline took place than had been observed at any time since research started 15 years earlier. This decline occurred irrespective of the quality of the moor and its management, and involved in turn: abnormally high levels of spring emigration and numbers of unmated territorial cocks, a low proportion of young cocks establishing territory, summer emigration of adults with very young chicks (of which at least some were reared, and subsequently established territory, in another area), and finally a high late-winter loss of territorial cocks. The proportion of inherently aggressive cocks was higher after the decline than during years of peak population density, suggesting that fluctuations in genetically determined behaviour patterns, which affect territorial spacing, are associated with cyclic fluctuations in numbers (Watson & Moss 1980).

Another factor involved in the pronounced long-term decline in grouse numbers, over big tracts in Moray, Banff and Perth and smaller areas elsewhere, has been an increase in the incidence of louping-ill, a tick-borne disease usually fatal to grouse. This has so reduced stocks in these areas that, with bags perhaps only 10% of what they were early in the century (A. Watson in Winter Atlas), grouse-shooting has ceased to be economic and many once-good moors have been turned over to forestry. Changes in distribution are consequently taking place, and may be expected to become increasingly extensive as afforestation and the conversion of part-heather moorland to reseeded grassland continue. Even in the better grouse areas moor management is often much less efficient today than it was pre-war, owing to the high cost of the labour required to carry out a proper programme of heather-burning. No detailed information is available but marked decreases due to afforestation have been reported in the southwest and have certainly occurred also in Argyll, parts of Perthshire and the northeast. Further north and in the Hebrides uncontrolled and ill-timed muirburn reduces numbers, while in Orkney the area suitable for grouse is steadily shrinking as more moorland is brought into cultivation.

Ptarmigan *Lagopus mutus*

Resident, breeding on much of the high ground north of the Highland Boundary Fault and on Skye and Mull, but occurring only locally or sporadically in a few other areas. Numbers fluctuate considerably and there has been some contraction of the range this century.

Long-term studies of Ptarmigan have been carried out by Adam Watson (1965a,b,c, 1972, 1979, 1981), upon whose work this account is largely based. This sedentary species is confined to the arctic-alpine zone, where it favours mixed heathland with abundant crowberry and blaeberry and only rarely breeds in areas dominated by grasses or heather. Although nesting mainly above 800 m asl, and at over 1200 m in the Cairngorms, it occurs down to 200 m asl in northern Sutherland. Studies from the 1940s to the early 1960s showed that spring densities in the Cairngorms ranged from 5 to 17 birds/100 ha. Predation on eggs and chicks did not appreciably depress the population, nor were breeding numbers limited by adult mortality due to predation. Flocks form in autumn, when counts in different years ranged from 8 to 46 birds/100 ha, and numbers remain fairly constant through the winter until the flocks split up in March/April.

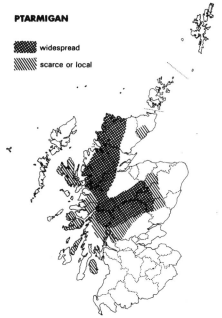

PTARMIGAN

▓▓ widespread

░░ scarce or local

has taken place; in 1977 Ptarmigan bred there for the first time this century, and have done so each year since. In Stirling and Dunbarton breeding is possibly only sporadic. No Ptarmigan were found in Kincardine in 1968–72, although Mount Battock, near Banchory, used to be occupied in most years (Watson 1965a), and there have been only occasional recent reports from the outlying Banff site of Ben Rinnes. The 'probable breeding' shown in the Atlas just east of the inner Moray Firth probably involved only stragglers, which occasionally turn up well outside the normal range, as for example the bird seen on Bennachie (Aberdeen) in January 1975.

Ptarmigan formerly bred in southwest Scotland but became extinct there during the 19th century (B&R). In the late 1960s and mid 1970s attempts were made to reintroduce them in the Leadhills area of south Lanark, using eggs collected near Braemar, but these were not successful. A few birds moved east into the Tweedsmuir Hills, however, and bred there in the late 1970s. Twenty were seen in the winter of 1977/78 and, although several were shot in 1978, there were still a few present in autumn 1980 (R. D. Murray) and one was seen in July 1984 (per J. G. Young). It is probable that in the peripheral parts of the range, where the amount of suitable habitat is very limited, the population is too small to be self-sustaining through more than a few consecutive seasons.

Major fluctuations in population size apparently occur about once in every ten years.

Since the 1960s ski developments on Cairn Gorm and the Cairnwell have brought increasing numbers of visitors to these areas, in summer as well as winter since the chairlifts provide easy access to the high ground. Many Ptarmigan on Cairn Gorm are killed by flying into ski-lift wires, and in May 1981 none were found alive in the developed ski grounds in Coire Cas, where the length of wire/unit area is now higher than anywhere else, and the heavy usage of the area by skiers increases the risk of the birds being flushed and hitting the wires. In addition to adult mortality, numbers on Cairn Gorm are affected by crow predation of eggs, which occurs both near the ski developments and in relatively undisturbed areas nearby; the numbers of crows and gulls on this high ground have increased greatly in the last 10–15 years, attracted by discarded food scraps. As a result, breeding success on Cairn Gorm is now consistently poor, with no young at all reared some years, and an average production of only one or two young/10 adults over a ten-year period. Both average population density and breeding success are higher in the Cairnwell hills, in association with more base-rich underlying rocks and plants of higher nutritive value.

Recent changes in distribution have been recorded only on the periphery of the breeding range. Ptarmigan bred in the Outer Hebrides until about 1924, but none has been recorded there since 1950 (Cunningham 1983). Only stragglers have been seen recently on Rhum, where breeding formerly occurred; these are possibly wanderers from the small Skye population, which is currently decreasing (A. Currie). In 1981 there were reports from three mountains in Mull, where small numbers still breed, but there are few recent records from Jura. On Arran, however, recolonisation

Black Grouse *Tetrao tetrix*

Resident, breeding in most mainland areas below c400 m asl where there is a mixture of heather moor – often with rough pasture – and woodland, but scarce in the northwest and absent from most of the islands.

The Black Grouse, a sedentary species, undergoes both short-term and longer-term variations in abundance (Mackenzie 1952). These, together with distribution changes associated with afforestation and woodland clearance, make it difficult to identify population trends with any certainty. A considerable decrease occurred in some areas during the first half of this century (Parslow) but since the 1950s the Black Grouse has shown local increases. Recolonisation is particularly apparent in recently afforested areas, but these

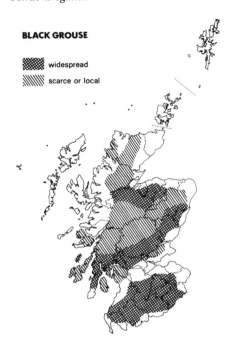

BLACK GROUSE

▨ widespread

▨ scarce or local

Jura small numbers survive after restocking (Reed *et al* 1983). In the Outer Hebrides attempts at introduction were made last century but were unsuccessful, and the only recent record is of a cock in North Uist in May 1978 (Cunningham 1983). An unsuccessful introduction was made in Orkney last century; the Black Grouse occurs there only as a rare vagrant. There are no records for Shetland.

Capercaillie *Tetrao urogallus*

Resident, currently confined to conifer woodlands between the Forth and southeast Sutherland, with the greatest concentration in the eastern Highlands. Although colonisation of plantations has permitted slow expansion, there is evidence of a recent decrease in range and numbers.

are deserted when the trees close canopy. Hardwood scrub, particularly of birch and hazel, and open mature pinewoods provide food and shelter in winter and may be important factors affecting distribution. There is little information on densities. On a study area in the Forest of Birse (Kincardine) there was one cock / 87 ha of suitable habitat in spring 1983, but densities in the old Caledonian pine forests of Glen Tanar are greater than this, though still liable to fluctuations (N. Picozzi).

Leks of 15–25 cocks were reported during the 1970s in Kirkcudbright, Dumfries, southern Ayr, the Ochils, central and west Perth, Aberdeen and Speyside; similar numbers doubtless occur in other areas too. In Sutherland there has been an extension of range as new conifer plantations are colonised, and in parts of central Ross (around Achnasheen) and northwest Inverness numbers are thought to be increasing. Elsewhere on the mainland there is little evidence of recent change other than local declines or increases. Game records for the early 1900s show that in some areas more than 200 birds were occasionally shot in a single day. Comparable numbers could not be shot anywhere now, although an exceptional bag of over 100 was obtained in Dumfries in 1970.

On the islands, where this species is usually scarce or absent, status changes are easier to detect. On Bute, where no Black Grouse were seen in 1968–72, there has been an increase since the early 1970s and flocks of up to 12 birds are now occasionally reported (Gibson *et al* 1980). On Arran numbers remain very small, but the expanding forests may lead to an increase in the fairly near future. Although possibly scarcer than in the past, the species is still present in suitable habitat on Mull, and breeding probably occurs regularly there. On Skye and Raasay, where it was once plentiful (B&R), it has recently become extinct but on Colonsay and

The Capercaillie, which is the world's largest grouse, is native to Scotland but became extinct in the 18th century; its decline is poorly documented but is usually attributed to extensive felling of pinewoods. Swedish birds were imported on several occasions in an attempt to re-establish the species, the largest and most successful release being at Taymouth Castle (Perth) in 1837–38; subsequent introductions were made in Angus, Kincardine, Moray, Inverness and Fife (B&R). From the successful introductions the Capercaillie spread out along the valleys and colonised most suitable woodlands. Extensive felling during the two world wars resulted in a decline in the population, but some recovery has already taken place, though numbers in east Scotland are currently lower than in 1968–72 (R. Moss). A detailed account of the history and distribution of this spectacular bird is given by Pennie (1950).

B&R knew of Capercaillie breeding in Lanark and Peebles but there have been no reports from these counties for many years, nor any recent ones from Kinross, where nesting was proved in 1968–72. Increasing numbers have been seen in Tentsmuir Forest (Fife) since the early 1970s and breeding is probably now taking place there. The extensive forests of Argyll have not yet been colonised, a small area on the east side of the Cowal peninsula comprising the only occupied site reported to date. Breeding occurs regularly on some

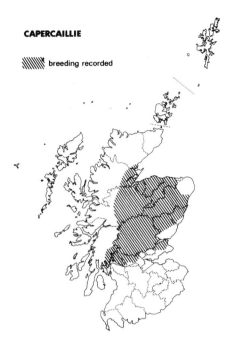

CAPERCAILLIE

\\\\\\ breeding recorded

difference in size between the cock and his harem of soliciting hens is striking – so great in fact that it led one researcher to ask 'Why are Capercaillie cocks so big?' (Moss 1980). The tentative conclusion reached was that size has conferred an advantage in fighting, and that hens prefer the best fighters – but this hypothesis has yet to be satisfactorily proved.

[Northern Bobwhite *Colinus virginianus*]

Escape / vagrant (Category D)

Reports of this American gamebird calling near Dunblane in 1974 undoubtedly referred to escapes or releases.

Red-legged Partridge *Alectoris rufa*

Feral breeder (Category C)

This widely-introduced gamebird is native to southwest Europe. In the last 20 years considerable numbers have been hand-reared and released in many parts of Scotland and breeding in the wild has been recorded in several counties. As releases continue to take place it is not possible to be certain whether, or to what extent, self-supporting feral populations may have become established. The Red-legged Partridge hybridizes freely with the Chukar *Alectoris chukar* and the Rock Partridge *Alectoris graeca*, both of more easterly origin, which have also been introduced, though on a more limited scale. Some of the records probably refer to hybrids, which are difficult to distinguish.

Since 1968 releases have definitely taken place in Kirkcudbright, Roxburgh, Berwick, Midlothian, Stirling, Dunbarton, Perth, Angus, Aberdeen, Banff, Moray, Caithness and Lewis. Breeding in the wild has been recorded in Dumfries, Midlothian (including one brood at 300 m asl in the Moorfoots), Perth, Aberdeen, Banff, Moray, Caithness and the Borders. Red-legged Partridges or hybrids have also been seen in Wigtown, Peebles, East Lothian, Fife, Stirling, Argyll, Kincardine, Easter Ross and Sutherland.

of the Loch Lomond islands, and up to 26 birds have been seen together on Inchmoan in recent years.

When range expansion is taking place it is normally the female that first occupies a new area. In the absence of male Capercaillies such birds may mate with Black Grouse, producing offspring which are intermediate in character (see SB 4 plate 9). The sighting of single females in Caithness in 1972, Ayr in 1977 and 1982, and Kirkcudbright in 1977, all well outside the present range, suggests that hybrids might be found in these areas and that colonisation may eventually take place.

The natural habitat of the Capercaillie is open woodland of Scots pine, with an abundant field layer including blaeberry and heather and a good number of mature 'grannie' trees; the latter are particularly important for winter feeding (Zwickel 1966). Such habitat, which occurs mainly in the remaining remnants of the native pinewoods, supports greater densities and more successful breeding than does planted forest (Moss *et al* 1979). Winter densities have been found to range from 5 to 36 birds / km². As is the case with related species, such as Red Grouse and Ptarmigan, numbers fluctuate between years, though not so rapidly nor in the same cyclic pattern. Where Capercaillies move into plantations they favour those with a good proportion of Scots pine, but will also use larch, spruce and fir. Near Perth they can quite often be seen in oak trees or foraging at dawn in stubble fields; in the days before combine harvesters appeared on the scene I have watched cocks feeding in winter on the tops of unthatched stacks.

Capercaillies are seldom easy to observe, as the first indication of their presence is often a crashing departure through the branches. Some cocks are peculiarly aggressive, however, and will attack both humans and vehicles in defence of their display ground. When observation at a lek is possible, the

Grey Partridge *Perdix perdix*

Resident, widespread in lowland agricultural areas, scarce or absent over much of the Highlands, and absent from most of the islands. Numbers have been affected by agricultural developments and in some areas are much lower than formerly.

The Grey Partridge feeds largely over farmland and nests under hedge or shrub cover, or in other tall ground vegetation; these requirements, together with the need for an adequate supply of suitable insect food for chick-rearing,

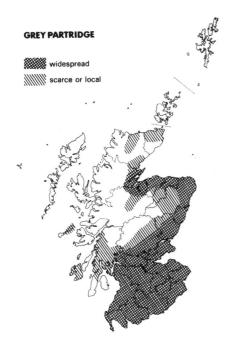

GREY PARTRIDGE

▓▓ widespread

░░ scarce or local

determine not only the species' natural distribution but also survival prospects for birds introduced to other areas. Nesting occurs up to about 500 m asl and there are few, if any, 10 km squares south of the Highland Boundary Fault that do not hold at least a few pairs. In the northeast the Grey Partridge is abundant (5–15 pairs/km², G. R. Potts *in litt.*) in Aberdeen and on the coastal strip of Banff, Moray, Nairn, east Inverness and Easter Ross; in southeast Sutherland and Caithness it is local only and numbers are relatively small. Further west, a few pairs breed at Altnaharra and Melvich (Angus 1983), but none are known to be regularly present elsewhere in Sutherland, in Wester Ross, or in west Inverness. The situation in Argyll is uncertain; the Atlas shows only one 'probable' breeding, but a recent assessment suggests a population of over 100 pairs (M. J. Gregory). The most marked recent decline has taken place in East Lothian, where a large proportion of the low ground is intensively farmed; densities have dropped from about 25 pairs/km² in the early part of the century to about 5 pairs/km² now. The glens of Angus, Kincardine and Aberdeen, in contrast, remain important local strongholds (G. R. Potts).

Grey Partridges have been introduced to many of the larger islands but now breed only on Islay, Bute and possibly Coll. There have been frequent releases on Islay, the most recent in the 1970s; breeding took place in 1969 (not shown in the Atlas) and a covey of 15 was seen in spring 1972, but few were left by 1974 (Booth 1981). On Coll, Grey Partridges were said in 1937 to be breeding commonly on the sandhills (B&R) – as they also do on the Findhorn dunes. B&R record introductions on Gigha, Colonsay, Tiree, Eigg, Rhum, Canna, Skye and Raasay; in every case the birds died out within a short time. The species formerly bred on Mull, but has recently become extinct there (Reed *et al* 1983), and on Arran, where there has been no recent record. On Bute, where some 9,000 were shot between 1900 and 1927,

it is now scarce. There have been several unsuccessful attempts to introduce Grey Partridges into Orkney and Shetland.

Following a drastic decline in numbers during the early 1960s, over much of England and also on the Continent, intensive studies of Grey Partridge chick feeding and survival were initiated by the Game Conservancy. These showed that modern farming techniques were seriously affecting chick survival rates (Potts 1970, 1980). Among the factors involved are increased use of herbicides and fungicides on cereal crops, with consequent reduction in the variety and number of weed species and insects present, and the removal of hedges and other areas of permanent vegetation important as nesting cover safe from predators. Chick survival rates have declined largely as a result of the decreasing availability of four insect groups which were formerly a main food source: plant bugs, caterpillars (of *Lepidoptera* and sawflies), weevils and leaf beetles. The eradication of knotgrass and mayweeds has been important in reducing the numbers of these insects (Potts 1983). Preliminary trials have recently indicated that survival rates are significantly improved if a strip about 3 m wide round the edge of a cereal field is left unsprayed.

While conditions in much of Scotland differ considerably from those in southern England, where this work was carried out, similar agricultural changes have certainly been introduced in the arable eastern counties and are continuing. Predation by crows is probably a limiting factor in areas where gamekeepering effort is now low. Overall, however, the decline of the Grey Partridge in Scotland has been much less severe than in parts of England and in other European countries (G. R. Potts).

Quail *Coturnix coturnix* [1]

Summer visitor, in very variable numbers, and sporadic breeder.

The Quail is a secretive bird, which generally remains hidden in long grass or corn and is identified only by its distinctive call. The presence of calling males is not, however, a reliable indication of breeding, and calling may cease once pairing has taken place (Atlas). In most cases breeding is confirmed only when chicks are sighted. Although birds sometimes arrive in April, the main influx is generally during May – in some years even later. Most have left by the end of August, but there are a few records for September/November and one mid-winter report of two at Newburgh, Aberdeen, in January 1976. From 1958 to 1983 inclusive, Quails were reported in every county except Peebles, Kinross, Clackmannan and Nairn, and breeding was confirmed in Berwick, Roxburgh, Fife, Aberdeen, Caithness, Orkney and Fair Isle. Although so difficult to prove, breeding is known to have occurred in most counties at least once this century.

Numbers fluctuate markedly between years, probably in response to weather conditions at the time of the spring migration. A warm spring with southeast winds apparently stimulates more birds to move towards the limits of the breeding range, where nesting is often only sporadic (BWP). Over much of the western European range, which extends from southern Fenno-Scandia to the Mediterranean (and into northwest Africa), marked population fluctuations occur. The Quail breeds annually in southeast England but only irregularly further north in Britain. In Scotland a 'good' Quail year produces widespread reports of calling birds; in 1976, 1981 and 1982, for example, there were records from 17–18 counties.

Quails winter in Africa, probably mainly south of the Sahara. As the only European migratory gamebirds, and with a good market value, they are heavily shot and trapped during migration, especially around the Mediterranean. Long-term declines are thought to have taken place in some parts of the range, but knowledge of the species is so limited, and so difficult to acquire, that neither the causes nor the overall effects of local population fluctuations are fully understood.

Pheasant *Phasianus colchicus*

Introduced resident (Category C); widespread and abundant throughout the lowlands but very local in the northwest Highlands. As a result of recent introductions, now also present on many of the larger islands.

In most areas the distribution of the Pheasant is determined by the availability of sufficient woodland or scrub cover for roosting and breeding, and of farmland or open deciduous woodland for feeding. On the west coast of Jura, however, a thriving feral population exists in an area virtually devoid of both deciduous woodland and cropping. (The original release was at Ardfin, but Pheasants are now common north to Loch Tarbert.) Effective control of crow predation on clutches can be a vital factor in ensuring the survival of wild populations in such marginal habitats (J. Phillips).

Introduction is said to have started by the late 16th century, and hand-rearing continues today on many lowland estates, often on a large scale and as an economically important enterprise. In most non-mountainous mainland areas there is also a well-established feral population, and Pheasants are scarce and local only in northern Argyll, west Inverness, Wester Ross, Sutherland and Caithness. There is little evidence of change since the Atlas survey, apart from a slight expansion in Sutherland and Caithness, which may be expected to continue as the area under afforestation increases.

More detailed information is available for many of the islands, where introductions seldom took place before the 19th century. B&R record that the Pheasant was introduced to Bute, Arran, Islay, Jura, Mull and Skye in the 19th century, and to Coll, Eigg, Rhum, Canna and Raasay in the late 19th or early 20th century. It became established on most of these islands, with wild breeding presumably taking place fairly regularly, and now occurs also on Colonsay and Gigha. On Muck and Raasay it has long been extinct and it has recently died out on Tiree and Rhum (Reed *et al* 1983). Late 19th century introductions in the Outer Hebrides survived only in the Stornoway Woods and vanished altogether during the 1939–45 war. Several small-scale introductions have since taken place in Harris, where they did not survive

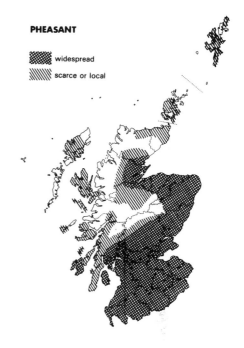

PHEASANT
widespread
scarce or local

long, and Lewis, and probably also in the Uists, where birds are occasionally reported. The most recent and important introduction to the Outer Hebrides was in the early 1980s, when over 1,000 poults were released in the Garynahine forestry plantations in Lewis; the prospects of the birds establishing themselves there should be quite good. In Skye too the establishment of forests in formerly treeless areas is likely to result in an extension of the species' distribution.

There have also been several introductions in Orkney and Shetland. The early ones were unsuccessful but this century they have generally resulted in survival for a few years. Venables & Venables (1955) record releases at four sites between 1900 and 1950; those on Vaila and Fetlar died off quickly, but survival was better at Kergord and Burrastow on the Mainland, where more cover is available. Although the Atlas shows no breeding in Shetland, Pheasants were said to be still breeding there in 1969 (SBR) and are seen quite frequently at Kergord; small numbers have since been released in Unst and at Tresta on the Mainland. In Orkney the success rate has been rather better. Wild breeding has been recorded on Hoy, where 15 were at South Walls in November 1981; there are regular sightings on South Ronaldsay and at several locations on Mainland; and Pheasants are occasionally reported on Rousay. Small-scale introductions still take place at intervals.

[Pheasant spp.]

Introduced (Category C or D)

Over the last 30 years or so several decorative species of pheasant have been deliberately released or have escaped. So far only one of these, the Golden Pheasant *Chrysolophus pictus* has established a self-supporting feral population; it has bred freely for several decades in Kirroughtree Forest, Kirkcudbright/Wigtown. Some 250 birds were thought to be present in 1974 but numbers were down by the early 1980s. Occasional individuals have also been reported in Peebles, West Lothian, Fife, Perth, Angus, Argyll (Kintyre), and Caithness.

Reeves' Pheasants *Syrmaticus reevesii* released in Speyside in the early 1970s were still present in 1974, so were probably breeding in the wild at that time. There have been no subsequent reports from there but single birds have been reported in the last few years in Perth, Ayr and Peebles.

Lady Amherst's Pheasant *Chrysolophus amherstiae* is reported to be hybridising with the Golden Pheasants in Kirkcudbright, and a bird of this species was seen on Arran in 1977. Silver Pheasants *Lophura nycthemera* have been recorded in Kirkcudbright, Roxburgh, Peebles and Angus.

As birds of these species are so obviously exotic many doubtless go unreported and this account is therefore almost certainly incomplete.

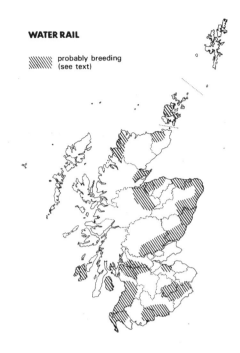

WATER RAIL

\\\\\\ probably breeding
      (see text)

[Helmeted Guineafowl *Numida meleagris*]

Widespread escapes are reported to have occurred in Angus since this species was introduced in 1980 (N. K. Atkinson).

Water Rail *Rallus aquaticus*

Local resident, requiring fairly extensive areas of dense aquatic vegetation for breeding and consequently absent from much of Scotland during the summer. In winter, when many immigrants are present, it is less selective and occurs also in more open situations, such as ditches and along river banks.

Water Rails are extremely secretive and are much more often heard than seen, except in hard weather when they may be forced to forage in the open. During the breeding season they are most likely to be located in the evening or at night, when weird squeals and grunts reveal their presence; tape-lures have recently been used successfully to stimulate 'song'. Little information is available on numbers but it is probable that the species is considerably underrecorded.

During 1968–72 proved or probable breeding was reported in Caithness, west Sutherland, east Ross, Inverness, Nairn, Aberdeen, Kincardine, Fife, Dunbarton, Lanark, Ayr, Wigtown, Kirkcudbright, Dumfries, Selkirk and Berwick, and on Orkney (Mainland), Islay and Arran. Breeding has since been proved in Aberdeen, Angus, Perth, west Stirling (Loch Lomond), and Renfrew. In Caithness nesting took place in 1972 and 1973 but not again; in Orkney it has not occurred since 1968; and there have been no further reports of 'singing' birds from west Sutherland or east Ross. B&R give old breeding records for Bute, Tiree, and the Uists, but the only recent summer reports from these islands are of calling birds on North Uist in 1971 and on Tiree in

1977; a pair possibly bred on Colonsay in 1983. Among the sites where breeding is known to occur regularly there are population estimates of 10+ pairs on Speyside, where Water Rails have nested on five different marshes (R. H. Dennis), 15–20 pairs in Ayr (Hogg 1983), and 10+ pairs in the Borders (1982 Borders BR). Using tape-lures Alan Wood located c25 pairs in the River Kelvin area in 1982 and c35 pairs in 1983; in 1984 two sections of his study area were drained during the drought period, with a loss of four pairs of Water Rails. The use of tape-lures in other areas could add greatly to our knowledge of the species' breeding population.

The Water Rail is a regular passage migrant through the Northern Isles, in both spring and autumn, and a few usually remain there throughout the winter. Passage is mainly from late March to early May and late September/November; the autumn movement is sometimes quite strong, as in 1975, when 17 birds were ringed on Fair Isle. This species is generally absent from Shetland in summer, but in 1973 one was heard in Unst in July. Passage and/or wintering birds also occur on the Outer Hebrides (including St Kilda) and Inner Hebrides, the Clyde Islands and, occasionally, the Isle of May. On the mainland, winter records are widespread, especially south and east of the Great Glen. Little information on numbers is available but there have been counts of 10–12 birds at Invergowrie Bay, Barr Loch and Lochar Moss (Dumfries).

There is no evidence that native birds migrate, although presumably dispersal takes place and local movements may result from particularly severe weather conditions. Birds of the Icelandic race *R.a. hibernaus*, have been identified in northern Scotland (BOU), but most of the winter immigrants probably belong to the nominate race *R.a.aquaticus*, which breeds throughout the remainder of the European range.

The recovery in Ireland in January of a bird ringed on Fair Isle the previous September indicates that at least some of the immigrant population moves further west in mid winter. The only evidence of the origin of Scottish-wintering immigrants comes from the recovery of a Netherlands-ringed juvenile at Loch Insh in December, and from one caught on Fair Isle in November and found dead in Sweden the following August. The ringing recovery data for Britain as a whole suggest that most of the immigrants arrive on the east coast but leave by a more southerly route (Flegg & Glue 1973).

So little is known about either past or present breeding populations of the Water Rail that it is impossible to assess whether any change in distribution or numbers is taking place. Many of the areas apparently most favoured are either so wet and prone to flooding that draining for agriculture is unlikely to be attempted, or else they have already been established as reserves. Recent experience along the River Kelvin has shown, however, that local drainage can result in a rapid decline in breeding numbers.

Spotted Crake *Porzana porzana* [1]

Irregular visitor, mainly in summer; annual since 1960 but probably breeding only sporadically.

The Spotted Crake is a very secretive species, most likely to be detected in potential breeding areas by its characteristic call. It is reported most often in May/June, although there are several April records (the earliest on the 2nd), and many of the calling birds remain in their chosen marsh well into July. It favours densely vegetated marshy ground, rather than reedbeds, and is consequently able to find suitable habitat in almost any part of Scotland. Reports of possible breeding, based on the presence of calling birds at the same site for several weeks, come from widely scattered localities on the mainland and from Shetland. In Sutherland there were three or four Spotted Crakes near Scourie in 1966 and the same site was occupied each summer, by at least one bird, from then until August 1970. In Inverness calling was heard in the Insh area in eight seasons between 1969 and 1980, in Dumfries there were records from several 10 km squares in 1968–72, in Perth birds were present in suitable habitat in four years between 1971 and 1976 (but not since), in

Shetland there were reports of calling birds from Unst, Fetlar and Mainland in the 1970s, and in Argyll one was heard for two weeks in June 1982.

The records suggest that the number of Spotted Crakes reaching this country in spring varies greatly between years; there were five early summer reports in 1978 but not more than three in any season from 1979 to 1983 inclusive. Autumn passage birds, presumably originating from the limited Scandinavian breeding grounds or the western edge of the main Eurasian range, occur mainly in August/September but have occasionally been recorded well into winter (the latest date is 31 December – on Islay in 1973). This species probably bred regularly last century, especially in the southwest; in the early 1900s breeding was proved once, in Roxburgh, and there were many passage records.

Sora *Porzana carolina*

Vagrant (N America) – three records, 1901–83.

There are two records from early this century, involving shot birds, and one recent sight record: from Foula on 30 October 1982 (BB 77:520). The earlier reports were from Tiree (October 1901) and Lewis (November 1913); both birds were immature males. With a secretive species such as this the chances of detecting the occasional wanderer must be minimal now that rails are afforded full protection – although a Sora on Bardsey obligingly trapped itself and so made identification simple!

Little Crake *Porzana parva*

Vagrant (E Europe/Asia) – five records, 1852–1983.

The only post-1950 records of the Little Crake are of one found dead on Unst in April 1959 and one caught on Fair Isle in May 1970; sadly, the latter was killed by a cat soon after it was released. Both were males. Birds believed to be of this species have been seen twice at Duddingston – in November/December 1952 and December 1962. The older records are from Banff (March 1852), Ayr (March 1909), and mainland Argyll (September 1911). The Little Crake, which breeds from Germany eastwards and sporadically in western Europe, has been recorded in Britain only about 100 times.

Baillon's Crake *Porzana pusilla*

Vagrant (Europe/Asia) – two records early this century and several older ones.

The most recent record of this small rail, which breeds only locally in western Europe, is of a female shot on Fair Isle in May 1929. B&R give one other 20th century record, from Caithness in August 1910, and several 19th century ones, scattered from Wigtown to Sutherland; where dates are given they are all between May and September. As with other skulking species, the decline in records this century may be as much a reflection of changing shooting practices as of decreasing numbers.

Corncrake *Crex crex*   [1]

Local migratory breeder which has been declining steadily for many years. The Hebrides now hold most of the small relict British breeding population, whose decline is associated with changing agricultural practices.

The Corncrake was once a widespread and common bird of farmland, favouring meadowland as nesting habitat, but by 1938–39 it was described as local and declining over much of Scotland and most of England and Wales (Norris 1947). The speed with which it has since virtually vanished from most of mainland Britain has been both spectacular and reasonably well documented. In 1968–72 breeding was recorded as confirmed or probable in 341 10 km squares in Scotland and 187 in England and Wales; by 1978–79 there had been an overall reduction in occupied squares of about 56% in Scotland and 95% south of the Border (Cadbury 1980b). This decline was most marked on the mainland, especially in the east and from the Clyde southwards on the west (Table a).

In 1968–72 Corncrakes were noted as at least probably breeding in all counties except Nairn, Banff, Kinross, Selkirk and Berwick; in 1978–79 the only mainland counties south of the Great Glen holding regularly-calling birds (as opposed to single reports, probably referring to passage migrants) were Argyll, Stirling, Perth, Renfrew, Ayr, Kirkcudbright, Dumfries and Roxburgh. West Sutherland was the only mainland area with more than a few pairs left (Table b). By far the largest numbers are now found in the southern Outer Hebrides and Tiree, although regular breeding still occurs in Lewis, Skye, Islay and Colonsay. In Orkney no calling birds were recorded on Mainland in 1984, where 38 were present in 1978–79, but a few pairs probably still breed on the other islands.

Numbers have declined substantially in many areas even where the breeding distribution has not contracted to any extent. Most of the areas surveyed in 1978–79 held only

one to five calling birds / 10 km²; 15 or more were recorded only in North Uist (in three 10 km squares), South Uist and Tiree (two squares each), and Benbecula, Colonsay and Iona (one square each). The maximum density found was on North Uist, where there were 35 birds calling in a single 10 km square (Cadbury 1980b).

The agricultural change believed to be the major factor responsible for the decline in Corncrake numbers is the replacement of semi-natural meadows by faster-growing leys cut earlier in the season, often when the birds are at the most vulnerable stage of their breeding cycle in late May and early June. Fast modern machinery is liable to destroy not only nests but also young birds unable to get out of the way in time. There are records from Wigtown of three nests lost during silage cutting in one season, and of two broods killed by a forage harvester in another, while in Stirling four of a brood of seven well-grown chicks (38 days old) were killed by a mower. Over the last 20–30 years silage making has spread to most parts of Scotland, including Orkney, but it is still unusual in the Hebrides, Shetland and the far northwest.

No attempt to investigate the Corncrake's habitat preferences was made until 1978–79, when James Cadbury carried out a survey in Orkney and the Hebrides. His findings (Cadbury 1980b) suggested that factors other than just the physical damage resulting from grass-cutting may be involved in the progressive decline of this species. When the birds arrive on the breeding grounds in May they tend to occupy the tallest available herb-vegetation. In the past this would usually have been marshy areas with tall herbs such as iris and meadowsweet; with improving grassland management, leys now frequently offer the best cover. More than 60% of calling birds in Britain as a whole were found in such grassland, which is intended for early cutting, but in the Outer Hebrides 50% were in fields with irises and in other

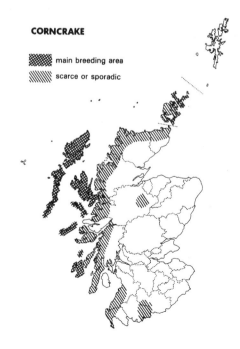

CORNCRAKE

▓▓ main breeding area

▨▨ scarce or sporadic

(a) *The recent decrease in the Corncrake population (Data from Cadbury 1980b)*

	No. of 10 km squares in which breeding confirmed or probable 1968/72	1978/79	% reduction in occupied squares
E Mainland (inc. Stirling)	88	10	89
SW Mainland (S of Clyde)	67	20	70
NW Mainland (Argyll to Sutherland)	61	24	61
Inner Hebrides	54	36	33
Outer Hebrides	32	31	3
Orkney	30	27	10
Shetland	9	2	(78)
Total	341	150	56

(b) *Numbers and distribution of regularly-calling Corncrakes recorded in 1978–79 (Data from Cadbury 1980b)*

Outer Hebrides:	total 260	Orkney:	total 102–104
S. Uist	83	Mainland	38
N. Uist	75	Westray	11–12
Benbecula	33	S. isles	25–27
other S isles	26	N. isles	28
Lewis	31	(exc. Westray)	
Harris/Berneray	12		

Inner Hebrides:	total 235–242	Other areas:	total c91–93
Tiree	85	Shetland Mainland	3
Skye	31–34	Clyde Is.	4+
Coll	28	W Sutherland	38–40
Islay	22–24	W Ross/W Inverness	5
Colonsay	20	Argyll	12
Small Isles	18–20	other W coast counties	15
other islands	31	Caithness/E Ross	7
		E Inv/Perth/Stirling	6
		Roxburgh	1

Scottish Total: 688–700 (+25 recorded on a single occasion and possibly breeding)

marshy areas. Alternative sources of tall cover, such as patches of nettles, docks, hogweed and other 'weeds' along field margins or in odd corners, have become progressively scarcer in most arable areas, but are still not uncommon in the islands. Leys and improved grasslands offer better cover for the birds but are probably poorer as breeding habitats than semi-natural meadows. They contain a smaller range of plant species and consequently may support a more limited range of invertebrates, and the lush growth of the grass makes movement difficult for the chicks, especially in wet weather. A study by Andrew Henderson (1983) found that in the Uists recent leys were the preferred habitat throughout most of the breeding season. Stands of iris were favoured by calling birds early in the season but after mid June there was a tendency to move into semi-natural meadows and arable crops, which by then afforded adequate cover.

Cadbury concluded that among the chief reasons why the Corncrake had so far maintained its numbers fairly well

in the Outer Hebrides were: the variety of habitats within a potential territory offered by the pattern of very small fields, the fact that many semi-natural meadows are still not mown until July, and the alternative cover and feeding sites provided by weedy crops, wide field margins, and marshy ground. He noted, however, that ewes and lambs were increasingly being spring-grazed on former meadowland around the crofts, with resultant loss of Corncrake habitat. The proposals for agricultural improvement since included in the Integrated Development Programme for the Western Isles aroused renewed concern for the future of this species, and stimulated a more intensive survey in the Uists and Benbecula in 1983. Only 167 calling birds were found where there had been 191 in June 1978. This represented a decrease of at least 12.6%, since it is likely that multiple visits in 1983 detected a higher proportion of calling birds than did the single June visit made to most areas in 1978 (Henderson 1983). On Tiree 96 calling birds were located in June 1983, a similar total to those recorded in 1975 and 1977 (J. Cadbury).

Elsewhere in Europe numbers have also been declining, and the Scottish population of c700 pairs, together with those in Ireland (estimated as c1,200–1,500 pairs in 1978), represents a substantial proportion of the total remaining in western Europe. As Cadbury comments, the future of the Corncrake appears bleak. The species shows no signs of being able to adapt to the environment created by more intensive farming methods, and its continued existence in western Europe may soon depend upon steps taken to preserve outmoded farm practices on reserves.

Those Corncrakes that do survive leave Scotland between late July and September; occasional October records may relate to passage migrants from Scandinavia. B&R describe the species as occasionally wintering, but the latest recent record has been 3 November. Corncrakes are vulnerable on migration too; they quite often collide with overhead wires, and of five overseas recoveries of Scottish-ringed birds, all in their first year, four were shot in France and Spain during September. The fifth was in the Congo in January. Little is known about the wintering range of this species but it is thought that the main wintering quarters are in eastern Africa (BWP).

Moorhen *Gallinula chloropus*

Resident, widespread and abundant except in the northwest Highlands and the islands. Native birds are sedentary but some immigration takes place in winter.

Moorhens are catholic in their habitat requirements, breeding on waters which range in size from small pools to the largest lochs, and beside both muddy ditches and large rivers. They frequently feed in the open, and are often at risk when crossing roads. The breeding season is unusually prolonged and downy chicks have been seen as early as January. Studies of the breeding biology of the Moorhen, and of a method of sexing this species in the hand, have been carried out at Culterty Field Station (Anderson 1965, 1975). Little information on breeding densities

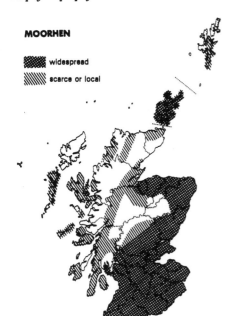

MOORHEN

▓ widespread

▨ scarce or local

is available, but about 100 pairs nested on St Serf's Island, Loch Leven, in 1974, and 64 pairs at Loch of Kinnordy in 1980.

As with many of the commoner species, there is little reliable evidence of change in distribution or numbers, and what there is relates to the fringes of the range. B&R considered that a decrease occurred following hard winters in the early 1940s, but there are no figures to substantiate this. However, the capture of over 2,000 in traps set for musk-rats in lowland Perth in 1933–34 might well have reduced the local population, at least temporarily. More recently mink have been suspected of causing local decreases (A. Anderson).

In the Outer Hebrides, where numbers have never been large, there has certainly been a decline, and no Moorhens now breed on Loch Stiapavat, Lewis, which once held six to eight pairs. It is doubtful if any currently breed in Lewis and Harris, although a few pairs nest on the machair lochs of the Uists and Benbecula; there is no record for Barra (Cunningham 1983). In the Inner Hebrides small numbers breed regularly on Islay, Tiree, Coll, Skye and Raasay, and sporadically on Eigg and Muck, and possibly Mull. The Moorhen has recently ceased to nest on Colonsay and has long been extinct on Rhum (Reed *et al* 1983); it is scarce on Arran but breeds regularly on Bute and Great Cumbrae. In Orkney it is widespread, occurring on most of the islands, but in Shetland it is a scarce and sporadic breeder. In Sutherland, Wester Ross, and much of Inverness, Argyll and Caithness Moorhens are uncommon and very local.

Unlike the Coot, this species seldom occurs in large concentrations on the water and counts of more than 50 at one site are relatively unusual, probably because many of the birds present remain unseen among vegetation. During periods of hard frost, however, they are forced out into the open to forage, and at such times 50–80 may be seen together.

Ice-cover on the lochs and marshes may also drive Moorhens to roost in trees, and I have seen nearly 40 perched in scrub willows at Stormont Loch under such conditions. In the north numbers are much smaller and counts seldom exceed 20–30 in Caithness and Orkney, or ten in Shetland.

Ringing recoveries indicate that native birds are largely sedentary; more than 70% of recoveries are within 10 km of the ringing site, and more than 85% within 100 km. All distant recoveries to date have been of birds ringed on passage at Fair Isle or in winter on the mainland; they came from, or were found in, Norway, Denmark, the Netherlands and France. Birds from much of the northwest segment of the species' European range, whch extends from southern Scandinavia, Britain, France and Iberia eastwards, are believed to winter in Britain (BWP), but little is known about the numbers involved.

[Gallinule spp.]

The record of a Gallinule of indeterminate race in Lewis in September/October 1964 (Cunningham 1983) is not accepted by the BBRC or BOU. An Indian Gallinule *Porphyrio porphyrio* found dead in Aberdeen in February 1965 was undoubtedly an escape. This species, which can be confused with the American Purple Gallinule *Porphyrula martinicia*, was formerly imported into this country (SB 3:231).

Coot *Fulica atra*

Resident, abundant and widely distributed in the lowlands but absent as a breeder from much of the Highlands and many of the islands. Immigrants increase the winter population, especially in the east.

The breeding distribution of the Coot is determined by its need for relatively shallow, eutrophic waters with abundant bottom vegetation for feeding and sufficient emergent vegetation for nest anchorage and concealment. Most such waters are at low altitudes, but B&R record nesting at more than 500 m asl. Although mainly vegetarian the Coot also eats insects and a variety of other foods, including eggs and chicks; one killed and ate a two-day old Slavonian Grebe at the Loch of the Lowes (Perth) in 1973.

No attempt has been made to assess the size of the breeding population and little information is available on densities. A count of 92 pairs on the Loch of Kinnordy in 1980 probably represents the higher end of the density range, while many larger waters hold no more than one or two pairs. The extent to which the population may fluctuate between years is apparent from counts at Lochwinnoch, where 17 pairs were present in 1980 but just three pairs in 1981. Only on the fringes of the range, where Coots are scarce enough to be noteworthy and easily countable, is there any record of recent population changes. It has been suggested (Atlas) that the apparent slight southerly contraction is associated with declining mean summer temperatures.

In Shetland small numbers bred in Dunrossness, and occasionally elsewhere, until the mid 1950s but no longer do so regularly; the Atlas shows no breeding record but nesting took place on Loch Asta in 1972, Sandwater in 1974, and Loch Tingwall in 1976 – these are the most recent reports. In Orkney the Mainland population decreased from the 1930s onward (Balfour 1972), but small numbers still breed both there and on several of the other islands. In the Outer Hebrides, 30 pairs nested on Loch Stiapavat, Lewis, in 1915; by 1932 the population was down to six pairs, and by 1974 breeding had ceased (Cunningham 1983). (The Atlas shows proved breeding in Harris, but according to Cunningham Loch Stiapavat is the only known site in Lewis and Harris.) Coots are quite numerous on the machair lochs of the southern Outer Hebrides and have bred on the Monach Isles. A few pairs nest on Islay and Colonsay but not on any of the other Inner Hebridean islands (Reed *et al* 1983). Nesting occurs on Bute but not on Arran.

Breeding numbers in Caithness are very small and the picture presented by the Atlas probably overstates the situation. In Sutherland only a few pairs nest regularly, all in the southeast, although sporadic breeding has been recorded in the west (Angus 1983). There are no breeding Coots in Wester Ross and few in west Inverness, though a temporary local increase was noted there during the 1970s (M. I. Harvey). This species is also scarce and local in Argyll, and relatively local throughout much of the southwest, southeast, and northeast.

After the breeding season large flocks of Coots gather in many areas and peak numbers are present in October (WWC 1983/84), when more than 4,000 are recorded in Scottish wildfowl counts. The relative proportions of native and immigrant birds in these gatherings is not known, but large influxes sometimes occur in the Northern Isles, and winter numbers in Caithness are much higher than can be accounted for by the breeding population. It seems likely,

therefore, that immigration occurs on a considerable scale. A particularly large movement was noted in the north in October/December 1976, when there were 180+ on Loch Spiggie, 470 on Loch Harray, 350 on Loch Ussie (Easter Ross), and 2,000 on the Loch of Strathbeg. In September 1981 Loch Leven held 2,400. Counts of 500 or more are recorded occasionally on many lowland lochs.

Numbers on waters in the east decline as winter progresses, but it is not clear whether this is due to continuing south-westerly movement, or simply to wider dispersal. The recovery in Ireland (in summer) of a Coot ringed as a chick in the Lothians shows that some long-distance movement of native birds occurs, but this is believed to be on only a very limited scale (BWP). All the overseas recoveries so far have been of birds ringed or recovered in Scotland in winter, and most involve movement from or through Holland, Denmark and Germany. The only breeding season recovery has been in Finland. Coots are migratory in the northeastern part of the range, which stretches from Britain, southern Fenno-Scandia, France and Iberia east through Europe and into Russia.

Even when waters are almost completely ice-bound Coots often remain in the area, feeding over nearby fields and marshes. Long-lasting ice cover is likely to result in the desertion of affected lochs, and movement to either running water or the coast. In hard weather considerable numbers are sometimes seen on the sea, for example 200 on Loch Ryan in January 1979. Stragglers and small flocks occur in the Hebrides more often in winter than at other times of year.

Crane *Grus grus*

Vagrant (Eurasia) – less than annual and in very small numbers.

Cranes have been reported with increasing frequency since the late 1960s and occur most often in spring and in the north and east. The maximum number reported in any year has been about ten (1969 & 1978), but as these very obvious birds sometimes move considerable distances in a short period it is not easy to be certain how many individuals are involved. The main arrival is in spring, starting in March and peaking in May, though 'new' birds may appear until the end of the year, some of these being immatures. Overwintering occurs occasionally, the longest 'resident' to date being one which stayed in the Ythan area from 26 April 1978 to 13 April 1979. Most reports are of single birds but small parties are sometimes seen, the largest so far being five at North Berwick in November 1976. In 1978 a displaying pair spent from early May until mid September in Orkney's east Mainland.

B&R give a few early 20th century records but only one (in Kirkcudbright) between 1910 and 1950. During much of this period Cranes were so scarce in Britain that the few isolated occurrences were thought to involve escapes; many are kept in collections throughout Britain. In the 1950s there were reports in only three years, in the 1960s in four years, and in the 1970s in every year except 1974. At least four Cranes arrived in 1980 and 1982 but there were no records

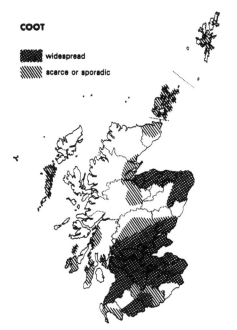

COOT

▓ widespread

▨ scarce or sporadic

in 1981. The four Northern Isles records in 1983 may have involved only one or two birds (one of those seen in Orkney was thought to be one of two adults which escaped from Kincraig Wildlife Park early in the year).

Cranes occur most frequently in Shetland, Orkney and Aberdeen but have been seen since 1950 in all east coast counties south to East Lothian, except Nairn, Banff, Angus and Midlothian. There are also records in this period from the Outer Hebrides (Uists & Benbecula), west Sutherland and Wester Ross, the Inner Hebrides (Mull), Argyll (Kintyre), Inverness, Lanark and Ayr.

The Crane breeds from Scandinavia, Denmark and north Germany eastwards through Russia and Siberia, and locally in the Middle East. It was formerly more widespread in Europe (it bred in east Anglia until about 1600) and the progressive contraction in range is probably due to loss of breeding habitat – it favours extensive wet meadows and bogs, often associated with birch scrub or mixed forest. Birds from the western part of the range winter mainly in Spain and the west Mediterranean area, but a regular wintering population has recently become established in France, c150 km east of Paris (BB 75:498). Migrating Cranes, which I have seen passing south over the Swedish island of Oland, give wonderful bugling calls and at a distance may both look and sound like flocks of geese.

Sandhill Crane *Grus canadensis*

Vagrant (N America) – one record, 1981.

The Sandhill Crane was first recorded in Scotland in 1981, when a first-summer bird was on Fair Isle on 26–27 April (BB 75:498). There had been an earlier occurrence in County Cork in September 1905. This species normally migrates west of the Mississipi when moving between its prairie breeding grounds and its wintering area on the coast of the Gulf of Mexico, so these two individuals had gone far astray.

[Demoiselle Crane *Anthropoides virgo*]

Escape. This decorative bird is often kept in collections, sometimes in a full-winged state. All British occurrences of the Demoiselle Crane, which breeds from the Black Sea eastwards and winters mainly in east Africa and the Indian region, are considered to involve escapes. The most recent Scottish reports are from Aberlady Bay (May 1972 and June 1977), Lanark (May 1972/March 1973) and Tiree (April 1973).

Little Bustard *Tetrax tetrax*

Vagrant (S Europe / Russia) – three records this century, the last in 1964.

The only recent record of the Little Bustard is of a male at Luce Bay, Wigtown, on 29 April 1964 (SB 3:253). B&R note two other 20th century occurrences: in Kincardine in 1912 and Aberdeen in 1935 (both during January), and several earlier. Although they refer to eastern and western races it is now accepted that this species is monotypic (BOU). The western part of the Little Bustard's range has contracted in the last 100 years and France is now the most northerly European breeding ground of this partly-migratory species.

Houbara Bustard *Chlamydotis undulata*

Vagrant (N Africa / Middle East) – one old record.

An immature female shot near St Fergus, Aberdeen, in October 1898 is the only Scottish record of this species, formerly known as Mcqueen's Bustard (ASNH 1899:51, 73). There have been fewer than ten British occurrences.

Great Bustard *Otis tarda*

Vagrant (S & E Europe) – three records this century, the last in 1970.

A bird of open country, this species formerly bred in many parts of England and is reputed to have done so in southern Scotland in the 16th century. There are several 19th century records and two between 1900 and 1950: in Orkney in January 1924 and Shetland in May 1936, both females. The most recent occurrence was on 11 January 1970, when an adult female appeared on Fair Isle; it was caught and kept for several weeks, then flown south to a zoo (BB 64:349). The Great Bustard breeds in Iberia and discontinuously in eastern Europe as far north as the Baltic. There have recently been attempts to reintroduce it on Salisbury Plain and also in West Germany.

Oystercatcher *Haematopus ostralegus*

Partial migrant, breeding in every county and island group; Scotland holds about 70% of the British breeding population. In autumn there is a marked southwest movement of local birds and an influx of immigrants; the largest gatherings of passage and wintering birds are on the Solway, Moray Basin and Forth.

Oystercatchers nest on all types of coast, along rivers and on loch shores far inland, and in arable and pasture fields. Inland nesting has been known since the 18th century in Scotland, although only a comparatively recent development further south (Buxton 1962). During the present century inland colonisation has continued and Atlas surveys recorded birds present in the breeding season in all but 25 10 km squares on mainland Scotland, most of the unoccupied squares being in either populous or steeply mountainous areas. Oystercatchers were not breeding inland in Clyde

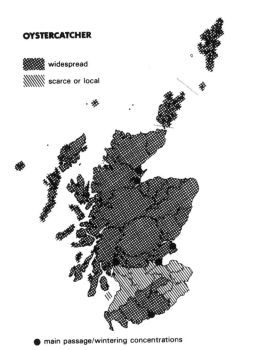

OYSTERCATCHER

▓ widespread

░ scarce or local

● main passage/wintering concentrations

in the late 1950s (Buxton 1962) but by the end of the 1960s much of Lanark had been colonised (Atlas). It has been suggested (Heppleston 1972) that a behavioural change enabled this species to exploit habitats which it formerly found unsuitable. Increasing numbers now nest on arable fields, but their successful breeding may be dependent upon the availability of pasture nearby, as ploughed ground offers inadequate food supplies for the chicks (Wilson 1978). Numbers have also increased in recent years at some, but not all, island sites. The Isle of May held 12 pairs in 1972 and 31 pairs in 1982, while the 1980 Balranald population of 110 pairs was more than double that of 1951. Fair Isle, in contrast, has shown little change, with 88 pairs in 1966 and in 1983, although numbers have fluctuated in between; the situation on Canna is similar, with the population fluctuating around 70 pairs (R. L. Swann).

There are few estimates of numbers, apart from those for discrete island populations, but Galbraith et al (1984) suggest that some 20,000 pairs breed on farmland and a further 2,000+ pairs on the Hebridean machair. On farmland below 300 m asl breeding density averages about 2 pairs/km^2; shingle areas may hold more than twice as many, while on machair the overall density is nearly 16 pairs/km^2. The Oystercatcher is clearly able to make use of a wide variety of habitats – as it does of nest sites. There are many records of nests in unusual locations, such as the hollow tops of posts and flat roofs, the latter situation proving satisfactory only because this is one of the few wader species that carry food to the young. Despite its adaptability it is wary and sensitive to disturbance; as a result many of those nesting in popular coastal areas probably fail to rear young.

Movement from breeding areas to the coast begins in July and numbers on the estuaries build up rapidly during August and September, when immigrants from Iceland, Faeroe and Norway start to arrive. At this time there is a marked overland movement from the Forth towards the Solway (Evans 1968), where the autumn peak is much more pronounced than on the east coast (Fig.). By mid winter numbers on the Solway are lower, as many birds move on to Morecambe Bay, into Ireland, or further south. On the east coast periodic influxes apparently continue into January, perhaps in response to severe weather on the Continent. Native birds return to the inland breeding areas during February and March, occasionally even in January, and mortality is sometimes high if severe weather follows an early mild spell. As they move inland they often gather in roosts of several hundred in major valleys before dispersing. By April/May most of the Oystercatchers remaining on the estuaries are non-breeders.

In 1969–75 the average November peak count on Scottish estuaries was nearly 50,000, representing 25% of the estimated British wintering population of 200,000 (Prater 1981). The largest concentrations in Scotland are on the North Solway, which regularly holds internationally important numbers (more than 7,500), the Moray Basin and the Forth (Table). The only other sites regularly holding nationally important flocks (more than 3,000) are the Eden and the Inner Clyde, but Loch Ryan, Wigtown Bay and Montrose Basin do so occasionally. Count coverage has been far from complete since the first phase of BoEE was concluded in 1975, but recent figures suggest that the Forth is increasing, and the Inner Clyde decreasing, in relative importance for

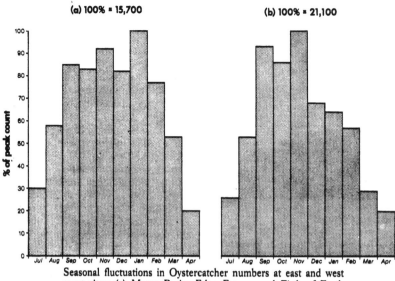

Seasonal fluctuations in Oystercatcher numbers at east and west
coast sites: (a) Moray Basin, Eden Estuary and Firth of Forth
(b) North Solway (based on highest average monthly counts 1969–75

this species. Oystercatcher numbers elsewhere on the mainland coast are much lower, but Loch Fleet, the Tay and
the Ayr/Ardrossan/Seamill coast frequently hold
1,000–2,000 (Prater 1981). Some 3,000–4,000 winter in the
Outer Hebrides (Buxton 1982a,b), more than 2,700 in
Orkney (Tay & Orkney R. G. 1984) and about 1,000 on Bute
(Rowe 1976). The wintering population in Shetland is pro-
bably fairly small, but preliminary results of the 1984/85
Winter Shorebird Survey suggest that some 10,500–14,500
Oystercatchers winter on Inner Hebridean and mainland
coasts from Kintyre to northern Sutherland inclusive (M.
E. Moser).

There have been several hundred recoveries of Oystercatchers ringed in Scotland and even a limited study of the
data reveals some interesting facts. Many native breeders
winter around the Irish Sea but some go south to Spain
and Portugal. Of 70 ringed as chicks and recovered more
than 250 km from the natal area in November/February of
their first winter: 28 were on the eastern shores of the Irish
Sea, 14 in Ireland, 15 in south Wales and southwest England,
four in France, two in Spain and only seven in Scotland.
Of 75 birds ringed in Morecambe Bay and on the Solway

during the winter and recovered in Scotland in the breeding
season, all were north of the Tay and more than half in
Aberdeen and the Northern Isles. Although a substantial
proportion of the Scottish Oystercatcher population clearly
moves further south and west, some remain to winter on
the Moray Basin and the Forth as well as on the Solway.
A high proportion of those wintering on the Moray Basin
are coastal breeders, and about 80% of them are adults,
suggesting that most Oystercatchers breeding inland in the
Highlands move right away to winter, as do many of the
young birds reared on the coastal strip (Swann 1985). Both
ringing recoveries and sightings of colour-marked individuals have confirmed that birds from Grampian and Shetland
are on the Forth in mid winter, and the same study has
shown that many return to the same area in successive winters
(Symonds *et al* 1984; Symonds & Langslow 1984). Immigrants come from a wide area in the western part of the species'
European range. There have been Scottish recoveries of
birds ringed as chicks in Iceland, Faeroe, Norway and Holland, while of three birds ringed on the Solway on the same
day in March, two were recovered in Faeroe and one in
south Norway.

There seems little likelihood of any serious threat within
Scotland to the Oystercatcher population, on either the
breeding or the wintering grounds. The varied nesting habitats currently occupied suggest that this species can adapt
to exploit a gradually changing environment, while the absence of shellfish operations on the principal wintering
estuaries means that the birds are not in conflict with commercial interests. This is not necessarily true of the major
wintering areas in England and Wales, where more than
25,000 Oystercatchers were shot between 1956 and 1974
in response to lobbying by cockle-fishers, who claimed that
the birds were having a significant impact on catches. Many
Scottish birds winter in these areas so this action, had it
been continued, might well have had an effect on the breeding population. It is now accepted, however, that other factors

*Peak numbers of Oystercatchers at major passage/wintering sites (Data
from Prater 1981 & M. Moser)*

	Highest av. monthly count 1969–75	Post-1980 counts Av. peak (n)	Range
North Solway	21,000	15,900 (4)	13,800–18,000
Moray Basin	8,000	6,400 (2)	5,000–7,800
Firth of Forth	5,500	6,800 (4)	5,700–8,300
Eden Estuary	4,100	3,800 (4)	3,600–3,900
Inner Clyde	3,500	4,600 (4)	4,200–5,000

are at least as important as the birds in determining cockle catches (Prater 1981). In Scotland the feeding ecology of Oystercatchers has been studied on the Ythan Estuary (Heppleston 1971).

Black-winged Stilt *Himantopus himantopus* [1]

Vagrant (Europe) – four records this century.

There were several old, and sometimes dubious, records of the Black-winged Stilt prior to 1900, but just one during the first half of this century, in Wigtown in 1920 (B&R). Since 1950 there have been three reports: from Erskine, Renfrew, in October 1958, Musselburgh in July 1975, and the Ythan area in October 1984. (A report from Moray in April 1953 (SN 1955:102) was not accepted.) The nearest regular breeding grounds are in southern France and Iberia but the Black-winged Stilt also nests sporadically in the Netherlands, and in 1945 and 1983 did so in England, where it occurs almost annually between April and November. Although some birds winter around the Mediterranean, most migrate further south.

Avocet *Recurvirostra avosetta* [1]

Vagrant (Europe, including England) – occurs less than annually and in very small numbers.

The Avocet is such a conspicuous bird that it is unlikely to pass unnoticed, so the records probably represent a fairly complete picture of its occurrence. From 1950 to 1983 inclusive it was recorded in 19 years, most often on the east coast from Shetland to East Lothian; 75% of reports were in March/June but there are records for all months except September, November and December. It is difficult to be certain how many birds are involved in any one year as they may move substantial distances in a short time; for example a bird seen flying west at Aberlady on 13 January 1968 is presumed to be the one that arrived at Skinflats later the same day, and may also be the one found dead at Fife Ness on 25 January. The maximum annual total to date has probably not exceeded ten. Most reports are of not more than two birds, but there were up to five at Aberlady between 6 and 11 April 1958.

Since 1950 Avocets have been recorded in Shetland, Orkney, Sutherland, Wester Ross, Inverness, Moray, Aberdeen, Fife, Midlothian, East Lothian, Stirling, Dunbarton, Ayr and Mull. The few records given by B&R include only two from an area for which there is no recent report; both relate to birds shot in Lewis at the end of last century.

The Avocet populations breeding in southeast England, the Netherlands and Denmark have increased substantially in the last 30 years. Small numbers winter in southwest England but most of the European breeding population migrates to coastal areas of southern France, Iberia and northern/ northwestern Africa.

Stone-curlew *Burhinus oedicnemus* [1]

Vagrant (Europe, including England) – nine records since 1950.

The post-1950 records of the Stone-curlew have been irregularly spaced and mostly from the east coast: Shetland (June 1955 & December 1981), Fair Isle (May 1963, May 1964 & May 1974), Fife (January 1965), Lanark (April 1966), Peebles (October 1971), and Angus (September 1974). B&R knew of five earlier occurrences, the first in 1858, when one was shot in Fife in January. The others were at Stirling (August 1897), Fair Isle (June 1913), and the Isle of May (April 1937 & May 1946). The 1981 Shetland record, of a bird which was found injured on 6 December and later died, is only the second winter occurrence this century.

The small breeding population in southeast England, estimated in 1972 as 300–500 pairs (Atlas), is at the extreme northwest of the Stone-curlew's range, which extends from France and Iberia eastwards but does not include Germany; breeding ceased in the Netherlands about 1958 (Parslow). Birds from the northern part of the European range winter in eastern Africa.

Cream-coloured Courser *Cursorius cursor*

Vagrant (Africa / SW Asia) – three occurrences, 1868–1983.

The first record of this species, which breeds in desert regions, was in 1868, when a male was shot near Lanark in October (B&R). The two subsequent records were also in autumn – as have been all the English occurrences to date. On 10 October 1949 three Cream-coloured Coursers were seen on ploughed land near Luggiebank, north Lanark (SB 5:28), and from 9 to 20 October 1965 a single bird frequented the beach and dunes around Gullane Point, East Lothian, where it was observed by large numbers of birdwatchers (SB 4:230). I spent some time there, crouched behind an oil-drum, filming the bird as it worked its way towards me along the tide-wrack, and was struck by its alternately 'dumpy' and 'neck-stretched' postures.

Collared Pratincole *Glareola pratincola*

Vagrant (S Europe / SW Asia) – five records 1950–83 and some older ones.

The post-1950 records of the Collared Pratincole, all between June and early October, are from South Ronaldsay, Orkney (1963), Fair Isle (1971), Slains, Aberdeen (1972), Mey, Caithness (1973), and Unst, Shetland (1974). B&R give six records between 1899 and 1935, several of them in summer, but it has been suggested (BOU) that some of the older reports of this species may have been due to misidentification of Black-winged Pratincoles. The nearest parts of the species' fairly extensive breeding range are along the Iberian and north African fringes of the Mediterranean. The Collared Pratincole winters in Africa south of the Sahara, generally near water in short-grassland or on areas of bare ground, and is said to be a wide-ranging species, which presumably explains its sporadic occurrences in Britain, where it had been recorded only about 70 times by the end of 1983.

Black-winged Pratincole *Glareola nordmanni*

Vagrant (E Europe) – two records, 1927 and 1976.

The first record was from Fair Isle, where a female was shot on 18 May 1927 (SN 1927:111). The only subsequent occurrence has been at the Loch of Strathbeg, Aberdeen, on 11 July 1976 (SB 10:314–15). (Note that in BB 71:498 and SBR 1976 the date is wrongly given as 15 July). All the other recent British records (only 19 by the end of 1983) are from the southern half of England and between June and the end of September. The Black-winged Pratincole breeds from Rumania eastwards through the Ukraine and winters in southern and eastern Africa. Like the previous species it favours wet areas surrounded by grassland.

Little Ringed Plover *Charadrius dubius*   [1]

Vagrant (Europe, including England) – occurs less than annually and in very small numbers. Has bred once, in 1968.

The Little Ringed Plover has occurred with increasing frequency since 1965, doubtless due to the expansion of the English breeding population; it first bred in Hertfordshire in 1938 and by 1972 had colonised gravel pits and waste-ground sites as far north as Northumberland (Atlas). Although now breeding so close to the Border, it is still a scarce visitor to Scotland. On the Continent it is widely distributed from southern Fenno-Scandia southwards; European birds winter in tropical Africa.

B&R give an old record for Aberdeen, an undated record of one shot in North Uist in 1908, and a somewhat strange report of three seen in Skye on 3 June 1949. There were no further records until 1965, when three, all thought to be young birds, appeared at Fair Isle (4 September), Whalsay (17–19 September) and Aberlady (12 October). From then until 1983 the Little Ringed Plover occurred annually except for 1966, 1972, 1973, 1978 and 1982. The maximum number of reports in any year has been three. Nearly two-thirds of occurrences are in spring, between mid April and mid May, and the remainder between 14 July and 12 October. There are records for Shetland, Fair Isle, Aberdeen, Midlothian, East Lothian, Berwick, Stirling, Clackmannan, the Lanark/Dunbarton border, Renfrew, Ayr, Dumfries and Islay.

On 1 July 1967 a Little Ringed Plover was seen on waste-ground near Hamilton, Lanark, and the following year a pair was in the same area from 8 May to 15 July. Their behaviour made it clear that they were nesting, and on 26 May 1968 a chick was seen (SB 5:282).

Ringed Plover *Charadrius hiaticula*

As a breeder widely distributed inland as well as on the coast; numbers are largest in the Outer Hebrides and Northern Isles. Passage movements peak in May and August/September, and there are important wintering concentrations in the Uists and Orkney and on some mainland estuaries.

Scotland holds about two-thirds of the British and Irish breeding population of Ringed Plovers, estimated in 1984 as c8,500 (A. J. Prater). Working from the Atlas data and sample counts in different areas, Prater (1976) assessed the Scottish total as c3,500 pairs, with by far the largest numbers in the north and west, where estimates were in many cases based on only a few records. Studies in the Hebrides and a more comprehensive breeding season survey in 1984 have shown that these figures were too low; the current estimate is c6,000 pairs (Table a). The only county for which there

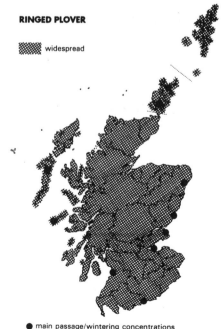

RINGED PLOVER

▨ widespread

● main passage/wintering concentrations

(a) Numbers and distribution of breeding Ringed Plovers in 1973–74 and 1983–84 (Data from Prater 1976 and A. J. Prater)

	No. of pairs 1973–74	No. of pairs 1983–84
Southeast:		
Midlothian, Peebles & Berwick	34	27
East Lothian	76	96
East-central:		
Stir., Fife, Angus, Kinc.	79	83
Perth	75	60
Southwest:		
Wigtown, Kirkcudbright & Dumfries	115*	154
Dumfries, Selkirk, inland	125*	40*
West-central:		
Ayr	105	100
Renfrew, Lanark, Clyde Is.	60	200
Argyll, mainland	300*	200*
Inner Hebrides	300*	550
Northeast:		
Aberdeen, Banff, Moray & Nairn	95*	120
North:		
Inverness	50*	60
Ross & Cromarty	205*	200*
Caithness & Sutherland	245	240
Orkney	350*	545
Shetland (inc. Fair Is)	359*	900
Outer Hebrides:		
S. Uist	450*	1,118
N. Uist, Benbecula	400*	718
Other islands	150*	430

* estimates based on small sample

Notes: (1) 1973–74 cover of inland areas, Outer Hebrides, Orkney and Shetland was very incomplete.
(2) In Ross & Cromarty and mainland Argyll cover was very poor in both surveys.
(3) 1983–84 estimates are preliminary, although considered reasonably accurate.

appears to be no recent breeding record is Clackmannan.

The density of breeding Ringed Plovers in the Uists and Benbecula is greater than in any area of comparable extent elsewhere in Britain and the islands support more than 35% of the Scottish population (Table a). On the Hebridean machairs densities range from less than one to nearly 40 pairs/km^2, with the average around 16 pairs/km^2 (Galbraith *et al* 1984). Numbers are also high on the sandy Inner Hebridean islands; Islay had 220 pairs in 1984 and Tiree 119 pairs (A. J. Prater). Elsewhere in Scotland coastal-breeding Ringed Plovers favour sand or shingle beaches, while inland they occur on river shingle-banks and the shores of lochs and reservoirs, sometimes as far from the sea as it is possible to go, eg by Loch Rannoch. They occasionally choose more unusual nesting sites, for example two or three pairs bred at Leith Docks in 1968, nests have been found on a concrete foundation in Wigtown and a forestry road in Stirling, and I once watched a bird return to a clutch neatly placed on a dried-up cowpat. B&R recorded nesting in ploughed fields and this habit has continued, although still relatively uncom-

mon except in the Uists. Breeding numbers vary considerably between years; there were 20 pairs inland in Midlothian in 1976 but only ten in 1978, and 43 pairs on Unst in 1973 but 64+ there in 1974. It seems likely that a slight overall increase has taken place during the last 30–40 years, with more inland breeding records and attempts to colonise new areas, as on the Isle of May where nesting (unsuccessful) was first recorded in 1977.

The Ringed Plover breeds from northeast Canada, Greenland and Iceland eastwards through northern Eurasia. Birds from the more northerly parts of the range are migratory, but many British-breeding birds remain to winter in this country (Prater 1981). The passage movements through Scotland in autumn and spring are known to involve birds from northeast Greenland and thought likely to include some from Iceland and Scandinavia. These winter as far south as West Africa; ringing recoveries include birds marked in Shetland and northeast Scotland on autumn passage and found in France, Spain, Portugal and Morocco between mid September and December; birds marked in southern Norway found in Ross and Kirkcudbright in October; and birds colour-marked in Greenland, seen in August/September the same year in Aberdeen, Angus and Kirkcudbright.

The first of the autumn immigrants pass through Fair Isle in late July but the main influx is during the second half of August and in September; thereafter numbers drop rapidly in most areas to a much lower wintering population. In autumn the Uists and Benbecula hold c1,500, of which about 66% of an August sample were juveniles (Summers & Buxton 1979). Flocks of 300 or more are frequent on the large east coast estuaries in autumn and also along the Ayr coast and on the Solway. Few regular counts are made on the Inner Hebrides, but there is an October record of 240 on Colonsay.

The Outer Hebrides hold what is apparently the largest wintering concentration in Britain; in February 1980 more than 1,800 were counted on the Uists and Benbecula, and the estimated total for the islands is 2,800–3,200 (Buxton 1982). This represents a significant portion of the Western European wintering population, estimated as 20,000–25,000 (Prater 1981). Little information is available on wintering numbers on the sandy Inner Hebridean islands, but c1,000 were on Tiree in mid winter 1984/85 (M. E. Moser). Orkney is now known to hold some 1,600, widely scattered through the islands (Tay & Orkney R. G. 1984). Comparatively few Ringed Plovers winter in Shetland but several hundreds are scattered along the beaches from Caithness to Easter Ross (Summers & Buxton 1983). On the Firth of Forth the December/February average peak count in 1976–81 was 570 (BoEE) but on the Solway numbers drop to 200–250 in mid winter; elsewhere around the coast they seldom exceed 100.

The large winter flocks on the Uists include a high proportion of local birds and some from the northern mainland (Summers & Buxton 1979; N. E. Buxton), while the recovery at Brora in December of a bird ringed at the same site as a breeding adult suggests that some Ringed Plovers are also resident throughout the year on other parts of the coast (Mainwood 1979). Ringing has shown, however, that some native birds move southwest to winter in Morecambe Bay and Ireland; there have been no recoveries of Scottish-bred birds further south than this.

Native birds start returning to their breeding grounds in mid-late February and in some seasons the first eggs are laid by the end of March; inland and in Shetland nesting may be several weeks later than on the mainland coast. By early April numbers around the estuaries are generally at their lowest but later in the month, and especially during May, many migrants pass through, and peak counts are recorded at most sites (Table b). The Solway has the largest concentrations at this time; an all-time peak of 3,400+ occurred in 1974, when there were 3,000 Ringed Plovers between Priestside and Powfoot. Nationally important numbers of passage birds (more than 300) also occur in May on most of the east coast estuaries. On the west coast the May passage is less pronounced but in Lewis an April movement has been recorded; it has been suggested that this involves Icelandic birds and that the May passage consists mainly of birds heading for Greenland, which overfly the Hebrides (Ferns 1980a). There have been several sightings of Ringed Plovers high in the Cairngorms at the time of the May migration.

The future of Scotland's breeding population of this species will clearly depend to a large extent upon developments in the Outer Hebrides. Any major change in the management of the machair, such as is likely to result from a subsidised programme of agricultural improvement, could have an adverse effect on breeding numbers and probably also on breeding success. The very high densities in some machair areas present an unusual opportunity for the conservation of a breeding population. Coastal breeding populations in many other parts of the country have already been affected by ever-increasing disturbance from recreational use of the beaches. Even if the gradual spread of inland breeding continues this would be unlikely to go very far towards compensating for the loss of even a relatively small proportion of the machair habitat. The passage population, frequenting sandy shores and often the less reclaimable parts of estuaries, does not appear to face any serious threat.

Killdeer *Charadrius vociferus*

Vagrant (N America) – four records, 1867–1984.

The earliest record of this plover, widely-distributed in America, is of one reputedly shot at Peterhead in 1867 (ASNH 1904:247); the specimen was not identified until 1904, when it was found (wrongly labelled as a Ringed Plover) in a museum drawer. This record should probably be regarded with some suspicion, although accepted by BOU and Sharrock & Sharrock (1976). There was no possible doubt about the identity of the Killdeer which appeared at Bo'ness, West Lothian, on 16 January 1983 and remained in the area – much observed and photographed by enthusiastic birdwatchers – until at least 17 March (BB 77:521). Later the same year another was reported, on South Uist, where it remained from 30 December until 7 January 1984. The most recent record is from Portencross, Ayr, on 21 January 1984.

(b) Peak numbers of Ringed Plovers at main passage and wintering sites (Data from BoEE, SBR, Buxton 1982, Summers & Buxton 1979, Tay & Orkney R. G. 1984, Moser 1984, BTO/WSG Winter Shorebird Count, preliminary figures per M. E. Moser)

Period	Site	Count (av. – if available, or max.)
Aug–Sept	Uist & Benbecula	1,500
	North Solway	200+ 1982–83
	Ayr coast & most east coast estuaries	300+ (each)
Nov–Mar	Outer Hebrides	c3,000
	(Uists & Benbecula	1,800+)
	Orkney	c1,600
	(Mainland	c500)
	Shetland	c600*
	Firth of Forth	550+ (av. 1976–81)
	Caithness	200+
	North Solway	c200
	'Strathclyde' (1)	2,400–3,000
	(Tiree	990)
	'Highland west coast' (2)	700–900
May	North Solway	1,350 (av. 1983–84)
	Eden Estuary	710 (1981)
	Montrose Basin	500+ (1975)
	Tyninghame	500 (1981)
	Aberlady Bay	400 (1981)
	Ythan Estuary	350 (1981)
	Tentsmuir	c300 (1981)
	Forth Estuary	300+ (1980)

*Preliminary figure from 1984/85 Tay Ringing Group/Shetland Bird Club Survey.
(1) 'Strathclyde' = coasts of Kintyre, all islands and mainland west coast to boundary with Highland Region.
(2) 'Highland west coast' = all mainland coasts of Highland Region from Strathclyde boundary to Strathy Point (Sutherland), Skye and the Small Isles.

Kentish Plover *Charadrius alexandrinus*　　　　[1]

Vagrant (Europe) – seven records, 1949–83.

Although formerly breeding in England, this species was not recorded in Scotland until 1949, when one was on Fair Isle on 14 May (SN 1950:24). There have been six subsequent records: at the Ythan Estuary (May 1962 & May 1981), Elie Bay, Fife (April 1966), Carnoustie, Angus (September 1974) and Culbin, Moray (June 1975). The Kentish Plover breeds on or near North Sea and Atlantic coasts, from Denmark to Gibraltar and southwards into western Africa, and also around the Mediterranean; further east, in Eurasia, it is typically an inland nester. It last bred in England (Sussex) in 1956, after numbers had declined progressively from around 40 pairs at the start of the century (Parslow 1973).

Greater Sand Plover *Charadrius leschenaultii*

Vagrant (Middle East/ W Asia) – two records, 1979 and 1982.

The first Scottish record was in 1979 (only six months after the first one in England), when a Greater Sand Plover was at Sandside Bay, Deerness, Orkney, from 9 June to at least 14 June (BB 73:568–73, 74:467). The second was at Aberlady on 24 June 1982. A detailed description, with comments on the problems of distinguishing this species from the Lesser Sand Plover, is given in Sharrock & Grant (1982). The Greater Sand Plover breeds on dry steppes in west and central Asia, Turkey and Jordan; its nearest wintering grounds are in the eastern Mediterranean, around the Red Sea and in eastern Africa. The increasing number of European records may indicate a recent expansion in the breeding range.

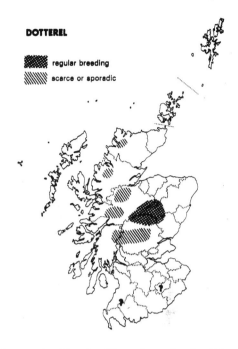

DOTTEREL

▓▓ regular breeding
▨▨ scarce or sporadic

Dotterel *Charadrius morinellus* [1]

Scarce and local migratory breeder, nesting mainly on the Cairngorm and Grampian plateaux. Occasional on passage, when often on coastal links and similar short-turf habitats.

Over most of its breeding range in Britain, Fenno-Scandia and southeast Europe the Dotterel is confined to the arctic-alpine zone, nesting either on exposed mountain summits or on tundra at lower altitudes. (A group which started to breed on newly-reclaimed polders in Holland during the 1960s is an interesting exception to this pattern.) It was formerly much more abundant in Scotland, both as a breeding bird and on passage, but suffered greatly from the attentions of both collectors and fowlers, to which its characteristic tameness made it particularly vulnerable (B&R). Scotland holds virtually all of the British breeding population which totals about 80+ pairs in a good year.

Long-term studies over the last 40–50 years suggest that the population has increased during this period, although breeding success fluctuates markedly in response to weather conditions. Nethersole-Thompson (1973) assessed the 1945–69 average population as c70 pairs, with perhaps 100 pairs in a good year; the comparable figure given for the 1930s is under 50 pairs. His estimate for the Cairngorms in 1945–69 was 20–25 pairs; 50+ pairs were thought to be present there in 1970–71, while in 1976–80 there was a larger average breeding population in the central Grampians than in any previous five-year period (Nethersole-Thompson & Watson 1981). Roy Dennis estimates the current Scottish total as 100–150 pairs, so there has clearly been some recent increase.

These high plateau areas hold by far the greatest concentrations of Dotterels (Table), which have a marked preference for hills of over 900 m asl with broad flat tops or long rounded ridges. Outwith the main breeding areas in Inverness, Banff, Aberdeen, Angus and Perth, Dotterels have bred occasionally this century in Sutherland (only one

Numbers and distribution of breeding Dotterels (Data from Nethersole-Thompson 1973)

	Est. no. of pairs breeding in most years 1945–69
Sutherland	0–1
Ross	3–8
Inverness – N of Great Glen	1–4
– Monadhliath	2–6
Cairngorms – east	8–15
– central	5–11
– west	7–10+
Grampians – east	10–18+
– central	10–23
– west	5–10
Perth – south & west	0–2
Argyll & S. Uplands	0–1

proved record, near Cape Wrath in 1967), north and west Inverness (1977), south and west Perth (from 1968), Kirkcudbright (1967), Selkirk/Peebles (1970), and Argyll (1982 & 1983), and more regularly in Wester Ross. A pair was seen in a suitable nesting area in Shetland on 10 July 1974. During the 1970s there was some evidence of extension of the breeding range and also of an increase in numbers, with more records from lower hills. Nethersole-Thompson (1973) has suggested that such an expansion may be related to climatic conditions, with the breeding population tending to spread and occupy lower sites during cold periods and to retreat from marginal habitats on the fringes of the range during warm summers.

Although the major breeding sites share the same physical features of altitude and landform, they include a variety of rock types. These influence the nature of the vegetation, which in turn determines the density of invertebrates present and consequently affects breeding success. The ratio of young:old birds is lowest on the barer granite summits, intermediate on the Moine gneisses, schists and quartzites of the central and western Grampians, and highest on the Dalradian schistose of the eastern Grampians, which carry a predominantly short-grassland vegetation (Nethersole-Thompson 1973). Dotterels generally nest on dry terrain, either on bare ground or in short vegetation, often among scattered boulders or grey hummocks of woolly fringe-moss, and usually exposed to the severe weather that these high tops experience even in mid summer. Not surprisingly the productivity of birds breeding under these conditions is not high and in years of snowfall in June/July few young are reared in the Cairngorms (Nethersole-Thompson & Watson 1981).

The increasing ease of access to the summit of Cairn Gorm has led to many expressions of concern regarding the adverse effects that more frequent disturbance might have on the breeding success of Dotterels and other montane species. There is good evidence that numbers of potential predators, notably crows and gulls, have increased in this area as a result of the easy pickings from discarded picnics (Watson 1979), and these present an added threat to eggs and young. Although individual birds suffer from egg-robbing, there is as yet no proof that Dotterel numbers have decreased and the most recent surveys show that the adult population on the Cairngorms is in fact very stable (Nethersole-Thompson & Watson 1981). It would nevertheless be foolish to assume that the species will not require continuing careful monitoring and protection from avoidable disturbance if it is to maintain its present numbers.

Dotterels winter on shores and semi-desert in North Africa and the Middle East; birds ringed as chicks in the central Highlands have been recovered in Algeria and Morocco. During migration they are most often reported from coastal areas, although there are occasional sightings of parties on inland hills in areas where they do not breed. Recent spring records cover the period from 11 April to 9 June, with over 80% in May; the earliest date noted by Nethersole-Thompson (1973) is 12 March, in the central Cairngorms, where breeding birds normally return during the first week of May. Autumn migration extends over a longer period but is most marked during August and September; there are several October records and one of a very late bird on Fair Isle on 6 November 1976. Birds on passage through the Northern Isles, and probably some of those seen further south, presumably come from Fenno-Scandia (the species does not breed in Iceland). Since 1968 there have been reports from most coastal counties, all island groups other than the Clyde Islands (for which B&R mention old records), the Isle of May and St Kilda. Most records are of single birds or parties of up to five, but a group of eight flew over Islay in September 1974 and flocks of twelve have been seen at Machrihanish and on Ben Lawers in May.

Lesser Golden Plover *Pluvialis dominica*

Vagrant (N America/Asia) – twelve records, 1883–1984.

Both races of this species – the American *P.d. dominica* breeding in Alaska and northern Canada, and the Asiatic *P.d. fulva* breeding in Siberia – have been recorded. B&R give one old record of each, the American in Perth in August 1883 and the Asiatic in Orkney in November 1887; both birds were shot. There were no further records until 1956, when a bird of the American race was on Fair Isle on 14–15 September.

During 1966 several sightings in Ireland were followed by an increase in reports of Lesser Golden Plovers in England, but no more were observed in Scotland until 23 November 1975, when a *dominica* was at Caerlaverock, where it remained until 26 February 1976. (A record from the same area for September/October 1976 was rejected by BBRC.) There were reports of Asiatic-race birds at Aberlady on 10–16 July 1976 and again (possibly the same bird) on 9 July 1977, and of American-race birds at West Fenton, near Gullane, on 13–14 November 1976, and at Aberlady on 16–20 September 1981. In 1977 adult American birds were seen at Kinneil, Stirling, on 14–22 August and at Musselburgh on 9–17 September; these reports could refer to the same bird. Recent records include an adult *dominica*, still showing some of its breeding plumage, on Fair Isle in late August 1983, and a first-winter bird, also there, in October 1984.

A paper by Pym (1982) on the racial identification of the Lesser Golden Plover includes a summary of distinguishing features at different seasons.

Golden Plover *Pluvialis apricaria*

Partial migrant, widely distributed in the breeding season on tree-less moorlands. In early autumn flocks gather on lowland farmland and estuaries, where some winter while others move further south. From August to April immigrants, mainly from Iceland, are also present.

The Golden Plover's breeding grounds include boglands and flows almost down to sea-level in the north, the rolling heather grouse moors of the central Highlands and the grassier Southern Uplands. Scotland holds about two-thirds of the British and Irish breeding population, estimated in 1969–74 as c30,000 pairs but probably now somewhat lower. Breeding densities vary considerably from one part of the country to another and even between apparently similar habitats; densities of $5+$ pairs/km² were present locally in the

Southern Uplands in the early 1980s (N. Easterbee), and 2 pairs/km² on Caithness flows (Reed *et al* 1983a) but many Scottish moors support less than one pair/km² (Ratcliffe 1976). Breeding numbers and success are influenced by weather, especially by severe conditions in late spring when the birds would normally be establishing territories, and by predation, with up to 75% of eggs sometimes being lost (Parr 1980).

In the first half of this century numbers declined in many areas, both in Britain and elsewhere in Europe. Possible reasons for the decrease in Britain, among them reduction in moorland carrying capacity and increase in predation, are discussed by Ratcliffe (1976) in his full account of the breeding of the Golden Plover. More recently local reductions in population size and distribution have resulted from widespread afforestation, as in Galloway, or from agricultural improvement, as in Orkney, where much moorland has been brought into cultivation in the last 10–15 years. Despite these reductions the breeding population as a whole is probably not yet seriously at risk, since much of the species' range is unsuitable for land-use changes of these kinds. There have been local increases in a few areas, with the first proved breeding on Canna in 1976, attempted breeding on Foula for the first time in 1970, and nesting on Fair Isle in 1970 after an interval of 26 years.

Golden Plovers start to leave the breeding grounds in early July and gather in flocks on low ground, where they often mingle with Lapwings. They seem to have a strong fidelity to particular areas, with flocks appearing annually at the same time in the same groups of fields, favouring those currently under grass or potatoes. As the winter progresses, inland areas are largely deserted and the birds move nearer the coast (Table). In the winter of 1977/78 all gatherings of more than 1,000 found between November and February were on the coast (Fuller and Lloyd 1981). Comparatively few Golden Plovers feed in the tidal zone, however, and grassland is by far the most important feeding habitat. Many roost on arable land, especially ploughed ground, but grassland and coastal areas are also used, the latter more often in Scotland than further south.

The first immigrants probably arrive in late July but little is known about the relative proportions of native and immigrant birds in the wintering flocks, or about their respective

GOLDEN PLOVER

▨ widespread

▧ scarce or sporadic

● main passage/wintering concentrations

Numbers of Golden Plovers at major coastal and near-coastal passage and wintering areas (Data from Fuller & Lloyd 1981, SBR, BoEE, Moser 1984 and preliminary figure from Tay Ringing Group/Shetland Bird Club 1984/85 survey)

	Count, year & season
North Solway	9,500 1977/78 c1,000 (estimate), 1982–84, mid-winter peak
Lothian Forth	8,300, 1977/78 2,250, av. peak 1979–84
Strathclyde Ayr coast	5,400, 1977/78 4,500, 1979 (SBR)
Grampian	6,300, 1977/78
Orkney	2,500, 1982–84, mid winter
Shetland	1,100 1984/85, mid winter

distributions (Prater 1981). Ringing recoveries suggest that the pattern of movements is complex, with birds from the same area dispersing in different directions, although there are as yet too few recoveries to give a comprehensive picture. Of four Sutherland-ringed chicks recovered within 12 months, one was in Lanark in September, two in Orkney in December, and the fourth in Portugal in February. Birds ringed in the breeding season in Shetland, Orkney, Perth and Rhum were wintering in Scotland when recovered in November/December but one marked in Kincardine was in Wales in February. Some movement to and from the Continent takes place – birds ringed in the Netherlands in March have been in Scotland in November/December in subsequent years. Birds from Iceland, which comprise the bulk of the Irish-wintering population (Prater 1981), have occurred in Orkney, the Inner Hebrides and the west and southwest, but not in eastern Scotland. Immigrants on the east coast are probably from the Fenno-Scandian population, which winters on southern North Sea coasts and south to the Mediterranean.

In the 1977/78 survey over 30,000 Golden Plovers were counted in Scotland in mid winter (Fuller & Lloyd 1981), all the nationally important flocks (more than 2,000) found being between the Ythan and the Lothians in the east and from the Clyde to the Solway in the west. Earlier in the autumn several thousand are often present in Caithness and around the Moray Basin, and 1,000+ in Shetland, but most apparently move out of these areas as winter sets in. Numbers on the Ythan have recently peaked at c3,000 in August/ September and on the Forth at 4,000–5,000 in mid winter, with an exceptional count of 6,000 at Queensferry in December 1975. In Orkney 4,500 were found in an incomplete November census (Booth 1979b) and more than 2,500 in mid winter (Tay & Orkney R.G. 1984).

The North Solway formerly held 5,000–6,000, generally peaking in January/February, but a decline has recently been recorded on the Inner Solway as a whole (ie including the south shore), with numbers dropping from 10,000+ in 1971–77 to under 2,000 in 1980–84; it is not clear whether this apparent decrease is due to changes in coverage of field habitats or to a real population decrease (Moser 1984). Around Loch Ryan and Stranraer there have been counts of between 4,000 and 7,000 in October/November, while a few thousand winter along the Ayr coast, with the largest numbers on the Bogside Flats at Irvine. On the Clyde up to 1,000 regularly commute between Glasgow Airport and the tidal flats (Gibson 1978) and an exceptional flock of 8,000+ was recorded at Hamilton in November/December 1974. Numbers on the Inner Hebrides are comparatively small, probably hundreds rather than thousands, but recent counts on the Outer Hebrides suggest a wintering population of 2,000–2,500 – mostly on Lewis (Buxton 1982a). Bad-weather movements may produce marked changes in distribution, with a concentration of birds in the southwest.

Native birds start to return to the breeding grounds as early as mid February, although this timing is much affected by weather conditions; most are on territory and in full breeding plumage by mid March (Parr 1980). At this season there is an obvious difference in plumage between most native birds and those of more northerly origin, many of which are still present on low ground in April and early May. The northern birds are both brighter in colour and more spangled in appearance. Opinions differ as to whether or not these northern birds, which are not distinguishable on the basis of measurements (Prater *et al* 1977), belong to a separate race *P.a. altifrons* or simply represent one end of a north-south cline (Parr 1980). As birds of the northern form are said to have bred on St Kilda and elsewhere in Scotland, and the proportion present in early May flocks (which are presumably heading for more northerly breeding grounds) has been noted as varying between 10% and 60%, it seems probable that the latter explanation is more likely to be correct.

Grey Plover *Pluvialis squatarola*

Passage and winter visitor to muddy estuaries in east-central Scotland and the Solway. Numbers are very variable and relatively small, but have recently shown a tendency to increase.

The first immigrants arrive in late July, and from August to October passage birds are seen on most coasts as far west as the outer Hebrides, and occasionally inland. In some years peak numbers are recorded in October but in others not until mid winter. Departure takes place in April/May but small numbers of immatures occasionally remain to summer on east coast estuaries and there are records of single adults in breeding plumage inland in the Grampians and Wester Ross in late April/May.

The Scottish wintering population is generally within the range 250–750 but may have come near to 1,000 in 1979–80. The only sites regularly holding flocks of over 100 birds, and thus qualifying as of national importance, are the Eden Estuary, the Firth of Forth, and the North Solway. Numbers on both the Eden and the Forth increased during the 1970s (Table) and the count of 700+ at the Eden in January 1980 is the highest yet recorded in Scotland. On the Solway more than 100 have wintered recently near Southerness and at Priestside, with peaks of c500 and 300 recorded in September and March respectively (Moser 1984). The only other sites known to have held more than 100 Grey Plovers are Culbin (130 in October 1978) and the Montrose Basin (100+ in October 1976). Flocks of 50–100 have been reported from the Tay, Rosehearty, the Dornoch Firth, North Uist and at Ardmore on the Clyde. Records from other mainland coastal counties (none for Berwick or Kincardine) and all the islands rarely involve totals exceeding 30 and often much smaller numbers. The few ringing recoveries include one

Numbers of Grey Plovers at main passage/wintering haunts (Data from Prater 1981 & M. Moser – figures rounded)

	Highest av. monthly count 1969–75	Post-1980 counts	
		Av. peak (n)	Range
Eden Estuary	270	640 (4)	580–740
Firth of Forth	140	390 (4)	310–580
North Solway	130	340 (4)	240–490

caught in Shetland in early October and shot four weeks later in Essex.

The 30,000–40,000 Grey Plovers wintering in Europe breed in western Siberia and follow a migration route through the Baltic to the North Sea (Branson & Minton 1976). About one third of this number winter in Britain, with the Wash holding by far the largest concentration. Recent counts suggest that the increase in wintering numbers observed elsewhere in Britain (Prater 1981) is now taking place in Scotland. Numbers certainly seem to have increased considerably over the last 30–40 years, as B&R give no records of flocks of more than 50–60 birds.

Lapwing *Vanellus vanellus*

Partial migrant, widely distributed as a breeder and scarce only in mountainous and moorland areas. Large flocks gather on low ground in late summer and many move to estuaries, or further south and west, to winter. During severe weather mortality is often high among those wintering in Scotland, which include immigrants.

Lapwings breed mainly on permanent grassland and arable fields and their breeding distribution and success is closely related to farming practices. Nesting densities are highest on machair and damp pasture and only slightly lower on arable land and rough grazing (Galbraith *et al* 1984). Eggs laid on arable land are liable to be destroyed during cultivations, while changes in the crops grown may influence numbers; in arable areas of Fife a decrease in breeding Lapwings has been attributed to a swing from cereal to vegetable growing (Oliver 1980). Damp grassland is an important feeding ground for the chicks (Fuller 1981) and its reduc-

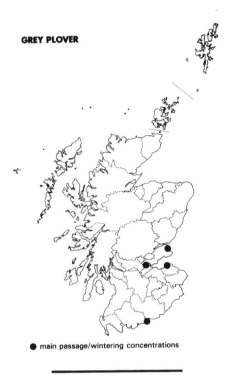

GREY PLOVER

● main passage/wintering concentrations

Sociable Plover *Chettusia gregaria*

Vagrant (Russia) – three records, 1926–83.

This species, whose nearest breeding ground is east of the Volga and whose wintering grounds are in Egypt and eastern Africa, was first recorded on 3 November 1926, when a first-winter female was shot on North Ronaldsay, Orkney (SN 1927:157). The only subsequent reports of the Sociable Plover are also from Orkney: one on Mainland in early December 1949 and the other on Eday, for about a week in mid January 1969 (SB 5:467–8). Most of the other British occurrences (21 to the end of 1983) have also been between August and January, but there are single April and July records.

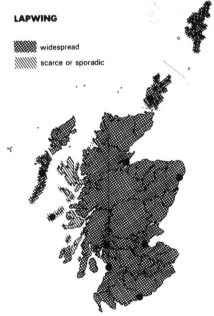

LAPWING

▨ widespread

▧ scarce or sporadic

● main passage/wintering concentrations

tion, or at worst elimination, through drainage is likely to have an adverse effect on the size and productivity of a local population. Lapwings nesting in sub-optimal habitats may need to move their broods to more suitable foraging grounds (Redfern 1982), and may breed only sporadically, as on Fair Isle, where the two pairs nesting in 1978 were the first for ten years, although 16 pairs had been present in 1961.

Using the density figures obtained in the 1982–83 survey of waders breeding on agricultural land, averaging 3+ pairs/km², Galbraith *et al* (1984) estimated that c64,000 pairs of Lapwings breed on farmland habitats in Scotland. Densities on the other habitats used, such as moorland and shingle, are on average much lower (eg Reed *et al* 1983a) but it seems possible that the total population is somewhere between 75,000 and 100,000 pairs. In the first half of this century there was some range expansion, both in northern Scotland, where several Shetland islands were colonised (Venables & Venables 1955), and elsewhere in Europe – Lapwings first nested in Iceland in 1922 (Voous 1960). This expansion has doubtless been helped by the lack of fidelity to the natal area shown by young Lapwings, while the reduction in egg collecting has doubtless helped to maintain breeding numbers. A decrease in the breeding population is generally apparent after a severe winter.

Flocking starts as early as mid June and by August there are groups of several thousands both around estuaries and inland, on pasture and later on potato fields. During the autumn there is a marked southwest movement of native birds and an influx of immigrants, while severe winter conditions result in extensive bad-weather movements, which probably also involve continental birds. It is not possible to trace the movements of these different populations with any certainty but ringing recoveries suggest that more than 75% of Scottish-reared birds winter well away from their area of origin. Of over 300 recoveries in November/February 50% were in Ireland, 13% in France, and 13% in Spain and Portugal. The proportion of first-winter birds among the Iberian recoveries was more than twice that for Ireland, where the ratio of first-winter:older recoveries is 1:2. Birds ringed as chicks in Norway, Denmark and East Germany have been found in Scotland, mostly in the east. Although some Lapwings return to breed in their natal area (Oliver 1980) there have also been summer recoveries of Scottish-bred birds in Iceland, Denmark and the Netherlands and a British-reared Lapwing has been found breeding 5,000 km away in Russia. A full analysis of the Scottish recovery data has yet to be made.

In 1969–75 the average January count of Lapwings on Scottish estuaries was 27,800 (Prater 1981), but this figure takes no account of the many flocks inland and total wintering numbers must be substantially higher. Attempts to detect population changes are further complicated by the fact that habitat usage varies not only at different seasons but also between years. The most important Scottish site, the Solway, regularly holds a nationally significant concentration, with 7,000–10,000 generally present from November to March and a peak count of 18,000 recorded during hard frost in December 1968. On the Inner Clyde autumn totals often exceed 5,000 (Halliday *et al* 1982); numbers there are fairly constant from mid November to late January, and recoveries have shown that many of the birds present at that time are

Strathclyde breeders (R. W. Furness). Large flocks build up on grassland around Glasgow Airport in autumn but some of these birds later move southwest to the Solway and Northern Ireland. On the Forth numbers are also at their highest in autumn; there were 12,000 in the Estuary in early October 1979, 5,000 of them at Skinflats. On the northern estuaries numbers seldom total more than 2,000 even during passage periods. Winter counts of 5,000+ are occasionally recorded in many areas and are probably often the result of bad-weather movements (Table). Mid winter numbers are low in Shetland, but higher in Orkney, where there were c3,700 in January 1983–84, more than 2,000 of them on Mainland (Tay & Orkney R.G. 1984). Some 5,000+ winter on the Outer Hebrides (Buxton 1982) and there have been counts of up to 3,000 in Islay and of several hundred in Bute.

Peak counts of Lapwings at major coastal and near-coastal sites; little information is available on inland flocks (Data from Prater 1981, SBR, Campbell 1978, Halliday et al 1982, BoEE/WWC, Moser 1984, Tay & Orkney R. G. 1984, BTO/WSG Winter Shorebird Count preliminary results, per M. E. Moser – figures rounded)

	Highest av. monthly count 1969–75	Post-1975 winter counts
North Solway	11,000	c3,000, November 1983
Inner Clyde	3,500	4,000–6,000
Firth of Forth	1,400	3,000–6,000
Ythan Estuary	1,000	7,500, February 1978
Wick	NA	5,000, December 1979
Hamilton	NA	8,000, Nov.–Dec. 1975
Loch Ryan/Urr	1,750	8,500, January 1977
Kirkcudbright Bay	2,250	5,000, January 1978
Irvine Flats, Ayr	2,030	4,600, January 1976
Orkney Mainland	NA	2,000+ January 1983–84
Tiree	NA	est. 5,000–10,000 mid winter 1984/85

When blizzard conditions strike in early or mid winter Lapwings can often be seen fleeing south ahead of the storm; those that do not leave in periods of prolonged frost are liable to starve. If they have already returned to the breeding area, from February onwards, they are likely to remain there even if the ground is snow-covered; in some years such conditions occur after the first eggs are laid and presumably result in considerable losses. Most Scottish birds are back on the breeding grounds by mid March, but passage – probably involving birds of Scandinavian origin – is evident in the Northern Isles throughout March and April and often well into May. Peak numbers (up to 1,000) are recorded on Fair Isle at this time.

Knot *Calidris canutus*

Passage and winter visitor, the main wintering concentrations being on east coast estuaries, especially the Firth of Forth, and on the Solway. Widespread as a passage migrant.

The Knot has a marked preference for sandy areas with some surface mud, but also occurs in considerable numbers on rocky coasts, for example between Berwick and Moray (Summers *et al* 1975). Immigrants start to arrive from mid July but numbers do not peak until mid winter. Small groups are occasionally seen inland and a few usually remain on the coast throughout the summer. Colour marking has shown that the Knot is a very mobile species which frequently changes its feeding area within an estuary or moves on to a different site; a lower proportion was recaptured on the Forth in successive years than was the case with most other waders studied there (Symonds *et al* 1984).

Knots wintering in Britain originate from northern Greenland and northeast Canada (Dick *et al* 1976); there

have been several ringing recoveries linking Scottish-wintering birds with Ellesmere Island and Baffinland. (Most of those breeding in western Siberia winter further south, to western Africa, though a few may reach Scotland with weather conditions which bring falls of Curlew Sandpipers and Little Stints.) Although some arrive from the northwest, perhaps after stopping-off in Iceland, many are known to come via southern Norway and the major moulting areas on the Wadden Sea and Wash. After moulting there is a marked dispersal northwards and westwards. Knots ringed between July and September in Lincoln and Norfolk have been caught in Sutherland and Ayr, and on the Solway, between October and February, and birds ringed at Teesmouth in November have been seen on the Forth in December/January (M. Pienkowski). In late winter there is movement back across the North Sea to the Wadden Sea, where the birds put on weight for their return journey to the breeding grounds (Prater 1981). Some Scottish Knots take part in this movement, but others apparently follow a more direct route, as indicated by a build-up of numbers at Findhorn Bay during April, and occurrences in Shetland in May and occasionally in early June.

Britain and Ireland support about 65% of the European and northwest African wintering population, estimated as c650,000. The major British sites are in England, where the Wash and the Irish Sea coast from Morecambe Bay to the Dee hold 75% of the population in autumn and spring (Prater 1981). In 1969–75 the highest average monthly count on Scottish estuaries (ie excluding birds on non-estuarine coasts) was around 40,000, in January. The Firth of Forth and the North Solway regularly hold flocks of international importance (more than 3,500), and some other east coast estuaries occasionally do so (Table). On the Forth the largest

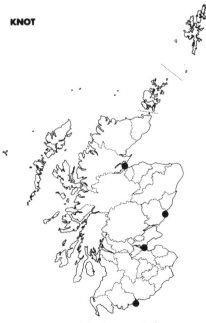

KNOT

● main passage/wintering concentrations

Peak counts of Knots at major wintering sites (Data from Prater 1981 & M. Moser – figures rounded)

	Highest av. monthly count 1969–75	Post-1980 counts Av. peak (n)	Range
Firth of Forth	16,000	9,600 (4)	8,100–11,500
North Solway	7,800	6,500 (4)	4,700–8,500
Moray Basin	4,000	2,100 (2)	1,100–3,000
Montrose Basin	4,000	1,600 (3)	750–3,000

concentrations are usually above the bridges; possibly exceptional flocks of 25,000 were recorded at Skinflats in 1971/72 and 1972/73 (Campbell 1978). Numbers on the Forth build up steadily from October to December and then remain relatively constant until February, thereafter dropping rapidly to under 1,000 by April. On the Solway the main influx occurs in November/December, numbers peak in January and most Knots have gone by the end of March; the main concentrations are in the Inner Firth, around Southerness (recent max. 6,700 in 1981/82) and Priestside (4,000 in 1982/83). Numbers wintering on the Inner Solway as a whole have decreased in the last decade, in line with the national and northwest European trends (Moser 1984).

In January 1972 the Moray Basin held more than 11,000

Knots, some 6,250 in the Moray Firth and around 4,750 in the Cromarty Firth. These numbers were probably unusually high (the average in 1969–74 was 5,500 for the whole area) as may have been the more recent record of c3,000 at Brora/Golspie in January/February 1979. Recent counts on the Tay have not exceeded 600, suggesting a decrease from the 'huge flocks' recorded there by B&R. Elsewhere around the coast numbers seldom exceed 200 at any one site and are much lower in the north and west. Few Knots occur in the Northern Isles and Hebrides outside migration periods; one on Fair Isle in January/March 1979 was the first winter record for the island.

The Knot is possibly the most variable of the wader species in terms of movement patterns within the general wintering area for the population, and this may in part explain the marked fluctuations in numbers between years. Weather conditions both during the breeding season and in winter probably have a considerable influence on the numbers reaching these shores. B&R thought that there had been an increase in the early part of this century, while there was a marked decrease between 1969 and 1974, attributed by Prater (1981) to a series of bad breeding seasons in the Canadian Arctic. In January 1979 the index for Britain was the highest since 1974, possibly due to cold-weather movement from the Wadden Sea (Marchant 1981); although there have since been quite wide fluctuations, the overall trend has been one of continuing decline (WWC 1983). As far as the future is concerned, reclamation of the mudflats of the Forth Estuary would appear to be the most serious potential threat to Scottish wintering Knots.

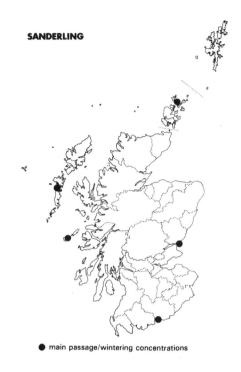

SANDERLING

● main passage/wintering concentrations

becula and Orkney's northern islands then hold by far the largest numbers (Table). Departure for the breeding grounds may start as early as March but the main return passage takes place in May. Apart from the recent discovery of the May concentration on the Solway there is no evidence of change in status or distribution this century.

Sanderling *Calidris alba*

Most abundant on passage, when it is widely, but in general thinly, distributed on sandy shores. Wintering numbers are much lower but there are important concentrations in the Outer Hebrides, Orkney and the Tay Estuary. Large numbers visit the North Solway briefly, and possibly only sporadically, on spring passage.

Autumn passage starts in mid July and Sanderling numbers peak in most areas towards the end of the month and in August. By the end of September only the comparatively small wintering population remains; the Uists and Ben-

Peak counts of Sanderlings at major passage/wintering sites (Data from SBR, Summers 1975, Summers & Buxton 1979, Buxton 1982, Clark et al 1982, Tay & Orkney R. G. 1984, BTO/WSG Winter Shorebird Count preliminary results, per M. E. Moser)

	Count	Period
Autumn passage		
Uists & Benbecula	400	August–September
Quendale (Shetland)⎫ N. Ronaldsay ⎪ Dunnet Bay ⎬ Aberlady Bay ⎪ Islay ⎭	230–260 each	July–September
Doonfoot (Ayr)⎫ Machrihanish ⎬ Achnahaird (W. Ross)⎭	130–180 each	July–September
Mid winter		
Uists & Benbecula	1,300	February
Orkney – north isles	860	January
Tentsmuir	c300	January–February
Tiree	400	January
Spring passage		
N. Solway	9,000–15,000	Late May
Barry–Buddon (Tay)	450	May
Rattray Head (Aberdeen)	130	June

The estimate by Prater & Davies (1978) of a Scottish wintering population totalling c1,175 was clearly much too low, in view of the figures obtained for the Outer Hebrides and Orkney in the early 1980s. Of the 1,200 Sanderlings counted in the southern Outer Hebrides in February 1980 more than 900 were on South Uist, with the biggest gathering – more than 300 – on the Howmore Estuary (Buxton 1982). (A count of 800 at Howmore in February 1975 (SBR) is far higher than any subsequent figure for the area). These islands are the second most important British wintering site for Sanderlings (the Ribble holds a few hundreds more in mid winter) and are of international significance for the species. In Orkney the largest numbers found in January 1983–84 were on Sanday (470), while Westray and Stronsay each held 170–180 (Tay & Orkney R.G. 1984). The wintering flock on North Ronaldsay, generally 50–100 birds, is the most northerly in the country. On the mainland the only site regularly holding important wintering numbers is the Firth of Tay, where there are high-tide roosts on both sides of the estuary, at Tentsmuir Point and the Dighty Mouth. Elsewhere winter numbers are generally very much lower but there are occasional reports of unusually large parties, eg 285 at Sands of Forvie in January 1980, possibly the result of birds concentrating in an area where feeding opportunities are good at the time.

Although Morecambe Bay regularly holds over 10,000 Sanderlings in late spring, no flocks of comparable size were recorded in Scotland until 1982. On 21 May that year R. T. Smith located a huge flock on the Solway shore near Annan Waterfoot. The birds remained in the area for at least two or three days and it was estimated that 9,000–15,000 were present on 22 May. More than 1,800 were caught on 22–23 May, including nine caught in the same area on spring passage in the previous three years and one ringed in Spain. These re-traps, together with the fact that there was nothing unusual about weather conditions at the time, suggested that this large influx might be a regular feature, previously unrecorded due to shortage of observer cover (Clark *et al* 1982). However, intensive searches in May 1983 and 1984 located maxima of only 700+ and 2,000+ respectively (M. Moser), so it seems that this site may be used only sporadically. Other May records include flocks of up to 450 on the Tay (Summers 1975) while two parties were seen departing at height from Aberlady on 1 June 1979 (Ferns 1980b). Most Sanderlings have left by mid June but a few often remain to summer and there are records of pairs in display flight over high ground in northwest Sutherland, and in full summer plumage in the Cairngorms.

Although there are records from the Clyde Islands and the Isle of May, the Sanderling is an uncommon visitor to these largely rocky islands. During autumn migration it occasionally occurs inland; such records generally involve single birds but there were ten at the Endrick Mouth in July 1977 and an exceptional 51 at Loch Leven on 25 October 1972.

Sanderlings breed in the high-Arctic, in a discontinuous range stretching from Alaska through Canada, Greenland and Spitsbergen to Siberia, and winter on temperate and tropical coasts in many parts of the world. The British passage and wintering populations include birds from both Siberia and Greenland/Canada. Recoveries of birds

marked on autumn passage through the Baltic, and presumed to be of Siberian origin, include one ringed at Ottenby, Sweden, in September and caught on the Tay the following January. Birds dye-marked in Greenland have been seen in Britain in August/September, and the pronounced passage up the west coast in May, when nearly half the western European population is concentrated between the Cheshire Dee and Morecambe Bay, is believed to involve birds from there and possibly also from northeast Canada (Prater & Davies 1978; Prater 1981). Recoveries of Scottish-ringed Sanderlings include four marked in Shetland on autumn passage and reported later the same winter: two were in Norfolk, one had reached the north Normandy coast 21 days after ringing, and the fourth was in the Netherlands in January.

[Semipalmated Sandpiper *Calidris pusilla*]

The report of a Semipalmated Sandpiper on the Isle of May in 1957 (SB 1:35 & 2:342) was subsequently rejected by the BBRC (along with several other records of this species – see BB 72:264–274), and there has not yet been an acceptable Scottish record. A discussion of the similarities between this species, the Little Stint and the Red-necked Stint is given in Sharrock & Grant (1982).

Western Sandpiper *Calidris mauri*

Vagrant (Alaska/NE Siberia) – one record, 1956.

The first British record of this small wader was from Fair Isle, where one was present from 27 May to 3 June 1956. It was at first thought to be a Semipalmated Sandpiper (SN 69:145–147), but was subsequently identified, from the detailed field notes, measurements, and plumage description submitted, as a Western Sandpiper (BB 56:55–58). There have been no further Scottish records, but two Irish and three English ones are listed by Sharrock & Grant (1982), who also give some useful hints on identification.

[Red-necked Stint *Calidris ruficollis*]

The report of a Red-necked Stint on Fair Isle in August 1982 (FIBOR 1982) was not accepted by BBRC. This small eastern Siberian wader closely resembles both the Little Stint and the Semipalmated Sandpiper; records are very carefully scrutinised by the Rarities Committee because of the risk of misidentification.

Little Stint *Calidris minuta*

Passage visitor, scarce in spring and occurring in very variable numbers in autumn.

The Little Stint is a high-arctic species, breeding from Lapland eastwards; small numbers winter in France and Iberia but most of the western population moves further south to Morocco and Mauritania. In Britain the largest numbers occur in eastern and southern England; Scotland is very much on the fringe of the migration route and in some years very few birds appear. Passage birds ringed in southwest Norway have been found in Shetland and Fife.

Between 1968 and 1983 the number of Little Stints reported in autumn ranged from ten to 700+, with under 100 in six seasons, 100–300 in five, and over 300 in four. The first arrivals may be in early July but in some years none appear until well into August. The timing of the main influx is also variable, sometimes occurring between mid August and mid September and sometimes up to a month later. Most have moved on by the end of October but there are a few November records and at least one in December. Although most reports are from the coast, inland occurrences are fairly common and birds are often seen well up the Forth Estuary.

Autumn passage is generally most pronounced in the east, from Aberdeen to the Forth, but in some seasons large numbers arrive further north and in such years birds are also likely to reach the Hebrides; in 1981, for example, there were high numbers north of Inverness throughout September and reports of 42 on Tiree and 15 on South Uist. In 1968–83 Little Stints were reported from all mainland counties other than Berwick and Selkirk, and from all the island groups except the Clyde Islands. Most records are of single birds or small groups but quite big flocks are occasionally seen, for example 200+ at Dornoch in August 1965. In 1978, when a total of 700+ birds was reported, numbers in the north were unusually high, with nearly 100 at Tarradale in Easter Ross and 80+ at Sumburgh.

Far fewer Little Stints are seen in spring and in some years none at all; the maximum recorded recently has been 15 in 1976, all on the east coast. There are a few April records but most spring sightings are between 8 May and 21 June. A June bird at St Cyrus in 1975 was heard singing during the night.

It seems unlikely that there has been any significant change in the status of the Little Stint during the last 30–40 years. B&R describe it as an uncommon passage migrant and refer to flocks of 40 and 60 as unusually large. They give records for Bute, Coll, Barra and North Rona, all areas for which there have been no recent reports.

Temminck's Stint *Calidris temminckii* [1]

An irregular passage visitor in small numbers and scarce migrant breeder.

Temminck's Stint has recently shown signs of establishing itself as a breeding bird here, although Scotland is considerably south and west of its regular breeding range in Fenno-Scandia and Siberia, where it normally nests in swampy tundra and birch scrub. Breeding was attempted in the Cairngorms in 1934, 1936 and 1956, with eggs laid on each occasion but failing to hatch (B&R and Parslow); birds were also present in the same area in 1935 and 1947. From 1969 onwards Temminck's Stints were seen displaying on low ground in Easter Ross, and breeding was proved for the first time in 1971, when two chicks were seen there (SB 7:94). Two pairs were present in 1972 and 1974, and three pairs in 1973, but breeding was not proved in any of these years. There were none at this site in 1975 and only single birds have been seen there since.

Two Temminck's Stints were found at a second site, in east Inverness, in May/June 1974. At least five were present in June/July 1975 and display was seen but breeding was not proved. In 1976 three displaying males were present, a clutch of four eggs was found on 27 June, and two chicks were seen on 18 July. At least four adults, and sometimes as many as eight, have been seen annually since then, with breeding confirmed in 1979, 1980, 1981 and 1982. In 1978 a clutch of four was discovered at another site, and by 1983 up to 16 birds had been recorded on territory in four different locations. The increasing frequency of June records from other areas suggests that it would be well worth while keeping a look-out for further possible breeding sites. It should be remembered, however, that this species is afforded special penalty protection under the 1981 Wildlife and Countryside Act.

As a passage migrant the Temminck's Stint was recorded only sporadically up to the mid 1960s. B&R give details of only four records but say that 'it has occurred several times on Fair Isle, both in autumn and spring'. There were only a few reports during the 1950s and not many more in the 1960s, but from about 1973 onwards numbers increased markedly. Since 1973 from two to 16 birds have been seen each year, fairly equally divided between spring and autumn. Almost 70% of the records are from east coast counties, from Shetland to East Lothian but there are also records from Renfrew, Dunbarton, Stirling, Perth, Kinross, Peebles, Ayr, Dumfries, Argyll (mainland), North Uist and St Kilda. The earliest date recorded is 9 May, and most spring occurrences are between 15 May and 20 June. Most autumn records are between 10 August and 21 September, and the latest date (a first-winter male shot in Dumfries in 1965) is 16 November.

Least Sandpiper *Calidris minutilla*

Vagrant (N America) – two records, 1955 and 1965.

The first record of this small stint was of one shot at Virkie, Shetland, on 14 August 1955; the skin is in the Royal Scottish Museum (SN 1957:170). The only subsequent record is from Cadder, near Lenzie, northwest Lanark, on 11–14 September 1965 (SB 4:506).

White-rumped Sandpiper *Calidris fuscicollis*

Vagrant (N America) – occurs irregularly, mainly in autumn and in very small numbers.

The first record of the White-rumped Sandpiper was in 1955, when one was at Gladhouse from 21 to 24 August (BB 49:39). In 1956–83 there were 22 further records, all but one involving single birds, and with a maximum of four in any one year (1980 & 1983). Most were, as might be expected, in the west and north: Orkney – seven, Fair Isle – two, Shetland – two, Caithness, Ayr, Lewis and South Uist – one each. The remainder were from the east coast: Midlothian – five (including the 1955 bird), East Lothian – two (including two adults at Aberlady in mid August 1983) and Sutherland one. Nearly all records are between late August and early November (one bird stayed on Fair Isle until 2 December) but there have been single occurrences in May/June and July. The White-rumped Sandpiper breeds in the Arctic and migrates along the Atlantic coast to winter in South America. Although scarce in Scotland, it is one of the commonest of the nearctic waders occurring regularly in Europe and has been suspected of overwintering in England.

Baird's Sandpiper *Calidris bairdii*

Vagrant (N America / Greenland) – ten records, 1911–84.

Baird's Sandpiper was first recorded in 1911, when an adult female was shot on St Kilda on 28 September (SN 1912:9). There were no further records for more than 50 years, by which time field identification of vagrant waders had become sufficiently competent to produce almost annual records of this species in England and Ireland. During the 1970s there were five occurrences: North Uist (September 1971), Gladhouse (September 1974 & September 1979), Aberlady Bay (August 1975), and Islay (3 June 1979 – one of very few summer records in Britain). In 1982 there were three further records: from Fair Isle and Rattray Head in September and Findhorn Bay in October, and there was one at Gladhouse in September 1984.

Pectoral Sandpiper *Calidris melanotos*

Vagrant (arctic N America / Siberia) – annual in very small numbers, mainly in autumn.

The Pectoral Sandpiper has been reported with increasing frequency since 1960 and annually since 1973. It was first recorded in Scotland on 3 September 1928, when an adult female was shot in Caithness (SN 1928:168); the only other record prior to 1950 was from Aberlady, on 10 August 1948. In the 1950s there were three records, each of a single bird, and in the 1960s 13 reports, involving 15 birds, and two blank years. From 1970 to 1980 inclusive there were 38 occurrences, totalling 40 birds, and only in 1972 was there a complete absence of records. 1981 produced a total of ten reports (the previous maximum in any one year was seven, in 1973) and there were six or seven in 1982 and about ten in 1983. Most reports are of single birds and no more than three have been seen together. Nearly 40% of occurrences are in the Northern Isles, the remainder being almost equally divided between east and west, with some well inland. Almost 75% of reports are between mid August and the end of the third week in October; the record of one in East Lothian on 19 November 1955 is more than three weeks later than any other. The comparatively few spring records are scattered from May (the earliest on 21st) to July. On 28–29 May 1974 a Pectoral Sandpiper was seen displaying by a Caithness loch.

Most, if not all, the Pectoral Sandpipers reaching Britain are believed to come from America, where autumn passage to the South American wintering grounds takes the birds well out over the Atlantic, making them vulnerable to sudden westerly storms.

Sharp-tailed Sandpiper *Calidris acuminata*

Vagrant (NE Siberia) – one record, 1956.

The sole Scottish record of the Sharp-tailed Sandpiper is of one at Hamilton, Lanark, on 13–21 October 1956 (SB 1:94). There have been only sixteen British occurrences, the most recent – in September 1983 – coming after five blank years. A paper by Britton (BB 73:333–345) gives guidance on the identification of this rare vagrant.

Curlew Sandpiper *Calidris ferruginea*

Passage visitor, occurring in very variable numbers in autumn, mainly in the east. Scarce in spring and occasional in winter,

Curlew Sandpipers breed in arctic Siberia and winter in Africa south of the Sahara. Although there is a small regular passage through northwest Europe, most follow a route which takes them over the Black Sea to the Mediterranean. Especially in seasons when breeding success has been high, weather conditions during migration sometimes result in the displacement westwards of large numbers of juveniles, producing occasional influxes into Britain (Wilson *et al* 1980). Apart from these irregular influxes the numbers recorded in Scotland are small.

Major autumn influxes occurred in four years between 1968 and 1983, involving more than 1,100 birds in 1969 (Stanley & Minton 1972) and several hundred in 1970, 1975 and 1978, while in six years fewer than 100 Curlew Sandpipers were recorded. Reports are most frequent, and numbers usually greatest, in the east, on estuaries from the Moray Basin to the Forth, but in some seasons there is a higher than usual proportion of occurrences in the north or the southwest. Inland sightings are not uncommon. There have been records during this period from all mainland counties other than Berwick, Selkirk, Banff and Inverness, and from all island groups except the Clyde Islands (for which B&R give old records).

In non-influx years most records are of single birds or groups of up to ten, but when the population is high much larger flocks are frequently reported. Recent maxima for the principal sites are: Tyninghame 100, Grangemouth (sewage works) 65, Ythan 35 – all in August/September 1969, and Invergowrie Bay 60, Skinflats 40, Eden Estuary 32, Aberlady Bay 24 – all in 1970. July 16 is the earliest autumn date but most passage takes place between mid August and the end of September, with small numbers present in October and sometimes a few stragglers in November. The few winter occurrences include three in December, one in January and two in February.

Spring passage, which does not occur every year, usually involves only very small numbers; the highest total recorded since 1968 has been 14, in 1971. There are a few March/April records but most passage takes place in May, with occasional birds still present in early June.

Purple Sandpiper *Calidris maritima*   [1]

Winter visitor, widely scattered along rocky coasts from July to May, especially in the north and east. Small numbers generally remain throughout the summer and breeding has occurred.

Scotland holds more than 75% of the Purple Sandpipers wintering in Britain, possibly substantially more than the 14,500–23,000 estimated by Atkinson *et al* (1978). Very much

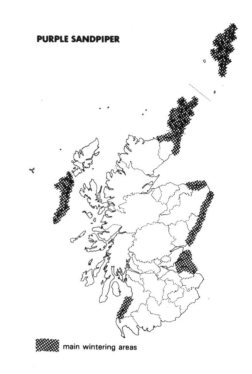

PURPLE SANDPIPER

▨ main wintering areas

larger numbers winter on the Norwegian coast and in Iceland but over much of the extensive European winter range, which is the most northerly of any wader, the Purple Sandpiper is relatively uncommon. Until recently little was known about the numbers and movements of this species but surveys and ringing programmes (eg Summers *et al* 1975; Atkinson *et al* 1978, 1981) during the last 15 years have added greatly to our knowledge.

The Purple Sandpiper has a discontinuously circumpolar distribution, the breeding grounds closest to Scotland being those in Iceland, Norway and Sweden. Ringing recoveries and colour-marking have shown that flocks wintering on the east coast, from Aberdeen to Northumberland, are largely of the short-billed type breeding in Norway, but birds from Greenland and probably Iceland may also occur. On the east coast the first immigrants arrive in July and numbers build up until November, then remain more or less static until April; virtually all have left by the end of May. Most first-year birds, which comprise only about 25% of the wintering population, do not reach east Scotland until mid September. In northern Scotland and the Outer Hebrides arrival does not start until October; a higher proportion of the birds in these areas are long-billed, possibly originating from Iceland or Canada (Tay & Orkney R.G. 1984). Purple Sandpipers are sedentary during the winter and show a high degree of fidelity to a particular stretch of coast in successive years. Of 210 marked on the east coast and recovered or re-trapped in later years only 16 were more than 20 km from the site of original capture, a dispersion of records no greater than movements recorded within a single winter (R. W. Summers).

The Purple Sandpiper has a marked preference for coasts with wide, flat shelves of tide-washed rock but also occurs where there are abundant not-too-steep rocks offshore. Such coastlines are not easy to count (Summers *et al* 1975; da Prato & da Prato 1979) and many of the figures given by Atkinson *et al* (1978) in their preliminary assessment of the population were consequently based on estimates. Subsequent surveys have produced revised, and generally higher, figures for several areas (Table).

Numbers and distribution of wintering Purple Sandpipers (Data from Atkinson et al *1978, Buxton 1982a,b, Summers & Buxton 1983, Tay & Orkney R. G. 1984 and R. W. Summers)*

County	Est. av. winter max. 1968–74	Post-1975 counts
Berwick	40	
E. Lothian	700	
Midlothian	100	
Fife	580	
Angus	330	
Kincardine	590	
Aberdeen	910	
Banff	600	
Moray	200	
Nairn/E. Inv.	10	
E. Ross	30	350 (1982)
E. Sutherland	240	70 (1982)
Caithness	600	1,440 (1982)
Orkney	1,200	5,700 (1983–84)
Shetland	3,500	1,540 (1984/85)
Outer Hebrides	730	c2,500 (1978–82)
W. Suth/W. Ross	50	
Argyll	120	
Renfrew/Bute	70	
Ayr	580	
Wigs/Kirkcud.	70	

Orkney, with its extensive tidal rock shelves, holds 5,000–6,000, much the largest total yet recorded in any district; most are on exposed stretches of coast on Mainland, Westray and Sanday, and comparatively few around Scapa Flow or the inter-island sounds (Tay & Orkney R.G. 1984). In Shetland, however, the Tay Ringing Group/Shetland Bird Club survey in December 1984/January 1985 found only about 1,500 Purple Sandpipers, a total well below the figure of 3,500 quoted by Prater (1981), which was based on two independent estimates of 2,000 and 5,000. The earlier estimate of 730 for the Outer Hebrides has, in contrast, proved too low. Although Fuller *et al* (1979) described the Purple Sandpiper as 'rare in winter', Buxton (1982) found more than 1,600 in the Uists and Benbecula, 350 in Lewis and Harris and 580 on the other islands in mid winter; he estimates the Outer Hebridean population as c2,500. There is a marked spring passage there, with mixed flocks of up to 500 Purple Sandpipers and Turnstones passing through, possibly *en route* to Iceland and Greenland. The Inner Hebridean population is very much smaller, with an estimated maximum of 150 (Reed *et al* 1983). Spring passage produces the highest counts on Fair Isle, where numbers

seldom exceed 100. On the Isle of May 200–400 may be present in autumn.

It seems unlikely that any significant change in the wintering population has occurred in the last 30 years, although numbers on the Ayr coast in mid winter 1980/81 were only one third of those recorded in 1969–75 (Hogg 1983). B&R give few figures for the Purple Sandpiper but imply that a flock of 100 is noteworthy. The 'immense numbers' they refer to as occasionally being seen on spring migration might represent the level of the May 1979 influx on Sanday, when 750 were recorded.

A few Purple Sandpipers regularly summer in Orkney and Shetland, and occasionally elsewhere on the coast. Breeding was suspected in the 19th century but it was not until 1978 that it was proved for the first time in Britain. That year a pair hatched three young; two broods were seen in 1979 and 1980, and four in 1981, but only one in 1982. In 1983 breeding was not proved, although at least one adult was present (Dennis 1983a). This species nests in exposed tundra-type habitat, often with boggy areas, moss hummocks and dwarf shrub vegetation, at altitudes ranging from near sea-level to mountain tops; such conditions are found in several parts of Scotland. In the interests of the birds attempting to establish themselves in this country, the breeding location is not being publicised.

Dunlin *Calidris alpina*

Migratory breeder, most abundant in the northern Highlands and the Outer Hebrides; elsewhere rather local or sporadic. Scottish breeders move south in autumn and the large wintering population consists of birds from northern Scandinavia and USSR. Spring and autumn passage involves birds from these areas and also from breeding grounds around the Baltic and as far west as Greenland.

The Dunlin nests at altitudes ranging from sea-level – as on the machair of the Hebrides – to about 1,000 m asl on the Cairngorms, and in a variety of habitats, including saltmarsh, heather moorland and sedge/moss mountain summits. Although occurring on moorland from Galloway to Shetland, Dunlins are often scarce and local, while in the Central Lowlands and the northeast breeding probably

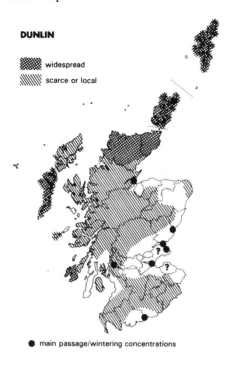

DUNLIN

▨ widespread

▧ scarce or local

● main passage/wintering concentrations

become possible to elucidate the rather complex movements of these three groups and much still remains to be learned.

Recent analyses (Hardy & Minton 1980; Ferns 1981) have shed light on the situation in Britain as a whole and present the following picture. Autumn passage starts in mid June with the southward movement of *schinzii*, which continues through July and August. Adults of this southern race usually do not moult in Britain and move straight on towards north-west Africa, up to a month ahead of the juveniles. During August this movement includes some *arctica* from northeast Greenland, which probably mix with *schinzii* when stopping-off in Iceland. Adult northern Dunlins, *alpina*, reach Britain between mid July and mid August and then moult, mostly on the Wash. The main arrival of juveniles occurs in early September, and from October a generally westward movement takes place; a massive influx into the Solway occurs in October/November (Moser 1984). The return movement of *alpina* starts in late February/March and continues into May, with movement to spring moult and staging sites (including the Wash and the Wadden Sea) where the birds remain until late April/May (Pienkowski & Pienkowski 1983); passage of this race is less marked in Scotland than in England. British-breeding *schinzii* return to their nesting grounds in April; Icelandic birds pass through, mainly on the west, in late April/May (over 90% of those caught on the Solway at this time in 1979 were considered to be of Icelandic origin); and Greenland birds, both *schinzii* and *arctica*, move up the west coast in the second half of May, when peak spring numbers occur in Scotland.

From July until May Dunlins are widely distributed around the coast and are most abundant where there are extensive mudflats. The population is at its peak in mid winter; in 1968–74 the highest average monthly total, c50,000, was in January. (In 1983/84 the total counted in Britain and Northern Ireland peaked at 412,000 in February; the population index was then at the lowest level yet recorded – WWC). All the principal estuaries hold nationally import-ant flocks during passage periods and/or in winter (Table), but internationally important numbers (more than 20,000) have been recorded only once, on the Firth of Forth, where there were nearly 40,000 in February 1977. This was excep-tional, however, and the Forth population is usually in the range 12,000–18,000. Although the birds are widely scat-tered within the Firth, and use all the available mudflats (Symonds *et al* 1984), the main feeding and roosting con-

occurs only irregularly. B&R describe the species as breed-ing in every mainland county and mention Aberlady, Tents-muir, St Fergus and Loch Leven as regular haunts. Since 1968 there has been only one report of breeding at Aberlady (1974) and none at St Fergus; the last nesting record for Tentsmuir was in 1937 and there have been none from Loch Leven since before 1950. It seems that there has been a decline in lowland nesting south and east of the Great Glen in the last 30–40 years.

On high ground in the north, and on the islands, numbers are larger and probably more stable, although there are few old figures with which to compare more recent ones. Dunlins breed locally on the Uists at densities of over 40 pairs/km² but the average density on the machair is much lower, at 15+ pairs/km² (Galbraith *et al* 1984), and lower still on peaty moorland (Reed *et al* 1983a) and acid grassland; the last two habitats probably seldom hold even one pair/km². Agricultural development of the machair clearly presents a potential threat to the very large breeding population in the Outer Hebrides, while afforestation of moorland must also be decreasing the extent of habitat suitable for this species.

Dunlins nesting in Britain belong to the southern race *C.a. schinzii*, which also breeds in southeast Greenland, Ice-land, the Faeroes and around the Baltic. Two other races occur in Britain: *C.a. alpina*, the northern Dunlin, breeds from northern Scandinavia eastwards through arctic Russia and winters in Britain and south to Morocco, and *C.a. arctica*, from northeast Greenland, occurs on passage and winters with *schinzii* in Morocco and Mauritania. These races differ sufficiently in measurements to allow the relative proportion of each present in a mixed catch to be estimated; they may also be distinguishable in spring plumage (Ferns 1981b). It has only recently, with the increase in wader-netting,

Peak numbers of Dunlins at major wintering sites (Data from Prater 1981, M. Moser – figures rounded)

	Highest av. monthly count 1969–75	Post-1980 counts Av. peak (n)	Range
Firth of Forth	12,000	6,300 (4)	3,300–8,400
North Solway	10,000	6,800 (4)	5,700–8,900
Moray Basin	6,000	5,200 (2)	4,700–5,700
Inner Clyde	4,000	1,600 (4)	1,100–2,700
Firth of Tay	4,000	3,900 (2)	3,600–4,100
Eden Estuary	3,500	2,300 (4)	1,600–3,300
Montrose Basin	1,500	1,600 (3)	1,600–1,700

centrations are in the Skinflats/Kinneil area. Dunlin numbers in Britain as a whole were much lower in 1978–84 than in 1973–77 and this decline has been very marked on the Forth, with numbers seldom exceeding 5,000 where 10,000 were previously often present. Dye-marking studies suggest that there is little turn-over of Dunlins on the Forth in mid winter (Pienkowski & Clark 1979).

During 1970–77 counts of 7,000–10,000 were recorded on the Tay, Inner Clyde and North Solway, but numbers have since been lower. On the Solway, where the main concentrations are around Priestside and Southerness, there is a large influx of *alpina* in October/November (Moser 1984). A similar mid winter influx occurs on the Moray Basin, though numbers vary greatly between years (R. L. Swann). Some 2,000+ winter in Orkney, mainly on Sanday (Tay & Orkney R.G. 1984) but peak numbers at the Ythan and other small east coast estuaries are generally under 1,000. In the west, more than 1,000 winter on the Ayrshire coast; north of the Clyde, around the Clyde Islands and on the Inner Hebrides, Dunlins occur only in hundreds. The wintering population in Shetland is probably small. Inland records are not uncommon during passage periods but are rare in winter and usually involve only small numbers.

Although many Dunlins have been marked in Scotland most have been caught on passage and there are few recoveries involving either native birds on their wintering grounds or Scottish-wintering birds on their breeding grounds. These few include birds breeding in the Uists and wintering in Mauritania, one ringed on the Forth in January and recovered near Archangel, USSR, in June, and one ringed in Fife in July and recovered in Greenland two years later. Birds ringed in winter (October/February) in central Scotland have occurred on passage, mostly in autumn, in the Netherlands, Norway and all the Baltic countries, and birds ringed in all these areas in autumn have occurred between late July and mid April from Shetland to the Solway, but not as yet in the Hebrides. Evidence of the post-moult dispersal within Britain includes records of Dunlins ringed in Norfolk and Lincoln in September and caught on the Clyde and the Moray Basin between early November and January. Examples of the north-south passage of *schinzii/arctica* through the country include a bird marked in Shetland in August and recovered five weeks later in Mauritania, and another marked on the Moray Basin in July and shot in Morocco six weeks later. Recovery details give an interesting indication of the fate that befalls some Dunlins: of 71 foreign rings, two were in owl pellets, one in a Peregrine pellet, one on a decapitated individual (presumably also a Peregrine victim) and one on a bird killed by a crow.

Broad-billed Sandpiper *Limicola falcinellus*

Vagrant (N Eurasia) – recorded in eight years, 1912–84.

All but one of the Broad-billed Sandpiper records, which involve a total of nine or ten individuals, are from the eastern half of the country, the exception being a bird at Stranraer in May 1983. The first two were from Fife: in 1912, when one was shot at Morton Lochs on 12 August (SN 1912:212),

and in 1946, when one was at the Eden Estuary in September. Between 1950 and 1982 there were reports from Skinflats (July 1967), Lossiemouth (August 1967), the mouth of the North Esk, Angus (May 1974) and Whalsay (November 1976). In 1983, after seven blank years, Broad-billed Sandpipers were reported at Stranraer, Aberlady Bay (two birds – one present 8–14 June and the other 8 June–14 July), and Tyninghame (15 June). The last could have been one of the Aberlady birds – but even if only three individuals were involved 1983 was a notable year for this species. In 1984 there were two records from Aberlady Bay, in late May/early June. The scatter of dates is surprisingly wide, from 17 May to 3 November.

The Fenno-Scandian segment of the population migrates along a southeasterly route to winter mainly on the coasts of the northern Indian Ocean (BWP). This pronounced easterly movement is presumably the explanation for the species' infrequent occurrence in Britain.

Stilt Sandpiper *Micropalama himantopus*

Vagrant (N America) – two records, 1970 and 1976.

The first occurrence of this long-necked and long-legged wader was at Dornoch, Sutherland, on 18 April 1970; this was the first British spring record and the bird was in a puzzling intermediate plumage (SB 6:280–281). The second report was from near Sumburgh, Shetland, on 11–18 September 1976. The Stilt Sandpiper usually travels to the west of the Mississippi when migrating between its arctic breeding grounds and South American wintering area, but it occurs more often on the Atlantic coast in autumn than in spring – hence the relatively high proportion of British records occurring in autumn. A very full field description is given in BB 48:18–20 and in Sharrock & Grant (1982).

Buff-breasted Sandpiper *Tryngites subruficollis*

Vagrant (arctic N America) – almost annual, in very small numbers, since 1970; most frequent in autumn.

First reliably recorded in 1957, when one was at Hamilton, Lanark, on 27–30 October (BB 51:193), the Buff-breasted Sandpiper has occurred (or been identified?) both in Scotland and in Britain as a whole with increasing frequency since the early 1970s. This increase is somewhat surprising in view of the fact that the species breeds in arctic Canada and Alaska and is described as 'rare' on the Atlantic coast when migrating to and from its South American wintering grounds.

Between 1957 and 1970 there were records in only three years – 1958, 1960 (two), and 1962, but since 1971 the Buff-breasted Sandpiper has occurred annually, with a maximum of eight reported in 1975. Most records are of single birds but there were three at the Loch of Strathbeg on 18 September 1977 and at Bornish, South Uist, on 18 September 1981, and two at Musselburgh in 1975, Crail in 1977, and

Caerlaverock in 1979. Twenty-seven of the 35 occurrences (to the end of 1983) have been in September, three in May, two each in June and August, and one in October. The proportion of east coast records is high for a bird of American origin, with occurrences in East Lothian, Midlothian, Fife, Aberdeen, Inverness and Sutherland. The other reports are from Shetland, Fair Isle, Orkney, the Outer Hebrides (St Kilda & South Uist), the Inner Hebrides (Islay, Mull & Coll), Argyll mainland, Ayr, Lanark and Dumfries.

Ruff *Philomachus pugnax* [1]

Variable numbers pass through Scotland in autumn and a few remain to winter. Spring passage is less marked but birds in breeding plumage are recorded most years in May/June and breeding has occurred.

Although the first immigrants may appear as early as 3 July, the main autumn arrival of Ruffs takes place in August/September and most have moved further south by mid November. The numbers involved vary greatly, with only a few hundreds recorded in some seasons and well over 1,000 in others, and the pattern of distribution is also variable. In 1973 nearly 80% of the 500+ birds reported in August/September were in the north and east; in 1974 about 70% of 1,100+ were in the north and east and 20% in the southwest; and in 1978 a major influx in early August was confined to Shetland. Though the largest flocks occur on the coast, inland records are quite common; between 1968 and 1983 Ruffs were reported from all mainland counties other than Nairn and Selkirk and from all island groups other than the Clyde Islands. Flocks seldom exceed 50 birds but there were 60 on Fair Isle and 84 at Sumburgh in September 1973, 63 at Cotehill Loch in October 1976, and 90 on Unst and 150 at Virkie in August 1978. Inland records of sizeable numbers include 41 at Linlithgow Loch and 27 at Forfar Loch.

Wintering birds occur most often on the Forth, at Aberlady Bay, where Ruffs have been recorded in every month. Numbers are never large, usually averaging less than ten in mid winter, but up to 26 wintered there in 1968/69. Forty were at Slains Lochs in January 1983, and a few are often present early in the year elsewhere on the east coast. Ruffs are scarce on the west in spring. Numbers increase slightly during May/June, when breeding-plumage birds have been noted displaying in areas as widely scattered as Dumfries, Argyll, Caithness, Shetland and the Hebrides.

This species breeds from eastern England, the Low Countries, and northern Scandinavia eastwards across Eurasia, and winters largely from the Mediterranean southwards; a bird caught on Fair Isle in September was shot the following autumn on the Guadalquivir in Spain. Ruffs started breeding in England in 1963, after an interval of more than 40 years, and in 1976 Sharrock (Atlas) commented 'It is noteworthy that there has never been any suspicion of nesting in Scotland, even though the bulk of the world's population breeds on marshy tundra in N Europe and Asia'. The following year breeding was suspected in the Inner Hebrides when an agitated female was seen on 17 July. Three years later,

in 1980, a nest with four eggs was found in Sutherland; although no male was seen the clutch did hatch (J. Massie). Breeding was also suspected on the Outer Hebrides during the early 1980s.

B&R regard the Ruff as a rather scarce passage migrant and give few records of more than a dozen birds together. It seems clear that, as in England (Prater 1981), Ruffs have increased in Scotland since the 1950s. The predominantly northern pattern of occurrences, and the recovery in Russia during the breeding season of one ringed at Montrose Basin in autumn, suggest that the Scottish population may come from Fenno-Scandia rather than the Netherlands, from which the English population is believed to originate.

Jack Snipe *Lymnocryptes minimus*

Passage and winter visitor to low-lying marshes; widespread but occurring only in small numbers, mainly between late September and April. Occasional in summer.

The Jack Snipe, a largely nocturnal feeder, is probably both more common and more widely distributed than recent published records suggest. B&R's graphic description explains why it can so easily be overlooked – 'We have watched it feeding by the side of the fresh-water pools on the top of the island (Isle of May); here we have several times been able to make observations on its habit of "freezing" – the pools lay in a little hollow and when we approached them suddenly the Jack, instead of flying away, remained absolutely still, sometimes with its bill buried in the oozy mud. If we did not look directly at it – in fact, catch its eye – it would stay thus until we were almost near enough to touch it'. This behaviour, together with its silent flight and non-flocking habits, makes the Jack Snipe one of the most difficult waders to assess.

Passage is most marked in autumn; early individuals arrive from mid August onwards, but the main movement is generally in September/October (Boyd 1956). Winter numbers are usually lower, and there may be a spring influx in March or April. May reports are mainly from Fair Isle and Shetland but include two displaying at the Endrick Mouth in 1975, and singles on North Uist in June 1978, Foula in July 1976 and Stronsay in July 1975. Records of ten or more together are unusual and come most often from the Northern Isles, the northeast, and the Clyde area; high counts outwith the autumn migration are almost exclusively from Renfrew and Ayr. Surprisingly, there are no significant records from the southwest and relatively few recent reports from the Hebrides. It is probable that this species is under-recorded in some districts.

Studies by Iain Gibson in the Clyde area – extending over some 10 years – have shown Jack Snipe to be widespread there, with concentrations at certain favoured marshes. More than 40 sites have been located, mainly in Renfrew, Ayr and around Cumbernauld, between them holding up to 200 birds during the autumn peak. The species' preferred habitat appears to be low-lying marsh with extensive areas of sedges, though a wide range of wetland types is used. Close observation at Paisley Moss has shown that certain areas are particularly favoured, with single birds often springing from the same patch day after day; trapping has revealed that during migration periods birds rising from the same spot on consecutive days may be different individuals. Gibson has recorded two fairly distinct autumn peaks at Paisley Moss, the first in early October and the second usually in early November. In mild winters most birds arriving in the second wave probably remain at the site and such wintering individuals appear to rest during daytime in an area little more than a few centimetres square, identifiable by the small patch of trampled vegetation, often with a few droppings. Mistnetting has demonstrated that Jack Snipe leave the marsh around dusk and return about an hour before sunrise. Their weight may increase by more than 50% between November and January and they can apparently tolerate severe frosts for longer than Snipe.

Jack Snipe breed from Fenno-Scandia and the eastern Baltic eastwards through Russia, and winter from Britain south through Europe to the Mediterranean and North Africa. There is believed to have been some contraction of the breeding range (Harrison 1982) so the decline hinted at by B&R may in fact be a real one. Ringing recoveries include birds caught on Fair Isle in September/October and shot in Orkney, Ireland and the south of France between November and January, and two ringed in winter in Wigtown and Shetland and recovered in the same areas, in November and March, one and three years later.

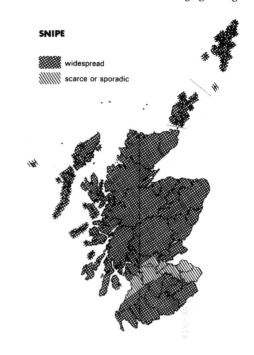

SNIPE

▓▓ widespread

▨▨ scarce or sporadic

Snipe *Gallinago gallinago*

Partial migrant, widely distributed as a breeder wherever there are marshy areas with some tall herbage; numbers are decreasing due to drainage. In winter many inland areas are deserted as native birds either leave the country or move to low-lying wetlands, where numbers are increased by the arrival of immigrants.

The drumming of displaying Snipe is a familiar sound in most parts of the country but there is currently concern for the species' future prospects in much of the agricultural lowlands. The wetter Hebridean machairs hold good populations, while rough grazing below 300 m asl and damp pastures are important breeding habitats. Marshy feeding grounds are essential and few Snipe nest on dry pasture or arable ground (Galbraith *et al* 1984). In England, where

the loss of damp grasslands has been extensive, the species is now scarce on lowland farmland (Smith 1983). Continuing drainage and fertilisation of rough grassland can be expected to decrease breeding numbers locally (Galbraith & Furness 1983), but the population as a whole is probably not under threat since many Snipe breed in bogs and on moorlands, though densities are apparently low even on wet moorland (Reed *et al* 1983a). Nests are not easy to find and censusing is difficult, despite the noisiness of breeding pairs; only the crudest of population estimates are consequently available. Parslow (1973) put the British and Irish population in the range 10,000–100,000 pairs and Sharrock (Atlas) considered it likely to be 80,000–110,000 pairs. Scotland probably now holds one-third to one-half of the total.

Snipe breed in every mainland county and all the island groups and there has apparently been little change in distribution since B&R wrote their detailed account. There are now, however, many breeding records for Fair Isle, where they knew of none. On the Isle of May the Snipe remains an occasional visitor only.

Assessment of wintering numbers is also difficult as Snipe seldom flock and often remain hidden in ditches and beside pools unless actually flushed. The recorded numbers can represent no more than a fraction of the birds present but they probably give a fair indication of the timing and scale of passage movements or influxes of immigrants. Concentration of local birds is presumably responsible for high counts in July – for example 100 at Gladhouse and 300 at New Cumnock – but from August onwards passage is evident, especially in Shetland, although numbers on Fair Isle may not peak until November. Mid-winter influxes sometimes occur, as in 1973 when there were 160+ in Unst on 11 January, but the largest autumn movements generally take place in September/October. B&R noted that 'enormous numbers sometimes winter in the Western Isles' and game records show that many still do so, although numbers may not be as high as in the past. On one South Uist estate the annual bag averaged 960+ in 1972–82 (per J. G. Young), while on Tiree the average number shot per gun-day dropped from 68 in 1911–14 to six in 1971–83 (per J. M. Boyd). Gatherings of 200–300 are not infrequent in autumn and there have been recent records of 200 or more from Shetland (Spiggie), Orkney (North Ronaldsay), Ayr (New Cumnock), near the town of Dumfries and at Loch Ken. Parties of under 50 are much more usual, however, and in mid winter and early spring large groups are seldom seen; 285 at Doonfoot, Ayr, in March 1977 was exceptional.

Ringing recoveries suggest a complex pattern of movements. More than half the native birds, and probably a higher proportion of those in their first winter, apparently move to Ireland. Birds ringed on passage at Fair Isle, St Kilda and Easter Ross have been recovered in Ireland and in Cornwall, while one marked on Foula was shot in Spain. Birds ringed in England in August have moved in the opposite direction and been shot on the Solway and in Ayr later the same year. The situation with regard to immigrants is equally complicated, with birds coming from many different areas. Snipe ringed as chicks in Iceland, Norway and Sweden have been recovered in winter in the Hebrides and Wester Ross, and recoveries in other parts of Scotland include birds caught on passage in Denmark, Finland, Poland, Germany and Czechoslovakia.

Most of the Snipe breeding in Scotland belong to the nominate race *G.g. gallinago*, which occurs throughout Europe and Russia, but in Shetland, Orkney and St Kilda this is replaced by *G.g. faeroeensis*, the race breeding in Iceland and the Faeroes. Birds of both races are represented among the winter visitors (BOU). There is a single old record of the American race *G.g. delicata*, a specimen of which was shot in South Uist in October 1920.

Great Snipe *Gallinago media*

Vagrant (Fenno-Scandia) – less than annual and in very small numbers, most often in autumn and in the Northern Isles.

There were only 14 records of the Great Snipe between 1958 and the end of 1984, although B&R knew of about 50 reports prior to 1950. The species is difficult to identify positively in the field and a high proportion of reports are rejected by BBRC – many of the earlier records refer, of course, to birds that were shot. There has been a marked decline in the southern and western part of the European range during the 19th and early 20th centuries, the cause of which is not fully understood (BWP); in the past breeding occurred in Germany and the Netherlands.

Ten of the 14 recent records are from Fair Isle: in 1960 (two), 1965 (two), 1967, 1969, 1977, 1981, 1983 & 1984; three are from Out Skerries (1973, 1974 & 1980), and one from Drumnadrochit, Inverness (April 1975). Most occurrences were between 10 September and 20 October, with a single record in December and two in spring (29 April and 1 May).

[Short-billed Dowitcher *Limnodromus griseus*]

The three old records of this N American species quoted by B&R (under the old name of Red-breasted Snipe) are not among those listed as acceptable in the BOU's *The Status of Birds in Britain and Ireland*.

Long-billed Dowitcher *Limnodromus scolopaceus*

Vagrant (N America/NE Siberia) — ten definite records.

The first recorded Long-billed Dowitcher was one (shot) in Fife in September 1867 and the second – more than 100 years later – was at Loch Lomond in May 1969 (SB 6:40). The other records are from Thurso (October/November 1975), Barra (August 1977), Meikle Loch of Slains (May 1980), Caerlaverock (November 1980/April 1981), North Uist (May/June 1983), Kirkwall (September 1983 and October 1984), Fair Isle (October 1983) and St Kilda (September 1984).

Since 1950 non-specifically identified Dowitchers have been recorded in Dumfries, Berwick, East Lothian, Angus, Perth, Dunbarton, Lanark, Argyll (mainland), Tiree, Orkney, Fair Isle and Shetland. Three of these records have been in April/May and the rest between August and November, mostly in September/October.

Woodcock *Scolopax rusticola*

Partial migrant, most abundant as a breeder in the lowlands, scarcer in the north and northwest Highlands and on the larger Inner Hebrides, and absent from the Outer Hebrides and Northern Isles. Some native birds move west or south in autumn, when there is an influx, sometimes very large, of immigrants from the Continent.

It was only in the 19th and early 20th century that the Woodcock became established as a breeding species throughout mainland Britain (Atlas); its colonisation of Scotland is documented in detail by B&R. By the 1940s it was breeding regularly on all the Clyde Islands, Islay, Colonsay, Jura,

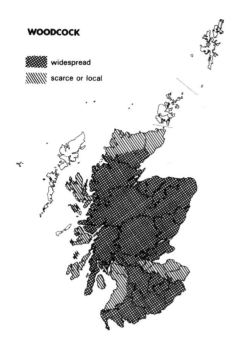

WOODCOCK

▓ widespread

▨ scarce or local

Gigha, Mull, Rhum, Skye and Raasay, and had nested sporadically in the Outer Hebrides, Orkney and Shetland. Atlas surveys produced little change in this general pattern of distribution but there have been no published breeding records from the Outer Hebrides or Orkney since 1900, and none from Shetland since 1905. The Woodcock continues to be a scarce breeding bird in Caithness and much of the western Highlands. In the southwest it is now abundant in conifer plantations, a habitat originally believed to be unsuitable for the species. Afforestation may eventually result in colonisation of Lewis and an increase in the Sutherland population.

Little information is available on numbers or breeding densities of this very secretive bird, but Shaw (1976) estimated 12–18 pairs/km² in the oakwoods of Glen Falloch and Gartfairn, Loch Lomond. Parslow (1973) put the British population in the range 10,000–100,000 pairs but Sharrock (Atlas) suggested that there are probably not more than 50,000 pairs in Britain and Ireland. Judging by the pattern of proved and probable breeding records in the Atlas, the Woodcock is more abundant in central, eastern and southern Scotland than in much of England and Wales; the Scottish population might therefore be of the order of 10,000–20,000 pairs – but this is a purely speculative total and the true figure may be very different.

In autumn immigrant Woodcocks arrive on the east coast; these influxes are most obvious, and consequently most fully documented, in places where the species does not normally occur. The records for Fair Isle show that the autumn arrival generally takes place in October/November; since 1968 the peak daily count there has only exceeded 100 in four seasons and a really large influx has occurred only once, in 1976, when more than 1,000 birds were present on 27–28 October – the most recorded in 30 years. Spring passage is usually slight but in late March 1980 there were arrivals on the east coast as far apart as St Abbs and the Out Skerries.

It is not clear whether the recent figures indicate a reduction in numbers since B&R's day or simply present a more comprehensive picture of the situation.

Because the Woodcock is a game species and extensive ringing was at one time carried out on Scottish estates, there have been many recoveries of marked birds; these shed some light on the rather complex pattern of movements. Of 315 recoveries, 231 were within 100 km of the ringing place, 140 of them shot in the period November/January, indicating that a substantial proportion of the Scottish population is resident throughout the year. Of 48 winter recoveries of birds which had travelled more than 250 km from their natal area, 36 were in Ireland, mostly in the south and west of Eire, and the other 12 in southern England, Belgium, Scilly, western France, northern Spain and Portugal. There are, however, exceptions to these general patterns: for example, a chick ringed in Kirkcudbright and shot near Moscow in May three years later, and one ringed in Perth and shot the following May in southern Norway. These records suggest that some interchange occurs between populations breeding in different parts of the species' very extensive range. Four Woodcocks ringed as chicks in Sweden have been shot in Scotland in mid winter and the only breeding-season recoveries of migrant Woodcocks ringed in Scotland (mostly on Fair Isle or the Isle of May) are from the same area.

Recent work by the Game Conservancy (Hirons 1981) has shown that, contrary to popular belief, male Woodcocks do not patrol exclusive territories but perform their roding display over large areas, often more than 100 ha, until attracted down by a receptive female. The pair remain together until the clutch is laid, when the male resumes his roding flights. At the start of the breeding season Woodcocks in Hirons' study area (in England) fed in the fields at night and roosted in cover during the day, but later on this pattern was reversed. From April onwards distribution in woodland corresponded closely with earthworm abundance.

Black-tailed Godwit *Limosa limosa*　　　[1]

Passage and winter visitor, widely though thinly scattered and in winter regular only at the Eden Estuary. A few pairs breed most years.

Black-tailed Godwits on passage, wintering and breeding in Scotland are all believed to belong to the Icelandic race *L.l. islandica*, as are most of the much larger numbers wintering in Ireland and England. Although there has been a big increase this century in the British wintering population, from under 100 in the 1930s to about 4,000 in the 1970s (Prater 1975), the numbers occurring in Scotland have increased only slightly and probably seldom exceed 400–500.

Autumn passage starts in mid July and continues until the end of September in the north and through October on the Solway. At the Eden there is sometimes a passage peak of over 100 birds in August or September but maximum numbers are often not reached until December/January, when 100–150 are usually present. On the Solway the highest counts are generally in September/October; in 1969–75

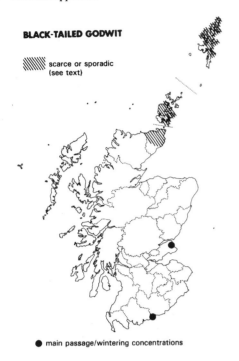

BLACK-TAILED GODWIT

⦀⦀⦀ scarce or sporadic
(see text)

● main passage/wintering concentrations

autumn numbers there peaked at 80–100, later dropping to five or fewer from December until early March. Movement northwards starts in March. The highest recent spring peaks on the Solway and Eden are c130 (April 1979) and c140 (April 1980) respectively. Most Black-tailed Godwits have left by mid May, although a few, mainly immatures, sometimes remain on the Eden and Solway throughout the summer; both these estuaries have stretches of the soft mud favoured by this species when away from the breeding grounds.

Elsewhere on the mainland parties are often of less than five birds, though flocks of up to 20 are occasionally seen. The figure of 125 given as the highest average monthly count for Brora/Golspie in Appendix 1 of *Estuary Birds* (Prater 1981) is an error and refers to Bar-tailed Godwit (M. Moser); the maximum group size recorded in Sutherland to date is only four (Angus 1983). Most reports are from coastal areas but there are also occasional inland records, including one of 16+ at the Endrick Mouth in April 1975. In the Hebrides this species occurs on both autumn and spring passage, mainly in July/August and April/May; it has been recorded a few times on Bute (B&R had no records for the Clyde Islands); and it is a scarce visitor to the Isle of May. In Orkney and Shetland it occurs both as a migrant and as a breeding bird.

Although breeding was first suspected (in Orkney) early this century, confirmation was not obtained until 1946, when a pair of Black-tailed Godwits nested in Caithness (B&R). There have been no subsequent records from that county. In Shetland breeding was first proved in 1949 and one or two pairs have nested fairly regularly ever since. A single pair bred on Sanday in 1956 and from 1973 onwards up to three pairs have held territory in Orkney, with two pairs

breeding successfully in separate localities in 1975. There are also breeding records from southeast Scotland, where one or two pairs were present annually from 1964–1971 at one site and a single pair at a second site in 1970–77, and southwest Scotland, where a pair reared two young in 1972. In 1979 there were eight birds in summer plumage at yet another mainland site in May/June but there was no proof that breeding took place. Details of these sites are being kept confidential in order to minimise the risk of disturbance.

In Scotland Black-tailed Godwits nest either in meadows or on damp moorland and blanket bog. It seems unlikely that there will be any substantial increase in the Scottish breeding population and it is consequently important that anyone coming across a breeding pair should avoid publicising their discovery. The birds are noisy, and very visible, on the breeding grounds but the disturbance caused by enthusiastic birdwatchers trying to find nests or young, both of which are very difficult to locate, has caused concern at some sites.

B&R noted that this species had increased as a passage migrant and winter visitor during the first half of this century and suggested that this increase (and also the species' re-establishment as a breeding bird in England) was due to expansion of the breeding population in Holland. It is now considered that all Black-tailed Godwits wintering in Scotland and Ireland (around 6,500) and also most of those wintering in England (a further 4,000+) are of Icelandic origin, while those breeding in East Anglia are of continental origin, belonging to the nominate race *L.l.limosa* (Prater 1981). On the face of it, it seems surprising that the population breeding in Iceland should have increased so greatly; although climatic amelioration may have had a beneficial effect (Prater 1975), agricultural improvement has been rapid and widespread in the more fertile areas of that country in recent years. If drainage and disturbance were indeed responsible for the loss of the Black-tailed Godwit as a breeding species in England during the early 19th century (Parslow 1973) one might expect that numbers in the more intensively-farmed areas of Iceland will also soon start to decline.

Bar-tailed Godwit *Limosa lapponica*

Passage and winter visitor, most abundant on the larger east coast estuaries, the Solway and the Outer Hebrides; widely distributed elsewhere on the coast, especially in autumn, though in much smaller numbers. Occasional in summer.

Scotland supports about 10% of the British and Irish wintering population of Bar-tailed Godwits, estimated as c58,000 or some 65% of the western European total (Prater 1981). No Scottish site regularly holds internationally important numbers (more than 5,500) but nationally important gatherings (more than 450) occur on the Uists, in Orkney and on all the principal Scottish estuaries except the Clyde (Table). This species has a preference for sandy estuaries and is highly mobile, often moving between, as well as within, wintering areas during a single season. There is no indication that any significant change in the status or distribution of this species has taken place during the present century.

Numbers of Bar-tailed Godwits at principal wintering sites (Data from Prater 1981 & M. Moser – figures rounded)

	Highest av. monthly count 1969–75	Post-1980 count Av. peak (n)	Range
Moray Basin	2,800	3,400 (2)	3,300–3,500
Firth of Forth	1,800	3,100 (4)	2,400–3,900
Eden Estuary	1,700	1,900 (4)	1,600–2,500
North Solway	1,600	4,300 (4)	2,200–7,100
Firth of Tay	1,100	1,300 (2)	1,100–1,400
Uists & Benbecula	NA	1,900 (1)	(Feb. 1980)
Orkney	NA	770 (1)	(Jan. 1983–84)

Immigration starts in mid July and numbers build up steadily during August. The first arrivals are adults which quickly commence their autumn moult; about 500 regularly moult in the Moray Firth near Ardersier. Immatures arrive from September onwards and passage through the Northern Isles is over by the end of October. Further south birds continue to arrive until January/February, when numbers are at their peak; colour-marking has shown that some of the mid-winter arrivals in the Dornoch Firth have come via the Wash (R. L. Swann). Some Bar-tailed Godwits leave during February but the main exodus takes place in late March and early April. There is no evidence in Scotland of a late April/early May movement corresponding to that recorded in the south of England (Prater 1981). Several hundred generally remain throughout the summer, and there are recent reports of up to 50 on Islay, 100+ at Aberlady, 200 at Findhorn Bay and on Sanday, and 300 at Tentsmuir;

most of these birds are immature. The Moray Basin holds the largest winter flocks, with numbers sometimes exceeding 5,000. Most are generally in the Moray and Cromarty Firths, where the highest average monthly counts in 1969–75 were 1,600+ and 1,000+ respectively, but flocks of 1,000 or more have also been seen at Dornoch, Udale Bay, Whiteness, and Brora. Ringing studies have shown that movement within the area results in fluctuations in numbers at different sites during the course of a season (Swann 1981).

Large numbers also regularly winter on the Firth of Forth, where the most important areas are Aberlady Bay, Musselburgh, and Burntisland Bay, and the birds move frequently between sites during the winter (Symonds *et al* 1984). On the Solway numbers peak earlier in the season than at other sites and drop markedly from January onwards, possibly due to movement across the firth or further south to Morecambe Bay. The Southerness area holds the largest gatherings, with a recent average peak of 3,900 and a maximum of 7,000 (during bad-weather movement in January 1982); as elsewhere, considerable interchange between sites occurs (Moser 1984). The only other estuaries regularly holding 1,000 or more Bar-tailed Godwits are the Eden and the Tay. Numbers seldom exceed 100 on Montrose Basin and are much smaller on the Ythan; few are recorded in the Clyde Estuary, although up to 250 occur on the Ayr coast at Hunterston.

By far the most important non-estuarine site is the southern Outer Hebrides; 1,900 were counted on the Uists and Benbecula in February 1980 – most of them on North Uist (Buxton 1982b), several hundreds are there in late August (Summers & Buxton 1979), and 800–1,000 are usually on South Uist from December to March. Islay is the only Inner Hebridean island to support a regular wintering flock; 300–400 are generally present in December/January (Booth 1981). Some 700 winter in Orkney, mainly on Sanday (Tay & Orkney R.G. 1984). Even during passage periods numbers in Shetland are small and there are few winter records. The Bar-tailed Godwit is a scarce visitor to the Isle of May, where it is recorded more often in autumn than spring (Eggeling 1974), and to the Clyde Islands. It seldom occurs inland.

Birds wintering in Britain belong to the nominate race *L.l.lapponica*, which breeds in the Arctic from northern Scandinavia to the Taymyr Peninsula in Siberia; those from the eastern part of the range, extending as far as northwest Alaska, winter in Australasia (Prater 1981). The possibility that birds from the middle of the range may migrate to either wintering area is suggested by the recovery in central Siberia, in early June, of a bird ringed on Fair Isle during autumn passage four years earlier. Other Scottish recoveries include birds ringed in south Norway while on migration, and subsequently found on South Uist, one marked on Fair Isle in late August and shot two weeks later in Denmark, and one caught on the Cromarty Firth in September and in Heligoland the following May. Many moult in the Wadden Sea, during September/October, before moving on to Britain.

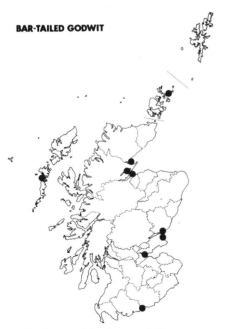

BAR-TAILED GODWIT

● main passage/wintering concentrations

Eskimo Curlew *Numenius borealis*

Vagrant (N America) – three old records.

The records of this North American wader – which is now verging on extinction – are all from the northeast and all in September. B&R give the first as in Aberdeen in 1878 and the others in Kincardine, in 1880 and 1885. However, the BOU's *The Status of Birds in Britain and Ireland* shows the first record as in Kincardine in 1855 and attributes the 1870 and 1880 occurrences to Aberdeen. The BOU statement is correct, as the Kincardine record was originally published in *Naturalist* 1855.

Whimbrel *Numenius phaeopus*   [1]

Local migratory breeder, regular in Shetland and sporadic, in small numbers, in Orkney and the Outer Hebrides and on the mainland. Widely distributed but nowhere abundant as a passage visitor.

Scotland supports the entire British breeding population of the Whimbrel. Numbers in Shetland alone have recently been estimated at around 300 pairs (Berry & Johnston 1980), well above the total of under 200 pairs suggested by Sharrock (Atlas). Even in Shetland Whimbrels have never been abundant and numbers declined markedly in the first half of this century (B&R); by the early 1950s there were only 50–55 breeding pairs, in four localities (Venables & Venables 1955). Since then the population has increased steadily, especially on Fetlar. In 1982 there were thought to be c105–135 pairs on Mainland, 70–75 pairs each on Fetlar and Unst, and 15–20 pairs on other islands (Herfst & Richardson 1982). A pair bred succesfully on Fair Isle for the first time in 1973, and again in 1974 but not since, and a pair summered for the first time on Foula in 1976.

B&R noted only two breeding records for Orkney from 1918 to 1950, and Balfour (1972) gave the most recent occurrences as three pairs on Eday in 1968 and one pair on Mainland in 1970. Whimbrels have since bred again on Eday and on at least one other northern island, with a minimum of five pairs in 1983. Older records detailed by B&R include breeding in North Uist, Argyll, north Sutherland and Inverness, and possible breeding on St Kilda, the Flannans and an island off Wester Ross. Breeding was first proved on St Kilda in 1964 and at least one pair has probably nested annually ever since; in 1979 two females laid in the same

nest (SB 11:164–166). The Whimbrel has bred sporadically on Lewis, most recently in 1971 (Cunningham 1983), but its status both there and elsewhere in the Outer Hebrides is unclear; SBR 1975, the Atlas and Fuller *et al* (1979) suggest that nesting took place in several areas in the mid 1970s, but Cunningham questions these reports. On the mainland there was a record of nesting in Sutherland in 1960 (SB 1:427) and a few pairs have bred in most years since 1973. There were up to three pairs at one undisclosed mainland site from 1974–77 and one or two pairs occasionally at a second site. Whimbrels have been noted song-flighting and apparently holding territory inland in Caithness since 1978.

In Scotland the breeding habitat is fairly-level open heather or grass moorland, often broken up by peaty channels. Studies on Unst indicate that the presence of such small-scale 'hills and valleys' is a desirable feature in a Whimbrel's territory, as it is important in providing ground cover for both adults and chicks (Herfst & Richardson 1982). Much Shetland moorland is currently being enclosed and reseeded, as common grazings are being apportioned to individual crofters, and this development is likely to make such areas less suitable as breeding grounds. On the mainland nesting has occurred at sea-level and within 100 m of the tideline (Headlam 1971).

Shetland-breeding birds start to leave the islands in late July and most have gone by mid August. Autumn passage elsewhere is from mid July (3 July is the earliest recorded date), generally peaking in late August/early September, occasionally earlier, and continuing well into October. Although most sightings are along the coast, Whimbrels can often be heard inland at night, especially in misty weather. Many of the passage records are from the Hebrides and southwest, but there is occasional heavy passage on the east, for example 100 at the Ythan in early August 1975, 70+ on the Moray/Inverness coast in 1979, and an exceptional 70 on the Isle of May on 4 August 1980.

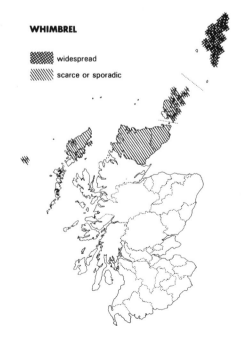

WHIMBREL

▓▓▓ widespread

▨▨▨ scarce or sporadic

Both summering birds and passage migrants belong to the race *N.p. phaeopus*, the breeding range of which extends from Iceland to western Russia. Most, possibly all, of the passage birds are probably of Icelandic origin, but as yet there are too few ringing recoveries to confirm this. At least some birds of continental origin are among those visiting England; Whimbrels ringed there in August have been recovered on their breeding grounds in Finland and USSR.

Whimbrels winter on coasts and estuaries south to South Africa; Shetland-ringed birds have been recovered in Ghana between October and February. Individuals occasionally winter in Scotland and there are records from as far north as the Cromarty Firth in December and January. Native breeders occupy territories in April and early May. Spring passage is mainly from early April to about the third week in May, but there are a few March records, especially from Islay, and birds are sometimes still passing through in June. High spring-passage counts include 130–150 on South Uist in May 1965 and 100+ at Leswalt (Wigtown) on 6 May 1982. Much larger numbers of migrants are recorded in south and east England than in Scotland; the peak count there, c2,500, occurs in May, when 1,500–2,000 may be present at the only important roost in Britain, on Steart Island in the Severn Estuary (Prater 1981).

The American race of the Whimbrel *N.p. hudsonicus*, has been recorded twice, on Fair Isle in late May 1955 (BB 48:379) and on Out Skerries from 24 July to 8 August 1974 (SB 8:427, see also SBR 1978:46). The only other occurrence of this subspecies, which is easily distinguished by the absence of white on the rump, has been in Ireland.

[Slender-billed Curlew *Numenius tenuirostris*]

The report of a Slender-billed Curlew at Avoch, Black Isle, in February 1960 (SB 1:255) was not accepted by the BBRC. There has as yet been no accepted British record of this rare and probably decreasing species, which breeds in eastern Russia and China and winters around the Mediterranean.

Curlew *Numenius arquata*

Partial migrant, widely distributed as a breeder wherever there is rough grassland or moorland; many native birds winter in Ireland. An abundant passage and winter visitor; the largest wintering concentrations are on the Solway and Forth and in Orkney, but in autumn sizeable flocks occur in many areas.

Originally a bird of damp moorland, breeding regularly at altitudes of 600 m or more asl, the Curlew has successfully colonised much of the predominantly agricultural lowlands, where it nests most often in rough grazing and damp permanent pasture but also quite often in arable fields – a habit noted by B&R during the 1940s when old grassland was brought into cultivation but the local birds remained faithful to the site. Breeding densities are apparently rather low, averaging from 2.9 pairs / km^2 on rough grazing over 300 m asl

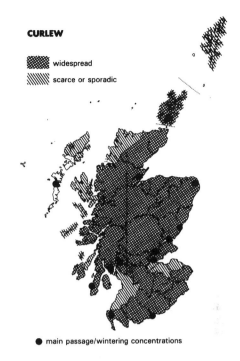

CURLEW

widespread

scarce or sporadic

● main passage/wintering concentrations

to less than 1 pair / km^2 on dry pasture, arable land and peat moorland (Galbraith *et al* 1984; Reed *et al* 1983a).

There is not enough information on past numbers and distribution to allow assessment of any trend in the breeding population, except on some of the islands. In the Outer Hebrides breeding was first recorded in 1965, on Lewis, but there has been only one subsequent confirmed nesting, also on Lewis, in 1980 (*contra* Fuller *et al* 1979, who claimed that about ten pairs nested on Lewis and that sporadic breeding might occur on other islands). On the Inner Hebrides numbers are apparently decreasing; the only island with a population of any size is Islay, while Tiree, Coll and Raasay no longer hold any breeding pairs (Reed *et al* 1983). The Shetland population is widely but rather thinly distributed; Berry & Johnston (1980) put it at 100–1,000 pairs but it seems likely that it falls in the lower half of this range. The first recorded breeding on Fair Isle was in 1968 (seven pairs in 1983) and on Foula in 1976; this suggests that some expansion may be taking place. Numbers in Orkney are higher, with 139 pairs recorded on Birsay Moor alone in 1974 and a suggested total of 1,000+ pairs (Heppleston 1981), but changing land-use patterns may be affecting distribution there, as they are in mainland areas where widespread afforestation is taking place.

Movement to low ground and towards the coast starts in late June; by early August the breeding grounds are largely deserted and big gatherings have built up on and near the coast. The first immigrants have already arrived by then – adults start to appear in June but juveniles come mainly in the second half of September (Prater 1981) – and it is not possible to be certain what proportions of native and immigrant birds are present. Ringing recoveries suggest that about two-thirds of the Scottish breeding population moves out to winter in Ireland (Bainbridge & Minton 1978); some young birds stay there throughout their first summer. Others, adults and immatures, winter on the Solway or on the west coast but some apparently remain within a short distance of their natal area unless driven away by severe weather. There are occasional exceptions to these general patterns of movement, for example a Shetland-reared bird found in Norway in early September the same year.

Curlews tend to use the same wintering area in successive years and the majority of immigrants appear to move directly to, and to remain within, a relatively limited area (Bainbridge & Minton 1978). Birds ringed as chicks in Norway (where the coastal breeding area is isolated from the main Eurasian range), Finland and Sweden have been recovered in Scotland; although numbers are small, the pattern suggests that Norwegian birds are most likely to winter north of the Tay and in the Hebrides, whereas those from Finland and Sweden are more widely scattered from Orkney to the Solway. It is presumably birds of Norwegian origin that form the small population wintering in Iceland, where the species does not breed.

In 1969–75 the largest estuarine counts of Curlews, over 15,000 on average, were in September (Prater 1981), but there are probably many more present at non-estuarine sites at that season and the total wintering population must be very much higher. Internationally important numbers (more than 3,000) winter in Orkney, where over 17,700 were counted in January 1983–84, more than 10,000 of them on Mainland (Tay & Orkney R.G. 1984). Presumably the combination of abundant good farmland (an important feeding ground for this species) near the coast and a relatively mild winter climate make this a favourable wintering area. Flocks of more than 3,000 also occur occasionally on the Firth of Forth and more regularly on the Solway and around Stranraer and Loch Ryan (Table).

On the east mainland flocks of more than 1,000 have been recorded recently at the Cromarty Firth, Strathbeg, Montrose Basin and Loch Leven, and in the west at Machrihanish and on the Ayr coast near Troon. The Ythan, Tay, Eden and Clyde seldom hold more than 500. In most areas numbers are highest in August/September and drop in October, but on the Solway over 2,000 are usually present

from July to February and on the Forth peak counts often occur in mid winter. Among the islands there are records of over 1,000 from Bute and Islay but on the other Clyde and Inner Hebridean islands numbers are much smaller. About 1,500–2,000 are in the Outer Hebrides in mid winter (Buxton 1982a,b) but only a few winter in Shetland. Although widely dispersed while feeding, Curlews roost communally, sometimes well away from the feeding area.

Native Curlews start to move back to their breeding grounds in February and in some years have penetrated far inland by the middle of the month; by early April most are on territory. The return movement of the immigrant birds is more protracted, with the first evidence of passage often in early March, a peak in late April, and birds still moving north through Fair Isle and Shetland during May. A small summering population remains around the estuaries and is believed to consist largely of first-year birds (Prater 1981).

Upland Sandpiper *Bartramia longicauda*

Vagrant (N America) – four records, 1933–83.

Known formerly as Bartram's Sandpiper (and in America as Upland Plover), this species was first recorded in 1933, when one was shot at Ruthwell, Dumfries, on 13 October (BB 27:205). The three subsequent records are from Fair Isle (October 1970 & September 1975) and St Kilda (April 1980). The St Kilda bird is the only British spring occurrence to date, all others (c30) being between September and December. The Upland Sandpiper breeds as far north as Quebec and migrates through the eastern USA to winter in South America.

Spotted Redshank *Tringa erythropus*

Scarce passage visitor, most regular on the east coast and in autumn. Occasional in winter and in summer.

Both the scale and the timing of the autumn passage of Spotted Redshanks are variable, with the numbers reported per year ranging from under 100 to over 300 and the peak occurring in either August or September. In most years immigrants start to arrive in early July, the largest numbers are present between mid August and mid September and along the eastern counties from the Moray Basin to the Forth, and nearly all have moved on by late October. Since 1968 the highest counts recorded were those in 1974, when the total was 300+ and more than 210 were reported in eastern Scotland. Most records are of single birds o very small groups but sizeable flocks occasionally occur, eg 38 at Morton Lochs and 25 on the Ythan Estuary, both in September 1974. Counts of 10–20 have also been recorded at Findhorn Bay, on the Tay and the North Solway, on Wigtown Merse and at Shewalton (Ayr).

Spring passage takes place from late March to early June, seldom producing more than 15 reports and often less than five. Birds in full breeding plumage have been recorded

Peak counts of Curlews at coastal passage/wintering sites (Data from Prater 1981, SBR, BoEE/WWC, Moser 1984, Tay & Orkney R. G. 1984 & BTO/WSG Winter Shorebird Count preliminary results, per M. E. Moser & Tay R. G./Shetland Bird Club count per R. W. Summers)

	Highest av. monthly count 1969–75	Post-1975 counts
Shetland	NA	c4,000 (1984/85)
Orkney	NA	17,700+ (1983–84)
N. Solway (Sept)	c3,000	Av. peak 1980–84 1,700; range 740–2,800
Firth of Forth	c2,100	2,000
L. Ryan/Urr	1,700	4,700 (September 1975)
'Strathclyde'*	NA	4,500–8,000 mid winter 1984/85

*'Strathclyde' = coasts of Kintyre, all islands and mainland west coast to boundary with Highland Region.

inland in May/June at Loch Ken and Harperrig (June 1974) and an Inverness marsh in May 1979 and 1980.

Although the largest numbers occur on the coast, inland records are quite frequent. Between 1968 and 1983 there were reports from all mainland counties except Banff, Nairn, Kincardine and Selkirk; relatively few were from the Highlands and the Borders. Spotted Redshanks occur fairly regularly in Shetland and Fair Isle on both spring and autumn passage, less often in Orkney, occasionally on the Isle of May, and only rarely in the Hebrides, where they have been reported since 1968 on Lewis, both Uists, Benbecula, Tiree and Islay. There has been no report from the Clyde Islands since the 1927 Bute record given by B&R, who knew of no occurrences in Shetland or the Inner Hebrides and only one, on North Rona, from the Outer Hebrides. As they point out, this species may well have been overlooked in the past.

The Spotted Redshank breeds from northern Scandinavia east across the USSR; those migrating through Scotland belong to the western population, which winters mainly in western Africa. During the period 1969 to 1975 there was an increase in the numbers recorded, especially in winter, in south and southeast England, where a few sites hold nationally important gatherings (over 50 birds). In Scotland up to five Spotted Redshanks have wintered almost annually since 1968. December/March records come most often from the Forth, Solway, and Ayr coasts but there are also reports from Renfrew, Fife, and the Moray Basin.

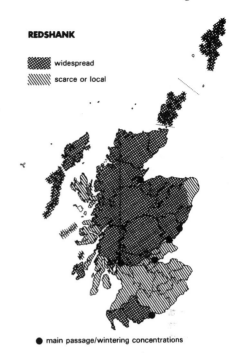

REDSHANK

▨ widespread

▧ scarce or local

● main passage/wintering concentrations

Redshank *Tringa totanus*

Partial migrant, widely distributed as a breeder wherever there is damp pasture or low moorland; numbers are currently declining as these habitats become scarcer. Many native birds leave in winter. An abundant passage and winter visitor on the coast, with the largest concentrations on muddy estuaries.

The breeding population of this species expanded during the early part of the 20th century (B&R) but is currently declining in many areas due to the drainage and improvement of what was formerly rough grassland. Once a familiar sight in most glens, the Redshank is now much scarcer and seems likely to decrease further. The SOC/WSG survey in 1982–83 (Galbraith *et al* 1984) demonstrated the extent to which this species is dependent upon marginal farmland.

On the mainland an average breeding density of 2 pairs/km² was found on damp pasture, whereas rough grazing below 300 m asl averaged only 0.7 pairs/km², with fewer still on dry pasture and arable land. Densities are low (0.2 pairs or less/km²) on the Caithness flows (Reed *et al* 1983a), and the population is sparse on the acid peatlands of the northwest and on many moorland areas further south. The Outer Hebridean machairs support much the highest densities, averaging about 15 pairs/km² and reaching more than 47 pairs/km² locally in small areas. Galbraith *et al* concluded from the findings of this extensive survey that some 5,000 pairs currently breed on farmland, of which 40% nest on the machair and a further 40% on damp pasture/rough grazing. Both these habitats are susceptible to agricultural development, which could pose a major threat to the future of this species. A comparable survey has shown that in England the Redshank is now an uncommon bird on lowland farmland (Smith 1983). It seems likely that numbers are now well below the 1968–72 estimate of 38,000–48,000 pairs for the British breeding population, of which Scotland must hold an increasingly large proportion.

As soon as the young have fledged the birds move to the coast and many of those breeding in the north and east probably later move southwestwards to the Solway or further. Some go only as far as Wales and southeast England, some to France and the Netherlands, and some cross to Ireland, with first-winter birds travelling further than adults (Prater 1981).

Immigrants from Iceland, belonging to the race *T.t. robusta*, start arriving in late July and account for a large part of the wintering population; several have also been recovered in Scotland during the breeding season (May-/June). British-breeding birds are intermediate in size

Peak counts of Redshanks at major passage/wintering sites (Data from Prater 1981 & M. Moser & Tay R. G. Shetland Bird Club prelim. fig. per R. W. Summers – figures rounded)

	Highest av. monthly count 1969–75	Post-1980 counts Av. peak (n)	Range
Inner Clyde	8,000	3,700 (4)	2,600–6,000
North Solway	4,400	850 (4)	680–1,100
Moray Basin	4,300	2,100 (2)	2,000–2,300
Firth of Forth	3,800	2,100 (4)	1,500–3,000
Montrose Basin	2,700	1,600 (3)	770–2,200
Firth of Tay	1,700	1,400 (2)	1,400
Orkney	NA	7,000 (1)	(Jan. 1983–84)
Shetland	NA	c2,000 (1)	(mid winter 1984/85)

between the Continental *T.t.totanus* (not known to occur in Scotland) and Icelandic races. Estimates of the proportion of Icelandic birds present, based on measurements, range from 50–80%, but may be subject to bias due to variations in the proportion of young birds (with shorter wings) in the sample; west coast estuaries appear to have consistently lower proportions of first-year Redshanks than those on the east (Furness & Baillie 1981).

All the major estuaries hold wintering flocks of international importance (more than 2,000), the main resorts being the Inner Clyde, North Solway, Firth of Forth and Montrose Basin (Table). In 1969–75 more than 25,000 Redshanks were counted during October on Scottish estuaries and the total must be well above this figure as the species is also widely scattered along rocky coasts, for example in Orkney. Over the country as a whole numbers are at their peak in autumn, but there are regional variations in timing, from September on the Tay and Forth to November or even later on the Moray Basin. Colour-dyeing has shown that some mid winter arrivals at the latter site have come from the Ythan, Eden and Wash (R. L. Swann). The largest concentrations are on the Clyde, where over 9,000 were present in autumn during most of the 1970s. Then in 1978 there was a rapid decrease after an October peak of 7,000; in 1979 the peak was only 4,200, and in 1980 numbers again dropped rapidly following a September count of 7,000. Furness & Galbraith (1980) have suggested that this situation may reflect a local deterioration in feeding conditions; it would be interesting to know how current distribution patterns compare with those found by Halliday *et al* (1982) in the period immediately before the decline.

Nationally important numbers also occur regularly on all the east coast estuaries from the Cromarty to the Forth, and on the Solway (Table). Elsewhere around the mainland coast, south of the Moray Basin and the Clyde, most muddy tidal sites hold several hundred Redshanks, at least in autumn if not all winter. Further north and west, and on most of the islands, numbers are in general smaller, for example a total of less than 1,000 on the Outer Hebrides (Buxton 1982), while flocks of over 100 are the exception rather than the rule in the other island groups except for Orkney, which has a wintering population of about 7,000 (Tay & Orkney R.G. 1984).

Recent studies of this species, involving cannon-netting and dye-marking, have shown that it is usually very loyal to its chosen feeding ground throughout the winter (which makes the desertion of the Inner Clyde the more surprising), and that it also tends to return to the same area in subsequent winters (Furness & Galbraith 1980; Nicoll & Summers 1980; Symonds *et al* 1984). A study has also been made of the effects of severe weather on body condition (Davidson & Evans 1982), using birds found dead during the prolonged period of severe weather in the winter of 1981/82, when the mortality rate was high (Davidson 1982). This suggested that death resulted from inability to utilise protein reserves fast enough, once fat reserves had been exhausted, rather than from very low temperatures.

Movement away from the wintering grounds starts in February and near the coast many Redshanks are on territory by mid March; the timing of their appearance inland is much influenced by the weather. Icelandic birds depart from late March onwards and passage is often marked during April and early May, with influxes of birds pausing briefly before moving on. There is a sizeable summering population on many estuaries, as numbers often nest in the vicinity.

Marsh Sandpiper *Tringa stagnatilis*

Vagrant (SE Europe) – five records, 1966–84.

The first record of this elegant wader, which breeds from Bulgaria and Rumania eastwards into Asia, was from Dunnet Bay, Caithness, on 3–5 September 1966 (SB 4:557). It has since been recorded in: Shetland (Mainland, 4–6 May 1969), Ayr (Doonfoot, 29 July 1979), Orkney (North Ronaldsay, 23 August 1979) and Moray (Lossiemouth 20–23 April 1984). The Marsh Sandpiper is a strongly migratory species; although it wanders quite far west during migration it winters south to South Africa and Indo-China. Its appearances in Britain, where it has been recorded less than 50 times, are very sporadic.

WN

Greenshank *Tringa nebularia*   [1]

Migratory breeder, confined to moorlands in the north and west. Scarce passage visitor, favouring river channels, pools and lake shores, both coastal and inland. Occasional in winter.

Breeding Greenshanks require an abundance of water-bodies, where the birds feed on insects living in the shallows and among waterside plants, and a large extent of gently undulating upland; this combination of features is found both on the Lewisian gneiss of the northwest and the glacial-drift of the eastern Highlands. The altitudinal range of the breeding grounds is from sea-level to 450 m asl in the northwest and from 210 m to 550 m, exceptionally 650 m, south of the Great Glen (Ratcliffe in Nethersole-Thompson & Nethersole-Thompson 1979). The highest density of breeding birds recorded is 20 pairs/km², in northwest Sutherland, but in most areas it is very much lower. Scotland holds the entire British Greenshank breeding population. The life-style of this bird on its breeding grounds has been studied

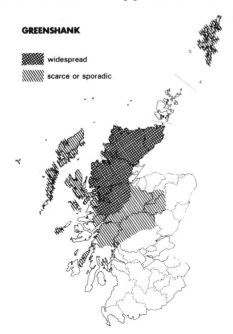

GREENSHANK

▨ widespread

▨ scarce or sporadic

over many years by the Nethersole-Thompson family, whose findings (Nethersole-Thompson 1951; Nethersole-Thompson & Nethersole-Thompson 1979) provide not only a great deal of information on distribution and behaviour but also an evocative account of the thrill of sharing Scotland's remoter moorlands with this and other 'wilderness' species.

Several changes in breeding distribution have occurred since 1950, the most notable being the recolonisation of the Outer Hebrides (from 1951) and the marked decrease in the upper Spey Valley, where overgrowth of vegetation, dry-ing-out of pools, afforestation, and increased human pres-sure have resulted in the loss of most of the small groups which formerly nested there in clearings and forest bogs (Nethersole-Thompson & Watson 1981).

Although breeding numbers are known to fluctuate mark-edly between years, there has apparently been some overall increase during the last 30 years. The most recent estimate of the population is 800–900 pairs (Nethersole-Thompson & Nethersole-Thompson 1979); this is considerably higher than the earlier estimates of 300–500 (Nethersole-Thompson 1951) and 400–750 pairs (Atlas). Most of the breeding popu-lation is concentrated in Sutherland, Wester Ross and main-land Inverness, with smaller numbers in Caithness, Aberdeen, Perth and Argyll and on the Inner and Outer Hebrides (Table). Sporadic breeding has also occurred in

Estimated breeding population of Greenshanks in 1970s (Data from Nethersole-Thompson & Nethersole-Thompson 1979)

County	No. of pairs
Sutherland	340–380
Ross & Cromarty	230–250
Inverness (mainland)	120–140
Outer Hebrides	50
Inner Hebrides	10–15
Caithness	25
Perth	10–15
Argyll	15–20
Aberdeen	5–10

Shetland, where a pair bred successfully on Mainland for the first time in 1980 (a nest on Unst in 1871 is the only earlier, but doubtful, record) and Orkney, where a pair with young was seen on Hoy in 1951. Nesting is said to have taken place in the Lammermuirs during the 1920s.

Most Scottish birds return to their breeding areas in March or early April but individuals may be inland as early as the last week of February. Movement back towards the coast starts in late June, the first birds to go often being females which have left their broods with the males. Autumn numbers are much increased by passage migrants, especially during August and September, and movement continues into October. Although the largest numbers are recorded at coast-al sites there is also a well-marked inland passage, with single birds or small groups visiting reservoirs, lochs and flood-waters. Even at peak migration, numbers are quite small and no Scottish site regularly holds more than 50 birds.

The highest recent count is only 49, on the Solway in September 1971, but there are records of 20–40 from the Inner Clyde, the Ythan, Montrose Basin, the Tay at Errol, and the Eden Estuary. Fair numbers are present on the Cromarty Firth in July but the figure of 45 given by Prater (1981) is perhaps misleadingly high.

Greenshanks winter from Britain south to southern Africa. In the past they were recorded only rarely in Scotland in winter but since the early 1970s wintering numbers have increased. In 1970 only 12 were reported but there were 25 records in 1972 and 90 in December 1975. This expansion has been attributed to a succession of mild winters during these years. More recently numbers have fluctuated between about 20 and 55. Although occasional birds occur on the east coast, the majority of wintering Greenshanks are in the west, mainly on the Clyde (up to 30 in 1976/77 but fewer since) and Solway (17 in 1979/80); a few also remain in the Outer and Inner Hebrides and around the sea-lochs from Argyll northwards.

It is not known what proportion of the Scottish population winters in Britain and Ireland, but Prater (1981) has suggested that 75% or more probably move further south, perhaps as far as northwest Africa. The few ringing recoveries include a Sutherland-ringed bird in Cork in its first October and one from Rannoch in France in September. The return passage in spring is much less marked than the autumn movement; although virtually complete for native birds by mid April it continues in the Northern Isles until late May/early June, when it presumably involves birds from Fenno-Scandia.

Greater Yellowlegs *Tringa melanoleuca*

Vagrant (N America) – three records, 1953–83.

The Greater Yellowlegs was first recorded in Shetland, at Dunrossness on 26–27 May 1953 (BB 48:363), one of comparatively few spring occurrences in Britain. The two subsequent reports are from the Ythan Estuary in October 1957 and South Uist in August 1978. This species, which breeds from Manitoba and the St Lawrence northwards and winters along the Atlantic coast of southern USA and in the Gulf of Mexico, is a much scarcer visitor to Britain than the Lesser Yellowlegs; only 26 had been recorded by the end of 1983.

Lesser Yellowlegs *Tringa flavipes*

Vagrant (N America) – recorded in ten years, 1950–84.

First recorded in 1910, when one was shot on Fair Isle in September (ASNH 1911:53), the Lesser Yellowlegs was very rare until the mid 1970s, since when there has been a marked increase in reports. As it passes through eastern North America *en route* from its Canadian breeding grounds to its wintering areas in South America, it is perhaps surprising that it is not recorded more often.

After a gap of 40 years, the Lesser Yellowlegs was recorded again in 1950, at Aberlady Bay. There have since been single records in 1951, 1953, 1976, 1978 and 1984, two each in 1974, 1975 and 1979, and four in 1980. Two Aberdeen reports in 1979, from Rattray Head on 25 September and the Ythan on 7 October, are assumed to refer to one bird, and a Lesser Yellowlegs at Cairnbulg, Aberdeen, from 17 January to 23 April 1980 may have been the same individual; there are a number of records of overwintering in Britain. Most occurrences have been between 6 August and mid October, but there are three May records and one in November. This species has been recorded in Shetland, Fair Isle, Caithness, Sutherland, Nairn/Inverness, Aberdeen, Angus, East Lothian, Argyll, Ayr and the Outer Hebrides (Monach Isles). It seems likely, in view of its American origin, that it occurs more often in the west than these records suggest.

Solitary Sandpiper *Tringa solitaria*

Vagrant (N America) – one old record.

B&R state that a Solitary Sandpiper was shot in Lanark 'before 1870'. Although vague, this record appears to be accepted and is quoted by BOU and Sharrock & Sharrock (1976). There have been only about 20 British occurrences, mainly in autumn and the most recent in 1983. The Solitary Sandpiper breeds in Canada and migrates through the eastern USA to winter in tropical America.

Green Sandpiper *Tringa ochropus*   [1]

Passage visitor, widespread in small numbers on autumn migration, when it is most abundant in the east and on the Northern Isles. Occasionally winters, is scarce on spring passage, and has bred.

The Green Sandpiper has a marked preference for freshwater and even on estuaries is more likely to be found beside streams or pools than in intertidal areas. It occurs frequently inland and, judging by reports from such areas as Crieff and the Inverness/Perth boundary, appears to make use of the passes through the hills as it travels south. A few birds often arrive in July, the main influx is in August/September, and stragglers are sometimes still passing through in November. The main autumn movement is generally well-marked, especially in Shetland and Fair Isle, where the Green Sandpiper is regular on passage. Although a large proportion of the records generally come from the east and north, in some seasons, eg 1975, up to one-third may be in the southwest. In 1968–83 inclusive there were reports from every mainland county other than Selkirk and Nairn, and from the Outer and Inner Hebrides and the Clyde Islands, where one on Arran in September 1979 was the first record. Numbers are never large, the annual total probably ranging from about 50 to 120, and most records are of single birds. The species' main wintering area is in the

Mediterranean Basin and Africa (BWP) but a passage bird ringed in September on Fair Isle was still on the west coast of France in December.

A few Green Sandpipers stay throughout the winter in most years, generally south of the Forth / Clyde but occasionally as far north as Easter Ross. Spring passage, between early April and the end of May, seldom produces more than 20–30 records and often far fewer. The bias towards the north, and to a lesser extent the east, is even more marked in spring than in autumn.

The Green Sandpiper, which breeds in marshy forested areas from southern Scandinavia and east Germany eastwards through Russia, has been suspected of nesting in Scotland since the 1930s. Displaying pairs were seen on Speyside in the 1930s and 1940s but it was not until 1959 that breeding was proved with the finding of a pair with one chick, on 29 May (BB 52:430). There have since been occasional reports of birds inland in suitable habitat during the summer but no proof that breeding has occurred again. The species' habit of using abandoned tree-top nests, eg of Mistle Thrush or crow, creates particular problems for the would-be nest finder – but it is, of course, given special protection under the 1981 Act.

Wood Sandpiper *Tringa glareola* [1]

Scarce passage visitor, occurring annually in spring and autumn. Since 1959 a few pairs have bred in most years in the Highlands.

It seems surprising that the Wood Sandpiper, which B&R regarded as an uncommon visitor and which still occurs in only small numbers on migration, has established even the somewhat tenuous foothold in Scotland that it has achieved since 1959. A singing cock was found in north Inverness in 1947 (D. Nethersole-Thompson) but breeding was not confirmed until 12 years later, when an agitated adult and accompanying juvenile were found in a north Sutherland marsh on 18 July 1959 (SB 1:150). Breeding has since been proved also in Caithness, Ross and Inverness (both west and east), and has almost certainly occurred in Perth and Argyll. Birds apparently holding territory have been reported from Aberdeen, Shetland and the Outer Hebrides. Many sites are remote and not all have been checked each year; numbers on those visited have fluctuated greatly, ranging from fewer than two pairs in 1972 to seven to ten pairs in 1980. The number actually known to have bred has not exceeded six pairs in any one year.

On its Scandinavian breeding grounds the Wood Sandpiper is most abundant in the forest-tundra transitional zone but in Scotland it also favours open boggy moorland with scattered lochans and marshy areas. It would be easily overlooked in such country were it not for its noisy behaviour. Both its distinctive song-flight and its agitated 'chipping' in defence of its young effectively draw attention to its presence where a more silent bird would often pass unnoticed.

Outwith the breeding season this species occurs both by coastal waters, where it is most likely to be found at the edges of pools and saltmarshes, and inland. Passage is not heavy and the numbers recorded seldom exceed 30 at either season, the total of 45–50 in autumn 1980 being the highest yet. Passage birds presumably come mainly from the western end of the species' continental range, which extends from Denmark and Fenno-Scandia east through Russia. There are no winter records; most Wood Sandpipers winter in the Tropics and sub-Tropics.

The Wood Sandpiper is most frequently reported in the north and east, occurring regularly on Fair Isle and in Shetland in both spring and autumn, and occasionally on the Isle of May. Between 1968 and 1983 it was recorded in every mainland county except Wigtown and Nairn, and from Lewis, North Uist, Islay and Rhum, while it occurred on St Kilda and Skye in the late 1950s. There are no records from the Clyde Islands. Spring passage takes place mainly between early May and mid June, with extreme dates of 8 April and 29 June, and autumn passage from mid July to mid October, the latest recorded date being 16 October. Most reports are of single birds but in 1970 there were eight at one site in Aberdeen and seven together in East Lothian, both occurrences being on 23 August.

Terek Sandpiper *Xenus cinereus*

Vagrant (NE Europe / Siberia) – two records, 1975 and 1977.

The Terek Sandpiper was first recorded in Shetland, on Whalsay on 20–21 June 1975 (BB 69:337), and the second occurrence was at Sandside, Caithness, on 5–12 June 1977. The British records, most of which are in spring, tend to occur in batches and it has been suggested by Sharrock & Sharrock (1976) that such irregular increases in vagrancy reflect periodic high population levels. Apart from an isolated group in eastern Finnmark (c30 pairs in 1980, BWP), most Terek Sandpipers breed in Siberia; they winter mainly in India and the Far East but some migrate westwards to Africa.

Common Sandpiper *Actitis hypoleucos*

Migratory breeder, widely distributed along upland streams, stony rivers and loch margins; local in the lowlands and scarce in the Northern Isles. Also occurs as a passage visitor, seldom in gatherings of more than a few birds.

In spring the first reports of Common Sandpipers are often during March (since 1968 the ealiest record has been 6 March) but the main arrival does not take place until late April / early May. This species is non-gregarious in its habits at all seasons and there is little visible evidence of passage on the coast in spring, when most reports are of birds moving inland. Spring migrants at Fair Isle in May are presumably Scandinavian birds. The Common Sandpiper breeds in all mainland counties and all island groups, but is absent from many of the smaller islands. Little is known about past populations or densities but there is some evidence that numbers declined over much of the breeding range between the 1930s and 1950s (Atlas). In Shetland this species has always been scarce and recent reports suggest a population of under ten pairs. B&R noted a decrease in Orkney, where the Common

Sandpiper was once common on Mainland; here too the current population is probably under ten pairs. On Arran numbers appear to be increasing, with counts of 14 pairs in 1975, 20 in 1978 and 34 in 1982. The comparatively few surveys of mainland rivers have produced densities ranging from 0.2–1.9 pairs / km (Cowper 1973; H. Galbraith & R. W. Furness), presumably depending upon both the proportion of suitable bank and the availability of food. The data currently available do not provide an adequate basis for assessment of the total population, put by Sharrock (Atlas) at not more than 50,000 pairs in Britain and Ireland.

As soon as the young have fledged, the birds start to move away from the breeding grounds, and in July fair numbers are often present at both inland and coastal sites with partly muddy and partly stony shores. Among the largest counts at this season in recent years are 60 at Glencaple on the Solway, 64 at the Endrick Mouth, 57 at Loch Insh and 44 at Gladhouse. Coastal passage in autumn, mainly in July and early August, is more marked than that in spring. At Findhorn in 1970 it was estimated that 350–500 Common Sandpipers, 70% of them juveniles, passed during this period (SBR). Movement continues on a smaller scale throughout September and by early October most birds have left, although there are occasional later stragglers and a very few December / February records. In the south of England small numbers now winter regularly (Prater 1981).

The Common Sandpiper breeds throughout Europe and Scandinavia but not in Faeroe or Iceland. Migrants passing through Scotland are probably of Scandinavian origin but there is as yet no ringing evidence of this. The few recoveries of Scottish-reared birds include one in north Portugal in its first September and others in western France in late April/ early May and in mid July. A bird marked in France in mid August was recovered in Dunbarton the following June. Although some birds winter in western France, Iberia and around the Mediterranean, most of the population goes much further south through Africa.

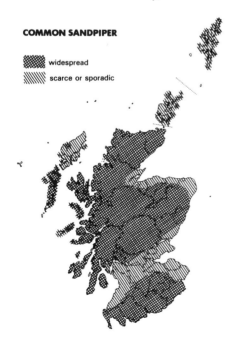

COMMON SANDPIPER

▨ widespread

▨ scarce or sporadic

Spotted Sandpiper *Actitis macularia*

Vagrant (N America) – six occurrences 1971–84, including attempted breeding in 1975.

The Spotted Sandpiper, an abundant bird in eastern North America, has been recorded more than 50 times in Britain. The first Scottish record was from Tyninghame, East Lothian, on 30 October 1971. The second, remarkably, involved a pair which attempted to breed on Skye in 1975 (Reed *et al* 1983), the first recorded breeding in Europe and the Palaearctic region. The birds were present at least from 15 June to 3 July and laid, but failed to hatch, a clutch of four eggs, two of which were fertile (BB 69:288; 70:12). There were single spotted sandpipers at North Berwick on 30 May 1976, on St Kilda on 22–23 May 1982, in South Uist from 31 July to 1 August 1983, and on Islay on 5 June 1984. The report of one at the Endrick Mouth in August 1977 (SBR 1978 – additions) was subsequently withdrawn (BB 74:473).

Turnstone *Arenaria interpres*

Passage and winter visitor, widely but fairly thinly distributed along all rocky coasts from August to May. Summers in small numbers; breeding has been suspected.

The first immigrants arrive in mid July with the main influx in August/September; thereafter numbers remain fairly constant until April. Like other widely dispersed species, Turnstones are difficult to census and there are few reliable estimates of numbers for much of the Scottish coastline. The most comprehensive surveys to date have been on the east coast and in Orkney (Table). Counts for other

Numbers of Turnstones counted on rocky shores (Data from Prater 1981, Summers et al 1975, Buxton 1982a, Summers & Buxton 1983, Tay & Orkney R. G. 1984, R. W. Summers, BTO/WSG Winter Shorebird Count 1984/85 – preliminary figures per M. E. Moser)

	Count/estimate	Date
Shetland	c5,500†	mid winter 1984/85
Outer Hebrides	3,200–3,400	1978–79 to 1981–82
Orkney	6,000	January 1983 & 1984
Caithness	1,050	1982
E. Ross & E. Sutherland	320	1982
Moray to Berwick	*7,300–8,200	1970–74
'Strathclyde' (1)	2,000–3,000	mid-winter 1984/85
(Tiree	990)	
'Highland west coast' (2)	500–800	mid-winter 1984/85

* = estimate   † = preliminary figure

(1) 'Strathclyde' = coasts of Kintyre, all islands and mainland west coast to boundary with Highland Region.
(2) 'Highland west coast' = all mainland coasts of Highland Region from Strathclyde boundary to Strathy Point (Sutherland), Skye and the Small Isles.

areas are very patchy and cannot be taken as a true indication of wintering numbers.

An estimated total of 7,500 has been suggested for Shetland (Prater 1981) but there are only scattered counts available to support this figure and it seems possible that it is on the high side. The largest wintering flocks recorded there in recent years have generally been in the range 70–100, while examples of island totals in December include 200 on Papa Stour, 150 on Bressay and 100 on Whalsay. In Orkney the Mainland and Sanday together held about half of the 6,000 Turnstones counted in January 1983 and 1984 (Tay & Orkney R.G. 1984). The Outer Hebridean population is estimated as 3,000+ (N. E. Buxton), with the largest numbers occurring on the Uists and Benbecula. The Argyll coast and Inner Hebrides held 2,000–3,000 in 1984/85 (M. E. Moser) and the Clyde Islands probably support a few hundred.

Numbers wintering along the northwest mainland coast, much of which is relatively inaccessible, are small. Further south, counts between the Clyde and Stranraer suggest a winter total of around 1,000–1,200 and rather more in September/October. In 1969–75 mid-winter totals on the

TURNSTONE

▓ main wintering areas

North Solway averaged 250 and on the Inner Clyde 150–200. There are occasional inland records, eg from Inverness, Dunbarton and Peebles, generally during the autumn migration period.

Turnstones tend to remain in one area throughout the winter, making only local movements of up to about 50 km, and many return to the same area in successive years (Summers et al 1975a; Symonds et al 1984). Although very much associated with seaweedy shores, Turnstones can also be found in somewhat surprising places. In Shetland they regularly forage in the fields during the winter, and in Aberdeen they have been seen in turnip fields and even feeding in a farmyard with finches and sparrows.

The Turnstone is one of the most northerly breeding species and a long-distance migrant which winters as far south as Australasia and southern South America. Spring passage occurs in waves during May, with pronounced movement along Scottish coasts; migration studies indicate a complex pattern of movements, still not fully understood (Branson *et al* 1978; Ferns 1981a). Although some Scandinavian birds occur in Britain, mainly on autumn passage, most of our wintering population originates from Greenland and arctic northeast Canada; there are several ringing records linking Scottish birds with northern Greenland and Ellesmere Island. Birds from that part of the breeding range winter from Britain south along the Atlantic coast to northwest Africa; those that moult on arrival in this country generally remain to winter here. Many of those going to west Greenland and Canada stop-off in Iceland before making the final stage of their journey to the Arctic (Wilson 1981).

Turnstones summer annually in Scotland, mostly in the Northern Isles. In recent years display has frequently been recorded in Shetland and Orkney and in 1978 a pair prepared a nest scrape on North Ronaldsay. The strongest suggestion of breeding has come from west Sutherland, where an agitated adult was seen on 9 August 1976; there was a downy chick nearby but this was unfortunately not identified. On their arctic breeding grounds Turnstones nest on pebbly shores or among tundra vegetation, often close to a rock or dwarf shrub.

Wilson's Phalarope *Phalaropus tricolor*

Vagrant (N America) – fifteen records, 1954–84.

Wilson's Phalarope was first recorded in 1954 at Rosyth, Fife, where one was present from 11 September until 5 October (SN 1954:188; BB 48:15). Occurrences have been almost annual since 1970 – a surprising situation in view of the fact that the species does not breed in eastern North America and is a rare, though regular, autumn migrant on the Atlantic coast. There are August/October records from Dunbarton (1962), Fife (1963), Angus (1970), Kirkcudbright (1972), Shetland (1973 & 1974), Ayr (1976 & 1978), Midlothian (1981), Orkney (1981), Lanark and Aberdeen (1984). The only spring occurrences have been in Shetland (9 May 1975 & 27 May to 1 June 1982) and Caithness (1–4 June 1980).

WN

Red-necked Phalarope *Phalaropus lobatus*   [1

Scarce and local migratory breeder; most of the population of under 50 pairs is on the islands, but breeding occasionally occurs on the mainland. Winters at sea, and outwith the breeding season is recorded most often on the coast and in autumn.

The history of this species has been documented by B&R and more recently by Everett (1971); both accounts depict a diminishing population much persecuted by collectors and disturbance. The main decline took place last century but there has been a continuing slow decrease in the number of breeding pairs since the early 1970s, although the population has remained relatively stable on Fetlar, which is now the main stronghold (Table). In Orkney, Tiree and North Uist breeding now occurs only irregularly. Nesting was recorded on St Kilda for the first time in 1972 but has not occurred again. There have been very small numbers at one or two mainland sites since 1977 (R. H. Dennis) and the species was recorded on Arran for the first time in 1983.

The Red-necked Phalarope has an unusually late breeding season, not arriving until mid May (11 May was the earliest recorded date 1968–83) and leaving again by the end of August. The favoured breeding areas are shallow freshwater lochans or bays, usually with emergent vegetation and adjacent marsh or rough grassland. In this species the sexual roles are reversed and the male undertakes responsibility for incubating the eggs and rearing the young; the brighter-coloured female is consequently the more likely to be seen as she feeds near the water's edge, spinning and pirouetting in pursuit of insects.

Over much of its circumpolar range the Red-necked Phalarope is abundant and large gatherings of females are a common sight. On its British breeding grounds, which

Numbers and distribution of breeding Red-necked Phalaropes (Data from RSPB & SBR)

	1975	1976	1977	1978	Number of breeding pairs 1979	1980	1981	1982	1983	1984
Shetland (inc. Fetlar)	19–20	20–23	NA	19–24	20–22	20–24	21–24	16–21	17–20	19–21
Fetlar	c13	16–19	19–22	17–21	20–22	20–24	21–24	16–21	16–19	17–18
Orkney	1	0	0	0	0	0	0	0	0	0
O. Hebrides	NA	2+	1+	5+	6–7	4–5	2–4	3–4	4	3
Mainland	0	0	0?	2	2–3	2–3	1–2	1	1?	1?

are at the southern limit of its range, it has probably never been present in comparable numbers and the colonies of up to 50 pairs recorded in Ireland in the past are unlikely to be seen again. Some former haunts have been affected by drainage or other agricultural activities but it is possible that climatic changes are as much to blame for the progressive decline of this species as are more tangible pressures. Despite the special protection afforded to it under the various Acts, and the establishment of reserves in areas where it might benefit, its future prospects do not look encouraging.

The wintering grounds of the British and Icelandic populations are not known, but many of those from Fenno-Scandia travel southeast overland to the Black and Caspian Seas and on to the Arabian Sea (BWP). Migrants are recorded most frequently on the coast in May and August, though numbers are always very small; there are occasional inland occurrences.

Grey Phalarope *Phalaropus fulicarius*

Passage visitor, occurring annually in very small numbers, mainly in autumn.

The Grey Phalarope has a circumpolar breeding distribution, within or close to the Arctic Circle, and winters at sea off the coasts of South America and western Africa. The origin of the birds recorded on passage off the Scottish coast is uncertain but they are probably from the Iceland, Greenland or Spitsbergen populations. Numbers vary markedly between years, presumably reflecting the severity and persistence of westerly winds in the Atlantic, which drive the birds inshore.

There is no regular pattern of occurrence; since 1969 Grey Phalaropes have been reported annually but during the 1960s there were four years without any records. From 1971 to 1983 annual totals ranged from one to 24 – in 1977, when there was a small wreck during September, with one in East Lothian, two in Islay, four in Shetland, and up to 15 on North Ronaldsay, Orkney. The only other records of more than three together are from Lewis, where there were four in October 1959 and up to seven in September 1961. (Scotland was virtually unaffected by the enormous influx of September/October 1960, when several thousand Grey Phalaropes reached southwest England and southern Ireland.) Although there are records for every month, about half the occurrences are in September/October and 80% between July and December. Females in full breeding plumage (living up to their American name of Red Phalarope) were seen between Sumburgh and Fair Isle on 25 May 1973 and at Kinnaber, on the Angus/Kincardine border, on 19 May 1975.

Since 1950 there have been records for most mainland counties with coastline exposed to the open sea and for Shetland, Orkney and the Outer and Inner Hebrides; there are no reports from the Clyde Islands. Birds are very occasionally seen inland, the only recent records being at Hamilton (Lanark) in October 1960 and Hule Moss (Berwick) in October 1976.

Pomarine Skua *Stercorarius pomarinus*

Passage visitor, abundant in spring off the Outer Hebrides but most widespread in autumn.

The Pomarine Skua breeds in the Arctic and winters at sea in the Tropics/sub-Tropics. During May it is apparently the most abundant of the skuas passing north off the Outer Hebrides (Davenport 1979; Thorpe 1981). Flocks of more than 40 occasionally travel together and from 19 to 22 May 1983 a record total of 786 were counted, most of them light phase adults (Cunningham 1984; SBR). These birds may have been driven closer inshore than usual by strong westerly winds in the Atlantic. An alternative possibility suggested by Blake *et al* (1984) is that this species uses the coast as a navigational aid during migration.

Annual totals vary widely, probably reflecting weather conditions at the time of the main passage, and since 1970 have ranged from under 50 to 1,000+. East coast occurrences are most frequent in autumn, when the peak may be at any time from late August to early October, and the biggest numbers anywhere from Shetland to Berwick. There was an unusually large autumn influx in 1976, when 186 passed into the Beauly Firth in two hours on 27 September and 110 were off Collieston (Aberdeen) on 2 October.

Since the early 1960s the Pomarine Skua has been reported from all coastal counties except Nairn and Renfrew, and from all the island groups. It is markedly scarcer in the southwest than elsewhere, and a flock of 21 well up the Solway Firth on 3 May 1982 was most unusual. Outwith the main passage period, it occurs fairly often in June/July and November and very occasionally in December/February, when small numbers have been seen in the Moray Firth. There are few recent inland records.

Arctic Skua *Stercorarius parasiticus*

Scotland supports the entire British breeding population, currently increasing, which disperses widely at sea in winter. Breeding numbers are highest in Shetland and Orkney. Also occurs as a passage visitor.

The Arctic Skua nests colonially and is noisy and demonstrative on its moorland breeding grounds. It is consequently easy to locate and comparatively simple to count, but information on numbers prior to 1969–70 is nevertheless sparse and patchy, except for a few well-recorded colonies such as those on Foula and Fair Isle. The figures available are sufficient to show, however, that the increase in numbers has accelerated during the last 10–15 years (Table a). The population is clearly still expanding.

(*a*) *Numbers and distribution of breeding Arctic Skuas. (Data from Everett 1982, Reed et al 1983b, Meek et al 1985, Argyll B R & local recorders)*

	Number of breeding pairs		
	1969–70	1974–75	Post-1975
Shetland	770	1,631+	
Orkney	230	716	1,034 in 1982
Outer Hebrides	40	37	80–94 in 1982
Caithness	20	28+	40+ in 1979–80
Sutherland	1	3	19 (Handa) in 1982
Argyll	25	26	36+ early 1980s
Total	1,086	2,441+	

Since 1950 colonisation of new sites has taken place in Shetland, Orkney, the Outer Hebrides, Handa and Caithness, among the most recently-established colonies being those on Handa (1968), Out Skerries (1969), North Uist (1970), North Ronaldsay (1975) and Stroma (1976). At many of the old-established colonies too there have been progressive increases, but at a few numbers have either remained fairly static or declined. In some cases, notably Hermaness and Noss, this is probably due to a local increase in the Great Skua population. The Great Skuas arrive earlier in spring and defend their already-established territories when the Arctic Skuas appear about two weeks later. When the Great Skua colony is an expanding one, with young birds staking claims around the fringe of the area, the effect on the nesting distribution of Arctic Skuas can be dramatic. On Foula Arctic Skuas have gradually been forced off the open moorland favoured by the Great Skuas and compressed into a much more confined area nearer to human habitation. Their overall numbers nevertheless continued to increase until very recently, although it had been suggested in the late 1940s that the population was being adversely affected by the growing population of Great Skuas (Furness 1983). A similar shift in distribution appears to have taken place on Unst, where few Arctic Skuas now nest on Hermaness (Table b) but numbers breeding elsewhere on the island have remained stable or increased (Bundy 1978a). There are no recent breeding records from Tiree or west Inverness, where B&R record nesting, nor from Wester Ross, where confirmed breeding is shown in the Atlas.

(*b*) *Numbers of breeding pairs of Arctic Skuas at some Shetland sites. (Data from Operation Seafarer raw data, FIBO Reports, Furness 1983 & Shetland Bird Reports)*

	Pre-1950	1969–70	1976	1982
Foula	1948: 125*	100	280	224
Fair Isle	1949: 15	180	136	100
Hermaness	1950: c70	40–50	13	5 (1980)

* individuals

Arctic Skuas are kleptoparasitic and much of their food is obtained by chasing other seabirds and forcing them to drop or disgorge their catch. The species parasitised varies according to local abundance; on Fair Isle Kittiwakes suffer most, and on Foula terns and Puffins. On the breeding ground this species is even more aggressive than the Great Skua, dive-bombing repeatedly at human or animal intruders as well as at other skuas. Arctic Skuas also indulge in elabor-

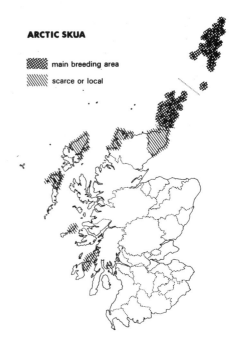

ARCTIC SKUA

main breeding area

scarce or local

ate distraction displays, squealing and trailing their wings in an effort to divert attention from their chicks, which leave the nest soon after hatching and move frequently thereafter, birds of the same brood often going in different directions.

The species occurs in several colour phases. Pale phase birds comprise about 25% of the population in Shetland whereas in the high-arctic part of the breeding range, in Jan Mayen and Greenland, the corresponding figure is 90–100%. The Scottish population is near the southern limit of the species' breeding range, which is Holarctic; in Europe nesting occurs also in Iceland, the Faeroes, Fenno-Scandia, Spitsbergen and northwest Russia.

The Fair Isle population has been studied for many years. The work carried out from 1948–62, by Ken Williamson and later Peter Davis, was concerned largely with the population's social structure and dynamics. During the 1970s the relationships between colour phase and mate selection, territory size and breeding success were examined in more detail (Berry & Davis 1970; O'Donald 1983); most of this work has been reported in specialist journals but a short account, with references, is given in FIBOR 1975:73–76. Dark and intermediate phase males are apparently sexually favoured over pale ones, but pale birds become sexually mature at an earlier age. In the Arctic the latter attribute is most advantageous, hence the predominance of pale birds. On Fair Isle continuing immigration is presumed to be responsible for maintaining the balance between the colour phases.

When they leave the breeding grounds in August young Arctic Skuas disperse widely; many – perhaps most – move rapidly southwards and are off Iberia or in the Mediterranean a few weeks later and off western Africa by mid winter. A few go in other directions, however, and first-year birds have been recovered in Norway and East Germany early in the winter. Most young birds probably do not return to British waters during their first summer and the record of one off New England in September suggests that they wander widely across the Atlantic. By their second winter some are well south of the Equator; the most distant records of this age group to date are from the coasts of Angola and Brazil. In their second summer some travel as far north as Greenland, where there have been several July/August recoveries. The few recoveries of adults indicate that they winter over much the same range as immatures.

Arctic Skuas are seen offshore fairly frequently in summer even in areas in which they do not breed, eg up to 40–50 together well up the Forth at the end of August. Passage is most obvious in autumn but under certain weather conditions marked spring movements are recorded off the west coast, eg 100+ off North Uist in early May. The occasional inland occurrences usually involve single immatures, but there is a record of 20 passing over Loch Garten in May. Few remain in British inshore waters in winter though there are occasional sightings in November and December.

In some breeding areas Arctic Skuas are persecuted because they cause disturbance to sheep – and are a nuisance to shepherds. On Fair Isle improvement of the moorland where Arctic Skuas were well-established increased the problem, because sheep attracted to the better grazing were constantly harassed by dive-bombing skuas. There is no evidence, however, that such harassment results in any increase in sheep losses.

Long-tailed Skua *Stercorarius longicaudus*

Scarce passage and occasional summer visitor. May have bred.

The growing interest in sea-watching has resulted in an increasing number of records of the Long-tailed Skua but it remains a rather scarce though regular passage migrant. It is now recorded annually in Shetland and the Outer Hebrides; there are also fairly frequent reports from the more exposed sections of the east coast as far south as Berwick but comparatively few from western inshore waters, and only a scattering of inland records.

In spring the main passage takes place off the west coast, and only a few stragglers occur on the east. Most are seen during May and early June but one was in Shetland on 30 April 1976. In autumn there is movement down both coasts but the majority of reports are from the east and between mid August and mid September; although there are quite often stragglers into October, two off Hound Point on 2 November 1982 were unusually late. The largest numbers together occur in the Outer Hebrides during spring passage; up to 271 have been counted passing north off Balranald in a single day in mid to late May. There have been up to 35 in a day off Orkney in August, while 11 off Fraserburgh in September is the peak east coast count so far. Since 1956 the spring and autumn totals have generally been roughly equal, but in 1983 exceptionally high numbers were recorded in spring, including 390 off North Uist between 18 and 25 May. A further 65 individuals were reported in autumn, making the 1983 total by far the highest yet.

On its arctic breeding grounds the Long-tailed Skua frequents the drier types of tundra and bare moorland. In this country Arctic Skuas occupy somewhat similar habitats and there are several June/July records of Long-tailed Skuas at Arctic Skua colonies in Orkney, Shetland and the Outer Hebrides. In 1980 an adult present in a mainland area from 2 June to 6 August was accompanied by an immature from 16 July onwards; the birds' behaviour suggested that breeding had taken place but this was not proved.

A ringed adult seen on the central Grampian plateau in June 1974 was suspected of predating Dotterel eggs, and the following June an adult was found freshly dead on a mountain road near Grantown-on-Spey.

Great Skua *Stercorarius skua*

The entire British population breeds in Scotland, with more than 90% nesting in Shetland and Orkney; birds are on the breeding grounds from April to August, when they disperse widely at sea. Both range and numbers have expanded in the last 30 years.

Although Great Skuas have bred in Shetland (where they are known as Bonxies) for many years, they were formerly much persecuted. After protection was introduced on Unst and Foula last century numbers gradually built up (B&R; Jackson 1966); a considerable population expansion took place during the first half of this century and this has continued in most areas since 1950 (Table). On Foula, however,

following an increase averaging 6.7% per annum from 1900–1977, there has recently been a slight decline, although no decrease in breeding success has been noted (Furness 1983). Bonxies first nested on the Scottish mainland and on several Orkney islands during the 1950s, and on the Outer Hebrides and islands off Sutherland in the 1960s. Breeding was recorded on the Summer Isles and Shiant Islands for the first time about 1980 and in 1982 respectively. The current population expansion seems likely to continue as there is abundant suitable habitat available and no apparent shortage of food. Elsewhere in the North Atlantic the Great Skua is well-established in Iceland and the Faeroes and has recently colonised Spitsbergen, Bear Island and Norway. The Scottish population probably represents about half the total breeding in the northern hemisphere.

No visitor to moorlands where Great Skuas are nesting can fail to be aware of their presence. Aggressively defensive of their eggs and young, they attack repeatedly, sometimes swooping low enough to strike the intruder's head with their feet. Their display too is attention-catching, as they stand calling with wings raised high above their backs. Despite their demonstrative territorial behaviour, skuas can be difficult to census accurately and estimates of breeding numbers made by different observers or at different stages of the breeding cycle may vary widely. Furness (1982) has suggested counts of 'apparently occupied territories', made from a distance and without disturbing the birds, as the best alternative to marking individual nests.

The Great Skua is very much a pirate among birds. Its aerial pursuit of other seabirds, to force them to disgorge fish, can be fascinating to watch; there is even a Shetland

Numbers and distribution of breeding Great Skuas by counties and principal colonies. (Data from Venables & Venables 1955, Cramp et al 1974, Everett 1982, Furness 1983, Meek et al 1985, FIBO Reports & local recorders – larger figs. rounded)

	Date of colonisation	Number of breeding pairs			
		Pre-1960	1969–70	1974–75	Post-1975
Shetland	?	NA	3,060	5,450+	NA
Foula	Pre-1800	300 in 1938	1,780	2,500	3,100 in 1977; 2,670 in 1980
Hermaness	Pre-1900	300+ in 1949	300	650	
Fetlar	Early 1900s	30 in 1946	275	237	
Noss	c1910	113 in 1946	210	260+	
Yell	Late 1800s	96 in 1946	125	?	
Fair Isle	c1921	4 in 1949	10	21	45 in 1982
Orkney			90	482+	1,652 in 1982
Hoy	Early 1900s	2 in 1915	72	462	1,570 in 1982
Outer Hebrides				24+	
Lewis	1945	?	12	11	20 in 1982
St K	1963	0	6	9	25 in 1978
N. Rona	1965	0	2	4+	?6+ in 1982
Sutherland			4	12	
Handa	1964	0	3–4	8	38 in 1982
Caithness	?1970s	0	0	1	?2+ in 1980
Total		1,000+	c3,200	c6,000	

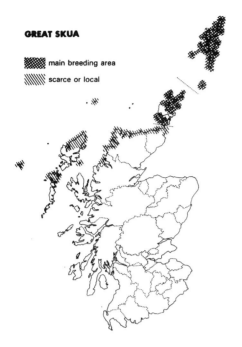

GREAT SKUA

▓▓▓ main breeding area

\\\\\\ scarce or local

record of a Bonxie forcing an Osprey to drop its catch. In addition to stealing fish, Great Skuas regularly kill other species and will attack, and sometimes drown, a bird as large as a Gannet. Feeding habits vary between colonies, with many Kittiwake eggs and nestlings but few fledglings being taken at some sites, and many fledglings but no eggs or nestlings at others. In Shetland as a whole a large proportion of the diet is either sandeels caught by the birds themselves or whitefish discarded by trawlers; the increasing abundance of these food supplies is probably responsible for the recent expansion of the Scottish population (Furness 1983). In the last few years rubbish tips have also been exploited as a food source (Furness *et al* 1981).

Well over 40,000 Great Skuas have been ringed and there have been some 1,300 recoveries; those up to 1974 have been analysed by Furness (1978). First-winter birds disperse widely, with recoveries in the Bay of Biscay, Iberia and the western Mediterranean, in USSR and Austria, near the Cape Verde Islands, and off the New England and Brazilian coasts. During the summer some immatures are in southern waters and others in the Arctic around Greenland and Spitsbergen. With increasing age the tendency to wander declines progressively and many birds over four years old winter in British waters, although from November to March they seldom come close enough inshore to be visible from the land. Along mainland coasts the largest numbers are seen in August / September; inland records are rare.

A few birds from Iceland and Faeroe have been reported in Scotland but never breeding. Nearly every British colony contains breeding birds ringed as chicks on Foula but the only Foula birds found nesting overseas have been in Norway and Spitsbergen. There is consequently no evidence of interchange between Scotland, Iceland and the Faeroes.

The breeding biology, population dynamics and feeding behaviour of the Foula colony have been the subject of inten-sive long-term studies by Bob Furness (1983), with much fieldwork assistance from Brathay Expeditions.

Mediterranean Gull *Larus melanocephalus*　　　　[1]

Vagrant (E Mediterranean) – annual in very small numbers.

The Mediterranean Gull has occurred in Britain with increasing frequency in the last twenty years and is now a regular visitor to some parts of the coast, especially in south and east England. It was first recorded in Scotland in 1957, when two individuals visited Fair Isle, a second-winter bird from 31 August to 2 September and a first-winter bird on 14 October (SB 1:117). The next report was from Aberdeen in 1972, and in 1975 there were records of singles from Aberdeen and Argyll. Since then the species has occurred annually, with a peak count of about ten in 1983 – though it is difficult to be certain of numbers as birds often remain in an area for several weeks and may also return to winter in the same area in successive years (BOU). Adults and immatures occur in roughly equal numbers and there have been occurrences in all months except June. About two-thirds of the records are from the east coast (Shetland, Sutherland, Banff, Aberdeen, Kincardine, Angus, Fife and West, East and Midlothian). West coast records are from Argyll, Dunbarton and Ayr, with inland occurrences in Kirkcudbright (Carlingwark) and Renfrew (Balgray Reservoir).

Although its main breeding area is around the Aegean and Black Seas, the Mediterranean Gull has bred in the Netherlands and in southern England, where it has nested annually since 1976, on one occasion hybridising successfully with a Black-headed Gull. In 1980 a male held territory in a Black-headed Gull colony in the west of Scotland and attempted coition, but no proof of breeding was obtained.

Laughing Gull *Larus atricilla*

Vagrant (N America) – seven records, 1968–83.

The first record, of an adult at the Endrick Mouth, Loch Lomond, on 2 April 1968, was not accepted by the BBRC until 1978, by which time there had been two further occurrences: on Islay on 21 April 1974 and Fair Isle on 13 September 1975. The Laughing Gull has since been reported at Loch Ken (September / December 1978), on St Kilda (June / July 1980), at Troon (January 1983) and on Fetlar (June 1983). All the records refer to adults or sub-adults. British occurrences totalled only 35 by the end of 1983.

Franklin's Gull *Larus pipixcan*

Vagrant (N America) – two records, 1980 and 1983.

The first record was of a sub-adult at Irvine, Ayr, on 2–6 July 1980 (SB 12:258–259). The following year a second-summer bird was seen on Canna on 5–10 July and subsequently picked up in a moribund condition; it is now in the Royal Scottish Museum. As Franklin's Gull breeds in the North American mid west, and occurs only irregularly on the western Atlantic coast, it seems surprising that it reaches this country at all – seven had been recorded in Britain by the end of 1983.

Little Gull *Larus minutus*   [1]

Passage visitor, most frequent in autumn and in east-central Scotland. Single birds or small groups occur sporadically throughout the country and at other seasons.

There has been a progressive increase since the early 1950s in the numbers of Little Gulls visiting east-central Scotland. B&R knew the species as only an occasional visitor; in 1953–56 Grierson (1961) recorded regular appearances on the Angus and Fife coasts, with county totals occasionally exceeding 100; and in 1968–83 maxima at any one site ranged from 100+ at Kilconquhar Loch in July 1968 to 700 at Westhaven in August 1981. Spring passage generally peaks in April/May and autumn passage in August/September, occasionally in late July. Grierson noted that in some seasons immatures outnumbered adults, but in recent years adults have always been very much in the majority. A few often remain throughout the year, and in June 1973 there were three, including one adult, at a gull colony in Fife.

Little Gulls favour freshwater pools and marshes during the breeding season but winter mainly off the coast, coming ashore to rest in bad weather and during migration. Passage birds visiting Scotland use both these habitats. At one time Morton Lochs was a regular resort and I have watched 50 or more hawking towards me over the northern loch and lifting at the road to circle and repeat the process. More recently Kilconquhar Loch has been the principal freshwater haunt. On the coast, this species' preference for fine-sand shores was noted by Grierson, who described the birds' feeding behaviour over the sea and adjoining moorland at Monifieth. Other coastal areas in Angus and Fife visited by sizeable flocks of Little Gulls since 1968 include Tentsmuir, Methil, Invergowrie Bay, Buddon and Westhaven. Both the totals involved and the numbers at individual sites vary greatly from year to year and peak counts are seldom high in both Angus and Fife in the same season. Little Gulls occur fairly frequently, though in much smaller numbers, in most east coast counties but much less often in the west. There are recent (post-1968) records from all island groups and nearly all mainland counties.

The Little Gull breeds locally in Europe, mainly east of the Baltic but with a few discrete groups further west, and in Siberia. One ringed as a chick in Finland was recovered in Fife the following year. Breeding is often sporadic away from the main range and the population in Sweden, Denmark and the Netherlands fluctuates markedly (BWP). Little Gulls attempted to breed in England in 1975 and 1978 but have not yet succeeded in raising young. Outwith the breeding season the species is widely distributed on southern North Sea and Atlantic coasts and around the Mediterranean. Its status in Britain and Ireland was reviewed in 1978 (Hutchinson & Neath 1978).

Sabine's Gull *Larus sabini*

Vagrant (Arctic/high-Arctic) – annual in small numbers since 1972.

B&R give a number of old records of Sabine's Gull, which winters off southwest Africa, but only seven in the first half of this century, the latest in 1925. There were two reports in the early 1950s, one in 1969, two in 1972, three (involving up to four birds) in 1973, one each in 1974 and 1975, three in 1976, and five to seven each year from 1977 to 1983. Occurrences are about equally divided between east and west coasts, and there have been two inland records, from Inverness (Loch Insh) and Roxburgh (Melrose). Since 1950 Sabine's Gull has been reported from Shetland, Fair Isle, Orkney, Ross and Cromarty, Aberdeen, Angus, Fife, the Isle of May, Midlothian, East Lothian, Berwick, Dumfries, Wigtown, Ayr, Argyll, Islay, Skye, Lewis, the Uists and Benbecula. More than three-quarters of sightings are in the period mid August to mid October, but there are records for all months except April and May. Slightly more than half the reports relate to immatures and most are of single birds.

Bonaparte's Gull *Larus philadelphia*

Vagrant (N America) – six records, 1850–1983.

Bonaparte's Gull was first recorded in 1850, when one was shot on Loch Lomond. There were no further reports until 1967, when one was seen at Kinlochbervie on 17 August (SB 5:175). There have since been records of an immature at Cupar, Fife (February 1972), a sub-adult at Scourie, Sutherland (June 1973), an adult on Islay (June & September 1975 – presumed to be only one bird involved), and an adult on Fetlar (June/July 1982). Although this species, which is the smallest American gull, breeds in northwest Canada, it winters off the east coast of North America.

Black-headed Gull *Larus ridibundus*

Widely distributed, breeding colonially on boggy ground, reed-fringed lochs and sand dunes, and dispersing in winter, when many move south or to Ireland. Immigrants from northern Europe winter, mainly on the coast.

The Black-headed Gull is a familiar sight almost everywhere; it follows the plough, hawks over grassland, scavenges around picnic sites – even on the summit of Cairn Gorm, feeds at estuaries, and visits gardens where bird-food is put out. It is a difficult species to census as its colonies are not only numerous and scattered, but are also liable to major fluctuations in size from one year to the next. In consequence little is known about the size of the breeding population. The most recent census, in 1958 (Hamilton 1962), was far from complete, but of the colonies counted 6% had over 1,000 pairs and more than 70% had fewer than 100 pairs. The largest was at Flanders Moss, with 10,000–12,000 pairs. In 1969–70 counts of coastal colonies produced an estimated total of 18,000+ pairs, by far the largest colony being at Tentsmuir, with c8,000 pairs. The changing fortunes of the Flanders Moss and Tentsmuir colonies illustrate the scale and speed of local fluctuations (Table); the decline at Flanders Moss has been paralleled at other sites in that area (Mitchell 1980). Decreases are sometimes attributed to predation by mink (Brown & Brown 1984b) and some are doubtless due to persecution, but habitat change (eg Tentsmuir) and competition from other gull species, perhaps associated with consequent habitat change (eg Flanders Moss), are other possible explanations. Black-headed Gulls are vulnerable to fluctuating water-levels and in some seasons flooding may be a contributory factor in the desertion of sites.

There has been some recent extension of range. B&R knew of no breeding records for East Lothian, Clackmannan, Kincardine, Banff, Arran, Mull or Skye; the Atlas survey confirmed breeding in all but Clackmannan and Skye. There has been a considerable recent expansion in the species' western European range; early this century colonisation of Norway and Iceland took place, and in the 1960s breeding occurred for the first time in Italy, Spain and Greenland. This expansion has probably been due to a reduction in persecution (large numbers of eggs were formerly taken for human food), an increase in the food supply resulting from changing agricultural practices, and possibly an amelioration of the climate; it is apparently still continuing (BWP).

Black-headed Gulls return to their colonies in March/April and form large pre-breeding roosts. Most nests are on the ground or on semi-floating vegetation but tree-nesting has been recorded. In some colonies the gulls nest in close association with other species, eg with Tufted Ducks on St Serf's Island (Loch Leven) and with Sandwich Terns at the Sands of Forvie. Presumably the potential threat presented by the gulls is more than outweighed by the protection they afford in keeping away more aggressive predators.

At the end of the breeding season both young and adults

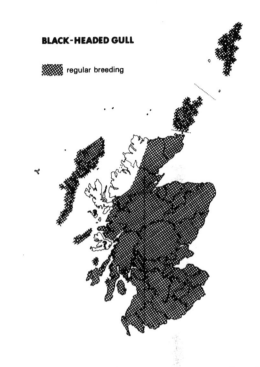

BLACK-HEADED GULL

▓▓ regular breeding

disperse rapidly. There is some movement south and west towards Ireland, especially of first-winter birds from northern Scotland, but comparatively few leave the British Isles (Flegg & Cox 1972). Immigrants from northern Europe arrive between August and December, mainly during September/October. Many come from Norway and most of the remainder from Iceland, Finland, Sweden and Denmark, but there have also been a few recoveries of birds from as far east as Latvian SSR and Poland. During the winter very large flocks gather both on the coast and inland; in mid January 1983 the biggest numbers recorded were around the Solway, which held 62,500 of the 175,900 counted in Scotland (Bowes *et al* 1984). Recent records include counts

Fluctuations in the numbers of Black-headed Gulls breeding at two colonies (Data from B & R, Grierson 1962, Hamilton 1962, Cramp et al 1974, SBR and Sandeman 1982)

	Estimated number of breeding pairs						
	1900	1936–38	1955–58	1969	1970	1974	1981
Tentsmuir (Fife)	6	3,500–4,000	c3,000	c8,000	0	0?	0?
Flanders Moss (Perth)	?	Several thousand	10,000–12,000	?	?	500	2

of 30,000 at Montrose Basin (in December 1978), 25,000+ on the Clyde, 20,000 at Skinflats and 10,000+ on the Ayr coast and in the Beauly Firth. The immigrants start returning to their breeding areas in late February/early March and most have gone by the end of April. At Fair Isle passage movements are noted in March/April and from early July onwards.

Ring-billed Gull *Larus delawarensis*

Vagrant (N America) – thirteen records, 1976–84.

The Ring-billed Gull was not recorded in Britain until 1973 and was first reported in Scotland in 1976, when a second-winter bird was seen on the Ythan on 14 February. The second record was an adult at Lossiemouth on 5 February 1979. Since 1981 there has been a rapid increase in the number of British – and to a lesser extent Scottish – records, and by 1983 the Ring-billed Gull had become the most abundant Nearctic visitor (Vinicombe 1985). There were two Scottish records in 1981 (South Uist 13–15 August and Lerwick 29 December to January 1982), four in 1982 (Scalloway 4 February to April; Golspie 17 February; Lewis 11–12 February; South Uist 17 July) and four in 1983 (Scalloway 13 January to 29 April; Tiree 27 January; Stromness 27 March; Aberlady Bay 2 October). Of these ten, five were second-winter birds, two first-winter (as was one at Aberlady in October 1984), two adults and one in its first summer. Two 1984 records, from Argyll and Ayr, are under consideration at the time of writing.

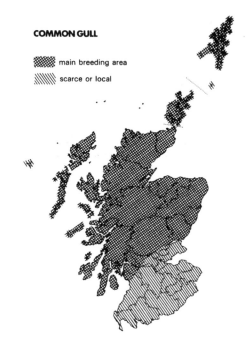

COMMON GULL

▨ main breeding area

▨ scarce or local

Common Gull *Larus canus*

Widely distributed resident, breeding both on the coast and inland, but absent from some areas of the Central Lowlands and southeast. In winter local birds disperse but few, if any, leave the British Isles. Also a winter visitor in large numbers.

Scotland probably holds 80–90% of the British and Irish population of this species, which is misnamed as it is by no means the commonest of our gulls. The Common Gull is more catholic in its choice of nesting sites than other gulls; moorland, bog, small islets, rocky shores and shingle banks are the most common situations but there are records of nests on the tops of posts, on the face of a dam, between pipeline joints, in a barley field, in trees, and even on a flat-topped hedge and a rooftop. Nesting occurs from sea-level up to about 1,000 m asl, sometimes singly, usually in small colonies of 20–50 pairs, and occasionally in much larger concentrations. By far the biggest colony is on the Correen Hills in Aberdeen; probably a long-established site, this held 2,000–3,000 pairs in 1972 (SB 8:75–76) and up to 5,000 in 1977 – one of the largest concentrations known in Europe. The biggest cohesive group within the colony numbered c800 pairs, many more than the largest coastal colony – of 320 pairs on Arran – recorded in Operation Seafarer. Another large colony at Corsemaul (between Dufftown and Huntly), held at least 1,000 pairs in 1978. The Correen Hills site is on raised heather moor at 450–550 m asl, and the birds feed over the surrounding rich farmland; numbers are being maintained, or perhaps increasing, despite shooting and insecticide poisoning (SB 10:50–53). The close proximity of good feeding grounds and suitable, relatively undisturbed, nesting areas presumably accounts for the great density of breeding pairs in this district, which probably holds a substantial proportion of the Scottish total. No reliable estimate of the total British and Irish population is currently available. Parslow (1973) suggested 10,000 pairs, the Operation Seafarer figure for the coastal population was 12,000+ pairs, most of them in Scotland, and Sharrock (Atlas) put the total at up to 50,000 pairs.

In most parts of the country the birds are more widely scattered and the average colony size is much lower than in the northeast, and relatively few breed south of the Forth/Clyde. In Unst the 500+ pairs are in scattered groups averaging 35 nests (Bundy 1978); in Lewis and Harris most Common Gulls nest singly – on islets in freshwater lochs (Cunningham 1983); in Ayr the maximum colony size is 30 pairs (Hogg 1983); and in Fife, the Lothians and the Borders numbers are very small indeed. The marked increase in the population that took place last century may still be continuing, with some southward expansion of the breeding range. B&R knew of no breeding records on the east side of the country south of Loch Leven (where this species last bred in 1935) nor in Stirling, Clackmannan or Dumfries. Common Gulls bred for the first time in Stirling in 1954, in Berwick in 1960, and in Dumfries in 1962 (Cramp *et al* 1974). Breeding was not proved during 1968–72 in Roxburgh, Midlothian, West Lothian, Clackmannan, Kinross or Fife, but has since been confirmed in Midlothian (1973), Clackmannan (20–30 pairs Upper Glen Devon Reservoir by early 1980s), and Fife (sporadic), while up to seven pairs have been recorded in Peebles and similar numbers in East Lothian. Though still small, populations on Foula and Fair Isle have been increasing steadily; Foula had one pair in 1948 and 10–12 pairs in 1975–80 and on Fair Isle, where breeding first took place in 1966, there were six pairs in 1983. Single pairs bred on St Kilda in 1963 and 1979. Scotland was formerly on the extreme western limit of the breeding range, which extends right across Eurasia (and into Alaska and northwest Canada), but Faeroe was colonised in the late 19th century and Iceland about 50 years later.

The breeding birds arrive back at their colonies from late-February/April; at higher-altitude sites laying may not take place until early June. After fledging the birds disperse, moving first to nearby farmland, where they feed over both grassland and arable ground. Most of those ringed as chicks in Britain either remain in the same general area or move southwards or southwestwards, with some young birds reaching Ireland (Radford 1960, Vernon 1969); few appear to winter in England but at least one Scottish-bred bird has been recovered in Portugal.

Immigration starts in late July and continues until mid November, occasionally later. During this period large numbers pass through the Northern Isles, with up to 4,000 being recorded on a single day on Fair Isle in mid September, and huge flocks build up in many areas. Peak counts since 1968 have included 6,000+ at Kinloss in late August, 10,000+ at Loch Leven and 19,000 on the Ayr coast in September, and 42,000+ at Loch of Skene in November. Visible evidence of passage to Ireland is provided by records of 700–800 passing southwest over Sanda in September, but it is not known whether these were local or immigrant birds. Nor is it known to what extent local birds contribute to the large mid-winter flocks, eg c20,000 roosting on St Serf's Island (Loch Leven), 10,000 in the Beauly Firth and 15,000 on Portmore Loch. Although coverage was by no means complete, the roost census in January 1983 indicated that in mid winter the largest numbers are in the Lothians and Grampian, which held 72,000 of the 128,000 counted (Bowes *et al* 1984).

Foreign-ringed Common Gulls recovered in Scotland include birds from Norway, northern Russia (Murmansk), Estonia, Finland, Sweden, Denmark and one from Iceland. Over 80% of recoveries involved immatures, most of them

in their first winter, and several had reached Scotland within six weeks of being ringed as chicks in Norway.

The spring departure takes place during April, when large numbers leave in a northeasterly direction. The scale and speed of this movement was first detected by radar observations (Bourne & Patterson 1962), which showed that the birds move across Scotland on a broad front and then converge on headlands before setting out across the North Sea, often too high to be seen. Departure is probably stimulated by the onset of fair weather and a favourable west wind, and the short, concentrated movement is in marked contrast to the slow and dispersed pattern of the autumn return. Large-scale passage is often visible at this time, for example on the Galloway/Solway coast, around the Forth, and in the Northern Isles. Adults and sub-adults apparently leave first, with the proportion of first-year birds increasing later on in April (Vernon 1969).

Lesser Black-backed Gull *Larus fuscus*

Migratory breeder, nesting colonially, both on the coast and inland; less abundant in Scotland than in the rest of Britain. Most migrate but increasing numbers now winter in this country. Also a passage visitor.

The Lesser Black-backed Gull is less of a scavenger than the closely-related Herring Gull and does not concentrate near towns and ports. It is less abundant in the north than in the central belt (Table), where the Forth and Clyde islands

Numbers and distribution of coastal breeding Lesser Black-backed Gulls in 1969–70. (Data from Cramp et al 1974)

County	Pairs	County	Pairs
Kirkcudbright	570	Caithness	16
Wigtown	5	E. Suth/Ross/Inv.	3
Ayr	430	Moray	50
Bute	2,280	Banff	14
Argyll	2,640	Aberdeen/Kinc.	5
W. Inverness	210	Angus	0
W. Ross	250	Fife	2,210
W. Sutherland	320	Midlothian	5
O. Hebrides	500	E. Lothian	330
Orkney	810	Berwick	6
Shetland	570	*Total*	11,224

hold several of the largest breeding colonies and numbers have been increasing. In Shetland and on some of the Hebrides there has been a decrease in the last few decades, possibly due to competition from, or predation by, Great Black-backed Gulls and Great Skuas. (Circumstantial evidence for this is provided by the fact that in 1980 this species ceased to breed on Foula, where the Great Skua population peaked at over 3,000 in 1977). There has also been a continuing slight expansion of the range, with breeding first recorded in Aberdeen in 1949, Kincardine in 1965, Banff in 1968, mainland Fife in 1972, Angus in 1975 and Sule Skerry in 1982. B&R had no records of breeding on Arran or Bute, where Lesser Black-backed Gulls were nesting in 1968–72.

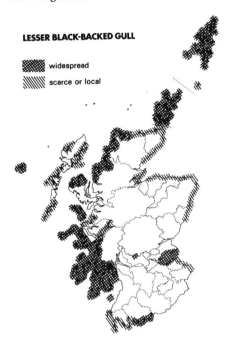

LESSER BLACK-BACKED GULL

▓▓ widespread

\\\\ scarce or local

a ninefold increase in the number of wintering Lesser Black-backed Gulls occurred between 1949–51 and 1959/60 (Barnes 1961). In Scotland no group of more than five was reported in 1959/60 and there were fewer than ten records in 1968/69. There has since been some increase; although numbers are still small, records are scattered north as far as Shetland instead of being confined mainly to the central belt and southwest. Since the mid 1970s wintering flocks – as opposed to very small groups – have been reported, eg December counts on Balgray Reservoir (Renfrew) of 100 in 1979, and 60 in 1980 and 1981, and 18 at Lanark in February 1981. In January 1983 a total of 146 were counted, the largest concentration (97) being in Dumfries and Galloway (Bowes *et al* 1984).

Lesser Black-backed Gulls from Iceland and Faeroe move through Scotland on both autumn and spring passage. Like our native birds, these belong to the race *L.f. graellsii*, which breeds also in France and northwest Spain. The small numbers of very dark-mantled Lesser Black-backs reported annually, usually singly and in spring or autumn but occasionally in summer, are probably *L.f. fuscus* from northern Norway and Sweden.

On the Isle of May the population increased at an average rate of 14.3% per annum from 1930, when breeding was first recorded, until 1972 (Duncan 1981). In that year culling was started (see Herring Gull) and some 1,700 birds out of a population of 2,500+ pairs were killed. In 1974 about a third of the breeding population was culled and by 1975 only 1,300 breeding and 70 non-breeding Lesser Black-backed Gulls were present. Culling continued at a lower rate but despite this the population had increased again to 1,870 breeding birds by 1979. The current policy is to limit the combined totals of Lesser Black-backed and Herring Gulls to a maximum of 3,000 pairs (NCC). Increases have also been noted on several other islands in the Forth, and on Horse Island, Little Cumbrae and Inchmarnock in the Clyde, but on Am Balg (off northwest Sutherland), the Flannan Isles and North Rona, where 'large numbers' had been recorded previously, this species appears to have been replaced by the Great Black-back, as none were found during Operation Seafarer. Between 200 and 300 pairs breed on St Kilda.

Most inland colonies are on moorland and bog. The largest is on Flanders Moss, where Lesser Black-backed Gulls have nested for many years. This colony held 150 pairs in 1951, c700 pairs in 1975, and c4,000 pairs in 1982 (A. Wood) – the figure of 8,000 pairs given in SBR 1980 is now considered an overestimate. Over the same period Herring Gulls have decreased at this site – perhaps they are being displaced by the Lesser Black-backs. Another colony, of 700–800 pairs, was found at 335 m asl near Braco in 1982 (SB 12:119–20), and there is one with 60–100 pairs on the Correen Hills.

After breeding the birds disperse and almost immediately start to move south, many reaching the Bay of Biscay by September, the western Mediterranean by October/November, and western Africa by December. In England

Herring Gull *Larus argentatus*

The most abundant of the large gulls, widely distributed as a breeder on all rocky coasts, with numbers highest in the east. In winter immigrants are also present and the main concentrations are near fishing ports or towns with extensive garbage tips.

In Britain as a whole the Herring Gull has increased markedly since the 1950s, probably due partly to the introduction of protection and partly to the species' growing habit of exploiting edible refuse. The main breeding concentrations are around the approaches to fishing ports, but the birds often commute considerable distances to feed. Scotland held about half the total of 330,000+ pairs counted in Operation Seafarer (Table). Although the population increase in Scotland has probably been smaller than in England and Wales, there has certainly been some expansion in range and marked local increases in breeding numbers, especially in the east. In 1968–72, coastal breeding was confirmed in Midlothian,

Fife, Nairn, east Inverness, Renfrew and Dunbarton, all counties for which B&R had no nesting records. Major population expansions have been noted on the Forth Islands, particularly on the Isle of May, where there was an average increase of 13% per annum between 1907 and 1972 (Duncan 1981). In the west the increase has not been so marked, averaging 7% per annum in recent years (P. Monaghan). Numbers have decreased in parts of Shetland, eg on Foula from a maximum of 40 pairs in 1960 to only five pairs in 1980 (Furness 1981b). The reason for such local decreases is not clear, but inability to compete with Great Skuas and Great Black-backs on the feeding grounds, or with Fulmars at nesting sites, have been suggested as possible explanations. In Caithness Herring Gulls, apparently displaced from cliffs by Fulmars, have nested in barley fields (SBR 1969).

Rapidly increasing populations at some sites have necessitated culling to protect both vegetation and other bird species. The most spectacular documented expansion has been on the Isle of May. From one pair in 1907, the colony there grew to c1,100 pairs in 1951 and c3,000 pairs in 1959; by 1972 a total of c16,850 pairs of Herring and Lesser Black-backed Gulls were nesting on the island. Terns ceased to breed in 1960, Eiders and other ground-nesting species were suffering badly from predation, soil erosion was occurring, and large areas of what was formerly a sward of sea pinks and sea campion had been replaced by coarse grasses. Control by egg-taking having failed, the NCC – at the request of the Isle of May Bird Observatory – embarked in 1972 upon a programme of culling, using bread baits containing narcotic placed at the nest. Culling was repeated up to and including 1975, and in 1976 the total population of Herring and Lesser Black-backed Gulls was down to c3,350 pairs. Small-scale culling has since continued at intervals to limit the population, which had risen again to 3,750 pairs by 1978; c3,000 pairs bred in 1983. The area occupied by the gulls was only marginally reduced following the cull, so the nesting density declined markedly (Coulson *et al* 1982).

Some Herring Gulls nest inland, often as solitary pairs or in very small groups, but colonies of 200 pairs or more have been recorded on the Correen Hills, among Common Gulls, and at Flanders Moss, where they are outnumbered by Lesser Black-backed Gulls. The Flanders Moss colony (where accurate assessment of the relative numbers of the two species is difficult) expanded from only two pairs in

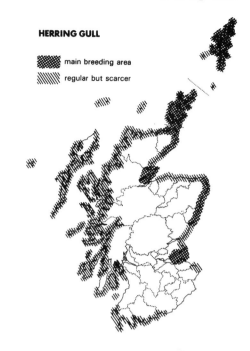

HERRING GULL

- main breeding area
- regular but scarcer

1952 to 50 pairs in 1974, 100 pairs in 1975, and 500 pairs in 1980; the increase seems now to have levelled off and numbers may be dropping, although the area occupied by the colony continues to extend. Ringing has shown that gulls from this colony forage over rubbish tips around Stirling and Cumbernauld (W. R. Brackenridge & A. Wood). In Lewis, where the non-coastal colonies are not so far from the sea, the combined effects of physical destruction of plants during nest construction and nutrient enrichment from droppings, corpses and discarded fish bones and other refuse, have greatly altered the vegetation and virtually killed off many of the typical moorland plants (Currie 1981). Similar changes in the vegetation can be seen at sites where gulls rest on moorland close to refuse tips; between Gairloch and Poolewe one such site is brilliant green against the surrounding heather. Rooftop nesting is now quite widespread on the east coast and has also been recorded in Dunbarton (P. Monaghan).

After breeding, Herring Gulls disperse, generally southwards and to well-defined wintering areas, often congregating near urban centres both inland and on the coast. More than 50% of those wintering in the Glasgow area originate from colonies in the north, northwest and west of Scotland, the remainder coming mainly from Ireland, the Solway and the Forth (P. Monaghan). Birds moving south from the Isle of May travel further than those heading west or north (Parsons & Duncan 1978). Second-year birds disperse furthest, on average 256 km from the island, while adults move less than 90 km away. Many Scottish immatures overwinter in northeast England. Isle of May birds have been recovered in Germany and Denmark, but only a small proportion of the population leaves the British Isles. Some young Herring Gulls settle to breed at colonies other than those at which they were reared – a fact which has important implications

Numbers and distribution of coastal breeding Herring Gulls in 1969–70.
(Data from Cramp et al 1974)

County	Pairs	County	Pairs
Kirkcudbright	5,200	E. Sutherland	180
Wigtown	1,070	E. Ross	10,100
Ayr	1,160	E. Inverness	3
Bute	2,860	Moray	1,160
Argyll	8,050	Banff	4,600
W. Inverness	4,600	Aberdeen	28,000
W. Ross	1,900	Kincardine	11,000
W. Sutherland	6,670	Angus	4,300
O. Hebrides	3,610	Fife	16,060
Orkney	7,800	Midlothian	30
Shetland	10,150	E. Lothian	7,300
Caithness	19,500	Berwick	1,500
		Total	c156,800

for gull control. Isle of May birds have been found breeding in colonies as far apart as Shetland, Sunderland, Cumbria and Copeland (Co. Down) (Duncan & Monaghan 1977).

During the autumn very large concentrations of Herring Gulls build up in many areas. Flocks of up to 8,000 shelter on Fair Isle in stormy weather from November onwards, and these include many adults of northern origin, noticeably bigger and darker than the local birds. Some 10,000–20,000 Herring Gulls regularly gather on Loch Leven from late summer, and there are records of 16,000 on the Tay in September, and c30,000 following sprats into the Beauly and Inner Moray Firths in December 1973. This species formed the bulk of the 50,000 large gulls on the Blackshaw Bank in the Solway in December 1969. Counts of 5,000–15,000 are usual in many areas.

Comparatively little is yet known about the relative proportions of homebred and immigrant birds in winter flocks. The Herring Gulls breeding in Britain belong to the race *L.a.argenteus*, whose range includes Iceland, Faeroe and the southern North Sea coast; the majority of immigrants are of the nominate race *L.a.argentatus*, which breeds from Denmark and Scandinavia to the Kola peninsula in northwest Russia. Both races have flesh-coloured legs but *argenteus* is smaller and paler than *argentatus* (BWP). Some Iceland/Faeroe birds winter in Scotland, while most adults around North Sea oil installations in winter are dark-backed birds, probably *L.a.argentatus* (W. R. P. Bourne). Coulson *et al* (1984) showed that the proportion of Scandinavian birds among Herring Gulls wintering in Scotland is much lower than in northeast England – only c2% in the Glasgow area and 5–10% in the east, compared with up to 30% in northeast England. Recoveries and sightings of marked birds indicated that most of these immigrants were from colonies in the far north of Norway.

Yellow-legged Herring Gulls are occasionally reported and these may belong to either *L.a.omissus*, which breeds from the eastern Baltic to northwest Russia, or *L.a.michahellis* from the Mediterranean, which has recently occurred with increasing frequency around the southern North Sea (BB 76:191–194).

The Herring Gull's scavenging habits have led to its being suspected of spreading disease, especially *Salmonella* infections, among domestic livestock, while roosting flocks have contaminated drinking water in storage reservoirs near Glasgow (Benton *et al* 1983). In some years large numbers die apparently of botulism, possibly also associated with scavenging; in 1975 more than 2,000 died in the Forth and smaller numbers on the Clyde and Moray Firth (Macdonald & Standring 1978). This species can justifiably be regarded as a pest which seems likely to cause increasing problems for both man and wildlife.

Iceland Gull *Larus glaucoides*

Scarce winter visitor, occurring annually. Most abundant in the north and from December until April. Immatures greatly outnumber adults.

This species breeds on low-arctic rocky coasts in Greenland and northeast Canada, and disperses southwards in winter. Most of the Iceland Gulls reaching Scotland are thought to belong to the east Greenland population, the majority of which are believed to winter in Iceland (BWP), An immature ringed in west Greenland has been recovered in northeast Scotland, however, and birds showing the characters of Kumlein's Gull *L.g. kumleini*, which breeds in Baffin Island, have been reported from Shetland (February 1983) and Orkney (January/May 1983).

Arrival is from mid September but the main influx of Iceland Gulls seldom occurs before mid winter. Individuals sometimes stay in one area for up to three months and a few, mostly immatures, remain through the summer. The annual total is generally in the range 50–100 but in 1981 around 200 were reported, including about 50 in Shetland. There was another large influx in early 1983, with at least 100 in Shetland between January and April and higher numbers than usual in many other areas eg 19 in Ayr and six on Mull. In February 1984 the largest groups yet recorded (45) were seen in Stornoway and Ullapool. Some 150–200 birds were reported from Shetland south to Dumfries between mid January and June. Occurrences are widely scattered around the coast, with a less marked northerly bias than in the Glaucous Gull; there are recent records for most coastal areas other than the Clyde Islands (for which B&R note occurrences) and Nairn (no records). Inland sightings are rare.

Glaucous Gull *Larus hyperboreus*

Winter visitor, occurring annually between September and May, in variable numbers.

The main influx of Glaucous Gulls occurs in November/December, sometimes later. Winter totals range from under 100 to 500+, with high numbers generally associated with persistent northerly gales in autumn. The lateness and coldness of the spring governs departure, but a few birds often remain through the summer. Immatures are generally more abundant than adults. Although primarily a marine gull, seen frequently near North Sea oil installations and recorded most regularly in the Northern Isles and Outer Hebrides, this species occasionally occurs inland and often scavenges with Herring and Great Black-backed Gulls.

In 1969 there was an exceptional arrival in Shetland, with about 300 on Fair Isle and 100 on Fetlar during the last week of November, but few further south. The proportion of adults to immatures was unusually high that year, on Fair Isle about 2:1. Numbers over the next decade were much lower, not exceeding 50 on Fair Isle, but in 1981 there was another big influx, with Glaucous Gulls recorded in most counties between January and March, a peak of 100 on Unst in late January, and at least eight as far south as Ayr in February.

This species has a circumpolar arctic distribution, nesting on rocky shores and islands. The Eurasian population disperses southwards in winter, when large numbers are in the Faeroes; those visiting Britain may come from the area of the Barents Sea, and perhaps also from east Greenland (Dean 1984). Birds from the Iceland breeding colonies, where the Glaucous Gull hybridises extensively with the Herring Gull, are thought to be largely sedentary. The only record of breeding in Britain is of a female mated with a male Herring Gull in Shetland in 1975–79. There have been several reports of apparent hybrids.

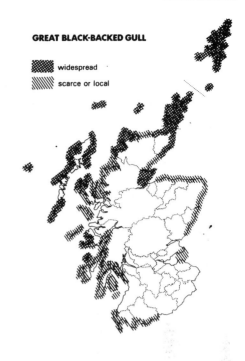

GREAT BLACK-BACKED GULL

■ widespread

▨ scarce or local

Great Black-backed Gull *Larus marinus*

Most abundant on the coasts of the north and northwest and relatively scarce as a breeder south of the Moray Basin and the Clyde. The breeding range has expanded and the population has increased this century. Winter visitor in fairly large numbers in the north.

Scotland held about 70% of the 22,300+ pairs of Great Black-backed Gulls counted in Britain and Ireland during Operation Seafarer, the majority of these being in the Northern Isles and Outer Hebrides (Table). This species prefers nesting sites which are safe from human disturbance. Many pairs nest solitarily on isolated stacks or rocky headlands but the main concentrations are on uninhabited or sparsely-inhabited islands, usually near fishing grounds or ports. In 1969–70 the largest colony (1,800 pairs) was on North Rona, while Hoy had 2,530 pairs dispersed over a much larger area. Some 750 pairs nested on the moorland interior of

Numbers and distribution of breeding Great Black-backed Gulls in 1969–70. (Data from Cramp et al 1974)

County	Pairs	County	Pairs
Kirkcudbright	60	Shetland	2,490
Wigtown/Ayr	45	Caithness	1,100
Bute	180	E. Ross	500
Argyll	540	E. Sutherland/	
W. Inverness	360	E. Inverness	4
W. Ross	180	Moray/Banff	38
W. Sutherland	1,960	Aberdeen/Kincardine	30
O. Hebrides	2,610	Angus/Fife/E. Lothian	6
Orkney	6,000	*Total*	c16,100

the Calf of Eday (Orkney), and several other islands held from 200–450 pairs, eg Fetlar, Noss, Stronsay and stacks off northwest Sutherland. Foula had 15 pairs in 1960 and 35 pairs in 1980, and Fair Isle had 55 pairs in 1969 and 41 pairs in 1975. The only mainland colony of any size is in Easter Ross, where some 400 pairs nest on Morrich More. This site differs from the majority in being unprotected by cliffs or sea – in this case the surrounding firing-range presumably ensures isolation. Inland nesting occurs in a few mainland areas but generally involves only small numbers.

B&R had no records of breeding on the east coast south of Moray, apart from an early 19th century report from the Bass Rock. Since 1950 Great Black-backed Gulls have bred for the first time in east Inverness (1970), Banff (1968), Aberdeen (1962), Kincardine (by 1980), Angus (1969), Isle of May (1962), mainland Fife (1972), East Lothian (1972 – Craigleith), and Renfrew (1970); birds have held territory in Berwick since 1978 but breeding has yet to be proved. In several of these counties numbers are still very small, but east Inverness had 40 pairs in 1978.

At the end of the breeding season the birds disperse and many of the big northern colonies are deserted. First-winter Great Black-backs travel on average c100 km southwards, while adults remain closer to their colonies (BWP). Some Scottish birds winter in Ireland but there have been only a small number of ringing recoveries outside Britain. The few records of migration include passage southwards over Papa Westray in mid August, and 80 passing down the Great Glen in four hours during September, 80% of them juveniles. Large numbers are present off the Hebrides and in the North Sea throughout the winter and in bad weather big flocks gather onshore, mainly on the coast but also on certain lochs. Counts of 500–1,000 are not uncommon between mid August and November; such numbers have been recorded at Loch Spiggie, the Longman refuse tip at Inverness, Lossiemouth, the Loch of Strathbeg (2,500 in September 1979), the Ythan, St Cyrus, Aberdeen (city), Kinnaber, the Isle of May and Aberlady. Inland roosting occurs on Loch Leven and on Loch Garten, where over 100 have been present in December.

The largest winter gatherings recorded have been in Shetland, where up to 5,000 have sheltered around Fair Isle in stormy weather, and 4,000 on the Out Skerries. These large flocks presumably include substantial numbers of immigrants. Shetland was not covered in the gull-roost survey of January 1983, when 10,000+ Great Black-backs were counted elsewhere in Scotland, more than 50% of them in Orkney and the Outer Hebrides (Bowes *et al* 1984). Ringing recoveries suggest that most immigrants come from north Norway, Murmansk and Iceland, with Icelandic birds most frequent in the northwest and west, and Norwegian/Russian birds in Shetland and the east. Blake *et al* (1984) suggested that winter distribution patterns in the North Sea may reflect the movement of immigrants, which are known to reach northeast England in large numbers in September and remain there until mid February, possibly moving north up the coast thereafter before returning to Norway.

The Great Black-backed Gull breeds on both sides of the North Atlantic, having colonised North America during the first half of this century. In northwest France the population is increasing steadily. Although this is the most coastal of the large gulls, it is visiting refuse tips and other sources of offal with growing frequency. The recent population expansion in Scotland is probably attributable to this habit and to the increasing availability of fish offal. In addition to taking the eggs and young of other species, Great Black-backed Gulls kill many full-grown seabirds, eg auks, and also rabbits.

Kittiwake *Rissa tridactyla*

Breeds on most coastal cliffs, with the largest concentrations in the Northern Isles and on the east coast; winters at sea. Numbers increased steadily earlier this century but the rate of increase has now slowed. Also occurs as a passage and winter visitor.

Scotland's cliffs support about 75% of the British and Irish breeding population of the Kittiwake, an oceanic gull which comes to land only during the breeding season. The majority nest on cliff ledges, but some use buildings – eg harbour-side warehouses at Dunbar – as substitute cliffs and on Foula a few build on boulders at the cliff foot (Furness 1983). Colonies often have traditional freshwater bathing places, which are visited during the nesting period by a continuous stream of birds. Distribution is determined largely by the availability of suitable nest sites. Many of the largest colonies are in Orkney and Caithness, where the horizon-

Ross's Gull *Rhodostethia rosea*

Vagrant (Arctic/high-Arctic) – about eleven occurrences, 1936–84.

Some Ross's Gulls are thought to winter in the north Norwegian and Barents Seas, but rather little is known about the post-breeding dispersal of this tern-like gull (Densley 1977). The species was first recorded in Scotland in 1936, when an immature was caught on Whalsay on 28 April (BB 35:276). All but four of the subsequent records are also from Shetland: one between Fetlar and Whalsay in October 1969, singles on Mainland in January 1972, January 1975, November 1977, January 1983 and two in January/February 1981, one on Yell in May 1979, one each on Unst and Whalsay in January 1981, and one wintering around Scalloway from December 1982 to February 1983 (last seen on 5th). An adult seen in Thurso on various dates between February 1983 and January 1985 might possibly have been the same individual. Three of the remaining occurrences were in 1976: adults at Thurso in January and the Monach Isles in May, and an immature on Islay in August. Only three of the records have involved immatures.

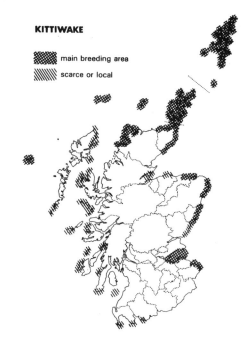

KITTIWAKE

main breeding area

scarce or local

tally-bedded rock provides ideal ledges; on the west coast there are few stretches of suitable cliff on the mainland and few colonies of more than 1,000 pairs. Some adults are at breeding colonies from early March until August/September; young birds disperse rapidly after fledging.

Kittiwake nests are fairly easy to count but many are beneath overhangs and in cave-mouths and can be seen only from the sea. Two census attempts have been made. In 1959 the estimated Scottish population was c100,000 pairs (Coulson 1963); this was undoubtedly a considerable underestimate. More comprehensive cover in 1969–70 resulted in a figure of 370,000+ pairs (Table a); however, this did not

(a) Numbers and distribution of breeding Kittiwakes in 1969–70. (Data from Cramp et al 1974)

County	No. of nests	County	No. of nests
Kirkcudbright	30	E. Ross	400
Wigtown	670	Moray	430
Ayr	7,740	Banff	11,320
Bute	0	Aberdeen	13,920
Argyll	3,990	Kincardine	38,060
W. Inverness	2,060	Angus	1,650
W. Ross	0	Fife	3,280
W. Sutherland	17,240	E. Lothian	1,070
O. Hebrides	25,380	Berwick	19,200
Orkney	128,680		
Shetland	42,770	*Total*	370,200
Caithness	52,340		

include an adequate assessment of the Orkney colonies and incorporated some counts made too late in the season to be reliable. There has not been a complete census since 1970, but counts made at several major colonies show an interesting contrast between the east coast, where numbers continue to increase, and the west, where at some sites there has apparently been a decline (Table b). Occasional counts

(b) Numbers of Kittiwakes breeding at selected sites. (Data from Operation Seafarer, FIBO Reports, SBR, Angus 1983, Coulson 1983, Harris & Galbraith 1983 and Swann & Ramsay 1984)

	Number of occupied nests		
	1969–70	During 1970s	Early 1980s
Ailsa Craig	7,700	1,400	?
Canna	c520	?	c980
Handa	8,000+	13,000	?
Fair Isle	12,000+	17,000+	?
Isle of May	3,100	3,000–4,000	6,000+

do not, however, provide a reliable measure of short-term population change, since wide variations in the rate of change occur between years and also in different parts of the same colony (Harris & Galbraith 1983).

In Britain as a whole, the Kittiwake population increased from the beginning of this century until 1969 at an average rate of 3–4% per annum, probably initially due to the cessation of persecution – this species was formerly shot for sport and to supply the millinery trade. In 1969–1979 the rate of increase dropped to 1–2% per annum (Coulson 1983). Although this population expansion has been almost as great

as that of the Fulmar, it has been less obvious as it has largely occurred in existing colonies, rather than through the establishment of new ones. Working from a partial survey, Coulson estimated that on the east coast of Scotland (excluding Orkney and Shetland) nesting pairs increased by 78% between 1959 and 1969 and by 37% between 1969 and 1979. In Shetland the population appears to be relatively stable; c31,700 apparently occupied nests were counted in a sea-based survey in 1981, as against c29,700 nests recorded in 1969–70 – neither figure included Foula or Fair Isle (Richardson 1985). On the west coast (excluding Ailsa Craig and St Kilda) an increase of 10% between 1959 and 1969 changed to a decrease of 15% between 1969 and 1979; the change has been even more marked on Ailsa Craig and St Kilda, with increases of 4% and 50% respectively between 1959 and 1969, and decreases of 80% and 61% between 1969 and 1979. The most rapid decline occurred in the mid 1970s and was too great to be attributable solely to poor survival of the young. In the absence of any evidence of toxic chemical or heavy metal accumulation, it is suspected that the decline, which occurred in a well-defined geographical area of west and south Britain and Ireland, was associated with a food shortage during the breeding season. As Coulson has pointed out, monitoring counts cannot explain population changes; complementary information on food availability, breeding success and survival rates is also needed to complete the picture. Such information is currently scanty, but a study of Kittiwakes on the Isle of May in 1982 showed that feeding frequency – presumably determined by the time needed to obtain an economic food-load – was unaffected by brood size and many of the youngest chicks in three-chick broods died of starvation that year (Galbraith 1983).

When the young birds leave the colonies many move first to the south-central North Sea, later dispersing very widely (Blake *et al* 1984). Coastal movements peak in autumn, eg 25,000–30,000 per hour passing Rattray Head (Elkins & Williams 1972). By December few young birds are in the North Sea and many have moved far across the north Atlantic; one Shetland-ringed chick reached southern Greenland less than six weeks after leaving the nest. But some wander south and east, to the Bay of Biscay and the Mediterranean. Immatures spend their first two years far from their natal area but in their third season return nearer to 'home' (BWP). Adults remain near their colonies until moulting is completed, then disperse, some leaving the North Sea and only returning at the start of the breeding season.

Kittiwakes breed on both sides of the Atlantic, well north of the Arctic Circle, and recoveries show that birds from Russian, Norwegian and Danish colonies visit Scottish waters. Inland occurrences are rare and usually follow onshore gales.

Ivory Gull *Pagophila eburnea*

Vagrant (high-Arctic) – recorded in thirteen seasons, 1950–83

Ivory Gulls winter near the edge of the pack ice and occur south of the Arctic Circle only as occasional stragglers, occurrences presumably being much influenced by climatic and ice conditions in autumn and winter. The records date back to 1822; although most are from the Northern Isles there have been reports as far south as Roxburgh and Dumfries (B&R). From 1950 to the end of 1983 there were 19 records, involving 19–20 birds. The most reported in any one year has been four, in 1980, when two were found dead in Shetland – in April and June – and an immature and an adult were there in November/December. Shetland accounts for about half the occurrences, the rest coming from Caithness (two), Fair Isle, Aberdeen, Angus, Perth, Argyll and Lewis. Immatures outnumber adults by about 2:1 and most sightings are in November/December.

Gull-billed Tern *Gelochelidon nilotica*

Vagrant (Europe) – seven occurrences, 1913–83.

The nearest breeding colony of the Gull-billed Tern, which is almost cosmopolitan in distribution but very local in western Europe, is in Denmark. The species has been recorded more than 200 times in Britain but is a very scarce visitor to Scotland. The first Scottish record was a male taken on the Pentland Skerries on 7 May 1913 and now in the Royal Scottish Museum (SN 1913:154). Five of the subsequent reports are from the Forth: Aberlady (September 1960 & March 1966), Dalmeny (September 1966), Skinflats (September 1969) and Bo'ness (May 1977), and there is one record from Fair Isle, in May 1971. Two adults were present at Skinflats but all other occurrences involved only single birds.

Caspian Tern *Sterna caspia*

Vagrant (Baltic) – fourteen records, 1968–83.

The Caspian Tern, cosmopolitan but very local in breeding distribution, was first recorded in Scotland in 1968, when one was at the Endrick Mouth, Loch Lomond, on 7 August (SB 5:390). There have since been records from Dunbarton/Stirling (August 1968 & July 1976), West Lothian (July 1981), East Lothian (June & July 1971), Aberdeen (August 1974, June 1975 & July 1976), Fair Isle (May 1978), Shetland (August 1976 & October 1979), Argyll (June 1981), Kirkcudbright (July 1982) and Barra (June 1982). With the exception of two at Aberlady on 1 July 1971, all records relate to single birds. They include a bird ringed as a chick near Stockholm in 1975 and found dead on Yell the following year.

Sandwich Tern *Sterna sandvicensis*

Migratory breeder, the most locally distributed of the commoner terns. Two east coast colonies – Sands of Forvie and Inchmickery – currently hold the bulk of the breeding population. There have been a few recent winter records.

The Sandwich Tern is an exclusively coastal breeder and seldom occurs inland. Although it generally favours sand or shingle for nesting, some of the largest concentrations in Scotland are on rocky islands and rough grazing. The total breeding population is around 2,000–2,500 pairs; there have been local fluctuations in numbers but little evidence of recent overall change (Table). By far the biggest long-established colony is at Sands of Forvie, where there has been a fairly steady increase over the last 30 years. In the Forth, Fidra has been deserted since 1974, while the colony on Inchmickery, which became firmly established during the 1920s, has gradually expanded.

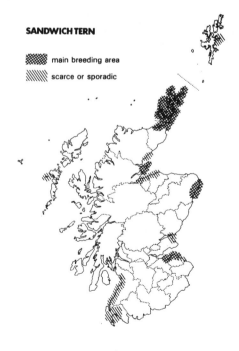

SANDWICH TERN

▓ main breeding area

▨ scarce or sporadic

The site at Tentsmuir, where 300–400 pairs bred earlier this century, was abandoned during the 1939–45 war and only sporadic breeding has since occurred; none have bred since 1968, when 500 pairs nested but all failed. Further north numbers nesting in Easter Ross have also dropped markedly. In Caithness, however, there has been a progressive increase since breeding was first recorded, on Stroma, in 1971; a mainland colony held 40 pairs in 1973. In Orkney nesting has occurred on Mainland, Rousay, Sanday, North Ronaldsay, Westray and Burray, and in Shetland on Whalsay between 1955 and 1960 but not again. A pair on the Monach Isles in 1978 is the only report of breeding in the Outer Hebrides since 1928. Breeding has not been recorded on the Inner Hebrides. The single record for Argyll, of a pair breeding at Loch Riddon in 1971 (SBR), is suspect (Gibson 1979), and the species is not known to have nested on the main Clyde Islands. In Ayr numbers have decreased on Horse Island, where the 11 pairs in 1982 all failed (Hogg 1983); Lady Isle, which held 94 nests in 1971, has been completely deserted; and nesting at Ballantrae is sporadic. The only other counties in which breeding has been recorded, often irregularly and usually only in small numbers, are East Lothian, Moray, Nairn (267 pairs on the Old Bar of Findhorn in 1958 – A. J. M. Smith), Inverness, east Sutherland and Wigtown.

Sandwich Terns usually nest in dense colonies and often in close association with other terns or gulls: on Stroma they are with Arctic Terns, on Inchmickery with Common Terns and at Sands of Forvie with other terns and Black-headed Gulls. They are very sensitive to disturbance by humans or ground predators, especially early in the season. Even after laying a colony may desert. Over 1,000 pairs nested on the Morrich More range in 1969, but suffered increasing disturbance from egg collectors and visitors (Bourne & Smith 1974). Many birds returned in 1971 but were later seen flying south towards Sands of Forvie, where numbers increased from 750 to 2,100 pairs; none bred at Morrich More. The Forvie colony has been the subject of a long-term breeding study by A. J. M. Smith, who found that kleptoparasitism by the gulls seriously affected chick survival only once in 16 years (Smith 1975). The gulls' value as deterrents to other potential predators presumably outweighs any damage they inflict. In 1973 and 1984 the Forvie colony suffered badly from predation by foxes; on the latter occasion breeding numbers dropped from more than 1,500 pairs in 1983 to only 315 in 1984 (A. J. M. Smith).

The first birds arrive in late March and the main influx occurs in the second or third week of April. By early May most have settled in and laying takes place, often almost synchronously. At Forvie the terns move into the gull colony at dusk, positioning themselves so that they are protected on all sides by gulls (Smith 1975). Sandwich Terns feed well away from a colony, which sometimes leads to unfounded reports of breeding elsewhere. At the end of the breeding season the birds disperse along the coast, in either direction, to areas with a plentiful food supply; many from the Farne Islands and as far south as Norfolk occur in Scottish waters at this time. During this period large flocks build up on the coast, for example 4,000 – of which 30% were juveniles – at Tentsmuir in early August 1973. Several hundred an hour are sometimes seen moving south.

Large numbers of Sandwich Terns have been ringed at Sands of Forvie and many, especially first-year birds, have been recovered on the African wintering grounds, where they are trapped for food. Some reach western Africa in September but the main arrival is during October. Most Scottish recoveries have been between Senegal and Angola, with a few south to Cape Province and Natal. Many immatures remain off Africa for at least 18 months, returning north in increasing numbers over the next two years and breeding in their third or fourth year (A. J. M. Smith).

Since the early 1970s Sandwich Terns have been reported occasionally in winter off the south coast of England, and in 1976 one was seen in early January at Dalgety Bay in Fife. There have been a few December/February records in each subsequent winter, mostly in the Forth (da Prato *et al* 1981) but also in Caithness and Ayr.

Britain and Ireland may hold about a third of the northwest European breeding population of this species, currently estimated at c41,000 pairs (Thomas 1982). Fair numbers breed on the coasts of East and West Germany, Denmark, the Netherlands and France, but there are few, if any, further north than those nesting in Orkney. The Dutch population was severely reduced by insecticide poisoning in the 1960s but has since partly recovered. Organochlorine residue levels in Scottish birds are low (W. R. P. Bourne).

Numbers of breeding pairs of Sandwich Terns (Data from: Cramp et al 1974, Thomas 1982, Bullock & Gomersall 1981, A. J. M. Smith & SBR.)

County/site	1969–70	1970–74 av.	1974–79 av.	range	1980–83 av.	range
Orkney	290	198	136	88–210	162	72–229
Caithness – Stroma	0	(130)	135	120–150	(645 in 1980)	
E. Ross – Morrich More	1,000	99	159	(3–250)	NA	
Aberdeen – Sands of Forvie	740*	1,056	1,062	825–1,194	1,216†	315–1,671†
Midlothian – Inchmickery	46	218	568	430–660	470	424–500
East Lothian – Fidra	60	109	0		0	
Ayr – Horse Island	108	25	0		0	
Total for Scotland	2,496	1,886	1,914	1,735–2,066	(NA)	(NA)

* 740 in 1969; 1,281 in 1970   † 1980–84   ( ) incomplete cover

Roseate Tern *Sterna dougallii*   [1]

Migratory breeder, regular only in the Forth, where the population is currently fewer than ten pairs. The scarcest of our breeding terns, it may become extinct in Scotland in the near future.

The Roseate Tern, an exclusively coastal nester, was first described from a specimen taken on Great Cumbrae in 1812. It has never been an abundant species in Scotland and its population has shown marked fluctuations this century. B&R note 19th century breeding records for Loch Lomond (suspect), islets off Carradale, and the Old Bar of Findhorn, but state that none were known in the early part of this century. In 1927 the species was nesting at Tentsmuir, in 1931 it bred for the first time in the Forth, and in 1963 one pair nested at Sands of Forvie. In 1943 it was rediscovered on the Clyde, where seven different sites have been occupied at one time or another, the most recent being Lady Isle and Horse Island. Lady Isle held 90 pairs in 1953 and eight pairs in 1968; two pairs attempted to nest in 1971 but none since. Horse Island held 65 pairs in 1958, 15 pairs in 1969–70 and an average of only eight pairs from 1971 to 1974, when the 11 pairs present all failed; none have nested since. Sporadic breeding has occurred in Angus (last record 1975, one pair – failed), Fife (up to 18 nests in 1956, when Leuchars runway was closed, and one pair, which reared at least one chick, in 1981), East Lothian (six pairs in 1969 and three pairs in 1970, when one chick was reared), Wigtown (one pair present 1970 but breeding not proved), and Orkney (three pairs displaying on Sanday 1969 but breeding not proved).

In the late 1950s and early 1960s Inchmickery had up to 450 pairs (Sandeman 1963), and sporadic nesting, mainly in small numbers, was recorded on the Isle of May, Fidra, Inchgarvie, Eyebroughty and Carr Craig. As the gull colony on Inchmickery expanded, Roseate Tern numbers decreased to under 50 pairs in 1969–70 and none in 1971. The Fidra colony remained at 50–100 pairs until 1971, since when there has been no further breeding record. In 1971 a cull of gulls was carried out on Inchmickery and in 1972 the Roseate Tern population was back up to 75 pairs. The subsequent picture has, however, been one of fairly continuous decline, becoming more rapid since 1977 (Table) – as it has throughout northwest Europe.

Although there are occasional late-April sightings, most Roseate Terns do not arrive until the second half of May. From May to July a few occur outside the normal range, and there are recent records from the Outer Hebrides, Inner Hebrides, Orkney and Shetland. South of the Moray Basin and the Clyde sightings are more frequent. The birds start to leave in August and all have gone by the end of September, moving south along the coast to reach their wintering grounds in western Africa by November. The fastest journey recorded for a Scottish-reared chick was by one which reached Senegal during the first week of October. First-year birds remain in Africa, not returning north until they are two years of age and breeding a year later (Langham 1971). Most of the Scottish recoveries have been in Ghana, where many terns are trapped for food.

The Roseate Tern has a fragmented distribution, breeding in widely scattered localities in every continent except Antarctica. The British and Irish population, which forms a substantial proportion of the European total, is thought to have peaked in the early 1960s and has since shown a marked decline, from c2,500 in 1969–70 to under 1,000 in 1979 (Thomas 1982). The future prospects for the species in Europe do not look good (Gochfeld 1983).

Common Tern *Sterna hirundo*

The most widely distributed tern, breeding in nearly all coastal counties and at many inland sites. Most colonies are small and only one regularly holds more than 500 pairs. Scotland supports about one quarter of the British and Irish total of c15,000 pairs. Also a passage visitor.

Common Terns nest in a variety of sites, most often on sand or shingle but also on rocky islands and sometimes on moorland, for example at over 300 m asl on the Correen

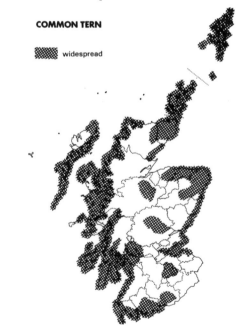

COMMON TERN

▓▓ widespread

Numbers of breeding pairs of Roseate Terns (Data from Thomas 1982 SBR & RSPB/G. Thomas)

	1969/70	1971*	1972	1973	1974	1975	1976	1977	1978	1979	1980	1981	1982	1983
Inchmickery	av. 46	0	75	50	80	61	100	54	51	40	21	2	14	6
Scottish total	126	88	100	52†	91	61	100	54	51	40+	21	3+	14+	6+

* gull cull carried out   † incomplete count

(a) *Estimated numbers of breeding pairs of Common Terns by counties*
(*Data from Cramp* et al *1974, Bullock & Gomersall 1981, Buxton 1985 & local recorders*)

County	1969–70	1975–83
Wigtown	50	C 200+ I c150
Kirkcudbright	350	C/I 250+
Dumfries	?	I c10
Ayr	160	C/I c100
Renfrew	?	I c70
Dunbarton	?	I c10
Stirling	?	C c50
Bute & Arran	150	C c20
Argyll	820	?
W. Inverness	70	C 50+
W. Ross	340	C/I 50+
O. Hebrides	76	C/I c600
W. Sutherland	70	C/I c50
Caithness	50	C/I c50
Orkney	200	C/I 231
Shetland	390	C/I 1,014
E. Sutherland	20	C/I c100
E. Ross	30	C/I c100
E. Inverness	110	C c20
Nairn	?	C c10
Moray	500	C/I ?
Banff	15	C/I 15–20
Aberdeen	490	C/I 200–300
Kincardine	5	C c10
Angus	4	C/I 150+
Perth	?	I 10–20
Kinross	?	I 5–15
Fife	260	C ?
Midlothian	100	C 670+ I c10
E. Lothian	420	C ?
Peebles	?	I 5–10

C = coastal; I = inland

Hills in Aberdeen. Man-made sites are occasionally occupied, as at Leith and Grangemouth docks, while near St Cyrus pea-fields have been used – with little success. During Operation Seafarer colonies were found in all coastal counties from Kirkcudbright to East Lothian (Table a), and the Atlas survey confirmed breeding well inland in all counties except Berwick, Roxburgh, Dumfries, Ayr, Lanark (which had been colonised in 1955 – Cramp *et al* 1974), West Lothian and Clackmannan. Inland nesting is most frequent in the eastern half of the country, possibly because river shingles and shallow waters for fishing are more widespread there. No full census has ever been attempted. Recent counts in Shetland and the Outer Hebrides indicate, however, that the populations there are substantially larger than was previously thought: 1,000+ and c600 pairs respectively (Bullock & Gomersall 1981; Buxton 1985). In Shetland about four times as many colonies were found in 1980 as in 1969–70 but the average size was only 12 pairs and none held more than 100 pairs; the most northerly site was on the Fidd, Unst. On Fair Isle Common Terns first bred in 1970 and c40 pairs nested in 1983.

Few west coast colonies hold more than 50 pairs, and in the east colonies of 100 pairs or more occur only between the Moray Basin and East Lothian. The largest include those on the Aberdeen and East Lothian coasts and the Forth Islands (Table b); regular counts at these illustrate the way local populations may fluctuate. At Sands of Forvie numbers peaked at 1,200 pairs in 1958, subsequently declining as the Black-headed Gull and Sandwich Tern colonies increased (A. J. M. Smith). The Isle of May also formerly held large numbers, peaking at an exceptional 5,000–6,000 pairs in 1946–47 (Cramp *et al* 1974), but there was a rapid decline in the early 1950s and by 1957 the species had ceased to breed there. In 1973, after the gull population had been reduced by culling, Common Terns again attempted to nest, but it was not until 1980 that a pair bred successfully. In 1983 about 30 pairs nested. Tentsmuir formerly held large numbers but was virtually abandoned in 1939–45, as a result of disturbance; although nesting has occurred since, numbers have never recovered and breeding success is often low due to flooding.

The main arrival of Common Terns is from mid April but a few individuals appear earlier, eg at Lossiemouth on 2 February 1980. After breeding the birds disperse along

(b) *Numbers of breeding pairs of Common Terns* at major east coast colonies (Data from Cramp* et al *1974, Thomas 1982, A. J. M. Smith, RSPB, NCC & local recorders*)

	1969–70	1970–74 av.	1975–79 av.	range	1980	1981	1982	1983
Strathbeg	NA	NA	100	70–121	99	100	170	138
Sands of Forvie	475†	351	247	133–320	120	150	200	465
Aberlady	75	291	245	217–318	(180)	(97)	(113)	(81)
Inchmickery	100	426	600	500–750	533	200+	415	365

* Since the late 1970s Arctic Terns have gradually replaced Common Terns at Aberlady Bay, with the proportion of Common Terns decreasing from <50% in 1980 to <10% in 1983; bracketed figures relate to 'commic' terns (P. R. Gordon)
† 1969

the coast. The extent of this dispersal is well illustrated by the recovery of two birds ringed within three days of each other on Carr Craig; about five weeks later one was in the North Sea some 65 miles off Spurn Point and the other on Sule Skerry. In August and September large flocks gather on the coast, often in company with Arctic Terns. These flocks often contain 2,000–3,000 birds, and may include some from the other side of the North Sea. Young birds from Norway and West Germany have been recovered on the east coast in July/September, one reaching Aberdeen little more than two weeks after ringing in Norway. During September movement southwards takes place and by November the majority of Common Terns have probably reached the main wintering ground off the west coast of Africa. Few remain in Scottish waters after early October, though there are a few November and December records. Most of the recoveries on the wintering grounds have been between Mauritania and Ghana but one bird had reached Angola by November, having travelled more than 7,200 km in about ten weeks. Immatures spend their first summer off Africa; although some return north at two years old, the general return to the nesting areas is not until a year later. Breeding takes place from three years of age.

Common Terns are very widely distributed in Europe and throughout most of the northern hemisphere. The British and Irish population, estimated at c15,000 pairs, forms only a small proportion of the northwest European total of well over 100,000 pairs (Thomas 1982).

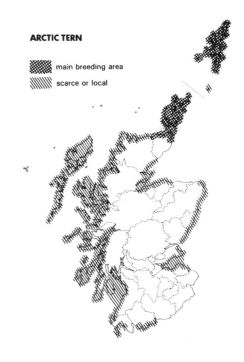

ARCTIC TERN

▓▓▓ main breeding area

╲╲╲╲ scarce or local

Arctic Tern *Sterna paradisaea*

By far the most abundant of our terns, breeding in nearly all coastal counties, often in large colonies and usually on rocky islands. Scotland supports over 90% of the British and Irish population. Also a passage visitor.

More than 90% of the Scottish Arctic Tern population is concentrated in the Northern Isles. In 1980 Papa Westray had the biggest single colony, of over 6,600 pairs, while the islands of Sanday, Rousay, Pentland Skerries, South Ronaldsay, Yell, Papa Stour and Fetlar each held more than 3,000 pairs. Six Orkney colonies and three in Shetland had more than 1,000 pairs, but the average colony size was only 158 pairs in Orkney and 80 pairs in Shetland (Bullock &

Gomersall 1981). In 1969 the Westray group had nearly 28,000 pairs but by 1974 had dropped to c15,500 pairs (Lloyd *et al* 1975).

No colonies of more than 300 pairs were found outwith Orkney and Shetland in 1969–70, but there were over 3,000 pairs on Stroma and 900 on the Monach Isles in 1979 (Thomas 1982), and 500 pairs at Kinloss in 1980 (SBR). On the mainland and the Outer Hebrides numbers are generally fairly small (Table a), with no significant change during the last 50 years. Arctic Terns now breed on Arran and Eigg (for which B&R had no records); one or two pairs nested in Midlothian at Leith docks during the 1970s; and breeding took place for the first time on Fair Isle in 1980. At Tentsmuir, which had a peak population of c500 pairs in 1953, few now breed, and on the Isle of May, which held 400–550 pairs in 1946, there were none from 1957 until 1984, when 19 pairs nested. The only coastal counties for which there are no breeding records are Banff and Berwick. In this country Arctic Terns usually nest within 100 m of the high-tide line, but they breed far inland in Iceland and have occasionally been recorded doing so here too. In 1966 I found three pairs incubating among Common Terns on a shingle island in the Tay, some 26 km upriver from the tidal limit, and a pair successfully reared two young at the Endrick Mouth in 1977.

Arctic Tern colonies are often short-lived and this, together with the sometimes large numbers involved and the need to avoid undue disturbance, makes censusing difficult and sporadic counts of little value. As Bullock & Gomersall (1981) put it 'The fact that Arctic Terns move unpredictably from site to site between seasons invalidates all but the most comprehensive of regular surveys. Any shift of colonies between seasons does not necessarily reflect a

(a) *Estimated numbers of breeding pairs of Arctic Terns, by counties (Data from Cramp et al 1974, Bullock & Gomersall 1982, A. J. M. Smith, RSPB/G. Thomas, & Buxton 1985)*

	1969/70*	1975	1976	1977	1978	1979	1980
Kirkcudbright	30	NA	NA	NA	NA	NA	NA
Wigtown	15	NA	NA	NA	NA	NA	NA
Ayr	60	71	75	76	10	26	NA
Bute	110	NA	NA	NA	NA	NA	NA
Argyll	1,700	NA	NA	NA	NA	NA	NA
W. Inverness	280	NA	NA	NA	NA	NA	NA
I. Hebrides	NA	276	NA	130	NA	816	NA
W. Ross	200	NA	NA	NA	NA	NA	NA
Sutherland	580	NA	NA	NA	84+	324+	NA
Caithness	60	NA	100	NA	NA	3,086	NA
O. Hebrides	1,200	200	250	300	470	900	2,800
Orkney	12,300	NA	NA	NA	NA	NA	33,000
Shetland	7,660	NA	NA	NA	NA	NA	32,000
E. Ross & E. Inv.	220	NA	NA	NA	NA	NA	NA
Moray	90	NA	NA	NA	NA	NA	NA
Aberdeen	320	121	132	136	162	172	120
Angus & Kinc.	c80	NA	NA	NA	NA	NA	NA
Fife & Lothians	60	NA	NA	NA	NA	NA	NA

* The only attempt at a comprehensive survey

population change; it may simply be a response to an unpredictable food source'. Foula is one of the few places where a relatively discrete local population of any size has been estimated regularly over a number of years. This population expanded rapidly during the early 1970s, to a peak of around 6,000 in 1975, but has since declined again (Table b). Coin-

(b) *Numbers of pairs of Arctic Terns nesting on Foula (from Furness 1983)*

1956	1964	1970	1971	1972	1973	1974	1975
250	200	500	930	1,100	750	1,800	6,000

1976	1977	1978	1979	1980	1981	1982
5,650	3,000	2,600	4,400	4,200	1,900	1,260

cident with the decline there was an increase in the nearest sizeable colony, on Papa Stour, where the population expanded from under 400 pairs in 1969–70 to 3,000 in 1978–80 and c10,000 in 1981 (Furness 1983). Large-scale mortality of chicks occurs in some seasons, with many abandoned before fledging. As such disasters sometimes happen in years of good weather (eg Ewins 1985), and when there has been no serious disturbance, they are probably attributable to a local failure of the food supply. At some sites trampling by cattle, or harrying by Arctic Skuas with consequent loss of eggs and young to predators, results in considerable losses.

Some Arctic Terns arrive at the breeding colonies during the second and third weeks in April, and the main influx in early May. Passage continues throughout May and often well into June in the north, where Icelandic birds are possibly involved. After breeding the birds disperse (as shown by recoveries of Farne Island chicks in Argyll a few weeks after ringing), but the majority probably head south fairly quickly.

During August and early September large mixed flocks of 'commic' terns gather on east coast beaches and these include passage birds from further east. Young birds from Denmark, West Germany, Finland and Estonia have been recovered in Scotland, mostly in July/September and on the east coast; one ringed as a chick in Finland was killed by a Great Skua on Foula four weeks later. Some young birds reach the Gulf of Guinea as early as August but, as with the previous species, the main arrival there is in October. Although most continue on their spectacular journey towards Antarctica, some are known to winter off the African coast. The most distant Scottish recoveries so far are birds which had reached Natal in November/December, having travelled more than 10,000 km in less than five months.

Arctic Terns have a circumpolar distribution, nesting far into the Arctic. The Icelandic population is enormous and considerable numbers breed around the Baltic and in Scandinavia. The estimated 77,000 pairs breeding in Britain and Ireland (Thomas 1982) are near the southern limit of the range.

Bridled Tern *Sterna anaethetus*

Vagrant (Caribbean / NW Africa) – one record, 1979.

The only Scottish record of this oceanic species is of a first-summer bird seen at Stromness, Orkney, on 6–7 August 1979 (BB 73:513). This was the first Bridled Tern to be seen alive in Britain, where only eight had been recorded up to the end of 1983.

Sooty Tern *Sterna fuscata*

Vagrant (Tropics) – two records, 1939 and 1954.

The first Scottish record of this oceanic species, which breeds on tropical islands right round the world, was in 1939, when one was found dead at Denny, Stirling, in May (BB 33:197). There has been only one report since, from the mainland of Orkney on 22 April 1954. Fewer than 30 Sooty Terns had been recorded in Britain by the end of 1983.

Little Tern *Sterna albifrons* [1]

Migratory breeder, nesting in small numbers on scattered beaches north to Caithness. It has suffered from increasing human use of beaches but protection has recently led to a slight recovery in the population.

Little Tern colonies are more stable than those of other terns, with the birds often returning even when they have been unsuccessful at a site the previous season, so this species is easier to census reliably. The first census was in 1967, when 172 pairs were located in Scotland, in 28 colonies and 13 counties (Norman & Saunders 1969). The largest colony had 30 pairs and 20 had fewer than five pairs; four colonies (15 pairs) were protected by wardens, and 12 (holding 69 pairs) were difficult of access. There were c250 pairs

in 1969–70; from then until 1979 the population fluctuated between that level and about 300 pairs (Thomas 1982).

The largest numbers are on the east coast south of the Loch of Strathbeg, and in the Hebrides (Table a). In 1969–70 Little Terns were found on the Outer Hebrides in Lewis, Harris, North Uist, Benbecula, South Uist, the Monach Isles and Barra, and on the Inner Hebrides in Coll, Tiree and Islay (B&R did not know of breeding on Islay). Few subsequent counts are available, but the Outer Hebridean population was c75 pairs in 1980; Islay held four pairs in 1973, eight in 1981 and 20+ in 1982; Tiree had 65 pairs in 1968, eight in 1969, 13 in 1973 and only six in 1983; and Coll had 18 pairs in 1969–70, only two in 1972, and about ten in 1983 (G. Thomas). A report of breeding in Orkney in the 1920s is considered to be unreliable (SB 5:102); Little Terns are now seldom seen in the islands. Breeding was first proved in Caithness in 1968, when seven pairs nested. Two pairs were in Inverness in 1972 but breeding was not proved, and in 1978 three pairs attempted, unsuccessfully, to breed in Nairn. In all other suitable east coast counties breeding occurs regularly, although numbers fluctuate markedly.

Regular counts at four reserves demonstrate the extent of the population fluctuations (Table b). Despite these colonies being largely protected from human disturbance, which is the commonest factor militating against successful breeding in this species (Norman & Saunders 1969), their success rate is seldom high. Sand-blow, high tides, Kestrels, rats and foxes have all been responsible for failures at some sites and in some years (Atkinson 1982). The Little Tern's

(a) Estimated numbers of breeding pairs of Little Terns, by counties (Data from Cramp et al 1974, Thomas 1982, RSPB/G. Thomas, Buxton 1985 & local recorders)

	1969–70	1975–79 av.	range	1980	1981	1982	1983
Kirkcudbright	2	?	?				
Wigtown	1	NA	NA	3–4	3	NA	NA
Ayr	6	15	5–30	4–6	6	4	5–6
Argyll	60	NA	NA	NA	NA	NA	NA
O. Hebrides	66	59+	18–84	c100	NA	35+	NA
Caithness & E. Suth.	9	c7	5–11	22–27	NA	NA	NA
Moray	10	NA	NA				
Aberdeen	24	75	51–146	NA	46+	54	NA
Kincardine	40	39	19–83	12	30	35	45
Angus	10	31	25–40	65			
Fife	10	14	0–51	NA	NA	NA	NA
E. Lothian	13	35	24–52	50	68	78	58

(b) Numbers of breeding pairs of Little Terns at the main east coast colonies (Data from Thomas 1982, A. J. M. Smith, NCC & local recorders)

	1971	1972	1973	1974	1975	1976	1977	1978	1979	1980	1981	1982	1983	1984
Sands of Forvie NNR	11	4	4	8	17	80	69	51	60	50	46	54	34	2*
St Cyrus NNR	75	106+	140+	158+	83	10	31	55	12+	12	14	6	16	60+
Tentsmuir NNR	10	NA	24	16	NA	18	10+	...See footnote † ...						
Aberlady LNR	16	NA	12	20	33	23	13	15	13	15	30	29	19	26

* predation by foxes
† Annual figures not available but range in this period 0–5 pairs

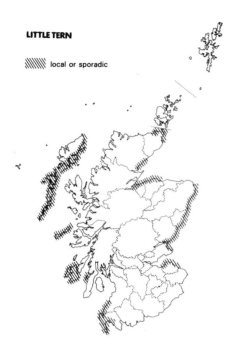

LITTLE TERN

\\\\\\\ local or sporadic

habit of nesting on bare ground within about 2 m of high water mark, and on the type of sandy beach that is most attractive to humans, makes its nest particularly vulnerable to abnormal weather and to accidental trampling (Knight & Haddon 1983).

Arrival at the breeding sites is from the third week in April, and the main influx is in early May. Birds are sometimes seen in areas where they do not breed, eg in Shetland and on Handa. Most Little Terns have left by mid September, but there are a few October records. A Scottish-ringed bird has been recovered in Morocco, and the main wintering area is believed to be the west coast of Africa (Cramp *et al* 1974).

The Little Tern's breeding distribution is almost worldwide. Britain and Ireland hold about a third of the northwest European population, with by far the largest numbers in England. Although the British population decreased earlier, largely due to disturbance, both it and the northwest European population now appear to be fairly stable, probably as a result of the increased protection given to the nesting birds.

Whiskered Tern *Chlidonias hybridus*

Vagrant (S Europe) – one old record, 1894.

An adult male Whiskered Tern was shot in Nithsdale, Dumfries, in May 1894 (ASNH 1894:179). The species breeds from southern France and Iberia eastwards and has been reported in Britain annually since the 1970s – with at least six records in May/June 1983 alone – so it seems surprising that there has been no Scottish occurrence this century.

Black Tern *Chlidonias niger*   [1]

Passage visitor, annual in small numbers.

The Black Tern breeds on freshwater marshes throughout western Eurasia, from southern Sweden and France eastwards, and winters in tropical Africa. It frequents coastal waters on passage and in winter.

There are recent records for all mainland counties other than Banff, Nairn and Clackmannan, and for the Outer and Inner Hebrides but not the Clyde Islands. Black Terns occur most regularly on the coast, especially between East Lothian and Aberdeen, and only rarely north and west of a line from the Clyde to the Moray Basin. They are more abundant in autumn than spring but numbers vary greatly between years, the annual total in the period 1969–83 ranging from 11 to c110, and the number of reports from one to 23 in spring and eight to 100 in autumn. Most records are of single birds or very small parties, the largest flock reported being 16 at Barnsness in August 1970. The main spring passage is generally in May/June, with 24 April the earliest date. Autumn passage sometimes starts in late July and is usually most marked between mid August and mid October. B&R give 26 November as the latest date but there has been no recent occurrence later than 8 November.

An adult Black Tern consorted with Little Gulls on Kilconquhar Loch from 9 to 24 July 1968, while at Sands of Forvie single birds were present for three weeks in May 1972 and 1974, and a pair in May 1973.

White-winged Black Tern *Chlidonias leucopterus*

Vagrant (SE Europe) – occurrences in ten years, 1964–83.

A bird thought to be of this species was seen off Arbroath in July 1932 (B&R) but the first positive identification was not until 1964, when one was on Benbecula, Outer Hebrides, on 23 May (SB 3:258). White-winged Black Terns have since been recorded in 1966, 1967 (four widely scattered occurrences in May/June), 1969, 1970 (two), 1973 (two), 1975, 1976 (two), 1979, 1982 (two) and 1983. All 17 reports have involved single birds. Most sightings have been in May/July but there are a few reports for August and September. There are records for Shetland, Fair Isle, Orkney, Caithness, Nairn, Angus, Midlothian, East Lothian, Renfrew and the Outer Hebrides. Although this is a bird of freshwater marshes, most records come from coastal areas, the only truly inland occurrences to date being at Eaglesham, Gladhouse and Forfar/Balgavies.

B

Guillemot *Uria aalge*

Breeds on most coasts with suitable nesting ledges; by far the biggest concentrations are in the north. Winters at sea, young birds dispersing further than adults; some adults visit colonies in fine weather during the winter. Also occurs as a winter visitor.

Scotland holds about 80% of the British and Irish Guillemot population (Cramp *et al* 1974), with the largest numbers in Sutherland, Caithness, the Northern Isles and the Outer Hebrides (Table a). Breeding density is much influenced by rock type and formation; the majority nest on cliff ledges or relatively flat-topped stacks, where they are often packed so tightly that they are difficult to count accurately, but in some places, notably Shetland and Canna, they occupy boulderfields and caves. Probably the most precarious recorded site is Rockall, where in 1946 Seton Gordon observed six apparently-brooding birds from the air; their prospects of success on this wave-washed rock must have been poor. The colonies on some of Orkney's horizontally-layered sandstone cliffs are very large, and present a scene of constant movement and clamour, not to mention smell, during the breeding season. The Noup (Westray) held an estimated 40,300 birds

(a) *Numbers and distribution of breeding Guillemots in 1969–70 (Data from Cramp* et al *1974)*

	No. of birds*	County	No. of birds*
Kirkcudbright	280	Caithness	62,200
Wigtown	1,900	E. Ross	750
Ayr	4,200	Banff	8,500
Argyll	4,500	Aberdeen	5,100
W. Inverness	2,400	Kincardine	33,000
W. Ross	0	Angus	1,800
W. Sutherland	49,300	Fife	9,000
O. Hebrides	65,000	E. Lothian	1,200
Orkney	129,800	Berwick	6,700
Shetland	77,400	*Total*	c461,400

*Note that counts were incorrectly expressed as 'pairs' in Table 22 of the Appendix in Cramp *et al* 1974).

in 1978 and Marwick Head c18,000 in 1981 (Stowe & Harris 1984). Regular counting of complete colonies is seldom practicable and the use of sample areas for long-term monitoring can result in misleading impressions of population change. This problem has recently been discussed, and a counting procedure recommended, by Harris *et al* (1983).

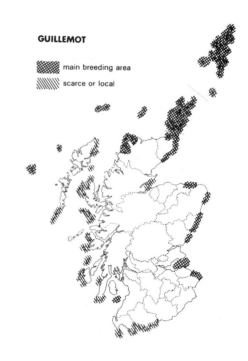

GUILLEMOT

▨ main breeding area

▨ scarce or local

Prior to 1969–70 few attempts had been made to count Guillemots but numbers appear to have increased fairly steadily over the last two or three decades and still to be doing so in most areas (Stowe & Harris 1984) (Table b). On the Isle of May the expansion since the late 1950s has been particularly marked, with an increase of about 5% per annum in 1969–81 and many new sites on the island colonised (Harris & Galbraith 1983). Smaller populations on the other Forth Islands have also increased in recent years, while a 5% per annum increase was recorded on the Berwick coast in the period 1957–78 (da Prato & da Prato 1980). Comparable rates of expansion have been noted on Canna (Swann & Ramsay 1984), Foula (Furness 1983) and Fair Isle. On Fair Isle the increase has been coincident with a trend towards earlier return to the colonies, many birds now returning in October. This trend may be due to growing competition for nest sites (Taylor & Reid 1981), or may perhaps be associated with an improvement in the food supply; it has not been apparent on Canna, where nesting starts 10–14 days later than at east coast colonies (R. L. Swann).

Most young Guillemots leave the ledges from mid June, before they can fly, and accompany their male parent out to sea; birds from different breeding areas subsequently disperse in different directions (Mead 1974). Many Shetland birds spend their first winter in Norwegian waters, travelling across the North Sea in early autumn and later moving mainly northwards along the coast. Young birds from east coast sites cross to the Kattegat and then go southwards as far as the English Channel, but many adults remain in British waters. Canna Guillemots are often in Scandinavian waters in their first autumn, and further south in the North Sea later in the winter. Most recoveries of adults from Canna have been from southwest England and western France (Swann & Ramsay 1983) but some Hebridean birds have reached Spain. There is also some movement between Scotland and the North Atlantic: Guillemots ringed in the Faeroes in June/July have been found in Shetland and on the east coast south to Berwick, while birds marked in Scotland have occurred in Faeroe and Iceland during their first winter. The pattern is thus one of dispersion rather than migratory movement, with many older birds not leaving home waters until deteriorating weather forces them to do so. Sizeable coasting movements sometimes take place then, with up to 3,000/hour recorded passing north off Girdleness in February, and c1,500/hour heading west off Faraid Head in April and south past Corsewall Point in late September.

Many of those recovered in Norwegian waters (about half the total recoveries) have been shot; in contrast oiling accounts for about 50% of British recoveries (Birkhead 1974). Guillemots are particularly vulnerable to oil pollution, eg nearly all of the 2,500+ Guillemots found on east coast beaches in January/February 1970 were oiled (Greenwood *et al* 1971) and 47% of those beached in Orkney in 1976–78 were affected (Hope Jones 1980). During the winter of 1980/81 immatures from Shetland, Orkney and the east coast were badly affected by severe oil pollution (Baillie & Mead 1982). On the west coast very large numbers died during the massive Irish Sea wreck in September 1969, in which both PCBs and gale-force winds were probably contributory factors. About 9,000 Guillemots were picked up between the Inner Solway and Loch Linnhe, many of them still alive but very weak and emaciated (Stewart 1970). Some 50% of the 10,000+ Guillemots found dead on the east coast of Britain during the wreck of February/March 1983 were first-winter birds and a further 22% immatures; ringing

(b) Counts of Guillemots at selected colonies (Data from Operation Seafarer and Stowe & Harris 1984)

	No. of birds in 1969–70	No. of birds in subsequent counts
Hermaness	15,990	22,760 in 1978
Noss	14,155	63,840 in 1981
Fair Isle	10,000	19,200 in 1975
Papa Westray	2,903	2,070 in 1981
Copinsay	9,000	23,640 in 1979
Wick-Lybster	10,600	37,430 in 1977
Inver Hill-Badbea	15,800	50,430 in 1977
Fowlsheugh	32,770	39,000 in 1982
Isle of May	9,000	16,310 in 1981
Craigleith	620	1,900 in 1982
Bass Rock	500	2,800 in 1980
St Abbs	6,730	14,790 in 1978
St Kilda	21,900	22,100 in 1981
Scar Rocks	1,200	1,200 in 1979

recoveries indicated that a large proportion were of Scottish origin (Underwood & Stowe 1984, Hudson & Mead 1984).

Stragglers are occasionally blown far inland by severe gales and these may even feed in freshwater. This is normally a strictly marine species, however, with few penetrating far up the estuaries. The influx of over 3,000 Guillemots feeding on sprats and herring well into the Beauly, in mid November/ December 1980, was a very unusual occurrence (SBR).

In the eastern Atlantic the Guillemot occurs from the Arctic Ocean south to Brittany and western Iberia. Three races have been recorded in Scotland. Most of the breeding population belongs to the nominate race *U.a. aalge*, which is blackish above; the browner southern race *U.a. albionis* breeds in the southwest and intergrades with more northerly populations. Birds of the arctic race *U.a. hyperborea*, from Bear Island and the Barents Sea, have been found dead on the east coast and may winter off Scottish shores quite commonly (Bourne 1968). The 'bridled' variant, with its distinctive white eye-ring, is most abundant in northern populations, the proportion present ranging from less than 1% in the south of England to 10+% in the Outer Hebrides and Orkney, and 20+% in Shetland. The proportion of bridled birds remained relatively unchanged from 1959–60 to 1981–82 (Birkhead 1984).

Brünnich's Guillemot *Uria lomvia*

Vagrant (Arctic/sub-Arctic) – less than annual and usually found dead.

Brünnich's Guillemot breeds no further south than Iceland and northern Norway and winters at sea in the same latitudes. The first acceptable record is of one found dead in East Lothian on 11 December 1908 (ASNH 1909:75). There were no further reports until 1968, when a freshly-dead specimen was picked up in Unst. By the end of 1984 Brünnich's Guillemot had been recorded in Shetland (1977, 1980 & 1983), Fair Isle (1980), Orkney (1981, 1982 & 1984), Sutherland (1982), Caithness (1976), Aberdeen (1979), Kincardine (1981), East Lothian (1980) and Argyll (1969). The 1980 Fair Isle bird, an adult in breeding plumage seen on 16–17 October, was the first live example of the species to be recorded in Britain; a second live bird was seen at sea in the Brent Oilfield the same year. All reports of Brünnich's Guillemots have been between October and March, with most in December/February. The skin of the Caithness bird is in the Royal Scottish Museum, that of the 1977 Shetland bird at Fair Isle Bird Observatory, and that of the Kincardine bird in Dundee Museum.

Razorbill *Alca torda*

Widely distributed breeder on coastal cliffs, in smaller and less dense colonies than the Guillemot. Sutherland, Caithness, the Outer Hebrides and the Northern Isles hold more than 75% of the Scottish population. Native birds disperse southwards in winter, immatures moving further than adults. Also occurs as a winter visitor.

The Razorbill's habit of nesting beneath overhangs, in narrow crevices and occasionally even in burrows means that colonies are both very dispersed and extremely difficult to count accurately. Little information on numbers was recorded prior to Operation Seafarer (Table), but neither

Numbers and distribution of breeding Razorbills in 1969–70. (Data from Cramp et al 1974)

County	No. of pairs	County	No. of pairs
Kirkcudbright	130	Caithness	19,000
Wigtown	150	E. Ross	60
Ayr	2,280	Banff	640
Argyll	2,190	Aberdeen	410
W. Inverness	1,170	Kincardine	5,800
W. Ross	3	Angus	130
W. Sutherland	14,200	Fife	350
O. Hebrides	22,200	E. Lothian	60
Orkney	8,500	Berwick	260
Shetland	8,900	*Total*	c86,500

the few earlier estimates available nor more recent counts give any indication of marked changes in the overall population (Stowe & Harris 1984). Since 1969–70, however, there

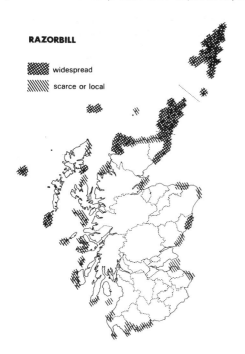

RAZORBILL

▓▓ widespread

░░ scarce or local

has been an increase of about 16% per annum on the Isle of May; this very rapid expansion may be partly due to immigration, since species laying only a single egg are unlikely to be able to increase by more than 10% per annum (Harris & Galbraith 1983). On Canna the increase averaged c9% per annum from 1974 to 1983 (Swann & Ramsay 1984). The Berwick population has also been expanding – at a rate of 4% per annum in 1957–78 (da Prato & da Prato 1980) – and numbers are increasing steadily on the Forth Islands (c410 pairs in 1969–70; c1,000 pairs in 1978) and more slowly on the Scar Rocks.

Since the 1950s Razorbills have been returning to east coast colonies six or seven weeks earlier than formerly, with many present in early February, though the trend towards earlier colony attendance is not so marked in this species as in the Guillemot (Taylor & Reid 1981). Razorbills have also been ashore on Fair Isle and the Isle of May in October/November.

The chicks leave in July, when about three weeks old and still unfledged, and are accompanied to sea by their male parents, which also become flightless as they moult at this time. It is not known how long the adult and chick stay together but ringing recoveries show that by September the young birds have largely separated from the adults. First-winter birds from northern colonies move east across the North Sea in autumn and then south in mid winter as far as Iberia and Morocco; those from west coast colonies appear seldom to enter the North Sea. Few adults are recovered in Norwegian waters and most are thought to winter off the east coast of Britain (Lloyd 1974; Mead 1974; Blake et al 1984); there is a marked southerly progression in the pattern of recoveries during the course of the winter. Interchange between colonies occurs (Steventon 1982) and Razorbills ringed as nestlings and as adults in Scotland have been recovered in summer in the Faeroes.

Sizeable coasting movements are occasionally recorded, with up to 1,000 birds an hour passing sea-watch points. Gales sometimes carry birds inland but major wrecks from this cause are rare; the February 1983 incident affected this species more than the other auks, however, presumably because a relatively large proportion of the population was in British inshore North Sea waters at that time (Blake et al 1984). Many of the ringed birds that died were from Orkney and Shetland, 80% of them being adults (Underwood & Stowe 1984; Hudson & Mead 1984). Oiling is also a potentially serious threat to this species, which spends much of the year in inshore waters (Blake et al 1984). More than 1,000 oiled birds were found on beaches from Kincardine to Fife in January/February 1970, while 51% of those beached in Orkney in 1976–78 were oiled. Adults from Orkney, Shetland and east coast colonies were the groups most affected by severe oil pollution during the winter of 1980/81 (Baillie & Mead 1982).

The Razorbill occurs only in the North Atlantic region, and the British Isles hold a substantial proportion of the world population. Two races are recognised: *A.t. islandica*, breeding in Iceland, Faeroe, Britain and south to Brittany, and *A.t. torda*, from North America, Greenland and continental Europe south to Scandinavia and the Baltic. *A.t. torda* has been recorded among beached birds in Scotland, including an immature ringed in Murmansk and found in East

Lothian, and this race may occur fairly regularly off our shores. Several first- and second-winter Icelandic birds were among the Razorbills wrecked on Scottish coasts in February/March 1983.

Great Auk *Pinguinus impennis*

Extinct. The last British specimen of the Great Auk was killed on St Kilda about 1840. An account of its earlier history is given by B&R.

Black Guillemot *Cepphus grylle*

Very sedentary; most abundant in the Northern Isles which hold more than half the British breeding population. Sparsely but widely distributed in the west, nesting mainly on offshore islands, but scarce in the east south of Caithness.

The Black Guillemot (or Tystie as it is called in the north) is our scarcest breeding auk. It nests in caves and crevices and among boulders near sea-level and is consequently a difficult species to census. Population estimates are generally based on the number of individuals sitting on the rocks or of pairs displaying offshore; since the numbers visible vary with time of year and day, many of the figures reported are unlikely to represent reliable assessments. In 1969–70 the Scottish total was estimated at 7,555 pairs (based on counts of single birds) (Cramp *et al* 1974). The figure of 2,330 pairs for Shetland was certainly too low, and even a subsequent estimate of 4,000 pairs (Kinnear, in Heubeck & Richardson 1980) is now considered to be minimal. The estimates for other areas were: Orkney – 2,240, Argyll – 710; Outer Hebrides – 530; west Sutherland, Wester Ross and west Inverness – each 420–440; Caithness – 370; Ayr – 60; Wigtown – 20; Bute, Kirkcudbright and Banff – not more than ten each.

The potential threat that oil-spills around Orkney and Shetland present to a large proportion of the population has recently led to intensive studies of Black Guillemots there. These have provided a more complete assessment of popula-

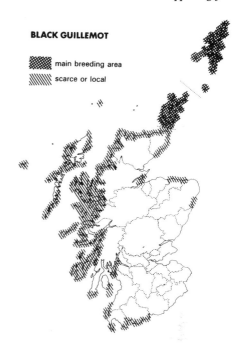

BLACK GUILLEMOT

▨ main breeding area

▩ scarce or local

tion size (Table) and information on the species' ecology and distribution outwith the breeding season (Ewins & Tasker, 1985; Ewins 1985a). In Orkney most Tysties were associated with open cliffs 11 m to 25 m high. Densities equivalent to at least 500 birds/km of coastline were found on the Holm of Papa Westray and 900 birds/km at Lyness Pier on Hoy, but these are exceptional. The average density was 7.6 birds/km of coastline in Orkney and 7.5 in Shetland, and most colonies held less than 20 adults.

Little information is available on past numbers but recent counts on some of the smaller Shetland islands are similar to estimates made about 40 years ago (P. J. Ewins). Black Guillemots formerly bred in the southeast, ceasing to do so in the 19th century due to persecution. They bred in Kincardine (presumably at or near Fowlsheugh) at least once in the first half of this century (B&R) and in Aberdeen, near Muchalls, in the early 1980s (M. L. Tasker). The single pair recorded in Banff during Operation Seafarer was the first recent report from that area, where a few pairs continue to breed. On the west coast the population is apparently increasing slowly, with colonisation into the Solway in the 1960s and the Inner Clyde in 1979. In the latter case the birds nested in a hole in a building at Port Glasgow. Other 'artificial' nest sites used include crevices in the harbour wall at Portpatrick, a niche on an inter-island ferry in Shetland, and the pier at Lyness (Orkney).

Numbers of Black Guillemots (Tysties) counted during the breeding season in Orkney (1983–84) and Shetland (1982–84). (Data from Tasker 1984, Tasker & Webb 1984, P. J. Ewins)

	Orkney	Shetland
Length of coastline surveyed (km)	495	1,601
Number of individuals counted	6,883	12,008

Black Guillemots seldom move far from their breeding areas; of 42 ringing recoveries, 37 were less than 100 km from the ringing site and only two, both Fair Isle birds, had moved more than 250 km: a first-winter bird in Essex in November and an adult in Yorkshire in February. Small numbers occur fairly frequently off the east coast as far south as East Lothian; most reports there are for September/May, when up to 12 have been seen around the Isle of May. A few have summered on the Kincardine and East Lothian coasts in recent years and one has been ashore with Puffins on the Isle of May (M. P. Harris). During the winter of 1978/79 unusually large numbers were off Aberdeen, with up to 44 counted in half an hour in January. Very large concentrations occur at times in Shetland, eg 1,100 in Colgrave Sound, between Yell and Fetlar, in March/April 1974. Elsewhere coastal counts seldom exceed 100–200 and in most areas are much lower.

Before oil operations started in the Northern Isles few Black Guillemots were affected by oil-spills. Only 3% of those beached in Orkney in 1976–78 were oiled but in 1978–80 the figure rose to 35%, though it has since fallen again (E. R. Meek). The *Esso Bernicia* spill at Sullom Voe killed more than 600 birds in early 1979 (Heubeck & Richardson 1980) and numbers breeding on the Yell Sound Islands dropped markedly, with only one pair in 1979 and 1980. Not all local declines can be attributed to such obvious causes, however, as the Foula population has shown marked fluctuations over the years and dropped suddenly in summer 1978, before the Sullom oiling (Furness 1983). Rats, otters and feral cats are potential predators, especially on inhabited islands, and the abandonment of some sites has been attributed to colonisation by rats (Ewins & Tasker 1985). Neither Orkney nor Shetland has yet been infested by feral mink, which have affected breeding distribution and numbers in other parts of the Black Guillemot's range. The results of long-term studies of the breeding biology and ecology of Tysties on Fair Isle have recently been summarised by Broad & Ewins (1984), who conclude that this species – here at the extreme southern limit of its circumpolar distribution – remains our least known auk.

Little Auk *Alle alle*

Winter visitor in very variable numbers, most abundant in the north.

A high-arctic species, breeding south to the northern tip of Iceland and wintering at sea, the Little Auk is a regular visitor to Shetland waters but becomes progressively scarcer inshore further south on the east coast and is much less frequently recorded on the west. There are recent records from all coastal counties except Nairn and from all island groups; from Orkney to Berwick occurrences are almost annual on exposed coasts. Little detailed information is available on numbers but it is clear that the totals wintering in Shetland waters fluctuate widely. January counts on the Fair Isle to Grutness crossing have ranged from under 50 in 1969 to 'thousands' in 1973. Off Foula more than 2,000 per hour were recorded passing south in a steady stream

on 26 December 1966 (Furness 1983) and numbers were high in Shetland in mid winter 1982/83. Most reports are for November/March but there are scattered records throughout the year, including several of birds in breeding plumage in July/August.

Little Auks are vulnerable to adverse weather conditions, which sometimes result in quite large wrecks on the coast as well as scattered records far inland. This situation appears to arise more often in the relatively confined waters of the North Sea – where the species is quite widely distributed (Blake *et al* 1984) – than in the eastern Atlantic, though the more frequent east coast records may simply reflect better coverage in the Beached Bird Survey. In the most recent wreck, in February 1983, counts of beached birds were highest in the southeast, where 184 were found on the Berwick coast. Oiling is also a potential threat and Little Auks suffered badly in 1970, when there was a major incident in the North Sea. Dead oiled birds were found on beaches from Shetland to Berwick, the largest numbers recorded being in Aberdeen (428), Angus/Kincardine (254) and Fife (132).

Most of the Little Auks visiting the North Sea are believed to belong to the nominate race *A.a. alle* which breeds throughout the species' range except in Franz Jozef Land, where it is replaced by the high-arctic form *A.a. polaris*. A few individuals of the latter race have been found among birds beached in Shetland, Fife and the Forth (W. R. P. Bourne).

Puffin *Fratercula arctica*

Breeds most abundantly on islands in the north and west, though smaller colonies are scattered around much of the coast. Puffins come ashore at the breeding sites from late February to mid April and depart in August to winter far at sea.

Scotland holds about 90% of the British and Irish breeding population, estimated to be around 700,000 pairs (Harris 1984b) (Table). The distribution of Puffin colonies is determined largely by the species' preference for steep grassy slopes with an adequate depth of soil for burrowing in, easy access to good fishing grounds and absence of ground predators; the major colonies are all on remote islands or near-

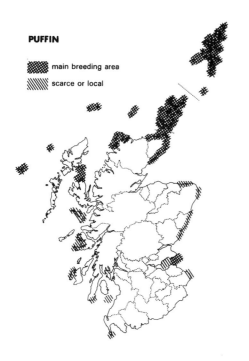

PUFFIN

▓▓▓ main breeding area

░░░ scarce or local

inaccessible cliffs. This is one of the most difficult of the seabirds to census as not only are the birds burrow-nesters, but also many non-breeders are often present at the colonies, and the numbers at burrow-mouths and offshore fluctuate widely according to time of day and weather conditions. Estimates of numbers, especially the earlier ones, should be viewed with some suspicion and taken simply as indicators of the relative size of local populations. Recent studies have shown that temporary failure to breed may result in numbers at a colony decreasing by up to 30% from one season to the next and recovering again the following year (Harris & Murray 1981); careful long-term monitoring is therefore generally necessary to detect population changes. The monitoring of sample areas at several Scottish colonies, notably St Kilda and the Isle of May, in the last decade has provided the most reliable measure of population changes yet available.

In his recent monograph Mike Harris (1984b) has summarised the past history – in so far as it is known – of the principal colonies. By far the largest concentration of breeding Puffins is that at St Kilda, currently estimated to hold some 300,000 pairs. The population there declined around the early 1950s, when most of those nesting on Hirta disappeared, leaving large areas of burrow-riddled ground completely deserted. After a period of stability, it has been increasing again – at an average rate of c4% per annum (*BTO News* 134:2). The colonies on the Shiants, Foula, Her-

maness and Clo Mor (Cape Wrath – the largest mainland site) each hold more than 50,000 pairs and those on Sule Skerry, Fair Isle and the Isle of May between 10,000 and 50,000 pairs. There are many in the 1,000–10,000 range, nearly all of them in Caithness and the Northern Isles. Off the east coast the only large colonies are on the Isle of May, where numbers have been increasing steadily, and Craigleith, which had 1,500 burrows in 1978. In the west few sites south of Skye hold more than two or three hundred pairs.

The Isle of May colony has shown a spectacular and accelerating increase since 1950. From under ten pairs that year it expanded to c2,000 in 1970, over 3,000 in 1976, 8,500 in 1980 and c12,000 in 1984; new areas were still being colonised in 1984 (M. P. Harris). From 1969–80 the average annual increase was 22%, this very high growth-rate being partly due to immigration from the Farne Islands, where numbers have been increasing only slowly (Harris & Galbraith 1983). In addition to increases on the Isle of May and Craigleith, there has been colonisation of other islands in the Forth, with breeding first recorded on Inchkeith in 1965 and on Fidra in 1967.

The reasons for marked fluctuations in the size of a Puffin population are often far from clear. Land-slides occasionally damage sites, as happened on Foula in 1978–79 when a rock fall destroyed c5,000 burrows on the Kame (Furness 1983). Changes in the marine environment which affect the available food supply are obviously important; a shortage of sprats and sandeels of the right size for chick-rearing can result in very heavy mortality among well-grown chicks. For example, in July 1959 at least 8,000 chicks 4–6 weeks old were estimated to have died of starvation on St Kilda (Boddington 1960) and in the 1970s and early 1980s many thousands starved on the Lofoten Islands (Harris 1984b). Predation by rats may be a locally important factor, as may pollution, but most fluctuations are believed to be due to natural causes. Mike Harris's view of the Puffin's future is optimistic; after many years' study he concludes that 'the general state of Puffindom is far better than at any time this century'.

Most Puffins leave the colonies during August and disperse at sea, generally remaining far from shore until February/March. Harris (1984a) has analysed the limited ringing recovery data and concluded that different North Atlantic populations winter in different areas. No clear pattern of movements is apparent but the majority seem to head south; more than half the overseas recoveries are from France and about 30% from Iberia, the western Mediterranean and Madeira/Canaries. Some go in the opposite direction, however, to Norway, Faeroe, Iceland and even to the western Atlantic. The wide winter range is demonstrated by two ringed on Sule Skerry on consecutive days and recovered in their first December, one in Newfoundland and the other in Tenerife. West coast birds apparently move furthest, but only rarely enter the North Sea, while east coast adults are more sedentary, seldom if ever travelling further south or west than the Straits of Dover.

Breeding adult Puffins are faithful to their natal colonies, but a fair amount of movement takes place among immatures at east coast sites. The significant difference in size between Isle of May birds and those on St Kilda suggests that there is little interchange between east and west coast populations

Estimated Puffin populations in 1982. (Data from Harris 1984b)

	No. of pairs		No. of pairs
Outer Hebrides	400,000	North and east	20,000
Shetland	125,000	Inner Hebrides	2,500
Orkney	45,000	Firth of Clyde area	500
Northwest	35,000	*Total*	625,000

(Harris 1977). However, such movements do take place occasionally, as an Isle of May-ringed bird was found breeding on Sanda, Argyll. East and west coast birds differ in the timing of moult, spring return to the colony and laying; in each case the east is three to four weeks ahead of the west (Harris 1982).

During the winter Puffins are very seldom seen close inshore and few are found oiled on the beaches. Even the *Esso Bernicia* spill in Shetland affected only a small number, although oiled birds were still being washed up in March. Some Norwegian birds winter in the North Sea and are occasionally cast up on the east coast; those breeding from about Bergen northwards and in Iceland are *F.a. arctica*, and those in southern Scandinavia, the Faeroes, Britain and France *F.a. grabae*.

Many aspects of the Puffin's ecology and breeding biology are discussed in detail in Mike Harris's monograph (1984b), which includes a chapter by Kenneth Taylor on behaviour and social life – aspects which make this a particularly fascinating and entertaining bird to watch.

Pallas's Sandgrouse *Syrrhaptes paradoxus*

Vagrant (east of the Caspian) – two records this century, 1969 and 1975. Has bred.

Pallas's Sandgrouse is well-known for its sporadic eruptions, which sometimes involve thousands of birds. Several such movements were recorded towards the end of last century, the most notable being that of May 1888, when an estimated 2,000 or more reached Scotland, penetrating as far west as the Outer Hebrides. Breeding took place at Culbin both that summer and the following year (B&R). No large-scale irruptions have occurred recently and the only records since 1900 are of a single bird on Foula in May 1969 and two on the Isle of May in May 1975. Pallas's Sandgrouse breeds on steppelands eastwards from the Caspian Sea and must cover enormous distances during its eruptive movements, the reasons for which are not understood.

Rock Dove (& Feral Pigeon) *Columba livia*

Resident, breeding in coastal caves from the Clyde to Caithness, on both mainland and islands. Elsewhere around the coast flocks of doves consist largely of feral pigeons or hybrids between these and Rock Doves.

The true Rock Dove is now restricted to the coastal west and north. Inter-breeding between this species and domesticated pigeons dates back several hundred years, to the days when dovecots provided an important source of fresh meat in many lowland areas. Feral birds of this origin, and later lost homing pigeons, settled in coastal areas and mixed so thoroughly with the resident dove population that flocks consisting solely of typical Rock Doves now occur only in the remoter areas, where domestic pigeons have never been common. In 1965 Hewson (1967) found 100% pure Rock Dove populations in Bute, Argyll, west and north Sutherland, the Inner and Outer Hebrides, Shetland and parts of Orkney.

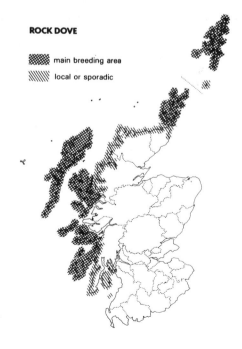

ROCK DOVE

main breeding area

local or sporadic

Some hybrids (under 20%) were present in Wester Ross, Caithness, east Sutherland and parts of Orkney, notably Eynhallow. Hewson concluded that factors common to the areas still holding a pure stock were: lack of colonisation by Stock Doves or Woodpigeons (attributable at least in part to the fact that arable land tends to be restricted to a narrow, relatively treeless, coastal strip) and rocky coasts with an abundance of caves for nesting and roosting.

The situation has probably changed little since 1965, although the numbers of apparently pure Rock Doves breeding on the Clyde Islands have almost certainly decreased. There are scattered small colonies in Argyll, giving a total population estimated at a few hundred (M. J. P. Gregory), apparently very few – if any – on the west coast of Inverness, and relatively small numbers in Wester Ross. The Rock Dove is more abundant in both the Inner and the Outer Hebrides, breeding on all islands of any size, on the west and north coasts of Sutherland (where there is at least one inland breeding record and nesting took place in Smoo Cave for the first time in 1979), and in Caithness. It is also widely distributed throughout Orkney and Shetland (*contra* Atlas, which shows no breeding north of Orkney's south isles). On Fair Isle this species became extinct around 1895, but recolonisation had taken place by the middle of this century; a few pairs still breed although no recent figure is available.

Rock Doves are largely sedentary, with most recoveries less than 10 km from the ringing site, but some movement does take place and numbers on Fair Isle increase in the autumn; a count of 85 there in November 1983 was the highest yet recorded. Sizeable flocks build up in many areas during the autumn and recent peak counts include: 500–700 on Orkney Mainland, 200–300 on Shetland Mainland, 400+ on Islay, 100 each on South Uist and Mull, and 140 at Noss Head in Caithness.

Stock Dove *Columba oenas*

A fairly common resident on low ground south of the Highland Boundary Fault, scarce in the northeast and around the Moray Basin and absent from most of the Highlands and islands. Small numbers occur on passage.

The Stock Dove was not recorded breeding in Scotland until 1866 but within 100 years had become a common bird over much of its present range. Its rapid spread northwards and westwards last century has been attributed to the increase in arable farming at that time (Parslow). B&R considered that the colonisation of Scotland involved movement both inland up the valleys and northwards up the coast, and that in the Moray Basin a separate nucleus was involved. They noted some decrease during the late 1940s, which they thought due to food shortage, while in the 1950s a population crash in England was attributed to the use of pesticides. The English population has since recovered (O'Connor & Mead 1984) but numbers in Scotland still seem to be lower than in the early part of this century, suggesting that factors other than pollution have been involved here. It is possible, however, that Stock Doves are under-recorded in some areas.

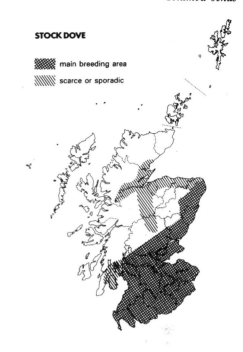

STOCK DOVE

▨ main breeding area

▨ scarce or sporadic

Although the current breeding range is virtually unchanged from that recorded by B&R, ie nesting in every county north to Argyll on the west and Sutherland on the east, recent reports suggest that there are now fewer in the southwest, Argyll and around the Moray Basin. The Stock Dove was described as 'swarming' among the Moray sand-hills (it is normally a hole-nester) and also occurring well inland in the late 19th century, and still common in the 1930s; today it is a scarce breeder along the coastal strip (although probably not as rare in Banff as the Atlas map suggests). Numbers are thought to have decreased recently in east Inverness and Easter Ross (M. I. Harvey), but a few still breed at least sporadically in southeast Sutherland. There have been few recent inland records in the Highlands away from the Great Glen and the Spey/Tummel/Tay valleys, and few even in coastal areas of Argyll. On Arran, where 30 pairs once bred on ivy-covered cliffs, only one pair was known to be present in 1982, and on Bute, which held a few pairs during much of the early part of this century, breeding has not taken place since 1969 (Gibson *et al* 1980); a single bird was seen in suitable breeding habitat on Little Cumbrae in summer 1984 (R. Broad). The only other islands on which breeding has ever been recorded are in the Firth of Forth. B&R found Stock Doves nesting in clefts of the cliffs on Fidra in the 1920s and a single pair was there in 1979, when a pair was also present on Craigleith. In 1981 breeding occurred for the first time on the Isle of May, where the species reappeared in 1978 after being entirely absent for 16 years.

Little information is available on Stock Dove movements and numbers, but the December recovery in Cambridge of one ringed as a chick in Lanark the previous summer suggests that some southward movement of native birds takes place in winter. What passage does occur in Shetland and Fair Isle is both irregular and on a very small scale, seldom involving more than five birds in either spring or autumn; stragglers occasionally reach the Outer Hebrides, including St Kilda, and flocks of up to 20 have been recorded in late autumn on Islay. Birds from the northern part of the species' European range, which extends from central Fenno-Scandia to the Mediterranean, probably do occur occasionally but there is little evidence of regular immigration (BOU). Winter flocks must therefore be assumed to consist largely, if not entirely, of native birds. Among the largest flocks reported since 1970 have been 170+ in Perth (Kingoodie – March 1974), 95 in Moray (November 1975), 100 in East Lothian (November 1982), and 105 in West Lothian (November 1981).

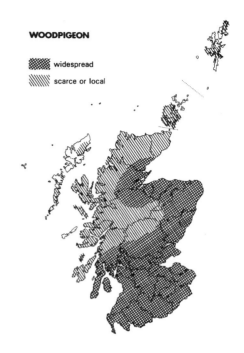

WOODPIGEON

widespread

scarce or local

Woodpigeon *Columba palumbus*

Widespread resident, breeding in all mainland counties and at least sporadically in every island group. Most abundant in districts with extensive arable feeding grounds, where it frequently reaches pest proportions. Also occurs as a passage and winter visitor.

Woodpigeons only became numerous in Scotland quite recently, following a marked expansion in the 19th century (Parslow). Habitat changes this century, including more extensive cultivation of cereals and brassicas, and the establishment of conifer plantations close to arable land, have allowed the increase to continue. The Woodpigeon's prolonged breeding season (well-incubated eggs have been found in January and unfledged young in November) must assist rapid population expansion. Flocks of 5,000–10,000 are now regularly reported in the central and southern lowlands in the winter months and considerable damage to crops occurs, especially during hard weather. The Woodpigeon is classified as a pest species under the 1981 Act and can be killed or taken by an authorised person at any time; it is, however, extremely difficult to control effectively (Murton 1965).

Breeding densities decrease towards the northwest, where only small areas of suitable habitat exist, and in some parts of the Highlands Woodpigeons are scarce or absent in winter. The Clyde Islands had been colonised by the early 1800s and by the end of the century the species was breeding on Mull, Islay, Jura, Eigg and Skye (B&R). Nesting now also occurs regularly on Colonsay, Jura and Raasay, and at least sporadically on Canna and Rhum; there is apparently no breeding record for Tiree or Coll. Increasing afforestation on several of these islands may be expected to result in expanding populations. Nesting first occurred in Stornoway Woods early this century and some 30–40 pairs probably now breed there. The only other Outer Hebridean site is South Uist, where one or two pairs have bred at Grogarry since 1979. Orkney was colonised in the late 19th century and Woodpigeons now breed in some numbers on Mainland and the south isles but are scarce in the north isles; nests are quite often on the ground among heather or rushes. Although breeding was first recorded in Shetland about 1939 (Venables & Venables 1955), none was reported in 1968–72 and there have been only a few records since, most often from Kergord plantation, but occasionally elsewhere, including Unst.

Passage birds occur in both spring and autumn on Fair Isle; numbers are generally small but in October 1976 about 150 were present after an unusually large influx. Numbers were higher than normal that autumn elsewhere in Shetland, and passage was noted on the east coast further south. Although it is impossible to be certain whether large-scale passage indicates immigration or simply bad-weather movements of native birds, it seems likely that some of the large flocks seen on the east coast in winter have come across the North Sea; Woodpigeons are quite often sighted from oil installations, though seldom in any numbers. Among the largest movements recorded recently on the east coast are 2,000 in Easter Ross in late December 1971, 4,600 passing northwards over Girdleness in $2\frac{1}{4}$ hours in mid February 1977, and 10,000 travelling southeast at Gullane in mid February 1978. Inland and on the west there are recent winter counts of 5,000–8,000 in Selkirk, Midlothian, West Lothian, Renfrew and Dumfries – and there have doubtless been many more vast flocks that nobody has attempted to count.

Although ringing has shown that native Woodpigeons are largely sedentary, with few recoveries more than 100 km from the nest site, some long-distance movements do occur, as evidenced by an Aberdeen-reared bird shot 14 years later in Middlesex. Immigrants are most likely to come from the northern part of the species' range, which extends from near the Arctic Circle in Scandinavia to the Mediterranean, but the only evidence of their possible origin comes from the May recovery in West Germany of one ringed on Fair Isle the previous July.

[Barbary Dove *Streptopelia 'risoria'*]

There are occasional reports of this domesticated variant of the African Collared Dove *S.risogrisea* in an apparently wild state, for example one arriving in Shetland off the sea in May 1982.

Collared Dove *Streptopelia decaocto*

Recent coloniser, now breeding in every mainland county and on all the larger islands, generally in close association with human habitation. Most abundant in the Central Lowlands and along the coastal fringe.

The Collared Dove's speedy and complete colonisation of Britain since 1955 is one of the most remarkable ornithological events of this century. After spreading rapidly from Turkey northwest across Europe between 1930 and 1950,

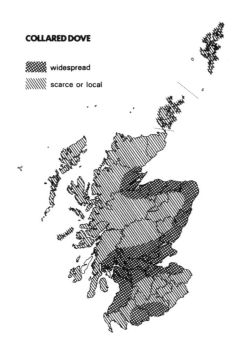

The spread of the Collared Dove (Data from Macmillan 1965, Hudson 1972 & SBR) Counties/islands in which recorded for the first time (first breeding record)

1957	Moray (1957)
1959	Ayr (1959), Fife (1964)
1960	Midlothian (1961), W. Lothian (1960), Perth (1962), Aberdeen (1963), Banff (1961), O. Hebrides (Lewis 1962), Fair Isle (–), ?Mull
1961	Dumfries (1962), Berwick (1961), E. Lothian (1963), Kincardine (1963), Nairn (?), Inverness (1964), Shetland (Mainland 1965)
1962	Angus (1962), Ross (1962), Orkney (1964)
1963	Wigtown (1963), Kirkcudbright (1969), Roxburgh (?), Renfrew (c1965), Lanark (1964), Stirling (c1969), Argyll (1963), Sutherland (1966), Skye (?), Islay (1963), Isle of May (–), Rhum (1968)
1964	Clackmannan (c1967), Dunbarton (1965), Caithness (1964), (Unst 1970)
1965	St Kilda (–)
1966	Arran (?), Bute (?)
1968	Canna (1968), *Selkirk (1968)
1969	Coll (1969), Tiree (1969), *Peebles (1969)
1974	Ailsa Craig (–)

* birds probably present several years earlier

the first birds reached England in 1955 and Scotland in 1957. They are now such a familiar sight – and sound – that it is hard to believe that in 1957 many ornithologists journeyed to Covesea, Moray, specially to see them. The species' subsequent spread has been documented by Macmillan (1965) and Hudson (1972); by the mid 1960s breeding had occurred as far north as Shetland and as far west as the Outer Hebrides (Table).

One of the reasons for this species' success must be the fact that it will attempt, often successfully, to breed virtually throughout the year, and consequently has the opportunity to raise several broods each season. Collared Doves have been recorded mating in January and with unfledged young at the end of October. They can be remarkably stupid in their choice of nest site, however; in Wigtown one pair tried to nest on the crossbar of a telegraph pole, and in Lerwick a pair spent many days attempting to build on a fire-escape ladder, off which every twig or straw blew as soon as it was laid down. Their tameness and willingness to feed at bird-tables, among poultry, and around distilleries must also be important in aiding survival in severe weather, while the fact that they cause relatively little damage to crops is also in their favour. In addition to the rapid expansion of colonies (there were more than 200 Collared Doves at Covesea by 1964), numbers were supplemented throughout the 1960s and 1970s by continued immigration, though the scale of this is not known. Ringing recoveries certainly suggest that it was substantial, with 19 English-ringed birds and one from Belgium recovered in Scotland between 1966 and 1974. Groups have been seen flying north over Drumochter Pass in June, and there are occasional sightings at North Sea oil installations. The few Scottish-ringed recoveries suggest random dispersal, with a bird ringed in Perth moving to the Moray Firth while an Orkney-ringed bird was in Kent two years later.

Within 15 years of colonisation flocks of 80 or more had been reported from places as far apart as the Mull of Galloway and Thurso, and there were roosts of over 200 in Perth and Glasgow. In the islands numbers have seldom been so large, but 200 were counted in Stornoway in June 1977. By that time the population was apparently levelling off, with reports of decreases in some counties and increases in others. Local populations are probably still expanding in the more recently-settled areas, such as the Clyde Islands, but it is just as well that the 50% per annum expansion of the early 1970s is no longer continuing.

When the Collared Dove first bred in this country it was put on Schedule 1 of the Protection of Birds Act 1954, and in 1963 a boy was fined for shooting one in Gullane. In 1966 it was taken off the specially protected list, and by 1971 there were reports of flocks being reduced by shooting. Under the Wildlife and Countryside Act 1981 this species can be killed or taken by authorised persons at all times – in other words it is now regarded as a pest species; this remarkably rapid transition from rarity to pest was anticipated by Andrew Macmillan as long ago as 1966 (SB 4:270).

Turtle Dove *Streptopelia turtur*

Passage visitor, recorded annually in both spring and autumn but rather scarce. A few summer most years and sporadic breeding has occurred.

The Turtle Dove has been recorded in most mainland counties and all island groups, but is most frequent and abundant as a migrant in the northern half of the country. Numbers reported in any one season have not exceeded 100. The spring influx takes place from late April or early May, with passage continuing into June in the Northern Isles; these birds have presumably overflown the English breeding range, mainly in the Midlands, south and east. Autumn passage is between early August and late October (6 November is the latest date) and is likely to involve birds from the northern part of the continental range, which extends to the southern edge of the Baltic. The species is migratory throughout its European range, wintering mainly in Africa, but there have been a few recent reports of individuals remaining in Britain, often in company with Collared Doves; in 1973/74 a Turtle Dove overwintered in Inverness. One ringed on Fair Isle in mid May was in Spain by September.

Turtle Doves are lowland birds, requiring woodland or shrubs for nesting and farmland for feeding. One of their most important food plants is fumitory, *Fumaria officinalis*, a common weed of arable land on lightish soils (Atlas). Agricultural developments during the 19th century led to some expansion of range in Britain, but in the last 50 years there has been only a slight northward extension. In Scotland breeding was first recorded in 1946, in Berwick; it was again proved in that county during the Atlas survey but not since, although nesting possibly occurred in 1982. In 1951 Turtle Doves bred in Roxburgh and probably did so in 1968–72,

but there has been no subsequent report. In 1958 a nest was found near Longniddry, East Lothian (SB 1:120); breeding occurred in the same area in 1960, 1964, 1965 and 1966 but has not been reported since. These are the only confirmed records of breeding in Scotland.

Rufous Turtle Dove *Streptopelia orientalis*

Vagrant (Asia) – one record, 1974.

The only record of the Rufous Turtle Dove is of a first-year bird on Fair Isle from 31 October to 1 November 1974 (SB 10:55), only the fifth British occurrence of this close relative of the familiar Collared Dove. This record is currently under review, however, and may ultimately be rejected.

[Mourning Dove *Zenaida macroura*]

In September 1983 the nest and eggs of this North American dove were found on board a ship at Montrose; one of the eggs contained a half-grown embryo. The nest was on a steel structure originally shipped from Texas (SB 13:51–52).

[Ring-necked Parakeet *Psittacula krameri*]

Escape / vagrant (Category D).

Considerable numbers of this African parrot (now known as the Rose-ringed Parakeet) were released, or escaped, in southeast England in the late 1960s, and feral populations are thought to have become established in several areas. In 1980–83 there were records from Dumfries, East Lothian, Lanark, Dunbarton, Aberdeen and Caithness, and there may well have been other occurrences not reported as the species is so obviously an exotic one.

Great Spotted Cuckoo *Clamator glandarius*

Vagrant (S Europe) – one record, 1959.

The only record of this large and striking cuckoo, which breeds east and south from Iberia and southern France and parasitises members of the crow family, is of an immature on the mainland of Orkney from 14 to 30 August 1959 (SB 1:152). The species has occurred more than 20 times in England, mainly during passage periods.

Cuckoo *Cuculus canorus*

Migratory breeder, widely distributed over most of the mainland and in the Hebrides but scarce and irregular in the Northern Isles. Also a passage visitor.

Because this species makes its presence so audible its spring arrival is widely reported in the national press as well as in ornithological publications; in recent years there have undoubtedly been occasions when a Collared Dove rather than a Cuckoo was responsible for the earliest calls reported. In most years the first birds arrive at the beginning of April (2nd is the earliest date since 1970), and the main influx in late April and early May. In cold springs arrival may be considerably later, as in 1977 when the first report was not until 26 April. Passage through Fair Isle and Shetland continues well into June.

The Atlas survey showed that in summer Cuckoos were absent from only a few 10 km squares in mainland Scotland, most of these being in the arable areas of Berwick, Fife and Buchan. In the Hebrides they are widespread and com-

mon. Breeding occurs up to altitudes of well over 300 m asl and the Meadow Pipit and Dunnock are probably the commonest host species in moorland and farmland areas respectively. Little information is available on numbers, but Yapp (1974) ranked the Cuckoo 11th of the 12 most abundant species in northwest Highland birchwoods, while Henty (1975) concluded that in Strathbraan, Perth, it was appreciably scarcer than it had been early this century. CBC results for Britain as a whole indicate little change in the population between years, but subjective comments by local recorders suggest that in Scotland numbers may vary considerably, perhaps being affected by spring weather conditions and consequent arrival dates.

Most adult Cuckoos depart during the second half of July or early August and the young birds two or three weeks later; stragglers are often still present in late September and occasionally in October, the latest in recent years being a passage bird in Shetland on 27 October 1975.

Ringing recoveries show that some young birds move well to the east on leaving this country; a chick ringed in Ayr was shot four weeks later in West Germany, and one marked in Lanark had reached Italy by early September. Other recoveries involve a bird which crossed from Fair Isle to Bergen within a week, one ringed on the Isle of May and shot in Spain, and a Netherlands-ringed bird found dead the following summer on Fetlar. The Cuckoo is very widely distributed across Eurasia, occupying a variety of habitats and breeding well beyond the Arctic Circle in Scandinavia; it winters in Africa south of the Sahara.

Black-billed Cuckoo *Coccyzus erythrophthalmus*

Vagrant (N America) – two records, 1950 and 1953.

The first Black-billed Cuckoo, an immature, was found dead in Kintyre early in November 1950 (SN 1951:131); the second was dying when found on Foula in mid October 1953 (SN 1953:196). Although occupying much the same range as the closely related Yellow-billed Cuckoo, this species appears rather less liable to wander across the Atlantic and had been reported in Britain and Ireland only eleven times by the end of 1983 – in most cases dead or dying when found.

Yellow-billed Cuckoo *Coccyzus americanus*

Vagrant (N America) – six records, 1904–83.

The Yellow-billed Cuckoo was first recorded in November 1904, when one was found dead on Colonsay, Inner Hebrides (ASNH 1910:184). It occurred again in 1936, and four times between 1952 and 1970 (inclusive) but not since. The records have all been in late September/ early November (there have not yet been any spring occurrences in Britain) and all involved dead or dying birds. In 1953 there was a remarkable trio of reports in early October – when a Black-billed Cuckoo also occurred – from Muck,

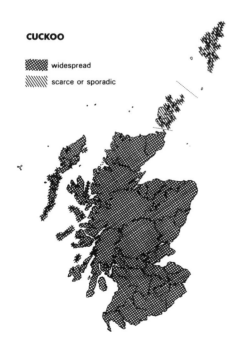

CUCKOO

▓▓▓ widespread

▨▨▨ scarce or sporadic

Nairn and Montrose. The other records are from Orkney Mainland (1936 & 1956), Shetland Mainland (1952), mainland Argyll (1969) and Caithness (1970).

Barn Owl *Tyto alba*   [1]

Resident, most abundant south of the Highlands and especially in the southwest. Very local, and apparently scarce, in the northeast and around the Moray Basin, and absent from the Outer Hebrides, the Northern Isles, and most of the Highlands.

Barn Owls can be difficult to locate when nesting in tree-holes and cliffs, and even sometimes in buildings. As a result they are probably under-recorded in most of the range, and little detailed information on numbers is available. Large natural short-term population fluctuations occur and casual estimates of abundance are consequently unlikely to be reliable; intensive work over a number of years is necessary in order to produce realistic estimates of population size or trends. One such study, carried out by Iain Taylor in southern Scotland from 1979 to 1984, showed that the number of breeding pairs was related to the abundance of the short-tailed vole (the principal prey species) and to winter severity. On low-lying arable farmland Barn Owl numbers remained stable from year to year, but on sheepwalk and in young forestry plantations there was about a sevenfold difference between the minimum and maximum number of pairs present during a single vole population cycle. Heavy winter mortality occurred when there was continuous blanket snow-cover for more than about seven to ten days. Mortality was thus related to altitude, and higher altitude nests were occupied least consistently. It is probably because of this sensitivity to prolonged snow that the species' stronghold is in the mild southwest and that it is seldom found nesting more than 300 m asl.

In Dumfries, Kirkcudbright, and Wigtown the Barn Owl

is not uncommon; in 1968–72 every 10 km square in the first two counties was occupied, with breeding proved in most. In 1981 about one third of the area was searched thoroughly and 118 pairs were located; this was a peak vole year so the owl population would be higher than the average for the area, which might be around 80 pairs (I. R. Taylor). In Ayr in 1979 some 36 territories were known, but in 1984, despite intensive searching, only 15 pairs were found (I. Leach). In Lanark, Renfrew, Dunbarton, Stirling and Clackmannan Barn Owls are much scarcer. Argyll is a particularly difficult county to cover because of the abundance of potential nesting places in cliffs and trees. There are many areas of suitable hunting habitat and winters are generally mild, so one might expect a population higher than the 10+ pairs M. J. P. Gregory believed to be present in the early 1980s.

Further east and north numbers seem generally to be lower. The Barn Owl occurs throughout the four Border counties; Roxburgh probably has the highest density with the number of known breeding pairs varying from eight to 15 between 1979 and 1984. There are very few recent authenticated records of breeding in the Lothians and Fife and none for Kinross. These areas have many active observers and serious under-recording seems unlikely. In Perth there were no positive breeding records between 1977 and 1983, although road-side casualties indicate that a few pairs may still be present. Barn Owls occur along the east coast from Angus (where nine or ten pairs were located in 1983) to Nairn but apparently only in small numbers. In the Black Isle and lowland Easter Ross there is evidence of a rather higher density but no estimates of actual numbers have been made. In southeast Sutherland, where breeding took place in a cave in 1961 and 1962, the Barn Owl now seems to be only an occasional visitor, and in the central Highlands, Wester Ross, north and west Sutherland and Caithness it is extremely rare or absent.

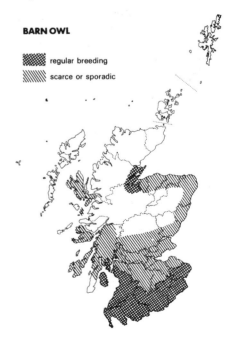

BARN OWL

▓▓▓ regular breeding

▨▨▨ scarce or sporadic

Barn Owls breed regularly on Arran, where eight territories were known in 1982, and on Bute, Inchmarnock and the Cumbraes. On the Inner Hebrides there are no recent records from Eigg, Muck, Rhum or Canna and no information from Tiree or Coll, but breeding occurs fairly regularly on Islay, Jura and Mull (Reed *et al* 1983). Skye holds a few pairs and possible breeding was recorded in Raasay Forest in 1982 (A. Currie). In the Outer Hebrides and Northern Isles the Barn Owl is only an occasional visitor. Most such vagrants are of the British race *T.a. alba*, but there have been a few Shetland records of the dark-breasted *T.a. guttata*, which breeds from Germany and Denmark eastwards. In Scotland the Barn Owl, a cosmopolitan species, reaches the northern limit not only of its European but of its world range.

Although largely nocturnal, Barn Owls also hunt in daylight, especially when feeding young or during severe winter conditions. They hunt mainly over open ground, often perch-hopping along the edges of woodland in arable areas. In Scotland they are normally sedentary; in the southwest about 95% of breeding birds move less than 20 km from their natal area and once established remain faithful to their nesting site. There is evidence that established adults will not move to avoid severe winter conditions, although pre-oreeding young birds may do so (I. R. Taylor).

Changes in the status of this species are particularly difficult to assess because of the paucity of reliable data. B&R documented a major decline in the late 1800s and early years of this century; by 1910 the Barn Owl was said to be almost extinct in many areas, including Dumfries. A recovery subsequently took place, but Prestt (1965) reported another decrease between 1953 and 1963, and implicated the use of pesticides. However, the very severe winter of 1962/63 was probably also responsible for a reduction in numbers, so the overall picture was far from clear (Bunn *et al* 1982). Changing agricultural practices after the 1939–45 war must have reduced the suitability of farmland for this species. Little hay is now stored in lofts, with consequent loss of suitable nest sites, and there are no longer stackyards populated by rodents, House Sparrows and finches to provide a valuable food source for owls during the critical winter period. On the other hand, much valuable new Barn Owl habitat has been created in the form of young forestry plantations with potential nest sites in abandoned and derelict farm buildings. Habitat changes obviously vary from area to area and the limited evidence available suggests a progressive decline in Barn Owls in the more intensively farmed eastern and central regions and much less of a change in the more varied southwest.

Scops Owl *Otus scops*

Vagrant (S Europe) – four records since 1950 and several older ones.

Most records of the Scops Owl are from the Northern Isles or the Hebrides and are of dead or captured birds. The post-1950 reports are from Shetland (June 1953), Orkney (June 1965 & November 1970) and Barra (June 1980), while the most recent mainland record is of one caught in

Kirkcudbright in April 1944. On its breeding grounds the Scops Owl favours open woodland, groves and clusters of trees near habitations; it is most likely to be detected by its distinctive and monotonous call.

Eagle Owl *Bubo bubo*

Escape/vagrant

There are four accepted 19th century records of this magnificent bird, from Shetland (two), Orkney and Argyll, but all reports since 1883 are considered to refer to escapes (BOU; Sharrock & Sharrock 1976). The Eagle Owl is a sedentary species, breeding in little-disturbed rocky or wooded areas of Europe and eastwards into Russia; it has decreased recently in many parts of its range. All the old records are being reviewed by the BOU to determine whether this species can justifiably be included in the British list. Several birds escaped, or were released, in Scotland between 1975 and 1985, and breeding is known to have occurred on at least one occasion (R. H. Dennis).

Snowy Owl *Nyctea scandiaca*   [1]

Irruptive vagrant (Arctic/sub-Arctic) – annual in very small numbers. Bred on Fetlar, Shetland, 1967–75.

The Snowy Owl has a circumpolar distribution, its nearest breeding grounds to Britain being in Iceland and the Scandinavian mountains. Over much of the breeding range it is dependent upon mammalian prey, such as lemmings, which fluctuate greatly in abundance; in years of food scarcity dispersive eruptions occur, with birds appearing far to the south of the normal range.

Snowy Owls have been known to visit Scotland, in variable numbers and at irregular intervals, since the early 19th century. During the first half of this century they were rather scarce and the few recorded in 1950–62 were nearly all in the Grampians and Cairngorms. In 1963 numbers suddenly increased and over the next two or three years there were many reports, the majority in Shetland but a few in Orkney and south as far as the Moorfoots. During the winter of 1965/66 an adult female and one or two adult males were regularly seen on Fetlar, and in 1967 ornithological history was made when breeding was proved for the first time. Seven eggs were laid, six young hatched, and five were successfully reared. As soon as the nest was discovered a round-the-clock watch was mounted by the RSPB; I was one of those privileged to help with the security watch and was on Fetlar when hatching started – a most exciting moment. Bobby Tulloch (1968, 1969) has provided full and fascinating accounts of events in 1967 and the following year.

What was taken to be the same pair bred successfully again on Fetlar in 1969, 1970 and 1971, but failed to rear any young in 1972. From 1973 to 1975 two females attempted to breed but only one male was present; as he could not supply both incubating birds with adequate food, one nest was deserted each season. By the spring of 1976 the male had vanished. At least two, and up to four, females have been on the island each year since, but no replacement male has appeared and there has consequently been no further breeding, although infertile eggs have been laid several times. The presence of at least one, and possibly two, unringed immature females among the four on Fetlar in November 1983, demonstrates that further immigration has taken place (R. J. Tulloch).

Twenty young were raised in 1967–75 (14 females and six males) and the many records in Shetland during this period suggest that they scattered widely through the islands; all were ringed but none has been recovered outside Shetland. One of the 1975 chicks was found, long-dead, on Unst in May 1982. Individuals have occasionally become entangled with barbed wire while hunting and required medical treatment; in all cases this proved successful.

Since the early 1970s Snowy Owls have also been reported in summer from Fair Isle, Orkney, St Kilda, the Outer Hebrides and the Cairngorms. They are most often seen on moorlands or mountains with boulders scattered among short tundra-type vegetation. In Scotland they feed largely on rabbits (Tulloch 1969, SB11:56–7), but following an outbreak of myxomatosis the Fetlar pair was forced to depend largely upon bird prey. During the period of rabbit shortage there, 1971–74, breeding success was low, with less than six young surviving into their first winter; by 1975 the rabbit population had recovered and the owls reared four young that year (M. Robinson). A Snowy Owl on St Kilda in 1968 fed mainly on mice.

In 1981 four females were seen on Fetlar, one or two on Unst, and one each on North Ronaldsay and Papa Westray, but the only male reported in the Northern Isles was a first-year bird on North Ronaldsay in June. It thus appears that females currently outnumber males by a considerable margin and that prospects of breeding in the future will be largely dependent upon a fresh influx of birds from elsewhere. It is not known where the Snowy Owls reaching Scotland come from, but the Icelandic and Norwegian populations are probably fairly small so it seems likely that some may come from much further north. In February 1971 an adult female caught on a trawler off Spitsbergen was released on Fetlar, but the only ringing evidence of long-

distance movement is from a bird (not Fetlar-reared) ringed on Fair Isle and found, long-dead, in Lewis more than two years later.

Hawk Owl *Surnia ulula*

Vagrant (Fenno-Scandia/Siberia & N America) – one recent record, 1983.

The only 20th century record of the Hawk Owl in Scotland is of one in Shetland – seen first near Lerwick and later on Bressay – in mid September 1983 (BB 77:538). Of the four old records, a bird in Lanark in 1863 was considered to be of the American race *S.u. caparoch*, and one in Aberdeen in 1898 to be of the Eurasian race *S.u. ulula*; the others, from Shetland and Renfrew, were not subspecifically identified. A bird of open coniferous and mixed forest, the Hawk Owl is erratically eruptive when food is scarce.

Little Owl *Athene noctua*

Scarce and local resident breeder south of the Forth/Clyde Valley. Breeding has been satisfactorily confirmed only in Berwick, Midlothian, East Lothian and Dumfries, but may also have taken place in Roxburgh and Kirkcudbright.

The Little Owl is widely distributed in Europe south of the Baltic but is not indigenous to Britain. Numbers were released in southern England on several occasions during the 19th century and, once established, the population rapidly expanded northwards. By 1930 the species was breeding over much of the country north to the Humber, favouring farming countryside with abundant hedgerow trees. After 1930 the rate of spread became slower, and there was some evidence of decline in the south, possibly due to pesticides, but by 1950 breeding had been recorded in all the northern English counties. The Little Owl was reported in Scotland from 1925 onwards, and B&R predicted that it would shortly be breeding. At least occasional northward movement of young birds is demonstrated by the recovery at Musselburgh in October 1962 of a Little Owl ringed as a chick in Northumberland earlier that year.

The first evidence of breeding north of the Border came in 1958, when a pair nested near Edrom, Berwick (SB 1:37). Little Owls bred there again in 1960 but have not been confirmed as doing so since, although birds were present all summer in 1980. Ten years later a pair bred near Borthwick, Midlothian; this site was occupied from 1968 to 1971 and again in 1975, but not subsequently. In Dumfries, following sightings in several years, breeding was confirmed in 1979. At least three pairs are now thought to be present in the Glencaple / Caerlaverock, Langholm and Annan areas. Recent sightings in the Lauriston area of Kirkcudbright suggest that breeding may now also be taking place there (A. D. Watson). Finally, in 1981 and 1982 a pair bred in East Lothian, at a site first occupied during 1979; there was no report in 1983, however.

In the Atlas confirmed breeding is also shown in Roxburgh (apparently on the evidence of calling birds in April) and in Perth, where a dead juvenile was found on the A9 north of Perth city in May. The first of these records is well within the apparent breeding range of this species at the present time and is probably acceptable, but the Perth record must be open to doubt as there has been no subsequent report of Little Owls in the county and the corpse found could have fallen from a vehicle.

Away from the areas where breeding has been proved or suspected, Little Owls have been reported in Wigtown, Lanark, Renfrew and Kinross but these have been only sporadic records. In view of the species' habit of using prominent perches during daytime it seems unlikely that many resident pairs would be overlooked. One must conclude, therefore, that colonisation of Scotland cannot yet be said to have taken place and that there are still very few established breeding Little Owls. Nor does the evidence suggest that any appreciable change in the situation is likely to occur before the end of the century.

Tawny Owl *Strix aluco*

Resident, breeding in all mainland areas with suitable woodlands but very local in the northwest Highlands, absent from the Northern Isles and Outer Hebrides and relatively scarce on the Inner Hebrides and Clyde Islands.

The Tawny Owl is the most abundant and widespread of our breeding owls. It nests at highest densities in deciduous or mixed woodland but also occurs among scattered trees in built-up areas and is widespread in the larger mature conifer plantations, sometimes nesting on the ground, on crags or in the tree nests of other species. Normally a hole-nester, it takes readily to nest-boxes; the provision of boxes in 7,800 ha of spruce plantation in Northumberland resulted in a population increase of about 50%, with 90% of pairs nesting in boxes (Petty 1983). A study of the species' breeding biology in Aberdeen woodlands, in progress since 1978, has demonstrated the extent to which a population may fluctuate in response to weather conditions and food availability (Massie 1984). In 1984 the study area held a peak of 42 nesting pairs (39 using boxes), whereas in the late spring of 1982 only 15 nests were found (13 in boxes). The number of

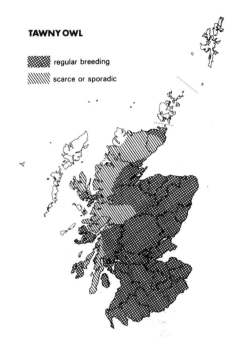

TAWNY OWL

▨ regular breeding

▧ scarce or sporadic

chicks fledged per nest ranged from 0.68 (1983) to 1.52 (1981). Little information on breeding numbers is available, though the relative frequency of proved breeding records in 1968–72 gives some indication of the way densities vary in different parts of the country. Breeding densities of 0.17 pairs / km^2 have been recorded on farmland in Aberdeen (Hardy 1977) and 2.2 pairs / km^2 in upland spruce forest in Dumfries (Hirons 1976). A general increase believed to have taken place in the first half of this century, especially in southern Scotland (Parslow), may have been due partly to a reduction in persecution and partly to increasing afforestation. What little evidence there is of more recent change relates mostly to the islands.

B&R described the Tawny Owl as a common resident on Bute and Arran, but breeding was not proved at all on Arran in 1968–72; in 1982 it was described as 'spreading', with five known territories. This suggests that either the original description was generous, or there was a decrease around the middle of this century. On Islay, where B&R knew of no confirmed nesting, Tawny Owls now breed in at least three 10 km squares (Booth 1981). Breeding also occurs regularly on Mull, Raasay and Skye, and probably at least sporadically on Eigg, where this species is a recent arrival (Reed *et al* 1983); on Rhum it occurs only as a vagrant. Comparison of B&R's records with those in the Atlas suggest that a substantial increase has taken place on Skye where, as on Mull, Arran, Islay and Raasay, there has been extensive afforestation over the last 30–40 years. Wandering Tawny Owls occasionally reach the Outer Hebrides but there is no evidence that they have ever bred there, although suitable habitat is available in the Stornoway Woods; most of those recorded in Lewis recently have been dead or dying when found (Cunningham 1983). There are no acceptable records for Orkney or Shetland, nor for Fair Isle or the Isle of May. The large areas afforested during the last 20–30 years will provide considerable potential for this species to extend its range, albeit slowly due to its very sedentary habits; more than 75% of recoveries are within 10 km of the ringing site.

Long-eared Owl *Asio otus*

Resident, breeding locally in all mainland counties and at least sporadically in all island groups; least abundant in the north and northwest. Also occurs as a passage and winter visitor, especially in the Northern Isles.

The Long-eared Owl is largely nocturnal and almost certainly under-recorded; very little is known about the size and distribution of the breeding population, and the Atlas map probably under-represents the species' true status. Pairs typically occupy small stands of conifers surrounded by open hunting habitat. They are usually absent from, or occur only at low densities in, the larger, mature conifer plantations where Tawny Owls are well established. Detection in either of these situations is largely a matter of luck. The adults are not particularly vocal, incubating females tend to sit tight, and the eggs are usually laid in an old crow nest or a stick nest of similar size. The calls of hungry young are often the first indication that a breeding pair is present. In Shetland and Sutherland nesting in heather on open hill ground has been recorded; on at least one occasion in Shetland the nest was destroyed by muirburning. In the southwest the provision of artificial stick nests has met with some success (Village 1981).

The dearth of reliable figures makes it impossible to say whether any real changes in status have taken place this century. In some areas, notably the southwest, numbers are thought to be considerably lower than in the 1920s, but this is a largely subjective assessment. The changing pattern of afforestation, especially the introduction of shelterbelts and small conifer plantations in formerly treeless areas, seems likely to result in changes in distribution, if not in overall numbers. Only for the islands, where the extent of suitable habitat is in many cases very limited, is there any firm indication of numbers or of status changes.

Gibson *et al* (1980) recorded that in Bute, where Long-eared Owls formerly nested fairly regularly, no evidence of breeding had been found for at least ten years, despite intensive searching. Not even possible breeding was reported on Skye in 1968–72, but there have since been at least two positive records and it is suspected that several pairs may be present (A. Currie). Elsewhere in the Inner Hebrides intermittent breeding occurs on Colonsay, Mull, Eigg, Rhum and Raasay, and more regular nesting on Islay and Muck (Reed *et al* 1983). In the Outer Hebrides the Long-eared Owl breeds regularly in Stornoway Woods and on the Loch Druidibeg NNR, and has done so at least once in the conifer plantation at Balallan in Lewis (Cunningham 1983). Although confirmed or probable breeding is shown in the Atlas for three 10 km squares in Orkney, one of them in the north isles, this was possibly an atypical situation; from 1974 to 1983 there was no evidence of breeding, and few summer records. In Shetland breeding is also apparently

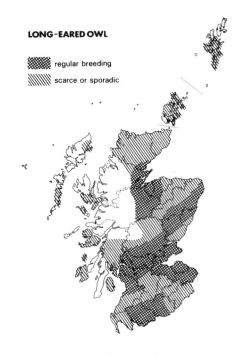

LONG-EARED OWL

▓ regular breeding

▨ scarce or sporadic

sporadic, with up to three pairs recorded in 1973 but none from 1974 to 1983.

Native Long-eared Owls are largely sedentary, with few recoveries more than 100 km from the nest site and none outwith Scotland. Those breeding in the northern part of the extensive European range are migratory and passage birds occur annually in the Northern Isles and occasionally elsewhere; passage is most marked in autumn and numbers are very variable. The most recent large influx was in 1978/79, with totals of 61 and 45 wintering, and maximum roost counts of 24 and 17, in Orkney and Shetland respectively (Davenport 1982). Daily totals of ten and six, the highest ever, were recorded on Fair Isle and the Isle of May respectively in mid October 1978. Migrants ringed in the Northern Isles and Aberdeen between October and April have been recovered in Ireland, Norway, Sweden, the Netherlands, Belgium and West Germany. One marked on Fair Isle had travelled nearly 900 km to reach the Netherlands a week later.

In most years at least small numbers of Long-eared Owls winter in Shetland and Orkney, roosting together in the few sizeable woodlands. Pellets collected at these roosts have shown that in Orkney the very local Orkney race of vole is an important prey item (Orkney BR 1981), while in Shetland, where there are no voles, the principal prey species is the field mouse (Okill & Ewins 1977). Breeding season studies by Andrew Village (1981) in Eskdalemuir in 1975–79 – when much of the recent afforestation had not yet closed canopy and consequently provided good vole habitat – indicated that Long-eared Owls fed preferentially on short-tailed voles. Although breeding density and the proportion of pairs attempting to breed were highest in years when the vole population was high, the percentage of hatched young fledging was highest in years when vole numbers were low. Clutch desertion and loss of broods were not correlated with vole numbers, and Village suggested that this was possibly because only the 'better' owl pairs bred in poor vole years.

Short-eared Owl *Asio flammeus*

Breeds in most mainland counties, in Orkney and on many of the other large islands but not in Shetland, Lewis or Harris; most native birds disperse in winter and some emigrate. Also occurs as a passage and winter visitor.

For successful breeding the Short-eared Owl requires an extensive area of open ground supporting a high population of small rodents. A diurnal hunter, it feeds largely on short-tailed voles, which are often particularly abundant in recently-afforested areas. It is most widespread in the Southern Uplands and on the foothills to the south and east of

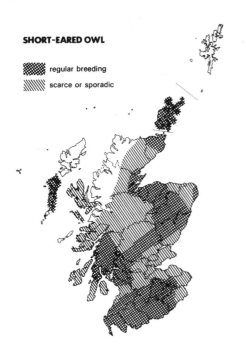

SHORT-EARED OWL

▓ regular breeding

▨ scarce or sporadic

the Grampians and Cairngorms, but its distribution and numbers vary markedly in response to habitat changes and prey availability. When the vole population in an area is very high the territory of a pair of Short-eared Owls may be as small as 16 ha; when vole numbers return to normal, territory size increases to 120 ha or more (Lockie 1955). However, at Eskdalemuir in 1970, although Short-eared Owl numbers were very high, investigation showed that the vole population was no higher than would be expected in a recently-planted area (Picozzi & Hewson 1970). During this study it was found that the birds were regularly roosting in road-side quarries, especially in those where an overhanging mat of turf offered shelter.

The Short-eared Owl is very local in predominantly arable areas, such as Berwick, East Lothian and Fife, and in the mountainous north and west Highlands. Afforestation since the early 1970s has probably resulted in considerable changes in the breeding distribution shown in the Atlas. Where planting has recently taken place on land formerly heavily grazed, eg in Lanark, the area of suitable habitat will have increased and a consequent expansion in range and increase in numbers can be expected to have taken place. In Orkney, on the other hand, the situation is reversed, with the extent of suitable moorland progressively decreasing; the population appears to be remaining relatively stable, however, and in 1982 breeding was recorded on Eday, Hoy, Mainland, Sanday (where one nest was in a silage field) and South Ronaldsay. An estimated 15 pairs breed in the Uists and Benbecula, all leaving to winter elsewhere (C. Spray). In the Inner Hebrides breeding occurs regularly on Skye (where numbers are probably increasing with the extension of afforestation), Mull and Jura, but only intermittently on Islay, Rhum, Eigg and Raasay (Reed *et al* 1983). B&R had no record for Bute, but breeding has since occurred on at least three occasions, the most recent in 1969 (Gibson *et al* 1980). Seven territories were occupied on Arran in 1978 and at least four in 1982.

Many native birds apparently leave Scotland in winter, immatures travelling further south than adults. Of four recovered in September/November of their first winter, one (an Orkney bird) was in Dumfries, one in Eire and two in Spain; of five winter-recoveries of older birds three were in Eire, one in Lincoln and one in Wales. It is not possible to be certain to what extent those seen on low ground and near the coast in winter include native birds as well as immigrants, varying numbers of which arrive every autumn.

In Shetland and Fair Isle, where passage is most obvious as this species is not resident, the first birds usually start arriving in mid September and immigration may continue into December. Numbers are often small, with a total of only 10–15 recorded during the season on Fair Isle, but much larger influxes occasionally occur, as in 1974 when 14 were present on 15 November. An exceptional arrival took place in 1978 and extended much further south than usual: Fair Isle peaked at 30 on 15 October, 32+ were reported on the Aberdeen coast the same day and a record nine on the Isle of May, and numbers were high as far south as Berwick. Many of the birds involved in this influx were in an exhausted or dying condition. Short-eared Owls are widely distributed across northern Eurasia, breeding well beyond the Arctic Circle in Scandinavia and also occurring in Iceland. Those in the northern part of the range are migratory, the extent of their movement – in terms of both numbers and distance – presumably being governed by the availability of prey. Birds ringed as chicks in Iceland and Finland have been recovered in Scotland, and immigrants from Norway and Sweden probably also occur here.

During the autumn and winter months roosts are established in some areas, the largest reported to date being in Mull, where 20 were seen together on 1 January 1981. Native birds return early to the breeding grounds, with display reported in Orkney before the end of February but inland sites, for example in Speyside, sometimes not being occupied until early April. Spring passage through Fair Isle and Shetland, where a few occasionally winter, involves only small numbers and takes place mainly in May / June.

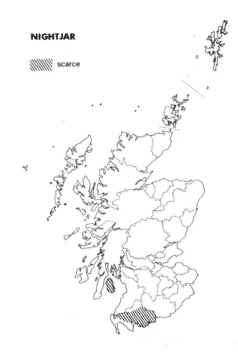

NIGHTJAR

░░░░ scarce

Tengmalm's Owl *Aegolius funereus*

Vagrant (Fenno-Scandia / central Europe) – four records since 1950.

B&R give four 19th century records of this small owl and four in the period 1900–1915; half of these old records are from the Scottish mainland. Since 1950 Tengmalm's Owl has been reported only from Orkney: in 1959 at Firth from 26 December to 1 January; in 1961 at Stromness on 1 May; and in 1980 two at Finstown, on 13–20 October and 18 November (both trapped). This species is strictly nocturnal and normally sedentary but occasionally makes dispersive or eruptive movements, which account for its occurrences in this country.

Nightjar *Caprimulgus europaeus*

Scarce and local migratory breeder, probably now regular only in Arran and Galloway. Numbers have decreased steadily this century and the breeding range has contracted southwards.

At one time the Nightjar was widespread and fairly common in open woodlands throughout much of Scotland; B&R described it as still breeding locally north to Caithness, but

noted that in Strathspey numbers were declining in the 1930s. By the mid 1960s it was very scarce in the north and west Highlands (Parslow), and in 1968–72 proved or probable breeding was recorded on the mainland only in southeast Sutherland, Easter Ross, Nairn, Moray, Perth, Stirling, west Inverness, Argyll, Ayr, Kirkcudbright, Wigtown and Dumfries, and on Arran, Bute and Mull. In 1981 a national survey produced a total of only 35 singing males, some of which were probably passage birds, at 25 sites in Scotland (Gribble 1983). The principal concentrations were in Kirkcudbright and Arran (Table a). Not all areas were well searched nor all Nightjars reported, however. For example no information was obtained for the Kilcreggan peninsula, which had held up to seven singing males in 1977; a dead juvenile was found in Dumfries, where no churring birds had been recorded in the survey; and on Bute two males were heard (SBR). It is possible that the population lies in the range 50–100 pairs (F. C. Gribble); the British total in 1981 was c1,800 singing males, nearly half of which were in southeast England. Although it is stated in the

(a) Numbers and distribution of singing Nightjars in 1981 (Data from F. C. Gribble, A. D. Watson & SBR)

County	No. of singing males located	No. of 10 km sq occupied	Males/10 km sq 1–5	6–10
Kirkcudbright	22	9	8	1
Wigtown	1	1	1	0
Arran	8	2	1	1
S. Argyll	2	2	2	0
Lanark	1	1	1	0
Sutherland	1	1	1	0
Total	35	16	14	2

Atlas that Galloway, Ayr, Bute and south Argyll were all recolonised between 1957–58 and 1968–72 it is much more likely that information was simply not available for 1957–58; this is certainly true of Galloway (F. C. Gribble). The Atlas records for Ayr probably give an exaggerated picture of the situation in 1968–72; by 1963 Nightjars were being regularly recorded only in the Glen App area (Hogg 1983) and since 1972 only passage birds have been reported there.

Nightjars are crepuscular and nocturnal and are consequently difficult to census, although the fact that they sing from the time of their arrival in May until early August means that an extended fieldwork period is available. Study has shown that the optimum times for detection of the churring song are the hour starting 30 minutes before sunset and the hour starting 30 minutes before sunrise (Cadbury 1981). The song carries well and can be heard over a considerable distance under calm conditions, but both adults and young are so well camouflaged that nests and broods are very difficult to find. Most of those recorded in the 1981 survey were at altitudes below 125 m asl and in young conifer plantations, some of which had extensive, relatively open, bracken-covered areas (Table b). Strong growth of rhododendron in areas on Arran which were formerly good for Nightjars may be responsible for decreasing numbers there, because of the dense ground-cover it forms and the paucity of insect life associated with it.

(b) Habitats occupied by Nightjars in 1981 – numbers of occupied sites (Data from Gribble 1983)

Even-aged conifer plantation	10
Bracken, extensive open areas	9
Upland heather/moor	3
Upland grassland	3
Wet heath/heath bog	3
Broad-leaved woodland	2
Mixed woodland	2
Lowland dry heath	2
Raised bog/mosses	1

In his discussion of the continuing decline of the Nightjar population in Britain as a whole, Gribble (1983) concludes that the main cause is climatic change, resulting in poorer breeding success; both lower spring temperatures with consequent later arrival, and wetter summers, with reduced availability of night-flying insects, especially moths, may be involved. The areas still occupied in Scotland are all at low altitude and in the drier coastal fringe; apparently-suitable habitat in the new forests of Dumfries and the Borders has not been colonised, possibly because these areas are higher and wetter. Habitat loss and disturbance are contributory factors to the decline in England but are less likely to be significant in Scotland. At the local level, population fluctuations may result from afforestation, with an increase during the early years after planting followed by a decline as the plantations mature. Provided that sequential felling and planting of not too large areas takes place in the Nightjar's Galloway stronghold there is reason to hope that the population there may be maintained. On Arran, where the other

principal concentration occurs, active management to control the spread of rhododendron might be worth attempting in the most favoured areas.

The Nightjar is widely distributed across Eurasia, breeding from southeast Norway and southern Sweden and Finland to northwest Africa. It occurs as a scarce and irregular migrant on the Northern Isles, the Inner and Outer Hebrides and the Isle of May, and is recorded occasionally in most mainland counties. Arrival of breeding birds takes place in May (3rd is the earliest recent record), migrants may appear at any time from May to October, and the latest recent autumn report is 29 October. During the mild winter of 1973/74 a Nightjar was on Mull on 15 January; this was the first acceptable British winter record. The species normally winters in Africa south of the Sahara.

Common Nighthawk *Chordeiles minor*

Vagrant (N America) – one record, 1978.

The only Scottish record is of one trapped near Kirkwall, Orkney, on 12 September 1978 (SB 11:85). The Common Nighthawk, a widely distributed North American species which winters in South America, had occurred in Britain on only ten other occasions by the end of 1983, always in autumn and most often in Scilly.

Needle-tailed Swift *Hirundapus caudacutus*

Vagrant (W Siberia) – two records, 1983 and 1984.

As the earlier report of a Needle-tailed Swift on Fair Isle in August 1931 (SN 1932:38 and B&R) is no longer accepted (BB 26:27), the bird seen on South Ronaldsay, Orkney, on 11–12 June 1983 is the first record of the species in Scotland (BB 77:539). Astonishingly, a second appeared less than 12 months later, and spent several weeks around Hillwell, Shetland, in late May and early June 1984.

Swift *Apus apus*

Migratory breeder, abundant in central Scotland, widespread but less common in the Borders, southwest, northeast and Moray Basin, very local in the west from the Clyde northwards and in the Highlands, and absent from most of the islands. Also occurs as a passage visitor.

Swifts, which are totally dependent upon flying insects, are among the shortest-staying of our summer visitors. Few reach Scotland before the end of April, the main arrival is during the first half of May, and most breeding birds have gone by mid August, although there are occasional passage stragglers in October and even one recent record as late as 8 November. The timing of the main influx is much affected by spring weather conditions, varying from

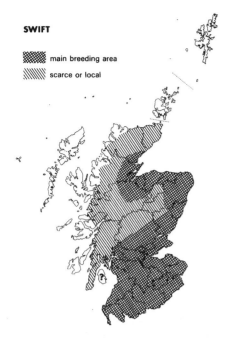

SWIFT

▓ main breeding area

▨ scarce or local

the first few days of May to almost the middle of the month; arrival in northern Scotland is often not until late May and numbers there are probably more variable than further south. Passage through Fair Isle and Shetland is most marked from the last week of May until about the third week of June. Numbers seen on the islands off the west coast are small, but there are a few records from St Kilda.

The great majority of Swifts nest in buildings, and densities are greatest in the larger towns, where 'screaming parties' are a familiar sight – and sound – in summer. Nesting in old woodpecker-holes has been recorded in Strathspey (R. Broad), and small numbers breed in crevices on cliffs. B&R give old records of cliff-nesting on Eigg and also record breeding on Islay and Arran, where nesting no longer occurs. About six pairs were nesting on the Mull of Kintyre in 1981 but the record of proved breeding on North Rona in the Atlas seems likely to be an error. The most northerly regular breeding site is at Thurso; on the west coast there are few north of Fort William. Virtually nothing is known about the overall size of the population.

During fine weather Swifts are often seen hawking over mountain summits. According to B&R they may have nested at over 1,000 m asl on Lochnagar, where they were seen 'going in and out of crevices in the crags' – but this conclusion should perhaps be regarded with some suspicion. Very large gatherings are frequently reported over both lochs and moorland areas during June and July, eg 2,000+ over Loch Leven, 2,200 over the Loch of Skene, and 1,500 over the Lowther Hills. Large-scale movements are also sometimes noted near the coast, as on 19 July 1979 when 4,500 were counted in three hours at East Fenton in East Lothian.

Once autumn passage has begun the numbers moving as a group tend to be much smaller; peak daily counts at Fair Isle and the Isle of May occur in July and seldom reach 150. Most of those passing through the Northern Isles, and

perhaps also the smaller numbers noted in the Hebrides, probably come from Scandinavia, where Swifts breed well beyond the Arctic Circle in Sweden and Finland. A migrant ringed on Fair Isle at the end of April and recovered six weeks later in Sweden was presumably then in its breeding area. Recoveries at or near the ringing site up to five years later indicate the fidelity of most Swifts to their natal area. This species, which is widely distributed in Eurasia and northwest Africa, winters in southern Africa.

Alpine Swift *Apus melba*

Vagrant (S Europe) – recorded in 13 years, 1950–84.

Since 1950 this supreme aeronaut has been reported at localities scattered from the Borders to Shetland and the Outer Hebrides. Most occurrences have been in May/June (those for Britain as a whole peak in September/October) but there are records for all months from April to November. Apart from a bird at Hawick in August 1951, all the records are for coastal areas, and more than half of them from the Northern Isles. The Alpine Swift has been recorded in Shetland, Fair Isle, Orkney, St Kilda, Benbecula, Banff, Aberdeen/Kincardine, East Lothian, Midlothian, Roxburgh, Ayr and mainland Argyll. The three definite older records given by B&R include 'a small party' in Wigtown in 1923, and a single bird in Ayr, the first recorded in Scotland, in 1892. All the recent reports have been of single birds and the highest total in any one year has been only three, in 1972. Alpine Swifts breed in mountainous or cliff-bound areas from Iberia, southern Europe and northwest Africa eastwards, and winter in the Tropics.

Kingfisher *Alcedo atthis*   [1]

Resident, breeding locally in small numbers from lowland Perth southwards. Sporadic breeding occurs further north and there are occasional sightings throughout the country. Numbers fluctuate in response to winter severity.

The Kingfisher favours slow-flowing water and is consequently absent from the Highlands and many of the islands.

KINGFISHER

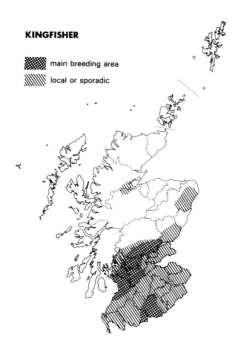

▨ main breeding area

▧ local or sporadic

Breeding occurs most regularly, and numbers are largest, on the west between the Clyde and the Solway. Away from the southwest the population thins out and breeding is often sporadic, but the species has been recorded at least once since 1968 in every mainland county. In 1970 a pair bred in Easter Ross but the unfledged young were destroyed by a flood; this is the most northerly nesting record. Kingfishers occur only rarely in the Northern Isles and Outer Hebrides but more often in the Inner Hebrides (Islay & Mull only) and Clyde Islands. The frequency of recent records from Islay and the presence of three pairs on Arran in 1977 suggest that breeding may take place at least sporadically on these islands. A brood was found on Bute in 1976, but there have been few other sightings on the island (Gibson *et al* 1980). There is one record for the Isle of May, in September 1969, but none for Fair Isle.

Kingfishers suffer heavy mortality in severe winters, but their ability to raise more than one brood in a season helps them to recover from such natural disasters. The rearing of three broods by pairs in both Lanark and Kirkcudbright was recorded in 1973; in the latter case a total of 18 young fledged. Nests are often flooded out, however, and the mortality rate among newly independent young is very high (Morgan & Glue 1977). After the very severe winter of 1962/63 there was a marked decline in numbers, which then increased fairly steadily until again affected by hard weather in 1978/79. The population was still low in 1980 and suffered badly in the prolonged frosts of 1981/82, when WBS figures showed a 64% decline in England – the largest annual decrease yet recorded for any species monitored by the scheme (Taylor & Marchant 1983). Unfortunately too few regular surveys of Kingfisher haunts are carried out in Scotland to permit assessment of the extent to which birds here may

be more seriously and lastingly affected by hard winters than those further south.

Being at the end of a food chain, the Kingfisher is very vulnerable to pollution, through the accumulation of toxic chemicals in its body. It may also suffer from reduction of its food supply due to other forms of water contamination, for example fine silt causing fish deaths. It seems likely that pollution was responsible for the decline which apparently occurred in some districts of central Scotland during the late 1950s and 1960s, when many waters became heavily polluted with industrial and/or agricultural effluents. Little is known about past numbers in most areas, but studies carried out on the White Cart in Renfrew in 1933 (Brown 1934) revealed 21 nests between Waterfoot and the outskirts of Paisley, a distance of only 22.5 km of which some 3 km was described as 'in Glasgow' and unsuitable for Kingfishers. The following year 33 nests (not 33 breeding pairs as stated in SB 8:33) were found between Waterfoot and Hawkhead (Clancey 1935), several of these being second clutches and one a third. Action by the Clyde River Purification Board during the 1970s led to a progressive improvement in the cleanliness of the Clyde and its tributaries, with the result that fish now flourish where at one time there were none, and Kingfishers can quite often be seen at the Falls of Clyde Reserve, New Lanark. In 1984 a minimum of seven pairs, and probably at least ten pairs, held territory on the White Cart Water (I. P Gibson *et al*), while an intensive survey of the River Kelvin and its tributaries, by A. D. Wood, D. H. McEwen and A. J. Young, located ten territories in 25 km, an average breeding density of four pairs/10 km.

The area where this species has bred most regularly in the last 30 years is the southwest, which has suffered little, if at all, from industrial pollution. Population fluctuations there are presumably almost entirely attributable to variations in winter mortality and breeding success, and this should eventually become the case in other areas too, as waters are gradually cleaned up. Unfortunately, no records of breeding numbers are available for the southwest.

Kingfishers are usually very sedentary, and to date no Scottish-ringed bird has been recovered more than 31 km from its nest site. There is some dispersal in autumn, however, and relatively long-distance movements have been recorded in some continental populations (Morgan & Glue 1977). Birds from the northern limits of the breeding range, which extends from the Mediterranean to southern Sweden and Finland, move south in winter to areas in which waters remain unfrozen. Birds of continental origin are perhaps more likely to be involved in the occasional records from the Northern Isles, Outer Hebrides, and northern Highlands, than are members of the resident population.

Problems occasionally arise locally when Kingfishers gain access to tanks or ponds on fish-farms and take advantage of the easy pickings afforded by the concentrations of small fish these contain.

Bee-eater *Merops apiaster* [1]

Vagrant (S Europe) – recorded in 13 years, 1950–83. Has attempted to breed.

Both its beautiful colouring and its habit of perching on wires make the Bee-eater likely to be seen and reported whenever it occurs in populated areas. The records therefore probably present a fairly true reflection of the irregularity of its occurrence in Scotland. Since 1950 Bee-eaters have been recorded most often in May/June but also occasionally in July/August and once in October. More than three-quarters of the records are from the Northern Isles; most refer to single birds but there were three together, a male and two females, on Fair Isle in July 1979, and two on Islay in June 1981. Some Bee-eaters remain in one area for several weeks, eg one on Fair Isle from 30 June to 18 August 1971, while others move on rapidly. It is consequently not possible to be sure whether successive reports – from, for example, Orkney, Fair Isle and mainland Shetland – refer to the same individual or to different birds. The absolute maximum recorded in any one year is only five.

The post-1950 records are from Shetland, Fair Isle, Orkney, Sutherland, Islay and Ayr. B&R note older records for Caithness, Inverness, Moray, Aberdeen, Renfrew and Wigtown as well as the remarkable attempt at breeding in 1920, when a pair settled at a sand-bank by the River Esk in Musselburgh, Midlothian. Unfortunately the female was injured, but before dying she laid an almost fully-developed egg. The nearest regular breeding colonies of Bee-eaters are in Iberia and southern France, but the species has bred sporadically in England, where it occurs annually in spring.

Roller *Coracias garrulus*

Vagrant (S & E Europe) – recorded in 18 years, 1950–83.

This conspicuous Jay-like bird has been recorded more often in recent years than formerly but is still only an irregular visitor. All post-1950 records of Rollers have involved single birds, but there were three separate reports in 1969 and two each in 1952, 1973, 1975, 1979 and 1983. The records cover the period May to October, with peaks in June and September; spring occurrences are about twice as numerous as autumn ones. The geographical scatter is wide: from Kirkcudbright and Berwick to Sutherland and Shetland, with more than half the post-1950 reports coming from the north. Individuals often remain in the same area for at least a week and in 1958 one spent nearly two months in Orkney. Since the earliest known record, given by B&R as 1700, the Roller has been reported in all mainland counties other than Nairn, Clackmannan, Stirling, Dunbarton, Kinross, West Lothian and Wigtown, and on Islay but not on the Clyde Islands or the Outer Hebrides (an old record from St Kilda is considered unreliable).

[Abyssinian Roller *Coracias abyssinicus*]

B&R quote an old and unaccepted report of two shot near Glasgow in 1857.

Hoopoe *Upupa epops* [1]

Vagrant (Europe) – occurs annually between April and October, in very small numbers.

The Hoopoe is such an unusual-looking bird that it attracts the attention of even the most casual of observers, so the records probably represent a fair indication of the frequency of its occurrence. On its breeding grounds, which extend over most of Europe south of the Baltic (and further south and east), the Hoopoe favours woodland edges but also frequents more open and arid areas. In Scotland it occasionally turns up in odd places; my own first sight of one was on a derelict army camp-site in Shetland, and in the early 1970s Hoopoes were seen in two successive years at Fealar Lodge, deep in the Grampians near the Perth/Inverness county boundary. B&R noted records for all mainland counties except Selkirk, Roxburgh, Kinross and Nairn. There have since been reports from the first three, and there are also recent records from the Outer Hebrides (St Kilda, Lewis & North Uist), the Inner Hebrides (Skye, Eigg, Islay & Iona), the Clyde Islands (Arran) and the Isle of May. Reports are rather more frequent in Shetland than elsewhere, but there is otherwise no particular pattern to the occurrences.

About 50% of reports in the last 25 years have been between mid April (5th is the earliest) and the end of May, and nearly 40% in September/October. This is in contrast to B&R's statement that 'by far the largest numbers occur in autumn'. A few Hoopoes appear in June/August and November, and one was in Perth from 13 to 20 December 1978. Most records refer to single birds, but up to three have been seen together. Individuals appear often to move around quite rapidly, making it difficult to be certain how many birds are present in any season, but annual totals in 1971–83 ranged from only two to about 21 (1983) and were usually between eight and twelve.

Overflying spring migrants are often numerous further south, and in warm summers some of these occasionally remain to breed in England. The Hoopoes reaching Scotland in autumn have presumably been wind-drifted westwards while moving from the northern parts of the breeding range to their African wintering grounds.

Wryneck *Jynx torquilla* [1]

Scarce but annual passage visitor, mainly on the east coast and in the north. Very local, and probably sporadic, breeder.

The changing fortunes of the Wryneck in different parts of Britain present an intriguing picture. In Scotland it was until recently known only as an irregular passage migrant; breeding was first recorded in 1969 and has probably occurred in the Highlands in most subsequent years, although not always confirmed. In England it was widely distributed as a breeding bird last century but during the early 1900s its breeding range contracted steadily and by the mid 1950s there were few Wrynecks nesting outwith southeast England. The decline was even more rapid in the 1960s, by the end of which regular nesting was probably confined to Kent and possibly Surrey (Parslow 1973). Since 1972 not more than five occupied sites have been reported in England, and in 1980 there was only one.

While the species was apparently approaching extinction south of the Border, increasing numbers of singing Wrynecks were being reported in the Highlands – in Strathfarrar (1951), Speyside (1952, 1961, 1962, 1965, 1968) and Drumnadrochit (1952, 1965) – but it was not until 1969 that breeding was eventually confirmed on Speyside. That season there was a large influx of migrant Wrynecks in early May, and calling birds were recorded during the summer in Argyll, Kincardine, and Easter Ross as well as Inverness. Burton *et al* (1970) describe the nesting of at least three pairs on Speyside in 1969. They suggest that these birds were of Scandinavian origin, on the grounds that unusual numbers of migrants were recorded that spring and that breeding occurred in, and most other summering records were from, semi-natural Scots pine forest similar to that widely used by Wrynecks in Scandinavia. Such forests are a rich source of the ants upon which the species largely feeds.

During the 1970s breeding was confirmed several times in Inverness (Speyside and elsewhere) and once in Aberdeen/Kincardine; a pair occupied a nest-box in Easter Ross in 1974, but no eggs were laid. Summering birds have also been recorded since 1970 in Sutherland and Perth. The number of sites known to be occupied in any one year has ranged from one to a maximum of 18 (in 1978). Only one pair was proved to breed in 1980 – the Aberdeen/Kincardine birds, which used a nest-box and were believed to fledge seven young – but at least ten sites were occupied. In 1982 birds were present at five sites on Speyside and one in Perth. One nest-hole has been used in three consecutive years, suggesting that at least some birds are returning deliberately and are not just lost migrants (R. H. Dennis). The Scottish breeding population is, however, so small that its future will almost certainly depend upon further periodic influxes of migrants.

The Wrynecks passing through Scotland, and presumed to be responsible for the recent colonisation, are likely to be from the northern part of the Eurasian breeding range, which extends from north of the Arctic Circle in Fenno-Scandia south to the Mediterranean. Most are migratory, wintering south to tropical Africa. Spring passage generally starts about the third week in April (1st is the earliest record), though sometimes not until early May, and is usually over by mid June. The return movement takes place between mid August and the end of September, with occasional stragglers to 20 October. Numbers vary greatly between years; during the massive arrival of early May 1969 there were 45 Wrynecks on Fair Isle at one time, many more than the total recorded in Scotland in most years. The great majority of records are from the Northern Isles, east coast headlands, and the Isle of May – all places with comparatively little cover, where even a well-camouflaged and skulking bird can easily be seen. The facts that there are occasional reports from as far west as St Kilda, and that those recorded away from the east coast are often found dead, suggest that passage birds occur inland more frequently than the records indicate, but remain undetected. Observations on the Isle of May (Langslow 1977) have shown that migrant Wrynecks may gain weight with remarkable speed during a stay of only a few days, indicating that they have probably used up a considerable proportion of their fat reserves in transit. Many of those that reached Fair Isle in the fall of May 1969 had gone beyond the point of potential recovery and were found dead on the isle.

[Northern Flicker *Colaptes auratus*]

Vagrant (N America) (Category D)

The corpse of a Northern (or Yellow-shafted) Flicker was found in Caithness in July 1981, having apparently died on board ship and been brought ashore in or on a goods container (BB 75:531). The species was already known to make use of 'assisted passage' across the Atlantic; in 1962 ten landed on the RMS Mauretania and one flew ashore on reaching Co. Cork.

Green Woodpecker *Picus viridis*

Recent coloniser, now breeding in lowland woodlands north to the Moray Basin. Its gradual spread is continuing despite periodic set-backs due to severe winters.

Although the Green Woodpecker was known as an occasional visitor from the late 19th century, and was recorded from widely scattered areas in the 1920s and 1930s, it was not until the 1940s that it was reported with any frequency. The first proved breeding was in Selkirk in 1951 – just in time for B&R to include it in an Appendix to *The Birds*

The spread of the Green Woodpecker and the estimated breeding population in the early 1980s (Data from G. Waterston's enquiry (lodged in Waterston Library), SBR and local recorders)

County	First record	First proved breeding	Est. no of pairs early 1980s
Wigtown	1968	1975	not >5
Kirkcudbright	1955	early 1960s	10–20
Dumfries	1946	1953	10–20
Berwick	1948	1951	10–20
Roxburgh	1939	1952	20–25
Selkirk	1950	1951	5–10
Peebles	1953	1960s	5–10
Ayr	1931	late 1960s	not >10
Renfrew	1943	late 1960s	c10
Lanark	1938	1969	<5
E. Lothian	1941	prob. 1956	5–10
Midlothian	1951	1955	5–10
W. Lothian	1934	?	c5
Fife	c1925	1970	10–20
Kinross	1968	1973	not >5
Clackmannan	1951	1965	10–20
Stirling	1930s	?	10–20
Dunbarton	?	1972	5–10
Argyll	1969	?	<5
Perth	1920s	1967	20+
Angus	1968	1972	5–10
Kincardine	?	1973	<5
Aberdeen	1950	1981	20+ *
Banff	1982	prob. 1982	?
Moray	1977	?	1–2
Inverness	1931	1981	10–15
Ross & Crom.	1930	?	?
Islands			
Bute	1970		
Islay	1978		
Mull	1979		
Coll	1982		

*by the end of 1984 breeding had been confirmed in three 10 km squares and 'probable' breeding recorded in a further 16 squares (NE Scotland Atlas/S. Buckland)

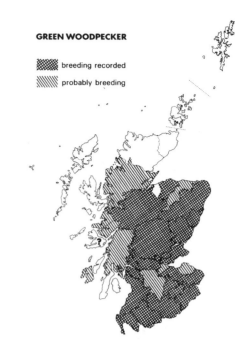

GREEN WOODPECKER

▨ breeding recorded

▨ probably breeding

of Scotland. Within five years it had been reported, and was probably breeding, in all four Border counties, Dumfries and the Lothians; many of these early records were collected by George Waterston, who organised a Green Woodpecker Enquiry in the late 1950s. The range has continued to spread both westwards, where Mull was reached in 1979, Cairnryan in 1980, and Coll in 1982, and northwards, with records from Easter and Wester Ross since 1976 (Table). In some cases this spread has involved the crossing of extensive tracts of inhospitable terrain; movement into Speyside must surely have been from north Perth over Drumochter, and into Deeside (where the first recorded breeding was at Banchory) possibly from the Angus glens rather than via the coast. The attempt at colonisation of Mull has apparently failed (R. Coomber) and there have not yet been any records from Nairn – where birds seem likely to appear soon if not already present – nor from Sutherland, Caithness or the Outer Hebrides. The only reports from Orkney and Shetland are pre-1950; they might possibly have involved birds from further north in the species' range, which extends from southern Scandinavia south to the Mediterranean and Iberia, rather than from the south.

Green Woodpeckers occur most often in areas with both mature deciduous trees and open ground. They appear to roam over quite large areas, often feeding on the ground and being especially partial to ants; estate 'policy parks' provide just the right combination of habitats. Their 'yaffling' calls proclaim their presence in an area but proof of breeding is generally difficult to obtain, and long intervals often elapse between the first report and confirmation that nesting has occurred. It is also difficult to be certain how many pairs are present, but there is little doubt that numbers are considerably reduced by severe winters. Recovery can be rapid, however, as CBC figures for England show that an increase of around 25% per annum is possible in the years following a population crash. From the estimates provided in 1983 by local recorders and others, it seems likely that the population was then somewhere in the range 200–350 pairs; a decrease would be expected in 1984 following the prolonged heavy snowfall in January that year.

Great Spotted Woodpecker *Dendrocopos major*

Resident, breeding in every mainland county and on a few of the well-wooded inshore islands. Also occurs almost annually as a passage and winter visitor, in very variable numbers.

The Great Spotted Woodpecker has undergone remarkable changes in status over the last 200 years. From being widespread in the 18th century, it decreased as forests were cleared until by the middle of the 19th century it was possibly extinct. By the late 1800s it had recolonised much of lowland and central Scotland, and during the first half of this century

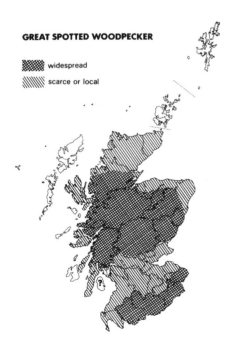

it spread north to Sutherland and west to Ardnamurchan and Mull (B&R). Though the population has fluctuated since then, the only real changes in distribution have been on the islands. Skye and Jura have been colonised since 1953 (exact dates unknown) and small numbers now breed there (Reed *et al* 1983). Single birds have been seen several times in summer on Islay and Rhum but there has not yet been any evidence of breeding. Nesting has occurred on Arran (Gibson 1981) but has not been recorded since 1972, although birds are still occasionally seen, while on Bute there has been a marked decrease since the mid 1940s (Gibson *et al* 1980). On Mull the species is more widespread than the Atlas records suggest (R. F. Coomber).

The Great Spotted Woodpecker is a strictly woodland species, frequenting both deciduous trees and conifers. On the mainland it is currently present wherever there are adequate woodlands to provide invertebrate food throughout the year, and occurs locally even in Caithness and northern Sutherland. Its distribution in Wigtown is wider than shown in the Atlas and it is believed currently to be breeding at several sites in the county (A. D. Watson). The spread of afforestation in many areas may be expected to result in future distributional changes, although colonisation may well be restricted by lack of suitable dead trees for nesting; however, the after-effects of the outbreak of Dutch elm disase in the late 1970s/early 1980s may help to make good such deficiencies in some districts. Little information on breeding densities is available, and what there is relates mostly to relatively small areas of species-rich woodland. Numbers usually decline noticeably following a severe winter.

Native birds belong to the race *D.m. anglicus*, and are sedentary. The larger northern race *D.m. major*, which breeds in Scandinavia and eastern Europe, intergrading with other races further south, apparently migrates in small numbers fairly regularly and occasionally makes major eruptive movements in late summer and autumn. Spring occurrences of the northern race are rare; there have been only a few records from the Northern Isles, the most recent a male on Fair Isle in April/May 1976, and one in Orkney on 30 March 1980. In most seasons a few migrant Great Spotted Woodpeckers arrive in Shetland and Orkney between August and November, and even smaller numbers are recorded further south. When an irruption occurs arrival starts earlier and records are much more widespread, with stragglers reaching the Outer Hebrides. In the invasion of 1974 the first bird appeared in Shetland in mid June, but both that year and in 1968 the main influx occurred from early September onwards. A hundred were on Yell on 30 September 1968 and there was a steady trickle of birds through Fair Isle from 7 September to mid October, nearly all of them immatures. The 1974 movement was less concentrated, with peak counts of only 16 in Shetland, but records there spread from 15 June to 14 November, and there were also a number of reports from Orkney and Caithness. After such invasions a few individuals sometimes overwinter in the Northern Isles. Further south movement is seldom sufficiently large-scale to be obvious, but birds of the northern race have been trapped occasionally on the Isle of May and seen at Fife Ness, where on 24 July 1972 one was watched coming in off the sea.

Lesser Spotted Woodpecker *Dendrocopos minor*

Status uncertain – probably vagrant; may have bred.

Although there are several earlier unconfirmed reports, there was no accepted record of the Lesser Spotted Woodpecker in Scotland until 1968. Two birds were seen in September that year at a locality on the Stirling/Perth county boundary, where their presence had been suspected during the previous two years (SB 6:210–212). They were not seen in 1969 but in early January 1970 three were watched in the same area; there are no subsequent reports from this site. The Atlas shows 'presence' in Kinross and central Perth but reports of these occurrences have not been published elsewhere and must be open to doubt. More recently a Lesser Spotted Woodpecker was both seen and heard on Speyside by J. P. Grant, who is familiar with the species, on 19 May and 4 June 1980 (SBR 1980).

The Lesser Spotted Woodpecker is quite widely distributed in England and Wales, becoming scarcer north of the Humber/Wash, but is normally very sedentary. However, in the northern part of its extensive Eurasian range it is a partial migrant, and those occurring in Scotland may originate from the Scandinavian population (SB 6:212). Broadleaf or mixed forest is the species' preferred habitat but it also occurs in patches of deciduous woodland surrounded by conifers.

[Three-toed Woodpecker *Picoides tridactylus*]

B&R quote an unsubstantiated record of one shot in 1809.

Calandra Lark *Melanocorypha calandra*

Vagrant (Iberia / Mediterranean) – one record, 1978.

The only record is from Fair Isle, on 28 April 1978 (BB72:530) – the second British occurrence. A bird of lowland steppes, the Calandra Lark breeds from Iberia and Morocco eastwards through the Mediterranean to Afghanistan and is normally sedentary in the western part of its range. A full description is given in Sharrock & Grant (1982).

Bimaculated Lark *Melanocorypha bimaculata*

Vagrant (SW Asia) – one record, 1976.

A Bimaculated Lark on Fair Isle on 8 June 1976 (BB 72:462–463) was the third British occurrence of the species, which breeds in dry heathlands in southwest Asia. A full description, including a comparison with the somewhat similar Calandra Lark, is given in Sharrock & Grant (1982).

[White-winged Lark *Melanocorypha leucoptera*]

The record of one seen in Caithness on 8 June 1958 (BB 51:320) was not accepted.

Short-toed Lark *Calandrella brachydactyla*

Vagrant (Europe) – annual since 1968, in very small numbers.

The earliest records of the Short-toed Lark date from the beginning of this century, when the reddish western form was first reported in 1904 on the Flannan Isles, and the greyish eastern form in 1907 on Fair Isle (B&R). Since 1968 the species has occurred annually on Fair Isle, whence the bulk of the records come, and almost as regularly elsewhere in Shetland. Outwith these areas there are records only from St Kilda (1957), Midlothian (1973), the Isle of May (1977) and Orkney (1983). Numbers are never large and the maximum recorded in any year is only ten. Nearly 60% of occurrences are in autumn, mostly in September/October but with extreme dates of 9 August and 19 November. Most spring sightings are in May, but there have been records as early as 20 April and as late as 5 June. In autumn the majority are of the eastern form; all racially identified spring birds have been of the western form. The Short-toed Lark breeds in open steppes and semi-desert areas, from Southern France, Iberia and the Mediterranean Basin east through southern Russia to Manchuria. Birds from the western part of the range winter in Africa south of the Sahara.

Crested Lark *Galerida cristata*

Vagrant (Europe) – one record, 1952.

The Crested Lark frequents flat and sparsely vegetated country and is widely distributed throughout Europe but is very sedentary and seldom crosses the England Channel. It has been recorded about 20 times in England but only once in Scotland: on Fair Isle on 2 November 1952 (BB 46:211).

Woodlark *Lullula arborea* [1]

Vagrant (Europe, including England) – occurs less than annually and in very small numbers.

This species was regarded by B&R as a fairly regular autumn visitor to Fair Isle but uncommon elsewhere. With only 12 records in 1971–83 it must clearly now be classed as no more than a vagrant. The Woodlark breeds in England but its range there has shown a marked southward contraction since the early 1950s (Parslow). In Europe it is widely distributed from southern Scandinavia southward; only those in the northern part of the range are migratory, moving no further than the Mediterranean area in winter. There is no evidence of the origin of the Woodlarks occurring in Scotland, but it seems probable that they come from the Continent rather than England. Between 1968 and 1983 the maximum recorded in any year has been eight (in 1969) and there were none at all in 1970, 1974–77 and 1981. Most records are from Fair Isle, but there have also been reports in this period from the Isle of May, Whalsay, Out Skerries, Rhum, Angus and Berwick. Spring occurrences are slightly more frequent than autumn ones and there is a greater time-spread at that season, with records from 20 March to 27 May; autumn birds nearly all appear in a five-week period from early October to mid November. In 1968/69 a Woodlark overwintered on Fair Isle and was heard singing from 23 January.

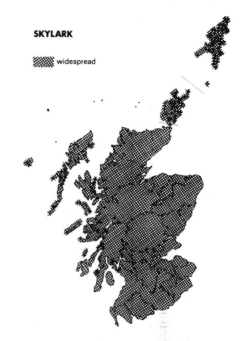

SKYLARK

▨ widespread

Skylark *Alauda arvensis*

The most widely distributed British species, breeding in open country from sea-level to over 1,000 m asl, throughout the mainland and on most islands. Native birds move towards the coast in autumn, when large numbers of immigrants arrive; some emigrate and 'weather movements' are frequent in winter.

The Skylark's basic habitat requirement is an area of short vegetation in which to feed; it is catholic in its choice of vegetation type and occurs on rough grassland, arable fields, saltmarsh, machair, heather moorland and mountain summits. Undisturbed ground is necessary for successful breeding, and those nesting in arable fields are at risk from operations such as harrowing, rolling and spraying. Breeding densities vary widely according to habitat: up to 90 pairs/km² have been recorded in some areas (Atlas), and the farmland average for Britain as a whole was c18 pairs/km² in 1972. In a census of c1,700 ha of mainly cereal fields in the Lothians in 1984 Stan da Prato (1985) found Skylarks at an overall density of only 11 pairs/km², whereas a few kilometres away on coastal grassland and saltmarsh at Aberlady Bay densities equivalent to 53–71 territories/km² were recorded in 1981–83 (P. R. Gordon). Few figures are available for sparsely populated moorland areas and mountain tops.

The Skylark is among the first species to abandon afforested land, so local changes in distribution and abundance must clearly be taking place in many areas. There is no evidence, however, of any long-term change and CBC results show that only the most severe of winters result in significant mortality. The CBC index remained remarkably stable from the mid 1960s until 1981, but dropped by 35% after the prolonged frost of 1981/82.

This species moves away from many of its breeding areas in winter but there is not yet sufficient ringing evidence to show whether native birds are mainly migratory or simply dispersive. Skylarks desert their Shetland and Fair Isle breeding grounds in autumn, while those nesting on the hills move to lower ground and nearer the coast. In mid winter the largest concentrations are in the Hebrides and Central Lowlands, with fair numbers also present in Orkney and near the east coast (Winter Atlas). Visible movement towards Ireland from Crammag Head (Wigtown) in late November suggests that some Scottish birds may winter there, while the November recovery in Portugal of a Skylark ringed as a chick in West Lothian shows that some travel much further south. Spells of bad weather produce very obvious movements, usually to the south and west and often involving thousands of birds. Recent records of such movements include 8,300+ in two hours in Aberdeen on 24 January 1976, 3,000+ in Wigtown on 2 February 1979, and c3,000 in Ayr after a snowstorm on 21 March 1980 – all were moving south.

On the Continent the Skylark is widely distributed to beyond the Arctic Circle; most of those breeding north of the Baltic are migratory, wintering further south in the range. Immigrants reach Scotland mainly between mid September and late November, and return northwards between late February and April. Numbers passing through the Northern Isles are rather variable and seldom very large; an autumn peak of 1,650 on Fair Isle in early October 1980 was unusually high. A few sometimes winter in Shetland. Movements at the Isle of May, where counts only occasionally exceed 100, are as likely to involve native birds as immigrants.

Shore Lark *Eremophila alpestris*   [1]

Passage and winter visitor, annual in variable but generally small numbers. Has bred.

The Shore Lark was first recorded in January 1859 at Tynemouth, East Lothian, and the Tyninghame shore is still the locality at which it occurs most regularly. Most reports are from the east coast, from Shetland to Berwick (no record yet for Kincardine). On the west there have been only a few scattered records – from Dumfries, Kirkcudbright, Wigtown, Ayr, west Inverness, Sutherland, Islay, Eigg, Skye, South Uist and North Rona. The main arrival usually takes place in October/November and some birds often remain throughout the winter. There have been sightings in every month and some spring passage occurs in late April and May. In many years fewer than ten Shore Larks are reported, but occasionally much larger influxes occur. In 1976, for example, there were reports from 11 counties and flocks of c60 on North Ronaldsay in October and 70 at Tyninghame in November/December.

An adult male in song on 10 July 1972, in the Highlands, was the first summer record. A pair was in the same area in July/August the following year, and possibly bred (BB 66:505–508), but none were seen there in 1974 and only single birds in 1975 and 1976. In 1977, following exceptional passage and wintering numbers in 1976/77, two males were singing at the same site, a nest with three eggs was found on 25 June, and a juvenile was present in August/September. This remains the only proved breeding record. A second site also held a singing male in June 1977, but there have been no subsequent summer sightings.

The Shore Lark has an unusual and very discontinuous distribution. In northern Eurasia it breeds on tundra north of the Arctic Circle, from Fenno-Scandia eastwards, and also on mountains in southern Norway. It occurs in mountainous areas in northwest Africa and from the Balkans east across Asia, and it is widely distributed in North America, where it is known as the Horned Lark. Birds from the northwest European part of the range winter from Britain to the Mediterranean. The only example of the American race *E.a. alpestris* (breeding in Newfoundland and eastern Canada) recorded in Britain is a male on South Uist in September 1953. The two or three birds on North Rona at the time of a sizeable influx in autumn 1981 probably belonged, like most of those wintering here, to the northern Eurasian race *E.a. flava*. This race has spread southwards into the Norwegian mountains during the last 200 years.

Sand Martin *Riparia riparia*

Migratory breeder; widely but thinly distributed on the mainland, becoming scarce in the northwest. Breeds on the larger Clyde Islands and Inner Hebrides, but is absent from the smaller islands, the Outer Hebrides and the Northern Isles. Mass movements of native birds take place in autumn. Also a scarce passage visitor.

The Sand Martin, a colonial breeder, is an opportunist, varying its nest site according to what is currently available. River banks, sand and gravel pits, sandy faces exposed during roadworks, drainage pipes below railway platforms and coastal cliffs are among the sites used. Many such sites are ephemeral in character, being removed or becoming vegetated (and consequently unsuitable) within a year or two of being colonised. Sand Martins are very faithful to their natal colonies, and when compelled to abandon them usually settle not more than 10 km away (Mead 1979). Because colonies are variable in both location and size, it is not easy to assess trends, but there is good evidence that mortality

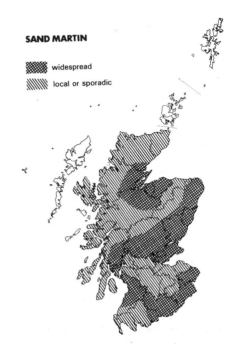

SAND MARTIN

▨ widespread

▧ local or sporadic

was abnormally high in the winter of 1968/69; this has been attributed to drought on the wintering grounds along the southern edge of the Sahara (Cowley 1979). A succession of late, cold springs may also have contributed to the decrease, or at least slowed the rate of recovery, by reducing breeding success. There are strong indications that a further decline took place in winter/spring 1983/84 and that the current population may be as little as 10% of that in the mid 1960s (C. J. Mead). Mink predation has been suggested as a possible cause of decrease in Kirkcudbright (G. Shaw) and some colonies have been deserted due to flooding (C. J. Mead).

Both distribution and numbers are determined largely by the availability of nesting sites; vertical faces of some height, or over water, are required to deter ground predators. Districts with extensive glacial deposits can usually offer both riverside and sand quarry sites, though the latter are often available for only a limited period, but in barer rocky terrain there may be only short lengths of suitable river bank. Although there are recent breeding records for all mainland counties, the Sand Martin is scarce and very local over much of the Highlands, and also in some parts of the Central Lowlands; surprisingly, there was no known colony in Kinross in 1983. Most of the large colonies are south of the Grampians, and many river banks hold only scattered pairs. The biggest colony reported in the last decade was at Barbush, Dunblane; this held about 920 nests in 1982 but has since declined (D. M. Bryant). From 1982 to 1984 estimates were made of nesting numbers at most colonies of more than 50 pairs in the Stirling area. In 1982 4,000+ pairs attempted first broods, 81% nesting in sand and gravel quarries, 18% in river banks and 1% in a glacial esker. In 1983 and 1984 numbers in 'artificial' (ie quarry) sites decreased to 76.5% and 16.6% respectively of the 1982 total. The decline was less marked at natural sites, but there was nevertheless an overall drop of c80% between 1982 and 1984 (Jones 1985).

Sand Martins breed regularly, though in small numbers, on Islay, Mull and Skye, and possibly sporadically on Jura, but are absent from the other Inner Hebridean islands. They are said to have bred in the past on the dunes of the Outer Hebrides but had ceased to do so by the beginning of this century. The last recorded nesting in Orkney was in 1918 and the only report of nesting in Shetland was last century (B&R).

This is usually the first of the hirundines to arrive in spring. 'First dates' vary widely according to the weather, ranging in the last two decades from 12 March to 14 April, and the main influx takes place in late April/early May. A spell of cold, wet weather soon after the birds arrive often results in concentrations – in the past sometimes numbering several thousands – over lochs and reservoirs, where flying insects are most abundant. By mid July the juvenile Sand Martins are starting to flock away from their natal colonies and forming communal roosts in areas such as reedbeds. The largest gatherings reported have been in the Tay reedbeds, where 1,000+ are often present between July and September and a maximum of 8,000 was counted near the Earnmouth in mid July 1975. Most have gone by the end of September but there are usually stragglers in October and there is one recent record on 10 November. The small numbers of passage birds recorded in the Northern Isles probably originate from the northern part of the species' continental range, which extends to northern Fenno-Scandia.

Since mist-netting was introduced, large numbers of Sand Martins have been ringed both at colonies and at their autumn roosts. Analyses of recoveries indicate that during autumn migration birds break their journey at preferred sites, which vary for populations from different areas. Scottish birds show a marked preference for roost sites in Yorkshire/Lancashire/Shropshire and around the Humber and Wash (where the principal Fenland roost has held about two million birds) but generally overfly those in East Anglia and Sussex (Mead & Harrison 1979a). On leaving Britain they travel mainly through western France on their way south, but in spring many take a more easterly route (Mead & Harrison 1979b). The most southerly recovery of a Scottish-ringed bird is from Senegal, while the speed with which some Sand Martins migrate is indicated by the capture in Sussex of one ringed seven days earlier in Stirling.

Analyses of nest record cards, and the results of Scottish studies of Sand Martin growth (Turner & Bryant 1979) and breeding season diet (Waugh 1979), are included in *Bird Study* 26(2), an issue devoted entirely to this species.

Swallow *Hirundo rustica*

Migratory breeder, abundant in agricultural areas of the mainland and inner islands, but scarce and local in mountainous areas and on the outer islands. Large pre-migration flocks gather in autumn and roost communally. Passage visitor in small numbers.

By tradition the harbinger of spring, the Swallow's arrival varies with the prevailing weather conditions. A few birds sometimes appear in March (7th is the earliest date), but the first usually arrive early in April and the main influx during the second half of the month. In cold springs arrival is delayed; in 1981, for example, when there was a late April blizzard, most did not arrive until early May. Breeding numbers fluctuate considerably, presumably partly reflecting breeding success the previous year and partly in response to the conditions prevailing either on the species' African wintering grounds or along its migration route. CBC results showed a marked decline during the early 1970s, a progressive recovery from 1974 to 1978, and then a further decline; in 1982 the index was the lowest since 1974.

There is evidence of a slight recent increase on the Outer Hebrides and Northern Isles. B&R had few records for the Outer Hebrides and breeding apparently ceased altogether for a period prior to 1971; since then Swallows have nested annually around Stornoway (Cunningham 1983) and also in South Uist, where numbers increased from two or three pairs in the early 1970s to about ten pairs in 1981 (C. Spray). In Orkney, where B&R (quoting Lack) described the species as sporadic, nesting now occurs regularly; a survey in 1982 located 25 pairs in Mainland alone, while pairs have also been recorded on Hoy, South Ronaldsay, Flotta, Rousay and Sanday since 1980. Many of the nests on Mainland were in old air raid shelters. An increase has occurred in Shetland too, with from four to eight pairs nesting annually

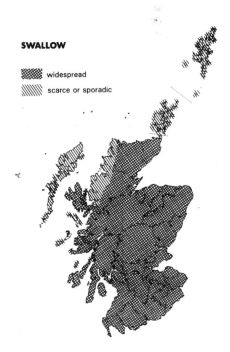

SWALLOW

▓▓▓▓ widespread

\\\\\ scarce or sporadic

and several records for Fair Isle and Foula in the last decade, whereas there were only sporadic reports prior to 1950. Swallows now nest annually on the Isle of May.

Changes in agricultural practices are likely to have affected status and distribution locally on the mainland. Modern farm buildings offer a more limited supply of suitable nest sites than the abundantly-raftered steadings and open-fronted cartsheds of the past, while modern methods of manure-handling, with slurry silos replacing open dung-heaps, must have greatly reduced the fly population on many farms. Little information is available on the extent to which such changes may have reduced local populations or may affect breeding success, but 24 farms in the Stirling area surveyed in 1980 held an average of three nests/farm (range 1–8) and one had nine nests in 1983 (D. M. Bryant).

Flocking preparatory to migration starts in August, and by the end of the month roosts of several thousand birds gather at many sites. Among the largest recorded in the last decade have been 10,000+ (a mixed flock including Sand Martins – Lynch 1984) in the Tay reedbeds, 3,500 at Condorrat (Dunbarton), 3,000+ near Loch Leven, and 3,000 at Duddingston. Of 297 ringed at a roost of 1,500 near Ayr in August 1978, 94% were juveniles (Hogg 1983). Most Swallows leave during September, but there are often a good many still passing through in October and stragglers in November; the latest recent record is 15 December. Such late records probably involve birds from the northern part of the continental breeding range, which extends quite far north in Fenno-Scandia.

Many, possibly most, British Swallows winter in southern Africa. There have been November/March recoveries of Scottish-ringed birds in the Orange Free State, Transvaal, Cape Province and Natal, and breeding season recoveries here of birds ringed in the same areas. A Swallow ringed in Cape Province on 17 March 1967 was caught in Moray on 23 March 1968, and one ringed as a chick in Aberdeen had travelled more than 10,400 km by 1 December the same year. Although normally showing great fidelity to their natal area, and even to the actual nest site, Swallows do occasionally go astray, as evidenced by one ringed in the nest in Ross and found dead in Norway the following May.

Red-rumped Swallow *Hirundo daurica*

Vagrant (S Europe) – eight records, 1906–84.

The Red-rumped Swallow was first recorded on 3 June 1906, when one of three birds on Fair Isle was shot (ASNH 1906:205, 1908:83). There were no further records until the 1970s, when singles in Shetland on 23–25 September 1971 (Whalsay) and 29 May 1972 (Mainland) were followed by three reports in 1976: from Fair Isle on 9–11 May and 3 June, and North Ronaldsay on 7 October. The only subsequent records are of one on Shetland Mainland on 15 May 1980 and one on the Isle of May on 23–24 May 1984. The Red-rumped Swallow can be difficult to identify with certainty and may well occur more frequently than the records suggest. It breeds in Iberia, southern France, northwest Africa and eastwards from the Balkans, and winters south of the Sahara. In England it is recorded almost annually most often between mid April and mid June, making it some-what surprising that most of the Scottish records have been from the Northern Isles.

House Martin *Delichon urbica*

Migratory breeder, most abundant south of the Grampians, more local north to Easter Ross, and scarce in the northwest mainland. Breeds regularly on the largest inshore islands, but only rarely in the Outer Hebrides and Northern Isles. Rather scarce as a passage visitor.

Although originally a cliff-nester, the House Martin is now closely associated with man, the majority nesting under the eaves of buildings; colonisation of new housing schemes sometimes takes place almost as soon as building is completed. Since mud is essential for nest construction, breeding can be much delayed in a dry spring. Cliff-nesting still occurs in many coastal counties but is possibly decreasing; in Caithness the House Martin is less common on the cliffs than in the past but becoming more abundant and widespread on buildings (P. M. Collett). There were colonies of more than 100 pairs at Tantallon (East Lothian) and on the Arbroath cliffs in the mid 1970s – and House Martins still nest at both these sites – but many cliff colonies are much smaller. Past records of cliff-nesting collated by Clark & McNeil (1980) include a number of inland records but there have been few recent reports from such sites. One site near Dalkeith has been lost through open-cast mining and others may also have been destroyed. A pair of House Martins nested under a Golden Eagle's eyrie in 1970 (SB 6:336).

Little is known about House Martin numbers or breeding densities, but, apart from local changes of the kind referred to above, there is little indication of any significant recent change in status or distribution. Some observers suspect that there has been a progressive decline but firm evidence is lacking; as this species is not adequately covered by the CBC no annual index is available. There is also little published information on colony sizes – and clearly much scope for further study. Breeding still occurs only very irregularly in the Northern Isles, where the most recent records are: Shetland, 1971 on Unst and 1975 in Lerwick; Orkney, 1969 and possibly 1983 on Mainland and 1977 on Westray. House Martins nested in the Outer Hebrides (Stornoway) for the first time in 1974, and again in 1975, but have not done so since. Small numbers breed regularly on Skye, Mull and Islay, but there have been no recent reports from other Inner Hebridean islands.

The first spring arrivals usually appear in early April, very occasionally in March (16th is the earliest recent date), and the main influx towards the end of April or in early May. Young birds prospecting for suitable sites continue to appear for a further few weeks and passage birds trickle through the Northern Isles throughout May and into June. Unlike the other hirundines, House Martins do not congregate at large pre-migration roosts. Autumn passage, which must involve large numbers, is most obvious on the coast, where successive small groups can often be seen flying south in

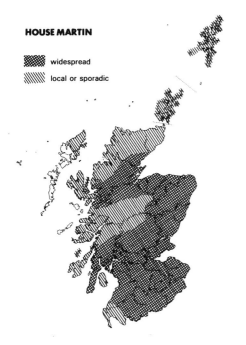

HOUSE MARTIN

▓ widespread

░ local or sporadic

September / October; large movements down inland glens also sometimes occur at this time. Latest dates are frequently well into November and even, rarely, in early December.

House Martins from the western part of the European range winter mainly in southeast Africa, but there have not yet been any mid-winter recoveries of Scottish birds. Recoveries have shown that the species is very faithful to its natal area; adults invariably return to the site where they first bred and even young birds generally settle to breed near where they were reared (D. M. Bryant). Long-term studies of various aspects of the species' breeding biology have been carried out by David Bryant and others working from Stirling University (Bryant 1978b, 1979).

Richard's Pipit *Anthus novaeseelandiae*

Vagrant (W Siberia) – annual in autumn, in very small numbers.

Richard's Pipit occurs annually on Fair Isle and nearly as regularly in other parts of Shetland, but has been recorded only occasionally elsewhere. Numbers are never large, the highest total since 1968 being only c17 (in 1970), and most records are of single birds. Nearly all occurrences are in autumn, the majority in September / October, with a few stragglers to mid November and a single August record. In some seasons there is a wide spread of dates and in others the birds all apparently arrive within a short period, most often in late September but sometimes not until October. In 1967–83 there were autumn records from Orkney, Aberdeen, Kincardine, the Isle of May, East Lothian and Islay, as well as from Fair Isle and Shetland. The only recent spring records were in 1967, when there was one in Orkney on 13 April, and in 1975, when there were three in Shetland

between 6 May and 20 June. In 1972 a Richard's Pipit was found at Whiteness Head near Nairn on 16 January. There have been a number of winter records in England and it has been suggested that wintering there may occur more regularly than is generally supposed (BB 75:514). There is little to indicate any change in status from the rather vague account given by B&R. This bird of damp lowland grasslands breeds from western Siberia to Mongolia and south into Australasia, and winters in Indo-China and Africa.

Tawny Pipit *Anthus campestris*

Vagrant (Europe) – less than annual, in very small numbers.

Although recorded with increasing frequency since 1970 (probably as a result of greater observer-activity rather than an actual increase in the numbers present) the Tawny Pipit remains a very scarce and irregular visitor. It was first recorded in 1933, when one was shot on Fair Isle, and there had been only two other reports by 1950. Following single records in 1951, 1952, 1961, 1962, 1963 and 1964, there were three sightings in 1970 and there have been reports almost annually since. About 60% of sightings are in spring, between 5 May and 10 June. In autumn the spread of dates is wider, with extremes of 15 September and 24 October. Somewhat surprisingly there have also been two July occurrences. To date this species has never appeared in both spring and autumn of the same year. Most of the records are from Fair Isle, Shetland and the Isle of May, but there have also been reports from Fife, East Lothian, Midlothian, Lewis and St Kilda.

Tawny Pipits breed in open areas of sparse vegetation, from Denmark and the Baltic States south to Iberia, northwest Africa and the Mediterranean islands, but are absent from northern France and the Low Countries. They winter in Africa south of the Sahara and also further east. This species occurs much more frequently in southern and southeastern England (especially in autumn), than it does here, and the distribution of the records suggests that most birds arrive by a short crossing of the English Channel (Sharrock & Sharrock 1976). Nothing is known of the origins of those reaching Scotland, but it may be that they are from a more northerly part of the breeding range.

Olive-backed Pipit *Anthus hodgsoni*

Vagrant (NE Russia/central Asia) – 13 records, all on Fair Isle, 1964–84.

The first Olive-backed Pipit was recorded on Fair Isle on 17 October 1964 and the second on 29 September the following year (BB 60:161). By the end of 1984 there had been a further 11 records, all between 24 September and 26 November. The 1964 bird was thought to be the first recorded in Britain but a 1948 report from Skokholm was subsequently accepted (Sharrock & Grant 1982). There have now been more than 30 British occurrences, nearly all of them in autumn. In its native area this species frequents habitats similar to those favoured in this country by the Tree Pipit, to which it is closely related.

WN

Tree Pipit *Anthus trivialis*

Migratory breeder, nesting in every mainland county and on the largest of the inshore islands, usually in open woodland or where there are scattered tall trees; absent from the Outer Hebrides and Northern Isles. Also a passage visitor in small numbers.

Nineteenth century records suggest that at that time the Tree Pipit was scarce and local in Scotland, but by 1950 it had colonised every mainland county and was breeding regularly on the Clyde Islands, and in small numbers on Islay, Mull, Skye and Raasay (B&R). The general pattern of distribution remains the same today, although there have been local changes, for example in Wigtown, where a few pairs now breed around Cairnryan – a westward extension since the Atlas survey. Tree Pipits are only thinly scattered over much of the southern half of the country but more widespread and abundant north and west of the Grampians. They are scarce and very local in arable farming areas, such as Buchan, East Fife and East Lothian, and in many eastern counties are confined to the glens.

Tall trees from which to launch itself into song-flight are an essential element of the Tree Pipit's habitat. Perhaps most typically a bird of hillside birchwoods, this species also occurs in mixed woodlands and native pinewoods provided that these have open areas within them. It breeds in young plantations where some mature trees have been left standing (eg where birch has been underplanted with conifers) and has been found in older larch plantations in Kirkcudbright (A. D. Watson). Breeding density varies greatly with habitat. In oakwoods densities between 11 and 38 pairs/km[2] have been recorded in Argyll and Wester Ross, and 15–22 pairs/km[2] in Stirling (Williamson 1974). The Atlas suggests an average of 25–30 pairs per occupied 10 km[2], but as Tree Pipits are often confined to relatively small areas, overall densities are possibly below this level.

The first Tree Pipits usually appear around mid April (very occasionally in the first week) and the main arrival takes place in early May. On the Isle of May spring passage is usually slight, but on Fair Isle very large falls sometimes occur; as these are often associated with southeast winds it is likely that they involve birds heading for Fenno-Scandia which have been drifted across the North Sea (Riddiford & Findley 1981). An unusually massive influx occur ed in the Northern Isles in 1969, when there were 1,500 on Fair Isle on 3 May, and the following day 'thousands' at Sumburgh and 450 on Fetlar. Passage generally takes place within a fairly short period and few are seen in Shetland aft r the first week in June. Autumn migration starts in August, the main movement is in September, and there are few records after mid October (one on Skerries on 2 November 1974 is the latest). On Fair Isle numbers are smaller in autumn than spring, with peak counts seldom reaching 100; up to 500 a day are occasionally recorded in September on the Isle of May. There are very few records for the Outer Hebrides, where the species is probably best described as a vagrant.

Tree Pipits are absent from Ireland but widely distributed elsewhere in Europe, breeding south to northern Spain and Italy. They winter in northern tropical Africa. There have been no recoveries of Scottish-ringed birds.

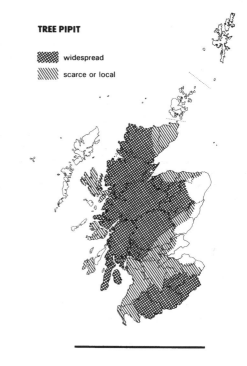

TREE PIPIT

▨ widespread

▧ scarce or local

Pechora Pipit *Anthus gustavi*

Vagrant (NE Russia / Siberia) – about 22 records, 1925–83.

All but one of the Pechora Pipit records are from Fair Isle, where it was first recorded in 1925; the odd one out is from Whalsay, in 1972. This species is unobtrusive and skulking in its behaviour and would be easy to overlook in areas with a fair amount of cover. (A field description is given in SB 7:263–4). B&R note nine or ten records prior to 1950 and there have been 12 since: two in 1958 but otherwise singles. Occurrences have been at very irregular intervals: in 1951, 1952, 1953 and 1958, then a gap until 1966 (the only record in the 1960s), 1971 and 1972, annually from 1975 to 1978, and no more to the end of 1983. All the post-1950 occurrences have been between 8 September and 16 October, but B&R refer to reports in late August and on 19 November.

Meadow Pipit *Anthus pratensis*

Widespread and abundant, breeding in open ground throughout the mainland and on all but the barest of the islands, and from sea-level to altitudes of over 1,000 m. Many native birds winter further south. Immigrants arrive in large numbers in autumn and some may remain throughout the winter.

Although often regarded as a bird of rough grassland, heather moor and mountain top, the Meadow Pipit also breeds in uncultivated areas on arable farms, and in young conifer plantations. The early stages of afforestation produce an increase in breeding densities. Williamson (1975a) found that deep ploughing of moorland prior to planting led to a doubling of breeding numbers, which then remained high until the young trees were around five years old, while Moss (1978b) recorded 82 pairs/km^2 on recently planted heather moorland compared with 52 pairs/km^2 on comparable unplanted ground, the corresponding figures for rough grassland being 55 pairs/km^2 after planting and 35 pairs/km^2 before. By the time the pre-thicket stage was reached, numbers in Moss's study area had dropped to 21 pairs/km^2, and the plantations were deserted before the trees formed a closed canopy. On Rhum, however, where the woodland areas were small and surrounded by open moor, Meadow Pipits were still nesting among dense woody growth in 15-year old conifer plots (Williamson 1975a).

Small islands and grassy mountain tops doubtless hold much smaller numbers, though few data are available for such habitats. Densities of less than one pair/km^2 (all in hay fields) have been found on predominantly arable farmland in the Lothians (da Prato 1985) and about 25–40 pairs/km^2 on rough grassland at Aberlady Bay (P. R. Gordon). CBC results indicate that in Britain as a whole the breeding population does not fluctuate greatly; only very severe winters, such as that of 1981/82, appear to result in a significant decrease.

Native birds usually return to their breeding areas around the end of March. The onset of a late spell of hard weather may cause flocks that have arrived on high ground to return temporarily to lower altitudes, while in years when snow lies long, breeding may not occur at all on the high plateaux. Spring passage of migrants continues throughout April and often well into May, with the largest numbers generally recorded in April.

Daily peaks at the observatories are seldom high in spring, but an exceptional 1,000 or so were on the Isle of May on

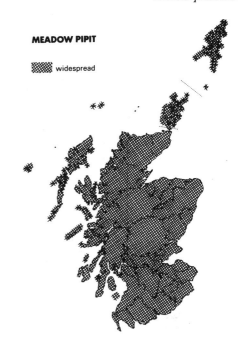

MEADOW PIPIT

▨ widespread

16 April 1950, while 2,000/hour were counted passing north at St Andrews on 20 April 1975. Autumn passage, which takes place mainly between late August and mid October, regularly involves much larger numbers and is often obvious on the coast, for example the 'thousands' seen at Kinloss in mid August 1970 and 1,000 at the Ord of Caithness in early September 1979. Recent peak daily counts on Fair Isle and the Isle of May have been 1,500 (8 September 1981) and 6,000 (6 September 1984) respectively. The few Meadow Pipits wintering in the Northern Isles are almost certainly immigrants, as the local breeding birds have generally left by the end of September (Davis, in Williamson 1965). In a hard winter such as 1981/82 the largest numbers are in the Central Lowlands, the Inner Hebrides and the southern Outer Hebrides, but in a more open season overwintering Meadow Pipits are much more widely scattered, with fair numbers in the northeast and Orkney (Winter Atlas).

Ringing recoveries suggest that many Scottish Meadow Pipits winter in Iberia; none has been recovered in Scotland between December and March. Most of those recovered have been shot – on passage through France in October/November, in Spain and Portugal from October to February, and, less often, in Morocco between November and January. There is no ringing evidence that passage visitors include birds from the northern part of the continental range, which extends to northernmost Fenno-Scandia, but there have been recoveries here of birds ringed in Iceland and *vice versa*.

Red-throated Pipit *Anthus cervinus*

Vagrant (N Eurasia) – less than annual, in very small numbers.

The Red-throated Pipit is recorded most often on Fair Isle, occasionally elsewhere in Shetland, and only rarely further south or west. B&R give the earliest record as 1908, on Fair Isle, and refer also to reports from St Kilda, Auskerry and Unst. Between 1950 and the end of 1983 there were records from Unst, Out Skerries, Whalsay, Fetlar, Shetland Mainland, Papa Westray, the Isle of May, Mull, Midlothian, Wigtown, and at sea in the Beryl A oilfield. Apart from 1974, which was a blank year, there has been at least one report annually since 1965. Three were on Fair Isle at the same time in spring 1975, but the maximum in a season has been only 12–13, also in 1975. Records are about equally divided between spring (mostly between 7 May and 11 June, but a very early bird at Whithorn, Wigtown, on 19 April 1983) and autumn (between 8 September and 22 October). In some years birds appear at both seasons. The Red-throated Pipit has a distinctive call which helps to draw attention to it. It breeds in the tundra zone of northern Eurasia and winters in the Tropics.

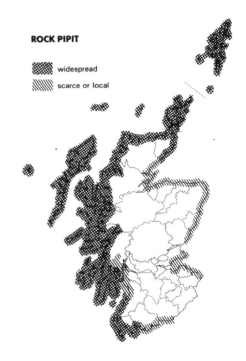

ROCK PIPIT

░░ widespread

▨ scarce or local

Rock Pipit *Anthus spinoletta*

Resident, breeding virtually right round the coast and on all the islands; abundant in the north and northwest, but relatively scarce or local in the west south of the Clyde and in the east. Also a scarce passage and winter visitor.

In this country the Rock Pipit is typically found on rocky shores, where it feeds among the debris along the tideline. Its distribution is consequently governed not only by the presence or absence of rocky shore but also by the extent of intertidal feeding ground available; cliff-girt coastline holds relatively few, unless it is well dissected by inlets, while low, hummocky coasts with many small islets hold much higher numbers. Little information on breeding densities

is available, but records suggest that numbers have declined, at least locally, on the east coast in the last decade or so.

Although this species is generally regarded as occurring inland only rarely in Britain, Rock Pipits (not assigned to any particular race) have been reliably reported at some altitude inland in recent years. They have been seen well inland in the breeding season in the Grampians, and apparently breeding at 300 m asl on Skye and over 500 m on Mull (R. Broad).

The native breeding stock is largely sedentary though some dispersal, probably mainly of young birds, takes place in early autumn. Only c30% of recoveries are more than 100 km from the ringing site, but first-winter Fair Isle birds have been found in the Netherlands, Northumberland, and at sea more than 400 km WSW of the island.

Rock Pipits breeding in Britain belong to the race *A.s. petrosus*, which also occurs in coastal northwest France. Elsewhere in Europe *A.s. littoralis* breeds on the coasts and islands of the Baltic and western Norway, and *A.s. spinoletta*, the Water Pipit, breeds in the mountains of Spain and France and disperses to low ground in winter. Very small numbers of birds showing the characters of *littoralis* are recorded in most winters; usually only one or two birds are seen at any one site but there have been a few reports of up to five together (eg Caithness in March 1980). Most occurrences are between February and April and on the east coast, although there have been some west coast sightings. The recovery in Fife of a Rock Pipit ringed as a chick in Norway provides confirmation that this race occurs at least sporadically. The European Water Pipit *spinoletta*, which winters regularly in south and southwest England (usually near fresh water), also occurs occasionally; birds showing the characters of this race have been reported a number of times but there have been few satisfactorily confirmed records. The American Water Pipit *A.s. rubescens*, which breeds also in Greenland and north east Siberia, has been positively identified twice: on St Kilda in October 1910 and Fair Isle in September 1953. A bird showing the characters of this race was at Scatness, Shetland, on 13 May 1981; this record is still under consideration by BBRC.

Yellow Wagtail *Motacilla flava*

Very local migratory breeder, most abundant in Ayr and Lanark, less regular elsewhere south of the Tay / Clyde, and rare further north. Small numbers of the British race occur on passage, as do birds belonging to races breeding further east and north.

The Yellow Wagtail group is a complex one, with some fourteen recognised races, several of which have been reported in Scotland; hybridisation between races occurs commonly, making certain identification even more difficult. The races positively identified in Britain belong to four main groups: Yellow-headed *M.f. flavissima*, breeding in Britain and locally on the eastern North Sea coast; Blue-headed *M.f. flava*, breeding from southernmost Fenno-Scandia south to north Italy, France and Iberia and east to the Urals; Grey-headed *M.f. thunbergi*, breeding in Fenno-Scandia and Russia eastwards into Siberia; and Black-headed *M.f. feldegg*, breeding from the Balkans east to Afghanistan. Hybrids or mutant birds of the regularly-occurring races may resemble those belonging to more distant areas, so some reports of the latter are probably open to doubt. Among those breeding in Ayr, Iain P. Gibson has found birds resembling both Blue-headed and Sykes' (see below) Wagtails, and also males showing the characters of *M.f. lutea* (sometimes referred to as the Kirghiz Steppe Wagtail). He considers that such variants account for the occasional reports of Blue-headed Wagtails breeding in Scotland.

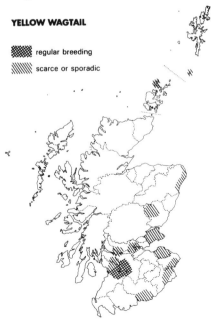

YELLOW WAGTAIL

▨ regular breeding

▧ scarce or sporadic

A marked decrease in breeding Yellow Wagtails, here at the northern limit of their range, took place during the first half of this century (Parslow). B&R detail many of the older records, which include a pre-1902 report of 250 in a flock in Lanark, and 50–100 together at Possil Marsh in 1913–14. By 1939 no Yellow Wagtails were coming to Possil Marsh in spring and only a few in autumn. Further north, the species bred regularly in many of the central lowland counties until the end of the last century, and not uncommonly near the east coast north to Aberdeen. It has also bred in the past on Islay and Raasay. B&R's conclusion that 'its headquarters continue to be in the Clyde area' is still true today.

Iain Gibson has studied Yellow Wagtails in Ayr and Lanark since 1978; his findings in respect of territory-holding pairs and breeding success are summarised in the Table.

Breeding Yellow Wagtails in Ayr and Lanark 1980–83 (Data from I. P. Gibson)

	Territorial males		Known successful pairs		Min. no. of young reared	
	Ayr	Lanark	Ayr	Lanark	Ayr	Lanark
1980	20	14	3	2+	4	2
1981	16	15	1	8+	4	c30
1982	22	22	11	9	30	22
1983	28	23	14	11	23	22

Note: the intensity of survey increased during the period, so the figures for territorial pairs do not necessarily reflect changes in population or in breeding success.

In Ayr most nests were in grass, where early cuts for silage often resulted in failure, the birds then moving to hay fields, where breeding pairs were sometimes already established. In Lanark barley fields were the preferred habitat. A colour-ringing programme was initiated in 1982 to facilitate study of the species' breeding biology, and has already produced the first foreign recoveries of Scottish birds: two on return migration through the Western Sahara in late April / early May 1983. A full report of this work is currently in preparation.

Outwith the main breeding area, nesting has been confirmed since 1968 in Dumfries, Renfrew, Stirling, Dunbarton, Roxburgh (possibly up to six pairs at times), Berwick, East Lothian, Fife, Aberdeen and Fair Isle (1981 – unsuccessful). Breeding occurred in Angus in 1962 but has not been confirmed since (although there was a 'probable' record during the Atlas survey); a pair probably nested on Papa Westray, Orkney, in 1981. There is no recent record for Arran, where the species formerly bred.

Small numbers of Yellow, Blue-headed and Grey-headed Wagtails occur occasionally in spring on the Isle of May and Fair Isle, most often in May and with Yellow the most usual; in autumn the species is even scarcer. These races occur occasionally in widely scattered parts of the mainland and in the Hebrides. Reports of other races are much scarcer and, as indicated above, may sometimes involve mutants or hybrids. The Black-headed *feldegg* was first reported in June 1925, in Dumfries (B&R), and there have been seven further reports, all males and all in May: from Whalsay in 1936, East Lothian in 1952, Out Skerries in 1969 and 1981, and the other three from Fair Isle, the most recent in 1974. There seems no reason to doubt the authenticity of these records, as this race is much more distinct than the others (BOU). A bird resembling Sykes' Wagtail *M.f. beema* (from southeast Russia) was shot on Fair Isle in May 1910, and specimens resembling the Eastern Blue-headed *M.f. simillima* were collected there in the autumn of 1908, 1909 and 1912 (Davis, in Williamson 1965). There have been no subsquent acceptable reports of these forms. There are a few sight records of birds showing the characters of the Ashy-headed *M.f. cinereocapilla*, which breeds in the Italian section of the Mediterranean, among the most recent being one on the Isle of May in May 1982. *Flava* wagtails resembling the form *iberiae* (which is not accorded subspecific status in the BOU's *The Status of Birds in Britain and Ireland*), have been reported from the Isle of May in 1975 (unpub. log), Whalsay in 1977 and Fair Isle in 1980 (SBR); all these sightings were in May.

Citrine Wagtail *Motacilla citreola*

Vagrant (E Russia / Siberia) – less than annual, in very small numbers.

Although not recognised in Britain until 1954, when two were trapped on Fair Isle (BB 48:26), the Citrine Wagtail is now recorded almost annually in autumn. This situation is likely to reflect increasing familiarity with the bird's field characters as well as an actual increase in numbers due to a westerly extension of the breeding range during the past decade. Most of the records are from Fair Isle but there are also reports from Out Skerries, Whalsay, Shetland Mainland, Inverness, the Isle of May and East Lothian. Not more than two have been seen in any one season. All arrival dates have been between 2 September and 17 October, and the majority from mid September onwards; several individuals have stayed in the same area for a week or longer. A possible hybrid bird was seen on Fair Isle in October 1980. The Citrine Wagtail breeds in marshy grasslands from Russia eastwards, and winters in India and southeast Asia. Many of those reaching Britain are immatures and might be confused with immature Yellow Wagtails (Sharrock & Grant 1982).

Grey Wagtail *Motacilla cinerea*

Breeds in all mainland counties, the Clyde Islands, and the larger Inner Hebridean islands, but only sporadically in the Outer Hebrides and Orkney; in Shetland little more than a casual visitor. Most move south or southwest in winter, though many remain in Britain. Suffers heavy mortality in severe weather, with consequent fluctuations in the population.

The Grey Wagtail's distribution is determined by the availability of its principal habitat requirement, which is fast-flowing water. Although often associated with hill streams, it also occurs on mill lades and fast stretches of lowland rivers, nesting regularly, for example, on the Clyde below Lanark and sporadically on the Brothock in Arbroath, on a stretch almost totally enclosed by buildings (M. Nicoll). It is rather local in many of the central lowland counties, and in Wigtown, Fife, Buchan, Caithness and parts of northwest Sutherland, but in favoured areas in Kirkcudbright

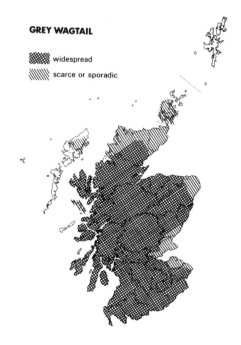

GREY WAGTAIL

▓▓ widespread

▨▨ scarce or sporadic

reaches densities much higher than the average of 10–20 pairs/10 km² quoted in the Atlas (A. D. Watson). WBS results for Britain as a whole show that between 50 m and 300 m asl densities increase with altitude, and are highest where the gradient is more than 2.5 m/km (Marchant & Hyde 1980b). In Angus and southeast Perth, Mike Nicoll has recorded densities ranging from 4–12 pairs/10 km on fast streams (eg the Monikie Burn, North Esk, Braan and Almond) to 1–5 pairs/10 km on slow streams (such as the Lunan Water) and on the headwaters in the glens. Breeding densities were apparently limited in some areas by a lack of suitable banks for nest sites.

Most Grey Wagtails leave Scotland in winter, the majority moving to southern Britain; only small numbers were recorded during the Winter Atlas surveys, mostly in the Central Lowlands. The few that remain usually feed around sewage works, by coastal streams, and on rooftops in urban areas. Mortality is high when sudden hard frosts occur. The WBS indices illustrate just how devastating to Grey Wagtails a period of prolonged severe frost can be; in Britain as a whole the hard weather of 1981/82 resulted in a 42% decrease, whereas the 1978/79 winter, which had more snow but less cumulative frost, produced a smaller, though still significant, decline of 33%. The number of pairs recorded in survey plots in 1982 was less than half the total present in 1977 (Taylor & Marchant 1983). Details of recent population fluctuations on sites in a wide area of Angus and southeast Perth are given in the Table.

Fluctuations in occupancy of 52 Grey Wagtail territories in Angus and southeast Perth checked annually in 1980–84 (Data from M. Nicoll)

	1980	1981	1982	1983	1984
No. of sites occupied by pairs	43	35	21	19	31
occupancy rate	82%	67%	40%	36%	60%
increase/decrease from previous year	?	−19%	−40%	−9%	+63%

There is no evidence of change in distribution since B&R's day, although the Grey Wagtail has now nested several times in the Outer Hebrides, for which they had no records; the first breeding record was near Stornoway in 1957 and three pairs were present in 1979. Nesting in Orkney continues sporadically, usually on Hoy; after a gap of many years breeding occurred in 1974 and has done so annually since 1976, with three pairs nesting that year and two pairs in 1981. There has been no further report from Fair Isle, where breeding occurred in 1950, and there is no record for Shetland.

Although widely distributed on the Continent, the Grey Wagtail is absent from much of the low-lying country bordering the North Sea, and from all but the extreme south of Fenno-Scandia. The fact that it is scarce and irregular on passage in the Northern Isles is consequently hardly surprising. On Fair Isle one or two birds are usually seen in spring, but sometimes none in autumn. Further south passage is more obvious, but still seldom involves large-scale movement; 145 passing south at Prestwick in one hour on 23 September 1981 is the largest gathering reported recently. Occasional birds winter as far north as Shetland, but ringing recoveries show that many are much further south between October and March (Tyler 1979). Most Scottish birds appear to winter around the Irish Sea and in southwest England, although two first-winter birds have been recovered in France. Most of the other winter recoveries have been in Devon and Cornwall, Lancashire, the Isle of Man and Ireland.

Pied Wagtail *Motacilla alba*

Widely distributed and abundant breeder throughout the mainland and on many of the inner islands, more local in the Outer Hebrides and Orkney, and scarce in Shetland. Many native birds move south in autumn and the large winter roosts which form in some areas presumably include immigrants. White Wagtails occur on passage in small numbers and occasionally nest.

The 'normal' Pied Wagtail is black-backed and belongs to the race *M.a. yarrelli*, which breeds in Britain and Ireland and very occasionally in adjacent areas of the Continent. The grey-backed White Wagtail breeds in Iceland and throughout continental Europe eastwards into Asia. Birds from Iceland and the northern part of the Continent are migratory, wintering in the south of the breeding range and in Africa; British birds are only partially migratory.

Pied Wagtails occupy a wide variety of habitats, wherever there is a good supply of insects and suitable cavities in which they can nest. They are common around farm steadings, and along roadsides, where they are often at risk from passing traffic as they hawk for flies. They regularly nest

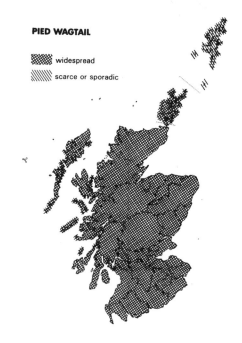

PIED WAGTAIL

▨ widespread
▨ scarce or sporadic

on the Forth Islands and have been recorded breeding at over 500 m asl in the Cairngorms. This is the most widespread of the riverine species censused in the WBS, which has shown that low-gradient rivers hold more territorial pairs per kilometre than steeper ones (Marchant & Hyde 1980). Riverside breeding densities recorded in Scotland range from three to 13 pairs / 10 km – but very many Pied Wagtails breed at a considerable distance from either rivers or large bodies of standing water. Despite the fact that a substantial proportion of the population migrates, severe winters cause a marked decline in breeding numbers: following the winters of 1978/79 and 1981/82 the CBC index dropped by 18% and 25% respectively. The wintering population, which may include a high proportion of immigrants, is concentrated largely in the Central Lowlands and northeast, with few remaining north of the Great Glen. The distribution pattern showed little difference between the hard winter of 1981/82 and the relatively mild following one (Winter Atlas).

In the 1970s and early 1980s Pied and/or White Wagtails have bred in Shetland or Fair Isle almost annually, though only in very small numbers; in 1975–76 up to six pairs of Pied and one or two pairs of White were recorded. Single pairs of White Wagtails also bred in this period in Arran (1972) and Perth (1981), and mixed pairs were recorded in North Uist, the Treshnish Isles, Skye, Islay, Caithness, Inverness and the Isle of May.

The White Wagtail is the more abundant and regular of the two races in Shetland, with passage occurring mainly in late April/early May and late August/early September. In spring daily counts seldom exceed ten but in autumn they are larger, with up to 200 occasionally present. Ringing recoveries show that at least some of these birds are of Icelandic origin; one ringed as a chick in northern Iceland on 18 July reached Fair Isle less than six weeks later. Some travel far south to winter; one ringed on the island was later shot in Mauritania. Pied Wagtails reared in Scotland are partial migrants; although some are still within 100 km of the ringing site in mid winter, the majority move south. November/February recoveries are widely scattered through England, France and Iberia (Galbraith 1977).

From late June until April Pied Wagtails gather at large roosts in a wide variety of situations. Many of the sites used throughout the season are in reedbeds, as on the Tay, or among willows, but in severe weather the birds move into more sheltered roosts. There are November/December records of 200+ around the churches in central Dundee, 500 in greenhouses near Prestwick, and c200 among beer bottle crates at an Edinburgh brewery (Dougall 1984), while a roost in the former Pullar's dyeworks in Perth was notably large. There R. L. McMillan, who studied and mist-netted at this roost for two seasons, counted c1,350 birds in late March 1978. Attendance at the roost was apparently related to overnight temperatures, and the birds crept in through a slatted wooden 'false roof' to perch on the (disused) heating pipes beneath. Some 63% of the birds using this roost were in their first winter. Analysis of recoveries of birds marked at the Perth roost showed that they gathered from up to 10 km away; on exceptionally cold days, however, smaller numbers were present, suggesting that birds from the periphery of the catchment area were reluctant to fly long distances to roost.

Waxwing *Bombycilla garrulus*

Irruptive winter visitor.

Waxwing numbers vary greatly between years. In a 'good' year several thousand birds may arrive and penetrate as far west as the Hebrides and Galloway, while in other winters there are only a few reports, all from the east side of the country. The scale of an irruption is determined largely by two factors: the species' breeding success in northern Fenno-Scandia and the size of the berry crop further south in Scandinavia. Although the first birds often arrive in September, and there is a late August record from Orkney, the main influx does not usually take place before mid October and occasionally not until December/January; the February//March arrival in the winter of 1956/57 was exceptionally late. In some years – presumably when the berry crop here is poor or has already been cleared by thrushes – the Waxwings move on quickly, but in other seasons some remain throughout the winter. This was the case in 1975, when display and courtship feeding were seen in Roxburgh on 1 April, a male was still in Abernethy Forest on 30 June, and a marked return passage was noted in April/May, with a few stragglers into June.

Waxwings occur most regularly in the Northern Isles and right down the east coast, but over the last two decades they have been recorded in all counties and island groups and on most of the principal islands. Since 1950 there have been only eight sizeable irruptions, irregularly spaced and varying greatly in scale, the largest being in 1956/57 (c7,000 birds), 1965/66 (c10,000) and 1970/71 (c5,000) (Macmillan 1964; Everett 1967; Lyster 1971). Hawthorn, rowan, rose hips, sea buckthorn and *Cotoneaster* are among the foods most regularly taken, but there are also records of Waxwings eating apples and toast and catching drips of birch sap (SB 9:350). As some *Cotoneaster* varieties are eaten by thrushes only when more favoured foods have been exhausted, berries often remain on these bushes well into winter and attract Waxwings close to houses; on occasion I have had these very beautiful birds feeding less than 2 m from my kitchen window.

Waxwings breed in thick coniferous or mixed forest with a berry-bearing shrub layer, from northern Fenno-Scandia east across northern Eurasia. The southern limit of the breeding range varies with the pattern of movement the previous winter, sometimes extending as far as southern Scandinavia. Birds from the extreme north of the range regularly move south in winter, but major eruptive movements take place only when the food supply in the normal wintering range is inadequate for the population. The 1965/66 eruption resulted from the combination of an exceptionally successful breeding season in northern Scandinavia and a poor berry crop there. Large numbers of Waxwings spread across Europe in October/November, reaching places as far apart as Iceland, Portugal, Italy and Turkey (Everett 1967).

Several birds ringed while on eruptive movements in Finland, Sweden and Norway have been recovered here; one found dead in Stranraer had travelled nearly 2,000 km in 18 days, an average of over 100 km a day. Evidence of onward movement is provided by recoveries – in Yorkshire, Lancashire, Wales and Germany – of birds ringed in Scotland earlier the same winter.

R.M

Dipper *Cinclus cinclus*

Breeds in all mainland areas with fast-flowing and unpolluted rivers and streams; most abundant in hilly districts. Occurs in the Clyde Islands, several of the Inner Hebrides, and locally on Lewis and Harris, and has bred in Orkney but not Shetland. The Scandinavian dark-bellied race occurs occasionally.

Dippers nest up to 600 m asl and are most abundant on streams with gradients ranging from about five to 25 m/km and at altitudes of from 50 m to 250 m asl (Shaw 1978). In the rather small Scottish WBS sample the number of territories/10 km ranges from nil to 8.8 on 'fast' rivers and nil to 3.7 on slow ones (Marchant & Hyde 1980b). The whole of the Angus South Esk (including its main tributaries) was surveyed in March 1984, when 67 territories were found on 101 km of river, giving an average of 6.6 territories/10 km (N. K. Atkinson). Breeding density and success is determined not only by gradient – and the associated frequency

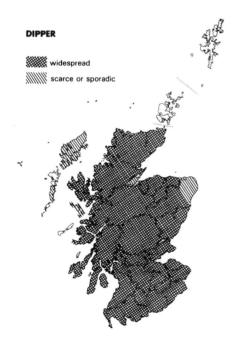

DIPPER

▓▓▓ widespread

░░░ scarce or sporadic

of falls, rapids and shallows – but also by stream productivity; da Prato & Langslow (1976) found that more second broods were produced on the lower reaches of the Midlothian South Esk than higher up, where the supply of large and easily caught prey items was less plentiful. In Grampian, no correlation was found between altitude and breeding success (Hardey *et al* 1978). Although many Dipper nests are under bridges or on other man-made structures, eg 70% in an analysis of over 1,300 Nest Record Cards covering Britain and Ireland (Shaw 1978) and 90% in Grampian (Hardey & Rae 1979), only 36% of those on the Angus South Esk in 1984 were in such sites, suggesting that the data collected by ringers and nest-recorders may be biased (N. K. Atkinson).

Studies of Dippers on the Midlothian Esk over a long period have shown that the population does not fluctuate greatly, although very severe winters may cause a short-term decrease. Cowper (1973) found that the Esk and its tributaries held 48–56 pairs of Dippers in the middle 1960s, with the South Esk system holding 24, 32 and 42 pairs respectively in 1964, 1965 and 1966. Studies by D. R. Langslow in 1974–79 showed the current population on the South Esk to average 40 pairs, equivalent to a density of about seven pairs/10 km.

Cowper's results suggested that Dipper numbers were much reduced by a severe winter, but the WBS data have shown that this is not so; only minor decreases resulted from the very prolonged frosts of 1981/82, which seriously affected many other resident waterbirds (Taylor & Marchant 1983). However, although the number of territory-holding pairs may not be affected, breeding success may be much reduced following a hard winter; in Dumfries & Galloway only one out of four pairs attempted to breed after the 1978/79 winter, and did so much later than usual (Shaw, in Marchant & Hyde 1980a). Dippers occupy winter territories from October, sometimes on loch shores, and have frequently been watched feeding beneath ice. In severe weather the habit of communal roosting under bridges, which provide protection from the wind, may have important survival value (Shaw 1979).

There is no evidence of any recent change in status or distribution. On the islands the Dipper is confined to those with hill ground and fast-flowing streams. It breeds regularly on Islay, Mull, Skye and Rhum, and probably also on Jura, and has done so at least sporadically on Raasay. In the Outer Hebrides it occurs on about ten streams in Lewis (Cunningham 1983), and very locally in Harris; it has also been reported from the South Uist hills and probably breeds there (C. Spray). Nesting was first recorded in Orkney, on Hoy, in 1919, and breeding apparently occurred fairly regularly until about 1940. Dippers reappeared in the late 1960s, and an empty nest was found in 1970 but breeding was not confirmed; the most recent record is of a pair on territory in 1974 (Booth *et al* 1984).

Dippers breeding in Scotland belong to the races *C.c. hibernicus* (Outer Hebrides, Kintyre, Clyde Islands and west coast) and *gularis*, and are sedentary, with young birds seldom dispersing further than 50 km from their natal area and many remaining within 5 km (Galbraith & Tyler 1982). Single birds of the dark-bellied Scandinavian race *C.c. cinclus* occur irregularly in Shetland and Fair Isle, between

November and April, and one suspected of belonging to this race was at Loch an Eilean, Inverness, in October 1978. One on Fair Isle in March/April 1983 is the most recent record from the Northern Isles.

Wren *Troglodytes troglodytes*

Resident breeder, abundant and widespread in a variety of habitats throughout the mainland and on most islands of any size. The population fluctuates markedly, as Wrens suffer heavy mortality in severe winters. Small numbers occur on passage.

The Wren is a highly sedentary bird, with few recoveries showing movements of more than 10 km from the ringing site, and distinguishable races have consequently developed on island groups well-separated from their nearest neighbours. These races vary in size, colour and song. Those currently recognised are: *T.t. zetlandicus* breeding in Shetland, *fridariensis* on Fair Isle, *hebridensis* on the Outer Hebrides, and *hirtensis* on St Kilda. *T.t. indigenus* breeds in all other parts of Britain and in Ireland, and *T.t. troglodytes* throughout most of continental Europe (BOU).

Wrens are very adaptable, nesting from near sea-level up to over 500 m asl, and in habitats ranging from boulder beaches, through gardens and woodland, to banks and gullies in moorland and mountainous areas. Both deciduous and coniferous woodlands are used, plantations being occupied at an early stage. Breeding densities vary widely with habitat. In western oakwoods Williamson (1974) found from 47 to 222 pairs/km²; numbers in Argyll and Wester Ross were only about half those in Stirling. The highest density among oaks occurred on Inchcailloch, Loch Lomond. Mixed deciduous woodland in Dumfries held even more, with up to 302 pairs/km² (Moss 1978a). In coniferous woodlands Moss found breeding density to be closely correlated with the amount of brash (trimmed branches) or other ground cover present. He recorded densities of 4–9 pairs/km² in Scots pine with no understorey, 116–121 pairs/km² in larch with a moderate understorey, and 71–173 pairs/km² in regenerating Scots pine with birch. The population in even-aged plantations varies only slightly with the stage of growth, being highest after thinning (Moss 1979). Farmland densities are very much lower. The Atlas suggests an average of

around 22 pairs/km², but the overall density on a large area of mainly arable land in the Lothians in 1984 was only 4 pairs/km² (da Prato 1985). Wrens were virtually absent from the hedgerows and were largely confined to woods and shelterbelts, where densities reached 127/km². In an area of scrub in the same district, censused over a seven-year period, the mean density was 58/km² – well below that of several other songbirds (S. R. D. da Prato).

The sensitivity of the Wren to winter severity is clearly demonstrated by the CBC results; population indices since 1964 have ranged from 47 to 351 on farmland and 57 to 274 in woodland, peak numbers in both habitats occurring in 1975. This is a greater fluctuation than that shown by any other common species. The effects of a severe winter are even more marked in Scotland than in Britain as a whole (see Chapter 8). In cold weather Wrens resort to communal roosting, which probably has considerable survival value; Joe Eggeling counted 43 emerging from House Martins' nests at his house near Dunkeld, and in East Lothian a few roosted in rat burrows among heated grain (SB 13:116). Population recovery can be rapid – an overall increase of more than 50% was recorded between the summers of 1979 and 1980 – but where numbers are small a succession of hard winters can result in a serious decline in the population. The island races are clearly most at risk. The Fair Isle population fluctuates between ten and 40 pairs and was estimated to include only ten singing males in 1980.

There is no evidence of any long-term change in status or distribution, but numbers may be expected to increase locally on islands and in moorland areas where afforestation has recently taken place. Although breeding has been recorded on Tiree, Ailsa Craig and the Isle of May it occurs only rarely on these islands. Nesting was recorded on North Ronaldsay for the first time in 1975.

Small numbers of migrant Wrens occasionally appear on

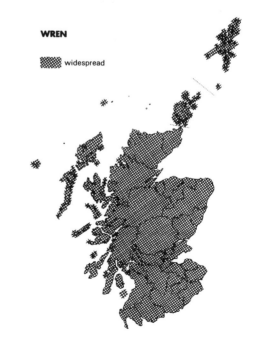

WREN

▓▓ widespread

the Northern Isles and the east coast. These are probably of Scandinavian origin, as birds in the northern part of the continental range are migratory. Movement most often takes place in September/November and February/April, and the numbers present at one time are never large. Peak counts in the last two decades have not exceeded 30 on the Isle of May.

Dunnock *Prunella modularis*

Resident, breeding throughout the mainland at up to c500 m asl and on many of the islands, but local in the Outer Hebrides, scarce in Orkney, and absent from Shetland. Also a passage and winter visitor in small numbers.

More than 30 years ago B&R expressed the view that the distribution of the 'Hebridean Hedge Sparrow' *P.m. hebridium* required further study before its limits could be definitely ascertained. No such study has yet been carried out, and subsequent publications have simply repeated the assessment given by B&R – as I do here. This race is thought to have spread to the outer islands, which then carried abundant scrub woodland, during the climatic optimum (Atlas), and is believed to breed in the Outer and Inner Hebrides, Argyll, Dunbarton, the Clyde Islands, Renfrew and Ayr. The main characteristic distinguishing it from *P.m. occidentalis*, the race breeding elsewhere in Scotland (and the rest of Britain), is apparently its ability to survive in areas virtually devoid of scrub, or with only bracken or long heather as cover. Although the Dunnock is an extremely sedentary species, with c95% of Scottish recoveries not more than 10 km from the ringing site, it seems unlikely that the two races remain separate on the mainland; presumably they intergrade, as *P.m. occidentalis* does with the continental race *P.m. modularis* in western France (BOU). There is clearly scope here for future study – even though it might result in the 'loss' of the Hebridean race.

The Dunnock is typically a bird of woodland with a shrubby understorey, gardens, and farmland with thick-bottomed hedges. It occurs in conifer plantations – but only at the thicket stage (Moss 1979), and is absent from, or very scarce in, grazed birchwoods and also many oakwoods (Williamson 1974). Dunnocks can be difficult to census due to their unusual breeding behaviour, which often involves two males occupying a territory holding only one female; in a study in Edinburgh 38% of territories had an extra male (Birkhead 1981). A census of more than 1,700 ha of predominantly agricultural land in East Lothian in 1984 showed the Dunnock to be one of the commonest songbirds present, at a mean density of 17 territories/km² but with densities approaching 200/km² locally in areas of scrub. The Dunnock was one of the few species to use severely-clipped hedges, though only at low densities (da Prato 1985).

On the Inner Hebrides the species is quite widespread, except on the barer islands such as Tiree, but on the Outer Hebrides it is very local and, according to Peter Cunningham (1983), 'not easily found'. In Orkney it breeds in small numbers and perhaps only sporadically in some areas. Berry & Johnston (1980) record that in 1965 eggs were laid in

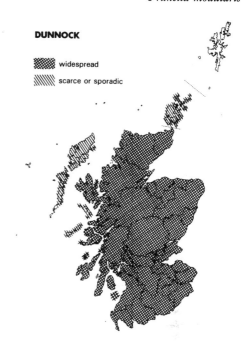

DUNNOCK

▓▓ widespread

▧▧ scarce or sporadic

Shetland but failed to hatch, while 'an old nest with four eggs', reputedly found on Fair Isle early in 1974, is claimed in the Atlas to be the first attempted breeding in Shetland. No details are given of the former record and the latter has not appeared in either the Fair Isle Report or the SBR; both records should probably be viewed with some doubt.

Continental birds are recorded regularly on passage on Fair Isle and occasionally on the Isle of May; they are likely to escape detection elsewhere, as the continental race is not distinguishable from the British in the field. As a passage visitor the Dunnock occurs more regularly in spring than autumn. Spring passage takes place between March and June, and the numbers involved are usually small, with seldom more than a few birds seen together and falls of any size occurring only rarely. Peak daily counts on the Isle of May and Fair Isle have exceeded 100 only a few times in the last two decades, maxima being 300 (April 1965) and 230 (March 1980) respectively. Autumn movement occurs between September and November; in some seasons very few birds appear, and even in the 'best' years for the species the maximum recorded on Fair Isle has been only 150 and on the Isle of May 50.

Small numbers of continental birds sometimes winter in Shetland, and even on Fair Isle, especially in years when numbers have been relatively high in autumn. Most move on further south, some possibly as far as the Mediterranean; the most distant winter recovery of a Dunnock ringed on passage through Scotland has been in France. Immigrants presumably originate from the northern part of the breeding range, which extends from Fenno-Scandia nearly to the Mediterranean; Scottish-ringed migrants have been recovered in Norway in the breeding season. Dunnocks breeding north and east of the Baltic are almost all migratory.

Alpine Accentor *Prunella collaris*

Vagrant (S Europe) – two records, 1908 and 1959.

This rather sedentary species breeds in mountainous areas from Iberia eastwards through southern Europe and normally makes only altitudinal movements. Both the Scottish records are from Fair Isle: on 6 October 1908 (B&R) and 27–28 June 1959. The Alpine Accentor has occurred about 30 times in Britain, most often in southern England and in August/January.

[Rufous Bush Robin *Cercotrichas galactotes*]

The report of a Rufous Bush Robin on the Isle of May during the October 1982 fall of Siberian migrants (SB 12:218) was not accepted by the BBRC. There had been eleven British records of this Mediterranean/African scrubland bird by the end of 1983, one each in April and August and the rest in September/October.

Robin *Erithacus rubecula*

Widespread as a breeder on the mainland and most of the inner islands, but scarce and local in the Outer Hebrides and Orkney and absent from Shetland. Passage sometimes involves large numbers; most autumn immigrants move on further south but some remain to winter.

Originally a forest bird, the familiar Robin has adapted to life in a wide variety of habitats, from suburban gardens to conifer plantations, and from sea-level up to at least 500 m asl. In some woodland types it is so successful that it rivals the ubiquitous Chaffinch in dominance ranking, and in many it is among the five most abundant species. Woodland breeding densities vary widely, ranging from the equivalent of more than 170 pairs/km² in a Stirling oakwood (Williamson 1974) to less than 10 pairs/km² in a Scots pine plantation without undergrowth (Moss 1978a). The importance to this species of the cover afforded by a dense shrub layer is demonstrated by the fact that Robins increase in abundance in well-grown conifer plantations when wind-blow produces an 'artificial' understorey (Moss 1978a). The Atlas quotes an average of 20 pairs/km² on farmland, while in suburban gardens and in estate policy woodlands, which so often have a good mix of mature deciduous trees and shrubs, densities equivalent to 300 breeding pairs/km² have been found on census plots. Farmland densities vary according to the amount of cover available and the general population level, since Robins tend to colonise hedgerows when numbers are high in adjacent woodlands. Census work in the Lothians in 1984 showed that Robins occurred at densities equivalent to from 60 to 159 pairs/km² in small woods and shelterbelts, but were absent from most hedges (da Prato 1985).

In autumn, Robins take up territories which, as they hold only single birds, tend to be smaller than breeding territories. Observations of colour-ringed birds in Midlothian over two

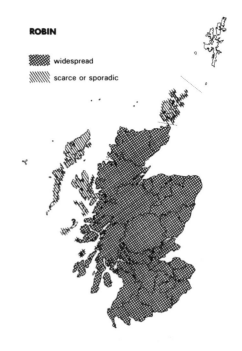

ROBIN

▨ widespread

░ scarce or sporadic

winters showed that in an area with a breeding season density of 115/km², autumn territory density reached 173/km², dropping to 58/km² in mid winter under normal conditions, and to zero during prolonged snow and frost, when many Robins moved into gardens (S. R. D. da Prato). In Aberdeen some Robins roost communally in winter (usually with Blackbirds), dispersing during the day to feed in suburban gardens (Swann 1975). Ringing recoveries indicated that the birds using such roosts returned to them in successive winters and came to them from considerable distances; one was recovered 193 km from the roost. The native population is largely sedentary; about 70% of 341 recoveries of Scottish-reared Robins were within 10 km of the ringing site. The remainder disperse, but none has been recovered more than 250 km away from its natal area. Robins breeding in upland woods are perhaps more likely to move away in winter than those nesting in areas where winters are less harsh; more ringing results would, however, be required to confirm whether this is so.

CBC results for Britain as a whole suggest that the breeding population is fairly stable, fluctuating over a comparatively small range and suffering a significant reduction only after protracted periods of very severe weather. The continuous frosts of 1981/82, for example, produced declines of 25% and 29% respectively in woodland and farmland indices, whereas the corresponding decreases were only 7% and 12% following the winter of 1978/79, which took a heavy toll of some other small resident passerines. The rather small Scottish sample showed a more marked weather effect, with declines of 27% in woodland and 50% on farmland between 1981 and 1982. Presumably the species' varied diet and willingness to live close to man – and consequently to supplementary sources of food – assist in its survival under severe conditions.

There is no evidence of any recent change in status or

distribution, apart from some slight increase in the Outer Hebrides. The Robin continues to be no more than a scarce and very local breeder in Orkney, and there is no recent nesting record for Coll, where small numbers were present in June 1984 (R. Broad), or Tiree. An increase may be expected to occur over the next few decades in areas where afforestation is taking place on what was previously heather moorland or rough grassland, for example on Lewis and Skye, and in parts of Sutherland and the Ochils.

The Robin breeds widely throughout Europe and western Russia, the continental race *E.r. rubecula* being distinct from that occurring in Britain *E.r. melophilus*. Those breeding north and east of the Baltic are migratory, wintering in central and southern parts of the range and southwards into northern Africa, and some southward movement probably also takes place in other parts of the Continent. Records of Robins marked or recovered in the breeding season indicate that most immigrants to Scotland are from Fenno-Scandia, Denmark and Germany. Although a few of those ringed on the Continent while on migration have been reported in Scotland in mid winter, the majority appear to move further south, with most October / February recoveries coming from France, Iberia and Algeria.

In autumn the first migrants sometimes reach Shetland in late August, the main movement is normally in October, and stragglers occur into November; a few occasionally remain to winter in Shetland. Peak counts at the observatories often do not exceed 50, but in some years easterly winds bring very large numbers across the North Sea. By far the biggest fall yet recorded was on 11 October 1982, when there were 4,000 Robins on the Isle of May (Ellis 1983) – four times the previous record for the island (1,000+ in late October 1976). The autumn 1976 peak of 600+ is the highest count for Fair Isle. Evidence of passage on the west coast, for example 80+ on Sanda and 70+ at Carsluith (Wigtown) in mid / late September, may reflect the dispersal of native birds, some of which are believed to winter in Ireland (BOU). Small numbers of native birds are recorded on passage at the Isle of May, but most migrants, and certainly all those involved in the occasional very large falls, are of continental origin. Individual birds sometimes return in successive years to winter in the same area, as evidenced by one that wintered on the Isle of May for at least three years (Eggeling 1960).

Spring passage starts in late February and continues into early June, with the main movement usually in April / early May. On Fair Isle passage is often slight at this season, with daily maxima of less than 50, and large influxes rarely occur; the highest recorded peaks are 500 in late March 1958 and 350+ in late April 1974.

Thrush Nightingale *Luscinia luscinia*

Vagrant (Europe / central Siberia) – less than annual, in very small numbers.

Since 1970 the Thrush Nightingale (sometimes known as the Sprosser) has occurred almost annually, with not more than three records in any year. Prior to 1970 occurrences were very irregular, with a gap of 46 years from the first record, on Fair Isle in May 1911 (SN 1912:9), to the second, in the same place and also in spring. Only one more Thrush Nightingale was reported in the 1950s and two in the 1960s – both on Fair Isle in 1965. Most sightings are in May, from 6th onwards, but there have been two reports in June, one in July, two in August, and a late bird on 3 October 1976. The bulk of the 26+ records are from Fair Isle but there are also reports from Whalsay, Out Skerries, Shetland Mainland, Orkney, St Kilda, Caithness, Aberdeen and the Isle of May. The Thrush Nightingale breeds in damp forest and shrubland, from Denmark, southern Sweden and the Baltic States east through Russia to southern-central Siberia, and winters mainly in East Africa. A field description is given in SB 6:283.

Nightingale *Luscinia megarhynchos*

Vagrant (Europe, including England) – annual in very small numbers.

B&R class the Nightingale as a very rare visitor and give only four records, the earliest in 1911 on the Isle of May (ASNH 1911:132). From 1950 to 1964 there were reports in seven years, and since 1965 the species has been recorded annually, most frequently from Fair Isle, Shetland and the Isle of May. Whether this represents an actual increase in occurrences or merely better coverage by observers it is impossible to say. In England there has been both a slight contraction in range and a considerable decrease in breeding numbers, the latter attributed to loss of the bird's favoured damp scrub habitat (Atlas).

The total recorded in Scotland in 1950–1983 inclusive was only 64, and six is the highest number reported in any one year (1973 & 1980). More than 80% of occurrences are in spring, with most birds appearing in May but a few records in April (the earliest is 19th) and June. The few autumn reports are scattered from 19 August to 30 October, the last a bird found dying on Fair Isle in 1971 which proved to be the first British occurrence of the large Eastern race *L.m. hafizi* (SB 8:195).

Most records come from islands with little cover, where small migrants are easily seen; in all probability many more Nightingales reach Scotland but escape detection. In addition to those from the bird observatories, Shetland and Orkney, there are records from St Kilda, Islay, Easter Ross, Kincardine, Fife, East Lothian, Berwick and Stirling – where a much-recorded bird sang near Stirling Castle from 14 May to 22 July 1952. More recently singing males have been reported in Wigtown (May 1980) and Unst (May 1982). Those arriving in Scotland in spring have presumably overflown the normal breeding range, which on the Continent extends from the Mediterranean and Iberia to the Low Countries and the southern Baltic. Nightingales winter in Africa south of the Sahara.

Siberian Rubythroat *Luscinia calliope*

Vagrant (Siberia) – one record, 1975.

A Siberian Rubythroat on Fair Isle on 9–11 October 1975 was the first recorded in Britain (BB 72:89–94). Birds of this species, and also of the similar White-tailed Rubythroat, are occasionally imported for the cage-bird trade, making it advisable to consider the possibility of escapes in the event of future occurrences (Sharrock & Grant 1982). The Fair Isle bird was thought to be a first-winter male – and hence unlikely to be an escape – and it arrived during a large influx of Siberian vagrants.

Bluethroat *Luscinia svecica* [1]

Scarce passage visitor, occurring annually in very variable numbers. Has attempted to breed.

The Bluethroat is most abundant in spring, when totals since 1968 have ranged from ten to about 275 (1981), and occurs most frequently in the Northern Isles and on the east coast. A record 100+ were on the Isle of May in mid May 1985. Autumn numbers seldom exceed ten. Most of those passing through Scotland belong to the red-spotted race *L.s. svecica* which breeds in the northern part of the range, from Scandinavia to Alaska, mainly north of about 60°N. Two young birds ringed on autumn passage through Sweden have been trapped here the following spring – one at Girdleness on 21 May and the other on Fair Isle on 22 June – while an adult ringed on Fair Isle in late May was at Ostend four days later. The few records of the white-spotted Bluethroat *L.s. cyanecula* have all been in spring, between 20 March and 29 May. B&R note several occurrences on Fair Isle and one on the Isle of May. From 1968 to 1983 inclusive there were only six reports: from Aberdeen and Out Skerries in 1969, Moray, Whalsay and Fair Isle in 1975, and Fife in 1982. This race breeds throughout much of Europe, north to Denmark and the Baltic and east into Russia, where it intergrades with the red-spotted form. Both races winter around the Mediterranean and in northern Africa.

The spring passage of the red-spotted race is comparatively late, with the main movement generally taking place in the second or third week of May; extreme dates since 1968 have been 4 May and 21 June, though B&R refer to mid-April records. The earliest autumn date is 29 August and the latest 13 November, but most passage occurs between mid September and mid October. In Shetland, Bluethroats occur annually in spring (there were at least 160 on the islands in the period 11–27 May 1981) and almost as regularly in autumn, but further south they are much less regular in autumn.

B&R knew of few mainland records but give old reports from Lanark and Midlothian. In recent years passage Bluethroats have been seen in Caithness, Moray, Banff, Aberdeen, Kincardine, Angus, Fife, East Lothian, Berwick, Dumfries, Ayr (Ailsa Craig), Stirling and Argyll (Kintyre). Single birds were on South Uist and North Rona in autumn 1981 and there are 19th century records from the Monach Isles.

In Scandinavia the breeding habitat generally includes marshy ground and low scrub. The only known British breeding attempt occurred in Inverness in 1968, when a female was found incubating a clutch of six eggs at the Insh Marshes. No male was seen and breeding was unsuccessful (BB 61:524–525). More recently a male in song was found in the Spey valley on 15 June 1980, but was not seen again. Injured birds occasionally summer on Fair Isle.

Red-flanked Bluetail *Tarsiger cyanurus*

Vagrant (Finland/Siberia) – six records, 1947–84.

The Red-flanked Bluetail, a bird of forested country, has spread westwards from Siberia to Finland but adhered to its southeast Asian wintering grounds. It was first recorded in 1947, when a first-winter bird was shot on Whalsay, Shetland, on 7 October (SN 1948:6). There had been five more records by the end of 1984, all but one in September/October; the exception was a nearly-adult male on Fetlar on 31 May to 1 June 1971. First-winter birds have been recorded in autumn on the Isle of May in 1975, at Fife Ness in 1976, and on Fair Isle in 1981 and 1984.

Black Redstart *Phoenicurus ochruros* [1]

Passage visitor, most regular on the Northern Isles and east coast; occasional in winter. Has attempted to breed.

The timing of the Black Redstart's migrations is rather variable; in some years spring passage starts in mid March and extends well into June, while in other seasons the first report is not until mid April and movement is completed by the end of May. Autumn passage is almost as irregular, starting any time from 9 August to early October and ending from late October to early December. In most recent seasons larger numbers have appeared in spring (range from 1968 to 1983, about ten to 70+) than in autumn, but in 1982 this situation was reversed, with an autumn total of 70+; B&R described the species as occurring most commonly in October and November. Most reports are of single birds or very small parties. Black Redstarts are recorded most regularly in Shetland, Fair Isle and the Isle of May; they occur occasionally elsewhere on the east coast but are much scarcer on the west and inland. There are records for the Outer Hebrides (Monach Isles, North Rona, South Uist & Lewis), the Inner Hebrides (Mull, Rhum, Islay & Tiree), and the Clyde Islands (Arran & Ailsa Craig), but there appear to be none for Nairn, Kinross, Clackmannan, West Lothian or Roxburgh. The only reports from Inverness and Dunbarton are prior to 1950.

Occasional individuals remain to winter, most often in the southwest, from Ayr to Galloway, though there have been mid-winter reports from as far north as Lerwick. Summering records are even scarcer. In 1973 an apparently unmated female on Copinsay was found to be incubating four infertile eggs (SB 8:80) and in 1976 breeding was suspected in Aberdeen, where a male in song was present from April until mid July and a probable female was seen. There is no other record of attempted breeding.

The Black Redstart is widely distributed in Europe from the Baltic to the Mediterranean, nesting in dry rocky areas or man-made substitutes such as ruined buildings, and wintering in similar habitats in the southern part of its range and in North Africa. During the last century it colonised southern Sweden and southern England, where it frequents both built-up areas and coastal cliffs. The British breeding population has fluctuated considerably since the species established itself on bombed sites during the 1940s, averaging around 30 pairs and probably never exceeding 100 pairs (Atlas).

The only Scottish ringing recovery is of a bird marked on the Isle of May in April and caught in East Germany in late June the same year.

Redstart *Phoenicurus phoenicurus*

Migratory breeder, nesting in every mainland county and on some of the larger inshore islands but not on the Outer Hebrides or Northern Isles; most abundant in the west mainland north of the Clyde. Regular on passage in both spring and autumn.

The Redstart is a hole-nesting species which favours relatively open woodlands, especially those containing oak and pine. In all counties south of the Clyde / Tay, with the possible exception of Dumfries, it is very local, occurring mainly in remnants of semi-natural oakwood or in estate policy woodlands. In Stirling, Dunbarton, Angus and Kincardine it breeds locally, most often in the glens; in Aberdeen it is common only on Deeside; and around the Moray Basin pairs are thinly scattered. Only in Perth, Argyll, Inverness, Ross and Cromarty and Sutherland can the Redstart be described as widespread, and even within these counties it is abundant only locally. On Arran, Bute, Mull, Islay, Jura, Skye and Raasay it breeds in very small numbers and in

REDSTART

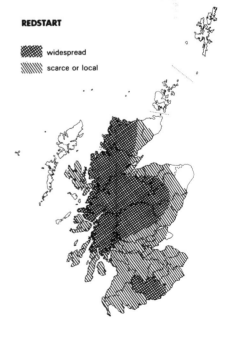

▨ widespread

▨ scarce or local

some cases probably only sporadically. Nesting occurred on Rhum in 1960–62 but has not been recorded since. B&R give reports of breeding on Lewis in 1914 and of an unsuccessful attempt in Shetland in 1901, but there has been no subsequent suggestion of nesting in either area, or in Orkney.

Breeding numbers vary markedly between years, probably due largely to conditions on the African wintering grounds. As with the Whitethroat, the CBC index dropped suddenly in 1969 and has since fluctuated between an all-time low of 23 in 1973 and a maximum of 90 in 1980. Few Scottish data are included in the CBC sample, and numbers here may vary even more widely than they do further south. These fluctuations in the Redstart population have been recognised for many years; as long ago as 1891 it was recorded that 'enormous numbers poured into the Spey Valley', to be followed in 1892 by a 'startling decrease' (B&R).

Because the size of the population as a whole fluctuates so much, figures for breeding densities obtained in different years are clearly not directly comparable. They can, however, serve to complement subjective assessments of relative abundance in various districts and habitat types. The densities found by Williamson (1974) in coastal oakwoods in the west are the highest so far recorded: nearly 50 pairs/km² in mid Argyll, 40 pairs/km² in Wester Ross, and 32 pairs/km² in north Argyll. In oakwoods around Loch Lomond numbers were much lower – at 8–21 pairs/km² slightly below the 23 pairs/km² recorded by Moss (1978a) in semi-natural mixed woodland. Redstarts are much scarcer than this in many areas, in at least some instances owing to a shortage of nesting holes – a deficiency which can be overcome by the provision of nest-boxes.

The Redstart is widely distributed across Eurasia, from northern Fenno-Scandia to the Mediterranean and east across Russia, but is almost entirely absent from Ireland. It winters in the southern Sahara. Breeding birds start to appear from early April, the main arrival is usually in early May, and passage continues through the Northern Isles well into June. On Fair Isle peak numbers in spring are often less than 50 but much larger falls very occasionally occur, as in 1970 when 700 were present on 9 May; such influxes are probably the result of birds being drifted across the North Sea, rather than moving north through Scotland (Hope Jones 1975; Riddiford & Findley 1981). The maximum recorded on the Isle of May at this season is only 50. In autumn the return movement takes place mainly between late August and mid October; again numbers are usually small in the Northern Isles, with large falls rarely occurring. The most recent big influxes on Fair Isle were in 1956 and 1957, when over 1,000 were present on 4th and 20th September respectively; daily peaks of less than 50 are the norm. Stragglers sometimes turn up in early November – the latest date is 11th – and individuals very occasionally reach the Outer Hebrides.

There have been no recoveries in Scotland of birds ringed on foreign breeding grounds, but passage birds presumably originate from the northern part of the range, probably in Fenno-Scandia. October/November recoveries in Spain, and April recoveries in Libya and Morocco, represent birds going to and returning from the wintering grounds, but a first-winter Redstart in Senegal in January was presumably

at the southern limit of its migration. A Redstart ringed as a chick in central Scotland was in Morocco the following March, while one marked in Kent on 19 May and subsequently caught on Fair Isle had travelled 970 km in only six days.

On two occasions male Redstarts showing the characters of Ehrenberg's Redstart *P.p. samamisicus* have been reported: the first on Fair Isle on 6 September 1948 and the second at Fife Ness on 23 September 1976. This eastern race breeds in southern Russia and Iran.

Whinchat *Saxicola rubetra*

Migratory breeder, widely distributed on the mainland, regular on the Clyde Islands and Inner Hebrides, more local in the Outer Hebrides, sporadic in Orkney and absent from Shetland. Passage is seldom conspicuous.

The Whinchat is typically a bird of rough ground with scattered gorse bushes; it breeds from sea-level up to about 500 m asl and is most abundant in the west. This is one of the species that benefits from afforestation, rapidly colonising young plantations in areas which previously lacked the prominent song-posts which are an essential feature of its breeding habitat. The fringes of upland plantations at the pre-thicket stage may hold the highest densities – A. D. Watson estimates c100 pairs/km² in parts of Kirkcudbright – but these sites remain suitable for only a few years and are abandoned when the trees meet in the rows. In most areas Whinchat densities are probably very much lower: the Atlas suggests an average of 10–20 pairs/10 km square, while in Ayrshire Phillips (1973) found c20 pairs/km² in Sitka plantation on grass but only 2.2 pairs/km² where the

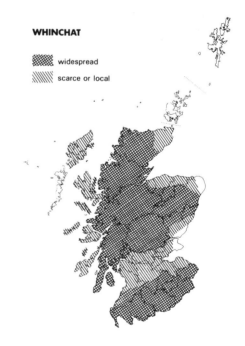

WHINCHAT

▨ widespread

▨ scarce or local

ground vegetation was heather. Little information on densities is available for the species' more traditional gorse/grassland habitats. Abandoned railway lines and similar derelict areas have been successfully colonised in some arable districts which otherwise lack suitable habitat.

In much of the eastern half of the mainland the Whinchat is local, while in intensively-farmed areas it is scarce. In 1984 none were found on c1,600 ha of farmland in East Lothian (da Prato 1985) and there are very few in east Fife and Buchan or along the fertile fringes of the Moray Basin. The species is also very local in Caithness, but may be expected to increase in areas currently being afforested. On the west it is local in Ayr, Renfrew and Lanark, but widely distributed and often abundant in all other western counties, over much of central Scotland and in many of the glens further north. It breeds regularly and in fair numbers on the Clyde Islands and the larger Inner Hebridean islands, but probably only sporadically on the Small Isles, Coll and Tiree. In the Outer Hebrides it is a regular though very local breeder in Lewis and Harris but nests only occasionally on the southern islands. There are old breeding records from Orkney but no recent confirmation of nesting, although territorial pairs have been seen on Hoy on several occasions since 1970.

The first Whinchats generally do not appear until about the middle of April, occasionally a week or so earlier or later; one in Peebles on 13 March 1983 was exceptionally early. Most are on territory by mid May. Passage through the Northern Isles often continues into June but is rather variable; on Fair Isle daily counts seldom exceed 100, and 350 in May 1970 is the highest recent count. On the Isle of May passage is more regular, in both spring and autumn, but there too numbers are generally small. The usual migration pattern for this species appears to be similar to that of the Wheatear, with birds travelling in small parties rather than large flocks and consequently little likelihood of major falls. The return movement starts in August and continues into October, usually peaking on Fair Isle in late August/early September. Again numbers are relatively low, the maximum count recorded on Fair Isle being 500, in 1956. Occasional stragglers occur in November, and in 1980 three were reported in January and one in December.

The Whinchat breeds throughout most of Europe, from Fenno-Scandia south to northern Iberia and eastwards across Russia; only one race is involved. There have been no recoveries in Scotland of birds ringed on foreign breeding grounds but passage visitors presumably originate from the northern part of the range. The species' main wintering area is in tropical Africa. There have been no recoveries of Scottish-ringed Whinchats between November and March, but several from France and Iberia in September/October and from Morocco and Algeria in April.

Stonechat *Saxicola torquata*

Partial migrant, breeding locally in most mainland counties, the Inner and Outer Hebrides and Orkney, but only sporadically in Shetland. Most abundant in the west, where it occurs well inland; largely confined to the coast in the east. The resident portion of the population suffers badly in hard winters and is slow to recover. A scarce passage visitor.

The Stonechat frequents areas of rough grassland or heather with gorse or other shrubby growth to serve as song-posts, and will also breed in some types of young conifer plantation. Where Whinchats are also present there is considerable inter-specific competition; Stonechats, which begin to breed earlier, are generally dominant (Phillips 1970). In young Sitka plantations in Ayrshire, Phillips (1973) found this species more abundant than the Whinchat where heather

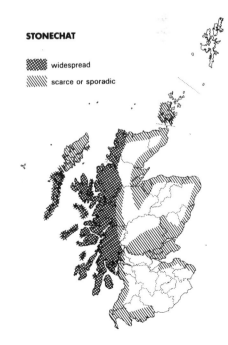

STONECHAT

widespread

scarce or sporadic

was the dominant ground vegetation, but the reverse situation where grass was dominant; Stonechat densities were about 9 pairs/km² with heather and less than 4 pairs/km² with grass. When the population is at a high level breeding densities in favoured west-coast areas, such as Kirkcudbright, may reach 50 pairs/km² locally (A. D. Watson).

Stonechats remaining in Scotland to winter are liable to suffer heavy mortality in severe weather. As this species is not adequately covered by the CBC, no figures are available for the decreases following the winters of 1978/79 and 1981/82, but from local recorders' comments it is clear that there has been a very marked decline since the comparatively mild period in 1968–72. The Atlas shows confirmed or probable breeding in every mainland county, with the possible exception of Selkirk (a 10 km square recorded as positive was on the Selkirk/Roxburgh boundary). The situation in the mid 1980s is very different, with Stonechats absent from Berwick (last bred 1981), Roxburgh, Selkirk, Peebles (last bred 1978), Dumfries (last bred 1978), East Lothian (last bred 1981) and West Lothian, and extremely scarce – if not absent – in Kinross, Perth, Clackmannan, Stirling, and Renfrew. In other areas, too, numbers are said to be much lower than in the mid 1970s. Some indication of the scale of the decline is given by the fact that in Renfrew 50+ pairs were located in 1972–75 but only one breeding pair in 1982, while Dunbarton probably had 20+ pairs up to 1978 but none were found there in 1982 (I. P. Gibson). The species is widespread on the Inner Hebrides and southern Outer Hebrides, but much scarcer on Lewis and Harris, and in Orkney. A pair bred on St Kilda in 1975.

B&R comment on similar declines during the 1940s, especially following the very hard winter of 1946/47, and the notorious snowstorms of 1962/63 also decimated the population. A quick recovery presumably takes place only when a series of mild winters and good breeding seasons follows the crash. Although the Stonechat often raises two, and sometimes three, broods in a season, nearly 50% of fledglings may die within three weeks of leaving the nest (Phillips 1976); when first-winter mortality is also high the rate of recruitment into the following season's breeding population is bound to be low.

Apart from weather-induced fluctuations, the only evidence of recent change in distribution comes from Shetland, where the Stonechat bred for the first time in 1961 (Magee 1965), and again in 1962. Nesting was not recorded again until 1975, when the population as a whole was at a high level; one nest was found and juveniles seen at three other sites had probably also been reared locally. A few birds overwintered in Shetland that year, and an unusually marked spring movement in late February/March 1976 was followed by confirmed breeding at four sites. Nesting occurred again in 1978 but has not been reported since.

In Scotland the Stonechat is at the northern limit of its European range; it is absent from Fenno-Scandia and Denmark but breeds throughout the rest of continental Europe and in North Africa. Some birds winter in their breeding area; some disperse, generally moving nearer to the coast; and some emigrate, going south as far as the Mediterranean. The majority of Scottish ringing recoveries involve first-winter birds, most of which were in the west and still within 100 km of the breeding area in December/January. More distant recoveries include birds in France and Spain.

Stonechats reaching Shetland in spring, or recorded elsewhere on spring passage, belong to the British-breeding race *S.t. hibernans* which also occurs in northwest France and coastal Portugal; there are no British records of *S.t. rubicola*, the race breeding in all other parts of Europe (BOU). Numbers seen on migration are usually very small (max. on the Isle of May only 16 and none at all in many years), but c100 at the Mull of Galloway in late September 1970 possibly represented a pre-migration gathering. In autumn birds belonging to one of the eastern races *maura/stejnegeri* occur occasionally; most reports are from the Northern Isles but there are also records from Aberdeen, Fife Ness and the Isle of May. All such sightings have been between 20 September and 23 November, and the largest number seen together has been three – on Fair Isle in mid October 1980.

Isabelline Wheatear *Oenanthe isabellina*

Vagrant (SE Europe) – one record, 1979.

A bird of dry steppes and semi-desert, breeding from southern Greece eastwards to Mongolia and wintering in northeast Africa, Arabia and Iran, the Isabelline Wheatear seldom strays as far west as Britain. The only Scottish record of this large sandy-coloured wheatear is of one in the Girdleness area of Aberdeen/Kincardine from 17 October to 10 November 1979 (BB 73:519). There are few other British records.

Wheatear *Oenanthe oenanthe*

Migratory breeder, widely distributed in open grassy country throughout the mainland and on many of the islands. Birds of both continental and Greenland races occur regularly on passage.

The Wheatear's principal habitat requirements are short-turf feeding areas and holes in which to nest. This combination of features occurs in many different situations, including cliff-tops, roadside verges, and steep mountain slopes, and Wheatears may consequently be found from sea-level up to 1,200 m or even higher. They are generally absent from extensive heather moorlands and from arable land. There has been little suggestion of any significant decline in the Scottish population as yet, but local decreases must be occurring in areas where previously-grazed grassland has been, or is currently being, afforested. Marked declines have already been noted in south and southeast England, where they have been attributed to a reduction in grazing pressure (by both sheep and rabbits) and the ploughing-up of marginal land during the 1939–45 war (Atlas). In Holyrood Park (Edinburgh), which formerly held five to ten pairs, numbers have dropped to only one or two pairs since grazing ceased in 1978 (L. Vick).

In 1977–83, on heavily sheep-grazed study areas c1,300 m and 1,100 m asl near Gladhouse Reservoir (Moorfoot Valley) and at Swanston (Midlothian), Lance Vick found no evi-

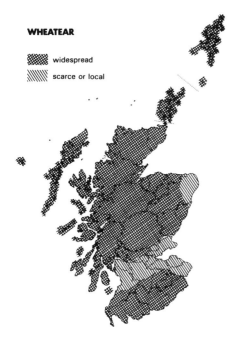

WHEATEAR

▓▓ widespread

░░ scarce or local

dence of double-brooding, which occurs regularly further south and at lower altitudes. Breeding numbers at Swanston were remarkably constant at 10–12 pairs on 0.8 km², whereas at Moorfoot Valley they ranged from 57 (1980) to 19 (1983), averaging around 40 pairs on 4.2 km² of suitable habitat. These figures suggest a maximum breeding density of 10–15 pairs/km², very similar to that quoted in the Atlas – up to 10 pairs/km² on favoured islands (including Fair Isle). Few other data on breeding densities are available.

The first Wheatears generally arrive during the second half of March, and the main influx of breeding birds two or three weeks later. Birds of the nominate race *O.o. oenanthe*, which breeds in Britain and throughout continental Europe but not in the Faeroes or Iceland, continue to pass up the east coast throughout April and sometimes later; numbers are usually small but the Isle of May sometimes has a peak of 100–200. Birds of the 'Greenland' race *O.o. leucorrhoa*, which also breeds in Iceland, Faeroe and northeast Canada, are recorded most often in the Northern Isles and on the west coast. The spring passage of these larger, brighter birds is slightly later, with most movement occurring in early May and stragglers quite frequent in early June. The two races sometimes occur together on passage, as in 1981, when about 100+ of the 450 Wheatears on Fair Isle on 12 May were considered to be Greenland birds.

Native breeding birds start to leave in July and the first south-bound migrants arrive in early August. Autumn passage numbers are very variable, with falls of several hundred continental birds arriving in some seasons and very few in others. Greenland Wheatears, most of which pass through in September/October, seldom occur in parties of more than 10–20 at this season and are recorded in very small numbers in widely scattered parts of the country. Most Wheatears have left by the end of November, but there have been a few December records in recent years and even a January

one – in Peebles in 1980. The only previous January record was in 1893.

The main wintering area for Wheatears breeding in western Europe and Iceland/Greenland is in Africa south of the Sahara. No Scottish-ringed birds have been recovered there, nor have any ringed on foreign breeding grounds been recovered in Scotland. Most recoveries to date have been in August/November and April/May in France, Iberia and Morocco, while the most distant was at an Algerian oasis 3,620 km south of the ringing site on Fair Isle.

Pied Wheatear *Oenanthe pleschanka*

Vagrant (SE Europe) – three records, 1909–83.

The first two Pied Wheatears recorded in Scotland, both females, were shot: on the Isle of May on 9 October 1909 (ASNH 1909:2) and on Swona, Orkney, on 1 November 1916. Had they not been shot the records would probably not have been considered acceptable; even today the identification of female and immature wheatears presents problems (BB 48:130, 49:317). The only recent record is of a male which frequented the mouth of the Don in Aberdeen from 26 October to 10 November 1976 and was trapped. There had been only eight records of the Pied Wheatear in Britain by the end of 1983, one in May and seven in autumn.

Black-eared Wheatear *Oenanthe hispanica*

Vagrant (S Europe) – eleven records, 1907–83.

The Black-eared Wheatear, the most frequently occurring of the vagrant wheatears, was first noted on Fair Isle in September 1907 (ASNH 1908:81) and has since been seen there in 1951, 1964 (spring and autumn), and 1979. The other records are from St Kilda (1911), the Isle of May (1949 & 1980), Caithness (1969) and Out Skerries (1981 & 1983). Six of the occurrences have been between 21 September and 13 November, the remainder between May and mid July. Four of the birds were adult males, two first-winter males and four females. All those subspecifically determined belonged to the nominate race *O.h. hispanica*, which breeds from southern Europe and northeast Africa eastwards and winters on the southern fringes of the Sahara. A field description is given in SB 6:214.

Desert Wheatear *Oenanthe deserti*

Vagrant (N Africa) – eight records, 1880–1984.

The first Desert Wheatear recorded in Scotland was a male near Alloa on 26 November 1880. There had been five more records by 1950 but only two since: males on Fair Isle (18 November 1970) and in Caithness (26 December 1984–10 January 1985). Three of the earlier occurrences were also on Fair Isle, in 1928, 1929 and 1940, and the others at Arbroath in 1887 and the Pentland Skerries in 1906. The Orkney record was on 2 June but the rest were all in autumn, between 6 October and 28 December. Three races of the Desert Wheatear are recognised: *O.d. homochroa*, breeding west of the Nile; *O.d. deserti*, breeding from east of the Nile to the Middle East; and *O.d. atrogularis*, breeding still further east. The Pentland Skerries bird and those on Fair Isle in 1928 and 1929 were ascribed to *atrogularis*, *deserti* and *homochroa* respectively; the others were not racially determined.

Black Wheatear *Oenanthe leucura*

Vagrant (W Mediterranean) – two records, 1912 and 1953.

Both the records of this distinctive bird are from Fair Isle. The first, a male, was seen on 28–30 September 1912 (SN 1913:26) and the second, probably a female, on 19 October 1953. The Black Wheatear breeds on rocky slopes around the western Mediterranean and is normally sedentary. There had been only five British records of this species by the end of 1983.

Rock Thrush *Monticola saxatilis*

Vagrant (S Europe) – five records, 1910–83.

This large, rufous-tailed chat was first recorded on 17 May 1910, when one of two males on the Pentland Skerries was shot (ASNH 1910:148). It has since occurred four times: on Fair Isle on 8 November 1931, 16 October 1936, and 30 June 1970, and on St Kilda on 17 June 1962. The most recent Fair Isle bird was a first-summer male (SB 6:336). The Rock Thrush, a bird of sparsely vegetated rocky slopes, breeds from Iberia and northwest Africa eastwards throughout southern Europe and winters in Africa.

[Blue Rock Thrush *Monticola solitarius*]

Escape/vagrant (Category D)

A male Blue Rock Thrush on North Ronaldsay from 29 August to 6 September 1966 is the only record (BB 59:302). This species, which breeds around the Mediterranean, is not strongly migratory and is quite often kept in captivity. This, together with the fact that the Orkney bird had damaged tail-feathers (often a sign that a bird has been caged), makes it probable that the record relates to an escape. (SB 4:451; BB 61:302).

White's Thrush *Zoothera dauma*

Vagrant (Siberia) – ten records, 1878–1983.

White's Thrush, bigger than a Mistle Thrush and very distinctively marked, was first recorded in 1878, when one was shot in Berwick. Three of the four other pre-1950 occurrences were on Fair Isle, in 1929, 1944 and 1948, and there have been three there subsequently, in 1958 (BB 53:412), 1971 and 1973. The other records are from Aberdeen (1913), Perth (1956), Whalsay (1975) and Lanark (1979). All occurrences have been in autumn/winter, the dates ranging from 24 September to 13 February, with most in October/November. White's Thrush is a forest bird, breeding from central Siberia to the Far East and wintering largely in Indo-China.

Siberian Thrush *Zoothera sibirica*

Vagrant (Siberia) – two records, 1954 and 1984.

An adult male on the Isle of May on 1–4 October 1954 (SB 8:114; BB 48:21) was the first British occurrence of the Siberian Thrush. There have since been two in England, in December 1976 and December 1977, and one on South Ronaldsay, Orkney, on 13 November 1984 – all males. Although it breeds no nearer than Siberia the species has been recorded in most European countries. The Isle of May record is currently under review as possibly relating to an escape.

Hermit Thrush *Catharus guttatus*

Vagrant (N America) – one record, 1975.

This species was first recorded in Britain on 2 June 1975, when one was found on Fair Isle (BB 72:414–417). There had been no further occurrences by the end of 1983. A wheatear-sized bird with a reddish tail which it cocks frequently, the Hermit Thrush breeds in mixed forest from Virginia north to Manitoba and Quebec, and winters in the southern USA.

Swainson's Thrush *Catharus ustulatus*

Vagrant (N America) – one record, 1980.

The only record of Swainson's Thrush (formerly known as the Olive-backed Thrush) in Scotland is of one at Scatness, Shetland, on 25–29 October 1980 (BB 74:484). There had been seven previous occurrences in Britain and Ireland; apart from the first, which was in May, all were in October (Sharrock & Grant 1982). Unlike the previous species, Swainson's Thrush is a long-distance migrant, breeding in spruce forests from Newfoundland to West Virginia and wintering in Central and South America.

Gray-cheeked Thrush *Catharus minimus*

Vagrant (N America) – five records, 1953–83.

All the records of this small thrush have been in autumn. The first British occurrence was on Fair Isle, on 5–6 October 1953 (BB 47:266–267). The subsequent Scottish records are from Fair Isle (October 1958), St Kilda (October 1965), Lossiemouth (November 1965) and Shetland Mainland (October 1982). Three were first-winter birds. The Gray-cheeked Thrush is a long-distance migrant, breeding in the spruce forests of northern North America and wintering in South America. There had been 16 British and Irish records by the end of 1983, five of them in October 1976. A summary of the characters distinguishing this species from Swainson's Thrush is given in Sharrock & Grant (1982).

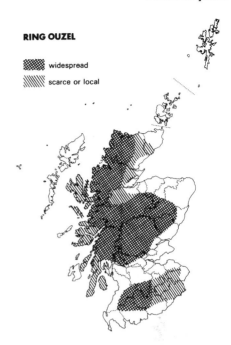

RING OUZEL

▓ widespread

▨ scarce or local

Ring Ouzel *Turdus torquatus*

Migratory breeder, widely but thinly distributed in most moorland and mountainous areas of the mainland, local on Arran, Mull and Skye, and sporadic in the Small Isles, Orkney and Shetland, but absent from the Outer Hebrides. Passage visitor in small numbers.

B&R noted that a 'serious decrease' in the breeding population of the Ring Ouzel had occurred during the first half of this century; previously a common bird in most hill areas it had become relatively scarce in many districts by 1950. Comments made by local recorders in the early 1980s indicate that numbers have continued to decline, although not to such an extent as to suggest that B&R's feared extinction of the species is likely to become a reality in the foreseeable future. An even more marked decline has occurred in Ireland, where it has been attributed partly to inter-specific competition with the Blackbird, which has been steadily expanding its range and increasing in numbers (Atlas). Evi-

dence of direct competition between Ring Ouzels and Blackbirds is hard to find and in one study aggressive encounters with Mistle Thrushes were far more frequent (Durman 1977). In Scotland afforestation has undoubtedly rendered some areas unsuitable for this bird of the open moors.

Most Ring Ouzels nest at altitudes between 250 and 500 m asl (Flegg & Glue 1975). Work by R. F. Durman (1977) in the Pentlands in the 1970s showed that they tend to concentrate in certain valleys and avoid open heather moor; one 2 km stretch of valley held 12 breeding pairs, with nests sometimes as little as 160 m apart. These favoured areas had steep rocky banks, where the nests were usually sited, and rich feeding areas nearby in the form of sheep-grazed grassland in the valley floor. Lance Vick found that c4.2 km² of the Moorfoot Valley (Midlothian) normally held five or six pairs, some of which probably raised two broods.

The Ring Ouzel has ceased to breed in Wigtown, while its recent decrease in Kirkcudbright and Peebles has been attributed to afforestation. It still occurs in fair numbers in the hills of south Ayr and Dumfries, and a few pairs nest in north Ayr glens, but there has been no breeding record for Renfrew since 1976. It is local in south Lanark and the Cheviots, Pentlands, Moorfoots and Lammermuirs, is locally common on the scarp faces of the Ochils and Gargunnock Hills, and has bred sporadically in the Lomonds. In the more mountainous regions from the Grampians north there is less evidence of decline and the Ring Ouzel continues to breed in many of the glens. Five territories were known on Arran in 1983 but there is no recent breeding record for Bute, and no confirmation of breeding on Jura since 1968–72. In Mull and Skye small numbers probably still breed regularly, and on the Small Isles nesting occurs at least sporadically. In Orkney Ring Ouzels bred on Hoy several times in the 1970s (and earlier) but there has been no confirmed breeding since 1977. In 1972 a pair nested in Shetland for the first (and so far only) time, on the Mainland.

In spring the first birds usually appear in March, sometimes as early as the first week, and most breeding pairs are on territory by early May. Passage through the Northern Isles often continues into June and seldom involves large numbers; peak counts on Fair Isle are usually under 50 but a record 300 were present on 3 May 1969. Autumn passage is more variable in both occurrence and extent; in some years only a few birds are seen, while in others falls of over 1,000 occur. Passage may start at the end of August, the main movement usually takes place in late September/early October, and stragglers often occur in November. By far the largest movement of Ring Ouzels recorded in the last two decades was in September 1976, when there were 80+ on Fair Isle on 26th and c1,000 at Kirkwall on 27th. During the following six weeks small parties were recorded as far west as the Outer Hebrides, where this species is a scarce and very irregular visitor. Most birds have left by the end of October but there have recently been occasional winter records, including a male in Shetland in January 1977 and one at Drumnadrochit in January 1980.

The Ring Ouzel has a fragmented distribution, breeding in Britain, west and north Scandinavia, and the mountain ranges of central and southern Europe. Birds passing through the Northern Isles, or arriving in numbers on the east coast, are assumed to be of Scandinavian origin, but there have not yet been any ringing recoveries confirming this, although one caught on the Isle of May in April was recovered (six years later) in Norway in October. The species' main wintering area is around the Mediterranean and in southern France and Iberia; birds ringed in Scotland have been recovered between October and April in all these areas, the most southerly records being first-winter birds in Algeria and Morocco.

Blackbird *Turdus merula*

Among the most widely distributed of our breeding birds, occurring throughout the mainland and on all the main island groups. Some native birds winter in Ireland and a few travel further south, but most are sedentary. Large numbers of immigrants arrive in autumn, many continuing south to winter in France and Iberia.

The Blackbird occurs in almost all habitats that include trees, but is scarce in the more open Highland birchwoods; where there are few trees it readily nests in buildings or clumps of shrubby vegetation. Breeding densities vary greatly with habitat. The Atlas quotes a figure of up to 250 pairs/km² in suburbia, where abundant supplementary food is usually available and where a patchwork of lawns, shrubs and hedges provides very favourable habitat; on an East Lothian housing estate densities ranged from 15 pairs/km² in gardens under five years old to 117 pairs/km² in 5–15-year old gardens nearby (S. R. D. da Prato). Woodland densities are in general lower: in predominantly oak woodland Williamson (1974) found 36–122 pairs/km² in Stirling, but far fewer in northern Argyll, and only ten pairs/km² in Wester Ross, while da Prato (1985) found densities equivalent to 126 and

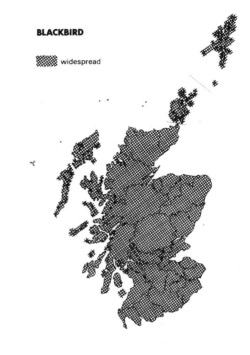

BLACKBIRD

▨ widespread

200 pairs/km² in small mixed woods and regenerating scrub respectively. Blackbirds are scarcer in large conifer plantations, where they tend to occur along the woodland edge (Moss 1978a). This species adapts better than most other woodland birds to living on farmland and is among the commonest passerines on farms with reasonable areas of cover, though in arable districts densities are relatively low; in 1984 there were fewer than eight pairs/km² on c1,600 ha of farmland with hard-trimmed hedges in East Lothian (da Prato 1985).

Blackbirds are less affected than Song Thrushes by severe weather, and CBC results for Britain as a whole show that the breeding population fluctuates to only a limited extent, with indices of abundance dropping by less than 15% even after winters as hard as those of 1978/79 and 1981/82. Communal roosting sometimes occurs. In Aberdeen suburbs juveniles started to move into roosts in June and more than 80% of the peak winter total was present by August/September; up to 100 birds occupied a single roost, often among rhododendrons (Swann 1975a). Although many of the Blackbirds using these roosts were local birds, some immigrants were also present; at a rural roost in Inverness a higher proportion were immigrants (Baillie & Swann 1980).

Well over 1,000 Blackbirds ringed in Scotland have been recovered. The records of birds ringed in the nest and recovered more than 250 km away indicate that many first- and some second-winter birds are in Ireland between November and February, with comparatively few going south into England and Wales or across to the Continent. Winter recoveries of immigrants ringed on their breeding grounds suggest that the majority come from Fenno-Scandia and Denmark; most of those originating south of the Baltic probably continue southwards to France and Spain and do not winter in Britain. There has been a recent marked increase in the numbers either ringed in Finland and recovered in

Scotland or *vice versa*, and this is believed to reflect the progressive northward expansion of the Finnish breeding population (Spencer 1975). Individuals appear sometimes to make navigational errors and travel in the 'wrong' direction: one adult caught on Fair Isle in mid October was shot nearly 500 km away in Norway two days later, and one ringed in Unst in September was in Iceland (where the species does not breed) 15 months later.

Immigration may start in early September, but the main passage period is usually between mid October and mid November, with later influxes in some years. Several hundred a day, and occasionally a thousand or more, are recorded on Fair Isle and the Isle of May at peak periods, while flocks can sometimes be seen arriving low over the sea at other points on the east coast. Spring passage takes place between March and early May and is most marked in late March and early April; the numbers involved are usually much lower than in autumn.

Although Blackbirds were nesting in all mainland counties and island groups by 1950, colonisation of many of the islands is comparatively recent. The species was relatively scarce in Orkney until the late 1800s, and in Shetland until the 1930s (B&R). It is now a common breeding bird in both areas and appears also to have increased in the Outer Hebrides.

Eye-browed Thrush *Turdus obscurus*

Vagrant (Siberia) – four records, 1964–84.

An Eye-browed Thrush on North Rona on 16 October 1964 was the first Scottish and second British record (SB 3:217; BB 61:218–223). There have been three subsequent Scottish records: at Lochwinnoch on 22 October 1978, at Newburgh (Aberdeen) on 27 May 1981, and on Orkney Mainland on 25–26 September 1984. The Eye-browed Thrush breeds in northeastern Siberia and winters south to Indonesia. A full field description is given in Sharrock & Grant (1982).

Dusky/Naumann's Thrush *Turdus naumanni*

Vagrant (Siberia) – three records, 1961–83.

All the Scottish records of the Dusky Thrush are from Shetland and in autumn. The first was on Fair Isle on 18–21 October 1961, the second on Whalsay on 24 September 1968 (SB 5:392–4), and the most recent on Shetland Mainland on 6–13 November 1975. There had been two earlier occurrences in England (BB 53:275) and there have been two since, in February/March 1979 and November 1983. This is a Far Eastern species which breeds from central and northern Siberia eastwards and winters from Manchuria south to Burma and Assam.

Black-throated/Red-throated Thrush *Turdus ruficollis*

Vagrant (central Asia) – six records, 1879–1983.

The first record of the Black-throated Thrush was in 1879, when an immature male was shot in Perth in February (Ibis 1889:579). There were no further records until 1957, since when there have been five, all in Shetland and in autumn/winter: on Fair Isle from 8 December 1957 to 22 January 1958 (BB 51:195) and on 17 October 1978, and on Shetland Mainland on 5–6 October 1974, 6–12 November 1977 and 7 December 1981. Both the Fair Isle birds were males, the first an adult and the second an immature, while those on Mainland included one female and one first-winter. There had been eleven British records by the end of 1983. The Black-throated Thrush *T.r. atrogularis*, similar to a Fieldfare in size and shape, breeds in lowland forests in central Asia and winters from Iran through India to Burma. The conspecific Red-throated form *T.r. ruficollis* has a more easterly distribution.

Fieldfare *Turdus pilaris* [1]

Abundant passage and winter visitor, many moving further south during the course of the winter. Scarce and sporadic breeder since 1967.

The Fieldfare has considerably expanded its breeding range in the last 150 years, spreading south of the Baltic from Fenno-Scandia; it now also nests from central France and Germany eastwards across Russia. Britain is on the western fringe of the current range, but in the 1930s Fieldfares bred on the other side of the Atlantic, in Greenland and North America. They do not nest in Iceland but have done so in the Faeroes. Ringing recoveries indicate that those visiting Scotland are from Fenno-Scandia.

The first arrivals sometimes appear in August and may consist largely of young birds. Numbers are seldom high until October, when the main influx generally occurs; this varies greatly between years, with successive waves of several thousands at a time moving across the country in some seasons and few flocks of more than a thousand reported in others. Peak counts at Fair Isle have ranged from 100 to

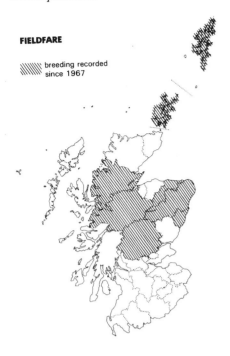

FIELDFARE

////// breeding recorded
\\\\\\\ since 1967

(Balfour 1968); Fieldfares bred there again in 1969 and 1974 and possibly in other years too. Nesting has been confirmed in Shetland (annually from 1968 to 1971 and in 1973, 1974 & 1982), Inverness (1970, 1981 & 1982), Banff (1972), Kincardine (1973), Aberdeen (1976 & 1977), and Perth (1979), and suspected in Caithness, Sutherland, Ross, Mull and Selkirk. Not more than two pairs have been confirmed as breeding in any one year. No sign of breeding was reported in 1978, 1980 or 1983. Most breeding records are from moorland valleys, hillside birchwoods, or plantation edge; nest sites have included elder bush, fir, sycamore and long heather.

Song Thrush *Turdus philomelos*

Mainly migratory breeder, widely distributed on the mainland, Clyde Islands and Inner Hebrides, regular but scarcer in the Outer Hebrides and Orkney, and sporadic in Shetland. Many native Song Thrushes winter in Ireland and some of the immigrants that arrive in autumn move on further south.

The Song Thrush occupies almost any habitat with trees or shrubs, but is least abundant where bushy cover is limited. In Orkney and the Outer Hebrides it often frequents ditches, dykes and areas of long heather. Breeding densities vary with woodland type and tend to be highest where tall scrub or young woodland is close to open feeding areas. In East Lothian da Prato (1985) found densities equivalent to 144 pairs/km^2 in 10 ha of regenerating scrub, 86 pairs/km^2 in three small mixed woods totalling 30 ha, and only 5 pairs/km^2 on c1,600 ha of farmland with hard-trimmed hedges and occasional shelterbelts. Blackbirds were nearly twice as numerous in the same areas. Although the CBC data for Scotland are very limited they suggest that farmland densities

8,000 (29 October 1979); on the Isle of May they seldom exceed 1,000 but reached a record 4,000 on 11 October 1982. Large numbers occur at times not only on the east coast (eg c33,000 near the Ythan Estuary on 26 October 1980) but also passing south inland, and in years of major invasions sizeable numbers reach the Hebrides, where this species is normally a much scarcer visitor than the Redwing. Severe weather in mid winter produces local movements – as on 17 December 1969 when some 2,500 an hour passed SSW over Caerlaverock in blizzard conditions – and may also result in further influxes from the Continent.

Fieldfares often occur in mixed flocks with Redwings, but tend to feed on the ground more frequently than they do. Although many apparently move on to Ireland or England, and some to France or south as far as the Mediterranean (da Prato *et al* 1980), a proportion usually stay throughout the winter. Scattered small flocks are seen in the Highlands most winters (R. L. Swann), a few occasionally remain in Shetland and Orkney, and there are wintering records from Speyside (eg in 1968/69, when there was a very heavy crop of juniper berries), but most are in the lowlands. Large numbers regularly gather to feed on sea buckthorn in the Gullane area; there were 5,000+ there in November/December 1975. Northward movement may be detectable from late March but is most obvious in April/May, when flocks of several thousands can be seen in many parts of the country. Recent high spring counts include 10,000 in Wigtown on 23 March 1976, 4,000 in Berwick on 24 April 1981, and 4,500 in Kincardine on 10 May 1981. The highest spring peaks recorded on Fair Isle and the Isle of May have been 5,000+ and 3,000+ respectively, but in some seasons passage birds are scarce, with numbers not exceeding 100 a day.

Breeding was first recorded in Britain in 1967, when a nest and three newly-fledged young were found in Orkney

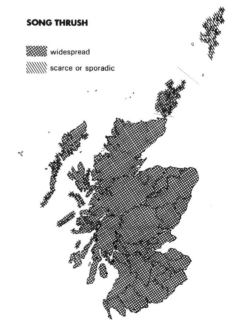

SONG THRUSH

▓▓▓ widespread

\\\\\\\ scarce or sporadic

are similar to those in other parts of Britain (Fuller *et al* 1985); the range of densities recorded in 1978 and 1983 for woodland census plots in southern England, northern England and Scotland also shows little sign of regional variation. CBC results for Britain as a whole suggest that woodland Song Thrush numbers have declined since the early 1970s, although recent fluctuations have been less marked in this segment of the population than among those breeding on farmland. Until the severe winter of 1946/47 a few pairs bred regularly in Shetland, but since then they have done so only sporadically. Apart from this decrease there is little evidence of any significant change in status in the last three or four decades.

Recoveries of Song Thrushes ringed as chicks in Scotland demonstrate that there is a marked southwesterly movement in autumn, with nearly all – both adults and immatures – deserting much of the Highlands. Many Scottish birds are in Ireland between October and January, while a few go south as far as Portugal. In mid winter Song Thrushes are thinly scattered over low-ground areas; slightly larger numbers were recorded in the Central Lowlands in the hard winter of 1981/82 than in 1982/83, presumably due to arrivals from the Continent (Winter Atlas).

The extent of immigration from the Continent varies considerably but seldom reaches anything like the scale of the other thrushes. Peak counts on Fair Isle rarely exceed 500, while the maximum recorded on the Isle of May – c3,000 in October 1966 – was about ten times the usual peak. Most autumn migrants probably belong to the continental race *T.p. philomelos*, which breeds from Fenno-Scandia and eastern Europe east across Siberia and almost entirely deserts its breeding areas in winter. Many of these birds move on further south and there have been winter recoveries in France, Iberia and Italy (but few from Ireland) of birds ringed while on passage. Immigrants sometimes start to arrive in late August, the main movement usually takes place in September/October, and later influxes occasionally occur in severe winters. Native breeding birds have returned to their territories by February/March, when small flocks are often to be found in highland glens, especially during cold spells (R. L. Swann). Spring movement through the Isle of May and Northern Isles often continues into May. Only a few continental birds have been racially identified on spring passage, when birds belonging to the British race *T.p. clarkei* also occur.

Redwing *Turdus iliacus*   [1]

Abundant passage and winter visitor, many moving on further south. Has bred annually in the Highlands since 1967.

Redwings of the nominate race *T.i. iliacus*, which breeds from Fenno-Scandia east through Siberia, and the Iceland/Faeroe race *T.i. coburni* are separable on plumage characteristics; both are present in the flocks which move south across Scotland in autumn. At that season birds of continental origin are in the majority and those trapped on Fair Isle include a high proportion of adults (ie experienced birds) suggesting that some Redwing populations cross the North Sea deliberately, rather than simply being wind-drifted across it as is the case with many other species (da Prato *et al* 1980). The evidence suggests that Icelandic birds occur mainly in the north and northwest, comparatively few reaching southeast Scotland. Flocks of Redwings sometimes travel with Field-

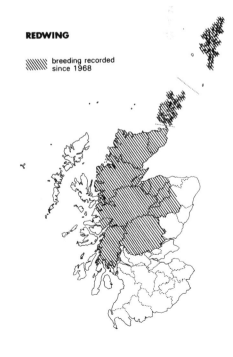

REDWING

⦀⦀⦀ breeding recorded
     since 1968

fares, stripping the berries from rowan and hawthorn as they go. In the north they appear to head for the inland glens, overflying the coast unless forced down by bad weather, and seldom remain in a glen for more than 24 hours. The size of the berry crop largely determines the numbers that stop to feed in an area (Swann 1983a).

Although the first arrivals are sometimes as early as mid August, the main influx occurs between late September and mid November, with the largest falls usually in October. Very large numbers are sometimes involved: recent peak counts (all in October) include c65,000 on Fair Isle and 10,000 on the Isle of May in 1979 (both all-time records), and 40,000–70,000 passing over Kirkwall in 1976, when 900+ were killed at the North Ronaldsay light during the same movement. In the west numbers are generally much lower, but there have been records of 5,000 on Mull (late September 1981) and 700 killed at the Rhinns light on Islay (October 1976); movements there may continue well into November and probably involve a higher proportion of Icelandic birds than do those in the east. Later influxes occur on the east coast when conditions are severe on the Continent (da Prato *et al* 1980) and even in the Northern Isles unusually large numbers are sometimes present in mid winter, as in 1978, when c12,000 were roosting in Berstane Wood, Orkney, in January.

Spring passage in March/April is regular in the Highlands, with small flocks (usually fewer than 100) occurring mainly on rough grazing along the moorland edge. The proportion of Icelandic birds among those trapped on Fair Isle is very much higher in spring, and it is thought that many continental Redwings return to their breeding grounds via the Continent, rather than through northern Britain (da Prato *et al* 1980).

Ringing recoveries indicate that Redwings passing through, or wintering in, Scotland originate from Iceland, Fenno-Scandia and USSR; a relatively high proportion appears to come from Finland and USSR (da Prato *et al* 1980). This species winters well south of its main breeding range – in Europe, north Africa and the Middle East. Most of those ringed on autumn passage and recovered in winter have been in France, Portugal and Spain, but there have been a few reported in November/February in Ireland, England, the Netherlands, Belgium, Germany, Italy, Greece, Turkey and Lebanon.

The first record of nesting was in Sutherland in 1925 and breeding was reported very sporadically over the next 40 years. Seven pairs were located in 1967 and intensive Atlas fieldwork the following season revealed 20 pairs in Wester Ross alone. Nesting has since occurred most regularly in Sutherland, Ross and Inverness, but has also been confirmed in Shetland, Orkney, Caithness, Nairn, Moray, Banff, Aberdeen, Perth and Argyll, and has possibly occurred in South Uist, Dunbarton, Stirling and Angus. The number of known territory-holding pairs reached a peak of c53 in the mid 1970s but then decreased considerably. Since 1981 numbers have risen again – although this apparent increase may be partly attributable to more intensive searching. Thirty pairs were confirmed as breeding in 1982, while in 1983 there were reports of 64 pairs in Sutherland/Ross/Inverness and a single pair nested, unsuccessfully, in Shetland.

Breeding has taken place in a variety of habitats, including birch scrub, oakwoods, gardens with shrubberies and pine forest (Sharrock 1972, Williamson 1973), and also in damp alder woods and young plantations with birches to act as song-posts. In 1935 a pair even nested on treeless Fair Isle; these birds were considered to belong to the Icelandic race, but those racially identified in recent years have been of continental origin (SB 5:342). The reasons for the apparently quite sudden colonisation of Scotland by the Redwing (and various other species) are not fully understood, but may be at least partly due to recent cold springs causing some south-westerly shift of northern species. 'Left-overs' from unusually large autumn influxes may be partly responsible too, while westerly drift of migrants moving north across the Continent in spring may also be involved (Williamson 1975).

Mistle Thrush *Turdus viscivorus*

Partial migrant, widely distributed throughout the mainland and on the Clyde Islands, but local on the Inner Hebrides and sporadic on the Outer Hebrides and Northern Isles. A scarce passage visitor.

The Mistle Thrush underwent a major range expansion last century; in the early 1800s it was found only locally in southern Scotland, but by 1950 it had colonised every mainland county (Parslow). A bird of woodland edge, it is equally at home in suburban parks and among scattered moorland trees up to at least 400 m asl; where suitable trees are scarce it occasionally nests among rocks. Although present in a wide range of habitats, it occurs at relatively low densities. About 9 pairs/km^2 have been found in oakwood in Argyll (Williamson 1974) and in some types of conifer plantation (Moss 1978a) but average breeding densities are

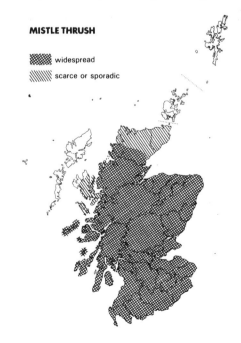

MISTLE THRUSH

▨ widespread

▧ scarce or sporadic

probably very much lower; figures of under 2 pairs/km^2 and under 5 pairs/km^2 are quoted for farmland and woodland respectively in the Atlas. Mistle Thrushes breed early in the year – first clutches are sometimes laid in March – and suffer badly in severe winters, especially during periods of prolonged hard frost such as occurred in 1962/63 and 1981/82. The CBC index showed a 26% decline in the farmland breeding population after the winter of 1981/82, and it is probable that the decrease was greater in Scotland than further south. Birds occupying woodland are less seriously affected.

On the mainland the Mistle Thrush is scarce only in parts of the north and west where woodland is absent, but it seems not yet to have colonised all the available habitat on the islands. Although quite abundant on the Clyde Islands and southern Inner Hebrides, it is very local in Skye and the Small Isles, on some of which it breeds only sporadically. Afforestation is progressively increasing the amount of suitable habitat on many of these islands, and numbers there may be expected to increase over the next few decades. Breeding has occurred sporadically in the Outer Hebrides in Stornoway Woods (last recorded in 1968), and in Orkney on Rousay (1850s), Shapinsay (1941) and Mainland (last recorded in 1971); nesting apparently occurred fairly regularly on Mainland earlier this century (Booth *et al* 1984). There is no record of even attempted nesting in Shetland.

Mistle Thrushes flock in autumn and counts of up to 300 are not unusual from late July on through the first half of the winter. Although many apparently remain in or near their breeding area throughout the winter, most upland sites are abandoned and some birds move south: there have been November/February recoveries, mostly of first-winter birds, in Lancashire, Ireland and France. B&R believed that it was those from high and exposed sites that emigrated, but there is little ringing evidence to support this view. Substantially more Mistle Thrushes were recorded during Winter Atlas survey work in 1982/83 than in the preceding, severer, winter; in both years few were noted north of the Great Glen.

As a passage bird through the Northern Isles this species is scarce and irregular. Peak daily counts on Fair Isle seldom exceed ten and are slightly higher in spring (February/May) than autumn (mainly September/October). Spring occurrences are probably due to birds overflying the Scottish mainland, while autumn migrants presumably originate from the northern part of the European range, which includes most of Sweden and Finland (but only southeast Norway) and extends south to the Mediterranean and east into Russia.

American Robin *Turdus migratorius*

Vagrant (N America) – six records, 1961–83.

The Blackbird-sized American Robin was first recorded in 1961, when one was on Orkney Mainland on 27 May (BB 55:577). There had been five more reports by the end of 1983: from Kirkcudbright (12 May 1966), Foula (11–16 November 1967 and 3–16 November 1982), St Kilda (14 January to 15 February 1975) and Caithness (5 November

1981). Most of the British occurrences have been between November and February. The American Robin is widely distributed in North America and is a partial migrant; it seems surprising that it is not carried across to this side of the Atlantic more often. Some are kept in captivity in this country so there is always the possibility of escapes.

[Cetti's Warbler *Cettia cetti*] [1]

The record of a Cetti's Warbler at Stirling on 11 July 1977 (SBR 1977) was subsequently rejected by the BBRC. There is, however, an increasing likelihood of this species occurring in Scotland as in recent years it has expanded its range in Europe and now breeds in southern and eastern England, although first recorded there only in 1961 (Sharrock & Grant 1982). It favours marshy scrubland.

Pallas's Grasshopper Warbler *Locustella certhiola*

Vagrant (Siberia) – five records, 1949–83.

Four of the records of this marshland warbler are from Fair Isle: on 8–9 October 1949 (BB 43:49), 2 October 1956, 20–24 September 1976 and 21 September 1981. The most recent occurrence was a first-winter bird on the Out Skerries, on 4–8 October 1983. Pallas's Grasshopper Warbler breeds from western Siberia eastwards and winters in Indo-China; vagrants occur sporadically in Europe but there have been few other British records.

Lanceolated Warbler *Locustella lanceolata*

Vagrant (east-central Russia) – thirty records, 1908–83.

This tiny, skulking bird, like a rather drab Grasshopper Warbler, could be easily overlooked and may occur more often than the records suggest. All but four of the records are from Fair Isle, the first in 1908, when an immature was shot on 9 September (ASNH 1911:71). The others are from Pentland Skerries (1910), Out Skerries (1973 & 1978), and an oil rig in the Forties Field (1978) – all sites with relatively little cover. Occurrences are sporadic; between 1950 and 1983 there were reports in only 12 years, seven of them in the 1970s. All but one of the records have been in autumn, mainly between 20 September and mid October but with extreme dates 8 September and 1 November. The 1953 record on 4 May should perhaps be regarded with suspicion as the only British spring record to date. In view of the species' rarity it is surprising that three should reach Fair Isle on the same day (11 October 1975); the island's annual total reached four that year and was nearly equalled in 1982, when three individuals were trapped between 13 September and 6 October. The Lanceolated Warbler breeds from east-central Russia to Japan, frequenting areas of tall herbage and scrub, and winters in Indo-China and Malaysia.

Grasshopper Warbler *Locustella naevia*

Migratory breeder, nesting at least sporadically in most mainland counties; most abundant in Galloway, the western counties north to Argyll, and the Clyde Islands. Scarce passage visitor.

Although the Grasshopper Warbler has such a distinctive song it may often be under-recorded; its high-pitched reeling is beyond the hearing range of many people, it sings most regularly at dawn and dusk but dull, cold weather may inhibit song, and it is skulking in habits. Confirmation of breeding is difficult, as nests are well concealed, and song may cease when a pair has mated. A further complication arises from the fact that sites may be occupied and deserted within a few years, as the habitat changes. This is especially true of recently afforested areas, which become suitable for occupation soon after planting, once thick ground cover has grown up, but are deserted again within ten years, as the young trees meet in the rows. This habitat is now probably as important for the Grasshopper Warbler as the damper vegetation formerly regarded as most favoured; in Kirkcud-

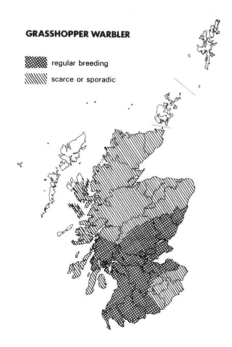

GRASSHOPPER WARBLER

▓ regular breeding

╱ scarce or sporadic

bright even young plantations near the upper limit of planting are colonised (A. D. Watson). Breeding probably occurs at least sporadically in all mainland counties, though there have been no recent reports from Berwick, Selkirk, Peebles or Clackmannan. Song is occasionally recorded in both Inner and Outer Hebrides but breeding has not been confirmed there or in the Northern Isles.

Little information is available on breeding densities but it is known that the population fluctuates markedly between years. At Lochwinnoch 14–20 singing males were present in 1968–72 but only about eight in 1975–82, and in east Dunbarton 21–23 males were located in 1981 but only four the following year. In the Lothians densities never exceeded one territory/km^2 on nearly 2,000 ha surveyed in 1979–82 (S. R. D. da Prato). In Scotland this species is at the northern fringe of its breeding range, and annual variations may to some extent reflect weather conditions during spring migration, with more birds reaching Scotland in a mild, early year than in a late one.

The Grasshopper Warbler's scarcity as a passage migrant is probably attributable to its breeding distribution. In Europe it is widespread from northern Spain to the southern shores of the Baltic but, in contrast to most of the warbler species that feature in falls at Fair Isle and the Isle of May, relatively few breed in Scandinavia. More Grasshopper Warblers are recorded in Wales and the west of England than in the east (presumably reflecting the breeding distribution in Scotland), but even there passage numbers declined in the 1970s (Riddiford 1983). European breeding birds winter in Africa, probably in the northern Tropics. The earliest spring arrival dates range from 12 to 29 April, and the main influx is usually in early May. In years when overall numbers are low, as in 1975, none at all may reach the Northern Isles in either spring or autumn, and even in a 'good' year daily maxima on Fair Isle and the Isle of May rarely exceed five. Most have left by mid October but there are a few records to 20 October. The larger numbers recorded on migration at observatories and lighthouses around the Irish Sea are presumed to involve British and Irish birds, whereas east coast arrivals have probably drifted from the Continent with southeast winds (BOU; Riddiford & Findley 1981). B&R recorded kills of this species at the Mull of Galloway and Little Ross lighthouses, but there have been no comparable recent reports from these sites.

River Warbler *Locustella fluviatilis*

Vagrant (Baltic States/Russia) – six records, 1961–84.

This rather large, fan-tailed warbler was first recorded in Britain in 1961, when one was on Fair Isle on 24–25 September (BB 55:137–8). All the subsequent Scottish records have also been there: in 1969 (16 September), 1981 (23–24 May), 1982 (22 & 24–26 September – both trapped) and 1984 (June). By the end of 1983 there had been a total of only ten British records. The 1981 Fair Isle bird was later found dead and its skin is in the Royal Scottish Museum. The River Warbler breeds from the Baltic States and east Germany eastwards through Russia, and winters in east Africa. Its breeding range is currently expanding westwards and it might therefore be expected to occur with increasing frequency in the future (Sharrock & Grant 1982).

Savi's Warbler *Locustella luscinioides* [1]

Vagrant (Europe, including England) – two occurrences, 1908 and 1981.

Both occurrences of Savi's Warbler have been on Fair Isle – but more than 70 years apart. In 1908 two were seen on 14 May (one – a female – was shot) and in 1981 a first-summer bird was trapped on 24 June. Savi's Warbler breeds in marshland throughout much of Europe and winters in east Africa south of the Sahara. A small breeding population has re-established itself since the mid 1960s in southeast England.

Aquatic Warbler *Acrocephalus paludicola*

Vagrant (Europe) – recorded in 17 years between 1950 and 1983.

A skulking little bird, streaked like a Sedge Warbler but sandier in colour, this species has been recorded only from sites where lack of cover has made it easy to see. It probably occurs more often than the records suggest, although undoubtedly much less regularly than in southern England. The Aquatic Warbler was first recorded in 1914 on Fair Isle (SN 1915:5), which has also provided most of the subsequent records. The other sightings have been on the Isle of May (1956, 1960, 1966 & 1967), Out Skerries (1969), St Abbs (1977) and Sumburgh (1980). Occurrences increased in frequency in the 1970s but there were no reports in 1980–83. Most records refer to single birds but there were three together on Fair Isle in August 1979; the maximum recorded in any year has been six, in 1969. Extreme dates are 9 August and 23 October, and most occurrences are in the second half of August. The Aquatic Warbler breeds locally, in marsh vegetation, from the Netherlands and Germany east through European Russia, and winters in Africa south of the Sahara.

Sedge Warbler *Acrocephalus schoenobaenus*

Migratory breeder, widely distributed throughout the mainland, but relatively scarce in the west and northwest Highlands. Breeds regularly on the Clyde Islands and many of the Inner Hebrides, more locally in the Outer Hebrides and Orkney, and not at all in Shetland. A scarce and irregular passage migrant.

Although often associated with reedbeds and other waterside vegetation, the Sedge Warbler is also frequently found in damp rough grassland, scrub and conifer plantations at the early thicket stage; it seldom breeds above 300m asl.

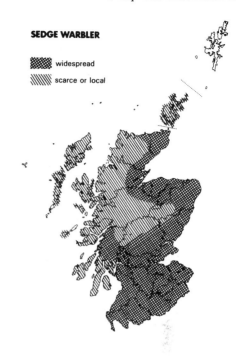

SEDGE WARBLER

▓▓▓ widespread

\\\\\ scarce or local

In favoured habitats densities may be very high, eg c260 in song at the Loch of Strathbeg in May 1981, the equivalent of 172 territories/km² in a mixture of scrub and perennial herbs in the Lothians (S. R. D. da Prato), and 54–124/km² in the Tay reedbeds, where numbers were not significantly affected by reed harvesting operations (McMillan 1979). The overall territory density on a large area of predominantly agricultural land in the Lothians was only 4/km² (S. R. D. da Prato). Numbers fluctuate markedly between years and the breeding population in Britain as a whole is now lower than it was prior to 1968, as this species was affected, like the Whitethroat, by drought conditions in Africa during the winter of 1968/69 (Atlas). Counts at Lochwinnoch have ranged from an average of 32 singing males in 1976–78 to a peak of 130 in 1980, dropping again to c90 in 1981–82.

The only evidence of change in status comes from the islands. Comparison of recent reports with the records given by B&R suggests that the Sedge Warbler has become more abundant in both the Inner and Outer Hebrides since the first half of the century. B&R give no records for Rhum, Raasay, Muck, Canna or Tiree, on all of which the species now breeds at least sporadically; few for the Uists and Benbecula, where it is now common around the machair lochs; and none for Lewis, where it now nests wherever there is suitable habitat – even in rhododendron on the fringes of Stornoway Woods (Cunningham 1983). There is less evidence of change in Orkney, although recent proved breeding on Rousay and singing birds on South Ronaldsay suggest a slight expansion since the early 1940s, when Sedge Warblers bred only on Mainland.

On the Continent this species occurs south to the Pyrenees and eastern Mediterranean but is absent from much of Scandinavia. British breeding birds winter mainly in tropical west Africa. The Sedge Warbler is a scarce and irregular passage migrant, possibly on account of its absence from so much of Scandinavia. Its arrival in spring is relatively late, with the first birds often not appearing until the last week of April. Through daily checking of colour-ringed breeding birds in Midlothian in 1979–82 it was found that males arrived between 9 and 16 May and females between 17 May and 2 June; the considerable differences between years were attributed to variations in the weather (S. R. D. da Prato). A trickle of migrants continues to pass through the Northern Isles until late June. Numbers recorded at the observatories are higher in spring than autumn but even so seldom exceed a daily maximum of ten; the highest-ever daily count for the Isle of May is 200 on 16 May 1970 and for Fair Isle 32 on 16 May 1980.

In Stan da Prato's Midlothian study area the breeding birds departed between 11 July and 19 August, little more than two months after they had arrived. At the end of the short breeding season the birds move into reedbeds to feed on aphids before emigrating (Bibby & Green 1981). Many Scottish birds seem to use reedbeds in southern England or northwest France for pre-migratory fattening. Birds from the Tay, Midlothian and Aberdeen have been caught at Radipole Lake (Dorset) in August/September and Scottish-ringed birds have been recovered in France on autumn passage. Most of the small autumn passage through the observatories takes place in August/September, but there are a few October records and one very late Fair Isle one on 11 November 1975. Evidence of the rapidity with which Sedge Warblers may move south comes from the capture in Devon of one ringed eight days previously in Fife, and the recovery (dead) in Cumbria of one ringed only two days earlier on the Isle of May. A Sedge Warbler ringed in August in the Tay reedbeds and caught in Mali the following April is the most distant Scottish recovery.

Paddyfield Warbler *Acrocephalus agricola*

Vagrant (Russia) – three records, 1925, 1953 and 1984.

All the records of this eastern species, which breeds from the Caspian and southern Russia eastwards to Mongolia and winters from Iran to Indo-China, are from Fair Isle. The first Scottish – and British – occurrence was in 1925, when a Paddyfield Warbler was on the island from 26 September until shot on 1 October; the second was on 16 September 1953 and the third on 30 May 1984. Up to 1969 there had been no other British records but there have since been six, all in autumn and all in England. The Paddyfield Warbler is a bird of shrubby grasslands and marshes during the breeding season; in winter it frequents reedbeds and the ricefields from which it gets its name.

Blyth's Reed Warbler *Acrocephalus dumetorum*

Vagrant (NE Europe) – four occurrences, 1910–83.

The three accepted pre-1950 records are all from Fair Isle and during September: a single bird in 1910 (the first

British record – ASNH 1911:70), four or five in 1912, and one in 1928. On each occasion at least one bird was shot. The only subsequent record is of one (trapped) in Orkney from 5 to 13 October 1979. A paper on the field identification of this difficult species appeared in BB 77:393–411. Although breeding as far west as Finland, Blyth's Reed Warbler winters in India; this markedly eastward migration presumably accounts for its scarcity as a vagrant.

Marsh Warbler *Acrocephalus palustris* [1]

Scarce passage visitor; annual in very small numbers, most frequent in June and on the Northern Isles.

The first record of the Marsh Warbler noted by B&R was in 1906; they give several Fair Isle occurrences and one on St Kilda in 1910. By 1967 there had been 19 spring and eight autumn reports from Fair Isle and more than half the subsequent records are from there. From 1968 to 1983 inclusive there were 75–85 reports, with total sightings in any one year ranging from one to 20 (in 1979). About two-thirds of the reports are in June; the earliest date is 22 May (a singing male in Wigtown in 1982) and the latest 14 October. There have been only seven recent autumn records. Away from Shetland and Fair Isle, the species has been reported only from Orkney, Caithness, Ross, Moray, Fife, the Isle of May and Wigtown.

There are small breeding populations of the Marsh Warbler in England, and in southern Sweden and Finland, but the main breeding range is from northeast France and northern Italy eastwards. Most winter in east Africa. A Marsh Warbler found dead on Copinsay in June 1979 had been ringed the previous August in Jutland; this was the first recovery of a Danish-ringed bird outwith southern Scandinavia.

Reed Warbler *Acrocephalus scirpaceus*

Scarce passage visitor; has bred.

Since 1968 the Reed Warbler has occurred annually in autumn and nearly as regularly in spring. More than 75% of reports are from the Northern Isles and nearly all the others from east coast counties; records from the west mainland and the Inner Hebrides are rare and there is none for the Outer Hebrides. Numbers vary greatly between years. In spring they are never large, seldom exceeding ten in total, but in autumn substantial falls occasionally occur, as in 1978, when there were 17 on Fair Isle on 29 September, and in 1981, when the 60+ recorded included c35 in Shetland from 11 to 16 September. Spring passage occurs mainly between mid May and late June, but sometimes earlier; in 1971, for example, all records were between 22 April and 23 May. The autumn movement extends from mid August into early October, with stragglers to 31 October. There are a few July/early August records, and several reports of song and display in May/June. To date the only breeding record is of a pair which nested, successfully, in a clump of snowberry in Unst in 1973.

The Reed Warbler is quite widely distributed as a breeding species in England north to Yorkshire, frequenting dense and extensive reedbeds. It is likely, however, that those visiting Scotland originate from further north in the breeding range, which extends throughout Europe to Finland, southern Sweden and Norway – where the species has bred only since 1947. The few ringing recoveries shed little light on the matter. They include a bird ringed on the Out Skerries in late September and controlled in Norway about three weeks later, one ringed on North Ronaldsay in late August and caught on Fair Isle two days later, and one ringed as a juvenile in Guernsey in July 1969 and found dead in a Glasgow ship repair yard in March 1971, having probably arrived there on board ship!

Great Reed Warbler *Acrocephalus arundinaceus*

Vagrant (Europe) – fourteen records, 1958–83.

This Skylark-sized warbler was first recorded in 1958, when one was singing at Loch Brow, Shetland, on 4–5 June (SB 1:254). It has since occurred in 11 years, with not more than two recorded in any one year and most reports coming from Fair Isle and Shetland. All but one of the sightings have been in spring, mostly in June but a few from 13 May onwards, and several have involved singing males; one bird remained on Unst for four weeks in 1975. The only autumn record is of one at St Abbs on 13–16 October 1979. Elsewhere in Scotland the Great Reed Warbler has been reported only in Inverness (a singing bird on the Insh Marshes in 1964), Fife (one singing at Kilconquhar Loch in 1970), and Caithness (at Wick in 1981). The species is widely distributed in marshy areas throughout Europe; it winters mainly in Africa south of the Sahara.

Thick-billed Warbler *Acrocephalus aedon*

Vagrant (Far East) – two records, 1955 and 1971.

The Thick-billed Warbler was recorded for the first time in Scotland, Britain and Europe on 6 October 1955, when one was caught on Fair Isle (BB 49:89–93). There has been only one subsequent British record, from Whalsay, Shetland, on 23 September 1971. As the species breeds from Siberia eastwards and winters in Indo-China it is hardly surprising that it is such a rare vagrant.

Olivaceous Warbler *Hippolais pallida*

Vagrant (S Europe) – one record, 1967.

The only record of this shrub-haunting species is of one first seen on the Isle of May on 24 September 1967 and killed two days later by a Great Grey Shrike (BB 61:350). It belonged to the eastern race *H.p. elaeica*, which breeds in southeast Europe and southwest Asia and winters in eastern tropical Africa. The Olivaceous Warbler has occurred about 12 times elsewhere in Britain and Ireland; in southern England birds of the western race *H.p. opaca*, which breeds in Iberia and northwest Africa, have been recorded.

Booted Warbler *Hippolais caligata*

Vagrant (NW Russia) – nine records, 1936–83.

The Booted Warbler was first recorded on Fair Isle, where one was shot on 3 September 1936 (BB 30:226), and occurred there again in 1959, 1966, 1968, 1976 and 1977. All these birds were first seen in late August and several remained on the island for a week or more. The other records are from the Isle of May (31 August 1975), Whalsay (26 September to 4 October 1977), and Out Skerries (11–16 September 1981). The Booted Warbler breeds from northwest Russia east to Siberia and south to Iran, and winters in India and Arabia; it is a bird of shrubby grasslands and forest clearings.

Icterine Warbler *Hippolais icterina*

Scarce passage visitor, annual in small numbers.

This species has a much more northwesterly distribution than the other *Hippolais* warblers occurring in Scotland but is nevertheless an uncommon visitor. In most years it occurs in both spring and autumn, but the annual total only occasionally exceeds 20–25. Numbers are usually larger in autumn than spring, with maxima of 35–38 (in 1980) and 14 (in 1982) respectively in the period 1968–83; in some years there are no spring records. Extreme dates for spring passage are 12 May and 7 July, and most movement takes place between mid May and late June. Some spring birds have been in song. The main autumn passage is in August/September but there are a few later records, to 20 October. Most reports are from Fair Isle, where the Icterine Warbler was first recorded in 1908, and Shetland, but there have also been occurrences in Orkney, Sutherland, Caithness, Aberdeen, Kincardine, Fife, East Lothian and Berwick, and on the Isle of May, Islay and North Uist. It seems surprising that this species, which breeds in open forest and shrubby woodland from Scandinavia and northern France east through Europe and Russia, and winters in Africa, does not occur here more frequently. Perhaps it does – but remains undetected at sites where no trapping is carried out.

Melodious Warbler *Hippolais polyglotta*

Vagrant (S Europe) – less than annual, in very small numbers.

B&R were responsible for the first record of a Melodious Warbler, a female 'got' on the Isle of May on 27 September 1913 (SN 1913:273). The next report was not until 1955 and only two were seen in the 1960s. There were sightings in six years during the 1970s and in 1980, 1981 and 1982 – but not in 1983. No more than three have been seen in any one year. Most occurrences have been in June and September, but there is a wide scatter of dates, from 18 May to 7 October. A bird singing on Fair Isle in June 1969 presumably lived up to its name and did so melodiously! Some two-thirds of the records are from Fair Isle and Shetland; the rest from Orkney, Caithness and the Isle of May. In the field this species is difficult to distinguish from the Icterine Warbler, and the scarcity of inland records anywhere in Britain is possibly a reflection of the problems of identification as much as of actual distribution. Nearly all the Scottish records are of trapped birds. The Melodious Warbler breeds in open forest and scrub from France and Italy south to Iberia and northwest Africa, and winters in western Africa. It occurs more frequently on the southern coasts of England and Ireland than in Scotland.

Dartford Warbler *Sylvia undata*   [1]

Vagrant (Europe, including England) – one record, 1983.

This normally very sedentary species is confined in Britain to the extreme south of England, where the fortunes of the small breeding population are much influenced by the severity of winter weather and by fires on its very localised scrub-heath habitat. It also breeds from northwest France to Iberia and the Mediterranean. Young birds are known to wander in autumn but the appearance of a Dartford Warbler at St Abb's Head, Berwick, on 18 May 1983, the first record for Scotland, is difficult to explain (SB 13:52–53). Presumably very abnormal weather conditions were responsible for this individual straying so far from its breeding range.

Spectacled Warbler *Sylvia conspicillata*

Vagrant (Mediterranean) – one record, 1979.

The only record of this Whitethroat-like warbler is of a male on Fair Isle on 4–5 June 1979 (BB 73:523). The Spectacled Warbler breeds in dry scrubland around the Mediterranean and winters slightly further south. Comments on the possibility of confusion with first-winter female Subalpine Warbler are given in Sharrock & Grant (1982). All other British records of this species were rejected during a recent review and the Fair Isle report is currently undergoing further scrutiny, which may result in it, too, being rejected.

Subalpine Warbler *Sylvia cantillans*

Vagrant (Mediterranean) – less than annual, in very small numbers.

The Subalpine Warbler was first reported in 1894, when one was shot on St Kilda, and there had been three more records by 1950. Two were seen in 1951 and three in 1958. A total of ten were recorded in five years during the 1960s and about 18 in eight years during the 1970s (none in 1973 and 1978). There were two in 1980, four in 1981, one in 1982 and three in 1983. More than 85% of the records are in spring, mostly in May/June but with dates widely scattered from 22 April to 16 July; the few autumn reports are spread from 11 August to 4 October. More than half the records are from Fair Isle and Shetland. Apart from two St Kilda occurrences, the others have all been on the east coast: Orkney, Caithness, Easter Ross (in 1935), Angus, the Isle of May and Berwick. Some individuals remain in the same area for a week or more, the longest stay recorded being a male at Sumburgh from 11 August to 30 September 1971. The distinctive males outnumber females by about two to one. Two races of the Subalpine Warbler are recognised: *S.c. cantillans*, breeding from Italy west to Iberia, and *S.c. albistriata*, breeding from Yugoslavia east to Syria and

Asia Minor. Few of those visiting Scotland are racially identified, but birds on St Kilda (June 1894) and Fair Isle (May 1966) belonged to the nominate race, while three others (Fair Isle – May 1951 & April 1964, St Kilda – June 1979) showed characters of the eastern race.

Sardinian Warbler *Sylvia melanocephala*

Vagrant (Mediterranean) – two records, 1967 and 1981.

Normally a very sedentary species, frequenting dry scrubland in the Mediterranean Basin, the Sardinian Warbler is readily identified by its distinctive red eye-ring. It has occurred only twice: on Fair Isle on 26–27 May 1967 (BB 60:483) and the Isle of May on 30 May 1981; both birds were males. There had been ten other British records by the end of 1983, several of them in autumn.

Rüppell's Warbler *Sylvia rueppelli*

Vagrant (E Mediterranean) – one record, 1977.

Somewhat similar in appearance to the Sardinian Warbler, but distinguishable by its white moustachial streaks and (in the male) black bib, this species is confined to the eastern Mediterranean, where it breeds on dry scrub-covered slopes. It winters in northeast Africa. The only Scottish record, of a male at Dunrossness, Shetland, from 13 August to 16 September 1977 (BB 74:279–283), was the first for Britain. There had been only one other by the end of 1983, in Devon in June 1979.

Orphean Warbler *Sylvia hortensis*

Vagrant (Mediterranean) – one record, 1982.

The first Scottish occurrence of this Blackcap-like warbler, which has a very distinctive eye-ring, was on 10 October 1982, when an immature was caught in an Aberdeen city park (BB 77:552). There had been only five previous British records, the first in 1948; two were in October, the others in July, August and September. The Orphean Warbler, a bird of open woodland and scrub, breeds in the Mediterranean Basin north to central France and northern Italy, and winters in Africa on the southern edge of the Sahara.

Barred Warbler *Sylvia nisoria*

Scarce passage visitor, annual in autumn but rare in spring.

Barred Warblers are reported most regularly from the Northern Isles and the Isle of May, but there have also been recent records from all the exposed-coast northern and eastern counties south to Berwick, and from the Argyll mainland, Skye and St Kilda. Passage usually extends from about mid August to mid October; recent extreme dates have been 1 August and 24 October, with one very late record, of an adult, at Wick on 15 November 1975. Numbers vary; on Fair Isle in a 'good' season 50+ may be recorded, with ten or more sometimes present at the same time, while in years when weather conditions do not encourage westward drift the season's total may be less than ten. Most of the Barred Warblers reaching Scotland are immatures and lack the distinctive breast-barring of the adults. The most recent of the few spring records of adults was one at Cruden Bay (Aberdeen) on 10 May 1970.

Barred Warblers breed in shrubby areas and open woodlands from southern Sweden, Denmark and northern Italy eastwards through central Russia, and winter in northeast Africa and southern Arabia. One ringed on Fair Isle in September was in Yugoslavia the following February, having managed to reorientate itself to some extent.

Lesser Whitethroat *Sylvia curruca*

Scarce migratory breeder, probably under-recorded owing to its very secretive habits. On passage more abundant in spring than autumn but never numerous.

After a period during which it apparently ceased to breed (BOU, Parslow 1973), the Lesser Whitethroat has been reported increasingly often in the last decade, with nesting recorded as far north as Aberdeen. Breeding was not confirmed in 1968–72, though noted as probable in East Lothian. Nesting has since been proved in Ayr, Berwick, Selkirk, East Lothian, Midlothian, Angus and Aberdeen and has probably occurred in Kirkcudbright, while singing birds

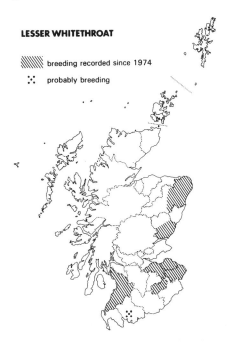

LESSER WHITETHROAT

▨ breeding recorded since 1974

∴ probably breeding

into Russia. Birds caught on autumn passage, which extends from mid August to late October and occasionally into November, include small numbers showing the characters of the Siberian race *S.c. blythi*, breeding from western Siberia east to Mongolia. The Lesser Whitethroat winters much further east than most British-breeding warblers, mainly east of the Nile – in northeast Africa, southern Iran and northwest India. A juvenile Lesser Whitethroat ringed in Midlothian on 3 July was in Derbyshire 13 days later, and one ringed as a juvenile near Doncaster in 1971 was found breeding in East Lothian in 1976. There have been no overseas recoveries of birds ringed in Scotland.

Whitethroat *Sylvia communis*

Migratory breeder, widely distributed in scrub and woodland edge throughout the mainland, Clyde Islands and Inner Hebrides. Most abundant in the Central Lowlands and southern counties, scarce and local over much of the Highlands, and sporadic in the Outer Hebrides, Orkney and Shetland. Passage numbers are generally small.

The Whitethroat is currently much less common than it was in the past – as a result of natural disaster rather than man's interference. In 1969 the CBC results drew attention to the fact that there had been a major crash in the breeding population, with a dramatic decrease of 77% from the previous year. Investigation showed that during the winter of 1968/69 severe drought conditions had prevailed at the species' wintering grounds on the southern fringe of the Sahara, and the resultant high mortality there was thought to be responsible for the population decline (Winstanley *et al* 1974). It was anticipated that the climatic change in that part of Africa was likely to be long-term and that the White-

have been reported in several other mainland counties and in Orkney, Skye and Islay.

The possibility of nesting Lesser Whitethroats escaping detection is evidently high. Even after mist-netting had led to the discovery of breeding birds in Midlothian, attempts to locate this species during CBC dawn and dusk censusing on the same site rarely resulted in either sightings or records of song (da Prato 1980). Breeding has been confirmed at this site nearly every year since 1974; the birds arrive from the first week of May and always take up territories on south-facing banks with thick shrubby cover of bramble, gorse, hawthorn and rose. On the basis of the numbers of young birds caught in the area, da Prato has speculated that the Lothians may hold an average 3–5 pairs/10 km² and a total population of at least 50–100 pairs. It would be interesting to know if similar densities might be revealed by mist-netting in other areas where breeding has been confirmed.

It is difficult to be certain just what, if any, real changes have occurred in the status of the Lesser Whitethroat. B&R recorded breeding in many of the southern counties in which nesting has recently been confirmed, and also scattered records further north, in Argyll, Inverness and Wester Ross. It is possible that sporadic breeding has occurred throughout this century, and that the study in the Lothians has simply directed attention to a situation which had existed for some time.

As a passage visitor the Lesser Whitethroat occurs in only small numbers, with daily maxima at Fair Isle and the Isle of May seldom exceeding ten in either spring or autumn. The larger falls recorded have all been in May. Old records suggest that migrant numbers were considerably higher around the turn of the century (B&R). Spring passage birds belong to the nominate race *S.c. curruca*, breeding from northern Fenno-Scandia, Britain and central France east

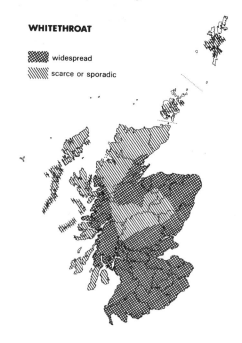

WHITETHROAT

▨ widespread

▨ scarce or sporadic

throat population would probably stabilise well below the pre-1968 level. This prediction has proved correct, with subsequent woodland CBC indices seldom rising above 50% of the 1960 figure, and farmland indices remaining even lower.

However, few Scottish figures are included in these indices, and from local recorders' comments it appears that the Whitethroat has made a good recovery in many districts, especially where there are large areas of young conifer plantation. Marked increases over the last decade have been noted in Kirkcudbright, Renfrew, Argyll, Aberdeen and around the fringes of the Moray Basin, but in the north and northwest mainland the species remains scarce and local. Breeding currently occurs regularly on the Clyde Islands and on all the main Inner Hebridean islands, and at least sporadically on the Small Isles, Coll, Tiree and Raasay. On the Outer Hebrides, where there were occasional nesting pairs prior to 1968, the only recent breeding has been in Stornoway Woods in 1981. There has been no recorded nesting in Orkney since 1941. In Shetland breeding was reported for the first time in 1974 and occurred again in 1982.

It has been suggested that hedges hold about 30% of Britain's Whitethroats (Pollard *et al* 1974), but in Scotland hedgerows are a relatively unimportant habitat for the species, probably mainly due to the fact that they are often severely trimmed and consequently provide insufficient cover. Stan da Prato (1985) found that on c1,700 ha in East Lothian densities ranged from only 3 territories/km² on farmland to the equivalent of 87/km² in scrub woodland; a herb-rich area of scrub at Cousland (Midlothian) had 158 territories/km².

In spring the first birds usually appear in mid April (10th is the earliest recent date) but the main influx, which is much influenced by weather, is often several weeks later and most British-breeding Whitethroats do not arrive until May (da Prato & da Prato 1983). Passage through the Northern Isles is heaviest in May and often continues throughout June; numbers are relatively small and daily peaks on Fair Isle seldom exceed 20. Spring passage is more marked further south, and peaks of 100 are not uncommon on the Isle of May. The return movement, which starts in late July or early August, is often protracted, with stragglers recorded as late as 10 November; most native birds leave in August and September. The numbers involved are even smaller than in spring, and there are no records of sizeable falls from either Fair Isle or the Isle of May. The fact that the Whitethroat is so much scarcer as a passage visitor than the Blackcap – which has a very similar European range, from central Fenno-Scandia to the Mediterranean – suggests that the two species have different migration strategies (da Prato & da Prato 1983). Most large falls of Blackcaps occur in October, by which time a large proportion of the west European Whitethroat population has probably reached Iberia and is consequently not vulnerable to drift across the North Sea. The few distant ringing recoveries relate to birds both marked and recovered while on passage (mostly in France and Portugal) and therefore give no information on origins or destinations; most British Whitethroats are thought to winter at the western edge of the Sahel zone south of the Sahara.

Garden Warbler *Sylvia borin*

Migratory breeder, regular – though often only local – in all counties south of the Grampians. Has recently expanded its range northwards and now nests sporadically in most of the northeast and highland counties, but is absent from the Northern Isles, the Outer Hebrides, and most of the Inner Hebrides, and has been decreasing on the Clyde Islands. Also a passage visitor.

The Garden Warbler requires areas of dense shrubby cover on its breeding ground, but is apparently less dependent than the Blackcap upon the presence of mature woodland and is consequently found in a wider range of habitats, including conifer plantations at the thicket stage. Most local recorders consider it currently to be the scarcer of the two species, whereas B&R believed it to be more plentiful than the Blackcap. Little information is available on Scottish breeding densities, but in England the average on CBC plots is 4.5 pairs/km² (Atlas), only a third that of the Blackcap. In the Lothians Stan da Prato found densities equivalent to 35 pairs/km² in tall scrub at Cousland (Midlothian), but only 5–10 pairs/km² in bigger patches of woodland in East Lothian, and less than one pair/km² on a large area of mainly agricultural land, where Garden Warblers occurred only in woods and shelterbelts (da Prato 1985).

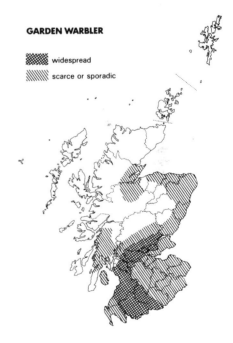

GARDEN WARBLER

▨ widespread

▧ scarce or sporadic

Although remaining scarce and local in most areas, the Garden Warbler has been expanding its breeding range northwards in the last few decades. B&R described it as a summer visitor 'south of the Grampians' and noted only one nesting record further north, in Easter Ross in 1919. In 1968–72 breeding was confirmed in Aberdeen, Moray and Inverness, and recorded as either proved or probable in eight 10 km squares in east Ross. Recent assessments by local recorders indicate that there has since been a further increase in Aberdeen, Inverness and Ross, and that nesting occurs sporadically in northern Moray and Banff. Singing birds are now present most years in Sutherland and Caithness but breeding has not yet been confirmed there, nor in Wester Ross or west Inverness. On the Clyde Islands the Garden Warbler continues to nest 'sparingly' on Arran, as it did in B&R's time, but no longer does so on Bute (Gibson *et al* 1980). There are no nesting records for the Northern Isles or the Outer Hebrides, though song has been heard in both areas. Singing males are occasionally reported from the Inner Hebrides but the only confirmed nesting has been on Colonsay in 1975.

The Garden Warbler breeds from northern Fenno-Scandia to Iberia and Italy, and eastwards into Russia; it winters in tropical Africa. A few birds often arrive in the second half of April, but the main influx of breeders does not take place until May, and small numbers of migrants continue to pass through the Northern Isles well into June. The return movement is from early August, occasionally late July, and may continue into November; late birds are probably not Scottish breeders. On Fair Isle and the Isle of May peak numbers generally occur in late August or September and daily maxima only occasionally exceed 100 and 50 respectively; 400 on Fair Isle on 21 September 1981 and 200 on the Isle of May on 3 September 1965 are the highest numbers on record. The latest autumn date in recent years has been 23 November. Most of the ringing recoveries relate to birds both marked and recovered while on passage. They include one caught on the Isle of May in October and shot 20 days later in Spain.

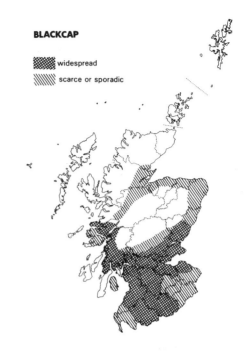

BLACKCAP

widespread

scarce or sporadic

CETK.

Blackcap *Sylvia atricapilla*

Migratory breeder, most abundant in the central and southern lowlands, scarce and probably irregular in most of the northeast and highland counties, and absent from nearly all the islands. Autumn passage numbers are variable but have shown a recent general increase; occasionally winters.

The distribution of the Blackcap is determined largely by the availability of mature deciduous or mixed woodland with good shrub cover, a combination typical of estate policies and parks. Where shrub growth is dense the singing birds can be difficult to see and breeding is not easy to confirm. Few data on Scottish breeding densities are available but in most areas these are likely to be well below the CBC woodland average of 12+ pairs/km^2 quoted in the Atlas. In scrub with some taller trees at Cousland (Midlothian) Stan da Prato found densities equivalent to 22 territories/km^2, but only 7–10/km^2 in more extensive woodland in East Lothian, and an average of less than one/km^2 over a large area of farmland, where the birds occurred only in shelterbelts and woods (da Prato 1985).

Judging by local recorders' comments, the Blackcap is very local in the four Border counties and in Fife, Kinross, Clackmannan and Stirling, but is much more abundant and widespread in western counties north to Argyll and in Perth and Angus. It is described as outnumbering the Garden Warbler in the Lothians, Fife, Argyll and Renfrew, but as the scarcer of the two in Kinross and Kirkcudbright; these are in most cases subjective assessments, however, unsupported by detailed survey counts. Further north the Blackcap becomes progressively less common, though there is thought to have been some recent increase in Aberdeen. Breeding occurs sporadically in Banff and Moray, but has not been recorded recently in Nairn. Small numbers breed most years in Inverness and Ross and Cromarty, mainly in the east. Singing males are often present in Sutherland and Caithness but there is as yet no confirmed breeding record for these counties. B&R noted nesting in Orkney, on Shapinsay in 1910 and 1949, but there has been no subsequent report, while in Shetland breeding has occurred at least twice, in 1948 (B&R) and 1974 (SBR). The Blackcap has not bred in the Outer Hebrides, nor does it do so regularly on the Inner Hebrides, although pairs may nest occasionally on Rhum (J. A. Love). A few breed on Arran but there has been no proof of nesting on Bute for many years (Gibson *et al* 1980).

Blackcaps breed from the Mediterranean north to central Scandinavia and eastwards into Russia. Expansion of the north European breeding range and of the scale of passage through Britain are probably responsible for the recent increase in wintering numbers here (Leach 1981). British-breeding birds winter mainly in Iberia and north Africa and birds overwintering in Britain are of continental origin (Langslow 1979). Blackcaps readily eat berries and come to bird tables in hard weather, and some are consequently able to survive even such a severe winter as 1978/79, as evidenced by the 10–20 reports in January and February 1979. Breeding birds generally start to arrive in the second half of April and the main influx takes place in early May; sightings earlier in the year almost certainly relate to over-wintering individuals. The return movement starts in August and passage generally continues well through November.

Passage numbers vary greatly between years, with much higher autumn peaks recorded in the last decade than previously (Langslow 1978). Writing in 1964/65 Peter Davis (in Williamson 1965) described the Fair Isle Blackcap peak as 'often up to ten and occasionally up to 30'; between 1968 and 1975 it was in the range 60–75 on three occasions; and from 1976–83 it failed to reach 50 in only two seasons, exceeded 100 in four seasons, and on 21 September 1981 reached a spectacular 1,000, five times the previous maximum for the island. Numbers on the Isle of May have also been higher recently, with an all-time record of 1,000 on 13 October 1982 (SB 12:220). The timing of the autumn passage is almost as variable as the numbers involved, with the main movement generally in September/October but large falls sometimes occurring in November; it is usually after such late autumn arrivals that wintering numbers are high. Spring passage is on a small scale, involving a trickle of birds from late April to early June but with daily peaks seldom exceeding ten on either Fair Isle or the Isle of May.

Blackcaps sometimes make surprisingly strong northward movements in autumn: one ringed in Kent on 25 September was caught in Sutherland on 9 October, and one ringed in Fife in late July was on Fair Isle on 10 September. Many of the recoveries of foreign-ringed birds result, as might be expected, from drift across the North Sea, with birds ringed in the Netherlands and Belgium turning up a few weeks later on the Northern Isles or elsewhere on the east coast. The fastest such journey recorded was a bird ringed in the Netherlands on 23 September and killed the following day in Shetland. There have been winter (October/March) recoveries in Eire, France, Spain, Italy, Algeria and Lebanon, of birds ringed on passage through Scotland.

Greenish Warbler *Phylloscopus trochiloides*

Vagrant (Eurasia) – less than annual, in very small numbers.

The Greenish Warbler was first recorded in 1945, when one was shot on Whalsay on 12 September (BB 39:153), and there was one on Fair Isle in 1949. There were reports in two years during the 1950s, four years in the 1960s, and

seven years in the 1970s, the maximum number in any year being only three. After a blank in 1980, four were seen in 1981, one in 1982 and five in 1983. Most sightings have been between mid August and late September, but there is a wide scatter of dates, from 21 May (singing male) to 15 October. More than half the reports are from Fair Isle and the remainder from Shetland, Orkney (three – all in 1981), Fife Ness, the Isle of May, East Lothian, Argyll and Perth. Records from St Abbs (1966) and Fair Isle (1961) were rejected. This is a difficult species to identify with certainty in the field and most accepted records are of trapped birds, on which the wing formula has been checked. In recent years the Greenish Warbler has been expanding its breeding range westwards in Europe, slowly but steadily. It now breeds from southern Finland and northern Germany eastward through Eurasia; it winters in India and Indo-China.

Arctic Warbler *Phylloscopus borealis*

Vagrant (N Eurasia) – almost annual in autumn, in very small numbers.

The Arctic (formerly known as Eversmann's) Warbler breeds in mixed forest and scrub, from northern Fenno-Scandia east through Eurasia and into Alaska. It migrates in a markedly southeast direction, to winter east of India – which explains why it occurs in such small numbers. It was first recorded in 1902, when one was killed at the lighthouse on Sule Skerry, and had been reported only ten times by 1950; from 1967 to 1983 it occurred annually. In most years fewer than five are seen, and the total has never exceeded eight. About 60% of occurrences are in September and the remainder mostly in late August and October. One on Fair Isle on 3 July 1982 was the earliest-ever date in Britain; the latest is 28 October. Most reports are from Fair Isle (15 in 1950–64; none 1965–6; c25, 1967–83) and Shetland (c22, 1967–82); there are also records from Orkney, Caithness, Aberdeen and the Isle of May.

Pallas's Warbler *Phylloscopus proregulus*

Vagrant (south-central Siberia) – recorded in ten years, 1966–83.

This tiny warbler was first recorded on 11 October 1966, on Fair Isle (BB 59:438). The second occurrence was in Aberdeen in 1968, after which there were no further reports until 1975; that year produced 29 British records, four of them in Scotland (Fair Isle, Shetland, the Isle of May and Perth). With the exception of 1980, there have been records each year since, with not more than five reports in any one season until 1982, when there was an unprecedented influx of at least 55–60 during the big October fall of Siberian migrants (SB 12:246–251). Apart from one late September record in 1976, and one trapped on the Isle of May on 14 November 1983 (the sole report for the year), all have been in October, and most between 10th and 25th. The Perth bird, which was at Rannoch, is the only inland record. In addition to the areas already mentioned there are records for Orkney, Caithness, Aberdeen, Kincardine, Fife and East Lothian – and a North Sea oil-rig 300 km east of Fife Ness. Pallas's Warbler, a bird of mountain forests, winters in Indo-China; its occurrence in this country is clearly dependent upon weather conditions at the time of its autumn migration.

occurred there again in 1968, on Whalsay in 1976, at Fife Ness and Cruden Bay in 1979, and in Orkney (two) and Shetland (one) and on the Isle of May (one) during the big fall of Siberian migrants in 1982. All the records are for October. This dark, ground-feeding warbler breeds from south-central Siberia eastwards and winters from Burma to Indo-China; it is a bird of forest edge.

Yellow-browed Warbler *Phylloscopus inornatus*

Scarce passage visitor in autumn.

This is a more northerly species than Pallas's Warbler and occurs with greater regularity. From 1968 to 1983 it was recorded annually in autumn, in numbers ranging from under ten to about 50 in any one year. The most in one place at one time has been 15 on Fair Isle (on 25 September 1973). Most passage takes place between mid September and early November; extreme dates since 1968 are 6 September and 19 November. About half the reports come from Fair Isle and Shetland and most of the remainder are from exposed east coast sites, from Orkney to Berwick (no Angus record so far), but there have also been recent reports from Ayr, Lanark, Kirkcudbright and the Outer Hebrides (Lewis). B&R give autumn records for the Inner Hebrides (Skerryvore), and refer to spring sightings in Dumfries, Lanark and Fair Isle. The latter, however, along with most other British spring reports, 'lack proper documentation' (BOU) and should therefore be regarded as unreliable. The Yellow-browed Warbler breeds in shrubby woodland and open forest right across northern Siberia, and winters in southeast Asia.

Dusky Warbler *Phylloscopus fuscatus*

Vagrant (central Asia) – five records, 1913–83.

The first British record of this very dark leaf warbler dates back to 1913, when one was shot on Auskerry, Orkney, on 3 October (SN 1913:271). There were no further Scottish records until 1961, since when there have been four, all between 26 September and 14 October: on Fair Isle in 1961 and 1974, and at St Abbs in 1976 and 1982. The Dusky Warbler is widely distributed in open woodlands from west-central Siberia eastwards; it winters in Indo-China. A full description is given in BB 65:497–501, in a paper on 'Field identification of Dusky and Pallas's Warblers'.

Bonelli's Warbler *Phylloscopus bonelli*

Vagrant (Europe) – eight records, 1961–84.

Bonelli's Warbler was first reported from Fair Isle, on 22 September 1961 (SB 2:343; BB 55:278) and has since occurred in Shetland Mainland (1974 & 1983), Whalsay (1979 & 1984), Out Skerries (1981), Islay (1976) and Aberdeen (1980). The Islay record – a singing male – was on 21–22 May, the others between 5 September and 3 October. This woodland warbler has been extending its range northwards during the present century, and now breeds in northern France and southern Germany; it winters in tropical Africa. There have been a number of spring records recently further south in Britain.

Radde's Warbler *Phylloscopus schwarzi*

Vagrant (central-east Asia) – recorded in five years, 1962–83.

Radde's Warbler was first recorded in 1962, when one was on the Isle of May on 8–10 October (SB 2:367–8). It

Wood Warbler *Phylloscopus sibilatrix*

Migratory breeder in deciduous woodland with sparse ground cover. Most abundant in western oakwoods, scarce and local in most eastern counties, and absent from all but a few of the islands. A scarce passage visitor.

B&R state that 'at the present time the Wood Warbler breeds in every county of Scotland', but it is questionable whether the species has ever been as widespread as their statement suggests. They make no reference to Orkney, where the Wood Warbler reputedly bred at Binscarth in 1914–15 (Balfour 1972) but has not done so again, nor to Shetland, for which there is no breeding record. In Caithness, Nairn, Moray, Banff, Aberdeen, Kincardine and Angus nesting apparently now occurs only sporadically – and there is little reliable evidence to suggest that the situation was significantly different in the first half of the century. Although B&R do not make specific mention of Fife and Kinross they imply that the Wood Warbler bred locally there at the time they were writing; there has been no record of nesting in these counties for at least ten years. Elsewhere in the east and in most of the Central Lowlands this species is a scarce and very local breeder.

In the southwest Wood Warblers are present in most suitable woodlands, but there is little information on numbers; the frequent presence of unmated singing males and the fact that the species is polygamous increase the difficulty of assessing breeding numbers. Kirkcudbright probably holds the bulk of the population in the southwest, while in the Wood of Cree (Wigtown), which includes much coppiced oak, this species is nearly twice as abundant as the Willow Warbler (Williamson 1976); 48 territories were occupied there in 1983 (SBR). Recent estimates for Ayr and Renfrew suggest totals of not more than ten and five singing males respectively (Hogg 1983; I. P. Gibson), and numbers are likely to be small in Lanark too. Breeding densities in the oakwoods of Argyll, Dunbarton, Stirling and west Perth are probably substantially higher than in most other parts of Scotland, but the highest recorded density was in a natural oakwood in Wester Ross. The estimated 90 territories / km^2 there compares with 23–43 / km^2 in Stirling, and 19–26 / km^2 in Argyll (Williamson 1974). Wood Warblers are widely scattered in central and north Perth, Inverness, east Ross and southeast Sutherland but totals in these areas are probably fairly small. Breeding occurs at least sporadically on Skye,

Raasay, Rhum and Mull, and possibly also on Islay and Jura, but there is no record for the Outer Hebrides, although singing males are sometimes present in the Stornoway Woods. A few pairs nest on Arran and Bute.

The paucity of information makes it impossible to be certain of current population trends, but it is suspected that loss of habitat is adversely affecting this species in some areas. This is likely to be the case wherever old oak or birch woods have been clear-felled or inter-planted with conifers, as they have been in parts of Perth. Deciduous woodlands are disappearing with even greater speed in south and east England, a situation which led to the mounting of a Wood Warbler survey in 1984–85. Only sample areas were surveyed in Scotland, but the 1984 results suggest that the population is probably well below the Atlas estimate, which was based on an average of 25–50 pairs/10 km^2 (C. Bibby).

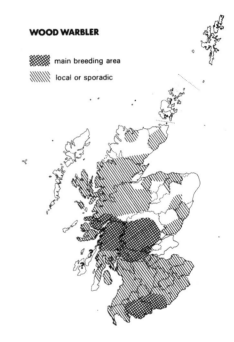

WOOD WARBLER

- main breeding area
- local or sporadic

The Wood Warbler breeds from southern Scandinavia to southern France and Italy, and east into Russia. It winters in tropical Africa. Breeding birds arrive from late April but the main influx does not take place until well into May. Few migrants reach the Northern Isles, and the daily maximum on Fair Isle, where passage may continue until the end of June, seldom exceeds five. Native birds start to move south in July and the main return movement is in August/ September, with stragglers sometimes occurring in October; the latest date is 25 October, in Shetland. Few Wood Warblers are ringed and even fewer recovered; the most interesting record so far is of one ringed on the Isle of May in mid July and shot the following April near Algiers.

Chiffchaff *Phylloscopus collybita*

Migratory breeder, most widespread and abundant in the south-west, very local north of the Grampians and absent from many of the islands. Summering and passage numbers are very variable; there has recently been an increase in wintering records.

The Chiffchaff favours sites where tall trees (broadleaf or conifer) are interspersed with shrubby growth, as for example in policy parklands with rhododendrons. Its distinctive song makes it easy to locate, but its nest is difficult to find and breeding is not often proved. The presence of a singing male is not a reliable indication of breeding, and many of the records in northern counties probably relate to unmated birds. Chiffchaffs are locally abundant from the Solway to Argyll, especially near the coast; local recorders' comments suggest that numbers there have increased in the last decade. In the eastern lowlands north to Perth and Angus the species is much more local but breeds fairly regularly, and in the northern half of the mainland nesting occurs at least occasionally. There are now confirmed breeding records for all mainland counties, whereas B&R did not know of any proved nesting north of Oban. Sporadic breeding has been recorded on Mull, Islay, Rhum, Eigg, Skye and Raasay, but has not been confirmed in the Outer Hebrides, though singing birds are present in Stornoway Woods most years (Cunningham 1983). The Atlas shows 'probable' breeding in Orkney, but nesting has not been proved in the Northern Isles.

The Chiffchaff is one of the earlier migrants; first arrivals often appear during March, and the main influx around mid April. Little information is available on breeding densities in Scotland, but even in the most favoured habitats numbers are not high, probably falling well below the average woodland density of 12.2 pairs/km² quoted in the Atlas.

Although proof is lacking, one gets the impression that the numbers reaching Scotland are to some extent dependent upon weather conditions during the migration period; in years with a mild early spring more males possibly penetrate further north than in cold late seasons. The CBC data for this species, which include few if any Scottish figures, show quite wide annual fluctuations. Breeding birds, belonging to the race *P.c. collybita*, start to move south in August and most have left by October; they are thought to winter in the Mediterranean region and tropical Africa.

Passage birds include representatives of three races: *P.c. collybita*, breeding in Europe north to Denmark and Poland, *P.c. abietinus*, breeding in Scandinavia and west Russia, and *P.c. tristis*, breeding in northern Siberia. Numbers of passage Chiffchaffs recorded at the observatories are relatively small in comparison with some other warbler species, and falls as large as the 200 on the Isle of May and 80 on Fair Isle on 11 October 1982 are unusual. Movement through Fair Isle may occur in any month from March to November (Riddiford & Findley 1981), but numbers are largest in autumn. Pale 'northern' birds, of the races *abietinus* and *tristis*, have been recorded on both spring and autumn passage, but those of Siberian origin occur more often in autumn than spring (Davis, in Williamson 1965). Most of the wintering records probably involve birds of continental origin which have arrived in Scotland relatively late in the season. In 1980, for example, there was a big October influx in the Northern Isles, and smaller numbers down the east coast to Berwick; there were subsequently 90+ reports in November (50+ in Shetland and Inverness), and 25+ in December. There have also been a few January/February records, mostly from the southwest, while in late December 1983 a Siberian *tristis* was trapped in Paisley.

The recovery rate of ringed Chiffchaffs is not high and there have been few distant recoveries of Scottish-marked birds. An indication of the origin and destination of some passage birds comes from the capture in Orkney in October of one ringed the previous month in Estonia, and the recovery in Sicily in February of one ringed on the Isle of May the previous October. Other recoveries of interest include one ringed on the Isle of May in April and found dead on Mallorca the following December, and one ringed on the Isle of Man and caught on Fair Isle a week later.

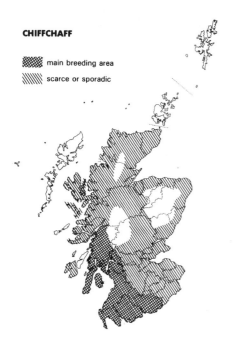

CHIFFCHAFF

▨ main breeding area

▨ scarce or sporadic

Willow Warbler *Phylloscopus trochilus*

Migratory breeder, the most abundant of the warblers. Widespread throughout the mainland and on the inner islands, in virtually all types of woodland other than well-grown conifer plantations; scarcer in the Outer Hebrides and Orkney and sporadic in Shetland. Passage is most pronounced in autumn.

The song of the Willow Warbler is almost synonymous with summer in the Highlands, as every little clump of birches, right up to the treeline (c700 m asl) appears to hold a pair. This is the commonest breeding bird in many birch-woods, where it may outnumber the ubiquitous Chaffinch by three to one (Yapp 1962); densities equivalent to 220–255 territories/km² have been recorded in birch and willow scrub at 250 m asl in the Pentlands (Keymer 1980), in herb-rich deciduous scrub in Midlothian (S. R. D. da Prato) and in oakwoods in Wester Ross (Williamson 1969). Williamson found fewer in oakwoods in Argyll and Stirling: c65 and 70–150 pairs/km² respectively, while on 1,700 ha of mainly agricultural land in East Lothian in 1984 the over-all density was only 11 territories/km² (da Prato 1985). Willow Warblers require a shrub-layer or its equivalent, and are consequently scarce or absent in mature beechwoods or similar woodlands with little secondary growth. Young conifer plantations hold quite good numbers; in study plots in southwest Scotland Moss (1978a) found 61–95 pairs/km² in the early years, dropping to 13/km² at the thicket stage and later to nil.

The only recent changes in the status of the Willow Warbler relate to the islands. Breeding was first recorded on the larger Inner Hebridean islands last century and on most of the Small Isles between 1900 and 1950 (B&R). Nesting is possibly still sporadic on the barer islands, but on those which are now more wooded breeding has become regular and numbers have increased. On the Outer Hebrides, where B&R had breeding records only for Storno-way Woods, Rodel (Harris), and Barra, most shelterbelts and plantations throughout the islands have now been col-onised (Cunningham 1983). In Orkney small numbers nest regularly in woods and gardens, and there have been reports of confirmed or probable breeding from Mainland, Hoy, Rousay, South Ronaldsay, Shapinsay and Sanday. Although the first record of breeding in Shetland was as long ago as 1901, there have since been only very infrequent reports of summering, the most recent in 1975, when a singing bird was in Kergord plantation during June but no evidence of breeding was obtained. Somewhat surprisingly, a pair bred on the Isle of May in 1922.

CBC results show that numbers do not fluctuate very widely between years (woodland indices ranged from 81 to 114 in 1964–82). Unlike the Whitethroat, the Willow Warbler, which winters in tropical and southern Africa, has clearly not suffered from any major disasters on its wintering grounds. The first birds usually arrive in early April – 26 March is the earliest date since 1968 – and the main influx from the middle of the month; arrival dates may vary by two weeks in response to weather conditions. At the end of the breeding season Willow Warblers start to drift south, appearing *en route* in areas, such as suburban gardens, where they do not breed and sometimes joining mixed flocks of tits and other warblers (da Prato 1981). The numbers of young birds moving around at this season are very large; by colour-marking wandering juveniles Stan da Prato esti-mated that at least 1,000 visited a 10 ha patch of scrub in July and August. Most have gone by the end of September; the occasional October records, and the very few in November and December, probably relate to continental migrants.

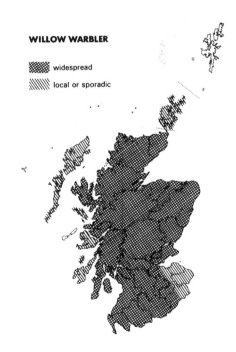

WILLOW WARBLER

▨ widespread

▧ local or sporadic

Passage through the Northern Isles involves birds of both the nominate race *P.t. trochilus*, which breeds from Britain, Denmark and northern France east to Poland and Rumania, and the 'northern' race *P.t. acredula*, breeding from Scandi-navia to Siberia. Some of the Willow Warblers breeding in Scotland resemble the northern birds, which are browner and less olive-and-yellow than the nominate race. On Fair Isle the main spring movement occurs in May (by which time British-breeding birds are already nesting), with stragglers into June; the daily peak at this season seldom exceeds 100 but there was an unusually big spring influx in 1970, when c1,000 were present on 8–9 May. Autumn passage is from late July, with the main movement in August/early September and small numbers still passing through in early October. The relatively early passage period means that Willow Warblers are less likely to be affected by bad weather than are some other warbler species; although overall autumn totals are high, major falls are infrequent. An influx of c1,000 on the Isle of May on 14 August 1978 was by far the biggest ever recorded there, and was also larger than any autumn peak on Fair Isle since the observatory was established.

Considerable numbers of Willow Warblers are ringed, mainly on passage, but the recovery rate is low (only 0.2% – Hickling 1983) and there have been few recoveries from either the wintering areas or the continental breeding grounds. Most of the foreign recoveries have been in August/October and April/May. The autumn ones are mainly from France, Spain, Portugal and Morocco, and there is a May recovery from Norway. A Willow Warbler ringed in Belgium in mid May was killed in Edinburgh only five days later, birds ringed in Sutherland and on the Isle of May in August/September were shot in Spain and Portugal respectively about five weeks later, and a juvenile ringed in Midlothian on 8 August was in Spain on 1 September.

[Ruby-crowned Kinglet *Regulus calendula*]

The record of two Ruby-crowned Kinglets shot near Loch Lomond in 1852 has never been formally accepted for inclusion in the British list, but the evidence has recently been reviewed by John Mitchell (1985), who concludes that reconsideration by the BOU Records Committee would be appropriate.

Goldcrest *Regulus regulus*

Resident breeder, widely distributed throughout the mainland; breeds or has bred in all the main island groups. Particularly vulnerable to severe winters. Variable numbers occur on passage, most abundantly in autumn and mainly in the north and east.

The Goldcrest, smallest of British birds, is typical of coniferous forest but also frequents mixed woods and other habitats, such as parks, which have scattered conifers. It breeds at altitudes of up to c400 m asl and occurs even in relatively isolated and quite small shelterbelts and upland conifer plantations. Breeding density varies with woodland type and is higher in spruce and larch plantations at or after the thicket stage than in mixed pine and birch woods (Moss 1978a,b); densities of more than 100 pairs/km² are apparently not uncommon in conifers. Western oakwoods studied by Williamson (1974) supported from nil (in Wester Ross) to 26 pairs/km², by far the highest numbers being found in the mixed woodlands of Glen Nant (Argyll). Overall numbers fluctuate greatly, however, and densities vary according to the severity of the preceding winter.

CBC results have shown that the Goldcrest's population fluctuations are the most marked among the resident species censused. A decrease of nearly 40% in the censused population followed the winter of 1975/76 and there had been no sign of recovery by 1978/79, when the prolonged hard

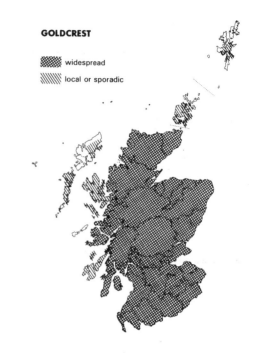

GOLDCREST

▨ widespread

▨ local or sporadic

weather caused a further drop of 40% to give an index of only 145 in 1979, compared with 469 in 1975. This time there was a rapid recovery, however, with the index increasing by more than 60% between 1979 and 1980, and much successful rearing of double-broods in Scotland in the latter year (SBR). The fact that the very severe frosts of 1981/82 apparently had no significant effect on the population suggests that the Goldcrest suffers from snow-cover rather than low temperatures; presumably wet, heavy or frozen snow adhering to conifer branches greatly reduces the availability of its insect foods.

There has been no recent major change in distribution, although many local changes in both range and numbers must be occurring in areas where afforestation has taken place on previously treeless ground. Such population expansion is likely to be most noticeable on the islands and in the extreme north, and it is indeed from the islands that the only evidence of recent colonisation has come. Numbers on Mull and Skye have been increasing with the spread of afforestation; on Rhum, where Goldcrests were first found breeding in 1933, there were c30 pairs in 1974; and in the Outer Hebrides, where the species was formerly confined to Stornoway Woods, there are now a few pairs scattered through the plantations on North and South Uist. In Orkney B&R recorded breeding at Binscarth (Mainland) in 1945 and nesting has since occurred sporadically both there and on Hoy and Eday; attempted breeding was recorded on Stronsay in 1975. The only confirmed records of breeding in Shetland are from Kergord plantation, in 1972 and 1976.

In the north of its range, which extends from Scandinavia to northern Spain and east into Asia, the Goldcrest is migratory, wintering in the southern part of the breeding range and south of it to the Mediterranean. Passage is recorded annually on the east coast and in the Northern Isles but numbers very greatly between years. The spring movement is generally small and takes place mainly in April, with a few birds sometimes appearing in March and stragglers into May. Peak daily counts on the Isle of May and Fair Isle at this season seldom exceed 20, but in 1983, following a big influx the previous autumn, spring passage was heavier than usual, with a peak of 170 recorded on Fair Isle in early April. Much larger numbers occur regularly in autumn, when passage may extend from August to November, with the biggest falls generally in October. In some years peak daily counts at the observatories are less than 100 but much larger influxes sometimes occur; by far the biggest yet recorded took place in October 1982, when an estimated 15,000 Goldcrests were on the Isle of May on 11 October (Ellis 1983).

Most of the Goldcrests ringed in Scotland are on passage and there is no ringing evidence to support the view that some native birds may emigrate (BOU). B&R believed that 'a considerable emigration takes place', on the basis of birds appearing at the lanterns of lighthouses in the Forth and on the Galloway coast in both autumn and spring. The possibility of those in the Forth being continental immigrants seems high, but recent reports of 100+ at the Mull of Galloway and 40–50 on Ailsa Craig in early April suggest that autumn dispersal of native birds may indeed involve some movement to Ireland. The comparatively small number of ringing recoveries include a bird marked in Finland on 27 September and caught on the Isle of May on 6 October, one ringed on Fair Isle and controlled in Orkney later the same day, and one ringed on the Calf of Man on 6 September and caught on Sanda (Argyll) the following day. It is likely that many immigrant Goldcrests come from central Europe (BOU).

Firecrest *Regulus ignicapillus*   [1]

Vagrant (Europe, including England) – almost annual, in very small numbers.

Since it was first recorded in 1959, when a male was on the Isle of May from 30 September to 3 October (SB 1:153), the Firecrest has occurred with increasing frequency. During the 1960s it was reported in five years, during the 1970s it was unrecorded only in 1978, and it has occurred annually since 1979. Reports are generally of single birds but on several occasions up to four have been seen together. The maximum number reported in any year has been 12 – in 1980 when there were records from Caithness and Kintyre in January, Sanda in May, the Bass Rock in June, St Abbs in August (three) and October, and Aberdeen (two), South Uist and Dumfries in November. Most first sightings are in September/November – the majority during October – but there are records for every month except July, and in 1983 four of the five reports were in April/May. Several of the birds first located in December are known to have remained in the same area for up to three months, so the early season reports may actually refer to birds which had arrived some time previously.

Firecrests are reported most frequently from the Isle of May and Shetland, but there have also been records from Fair Isle, Orkney, Caithness, Ross, Moray, Aberdeen, Fife, East Lothian, Peebles, Berwick, Dumfries, Ayr, Argyll and the Outer Hebrides. The chances of this species being overlooked, especially in woodland, must be high, and it seems likely that many more occur than are reported. Firecrests can be located in woodland by their song, which is harsher and less rhythmic than that of the Goldcrest; they breed both in the coniferous habitats favoured by Goldcrests and in mixed woodland.

This species has expanded its continental range to the north during the last 50 years and now breeds from the Mediterranean to the Netherlands (where nesting was first recorded in 1930), Denmark (1961) and north Germany. Since the 1950s it has also occurred more frequently in Britain as a passage migrant, and in 1962 breeding was confirmed in England for the first time. The current breeding distribution is wider than that shown in the Atlas map and, if the recent range expansion continues, colonisation of Scotland may take place in the fairly near future – a bird found dead in Caithness in February 1976 had been ringed in Norfolk the previous autumn.

Spotted Flycatcher *Muscicapa striata*

Migratory breeder, widely distributed in woodlands on the mainland and most of the inner islands; nests sporadically in the Outer Hebrides and Orkney, but has not bred in Shetland. Regular but relatively scarce on passage in the Northern Isles.

The Spotted Flycatcher is typically a bird of woodland edge or open glades but it will also breed close to habitations and is frequently found around farm steadings and in large gardens. Its principal requirements are a good supply of flying insects, suitable perches from which to hawk, and an alcove or ledge for nesting; it readily occupies open-sided nest-boxes. It is most abundant in lowland deciduous woodland near water but also occurs in relatively open native pinewoods. Little information on breeding densities is available but the 20 pairs/km² and 15 pairs/km² recorded in oakwoods at Badachro (Wester Ross) and Inchcailloch (Loch Lomond) respectively possibly represent the upper end of the range; other western oakwoods studied had fewer than

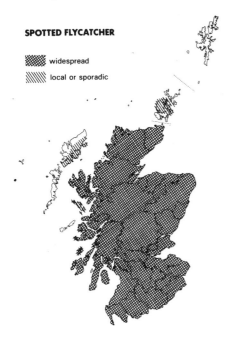

SPOTTED FLYCATCHER

▨ widespread

▨ local or sporadic

10 pairs / km² (Williamson 1974), while in highland birch-woods, Yapp (1974) found the Spotted Flycatcher less abundant than the Redstart and the three commoner tits. The CBC indices for Britain as a whole show that the breeding population fluctuates widely but give no indication of any long-term population trend.

Because it is so dependent upon flying insects, this species is among the last of the summer visitors to arrive. There are often a few reports in the last week of April (8th is the earliest date) but in some seasons no Spotted Flycatchers arrive until the end of the first week in May; the main influx generally takes place in mid May. On Fair Isle spring passage peaks during the third week of May and stragglers occur into early July (Riddiford & Findley 1981). Spotted Fly-catchers are often double-brooded further south, but evidence of the extent to which this may occur in Scotland is lacking. The return movement starts in late July and most breeding birds have left by late September. Small-scale passage continues to mid October, with occasional later stragglers; one at Berwick on 3 November 1968 is the latest recorded date. The daily maximum on Fair Isle seldom exceeds 15 in autumn but occasionally reaches 50 in May.

There is no evidence of any change in distribution during the last 30 years. Eigg and Muck are the only islands on which sporadic breeding occurs now but had not been reported prior to 1950, while on Rhum numbers have shown some increase, with about four pairs nesting in 1983 (J. A. Love). Nesting continues to be irregular in Orkney and the Outer Hebrides, with no more than two or three pairs present in either group in any one year.

The Spotted Flycatcher is widely distributed throughout Eurasia, breeding in Europe from arctic Scandinavia to Iberia. It winters in tropical and southern Africa. There have been recoveries of Scottish-ringed birds in September/October in France, Spain, Portugal and Italy, and one ringed

as a chick in Fife was recovered at sea off west Africa three months later. The probability that passage birds are from the Scandinavian breeding population is suggested by the late July recovery in Norway of a Spotted Flycatcher ringed on the Isle of May the previous August.

Red-breasted Flycatcher *Ficedula parva*

Passage visitor in very small numbers.

The Red-breasted Flycatcher should probably be regarded as a vagrant rather than a passage migrant, since Britain lies well to the west of its normal range. It breeds from northern Germany, south Sweden and the Balkan States eastwards into Russia (a different subspecies replaces it further east) and winters in western India, in coniferous and mixed forests, orchards and gardens. A bird caught on the Isle of May in mid October 1975 had been ringed five weeks earlier on the Aland Islands off southwest Finland.

Red-breasted Flycatchers occur annually in autumn and less regularly in spring, the total recorded in any one year seldom exceeding 25 and sometimes being less than ten. In 1968–83 inclusive the highest numbers occurred in 1981, when around 50 were seen, all but four of them in autumn. Most spring occurrences in this period were in the second half of May, extreme dates being 1 May and 15 June. The autumn records show a much wider spread, from 17 August to 15 November, with most between mid September and mid October. There is a single winter record, of one at Reay, Caithness, on 6 January 1981. Recent reports are widely scattered, with occurrences since 1968 in Shetland, Fair Isle, Orkney, Caithness, Sutherland, Ross, the Outer and Inner Hebrides, Aberdeen, Kincardine, Fife, the Isle of May, East Lothian, Berwick and Wigtown. As might be expected, the great majority of the records are from the Northern Isles and the east coast.

Collared Flycatcher *Ficedula albicollis*

Vagrant (central & SE Europe) – six records, 1947–83.

The first record of the Collared Flycatcher was in 1947, when an adult male was shot on Whalsay on 11 May (SN 1948:51). All the subsequent Scottish records have also been in May (there has been one autumn record in England) and from the Northern Isles: Harray 1963, Out Skerries 1975 and 1976, Bressay 1979, and Stronsay 1980. All but the 1976 bird were males. This species is less easily distinguished from the Pied Flycatcher in autumn and might consequently be overlooked at that season. The Collared Flycatcher breeds in mainly deciduous forests, from central and southeast Europe eastwards, and winters in Africa south of the Sahara; it is much scarcer here than the Red-breasted Flycatcher and had been recorded in Britain only 11 times by the end of 1983.

Pied Flycatcher *Ficedula hypoleuca*

Migratory breeder, most abundant in the southwest and the Trossachs; sporadic nesting takes place in many other mainland counties. Passage occurs regularly in both spring and autumn and is most marked in the Northern Isles.

The Pied Flycatcher is confined to deciduous woodland and occurs most often where oaks are present and there is water nearby. Old woodpecker holes and natural tree-cavities are the species' preferred nest sites and breeding numbers may be limited by shortage of suitable holes; in otherwise acceptable habitat the breeding population can be increased by the provision of nest-boxes. The first spring arrivals appear from mid April but the main influx does not take place until early May. Numbers appear to vary widely between years, as does the distribution of singing males outwith the main breeding areas.

The Atlas, which shows all possible, probable and confirmed breeding records for the four-year survey period, presents a somewhat misleading picture, as Pied Flycatchers breed only very irregularly in most parts of Scotland. Since 1968–72 there has been no further proved breeding in Sutherland, Ross and Cromarty, Lanark or West Lothian and there are no recent records from Moray, where sporadic breeding occurs (N. Elkins), or Clackmannan, where the species formerly nested in Dollar Glen (C. Henty). Probably not more than two pairs currently breed in Midlothian and there has been no recent confirmation of nesting in East Lothian, although one or two singing males are generally present in spring. Earlier changes in distribution were documented by Bruce Campbell (1954, 1965).

In the southeast Pied Flycatchers appear now to be confined to the Tweed Valley and some of the large estates, the total population for the four Border counties probably

not exceeding 10–15 pairs (R. D. Murray). These figures suggest a decrease, as B&R record 11 pairs in the Hirsel grounds alone in 1949, most of them in nest-boxes. In Ayr, where up to ten pairs nested in the early 1970s, not more than two pairs have been present in the last few years (Hogg 1983). In Renfrew a pair was seen in 1974, but breeding was not proved, while in Dunbarton nesting was first confirmed in 1979 at Luss and occurred in 1981 at Milngavie. In central and north Perth and in west Inverness breeding is probably sporadic, but on Speyside it occurs fairly regularly between Kingussie and Nethy Bridge, with up to ten singing males sometimes present and two or three pairs proved to breed (Dennis 1984). Pied Flycatchers bred for the first time in Aberdeen, on Deeside, in 1981 (SBR). There are no breeding records for Caithness, Nairn, Kincardine, Fife or Kinross. A 'probable' report from Skye is the only suggestion of nesting on any of the islands.

One of the most important concentrations is in Dumfries and Kirkcudbright, where the Pied Flycatcher is widespread in suitable habitat, for example the oakwoods in Glen Trool and the Glenkens. No detailed information on breeding numbers is available but the population in Kirkcudbright was estimated at 50–100 pairs in the early 1980s (A. D. Watson). The oakwoods of the Trossachs and Loch Lomondside have become increasingly important since nest-boxes were provided to facilitate a long-term study by Henry Robb, started in 1973. A single box was occupied that year, five pairs nested in 1974 and by 1984 numbers had built up to 46 pairs; in only two seasons has there been any evidence of breeding in natural sites. The easy access to birds nesting in boxes has made possible close study of breeding success as well as of fluctuations in numbers (Table), and ringing

Occupancy of nest-boxes by Pied Flycatchers in the Trossachs (data from H. Robb)

	No. of boxes	No. of clutches laid (range)	% occupancy (range)
1973–76	10 → 100	1–18	10–26
1977–80	127 → 190	21–38	17–20
1981–84	207 → 243	27–46	13–19

has demonstrated the fidelity of this species to its nesting site (H. Robb). Bad weather, which affects the availability of caterpillars, appears to have a significant effect on fledging rates; in 1980, when fledging success was particularly poor, the caterpillars were simply washed off the trees during prolonged rain (C. J. Mead). Grey squirrels and Great Spotted Woodpeckers occasionally take both eggs and young, and fledging success is also affected by the number of bigamous males present; some of these do not attend their second brood, which the female may be unable to rear by herself.

Most spring passage through the Northern Isles takes place in May, with a few arrivals from mid April and stragglers to the second half of June. The return passage is much more prolonged, starting in early August and continuing to late October, with occasional November records. Numbers are generally small, on Fair Isle seldom exceeding daily maxima of 15 in spring and 40 in autumn, but quite large influxes sometimes occur, as in 1970 when Fair Isle had a spring

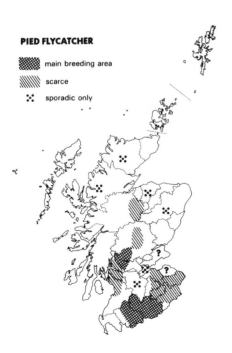

PIED FLYCATCHER

▓ main breeding area

╱ scarce

⋮ sporadic only

peak of 120, and 100+ were recorded on Unst. A few reach the Hebrides during migration periods, most often in autumn, but parties of any size occur only on the east. The largest recent non-island movement reported was at St Abbs, where 100 Pied Flycatchers were seen on 30 August 1980.

The breeding range of this species extends from northern Scandinavia south locally to France, Iberia and Italy and east into Russia. The wintering area is in tropical Africa. Birds ringed on passage and in the nest in Scotland have been recovered in Spain and Morocco, and adults ringed in August in Norway and Finland have been caught on the Isle of May.

Bearded Tit *Panurus biarmicus*   [1]

Vagrant (Europe, including England) – recorded in three years, 1972–83.

The Bearded Tit, a very local reedbed breeder, is normally sedentary but occasionally makes eruptive movements, when birds appear far from their regular breeding areas. It was during one such eruption in 1972 that the species was first recorded in Scotland; on 5 November two males and a female were seen at the Loch of Strathbeg and a male at Guardbridge. From that date to the end of the year a total of 17 were reported, from five counties: Aberdeen, Angus, Fife, East Lothian and Kirkcudbright. Two were still at Strathbeg in March 1973. There has been only one subsequent record, of a single bird at Strathbeg on 12 April 1976. The Bearded Tit founded new colonies in England (the most northerly at Leighton Moss, Lancashire), and in Denmark and Sweden, during the run of mild winters in the late 1960s and early 1970s. If the population was not too badly hit in 1981/82 and is still expanding, as appears to be the case at Leighton Moss, we may expect more occurrences in Scotland in the future.

Long-tailed Tit *Aegithalos caudatus*

Resident breeder, widespread in mainland woodlands and at least sporadic on most of the Inner Hebridean islands. Occurs as a vagrant in Orkney and the Outer Hebrides but has not been recorded in Shetland this century. The population fluctuates markedly in response to winter severity.

Although B&R record nesting at up to 500 m asl, most Long-tailed Tits breed in woodlands at much lower altitudes. They occur in a wide variety of woodland types, but appear to prefer fairly open oak, birch or mixed deciduous woods; in Scotland few nest in hedges, a site often favoured in England. Their remarkable skill in constructing their domed and elastic-sided nests of moss, lichen and spiders' webs, lined with feathers, has been described in detail by Perrins (1979). Many nests are predated, often at an early stage, and the proportion of pairs successfully rearing young in any year may be well below 50%; it is presumably because of this low success rate that the population may take several

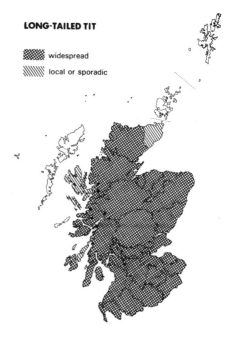

LONG-TAILED TIT

widespread

local or sporadic

years to recover from the losses sustained during a severe winter, especially in less than optimum habitats. The Long-tailed Tit's large brood size (ten or more eggs are usually laid) and habit of remaining in family or larger groups through much of the winter possibly give a misleading impression of abundance (Dougall & North 1983). The species is in fact by no means numerous in most areas; in terms of relative abundance it ranks well below Blue and Great Tits in all types of woodland (Yapp 1962, 1974). Little information is available on breeding numbers in Scotland, but densities recorded in oakwoods in western Scotland ranged from 26 pairs/km² in Stirling to 4 pairs/km² in north Argyll (Williamson 1974).

There has apparently been little recent change in distribution. B&R describe the Long-tailed Tit as breeding 'sparingly' on Mull, Skye and Raasay, and it is still not numerous, though quite widespread, on these islands. Nesting occurs sporadically on Jura and Eigg, and on Rhum, where the one or two families usually depart during the winter (J. A. Love). On Islay parties of up to 50 are present in summer, while a small group which bred on Colonsay in the 1970s is apparently now extinct (J. Clarke). Sutherland is the only mainland area in which there appears to have been an expansion of range; B&R had no records from the west, where breeding was proved in several 10 km squares in 1968–72.

After the breeding season Long-tailed Tits travel around in parties searching for food, and it is in autumn that they are most likely to appear in areas in which they do not breed. Most occurrences on the Isle of May and in Orkney and the Outer Hebrides are between September and November; some 20–30 were in Stornoway Woods in November 1973 but eight days later only two remained. Large parties sometimes gather in areas where food is abundant, especially at times when the population is high, the largest recent count being 100–150 feeding on aphids in an orchard near Melrose in mid April 1977.

Following the very severe winter of 1962/63 the population took a long time to recover, but by 1968–72 it had reached a fairly high level. A run of relatively mild winters during the 1970s allowed the increase to continue until 1975, but breeding numbers dropped quite markedly in 1976 and again as a result of the prolonged severe weather in January/February 1979, with the CBC index showing a decrease of more than 40% between 1978 and 1979. The two continuous spells of very hard frost in December 1981 and January 1982 affected farmland populations of this species much more than those in woodlands, which were not significantly reduced, presumably because insects remained available in sheltered thickets and similar areas.

Different races of the Long-tailed Tit occur: in Britain *A.c. rosaceus*, in Scandinavia *A.c. caudatus*, and in continental Europe *A.c. europaeus*; all are normally sedentary. Some of those occurring in the Northern Isles may belong to the white-headed Scandinavian form (BOU) but there have also been confirmed records of the British race from Orkney.

Marsh Tit *Parus palustris*

Very local resident, breeding only in the southeast, where the small population is slowly expanding.

The Marsh Tit is a recent coloniser, first recorded in 1921 at Duns, Berwick (B&R). Over the next 20 years numbers there increased but breeding was not proved in Scotland until 1945, when a nest with eight eggs was found near Coldstream. Marsh Tits have since spread into Roxburgh, where they were first reported in 1964 and breeding was confirmed in 1966, and Selkirk, where the first sighting was in 1970 but there has not yet been proof of breeding. The Atlas shows confirmed breeding in Peebles, but no details of this record have been published and there has since been only one report from the county, of a single bird in October 1979; this breeding record must therefore be considered of doubtful validity.

No systematic survey has been carried out but in 1980 the breeding population was estimated as about 50–100 pairs (Borders BR). Marsh Tits are most abundant in the deciduous woodlands of large estates in the Tweed and Teviot valleys, with smaller numbers present around Duns, Eyemouth, Coldingham and Cockburnspath. The only recent reports outwith these areas have been the single record from Peebles in 1979, several from East Lothian (two separate birds in 1966, two together at Gosford in September 1972 and at two sites in 1983), one at Langholm, Dumfries, in January 1974, and a male in song near Loch Ken, Kirkcudbright, in April the same year. An earlier report of two in Midlothian (SN 69:174) was subsequently rejected, as one of the birds was close-ringed (SB 1:118). The largest group yet seen together has been 12 at St Abbs in October 1981.

The Marsh Tit is widely distributed in England and Wales, becoming scarcer in the northwest; on the Continent

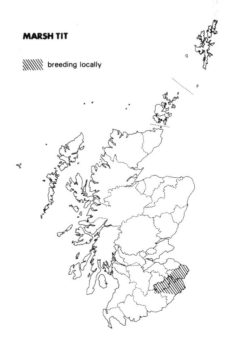

MARSH TIT

〰〰〰 breeding locally

it breeds from central Norway and southern Sweden to Italy and northern Spain. Although it is normally very sedentary some 'local wandering' does take place (Harrison 1982) and this presumably accounts for both the occasional records outwith the main breeding area and for the progressive spread westwards in the Borders.

Willow Tit *Parus montanus*

Local resident breeder, common only in the southwest, where numbers are largest in Dumfries and Kirkcudbright. The breeding range has contracted in the last 50 years.

As B&R point out, the possibility of confusion between Willow and Marsh Tits is considerable. They assumed that most old records of Marsh Tits actually referred to Willow Tits, but with the recent spread of the Marsh Tit it is no longer safe to do so. In the early 1900s the Willow Tit bred regularly in most counties north to Stirling, Perth and Angus, and sporadically elsewhere. Recent information on distribution and numbers is fragmentary, apart from a few well-watched areas in the southwest, and it is possible that breeding is going unrecorded in some districts. Willow Tits occur in most types of hardwood, and occasionally in conifers, but favour scrubby woodlands with alder and birch, often on water-logged soils near lochs and in river valleys. Dead stumps for nest-hole excavation are an important habitat requirement.

Comparing B&R's assessment of the Willow Tit's status 'today', ie during the first half of this century, with Atlas and subsequent records, it is apparent that there has been a marked contraction of the breeding range. Nesting has not been recorded for many years in Midlothian, West Lothian, Fife, Perth, Angus, Inverness or Ross – all counties in which breeding formerly occurred. There are recent records for Ayr, Lanark, Renfrew, Dunbarton and Stirling, and breeding is presumed still to take place regularly in most, if not all, of these counties, though the numbers involved are apparently smaller than in the past. The possibility of incipient colonisation from the currently expanding population in Northumberland is suggested by the increasing frequency of reports from Berwick during the early 1980s, and breeding was confirmed near Ayton in 1982 (Borders BR). At present, however, the Willow Tit seems to be flourishing only in the southwest, where it is currently widespread in Dumfries and Kirkcudbright wherever there is suitable habitat, and has recently been spreading westwards in Wigtown (A. D. Watson). Little information is available on numbers but in 1969 seven pairs were found in one Dumfries glen and up to ten nests around Kirkconnell Flow.

There is no obvious single reason why this species should have declined so markedly in so many areas. Severe winters in the late 1940s and early 1950s probably led to the extinction of the small populations formerly breeding in the Highlands (R. H. Dennis), while loss of habitat may well be responsible for most other local decreases. In view of the uncertainty regarding the present status of the Willow Tit it seems desirable that a comprehensive survey be carried out in the near future.

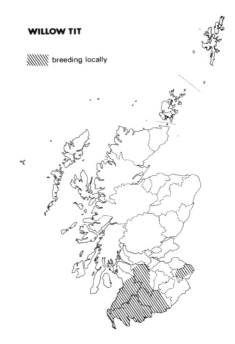

WILLOW TIT

▨▨▨ breeding locally

Willow Tits are widely distributed and quite abundant in England and Wales but, like the Marsh Tit, are absent from Ireland. On the Continent they breed from northern Fenno-Scandia and northwest Russia south to France, northern Italy and Greece. Although this is an essentially sedentary species, birds from the extreme north of the range move south in winter. The solitary record from Fair Isle, in November 1935, may have involved a bird of Scandinavian origin.

Crested Tit *Parus cristatus* [1]

Local resident breeder, confined to the Moray Basin catchment, where it now occurs in conifer plantations as well as the native pinewoods which are its traditional habitat. There has been some recent expansion of the range. Single birds and small parties are occasionally seen outwith the main range, and breeding may have taken place on Deeside.

The small isolated Scottish population of the Crested Tit *P.c. scoticus* belongs to a different race from those breeding on the Continent, where the species is widely distributed from Scandinavia to Iberia and eastward into Russia. The species nests regularly and in any numbers only in Inverness, Moray and east Ross, though breeding also occurs very locally, and in some areas probably sporadically, in Banff, Nairn, and southeast Sutherland. A breeding season survey was organised by Martin Cook in 1979–80 and the account which follows draws heavily upon the results of that study (Cook 1982).

The Crested Tit's habitat requirements apparently include dead stumps at least 15 cm in diameter for nesting in (or

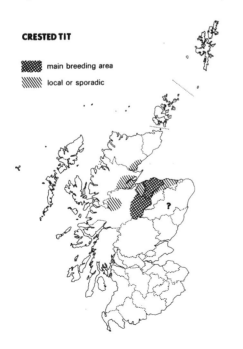

nest-boxes, which are readily used in some situations) and long bushy heather (which facilitates ground feeding during snowy weather), in addition to the presence of Scots pines. Every nest found in the 1979–80 survey was either in or near Scots pines; although three nest-boxes on Corsican pines were occupied, these were within 100 m of a stand of Scots pine. Slightly under half the sites for which habitat data were obtained were in predominantly native woodland, the remainder being in Scots pine plantations not less than 20 years old. At 75% of sites the lowest branches were at 2 m or less above ground level, and the birds appear to prefer rather open woodland with a well-developed shrub layer. Most of the Crested Tits found nesting in coastal Moray and the lower Spey valley were in plantations, most of those in Ross and west Inverness were in natural-looking forest, and in upper Speyside both habitats were occupied.

When Nethersole-Thompson studied the Crested Tit in Strathspey, before extensive afforestation took place in Glenmore, he found densities of 0.02–0.04 pairs / ha in open pinewood. Of the 324 nests he recorded, 239 were in pine stumps, 46 in dead pines and 8 in forks of living pines, the remainder being in alders, birches, a gatepost, woodpecker holes and a squirrel's drey. More than 90% were less than 4 m above the ground (Nethersole-Thompson & Watson 1981). Using estimates of breeding densities ranging from 0.01 pairs / ha in plantations and low-density natural pinewood to 0.15 pairs / ha in 'pure natural-looking pinewood', and applying these figures to the occupied areas of each habitat type, Cook assessed the total population in 1979–80 as in the region of 885 breeding pairs, slightly below the Atlas estimate of c1,000 pairs and appreciably higher than the 300–400 pairs suggested by Nethersole-Thompson in Darling & Boyd (1969). Numbers fluctuate markedly as the species suffers badly in hard winters (Campbell 1974). In good years there is sometimes a non-breeding surplus, which possibly helps the population to recover more rapidly after a severe winter. On CBC plots at the RSPB's Loch Garten reserve the breeding population has ranged from a peak of 29 pairs in 1978 to only 18 pairs in 1981 (Dennis 1984). The severe weather of 1981 / 82 apparently had little effect on the numbers present; in the same census area 18 pairs were present in 1981 and 19 pairs in 1982 (RSPB).

Crested Tits are usually sedentary, with young birds seldom dispersing further than 2 km from their nest site. As Cook points out, such sedentary habits mean that colonisation of new areas, for example plantations reaching an age suitable for occupation, can only occur if there is an existing population within the normal dispersal distance. Presumably when numbers are high, after a run of mild winters, some birds disperse further in order to find unoccupied territory, and at such times Crested Tits may appear in areas some distance from the nearest regular population. Grant (1984) recorded two at the Linn of Dee in 1950, and felt certain that they had travelled via the Lairig Ghru. There were no further authenticated reports from Deeside until the early 1970s, when breeding success in Inverness was good and numbers high, and parties were again seen on Deeside (Knox 1983). Between 1973 and 1978 there were sporadic sightings in Glen Quoich, near Aboyne, and in Glen Dye (Kincardine), but there have been none since the hard winter of 1978/79.

The exact timing and extent of the Crested Tit's spread this century is not clear, as earlier records are patchy and some of the published information is of questionable accuracy (eg statements that this species is an established breeder in Mar). Breeding was first recorded in east Ross prior to 1912, and it appears likely that the lower Findhorn valley in Moray and parts of Nairn and west Banff were colonised in the 1920s/1930s (B&R and Campbell 1974). The first published record for Sutherland was in 1956, and the first confirmed breeding there in Shin Forest in 1978 (the comment in *Sutherland Birds* (Angus 1983), that Crested Tits are believed to have been resident in pinewoods in the southeast of the county since at least 1952, refers to an area which was part of Ross and Cromarty until 1975). The most northerly sighting so far has been at Clynelish, near Brora, in December 1978. By 1968–72 there had apparently been some contraction, as there were no Atlas records from Sutherland, and only 'possible breeding' reports from Banff. It seems likely, however, that breeding continued at least sporadically in these counties; it was again confirmed in Banff in 1980, while nesting occurred in at least five woods in Sutherland between 1970 and 1983, with up to six pairs present in any one year (D. Nethersole-Thompson). Crested Tits were also found breeding in Strathconon, Ross, in 1980, but it is not known whether this was a new development or merely a previously overlooked population.

Outwith the areas already mentioned, there have been occasional reports from Glen Garry, Inverness. No confirmation of breeding there has ever been obtained, but there were sightings in the 1940s (B&R) and a Crested Tit was at the head of the pass between Loch Arkaig and Loch Garry in 1968 (SBR). There are pre-1950 records for Angus and Lanark, but the only recent reports south of the Grampians are of one in Stirling in February 1981 and seven at Rannoch, Perth, in autumn 1983 (SBR). With such isolated occurrences there is always the possibility of wanderers from the Continent, as there have been occasional records of continental birds in England (BOU).

A detailed account of the habits and breeding behaviour of the Crested Tit will be found in Nethersole-Thompson & Watson (1981).

Coal Tit *Parus ater*

Resident, widespread and abundant on the mainland, and increasing in many areas as a result of afforestation. In the Inner Hebrides breeds regularly on some islands but is a local and sporadic nester on others. Has bred in the Outer Hebrides but is absent from Orkney and Shetland.

The Coal Tit is much more strictly confined to woodland than are Blue and Great Tits and is most closely associated with conifers, but it is also widespread in birchwoods, where it is sometimes present in greater numbers than Blue and Great Tits together (Yapp 1962). It only rarely finds non-woodland conditions suitable for breeding, although even a few conifers in parkland may prove adequate. It is also less ubiquitous than the other *parus* tits, occurring in only 74% of all woodland CBC plots in 1973, as against 94%

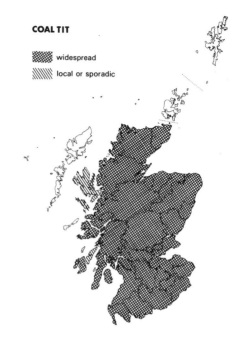

COAL TIT

▨ widespread

▨ local or sporadic

and 97%, respectively, in the case of Blue and Great Tits (BTO/J. Marchant). In favoured areas breeding densities may reach 100 pairs/km² (Atlas), but they are much lower in most mixed and deciduous woods. In conifer plantations numbers increase markedly when the thicket stage is reached.

In winter Coal Tits often form roving mixed flocks with Blue Tits, Goldcrests and Treecreepers. Many such flocks total fewer than 50 birds, but much larger numbers of tits occasionally gather where food is especially abundant, for example when feeding on beech mast. The CBC data suggest that Coal Tits are less affected by severe winters than other tits, but numbers nevertheless fluctuate considerably, presumably in response to food availability.

Coal Tits were already well-established by the first half of this century on the larger Inner Hebridean islands with suitable woodlands. B&R describe them as common locally on Mull, Islay, Jura, Gigha, Colonsay, Skye and Raasay and record breeding on Eigg in 1934 but make no mention of Rhum. At present the species is widespread but still not numerous on Skye, and common on Raasay (A. Currie). It was recorded on Eigg in 1968–72 and also on Rhum, where the population is only six to eight pairs and has not yet shown any sign of increasing (J. A. Love). Coal Tits bred in Stornoway Woods early this century but later died out; they were present again in the 1960s, and nested in 1965 and 1969 but have not done so since (Cunningham 1983). An increase can be expected to occur both on the islands and in the north mainland as extensive recently-planted forests reach a suitable stage for colonisation.

Native Coal Tits belong to the race *P.a. britannicus* and are sedentary; the few Scottish ringing recoveries are all within 10 km of the ringing site. The continental race *P.a. ater* is widespread right across Eurasia; birds breeding in the northern part of the range are partial migrants and irruptions from adjacent parts of the Continent are occasionally recorded in southeast England (BOU). Coal Tits are exceedingly rare visitors to the Northern Isles; the very few that reach the islands, and probably also some of the stragglers seen on the east coast, are likely to belong to the continental race, at least one example of which has been identified on the Isle of May.

Blue Tit *Parus caeruleus*

Resident, widespread and abundant in woods throughout the mainland, the Clyde Islands and the Inner Hebrides; has recently colonised the Outer Hebrides but does not breed in the Northern Isles. Immigrants from the Continent occasionally occur in small numbers, most often in early winter.

The Blue Tit is catholic in its choice of habitat, breeding in suburban gardens and conifer plantations as well as in the deciduous woodlands which were probably its original habitat and in which it reaches the greatest breeding densities. It readily nests in boxes, especially where the supply of natural holes is limited. Long-term studies in England have shown that the Blue Tit's breeding cycle and success are very closely related to the availability of caterpillars, the main food used in rearing the young (Perrins 1979), and work in the Carse of Gowrie has suggested that considerable variations in the timing and success of breeding can occur within even a comparatively small but non-homogeneous area of woodland (Greenwood & Hubbard 1979).

At the end of the breeding season Blue Tits move around in small groups, often with other tits and Treecreepers, and usually desert upland sites for woodlands lower in the glens. Large numbers sometimes visit the Tay reedbeds in winter; there were c100 feeding on the *Phragmites* there in December 1975. Although most are sedentary, with over 90% of ringing

recoveries within 10 km of the ringing site, some travel much longer distances, especially in years when the population is high. It is presumably as a result of such movements that colonisation of many of the islands has taken place.

Blue Tits have bred on most of the larger Inner Hebridean islands for many years (B&R) and nesting has recently been recorded on Eigg and Canna. On the smaller islands breeding is probably only sporadic, as is the case on Rhum and possibly Coll, where several were seen in summer 1984 (R. Broad). The survival of the tiny populations involved is likely to depend upon the periodic arrival of immigrants from the mainland (Love 1981). The first record of breeding in the Outer Hebrides was in 1963, in Stornoway Woods. Numbers there have increased fairly steadily and outside the breeding season birds are often seen in gardens. Nesting has recently been recorded at Keose, a few kilometres from Stornoway (Cunningham 1983), and in Harris (SBR 1981), and Blue Tits have been seen in suitable habitat in North Uist.

This is among the commonest and most widely distributed of woodland birds, occurring in more than 90% of the woods recorded in the BTO's Register of Ornithological Sites (Fuller 1982). Breeding densities are highest in mixed deciduous woodland and lowest in mature conifer plantations; the average over all CBC woodland plots in Britain is around 43 pairs/km² (Atlas), but locally may be very much higher. In East Lothian in 1984 Stan da Prato (1985) found 104 pairs/km² in three small mixed woods totalling 27 ha and surrounded by farmland. Jeremy Greenwood's study area near Perth, a 32 ha mixed wood containing 121 nest-boxes, holds the equivalent of 125–266 pairs/km². Numbers fluctuate in response to both summer food supply and winter severity; during wet, cold spells in the breeding season many young die in the nest, while some pairs may not even attempt to breed after a winter in which temperatures have been abnormally low (J. J. D. Greenwood, Love 1981). CBC results for Britain as a whole show that Blue Tit numbers were significantly reduced by the hard winters of 1978/79 and 1981/82, both on farmland and in the preferred woodland habitat, and that recovery takes place fairly quickly.

Native birds belong to the race *P.c. obscurus*, which is indistinguishable in the field from the continental race *P.c. caeruleus*. The latter breeds from southern Scandinavia across most of Europe, into North Africa and eastwards to central Russia. Those breeding in the north of the range are partial migrants, while eruptions occasionally take place further south, with considerable immigration noted in southeast England. Examples of the continental race have been identified on Fair Isle (Davis, in Williamson 1965) and it seems likely that the Blue Tits appearing irregularly in the Northern Isles originate from Scandinavia. The numbers involved are very small. The largest recent influx to Shetland, in mid October 1977, totalled 17+ birds; several remained throughout the winter, the last leaving in May.

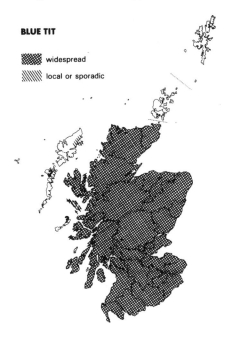

BLUE TIT

▓ widespread

▨ local or sporadic

Great Tit *Parus major*

Resident, breeding in woodland and scrub throughout the mainland and on several of the islands but scarce in the Outer Hebrides and absent from the Northern Isles. Much more abundant in deciduous trees than among conifers and seldom nests more than 500 m asl. Small numbers of immigrants occur irregularly.

The Great Tit extended its distribution in the north mainland during the first half of this century (Pennie 1962) and is still doing so in Caithness, where it was until recently restricted to a small area in the south. Although increasing afforestation may be partly responsible for this spread, it has been suggested that climatic factors, and especially milder winters, are also involved, as a comparable northward expansion has occurred in Norway, where woodland has long been present (Atlas). B&R recorded breeding on Islay, Jura, Mull, Skye and Raasay, but made no mention of Rhum or Eigg, on both of which nesting now occurs at least sporadically (Reed *et al* 1983; J. A. Love); no Great Tits were found on Rhum in 1968–72. Small numbers have been recorded recently on Coll and Colonsay. Following the first Outer Hebridean record of nesting, on Lewis in 1962, numbers increased until about 1970 but subsequently declined; only about six pairs were present in the early 1980s (Cunningham 1983).

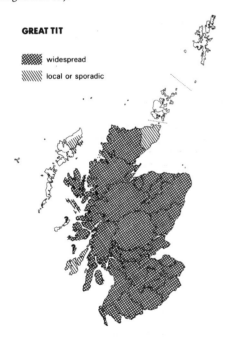

GREAT TIT

▨ widespread

▨ local or sporadic

Great Tits breed in parks and gardens and on farmland as well as in woods. In most habitats they are less abundant than Blue Tits, but in oakwoods in Wester Ross the situation is reversed (Williamson 1969). The Atlas suggests that woodland CBC densities for Britain as a whole often exceed 50 pairs/km² but seldom reach 100/km²; in a 32 ha mixed wood near Perth holding the equivalent of 19–59 pairs/km² many pairs occupy nest-boxes (J. J. D. Greenwood). Farmland densities are very much lower, for example only about one pair/km² on predominantly arable ground in East Lothian (da Prato 1985). In winter many upland woods are deserted as the birds move lower down the glens, sometimes gathering in large numbers where beech mast is plentiful. Great Tits appear to suffer less in hard weather than some of the other woodland species; even the severe winters of 1978/79 and 1981/82 produced decreases of only 10% or less among CBC woodland populations.

Native birds belong to the race *P.m. newtoni* and are sedentary, with 95% of recoveries less than 10 km from the ringing site. Small numbers of Great Tits occasionally appear in the Northern Isles and these are believed to belong to the continental race *P.m. major*, which is widely distributed across Eurasia and is a partial migrant in the northern part of its range. Eruptions of this race occur in some years, when the population is at a high level. At such times considerable numbers sometimes arrive in southeast England and some reach Scotland, as continental birds have been identified on the Isle of May in both spring and autumn (Eggeling 1974). The largest irruption recorded in Shetland for many years took place in 1977 and involved c17 birds, most of which arrived in mid October. Those reaching the Northern Isles in autumn often remain to winter. As it is impossible to distinguish the two races in the field, immigrants may be overlooked in parts of the country where Great Tits are normally resident; 54 at St Abbs in early October 1981 might well have been immigrants.

Nuthatch *Sitta europaea*

Vagrant (Europe, including England) – less than annual, in very small numbers.

Although the Nuthatch – a hole-nester which frequents mature deciduous woodland – is distributed widely in Wales and southern England, and more locally as far north as Durham and Northumberland, it has yet to be found breeding in Scotland. The first report of this species was of one 'said to have been caught on the nest' in Roxburgh in 1850, but as details are lacking this is not accepted as a confirmed breeding record. There were five further reports in the 19th century, but only three between 1900 and 1950, one of which involved a pair nest-building in Kirkcudbright in 1927; the others were in Berwick (1928) and Stirling (1945). The post-1950 published records are summarised below:

1963/64	Perth – Almondbank (November/February), feeding at a bird table
1966	Kirkcudbright – a few miles west of Dumfries (October/November)
1969	Kirkcudbright – same area (February)
	Kirkcudbright – Kirkconnell Flow (February/April)
1970	Perth – Invergowrie (May)
1971	Perth – Invergowrie (January)
1972	Perth – Strathardle (May)
	Perth – Glen Lyon (September)
1974	Perth – Invergowrie (March)
1975	Roxburgh – Yetholm (April)
	Argyll – Dalmally (June)

1976 Kincardine – Banchory (January)
   Argyll – Lochgilphead (January / February)
   Kirkcudbright – Glen Trool (August/December)
1977 Perth – Loch Earn (July)
   Wigtown – New Luce (October)
1978 Stirling – Loch Lomondside (March)
   Caithness – Wick (December)
1981 Angus – Montrose (March)
   Kirkcudbright – Glenkens (April), at a nest-box.
1983 Berwick – Ayton (April & July)
   Kirkcudbright – near Dalry (November)
   Kirkcudbright – Army Range near Dundrennan (December)

All these records relate to single birds.

The scatter of dates suggests that occasional irruptive movements may take place in autumn and winter, with some birds first being noticed when they come near habitations in hard weather. Although the white-breasted Siberian population of the Nuthatch *S.e. europaea* makes such movements and has spread westwards into Europe and Scandinavia, the buff-breasted British and western European race *S.e. caesia* is generally considered to be very sedentary. There have been no records of *S.e. europaea* in Britain – even the Caithness bird, which was the most likely to have originated from Scandinavia, was not white-breasted – so all Nuthatches wandering into Scotland presumably move north from England. Since the 1940s the English breeding range has been gradually extending northwards, into the Lake District on the west and to within 10 km of the Border on the east (Atlas). In view of this spread, and the increasing frequency of occurrences in Scotland, it seems quite likely that breeding will occur here in the future.

Treecreeper *Certhia familiaris*

Resident, widely distributed in woodlands throughout the mainland, the Clyde Islands and most of the Inner Hebrides; scarce and very local in the Outer Hebrides and absent from Orkney and Shetland. Vagrants from Scandinavia occur sporadically, most often in the Northern Isles.

This unobtrusive little bird occurs in all types of woodland but is more abundant in deciduous and mixed woods than in conifer plantations. It is present in 60% of all woodland CBC plots (in Britain as a whole), and an average density of 50–100 pairs/10 km² is suggested in the Atlas. The Treecreeper's requirement for a nesting site behind loose bark or in the crack of a broken branch probably acts as a limiting factor in some woodland types, especially immature conifer plantations. It will use artificial nest sites, including nest-boxes of suitable design, and a pair bred successfully for many years inside the hide at the Loch of the Lowes (Perth), treating the hide supports as tree-trunks (which they were) and often leaving the nest via the viewing window. Its habit of roosting in niches scraped out of the soft bark of Wellingtonias (*Sequoiadendron*) is well-known; the white streak of droppings below a niche identifies a roost currently or recently in use.

The Treecreeper is not an easy species to study and little fieldwork has been done on it in Scotland; a paper on its breeding biology by Flegg (1973), which covers weight, measurements and moult as well as breeding details, included the Scottish data then available. This is one of the few small passerines to be found, albeit in small numbers, in some upland woods in winter, when Treecreepers often join wandering tit flocks. Numbers are affected by prolonged spells of severe weather, during which insects become difficult to find. Woodland CBC indices showed declines of 18% and 20% respectively following the hard winters of 1978/79 and 1981/82, and losses were probably higher in Scotland than further south.

The only evidence of any change in status since 1950 comes from the islands, where breeding now occurs regularly on Rhum (up to six pairs in the early 1980s) and was recorded for the first time on Canna in 1976; B&R had no records for these islands. Colonisation of the Outer Hebrides took place during the 1950s, and up to ten pairs were present in the Stornoway Woods by 1966, but numbers have since

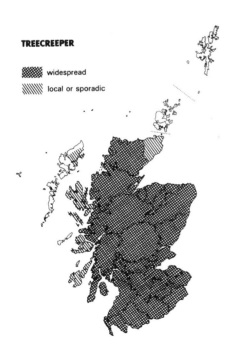

TREECREEPER

▨ widespread

▧ local or sporadic

decreased and not more than one or two pairs remained in 1983 (Cunningham 1983).

Native Treecreepers belong to the race *C.f. britannica* and are normally sedentary, as are those breeding elsewhere in the European range, which extends from central Scandinavia and France east across Eurasia. The latter belong to the so-called northern race *C.f. familiaris*; several of the stragglers that turn up in Shetland and Fair Isle, and single birds in Fife, Berwick and Harris, have been identified as belonging to this race, and it is probable that others reaching the Northern Isles also originated from across the North Sea (BOU). The occurrence of such vagrants is very irregular and numbers are extremely small; a total of four in Shetland in the autumn of 1980 was unprecedented.

Golden Oriole *Oriolus oriolus*   [1]

Vagrant (Europe, including England) – annual in very small numbers. Has bred.

Although occurring annually, the Golden Oriole reaches Scotland in such small numbers that it seems appropriate to class it as a vagrant. From 1968 to 1983 inclusive the maximum number recorded in any one season was about 15 (in 1978) and in some years only two or three birds were seen. Over 90% of sightings are between 8 May and early July, and individuals sometimes remain in the same area for several weeks. Autumn occurrences are in September/November, with 7 November (Fair Isle 1981) the latest date. Females/immatures and adult males are reported in roughly equal numbers and song is occasionally heard.

B&R state that 'the great majority of occurrences are south of the Grampians' but in recent years this has certainly not been the case. Most of the records since 1968 have been from Shetland, Fair Isle and Orkney. The remainder are widely scattered, from Berwick to the Outer Hebrides and from Caithness to Kirkcudbright, but there have been few recent reports from the southwest.

In 1968 song and display were noted in Kinross during June, and in 1973 there was a singing male in Fife in May. The following year a pair was seen feeding fledged young in the same area on 17 July. This remains the only Scottish breeding record. The small population of less than 30 pairs in south and southeast England is on the northwest corner of the Golden Oriole's breeding range, which extends from Iberia east to Iran and India, and from southern Sweden and Finland south to Asia Minor. The bulk of the population winters in Africa south of the Sahara.

Isabelline Shrike *Lanius isabellinus*

Vagrant (Iran/Mongolia) – four records, 1950–83.

This species, until 1978 classed as a race of the Red-backed Shrike *Lanius collurio*, was first recorded in Britain on 26 September 1950, when an adult male was seen on the Isle of May (BB 44:217–219). By the end of 1983 there

had been 16 more British records, the three Scottish ones all on Fair Isle – on 12 May 1960, 24 October 1979 (an immature), and 9–12 October 1981. Four races of this shrike are known, the most westerly being *L.i. phoenicuroides*, to which the Isle of May bird probably belonged. The 1981 Fair Isle bird showed characters considered to exclude the race *L.i. isabellinus* and was therefore either *speculigerus*, the race breeding in Mongolia, or a pale *phoenicuroides* (BB 76:519). The breeding range of the species as a whole extends from Iran to Mongolia but little is known about the overlap between the races (Sharrock & Grant 1982). A paper on the identification of the Isabelline Shrike appeared in BB 75:395–406.

Red-backed Shrike *Lanius collurio*   [1]

Scarce passage visitor, annual in Shetland but less regular further south on the east coast and only sporadic elsewhere. Numbers are largest in spring; most autumn birds are immatures. Has bred.

Spring and autumn passage totals of Red-backed Shrikes are generally less than 50 and 20 respectively but occasionally much larger, for example 185+ in May 1977 and over 100 in autumn 1981. In a 'good' year 10–15 birds may be on Fair Isle and other Shetland islands at one time. Most reports from other areas are of single birds. Spring passage generally starts during the first half of May and continues until late June, but there are a few April records (the earliest is 23rd) and a few in early July. The return passage is less consistent in its timing, with first reports ranging between 1 August and 3 September and last records between 8 September and 11 November; the main movement usually occurs between mid August and early October. There are records from most mainland counties, the Outer Hebrides (Lewis, South Uist and Benbecula) and Arran, but not from the Inner Hebrides. Outwith Shetland, Red-backed Shrikes are most frequently reported from Orkney, Caithness, Aberdeen, Fife, the Isle of May and East Lothian.

B&R recorded probable breeding in Midlothian in 1932, and a pair possibly nested in Orkney in 1970 (Balfour 1972), but it was not until 1977 that breeding was confirmed. That year three nests were found in the north and northeast, of which one was successful, one robbed by a predator, and one deserted. There was also a pair displaying and courtship feeding in Orkney and a pair on territory in Wester Ross. In 1978 two pairs nested in the north, both successfully, and five other males were present at suitable sites. In 1979 one pair bred in Aberdeen and males were on territory in Inverness, Perth, Angus and at a second Aberdeen site. Although in 1980, 1981 and 1982 males were seen inland in several areas during June, there has been no further evidence of breeding. Should another large-scale May influx occur, however, it would clearly be worthwhile keeping a look-out for breeding pairs.

There is a small and diminishing breeding population of the Red-backed Shrike in southeast England. Numbers there have been declining, and the range has been contracting, for more than 100 years; from 300 pairs in 1952 the population dropped to 80–90 pairs in 1971, and to under 40 pairs in 1981. A decline has also occurred elsewhere in western Europe and it has been speculated that this is linked with climatic deterioration; wet, cool summers reduce the activity, and probably also the abundance, of the insects upon which the species feeds. Most British pairs breed on dry, bushy heathland, often near water and where there is a mixture of gorse, heather and small trees (Atlas), but on the Continent they also occupy waste ground, hedges and orchards, and in Scandinavia forest clearings. Those breeding in Britain and Europe (from southern Sweden to Iberia and east to the Urals) belong to the nominate race *L. c. collurio*. Most of the passage migrants reaching Scotland probably originate in northern Europe, but there have not yet been any ringing recoveries to substantiate this.

Lesser Grey Shrike *Lanius minor*

Vagrant (Europe) – recorded in 11 years, 1965–84.

The Lesser Grey Shrike occurs only irregularly and in very small numbers; the most seen in any one year has been only four – in 1974, when there were three in Shetland/Fair Isle in June and one on Mull in September. Post-1974 records are from Out Skerries (May 1977) and Fair Isle (June 1984). Most reports are from Fair Isle, where the species was first recorded in June 1913 (B&R). There have been about 14 subsequent reports from there, including two birds on 19 September 1955, and nine reports since 1965 from Shetland. The Lesser Grey Shrike has also occurred in Orkney, Sutherland, Angus, Fife, East Lothian, Ayr, Argyll (mainland), and the Outer and Inner Hebrides. The records are about equally divided between spring and autumn, with most birds seen either between mid May and mid June or between mid September and mid October.

On its breeding grounds, which stretch from eastern France and southern Germany eastwards, this species frequents woodland-edge habitats. It winters in Africa south of the Sahara. Although not a large bird, the Lesser Grey Shrike occasionally tackles sizeable prey – or so it appears from the discovery of one drowned with a sparrow in a water barrel on North Ronaldsay.

Great Grey Shrike *Lanius excubitor*

Scarce and irregular passage and winter visitor, usually in only small numbers; occasional large autumn influxes are followed by higher wintering numbers. Occurs most often in the Northern Isles and the eastern half of the country, and very sporadically in the west.

The first autumn arrivals usually appear in mid September (13th is the earliest since 1968), but sometimes not until October. The main influx occurs in October/November, sometimes later, often with marked passage through the Northern Isles. In most years fewer than 40 are reported in autumn, but numbers are sometimes much larger, as in 1970 (140+), 1976 (100+) and 1982 (c100). Reports are most widespread, both geographically and through the winter, in seasons when autumn numbers are high; more than 70 wintered in 1970/71. Wintering birds often establish territories and favoured areas may be occupied in successive years. Spring passage is usually evident in the Northern Isles during April. Most birds have left by the middle of May, though occasional stragglers are still present in June; some spring migrants remain in one area for two or three weeks and occasionally sing. There are no July or August records. Since 1968 there have been reports from nearly all mainland counties and from the Outer Hebrides (North Uist and North Rona), but not from the Inner Hebrides (B&R give old records for Islay and Mull) or the Clyde Islands (B&R give an Arran record).

The feeding habits of wintering Great Grey Shrikes have been recorded by Hewson (1970) and Halliday (1970), who found that prey included voles, insects, birds (Chaffinch was identified) and a frog. Tulloch (1970) describes the importance of 'wedging-places' in enabling a shrike to feed from prey too large to swallow whole. Among the prey items recorded as taken by passage birds are lizard (SB 6:449–50) and the only Olivaceous Warbler yet recorded in Scotland (SB 10:24–25).

The Great Grey Shrike is widely distributed, throughout much of Eurasia, in Africa and in North America, breeding from the sub-Arctic to the Tropics, on forest edges or in more open areas with scattered trees and shrubs to serve as look-out posts. Most of those visiting Britain belong to the nominate race *L. e. excubitor* which breeds from Scandinavia and Iberia east to the Pacific; the Steppe Shrike *L. e. pallidirostris*, which breeds further south and east, has been identified twice on Fair Isle, in September 1956 and October 1964 (BOU). The April recovery in Norway of a shrike ringed on Fair Isle 17 months previously suggests that at least some of those reaching Scotland originate fairly far north.

Woodchat Shrike *Lanius senator*

Vagrant (Europe) – less than annual, in very small numbers.

B&R give only four records of the Woodchat Shrike, the first an immature killed at the Isle of May light on 19 October 1911 (SN 1912:10); this is still the latest autumn date. The highest annual total so far has been five, in 1965. Since 1968 the species has been recorded in 12 years, with not more than three birds in any one season and none at all in 1977–79 inclusive or in 1983. Some 70% of the records are between late April and June, mainly in May. Most of the autumn birds appear between mid August and the end of September, and many are immatures. More than half the records are from Fair Isle and Shetland but there have also been reports from Orkney, Aberdeen, Moray, Fife, the Isle of May, East Lothian, and the Outer Hebrides. In its breeding area, which extends from the Mediterranean Basin north into France and Germany and east to Iran, the Woodchat Shrike frequents dry fairly open ground with scattered scrub or trees for song-perches. It winters in Africa south of the Sahara and occurs more frequently in England than it does here.

Jay *Garrulus glandarius*

Resident breeder, widespread in the southwest, and from the Forth / Clyde to Aberdeen and north Argyll; outwith these areas the Jay is scarce and very local. Occasional irruptions of continental birds may occur in autumn.

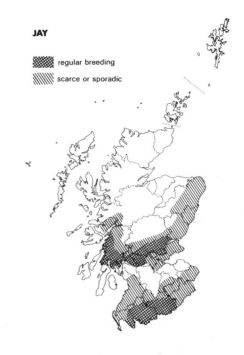

JAY

▓ regular breeding

▨ scarce or sporadic

Like most other members of the crow family, the Jay was much persecuted in the 19th century, on account of its habit of taking eggs and young gamebirds. An increase took place during the first half of this century and this population expansion appears to be continuing. Although primarily a bird of deciduous woodland, especially oakwoods, the Jay has adapted to life in, or at least on the fringes of, conifer plantations. As a result it has recently been expanding both its range and its population in areas of maturing conifers, especially in the southwest. Marked increases have been reported in Dumfries, where there were 120–150 pairs in the Forest of Ae in 1969, and in Kirkcudbright and Wigtown. Although still fairly scarce in west Wigtown, Jays were said to be increasing around Stranraer in 1975, while the expansion noted since the early 1970s elsewhere in the county is continuing. Further east there has also apparently been some increase, with up to ten pairs thought to be present in each of the four Border counties (none were recorded in Selkirk in 1968–72), although Jays are still persecuted on some estates in that area. In Ayrshire too there has been an expansion since the 1960s, starting with the establishment of a stronghold in the Girvan Valley – presumably originating with birds from the Kirkcudbright forests – and more recently spreading to well north of Kilmarnock (Hogg 1983). In Lanark Jays are scarce and their status uncertain, and there is no recent breeding record for Renfrew; if the current expansion in Ayr and Dumfries continues it seems likely to be only a matter of time before these two counties are colonised.

From the northern segment of the range expansion has been taking place both southwards and northwards. Breeding was recorded for the first time in West Lothian in 1980, a few pairs are now present in Kinross, and there have been several recent reports from west Fife. In Tentsmuir Forest, where nesting was confirmed in 1968–72, numbers have been

increasing steadily. Deeside was the only area of Aberdeen in which breeding was recorded during the Atlas survey and there were few other reports during that period, but by 1984 Jays were known to be definitely or probably breeding in 18 10 km squares and possibly doing so in a further six (NE Atlas). Argyll has long had a fair-sized and widely scattered population, currently estimated at over 100 pairs; there have been increasingly frequent reports from the north of the county in recent years, and also from southwest Inverness. It is clear that there has been a very substantial expansion of the breeding range since B&R described this species as confined to 'that part of Scotland which lies south of the Grampians'.

Although the Jay reputedly bred at one time on Arran (B&R), it is now recorded there only occasionally. It occurs as an irregular vagrant on Bute and has also been seen on Islay, but there are no records from the Outer Hebrides or Orkney, none this century from Shetland, and only one, in 1940, from Fair Isle. On the mainland north of the present breeding range there are sporadic reports from east and north Inverness, Ross – east and west, and Sutherland.

Immigrant Jays of continental origin are recorded in autumn fairly frequently, but in very variable numbers, in southeast England, and probably also occur elsewhere on the east coast in years when major irruptive movements take place; ringing recoveries have shown that at least some of these birds are from western Europe (BOU). Although there is no proof that such invasions occur in Scotland there is circumstantial evidence that they do. A small influx was noted in the Borders in October / early November 1975, but much more convincing was the report in October 1982 of a large roost of Jays in Midlothian, a county where none are known to breed. On 17 October some 320 were counted, assembling in a paddock near a mixed deciduous wood, in which they eventually roosted; 123 were present the following evening, but by 19 October the flock had dwindled to 18 birds (Young 1984). This influx, which involved by far the largest gathering of the species ever recorded in Scotland, was around the time of the massive fall of Scandinavian and Siberian immigrants all down the east coast of Scotland, and it seems likely that these birds were of Scandinavian origin. None were found at this site in October 1983, when an unprecedented invasion of continental birds occurred in southern England and numbers were unusually high in eastern Scotland.

Magpie *Pica pica*

Local resident, most abundant in the Forth/Clyde valley and the northeast, very local in the southern uplands and Fife/Perth/Angus, and absent from most of the Highlands and islands.

The Magpie was at one time much more widespread, breeding regularly north to Sutherland and Wester Ross, but during the 19th century it declined greatly due to persecution. By 1938 it was common only in the industrial belt, where it was described by Arthur Duncan – reporting on one of the earliest SOC enquiries – as abundant in Renfrew, north Lanark, West Lothian, Midlothian, Stirling and part of Dunbarton (quoted in B&R). In almost all other lowland areas it was then either extinct or very scarce and local. The decrease in gamekeeping during the two world wars resulted in some expansion of the population, and more recently colonisation of the increasing areas of upland forest has started. As the Magpie is sedentary, with little evidence of movements more than 10 km from the natal area, colonisation is a slow process.

Today the bulk of the population is still concentrated in the Central Lowlands, from north Ayr through north Lanark to West Lothian and Midlothian, and north to Dunbarton, Stirling, Clackmannan, west Fife and west Perth. Throughout this area, and also in Kincardine/Aberdeen, the Magpie is a common sight in parks, around the suburbs, and on farmland with abundant unkempt hedges and rough grassland. In the more intensively-farmed and keepered areas such as east Perth, Angus, Kinross, east Fife and East Lothian it is scarce and local, and is still persecuted on account of its habit of taking the eggs and young of game (and other)

birds. Further south too Magpies decrease in abundance from west to east. In Wigtown (where they bred at the Mull of Galloway for the first time in 1974) and in Kirkcudbright they are locally common; in Dumfries (where they nest in dense stands of Sitka – E. Fellowes) and in Roxburgh they are scarcer and more local. They are very uncommon in Berwick, and absent from Selkirk and Peebles, although possibly now beginning to move into the West Linton area from the north. In Argyll there are about ten pairs on the Cowal peninsula but none breed elsewhere in the county, nor in west Inverness, Wester Ross, or Sutherland. Magpies have been seen occasionally in the Rumster Forest area of Caithness, but there is not yet any record of breeding. Around 10–20 pairs are known in east Ross, and the same number in east Inverness and Strathspey, while in Nairn, Moray and Banff the small populations are widely scattered. The only recent island breeding records are from Skye, where Magpies nested at Braes in 1972 but have not done so again, and from Bute, where one or two pairs still survive despite persistent persecution. On Arran, where breeding formerly occurred, nesting has not been recorded for some years.

Outwith the breeding range, Magpies are occasionally recorded in all parts of the mainland, and have been seen at altitudes of up to 400 m asl. Unusually many were present in the north in January/April 1979, with eight sightings in Wester Ross; this influx was possibly due to the very severe mid-winter weather. Records from the islands are few. Although occurrences on Islay are relatively frequent, birds appearing there are liable to be shot with minimum delay, and there is little likelihood of colonisation. There are also records for Mull, though no very recent ones. B&R knew of no occurrences in Orkney, where there have been several sightings since the early 1970s, including one on North Ronaldsay in October 1976 – the most northerly to date. There are no records for Shetland or the Outer Hebrides.

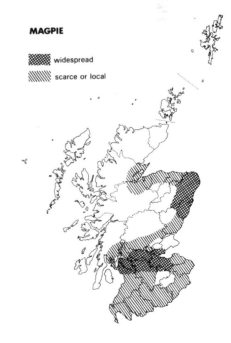

MAGPIE

▨ widespread

▧ scarce or local

The only systematic work on Magpies in Scotland is a limited study of breeding biology carried out in Aberdeen by Love & Summers (1973). In a study area with few hedges, but with small clumps of trees round most of the farms, they found one nest/4.5 km². This is a very much lower density than the CBC figures given in the Atlas: 2.3 pairs/km² on farmland and 2.8 pairs/km² in woodland. B&R quote reports of gatherings of over 100 birds, but there are few recent records of such large flocks. The absence of the Magpie from most of highland Scotland seems surprising in view of the fact that it breeds beyond the Arctic Circle in Scandinavia, while its scarcity in some other parts of the country seems unlikely to be entirely attributable to persecution. There is clearly scope for further study here.

Nutcracker *Nucifraga caryocatactes*

Vagrant (Eurasia) – two occurrences this century, 1968 and 1971.

The only recent records of this member of the crow family are of a single bird in Lerwick on 22–23 August 1968 and five at Eddleston, Peebles, on 28 August 1971. None were recorded in the first half of this century and only three prior to that, the first being one shot on Sanday, Orkney, in October 1868. The nearest breeding Nutcrackers are in southern Scandinavia and the mountains of central Europe; they belong to the thick-billed race *N.c. caryocatactes*, which is normally sedentary. Most of those reaching Britain belong to the more easterly slender-billed race *N.c. macrorhynchos*, which breeds in coniferous forests from northeast Russia to Manchuria and makes eruptive movements in years when there is a poor seed crop. The Shetland bird was the sole Scottish representative of a major irruption in autumn 1968, when more than 300 arrived in Britain.

Chough *Pyrrhocorax pyrrhocorax*   [1]

Scarce and local resident breeder, confined to Islay, Jura, Colonsay and Kintyre.

The British population of the Chough, the bulk of which is in Ireland and Wales, is far separated from the rest of the European breeding range, which lies from the Alps and Pyrenees south to North Africa. In Scotland, therefore, this sedentary species is at the extreme northern limit of its distribution. The history and present status of the Chough have recently been discussed in detail by Judith Warnes (1983); the short account given here draws heavily upon her work.

At one time Choughs occurred in many parts of the country, both inland and on the coast, but by the early 19th century most had vanished from inland areas and a decline was also becoming apparent on the coast. The east coast was deserted first, while on the west mainland small populations survived in Ayr until the late 1920s and in Wigtown until the 1930s. On the Inner Hebrides Choughs were once common on most of the islands but had gone from many by the end of last century. There is no satisfactorily confirmed proof this century of breeding in the Outer Hebrides, where the most recent record is of a single bird with crows on Barra in August 1963 (Cunningham 1983). By the early 1900s Islay was probably the only place where the species remained reasonably common.

Following several partial surveys, a breeding population census was carried out in 1982. This revealed a total of 171–211 birds, including 61–72 breeding pairs (Table). The great majority were on Islay, where the Rhinns and the Oa held the largest numbers. In Kintyre numbers appear to fluctuate and there is still not a firmly re-established breeding colony. Away from the known breeding areas the only recent

The population of the Chough in Scotland in 1982 (Data from Warnes 1983)

	Islay	Jura	Colonsay	Kintyre	Total no. of birds
Breeding pairs	53–61	6–8	1	1–2	122–144
Helpers*	3	0	0	0	3
Non-breeders	32–50	max. 7	0	max. 7	46–64
Total	141–175	19–23	2	9–11	171–211

* "Helpers" are birds present at a nest in addition to the breeding pair and known to assist in at least some stages of rearing the young.

reports have been from the Garvellachs (one in September 1977, SBR) and Ayr, where one was accidentally shot in November 1980 (Hogg 1983).

Choughs feed largely on the invertebrates present in the soil under unimproved grassland or associated with cattle dung, and it is likely that agricultural developments, such as reduction in permanent pasture area and changing methods of cattle management (especially in-wintering) have been detrimental to them (J. M. Warnes). Persecution, including egg collecting, was undoubtedly one of the factors contributing towards the earlier stages of the decline, while cold winters which reduce the availability of food are likely also to have had a depressing effect on the population. On Islay, where this normally cave-nesting species has recently taken to breeding in derelict buildings, the population is currently showing encouraging signs of increase.

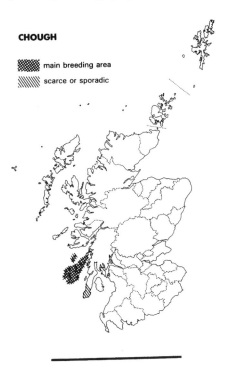

CHOUGH

- main breeding area
- scarce or sporadic

Jackdaw *Corvus monedula*

Resident breeder, abundant over much of the mainland, scarce and local in the northwest Highlands and absent from many of the islands. The autumn arrival of winter visitors is less obvious than their spring departure, especially in the north.

This is the most adaptable of the crows and is equally at home in the city, on farmland or on mountainsides; breeding has been recorded up to 400 m asl in Inverness (B&R). A social species, the Jackdaw is catholic in its nesting habits, occupying such varied sites as cliff crevices, holes in trees, rabbit burrows and buildings (especially chimneys), and forming large colonies where nest sites are abundant. Where colonies are close to concentrations of ground-nesting spec-

ies, Jackdaws may prey heavily on eggs, as on St Serf's Island in the early 1970s, when the 350+ nesting pairs took large numbers of duck eggs and control measures became necessary (Allison *et al* 1974). Egg predation is also common on seabird cliffs. Jackdaws and Rooks often share the same communal roost and feed together over farmland in mixed flocks, Jackdaws being the more opportunist in taking advantage of feed put out for domestic animals. In winter those breeding in the uplands move to lower ground.

A considerable increase and range expansion took place around the turn of the century and in the early 1900s (Parslow 1973), but there appears to have been little subsequent change. The Jackdaw remains scarce and very local in north and west Sutherland, Wester Ross and west Inverness. Although widespread and locally abundant on Islay, Colonsay, Eigg and Skye, it does not breed on Rhum, Coll, Tiree, Canna or Muck and is relatively scarce on Jura and Mull. The first record of nesting in the Outer Hebrides was in 1895 in Stornoway, where some 10–15 pairs still breed; there has been no expansion into other areas (Cunningham 1983). The arrival of 55 on Benbecula in November 1983 might lead to further colonisation. Orkney had been colonised by the 1880s and Jackdaws now breed regularly on most of the islands, usually on cliffs, and are plentiful in Kirkwall. Nesting was first recorded in Shetland about 1943, but breeding appears still to be only sporadic. The Kergord plantation is the most regularly used site and nesting has taken place on the Noss cliffs; not more than five pairs have been involved in any one year. There has been one attempt at nesting on Fair Isle; a pair remained after eight birds had wintered there in 1969/70 and built a nest but departed without laying.

Little is known about the numbers of immigrants visiting Scotland, and it is impossible to tell how many may be present in the occasionally huge flocks recorded in winter, recent

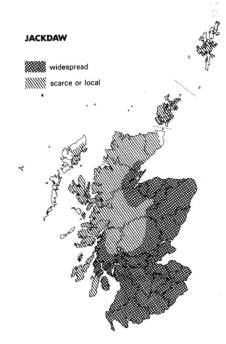

JACKDAW

- widespread
- scarce or local

examples of which are 4,750 at a roost near Cumbernauld in January 1980 and 1,800 on Arran in the winter of 1978/79. Small groups of migrants are recorded on Fair Isle and the Isle of May in October/November most years, but daily totals seldom exceed ten and 20 respectively. In 1979, however, there was an unusually heavy autumn passage in the north, with a peak of 270 on Fair Isle. In spring the movement through the Northern Isles is more obvious, with a fairly steady trickle of passage birds in March/May and occasionally much larger numbers, as in 1972 when there were 100 near Lerwick on 12 March and others on Whalsay, Unst and Foula. The few foreign ringing recoveries indicate that immigrant Jackdaws in Scotland come from Scandinavia, where the breeding stock belongs to the race *C.m. monedula*, and it seems possible that Shetland-nesting birds are of this race. Those breeding elsewhere in western Europe, including Britain, belong to the race *C.m. spermologus* (BOU). Most native birds are very sedentary, with few recoveries more than 100 km from the nesting site.

In the 1950s Lockie (1955) investigated the breeding and feeding of Jackdaws, while recent studies include work on dispersion and reproduction (including the effects of supplementary feeding on breeding success) by Hamadani (1978) in Aberdeen, and a behavioural study in the West Linton area in the early 1980s by Paul Green. The first showed that Jackdaws from the study site used different communal roosts in winter and summer, associating with Rooks at both; immediately after fledging, the young, guarded by adults, gathered at a roost near the colony which was not shared with Rooks. Green found that the Jackdaws in his study area followed a regular cycle of habitat use throughout the year, feeding mainly on grassland near the colony during the breeding season, moving into the hills in late summer, gleaning the stubbles in autumn and eating livestock feeding stuffs in winter.

Rook *Corvus frugilegus*

Resident; distribution is closely linked to that of agricultural land and Rooks are absent only from areas where such feeding grounds are scarce, ie much of the north and west Highlands and many of the islands. Immigration occurs, with numbers doubling in winter in some areas.

Rooks feed over farmland in large flocks, especially in winter when birds from several rookeries come together at huge roosts; they are consequently often suspected of being injurious to agriculture. Although local damage is undoubtedly caused at times, for example when flocks feed on newly-sown or lodged cereals, or dig up recently-planted potatoes, the Rook's diet is largely invertebrate and includes the larvae of various pest species as well as earthworms. Farmers' concern over Rook numbers led to a national survey of rookeries in 1944–46; 30 years later, when local studies had shown a major decline in some areas, a second survey was carried out. In Scotland there was 98% coverage in the 1975 survey, the detailed records of which are held in the SOC's Waterston Library. The account that follows is based largely on

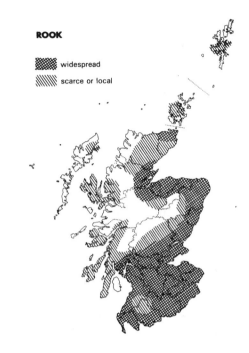

ROOK

▓▓ widespread

▨▨ scarce or local

reports of this survey by Castle (1977) and Sage & Vernon (1978).

In 1975 some 3,150+ rookeries were counted (Table a); these held around 250,000 nests, indicating a decrease of c36% from the estimated total in 1945–46, when the survey was much less complete. The decline in England, which had about twice as many nests as Scotland in 1975, was substantially greater at 45%. Decreases of more than 50% were found in Fife, Midlothian, Berwick and Roxburgh, of 40–50% in Kincardine, Kinross and West Lothian, and of 25–40% in Banff, Angus and East Lothian. Increases were recorded only in Caithness (14%), Sutherland (41%) and Orkney (64%).

Rookeries in Scotland are on average three times larger than those in England, with a mean of 79 nests as against c25. In 1975 thirty-six Scottish rookeries held more than 500 nests, 24 of them in Aberdeen and Banff and the rest widely scattered from Easter Ross to Lanark and Berwick. Rookery size ranged from an average of 138 nests/rookery in Aberdeen to only 31 in Bute, and the density of nests/km² from more than nine in Aberdeen and West Lothian to under 0.5 in Shetland, Sutherland, Inverness and Argyll. The largest single unit, with over 2,000 nests, was at Arnage Castle near Ellon, while the Hatton Castle complex of 16 groups, near Turriff, held over 2,600. Although only a few rookeries contained more than 200 nests the 9% that did accounted for nearly 40% of the total nests (Table b). Scots pines supported more than half the nests, with beech the next most popular choice at 19%, followed by sycamore with 12%. Only c20% of nests were at altitudes of more than 150 m asl, the highest being at around 350 m in Banff, Midlothian and Berwick. In 1975 no Rooks were found breeding on Arran (where breeding formerly occurred), Colonsay, Tiree, Coll, Eigg, Rhum, Barra, the Uists or Harris; in Shetland they were, and still are, confined to the Mainland. In Orkney breeding was recorded on Mainland, Rousay and Hoy, but in 1982, although the total population had increased from just under 1,000 pairs to c1,350, no nests were found on Hoy and only one on Rousay (Orkney BR 1982).

In the 1970s surveys were also made of winter roosts (Munro 1971, 1975). Sixty-one and 82 roosts were found

(b) Size distribution of rookeries in 1975 (Data from Castle 1977)

	<25	26–50	51–100	101–200	>200
% of total rookeries	33.8	20.3	21.7	15.1	9.1
% of total nests	5.2	9.4	19.6	26.2	39.6

(a) Distribution and densities of breeding Rooks in 1975 (Data from Castle 1977 – figures rounded)

County	Number of rookeries	nests	Av. no. of nests per rookery	km²
North & west				
Shetland	2	110	55	0.1
Orkney	15	960	64	1.0
Sutherland	30	2,060	69	0.4
Caithness	48	6,120	128	3.5
Ross & Cromarty	95	8,600	91	1.1
Inverness	59	4,450	75	0.4
Argyll	78	3,220	41	0.4
Bute	26	800	31	1.4
Northeast				
Nairn	21	2,000	95	4.7
Moray	77	7,390	96	6.0
Banff	103	14,120	137	8.7
Aberdeen	359	49,650	138	9.7
Kincardine	65	3,710	57	3.8
East-central				
Angus	134	8,420	63	3.7
Perth	147	6,780	46	1.1
Clackmannan	15	545	36	3.9
Kinross	35	1,510	43	7.1
Fife	149	7,400	50	5.7
Southeast				
West Lothian	47	2,900	62	9.3
Midlothian	99	4,900	50	5.2
East Lothian	48	3,490	73	5.0
Peebles	49	4,870	99	5.4
Selkirk	43	2,560	60	3.7
Roxburgh	127	8,170	64	4.8
Berwick	123	8,340	68	7.0
Southwest				
Stirling	113	5,640	50	4.8
Dunbarton	72	2,810	39	4.5
Renfrew	64	3,480	54	6.0
Lanark	154	13,250	86	5.7
Ayr	246	19,360	79	6.6
Dumfries	280	21,870	78	7.9
Kirkcudbright	155	12,300	79	5.3
Wigtown	119	10,580	89	4.4
Total & mean	3,197	252,340	79	3.3

respectively south and north of the Forth/Clyde. These roosts, many but not all of which are also breeding rookeries, were closely associated with good farming land, and the largest tended to be on or near the best land. This association, and the regularity of the spacing at 8–16 km apart in southern Scotland, made it possible to predict likely locations, and it was estimated that perhaps three remained unlocated in the south and about 18 in the north, where spacing was much wider in many areas. Many of the roosts held more than 10,000 birds (Jackdaws were present in all cases and could not be counted separately); the two largest were Hatton Castle (see Watson 1967) with c65,000 and Straloch (also in Aberdeen) with c49,000. Both deciduous trees and conifers were used, and there was a marked preference for small woods or strips of mature trees. Some of these winter roosts were known to have been in existence for 100 years or more, while others had been established within the previous five years, usually when a wood formerly used had been felled or windblown.

The very large Rook population in Aberdeen has been the subject of long-term ecological and behavioural studies by Aberdeen University (eg Dunnet & Patterson 1968; Feare 1978; McKilligan 1980). These have shown, *inter alia*, that the widespread practice of shooting young Rooks in May is not effective in controlling a local population, that a mix of grassland and arable provides the best year-round feeding opportunities, and that from late December onwards Rooks may make a daily round-trip of up to 90 km from the winter roost to visit their nesting site, regardless of prevailing weather conditions. Radio-telemetry has recently been used in Rook studies in the Borders (Green 1985).

Ringing recoveries show that native birds seldom travel more than 100 km from their home rookery, though young birds may disperse further during their first summer. Little information is available on the numbers of immigrant Rooks visiting Scotland as it is impossible to distinguish them from the resident population. In areas where resident Rooks are scarce, such as Shetland, only small numbers appear to be involved, though B&R mention records early this century of 'hundreds' on Fair Isle, where numbers currently seldom exceed 50. They also give a graphic description of the great invasion in October 1893 which resulted in the colonisation of Lewis. Very large numbers of Rooks reached the Outer Hebrides in late October/early November, and some 4,000 wintered in Stornoway Castle grounds; most left in the spring but about 200 remained and bred the following year. The few relevant ringing recoveries show movement between Scotland and the small and scattered breeding populations in Norway and Finland. In England, where a marked autumn arrival occurs along the east and southeast coasts, immigrants come from Germany, Denmark, the Netherlands and western Russia (BOU). The 1893 influx probably originated even further south as the first wave, several thousand strong, was recorded on Scilly.

Carrion / Hooded Crow *Corvus corone*

Resident; widespread and abundant throughout the mainland and the islands. Hooded Crows predominate in the north and west Highlands and the islands and Carrion Crows in the south and east. The broad band of hybridisation where the two subspecies overlap is moving progressively further north and west. Both forms occur as passage and winter visitors in small numbers.

Although Hooded *C.c. cornix* and Carrion *C.c. corone* Crows have such strikingly different plumages, they readily inter-breed and produce fertile offspring and are consequently regarded as belonging to the same species. Hybrids vary widely in colour, from only slightly darker than a pure Hooded Crow to nearly black, and birds of different plumages may occur in the same brood (SBR 1969). In Europe the Carrion Crow is the more limited in distribution, breeding only from Scotland (but not Ireland) south to Iberia and east to the Elbe. The Hooded Crow breeds in Ireland

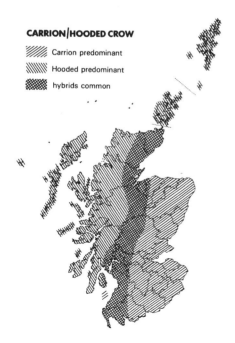

and the Isle of Man (but is absent from mainland England), in Faeroe and Scandinavia, and from the Elbe and northern Italy eastwards into Russia.

Crows, which are normally solitary breeders, occupy a wide variety of habitats, including moorland, farmland, coastal cliffs, rocky islands and city parks. Attempts are made to control their numbers in gamekeepered areas, as they take eggs and young birds, and on hill sheep ground, where they are often suspected of killing lambs. They sometimes feed extensively on carrion, and undoubtedly at times attack sick and dying ewes and lambs, but are really only a problem where ewes are in poor condition at lambing. Insects and grain are major food sources in summer and winter respectively (Houston 1977). Crows have been recorded preying on salmon fry (SB 1:268) and regularly scavenge on garbage tips in many areas; in some districts Carrion Crows are quite frequent visitors to town gardens. Breeding populations have been investigated in Argyll (Houston 1977; Hewson & Leitch 1982 – Hooded only) and in northeast Scotland (Picozzi 1975); territories were found to cover about 40 ha on lower ground but as much as 240 ha on heather moor. Nesting densities are highest in lowland arable areas; Picozzi found 3.3 pairs/km² in his study area, while in 1984 Rebecca (1985) located six pairs nesting on a 25 ha moss surrounded by arable land, and also recorded ground-nesting in a cornfield in Aberdeen. A. Wood found an average of 4.7 pairs/km² in Cumbernauld and Kilsyth District in 1983.

There has been a steady northward movement of the hybridisation zone this century, especially in the east. In 1928 the centre of the zone lay roughly along the Highland Boundary Fault east as far as Angus and then north along the edge of the hills to the coast just west of Fraserburgh. By 1974 it ran north through Perth and Speyside and in a semicircle along the hill edge inland from the Moray Basin, eventually turning east to the coast in southern Caithness (Cook 1975). The proportion of Carrion Crows in the Dornoch area rose from 15% in 1955 to 50% in 1974, while in Kincardine hybrids decreased from 28% in the late 1960s to nil in 1981 (Picozzi 1982). The relative proportions of apparently pure-bred birds and light and dark hybrids give a useful indication of the current situation in an area. The hybrid zone is still moving north, with inter-breeding now occurring in central Sutherland (Angus 1983), widespread in Caithness (P. M. Collett), and noted in Orkney in 1974. Hybridisation has not yet been recorded in Shetland or the Outer Hebrides, although hybrids are frequently seen and there have been reports of Carrion Crows there for several years. In the southwest, several pairs of pure Hooded Crows are now resident in west Wigtown, and hybrids are present in several areas, though inter-breeding has not yet been recorded in the county.

The factors determining the relative distribution of Carrion and Hooded Crows are not fully understood. In Arran, which is near the southern end of the hybridisation zone, Hooded Crows predominate on the high ground, Carrion Crows increasing in frequency with decreasing altitude. While it is possible that climatic amelioration is one of the factors involved in the northward expansion of the Carrion Crow, no direct correlation between climatic factors and the position of the hybrid zone has been proved (Cook 1975).

Breeding birds are largely sedentary but non-breeders,

CARRION/HOODED CROW

///// Carrion predominant

\\\\\ Hooded predominant

▓▓▓ hybrids common

and especially young birds, may move considerable distances and form large winter flocks sometimes numbering several hundred. More than 70% of recoveries are within 10 km of the ringing site. Little is known about the scale of immigration, but both Hooded and Carrion Crows are recorded annually on passage in the Northern Isles and small numbers of both occur well outside their respective breeding ranges during the winter months. Up to 40 hybrids have been recorded on Fair Isle, well north of the main zone of interbreeding, and it is possible that these originated from the limited area of hybridisation in Denmark. Around the Ythan the percentage of Hooded Crows increases in winter, presumably due to immigration. The few foreign-ringed Hooded Crows recovered in Scotland have been from Scandinavia and Russia. The size of the flocks occasionally reported on or near the east coast – eg 80 on the Isle of May in mid October 1974 and 50 coming in off the sea at North Berwick earlier that month – suggests that at least in some years sizeable numbers of immigrants arrive.

Where they are unmolested, crows often congregate in large winter roosts. Among the highest recent counts in contrasting habitats and geographical areas are 800 in Kergord plantation (Shetland), 240 at Loch Garten, 75 at Duddingston, and a total of c1,000 at five roosts in the Ken valley.

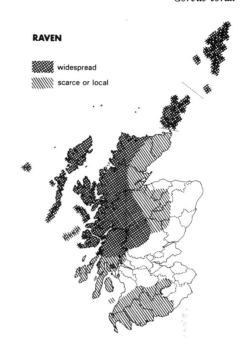

RAVEN

▨ widespread

▧ scarce or local

Raven *Corvus corax*

Resident; still widely distributed in the Highlands and islands but has decreased markedly in central Scotland and the Southern Uplands since the 1950s, especially in areas where there has been extensive afforestation.

Ravens suffered greatly from persecution around the turn of the century but recovered when gamekeepering was reduced during the wars. Between 1945 and 1961 there was an overall decline of not more than 14% in the British breeding population (Ratcliffe 1962); more recently land-use changes have adversely affected the population in some areas, while in others breeding success is low (Marquiss *et al* 1978). Although failure may at times be attributable to persecution this is not likely to be the sole factor involved. Once a familiar scavenger in towns, the Raven is now a bird of thinly-inhabited country, shunning close proximity to man. Most nest on coastal cliffs or inland crags up to about 600 m asl, but a few make use of trees, quarries or buildings. The limited data available on breeding densities indicate wide differences between regions, presumably due largely to variations in the carrion supply, although the availability of suitable nest sites

may also be a limiting factor. Ravens appear to be fairly sedentary, with young birds liable to move further than adults. Most ringing recoveries are within 100 km of the natal site but birds ringed as chicks in Orkney and Kintyre have been found in West Lothian and Northern Ireland respectively, and one from Anglesey reached East Lothian. There is no evidence of movement to the Continent or *vice versa*.

Non-breeding Ravens often remain in flocks until May, and big gatherings form soon after the young fledge (which may be as early as the end of April). Numbers are largest on the islands, with June 1981 records of up to 150 at Mossbank in the north of Shetland Mainland, and 200 near Stornoway, where the birds feed regularly on the garbage tip. There were counts during the 1970s of over 100 at winter roosts on Islay and Orkney Mainland, and of 40–70 on Mull and in Ross, Perth and Peebles. Reports from other areas usually relate to flocks of not more than 20–25.

B&R described the Raven as 'by no means uncommon' and said that it nested in every county except East and West Lothian, Clackmannan, Kinross, Fife, Kincardine, Aberdeen, Moray and Nairn. There have been recent (post-1970) records for Clackmannan, Aberdeen and Moray, and breeding is believed to occur at least sporadically in Nairn (N. Elkins), but Ravens had ceased to nest in Renfrew by 1968–72, and in Berwick, Roxburgh and Selkirk by 1981 (Mearns 1983a). A new nest found in Berwick in 1983 may indicate an attempt at recolonisation (R. Mearns). The last report of breeding in Midlothian was in 1976, and there seems to be no recent confirmation of nesting in Lanark. Breeding occurs regularly on all the principal islands in the north and west other than Bute.

Studies in Speyside and in southern and central Scotland have shown that numbers in these areas have declined markedly since the 1960s (Table). In Speyside poisoning (with

Occupation of Raven territories in three areas at different periods (Data from Weir 1978, Mitchell 1981, Mearns 1983a)

	Speyside		Central		South/southwest			
					Inland		Coastal	
No. of "traditional" territories known	22		30		81+†		29+†	
Years in which checked	a. 1964–68	b. 1977	a. 1970s	b. 1981	a. 1974–5	b. 1981	a. 1974–5	b. 1981
No. of territories occupied	16–17	5	24–25	15–17*	48	35*	19	12–13*
No. of pairs breeding	10–11	1–2	?	9–12	38	23	13+	7–10
Decline between period a. & period b.	c70%		c30%		c40%		c23%	

† excludes former sites known to be deserted prior to 1945–60
* includes sites where only single birds seen

meat baits) was suspected as being a major contributory factor and there is evidence from elsewhere that Ravens have been killed by Mevinphos, alphachloralose and strychnine (Cadbury 1980a). In the other two study areas extensive afforestation and improved sheep husbandry, with consequent reduction in the availability of sheep carrion, were considered to be the primary causes of the decrease. If the present rate of decline continues the Raven may shortly cease to breed inland in southern Scotland. The only other area in which fairly regular counts have been made over an extended period is Orkney where, in the period 1972–1982, the number of pairs attempting to nest ranged from 21 to 26 (Booth 1979a; Orkney BRs). Breeding success there is comparable to that among coastal-nesting pairs in the southwest, with an average brood size of just over three. In Orkney and Shetland there have been reports of failure to fledge due to oiling by Fulmars; in Fetlar in 1982 the failure of three out of seven pairs attempting to nest was attributed to this cause.

There seems little doubt that a poor standard of shepherding, such as occurs on many of the islands and on some of the difficult ground in western Inverness and Argyll, makes for good Raven country. The introduction of agricultural improvement schemes in the Outer Hebrides and Shetland may, in the long-term, cause a decrease in the currently large Raven populations in these areas. In southern Scotland the continuing improvement of land adjacent to the coast has already resulted in a marked reduction in the number of coastal-breeding pairs and may be expected to cause the species' disappearance from such areas in the fairly near future.

Starling *Sturnus vulgaris*

Ubiquitous, breeding in a wide variety of habitats throughout the mainland and islands, and absent only from the most mountainous and exposed parts of the north and west Highlands. Big flocks assemble after the breeding season and roost communally, and large numbers of immigrants winter here.

The Starling is now so familiar and abundant that it is hard to accept that it was scarce little more than 100 years ago. Around 1800 it was probably extinct in Scotland except in Shetland and the Outer Hebrides, where a small population survived and is now recognised as a separate subspecies, *S.v. zetlandicus*. Population expansion started in England in the early 1800s and had spread through much of Scotland by the end of the century; colonisation was virtually complete by the middle of the 20th century (B&R). The only recent

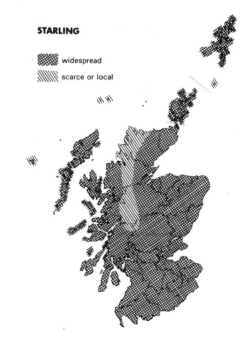

STARLING

widespread

scarce or local

'first' breeding records are from Handa (1973) and Sule Skerry (1981), and the only report of a decrease this century is from Ailsa Craig, where breeding no longer takes place. Among the more unusual nesting sites recorded is a barge associated with the Ekofisk oil installation in the North Sea.

A highly opportunist and adaptable bird, the Starling takes full advantage of man's activities in many ways, nesting in the roofs and walls of buildings, roosting in vast numbers along wires in city streets and under bridges, and feeding readily at bird-tables and on rubbish tips. Farming changes have probably also been beneficial to this species; winter cereals near large immigrant roosts sometimes suffer damage from feeding flocks and in England Starlings feed heavily on cattle food in intensive calf-rearing units (Feare 1984) – this behaviour has not yet been reported in Scotland. Little information is available on the size of the breeding population, which is largely sedentary, with few ringed birds recovered more than 250 km from the ringing site. No analysis of the 1,100+ recoveries of Scottish-ringed birds has been carried out, but there is presumably some movement into towns in autumn, while some first-winter birds certainly cross to Ireland.

Starlings breeding in the northern part of the Eurasian range are all migratory and winter further west or south, either within the breeding range or south of it around the Mediterranean. The majority of the immigrants arriving in autumn are from Scandinavia (c85% of foreign-ringed recoveries), the remainder coming from between the Netherlands and USSR. The biggest influxes occur from October onwards; in 1976 passage was notably heavy in late October with records of around 1,000 and 2,000 respectively on the Isle of May and Fair Isle. From autumn through to March vast flocks milling around in the air prior to roosting are a familiar sight in many areas; the various patterns of pre-roost assembly at a site where some 150,000 roosted have been described by Brodie (1974). Counting such huge numbers is no easy task but estimates of winter roosts have been made in various areas, among the highest totals being 500,000 at West Freugh near Stranraer, 175,000 at Berstane in Orkney, 100,000 in the city of Aberdeen, 50,000 in Cumnock, and 40,000 on the Tay railway bridge at Dundee. In addition to the vast gatherings in the Central Lowlands, there are also large mid-winter concentrations in Orkney and the Buchan area (Winter Atlas).

Nobody seems to have made any recent attempt to assess the numbers that pack together in Glasgow's city centre, causing a considerable public nuisance in terms of both dirt and noise. Many efforts have been made to dissuade Starlings from using such sites, among them the broadcasting of distress calls; this is the most effective method currently available for dispersing winter roosts but is ineffective in feeding areas (Feare 1984). Most roosts are in trees – which are gradually killed by the birds' droppings – but reedbeds and caves are also used, while at Stornoway Airport a large but little-used hangar accommodates some 7,000 Starlings (Cunningham 1983). Autumn roosts are often smaller and separate from the main winter roosts; a July roost of around 7,000 at West Water Reservoir, Peebles, consisted almost entirely of young birds.

Heavy mortality among Starlings due to natural or accidental causes is seldom recorded but in January 1976 some 430 were found dead on the beach at Stonehaven. The cause of death was not ascertained but the incident occurred during a period of easterly gales, which resulted in a small wreck of Little Auks in the area and may have caught a flock of Starlings on passage across the North Sea (NE Scotland BR 1976).

[White-shouldered Starling *Sturnus sinensis*]

One in Berwick in November/December 1981 was regarded as an escape (SBR 1981); the species is native to China.

Rose-coloured Starling *Sturnus roseus*

Vagrant (E Europe) – almost annual, in very small numbers.

B&R classed the Rose-coloured Starling as 'an occasional visitor', but this perhaps suggests rather more frequent occurrence than has been the case recently. From 1968 to 1983 inclusive the species was reported in all but two years, all records referring to single birds and not more than six individuals occurring in any one year (six in 1975). Most sightings are of adults and in June/August, but there are a few May records (the earliest on 4th) and a few in September/November (the latest a juvenile in Yell on 3–11 November 1977). In 1977 one stayed on Islay from 20 July until 10 September. Occurrences are most frequent in Shetland and Fair Isle, but there are also records from Orkney, Caithness, Sutherland, Inverness, Aberdeen, Angus, the Isle of May, Midlothian, Berwick, Kirkcudbright, Argyll and the Inner and Outer Hebrides. The Rose-coloured Starling breeds from Hungary and the Balkans east through southern Russia and Iran, and winters chiefly in western India. Large flocks quite often erupt westwards across Europe, and breeding has occurred sporadically in Italy.

House Sparrow *Passer domesticus*

Resident, breeding in every county and on most of the inhabited islands, but scarce and local where the human population is thinly scattered.

As its name suggests, the House Sparrow is very closely associated with man. The distribution shown in the Atlas is the best available information at the present time but may well give a somewhat misleading impression of the population. In 1968–72 breeding was confirmed in about 80% of 10 km squares; many squares probably held only a few pairs around a farm or small village, whereas in built-up areas each proved breeding record may have represented at least several hundred pairs. Although most often nesting in holes in buildings, the House Sparrow will occupy rock crevices and the bulky nest foundations of other species, and occasionally builds dome-shaped nests in trees and hedges.

The density figures quoted in the Atlas for Britain as a whole, ie 10–20 pairs/km², may be substantially too high to be applicable to most of northern Scotland and the islands. Few people attempt to count sparrows, however, and – apart from a count of 372 in Princes Street Gardens, Edinburgh, in January 1983 – the very limited information available on numbers relates solely to islands. In late June 1975 a census on Unst, where breeding has been recorded on the summit of Saxa Vord, produced a total of 637 birds; Out Skerries held 90 in August 1975; in 1981 the Fair Isle population's autumn peak was 210; and only 10–20 pairs currently breed on Rhum.

Ringing recoveries indicate that House Sparrows are very sedentary, with only c5% of recoveries more than 10 km from the ringing site. Local movements involve wandering individuals occasionally turning up on Stroma, Handa, and St Kilda, and on the Isle of May, where the species formerly bred. The only permanently inhabited island (other than lighthouse 'rock stations') on which breeding does not occur is St Kilda. B&R refer to about ten pairs nesting on Ailsa Craig, but none breed there now. Hybrids with the Tree Sparrow have been recorded on several occasions, especially on Fair Isle – where sparrows receive more attention than in most other areas.

HOUSE SPARROW

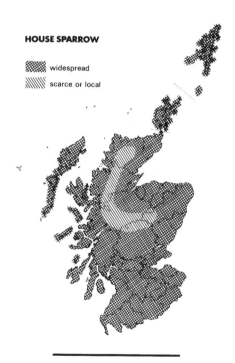

▨ widespread

▧ scarce or local

Tree Sparrow *Passer montanus*

Resident, relatively widespread south of the Highlands but very local further north and west. The population undergoes periodic fluctuations, the reasons for which are not understood. Passage and possibly winter visitor, in small numbers.

In Scotland the Tree Sparrow is near the northern limit of its European range, which extends from southern Norway and Sweden to the Mediterranean. Despite its names (English and scientific) it is not dependent upon trees and is certainly not confined to mountains, in fact it is entirely absent from the mountainous areas of Scotland. Primarily a hole-nesting bird frequenting cultivated land with patches of trees, it occasionally builds in hedges or bushes, nests readily in boxes, and will also use crevices in buildings and Sand Martin holes. It frequently nests colonially, when numbers can be difficult to assess, but also breeds as isolated pairs, which can easily be overlooked. Though nowhere abundant, it is quite likely to be under-recorded; Tree Sparrows can be surprisingly elusive in the breeding season.

There have been at least two major fluctuations in the Tree Sparrow population in the last 100 years. Towards the end of the 19th century breeding was widely recorded in the Hebrides and Clyde Islands and occurred at least sporadically in most counties, but by the 1930s numbers had declined and nesting had ceased on many of the islands (B&R). From about 1958 onwards an increase took place throughout Britain, and during 1961–67 recolonisation was reported from Kinross, Aberdeen, Orkney, Fair Isle, Shetland, St Kilda and some other Hebridean islands (Parslow 1973). The population was consequently high in 1968–72, when breeding was confirmed in all mainland counties except Argyll and Banff, and on Islay, Mull, Canna, Skye, Lewis,

St Kilda, North Rona, Eday, Papa Westray, Fair Isle, Mousa, Unst and possibly Foula (Atlas).

Local reports and recorders' assessments in the early 1980s suggest that Tree Sparrows are again declining, with low breeding numbers and few if any nesting in outlying areas. The only Outer Hebridean site still occupied is the Stornoway Woods, where a few pairs continue to breed. There were no nesting records from the Inner Hebrides, Orkney, Fair Isle or Shetland in 1980–83. Numbers in Caithness are down to two or three pairs, in Sutherland sporadic breeding now occurs only in the southeast, and in east Inverness and along the coastal plains of Nairn, Moray and Banff the population is very small and breeding possibly irregular. Larger numbers are apparently still present in Easter Ross and the Black Isle, although breeding is very local there. In Aberdeen Tree Sparrows are currently confined to the Buchan plain. Further south the comments 'decreasing' or 'scarce' have been used in respect of Kincardine, Perth and Kinross, but the species is still at least locally common in Fife and Renfrew, and in Ayr – where an increase was noted in the late 1970s (Hogg 1983). Censuses in the Lothians in 1982 produced totals of c25 pairs in West Lothian, c100 pairs in Midlothian, and c100 pairs in East Lothian; in view of the likelihood of under-recording the actual populations are probably substantially higher. In East Lothian in 1984 there was an average density of 2 pairs/km^2 on some 1,600 ha of farmland but numbers were much higher in small areas of scrub (da Prato 1985). In Berwick a few pairs still breed around Eyemouth and Ayton, in Roxburgh most of the small population is in the Tweed Valley, none now breed in Selkirk, and there are perhaps ten pairs in west Peebles. In the southwest Tree Sparrows remain scarce and very local in Wigtown, and are becoming increasingly so in Kirkcudbright and Dumfries.

The few ringing recoveries suggest that native birds commonly move up to 100 km from the breeding area, while the capture of a West Yorkshire bird in Inverness three months after ringing demonstrates that longer movements also occur. Passage migrants and winter visitors to Britain are presumed to originate from northern Europe (BOU). Small numbers of passage birds are recorded annually in Fair Isle and Shetland, and occasionally from other areas outwith the present breeding range; most are seen in April/June and a few in September/October. In the early 1970s more than 30 were sometimes on Fair Isle at one time, but since 1979 there have been few reports of more than five in a day. There were 100 on Foula in mid May 1971, but all recent Shetland records have involved very much smaller numbers. Winter Atlas surveys in 1981/82 and 1982/83 showed the main mid-winter concentrations to be in the east-central lowlands, from Angus to the Lothians and west to the Clyde valley, with smaller numbers in the northeast, around the Moray Basin and in the southwest. It is not known whether winter flocks on the mainland consist entirely of native birds or include some immigrants, but the fact that the largest flocks are most often reported from the east coast suggests that immigrants may be involved. Since 1972 there have been counts of 100–200 in Perth, Fife, East Lothian, Stirling, Dunbarton, Ayr, Selkirk and Berwick, and of 400 in Midlothian (1976), but the number of large flocks reported has been decreasing.

Chaffinch *Fringilla coelebs*

The most widespread and abundant of the finches, breeding in nearly every 10 km square on the mainland, on the Clyde Islands and on nearly all the Inner Hebridean islands, but scarce and local in the Outer Hebrides and Northern Isles. In autumn variable numbers of immigrants arrive and form large winter flocks, which may remain together until April.

Chaffinches breed in a wide variety of habitats, including gardens, farmland with hedges, conifer plantations, mature deciduous woods and hillside birchwoods, and have been found nesting up to an altitude of over 500 m asl (B&R). Their diet is almost as varied, including a wide range of weed seeds, beech mast, grain, insects (during the breeding season) and, wherever opportunity arises, picnic or other scraps provided by man. This adaptability makes the Chaffinch one of the most successful, and consequently most common, of our breeding birds. Farmland CBC data suggest that this species is more abundant in Scotland than in England (Fuller *et al* 1985), and this may also be true for woodlands. CBC breeding densities for Britain as a whole average 19 pairs/km^2 on farmland and about 37 pairs/km^2 in woodland (Atlas); the corresponding figure for largely arable East Lothian farmland in 1984 was only 2 pairs/km^2, but in woodland and scrub in the same district there were 149–183 pairs/km^2 (da Prato 1985).

Around 1960 Chaffinch numbers declined in most parts of Britain, especially in intensive arable areas, as a result of increasing use of highly toxic organochlorine seed-dressings; this species is particularly vulnerable as it regularly feeds on grain in recently cultivated soil (Newton 1972). The decline appears to have been short-lived, however, and

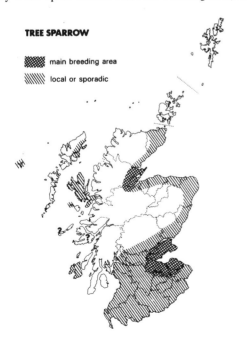

TREE SPARROW

▓ main breeding area

▨ local or sporadic

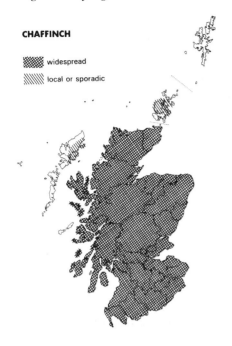

CHAFFINCH

▨ widespread

⧄ local or sporadic

the farmland CBC index has been fairly stable since the mid 1960s.

Chaffinches have spread into new forests wherever these exist on the mainland, and have also shown some slight expansion of island populations, which may be expected to continue as recently planted woodlands mature. Breeding now occurs regularly on all the larger Inner Hebridean islands except Coll and Tiree (Reed *et al* 1983); B&R had no record for Canna and make no mention of Muck. On Rhum c70 pairs nested in 1974 and numbers are increasing (J. A. Love). The Outer Hebridean population is concentrated in Stornoway, but sporadic breeding has been recorded elsewhere in the islands. In Orkney nesting has taken place recently on Mainland, Hoy, Rousay, Eday and possibly Shapinsay, but in Shetland there has been only one report of nesting since 1950, in Kergord plantation in 1973. The few previous records of breeding there were from the sycamore wood at Halligarth, Unst, in 1901 and 1930–33 (Venables & Venables 1955 – *contra* Atlas, which says no record for Shetland).

The essentially sedentary British Chaffinches belong to a different race *F.c. gengleri* from those breeding elsewhere in Europe *F.c. coelebs*, which tend to be larger and paler. Those breeding in the northern part of the range, which extends from near the forest limit in Fenno-Scandia to North Africa, move south to winter and variable numbers pass through and winter in Scotland. These immigrants behave differently from the native birds, forming large winter flocks which forage over open ground, concentrating on local abundances of food and roosting communally in woodland or scrub; native Chaffinches feed and roost singly or in small groups and remain near their nesting area all year. A further difference between residents and immigrants is that both sexes are present in roughly equal numbers among the former, whereas males markedly outnumber females in

flocks of northern Chaffinches in this country (Newton 1972). The very large flocks recorded during the winter may be assumed to consist largely of immigrants, although local birds may join them at times, especially in hard weather or where food is particularly abundant, for example in a good beech mast year. Flocks of 1,000 or more are by no means unusual and there have been recent counts of over 2,000 from places as far apart as Mull, Peebles, Ayr and Dunbarton.

Chaffinches ringed here in winter have been recovered during the breeding season in Norway, Sweden and Germany, and on passage in Holland and Belgium. Evidence of onward movement to Ireland is provided by birds ringed in autumn on the Isle of May and recovered in Ireland a few weeks later. One caught on the Isle of May on 18 April was found dead near Stockholm on 4 May, having covered 1,282 km in less than three weeks. No Scottish-reared Chaffinch has been recovered more than 250 km from the nest site.

Passage through the Northern Isles is most marked in March/April and late September/October; some usually remain in Shetland throughout the winter and a few sometimes summer there. Numbers have been fairly low in the last 10–15 years, with Fair Isle daily maxima often under 50 and more than 500 recorded only once, in late March 1980; spring and autumn peaks in 1948–64 were 1,000 in April 1958 and 500 in September 1961 (Davis, in Williamson 1965). The largest number recorded recently on the Isle of May has been 350, on 3 October 1976, a figure well above the average daily peak.

Brambling *Fringilla montifringilla* [1]

Winter and passage visitor in very variable numbers. Has bred.

Wintering numbers vary markedly between years; in some seasons most Bramblings appear to move on quickly, while in others many remain throughout the winter, often flocking with Chaffinches. Passage dates are most easily identified in the Northern Isles, where few Bramblings occur in winter. In a 'good' year the first birds may reach Fair Isle in August but numbers are usually small until mid September, the main arrival seldom takes place until October and is generally over by mid November, but stragglers may continue to appear into December. Several hundred are often on Fair Isle during October and in 1968 up to 1,300 were there in the middle of the month.

Further south and west the distribution of records and the maximum flock size give the best indications of the scale of the autumn influx and wintering population. In years when Bramblings are abundant there are records from every county and island group and many flocks of 1,000 or more are reported; in years when they are scarce the records come mainly from the eastern half of the country, from Aberdeen south, and there are few flocks of more than 500. Bramblings often form mixed flocks with Chaffinches and feed around woodland – especially beech – and in fields. During hard weather they occasionally come to feed in gardens; in early 1982, in conditions of deep snow, several fed under – but not on – my bird-table for a week or so. There was a striking difference in the numbers and distribution of Bramblings recorded in the first two years of the Winter Atlas survey. In 1981/82 reports were thinly scattered, with no counts of more than 30 north of the Central Lowlands. In 1982/83 many more were recorded throughout the eastern half of the country, from Easter Ross to the Borders (whereas numbers in England were very much lower than in the preceding winter). Spring passage through the Northern Isles sometimes starts in late March, though more often around mid April, and continues into early June; most have left by mid May.

A few summering birds are reported in most years, and among these have recently been increasing numbers of singing males. B&R accepted only one of the pre-1950 reports of breeding – a nest with seven eggs in Sutherland in 1920, which was robbed. Although pairs were subsequently seen in several areas, it was not until 1979 that breeding was again confirmed, with the finding in Grampian of a nest – later deserted – containing three eggs, in one of which there was a long-dead embryo; no male Brambling was seen in the area (BB 73:360–361). The first successful breeding was in 1982 in Inverness; a nest located on 6 June contained five downy young on 14th, and a fledged juvenile was nearby on 1 July (SB 12:191–193). A female and three fledged young were also seen and Bramblings were present at three other sites. Breeding occurred again in 1983, when two juveniles were noted in Inverness and adults were at several sites, including some in Sutherland. (R. H. Dennis).

Both the Grampian and Inverness nests were Chaffinch-like in structure and situated in the fork of a birch tree. In their note on the Grampian nest, Buckland & Knox (1980) point out that Bramblings frequently nest south of their normal breeding range in years when the wintering population in an area is unusually large, or when cold spring weather delays migration. As the numbers in northeast Scotland in early spring 1979 were high, and the spring was abnormally cold, they consider the breeding attempt in Grampian to be an isolated instance due to this combination of factors, rather than an indication of potential colonisation. Since in northern Scandinavia the zone of overlap between the Brambling and the Chaffinch (which replaces it to the south) has moved northward during the last 150 years (Newton 1972), the probability of casual breeding certainly seems more likely than that of impending colonisation.

The Brambling breeds in open coniferous and birch forest and shrub tundra right across northern Eurasia, in a broad band stretching from Norway to Kamchatka and locally in a few areas further south. It is one of the most migratory of the finches, wintering south of the breeding range, in Europe from southern Sweden to the Mediterranean. Ringing recoveries indicate that birds visiting Scotland come mainly from Scandinavia; one ringed in southern Norway in October was on North Ronaldsay three days later. There are several Scottish recoveries of individuals marked the previous winter on the Continent (the Netherlands, Belgium and Germany), and *vice versa*, while one ringed in Orkney in March was caught five weeks later on Texel. Newton (1972), commenting on the extent to which winter distribution is determined by the availability of beech mast (which is rarely plentiful in two consecutive years in the same area), notes that few, if any, Bramblings have been recovered in the same place in successive winters.

Serin *Serinus serinus* [1]

Vagrant (Europe, including England) – seven records, 1911–83.

This small Canary-like finch has considerably extended the northern limits of its range this century and started breeding sporadically in southern England in the late 1960s; up to seven pairs bred there in 1982. In Scotland it remains a scarce vagrant. The first record was of a male caught in Edinburgh in November 1911 (SN 1912:11). There were single Serins on Fair Isle during May in 1914, 1957 and 1964, on 1 October 1968 and on 30 September 1982, and one on Shetland Mainland on 17 November 1968.

The Serin now breeds from Denmark and southern Sweden to the Mediterranean. Its recent expansion of range possibly results from the way in which it has adapted from forest habitats to life in small wooded areas, such as orchards and gardens, in close association with man. It is unable to tolerate cold wet weather and the northern section of the population moves south to the Mediterranean in winter.

Greenfinch *Carduelis chloris*

Resident, widespread in lowland areas of the mainland, regular on the Clyde Islands and most of the Inner Hebrides but scarce and very local in the Outer Hebrides and Northern Isles. Small numbers occur on Fair Isle and in Shetland, but there is no evidence of regular migration.

This is one of the most familiar of the finches, occurring wherever there are hedges, gardens, parks, scrub woodlands or young conifer plantations. It is a frequent winter visitor to gardens, where it feeds avidly on peanuts. The tendency to congregate around towns and villages has become more marked since stackyards – formerly an important winter feeding ground – vanished from the farming scene, and in many places the Greenfinch is now most abundant close to built-up areas. Little information is available on breeding

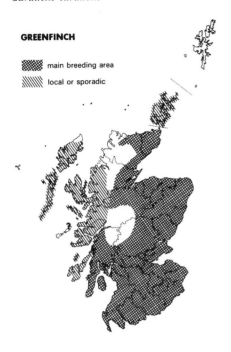

GREENFINCH

▨ main breeding area

▧ local or sporadic

south in the breeding range – which extends throughout most of continental Europe and into North Africa – but as yet there is no ringing evidence of passage through Scotland. The numbers reaching Fair Isle and Shetland (generally between October and May) are very small; flock counts seldom exceed ten birds, but there was an exceptional influx of 100 on Fair Isle in mid October 1983. Noteworthy recent counts on the Hebrides include 50+ on South Uist in March 1980 and 200 on Skye in December 1981. On the mainland autumn/winter flocks of 300–500 are not uncommon; a count of 1,500 near North Berwick in December 1981 was notably high, while 18 May was an unusually late date for a flock of 115 still to be together (at Aberlady).

Goldfinch *Carduelis carduelis*

Partial migrant, widespread though not abundant on low ground south of the Highlands, scarce and local in the northeast and around the Moray Basin, and absent from most of the north and northwest Highlands. Breeds regularly on the Clyde Islands and sporadically on Islay and Mull, but not in the Outer Hebrides or Northern Isles. In winter small groups appear well outside the breeding range.

numbers. CBC densities for Britain as a whole average 6 pairs/km^2 on farmland and 9.2 pairs/km^2 in woodland but are higher in suburban situations (Atlas). In Scotland densities in many areas are likely to be appreciably lower; in 1984 there was less than one territorial pair/km^2 on over 1,600 ha of East Lothian farmland, though the higher densities in woodland and scrub raised the overall average on 1,700 ha to 3 pairs/km^2 (da Prato 1985). In winter Greenfinches form large flocks, often joining Chaffinches and Yellowhammers, and forage widely over farmland, in woodland, and along coastal flats.

By the end of last century the species was apparently already established on Orkney Mainland, and early this century it was breeding on Islay, Jura, Gigha, Mull, Skye, Raasay and Lewis. It nested for the first time on Eigg in 1926 (B&R). Since 1950 it has bred, at least sporadically, on Canna, Muck, Rhum and Colonsay but its status on Coll and Tiree is uncertain. In the Outer Hebrides Greenfinches remain scarce and very local; their main stronghold is Stornoway Woods, but breeding has probably also taken place in young plantations at Horgabost and Borve in Harris, at Newton in North Uist and at Loch Druidibeg in South Uist (Cunningham 1983). They are also very local in Orkney, where breeding occurs regularly on Mainland and at least occasionally on Rousay and Shapinsay; there has been no recent record from Hoy, where two pairs were present in 1942. Breeding has not been recorded in Shetland.

Ringing recoveries indicate that most Scottish Greenfinches are very sedentary, with more than 80% of recoveries within 10 km of the ringing site. Some birds move much further, however, and it is possibly such dispersal, rather than any regular migratory movement, which accounts for the appearance of small groups on islands and in areas where the species does not breed. The Greenfinches breeding in coastal Norway are known to be migratory, wintering further

This very decorative little bird, formerly a prime target for birdcatchers, decreased during the 19th century; since trapping became illegal around 1880 it has spread and increased. Goldfinches feed mainly on the seeds of plants belonging to the *Compositae*, including dandelions, thistles, groundsel and ragwort, and the increasing use of herbicides in arable areas has probably affected local distribution in the last two decades. They are now generally most abundant

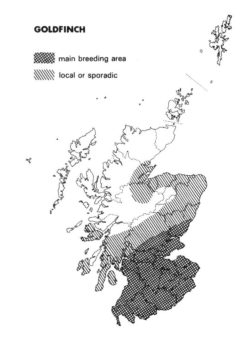

GOLDFINCH

▨ main breeding area

▧ local or sporadic

where there are patches of weedy rough grazing or waste ground with scattered trees or bushes.

It is not easy to be certain to what extent the distribution of the Goldfinch has changed in the last 30–40 years. B&R describe it as breeding 'with tolerable regularity' in Perth, Angus, Kincardine, Aberdeen and Inverness, but make no reference to 20th century records in the Lothians, Fife, Kinross, Clackmannan, Stirling and Dunbarton – in all of which the species now breeds regularly. They also refer to it as 'quite common' at Inverewe (Wester Ross) and say that it has bred on Skye and Raasay; there have been no recent reports from these areas. Nesting has, however, been reported in Caithness (1974) and southeast Sutherland (1979). It seems likely that no significant change in distribution has taken place since the early part of the century, and that apparent changes are largely due to fluctuations in the overall population, which increases when warm dry summers result in good seed crops. During the early 1970s numbers increased steadily and flocks of up to 40 were reported as far north as the Black Isle, but a general decrease followed the hard winter of 1978/79. The average breeding densities quoted in the Atlas are 2.2 pairs/km^2 on farmland and 5–10 pairs/km^2 locally in scrub; in East Lothian Stan da Prato (1985) found less than one territorial pair/km^2 on c1,600 ha of predominantly arable land but the equivalent of 19–22 pairs/km^2 in small areas of woodland and scrub.

From September to April flocks of Goldfinches are present on weedy open ground in many areas, often in largest numbers near the coast. Among the biggest flocks recorded since 1970 are 400+ at Eyemouth (October 1981), 200+ in Fife (December 1977), and 100+ in Midlothian, Ayr and Kirkcudbright. Winter Atlas records show that in mid winter the species is widely, though often thinly, scattered over most low-ground areas south of the Great Glen, but is very scarce further north and west. Small numbers occur irregularly in spring and autumn on the Isle of May, while in the Northern Isles and Outer Hebrides this species is a scarce and sporadic visitor.

Scotland is near the northern limit of the Goldfinch's breeding range, which extends from southern Fenno-Scandia to North Africa. The British population belongs to a different race *C.c. britannica* from those breeding on the Continent *C.c .carduelis* but the two races are not distinguishable in the field (Newton 1972). Although it is thought that continental birds winter in Britain there is not yet any ringing evidence to prove this. Nor is much information available on the scale and extent of movement within Britain, though it is known that most native birds move south in September to France and Spain, while some cross to Ireland.

Siskin *Carduelis spinus*

Partial migrant; has increased greatly since 1950 and now breeds in most mainland counties and on several of the islands. Passage and winter visitor in variable numbers.

The Siskin is perhaps the species which has benefited most from afforestation. Although in Scotland traditionally an inhabitant of the Caledonian pine forest, it breeds in other conifers over much of its range, which in Europe extends from central Fenno-Scandia and Germany south to the Alps and east into Russia. Spruce seed is its main summer food on the Continent, and in recent years it has adapted to life in, and rapidly colonised, the extensive plantations of Sitka spruce now present in many parts of Scotland. As recently as the mid 1960s the Siskin was described as 'distributed quite widely in northeast Scotland, rather locally elsewhere in that country' (Parslow 1973). By 1968–72 it was already widespread in west Inverness and in Argyll –

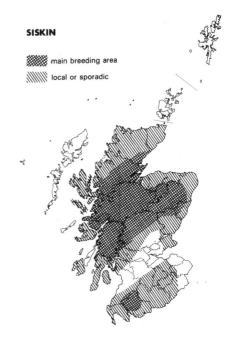

SISKIN

main breeding area

local or sporadic

where afforestation dates from the 1920s and some plantations had been colonised by the 1930s (D. Nethersole-Thompson) – and was becoming well-established in the southwest. Since the early 1970s there has been a further expansion in range, and this trend seems likely to continue as more recently planted forests reach cone-bearing age.

The only mainland counties in which breeding is not known to occur at present are East Lothian, Kinross and Clackmannan; it is probably only a matter of time until nesting takes place in the plantations there. Siskin numbers fluctuate in response to food supplies in the breeding season (Newton 1972) but many local recorders comment that a marked increase, involving an extension of range as well as expanding numbers, has taken place since about 1970. This has been particularly noticeable in the southwest, where probable breeding has now been recorded as far west as Dunragit (Wigtown). The Siskin is currently described as 'scarce' or 'uncommon' in Berwick, Renfrew, Lanark, West Lothian, Fife and Dunbarton, but in most of these counties it is increasing. In the north it remains a very local and sporadic breeder in Caithness, but here too there will probably be an increase as recent afforestation matures. In Sutherland nesting has been reported north to Borgie Forest near Bettyhill.

B&R had no records of breeding on any of the islands, but in 1968–72 Siskins were proved to be nesting on Arran, Islay, Mull, Rhum, Eigg and Skye. Nesting occurred on Canna in 1969 and 1970, but not since (Swann 1984). Breeding was first confirmed on Bute in 1977 (Gibson *et al* 1980) and a pair probably nested on Raasay in 1982 (A. Currie). There are no breeding records for the Outer Hebrides, Orkney or Shetland.

Little is known about breeding densities but numbers recorded in the CBC at Loch Garten have ranged from 25 to 45 pairs, while more than 100 pairs were thought to be nesting in the Carron Valley forest in 1977.

By late summer, when the supply of conifer seeds is exhausted, Siskins move into woods containing birch and alder; feeding on peanuts put out by householders has been reported with increasing frequency in hard weather since the early 1970s. Some birds move south to winter but there is no ringing evidence that any leave Britain. Nor is much known about the scale of immigration, beyond the fact that it is very variable, as in most irruptive species.

Few Siskins have been ringed or recovered in Scotland during the breeding season, but one ringed as a juvenile at Golspie was in Portugal by November. The limited data suggest that movements are irregular in both volume and direction. In the same months of successive (or later) years birds have been in West Germany and Easter Ross (July), and in Surrey/Middlesex/Hampshire and Inverness/Kincardine (March). Birds ringed in Italy (November) and West Germany (March) have been in Shetland and Perth the following June, while others ringed in USSR (September) and Holland (October) were in Easter Ross and Midlothian in January the same winter and two winters later respectively. Siskins ringed in the Moray Basin area in March/April have been recovered in Belgium the following winter.

Passage is obvious only in areas where Siskins are normally absent, such as the Northern Isles and the Isle of May.

Movement is recorded annually in both autumn and spring on Fair Isle, in widely varying numbers. Few pass through in April/May but in September/November the daily maximum ranges from less than ten to more than 100 in different years; the peak record so far is 300 on 23 September 1980. The Isle of May, where numbers are generally much smaller, had 80 one day earlier, and there were 140 at Walls in Shetland on 21 September 1980. This was the largest autumn movement recorded recently.

From October to March flocks of 100 or more Siskins are present in many areas. Among the largest recent counts have been 400–500 at Dunkeld, 350 at the north end of Loch Lomond, 280 at Dunblane, 250 roosting in reedbeds at Glencaple, 200 in Caithness, Sutherland, Argyll (mainland) and Stirling and c100 on Mull, in Upper Deeside and in the Trossachs.

Linnet *Carduelis cannabina*

Partial migrant, widespread but only locally abundant in lowland south, central and northeast Scotland, on the Clyde Islands and in Orkney. Scarce or local in the Highlands and Inner Hebrides and absent from the Outer Hebrides and Shetland. Although some move south in winter, many remain in Scotland. Possibly also a winter visitor.

In Scotland Linnets are most often found breeding among gorse bushes on rough ground near arable farmland, though some nest in hedges or young conifer plantations. They are heavily dependent upon weed seeds, hence the close association between their distribution and that of lowland cultivated ground. In some districts reclamation of moorland and improvement of rough grazing has resulted in loss of nesting

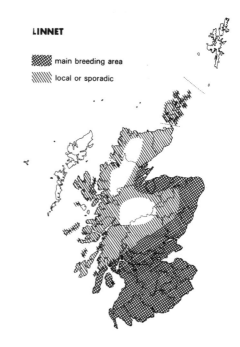

LINNET

▨ main breeding area

▨ local or sporadic

habitat, while in others this species will gradually be forced out as conifer forests mature and cease to offer suitable habitat. Little information is available on breeding numbers, but the greatest densities probably occur near the coast. In east Lothian in 1984 da Prato (1985) found only 3 pairs/km² on over 1,600 ha of farmland, but the numbers breeding in c150 ha of woodland, scrub and derelict land were high enough to give an overall density in his 1,700 ha study area of 6 pairs/km².

In the east mainland Linnets become scarcer north of Aberdeen and are very local in Caithness and north Sutherland. In the west they breed only locally in Argyll and are scarce or absent further north; where they do occur they usually move away in winter. Some of the Atlas records for the northwest Highlands and the Outer Hebrides should probably be treated with reservation, as this species is easily confused with the Twite, which replaces it in most of that area, though both species do occur in Orkney and Islay. The Linnet is quite widely distributed in Orkney, with recent 'probable breeding' reports from Mainland, Eday, Sanday, Rousay, South Ronaldsay and Hoy. It is also widespread and quite common on Islay, but breeds only sporadically on some of the other Inner Hebridean islands (Mull, Colonsay, Coll, Tiree, Canna, Skye, Raasay and possibly Jura) and not at all on Rhum or Eigg. Sporadic nesting has been recorded on the Isle of May and several of the other Forth Islands.

Because of the possibility of misidentification, it is difficult to be certain whether the reports of breeding in Shetland and Lewis quoted by B&R were reliable. They said the Linnet had colonised Shetland around 1934, but none were found there in 1945–53 (Venables & Venables 1955) and there have been no subsequent records of nesting, although a few birds are sometimes present in summer. Breeding reputedly took place in Lewis prior to 1915 but has not been recorded since, despite the fact that birds are occasionally seen in summer and suitable habitat is available (Cunningham 1983).

Flocking starts in July and from then until March large flocks are present in many areas. It is not known to what extent these include immigrants from the Continent, where the breeding range extends from the Mediterranean to southern Norway, Sweden and Finland, but the proportion of winter visitors is thought to be small (BOU). Passage certainly occurs annually in Fair Isle (daily peak usually less than 30) and Shetland, and occasional birds are recorded there throughout the year; there is no ringing evidence of where these passage birds come from. Many British Linnets winter on the Continent, mostly in Belgium, France and Spain (Newton 1972). There have been too few recoveries of Scottish-reared birds to demonstrate whether or not they regularly travel so far south, but one ringed in Edinburgh in August was shot in Spain the following January. Some of the flocks recorded in mid winter are very large, for example 1,700 at Gosford in November, 1,200 at Aberlady and 1,000 at Slains in December, and 1,400 at Inverness and 1,000 at Montrose in January. In the west flock sizes are generally smaller but there are recent records of 600 in Renfrew, 400 in Kirkcudbright, and 200+ in Wigtown, Argyll and Tiree.

Twite *Carduelis flavirostris*

Partial migrant, breeding on hill ground and coastal moorlands; most widespread and abundant in the islands and the northwest Highlands, much scarcer in the central and eastern Highlands, very local in the southwest, and now absent from the Southern Uplands. In winter large flocks gather on low ground, and these may include immigrants.

The Twite has a very limited European distribution, breeding only in Britain and Ireland and in coastal Norway and Finland; British breeders are *C.f. pipilans* and those in Scandinavia *C.f. flavirostris*. Elsewhere in Europe it is replaced by the Linnet, the two species overlapping only in Britain and southern Norway. Twites also breed in southwest and central Asia – at up to 3,500 m asl in Tibet – and it is supposed that the European relict populations became isolated during the retreat of the ice age (Voous 1960). A considerable contraction of the Scottish range has occurred this century and is apparently continuing. There has also been a recent decrease in numbers in some areas, such as Shetland, where the species was formerly abundant.

B&R had records of the Twite breeding at some time in every county except Clackmannan, Kinross, East and West Lothian and Wigtown, but noted that it was declining in many areas south of the Highlands. By 1968–72 it had vanished completely from Kincardine, Midlothian, Lanark, Dumfries and Roxburgh, and almost entirely from Peebles, Selkirk and Berwick; there have been no subsequent breeding records from any of these counties. A few pairs were then present in Ayr, both inland and on the coast, but these have now gone although a few pairs still nest on Ailsa Craig. The two Renfrew records in 1971 are the last known for that county, and in Stirling and Dunbarton the Twite has possibly also ceased to breed, or is at best down to a very few pairs. The one southern county in which a small but apparently thriving population still exists is Wigtown, where groups are present at the Mull of Galloway and on the southwest and south coasts. One or two pairs probably also continue to nest on the coast of Kirkcudbright, in the Port o' Warren area.

From Perth and Angus north to the Moray Basin the Twite breeds regularly in the glens, but little information is available on numbers; most comments suggest a declining population but it is possible that this unobtrusive little bird is under-recorded in some areas. A few pairs still nest on low ground at Morrich More, but there are none in the Black Isle and

TWITE

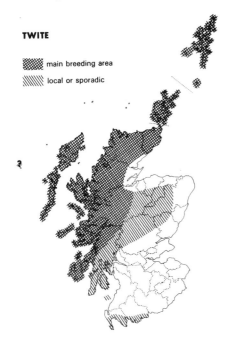

![legend] main breeding area

![legend] local or sporadic

reported during the 1970s and early 1980s in all the Highland counties and in Aberdeen, Angus, Perth (eg at 1,000 m asl on Ben Lawers), Stirling (at Skinflats), Ayr and Dumfries, and there were 600 in Ardwell Bay, Wigtown, in February 1983.

Redpoll *Carduelis flammea*

Partial migrant, widely distributed on the mainland and fairly regular on some Inner Hebridean islands but absent from, or very rare on, the Outer Hebrides and Northern Isles. Also a passage and winter visitor.

The British-breeding race *C.f. cabaret* is generally referred to as the Lesser Redpoll and is the smallest and darkest of the three subspecies. It occurs also in isolated pockets in Switzerland and other mountainous areas of

fewer in southeast Sutherland than in the past. In the western Highlands, from Argyll northwards, the Twite becomes progressively more widespread and locally common, and in Caithness it is abundant.

Fair numbers still breed on most of the Inner and Outer Hebrides, but nesting has ceased on St Kilda. There appear to be no recent breeding records for Arran or Bute. In Orkney and Shetland a marked decline has taken place within the last few years – particularly noticeable in Shetland about 1979; on Fair Isle numbers decreased in the 1960s. It has been suggested that in areas where Twites feed on farmland in spring they may have been affected by the increasing use of organochlorine seed-dressings but, as the decrease is so widespread and involves birds in moorland and hill areas as well as on crofting ground, it seems likely that other factors are also involved.

From late August onwards Twites gather in flocks on farmland and other open ground, such as coastal flats, where they do not breed. Little is known about the extent of these seasonal movements, many of which seem to be only local, but recoveries of Scottish-ringed birds in Donegal and on board ship in the North Sea indicate that some emigration occurs. Twites ringed elsewhere in Britain have been recovered in winter in Holland, Belgium, France and Italy (Newton 1972). Some immigration is also believed to take place, as Twites are sometimes seen arriving on the east coast in autumn and parties of up to 50 have been recorded on the Isle of May in both autumn and spring, but there is not yet any ringing evidence to show the origin of these birds.

Autumn flocks of 500+ are not uncommon on stubbles in the Outer Hebrides and Caithness, and 800 were in one flock in Argyll in February 1978. More than 800, mostly juveniles, were on Fair Isle in mid September 1977, dropping to 30 by November. Winter flocks of 100–250 or more were

REDPOLL

![legend] widespread

![legend] scarce or local

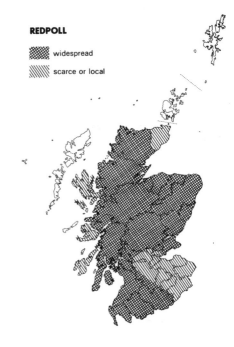

Central Europe. Most winter visitors are Mealy Redpolls *C.f.flammea* and are slightly larger, paler and greyer, with a more pronounced wing-bar; this race breeds from Scandinavia eastwards. The still larger Greenland Redpoll *C.f. rostrata*, which breeds also in Baffin Island, has darker flank stripes but is not easy to distinguish with certainty in the field; this race normally winters in Iceland but has been recorded in Shetland and the Hebrides. As in many seed-eating finches, the seasonal movements of Redpolls are affected by the availability of winter food, in this instance mainly birch seed; in years of poor seed crop in the breeding areas major eruptive movements occur.

During the first half of this century the Lesser Redpoll, here at the northern limit of its breeding range, increased in most parts of Scotland and colonised both birchwoods and young conifer plantations north to Sutherland (1928) and Caithness (1931) (B&R – cf. Parslow 1973 which says no breeding in Caithness). By 1950 nesting had been recorded on the Clyde Islands, Islay, Jura, Mull, Rhum and Skye, on all of which Redpolls still breed; sporadic nesting also occurs on Raasay and Canna. B&R give old records for Eigg and Barra but there have been none this century. The first proved nesting in Orkney was on Hoy in 1975, when two pairs were also present in the breeding season on Shapinsay; a pair bred again on Hoy in 1976 but not since. Mealy Redpolls summering in South Uist and Lewis in the early 1980s have been heard singing and may have nested there; a pair bred successfully at Voe, Shetland, in 1982, rearing three young (Shetland BR 1982).

In many counties the breeding Redpoll population has increased recently as a result of afforestation, and this trend may be expected to continue as more recently planted areas reach a suitable stage for occupation. Marked fluctuations in numbers occur and the CBC indices show a dramatic decline since 1968–72, considered partly attributable to unusually extensive emigration – in terms of both numbers and distance – in 1977/78. That winter British-ringed birds were recovered in Switzerland, Italy, Iberia and southern France, and the more distant emigrants probably did not return. The CBC figures almost certainly overstate the situation, however, since few census plots are in Scotland, which holds the bulk of the breeding population (M. Boddy in Winter Atlas). The only Scottish districts in which some decrease has been reported are Dumfries and the Black Isle. As conifer plantations mature they become less suitable for Redpolls, which will also nest in quite small patches of birch scrub and in suburban situations, where they may easily be overlooked by those not familiar with their distinctive flight call. Breeding densities of 25–50 pairs/km² are probably not uncommon in suitable woodland.

A substantial proportion of the native population moves south in winter, and there are a good many breeding season recoveries in Scotland of birds ringed in England between October and April. Some move on into Belgium, Holland, France and occasionally as far east as the Baltic States, the numbers that do so being greatest in years when the birch seed crop in England is poor. Those that remain in Scotland often frequent birch and alder scrub and sometimes join up with Siskins. Flocks are largest in August/October, when many hold 200 or more birds. In mid winter groups are usually much smaller though flocks of up to 150 are occas-

ionally reported in January/February. The Winter Atlas survey showed Redpolls to be more numerous and more widespread in mid winter 1982/83 than in the more severe preceding winter – but the records do not differentiate between the Lesser and Mealy races. The Lesser Redpoll is an uncommon passage bird on Fair Isle but a record 20 were present in early May 1981.

The autumn arrival of Mealy Redpolls takes place between September and December, the largest numbers generally being recorded in October. Numbers vary greatly between years, with daily maxima on Fair Isle of less than ten in some years and 500+ in others; there were notably large influxes in 1972 and 1975. In 1972 Mealy Redpolls were reported south to the Isle of May, which had c65 on 21 October, and 116 were trapped on Fair Isle during October, with c80 present on 28th. In 1975 the main influx took place around 15 October, when c500 were on Fair Isle. Smaller groups were reported in most east coast counties south to the Forth, where the Isle of May had a peak of 70 on 22 October. The Mealy Redpoll sometimes occurs further west on the mainland and in the Outer Hebrides, but numbers are generally small. Spring passage takes place between February and late May and seldom produces a daily maximum of more than five on Fair Isle.

Greenland Redpolls are also irregular in occurrence and are considerably scarcer. In some years none are reported and totals seldom exceed about a dozen; most of the records are from the Northern Isles but there are a few from the Outer Hebrides. The largest recent influx was in 1976, when the first birds reached Fair Isle on 3 September and 60 were present by 17th; there were also records from Orkney and Shetland that year. In 1974 up to seven Greenland Redpolls were in the Outer Hebrides between late August and December. The few recent spring records have been in May.

Arctic Redpoll *Carduelis hornemanni*

Vagrant (Arctic) – less than annual, in very small numbers.

There are two races of this species, which has a circumpolar distribution: *C.h. hornemanni* breeds in Baffin and Ellesmere Islands and northern Greenland, and *C.h. exilipes* over the rest of the range. The Arctic Redpoll is a partial migrant, moving southwards from its shrub tundra breeding grounds in winter, but only rarely reaching Britain. Because it is extremely difficult to distinguish from the Mealy Redpoll in the field, sight records are examined very critically by BBRC before acceptance. B&R, dealing largely with records of shot birds, note occurrences of both races (under their old names of Hornemann's and Coues' Redpolls), mostly on Fair Isle. From 1968 to 1983 inclusive there were accepted records of the Arctic Redpoll in only eight years, with a maximum of three birds in 1972, and sightings of 'probables' in another two years. Most occurrences have been in October but there are records for September (from 24th), November, December, February (and two 'probables' in March), and summer records – both in July – from Foula in 1965 and the Isle of May in 1982. Most of the confirmed records are from Shetland and Fair Isle but there have also been reports from Caithness and Orkney; some are still under consideration by BBRC. The latter will be included in a study of the records resulting from an influx in Autumn 1984. Two Arctic Redpolls on Fetlar on 13 October 1980 showed the characters of *C.h. hornemanni*, but the majority of those reaching the east coast of England are considered to be of Eurasian origin (Sharrock & Sharrock 1976).

Two-barred Crossbill *Loxia leucoptera* [1]

Vagrant (NW Europe/N America) – recorded in five years, 1950–83.

The Two-barred Crossbill is nomadic in winter and occasionally makes eruptive movements into western Europe, usually in company with other species of crossbill, in search of the larch seed which is its preferred food. B&R refer to it as an uncommon visitor but give few details. Since 1950 there have been reports in only five years: 1959 – two in August (Foula & Inverness); 1962 – one (Fair Isle) in late July; 1968 – a pair (Forest of Ae, Dumfries) in late February; 1972 – three single males (Fair Isle, Shetland Mainland and Whalsay) 6–19 July; and 1973 – a male (Dores, Moray) in early April. Despite its smaller size and prominent double wing-bars, the problems of distinguishing this species from the Crossbill *L. curvirostra* are sufficient to warrant study by the BBRC (BB 74:489). Only adult males can be sight-identified with certainty, and several of the records listed above may be rejected as a result of the review being carried out at the time of writing.

Crossbill/Scottish Crossbill *Loxia curvirostra/scotia* [1]

The Scottish Crossbill breeds from Argyll, Perth and probably Kincardine north to Sutherland, and the Common Crossbill locally from Sutherland to the Borders and Kirkcudbright. Some immigrant Common Crossbills from Fenno-Scandia and possibly USSR arrive in most seasons and major irruptions occur at irregular intervals.

The taxonomy of the Crossbills is both complicated and controversial. Some ornithologists consider that the Scottish

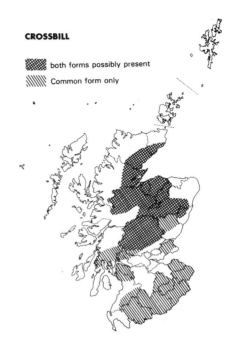

CROSSBILL

▓ both forms possibly present

▨ Common form only

Crossbill, which has a larger head and a heavier bill than the Common Crossbill, should be classed as a full species *L. scotica*, some that it is a race of the Parrot Crossbill *L. pytyopsittacus* – or *vice versa* – and some that it is a geographically isolated form of the widely distributed Common Crossbill *L.c. curvirostra*. (This problem is discussed by Knox, in Nethersole-Thompson 1975) There is some overlap in bill size between the two forms, making certain identification difficult in the field, and reports consequently often leave the question of race open. For this reason the two forms are considered together here, and the prefix Scottish or Common is used only when there is no doubt about identification.

It may be possible for observers sufficiently familiar with both forms to distinguish them by their calls, although much experience is necessary before this can be done with reasonable certainty (Nethersole-Thompson 1975). Perhaps the most practical clue to identification in most cases is the location and the type of conifer woodland in which the birds are found. Scottish Crossbills are closely associated with Scots pines and occur most often in old native pinewoods, although they do occasionally visit and nest in other conifer plantations containing a substantial proportion of pines. Common Crossbills feed principally on spruce seeds. Breeding seasons are synchronised respectively with the maturing of pine and spruce seeds but both races will take seeds of other tree species when their preferred food is in short supply. Numbers of Scottish Crossbills are generally small but very large numbers of Common Crossbills are sometimes present.

Crossbills have been the subject of long-term study by Desmond Nethersole-Thompson; most of the available information is recorded in his monograph *Pine Crossbills* (1975), upon which this account is largely based. The Scottish Crossbill is a difficult species to study; its population is small and widely scattered, its habit of feeding in the canopy makes it easily overlooked, and its numbers fluctuate and its distribution changes from one year to the next according to the pine seed crop. Nesting may take place at any time between late January and June, being earliest in years when pine seed is abundant; only one brood is normally raised. Common Crossbills may raise two or three broods in a year and in southern Scotland have been recorded nesting in all months from August to April (Newton 1972) and possibly May (SBR). Scottish Crossbill nests on Speyside range from 50 m to 1.65 km apart, and reach a maximum density of about 6 nests/10 ha (D. Nethersole-Thompson); Common Crossbill nests in Dumfries plantations are usually about 300 m apart. Scottish Crossbills are seldom seen in parties of more than 15–20 birds but flocks of Common Crossbills may number up to 80–100 or even more.

Because of the frequent uncertainty in identification, it is difficult to assess both the overall breeding numbers and the limits of distribution of the two forms. The fact that following an irruption Common Crossbills may remain to breed in areas where they are not normally resident further complicates the situation. In southern Scotland only Common Crossbills are present but over much of the Highlands both forms may occur. The breeding population of the Scottish Crossbill is probably in the region of 300–400 pairs; no corresponding estimate is available for the Common Crossbill, but several thousands may be present in the borders and southwest in a good year (I. Newton). Nesting has occurred on Arran and was recorded on Islay for the first time in 1982, but has not yet been proved on Mull, where birds are occasionally seen.

The Common Crossbill is widely distributed over much of Eurasia; those reaching Britain during irruption years probably originate from Norway east to the Urals (Newton 1972). Numbers on the breeding grounds fluctuate in any one area according to the spruce seed crop; when the cones fall the birds disperse, often for not more than a few hundred kilometres, until they find another good seed area, feeding on pine seeds until the new spruce crop is ready. Unlike truly migratory species, Crossbills make only one such major movement a year, generally during the summer (Parrot and Two-barred Crossbills behave in a similar way). In years when the cone crop is poor and the population large this movement starts earlier and ends later, movement is mainly in one direction and the birds travel much further; once on the move they may not stop even if they find an area of ample food supply. Both crop failures and high population level are probably involved in initiating eruptive movements, and it has been suggested that the birds reach a higher migratory state on such occasions (Newton 1972). Following an eruption, some birds return successfully to their area of origin but seldom do so in the same year; those which reach a suitable area of conifers may remain to breed, sometimes for a single season but occasionally establishing a more permanent colonisation, as in Dumfries and Kirkcudbright.

There were several Common Crossbill irruptions during the 1950s and 1960s but only one in the 1970s, in 1972, and there have been no major arrivals since. In July 1972 several hundred passed through Shetland, 50+ were on North Rona, and a few even reached St Kilda; other records were scattered down the east coast to Berwick and right across the country to Renfrew and Arran. On Fair Isle up to 54 were present at one time. All large arrivals there have occurred between late June and mid August, and in irruption years the totals involved have ranged from 50–300 birds. Crossbills ringed on the island have been recovered in northern Italy, three weeks later, and in Norway, two weeks after ringing. In most seasons at least a few immigrants appear on the mainland, generally in summer but occasionally in autumn, as in late October 1977, when birds were seen arriving over the sea in Moray. A comparatively small irruption in late September/October 1982 produced both a wider geographical spread of records and higher numbers in 'traditional' areas in 1982/83 than in the preceding winter (A. Knox in Winter Atlas).

Parrot Crossbill *Loxia pytyopsittacus*   [1]

Vagrant (N Europe) – recorded in five years, 1953–83.

The Parrot Crossbill, a resident of mature pinewoods in northern Europe, from Sweden east to northern Russia and south to Estonia, occasionally makes eruptive movements which bring it as far west as Britain. Although there were various old records B&R doubted their validity and the first report to be accepted was that of an adult female trapped on the Isle of May on 18 September 1953 (BOU); even this was regarded with some suspicion by the then Scottish Bird Records Committee (SB 3:168). There were no further reports until 1962, when an irruption occurred, with some 59 birds on Fair Isle alone and others in Shetland and the Outer Hebrides (BB 73:527). The first 20+ birds reached Fair Isle on 27 September and were mostly females and immatures. By 4 October 33 were present and on 11 October a further 20 or so, mostly adult males, arrived. Thirty-four were trapped and two found dead. There were also five or six in Shetland, two on North Rona and four in Lewis, all between 7 and 13 October. Apart from two on Fair Isle on 20 March 1963, and a male killed by a cat in Shetland on 22 October 1975, there were no further reports until the autumn of 1982, when there was another big influx. Again most were in the Northern Isles but there were also records from North Uist (a female and four immatures 21–22 October) and Wick (immature male 17 October). Fair Isle had one on 7 October and six on 8th, when a further 22 birds probably of this species were also present (FIBOR 1982). Four males, four females and a juvenile female were recorded in Shetland between 12 and 19 October. A total of 100+ Parrot Crossbills were reported in Britain during the winter of 1982/83, and in 1984 a pair bred in Norfolk (BB 77:557; Catley & Hursthouse 1985).

Trumpeter Finch *Bucanetes githagineus*

Vagrant (S Mediterranean) – two records, 1971 and 1981.

This normally sedentary species of desert and semi-desert areas in North Africa has recently established itself in southern Spain. The first two British records were both in the spring of 1971: in Suffolk on 30 May and, astonishingly, on Handa on 8–9 June. The Scottish bird was a female; it must have found conditions in Sutherland, even in June, very different from those of its native area. The only subsequent record is of one on Sanday, Orkney, on 25–29 May 1981. Trumpeter Finches are often kept in captivity and there has been extensive netting at the Spanish colonies. The BBRC has nevertheless accepted that these records relate to genuine vagrants (BB 70:45–49; BB 76:523).

Scarlet Rosefinch *Carpodacus erythrinus* [1]

Scarce passage visitor, annual in very small numbers. Has bred.

The Scarlet Rosefinch (also known as Scarlet Grosbeak or Common Rosefinch) breeds in forest edges and swampy clearings in mixed forest from Mongolia west to Scandinavia, where it has recently shown some extension of range (Voous 1960). It had been regarded as a potential future coloniser for some time before breeding in Scotland was finally proved in 1982 (BB 75:529). B&R described it as only an occasional passage migrant, not recognised in Scotland until 1906, but it now occurs annually in autumn and almost as regularly in spring, although numbers are still small. Peak totals to date have been 20 in the spring of 1981 and c50 in autumn 1980.

Most spring passage occurs between mid May and mid June, but there are occasional earlier records (from 2 April); singing males in breeding-plumage have been seen several times in May/July. The timing of the autumn passage is more consistent than in many other species, nearly always taking place between mid August and mid October – the latest sighting in recent years has been 29 October. Adult males comprise a substantial proportion of the birds seen in spring but very few are reported in autumn. Although most records of the Scarlet Rosefinch come from the Northern Isles, there have also been reports from Sutherland, Caithness, Inverness, Aberdeen, Angus, the Isle of May, Berwick, St Kilda, North Rona and Rhum.

In 1982 a pair was present throughout the summer in Highland Region and a nest with eggs was found on 23 June; it is not known whether any young fledged successfully (BB 77:133–135). No Scarlet Rosefinches were found when the site was revisited in June 1983, but a singing male was located in a different Highland area and another sang on territory on Fair Isle, where it remained from 31 May to 21 July.

Pine Grosbeak *Pinicola enucleator*

Vagrant (Fenno-Scandia) – one record, 1954.

A bird of coniferous forests, breeding from northern Sweden and Finland east through Siberia (and also in North America), the Pine Grosbeak winters regularly as far south as southern Norway and occasionally makes longer eruptive movements when food supplies are poor. It rarely reaches Britain, however, and there has been only one acceptable Scottish record: an adult female on the Isle of May on 8–9 November 1954 (BB 48:133). A report of four at Tomatin, Inverness, in May 1938 (SN 1938:115) was later rejected.

Bullfinch *Pyrrhula pyrrhula*

Resident, widely distributed and abundant in most wooded mainland areas, more local on the Clyde Islands and Inner Hebrides, and absent from the Outer Hebrides and Northern Isles. Recent population expansion has occurred in areas of extensive afforestation. Scarce irruptive visitor.

Originally a forest-dweller, the Bullfinch has adapted to life in a variety of habitats, ranging from town gardens to conifer plantations at altitudes of over 300 m and showing a preference for scrub and thick young woods. Its habit of feeding on the flower buds of fruit trees makes it unpopular in orchards, but in Scotland any such damage is of little commercial significance. It also takes a wide variety of seeds, including those of heather, on which it can be seen feeding in open moorland at up to 600 m asl.

There is no evidence of change in distribution over the last 50 years, but the population has shown a marked increase in the afforested areas of the southwest during the last decade. Plantations at the thicket stage with easy access to heather along rides or on adjoining moorland are especially favoured (A. D. Watson). Although not commented upon by local recorders, similar increases have probably also taken place in other afforested districts and may be expected to do so in the future in areas more recently planted, for example central Sutherland. Breeding now occurs more regularly than in the past on Rhum and small numbers nest annually on Skye, Raasay, Mull and Islay, but probably only sporadically on Eigg, Jura and Colonsay. Little information is available on breeding densities in Scotland, but in East Lothian in 1984 Stan da Prato (1985) found about 8 pairs/km^2 in 150 ha of scrub, woodland and derelict land and fewer than 0.5/km^2 on some 1,600 ha of arable farmland.

British Bullfinches, belonging to the race *P.p. pileata*, are smaller and darker than those breeding in northern Europe, and are very sedentary. There are few Scottish ringing recoveries but 85% of those in Britain as a whole are within 5 km of the ringing site (Newton 1972). The more brightly coloured immigrants belong to the race *P.p. pyrrhula*, which breeds from Scandinavia and France right across Eurasia. Birds from the northern part of the range are migratory, the scale of seasonal movement being influenced by food availability. In most seasons small numbers reach the Northern Isles in October/December, and occasionally there is

BULLFINCH

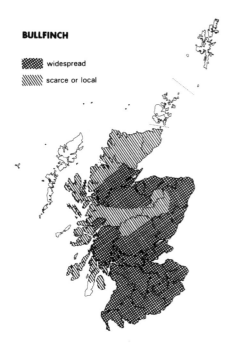

▓▓ widespread

▨▨ scarce or local

secretive behaviour and apparently erratic occupation of some sites make it easily overlooked. B&R give the first proved breeding record as in 1903. In England, where the species is more abundant, a rapid expansion of the breeding range is believed to have taken place during the 19th and early 20th centuries (Parslow); the spread in Scotland from 1903 onwards noted by B&R was presumably part of that expansion. Breeding currently occurs most regularly in Perth (in several areas from the Carse of Gowrie to Doune) and Midlothian (the Royal Botanic Gardens); the largest number seen together has been 13 at Scone (Perth) in December 1978. Since 1950 nesting has also been confirmed in Angus, Stirling, East Lothian, Peebles, Selkirk, Berwick, Roxburgh, Ayr and Dumfries, and has probably occurred in Kirkcudbright and Aberdeen. There are older breeding records for Fife and West Lothian. Sightings of pairs in Aberdeen suggest that breeding may have occurred there in the 1960s, and there have been sporadic reports from Kincardine, Renfrew and Dunbarton, all of which contain suitable habitat, but apparently none from Lanark. One on Arran in July 1983 was the first record for the Clyde Islands.

HAWFINCH

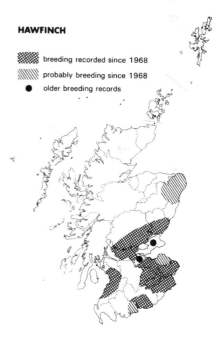

▓▓ breeding recorded since 1968

▨▨ probably breeding since 1968

● older breeding records

a slightly larger influx with stragglers as far west as the Outer Hebrides and south to Berwick. Daily totals on Fair Isle seldom exceed five in autumn. The return movement in spring is even smaller and less regular, with most records relating to single birds. A few 'northern' Bullfinches sometimes winter in Shetland.

Hawfinch *Coccothraustes coccothraustes*

Scarce and very local resident in deciduous woodland, from Perth (possibly Aberdeen) south to the Borders and Galloway. Extremely elusive and probably under-recorded. A passage visitor in very small numbers.

The Hawfinch's need for an adequate supply of fairly large seeds, such as those of cherry, blackthorn, beech and elm, limits its potential distribution in Scotland, while its

Further north the Hawfinch occurs only as an occasional straggler, mainly during passage periods. There are a few records from Argyll, Inverness, Ross and Cromarty, Sutherland and Caithness, mostly between March and September, also two from St Kilda and one from Tiree. In Orkney, Fair Isle and Shetland, passage is most often noted in May, but there have been occurrences in April, June/September and even December/January. The only Isle of May record is of a bird killed at the light in October 1937. The numbers seen on passage are very small, seldom more than two or three individuals, but in spring 1978 there was an unusually marked migration involving seven birds in Shetland, three or four on Fair Isle, two in Orkney and two in Caithness.

These migrants are well to the northwest of the Hawfinch's breeding range, which extends from Iberia, southern Britain, Denmark and south Sweden eastwards across Eurasia; the only indication of their origin so far is from the recovery in Shetland of a bird ringed in East Germany. Migration in this species is erratic and apparently to some extent linked with the availability of food (Harrison 1982).

Evening Grosbeak *Hesperiphona vespertina*

Vagrant (N America) – two records, 1969 and 1980.

A bird of the spruce belt of Canada and northwest America, the Evening Grosbeak winters erratically further east and south, occasionally appearing in eastern USA in large numbers. A particularly large movement of this kind took place in the winter of 1968/69 and on 26 March 1969 a male Evening Grosbeak was seen on St Kilda. This was the first British record and was accepted as being a genuine transatlantic vagrant (BB 64:189–194). The only subsequent report is of an adult female at Nethy Bridge, Inverness, from 10 to 25 March 1980. This species is sometimes kept in captivity, though rarely in Britain, so the possibility of escapes must be borne in mind.

Black-and-White Warbler *Mniotilta varia*

Vagrant (N America) – one record, 1936.

The only Scottish record of this North American woodland warbler, which breeds from the St Lawrence region south to the Gulf states, and winters from Florida to the Tropics, is of one found dead at Scalloway, Shetland, in mid October 1936 (BB 53:98), the first recorded in Britain. This is presumably the record which B&R included, without details, in their Appendix listing birds 'considered to be escapes'. There had been seven subsequent British and Irish occurrences by the end of 1983, four of them in September.

Tennessee Warbler *Vermivora peregrina*

Vagrant (N America) – three records, 1975–83.

This is a rather nondescript warbler which breeds in eastern Canada and USA and normally migrates well inland to its wintering areas in South and Central America. The first British record was from Fair Isle, on 6–18 September 1975, and the second in the same place about two weeks later, on 24 September; both birds were trapped and ringed (BB 74:90–94). A Tennessee Warbler on Orkney Mainland on 5 September 1982 (BB 77:160–164) is the only other British record to date.

Cape May Warbler *Dendroica tigrina*

Vagrant (N America) – one record, 1977.

The first Cape May Warbler recorded in Britain was a singing male in Paisley Glen, Renfrew, on 17 June 1977 (BB 73:2–5). This is one of the more vividly coloured American warblers, with tiger-striping (hence its scientific name) in black and yellow on its underparts and chestnut cheek-patches. It breeds in spruce forests in eastern Canada and USA and winters in the West Indies. The population fluctuates markedly in response to infestations of the spruce bud worm and a big increase took place in eastern USA in 1965–73.

Yellow-rumped (Myrtle) Warbler *Dendroica coronata*

Vagrant (N America) – two records, 1977 and 1982.

The first Scottish record was from Fair Isle, where a male was seen on 18 May 1977 (BB 71:526) and the only subsequent report is of one on North Uist on 22–23 October 1982 following severe westerly gales. Easily identified by its bright yellow rump, this species is widely distributed in the coniferous forests of Canada and the northern USA; it winters in the southern States. A detailed description of the first Yellow-rumped Warbler in Britain appeared in BB 48:204–207.

American Redstart *Setophaga ruticilla*

Vagrant (N America) – one record, 1982.

This very decorative wood-warbler was first recorded in Scotland on 1 November 1982, when a female or immature was found on Islay following an unusually marked autumn influx of American land-birds (BB 76:525). There had been only four other British and Irish records by the end of 1983, all in October. The American Redstart, a bird of deciduous woodland, is widely distributed across North America. It winters in Central and South America.

Ovenbird *Seiurus aurocapillus*

Vagrant (N America) – one record 1973.

The first acceptable British record of this secretive ground warbler was in 1973, when one was caught on Out Skerries, Shetland, on 7 October, the first to reach Europe alive (BB 68:453–455). There had been one previous 'Category D' record, based on a wing found on the Mersey tideline in January 1969, and there has been only one since, also of a dead bird, in County Mayo in December 1977. The Ovenbird breeds in deciduous woodland from northwest Canada south to Kansas and Georgia, and winters from Florida and Louisiana to the Tropics.

Common Yellowthroat *Geothlypis trichas*

Vagrant (N America) – one record, 1984.

The first Scottish record of this species was in 1984, when a male was on Fetlar on 7–11 June. This was only the second occurrence of the Common Yellowthroat, a widespread species in central-southern North America, since it was first recorded in Britain on Lundy in November 1954 (BB 48:145–147).

[Western Tanager *Piranga ludoviciana*]

A bird thought to be of this species was seen on the Isle of May in June 1973 (SB 8:45).

Song Sparrow *Zonotrichia melodia*

Vagrant (N America) – three records, 1959–83.

This heavily-streaked bunting was first recorded in Britain on 27 April 1959, when one was found on Fair Isle, where it stayed until 10 May (BB 52:419–421). In 1979, almost exactly 20 years later, two more occurred in Scotland: a male, which was trapped and ringed, on Fair Isle from 17 April to May, and an unringed bird at Sumburgh on 10 June. There had been only three other British and Irish records by the end of 1983: in 1964, 1970 and 1971 and all in April/May. The Song Sparrow, a partial migrant, is widely distributed in North America. Some comments on the differences between this species and similar 'sparrows' which might occur in future are given in Sharrock & Grant (1982).

White-crowned Sparrow *Zonotrichia leucophrys*

Vagrant (N America) – one record, 1977.

As has been the case with many other American vagrants, the first British record of the White-crowned Sparrow came from Fair Isle – on 15–16 May 1977 (BB 73:446). This remains the only Scottish record but a White-crowned Sparrow was found in England on 22 May the same year; there had been no further records by the end of 1983. This species is more migratory than the Song Sparrow, breeding much further north, near the tree limit in Canada, and usually travelling well inland on its way south to winter in the Gulf States.

White-throated Sparrow *Zonotrichia albicollis*

Vagrant (N America) – six records, 1909–83.

There is good evidence of 'assisted passage' across the Atlantic on board ship for at least some of the White-throated Sparrows reported in Britain (BB 58:230) and there must be an element of doubt as to whether any of the 10+ recorded so far got here entirely under their own power. The first Scottish record was in 1909, when a male was shot on the Flannan Isles (ASNH 1909:246) – B&R relegated this to their Appendix of birds 'considered to be escapes'. The other occurrences were in: May 1966 on Fair Isle; May/August 1970 near Thurso; November 1971 on Whalsay; May 1973 on Out Skerries; and June 1978 on Fair Isle. Apart from its black-and-white head and throat, the White-throated Sparrow looks – and behaves – rather like a Dunnock. It breeds in the North American spruce belt and winters in the southern States.

Dark-eyed (Slate-coloured) Junco *Junco hyemalis*

Vagrant (N America) – four records, 1966–83.

As this woodland bunting is sometimes kept as a cagebird the possibility of escapes cannot be entirely ruled out. The Scottish records have all been in May: the first two on Foula (1966 & 1967), the third on Out Skerries (1969) and the most recent in Glen Affric (1977). There had been earlier records in Ireland and in England, where three were reported in 1983. The Dark-eyed Junco is a common bird of coniferous forest, both breeding and wintering over much of North America.

Lapland Bunting *Calcarius lapponicus* [1]

Regular but scarce passage visitor, occasionally wintering; annual in Shetland on both spring and autumn passage, fairly regular in autumn on the east coast from Aberdeen to East Lothian and in the Outer Hebrides, and sporadic elsewhere. Has bred.

The Lapland Bunting has a circumpolar arctic/sub-arctic distribution, breeding in moss-shrub tundra north of the Arctic Circle and in the alpine zone on the Scandinavian and Alaskan mountains. The main wintering range of the Eurasian population is from the Black Sea eastwards, but there is also a regular wintering population along the coasts of the North Sea from Denmark to northern France; the latter presumably holds the Scandinavian breeding birds. The origin of those visiting Scotland is unknown, but in view of the frequency of their occurrence in Shetland and the Outer Hebrides it seems probable that many of them come from Greenland. Most mainland sightings of migrants and wintering birds are on or near beaches but the species also frequents grassy moorland.

On Fair Isle, Lapland Buntings start to arrive during the first half of September, building up to a peak towards the end of the month or in early October. The numbers involved are not large and there are seldom more than 20 birds on the island at once. Occasionally, however, the autumn immigration is more marked than usual, as in 1973 when there were 80 on Fair Isle on 16 September and well over 200 reported in Scotland as a whole. On that occasion the influx was concentrated largely in Shetland, but in some winters sizeable flocks (max. to date 30+) appear around the Ythan Estuary and on the Midlothian/East Lothian coasts; since 1979 up to 30 have wintered in the Musselburgh area.

Lapland Buntings have been recorded in all east coast mainland counties from Caithness to Berwick and are fairly regular on the Isle of May. In the west they have occurred on several Outer Hebridean islands and on Mull and Islay, but not on the Clyde Islands; on the mainland there are records only from Wester Ross, Argyll, Ayr, Dumfries and Kirkcudbright. This species is only rarely reported inland; winter records since 1968 have been from Angus (Loch of Kinnordy, November 1977), Midlothian (Silverknowes, February 1978), Perth (two near Crieff, December 1980) and Stirling (Campsie Fells, November 1982).

Spring occurrences are much scarcer and involve very few birds. It is noteworthy that in 1977, the year in which breeding was first confirmed, there were about twice as many spring reports as usual. Passage north through Fair Isle and Shetland starts about mid March and is generally over by mid June. The presence of a male at 500 m asl near Kingussie in April 1968 should perhaps be considered the earliest indication that breeding might occur. It was followed in 1974 by a report of an adult male in full breeding plumage in Caithness on 30 June. Three years later, after a cold late spring with above average snow cover on the hills, five occupied sites were located, with up to 16 pairs present. In July 1977 breeding was confirmed when two broods of four young were seen. In 1978 three sites held up to six pairs, two of which were proved to breed. In 1979, after another late cold spring, numbers increased again, with five sites holding up to 14 pairs, and no fewer than 11 pairs confirmed as breeding. At the main site, occupied by seven males and at least ten females, one male was thought to be polygamous with up to four females. This sudden colonisation was not maintained, however, and in 1980 only one pair bred and no other sites were occupied. In 1981 only a single bird was seen and none were reported in the summers of 1982 or 1983.

Snow Bunting *Plectrophenax nivalis*   [1]

Passage and winter visitor, most abundant in the east. A few pairs breed most years in the high corries of the Cairngorms and occasionally elsewhere in the Highlands.

In Scotland the Snow Bunting, which has a circumpolar arctic/sub-arctic distribution, is at the southern limit of its breeding range. It nests among the high tops, usually in barren areas of block scree and boulder fields, where snow often lies late; the craneflies and other insects associated with the snow-bed vegetation are an important source of food for the young. A particular corrie or slope may be favoured for several years and then abandoned, with an alternative site, previously unoccupied, becoming the main breeding area. Larger numbers of summering birds are often reported in years of late, heavy snowfall, but recent work in the Cairngorms has shown that there is no correlation between the number of cocks present and the extent of snow cover in late spring (Milsom & Watson 1984).

The two races of the Snow Bunting recorded in Britain are distinguishable by the amount of white on the rump. *P.n. nivalis*, with a predominantly white rump, breeds in Scandinavia, the Faeroes, Greenland and North America and is migratory, wintering in Britain, coastal Scandinavia and south to central Europe, and in the northern USA. *P.n. insulae*, with a much darker rump, breeds in Iceland and is believed to be almost entirely sedentary, merely moving to lower ground or coastal areas in winter. However, there are a few winter records of the Icelandic race in Scotland; one or two have been collected in South Uist and three or four on Fair Isle (Williamson, in Nethersole-Thompson 1966). In the Cairngorms both white- and dark-rumped birds have been found breeding, sometimes in the same year, and the relative proportion of the two types varies between years (Nethersole-Thompson 1976). The Scottish population is so very small – the number of pairs seldom enters double figures – that it is unlikely to be self-sustaining and probably depends upon fairly frequent additions of 'new blood' from among the wintering flocks. The movements of the small native population outwith the breeding season are not known.

Regular monitoring of the main breeding area in the Cairngorms has shown that marked fluctuations in the population occur – although, as Wynne-Edwards points out (in Nethersole-Thompson, 1976), Snow Buntings can easily escape detection, even when you know where to look and what to look and listen for. Between the late 1880s and about

1914, breeding was probably more frequent, and occurred over a wider area, than during the subsequent 30–40 years. In 1947, the best year in the period 1935–65, only three pairs and seven or eight unmated cocks were found in at least 2,870 ha of suitable habitat in the Cairngorms. The only proved breeding elsewhere during this period was a single record from Ben Nevis in 1954. Many more were recorded in 1966–75, with nests or broods reported from at least eleven different hills in Sutherland, Ross, Inverness, Banff and Aberdeen, and birds present in suitable habitat during the summer in 20 or more other locations, including Shetland, Caithness, Perth, Argyll, Skye and St Kilda (where

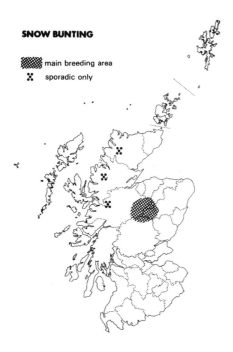

SNOW BUNTING

▨ main breeding area
✕ sporadic only

a pair had bred in 1913). In the Ben Nevis area in these years Snow Buntings bred definitely once, probably in a further two years, possibly in one year, and were absent in five years. In 1976–83 breeding has been proved in northwest Ross (1979 – in which year at least 14 young were raised in the central Highlands – and 1983); up to five pairs have been present in other areas outwith the Cairngorms and Grampians; and singing males have been recorded on several occasions in Shetland, Orkney and elsewhere. The central Highlands population fluctuated in this period between two and five pairs, with varying numbers of unmated cocks also present. 1983 was a very good year, with the highest numbers in the Cairngorms since 1974 (A. Watson).

Immigrants appear in the Northern Isles from the first week in September, but the main arrival is generally not until October/November. Birds from both Greenland and Scandinavia are involved, the former being the first to arrive (Williamson, in Nethersole-Thompson 1966); Snow Buntings from these two widely separated areas mix both on passage and in the wintering areas in Britain and on the Continent. In most seasons flocks passing through Shetland

and Fair Isle number hundreds rather than thousands, but much larger influxes sometimes occur. There was an exceptionally large and early arrival in 1970, with 2,000 each on Fair Isle and Foula and 1,000 on Fetlar, all on 18 September, and 3,000 on North Ronaldsay towards the end of the month. Most of the birds in these huge flocks were males. The only other recent major influx was in 1974 and was much later, with Fair Isle numbers peaking at 3,000+ on 8 November.

As the birds move south they apparently break up into smaller flocks. Counts of 500–1,000, and very occasionally more, have been reported from east coast counties south to the Lothians, but flocks of less than 250 are much more usual. In the west, including the Outer Hebrides, numbers are smaller, and counts of over 100 are exceptional in most areas, though increasing numbers have recently wintered on the hills south of New Cumnock, with 700 there in January 1982. On passage and in winter Snow Buntings can be found feeding on beaches among tide-wrack, on stubbles, and in the hills on grass and heather seeds – and in recent years on crumbs around skiers' car parks. Ringing recoveries show that passage continues south along the east coast, at least as far as Spurn Point and Humberside, that some cross to the Continent in mid winter, and that the return journey north from February onwards is by much the same route. The Snow Bunting is, in fact, one of the few passerines with a regular spring migration route through the Northern Isles, and fluctuations in numbers there are not associated with 'drift' conditions, as is the case with so many species (Williamson, in Nethersole-Thompson 1966). The main spring passage takes place in March and few are recorded on Fair Isle after mid April. One adult male ringed on Fair Isle in early April was shot in May the following year on Fogo Island, Newfoundland. It had presumably bred in southern Greenland, and moved southwest to winter in North America, as do most Snow Buntings from that area. An adult male ringed in summer in northeast Greenland and recovered on a boat off Foula in September probably represents a more typical movement.

Pine Bunting *Emberiza leucocephalos*

Vagrant (E Russia/Siberia) – six records, 1911–83.

All the Scottish records of this Yellowhammer-like bird are from the north and all but one in winter. As with many other vagrant buntings the possibility of escapes cannot be entirely ruled out. The Pine Bunting was first recorded in Britain on 30 October 1911, when a male was shot on Fair Isle (SN 1912:8). The subsequent records are from Papa Westray (October 1943), North Ronaldsay (August 1967), Golspie (January 1976), and Fair Isle (October & November 1980). The October Fair Isle bird was immature, possibly a female; all the others were males. There had been only one other British record by the end of 1983. On its breeding grounds the Pine Bunting frequents open forest; it winters in northern India and China.

Yellowhammer *Emberiza citrinella*

Resident, widely distributed on the mainland below about 300 m asl, regular on the Clyde Islands and some of the larger Inner Hebrides, but absent from the Outer Hebrides and Northern Isles. Small numbers occur on passage but little is known about the size of the immigrant population.

The Yellowhammer is widespread and relatively abundant in most mainland agricultural areas and up to the moorland edge. It requires fields or other open dry ground for feeding, and small trees, wires or similar prominent song-posts; it is often present near the edge of young conifer plantations. The breeding season is a long one, with unfledged young seen as late as the end of October in Easter Ross. There are few data on breeding densities in Scotland, but it seems unlikely, especially in intensive cereal-growing areas, that they often reach the figure of 9.5 pairs/km^2 quoted in the Atlas as the CBC average for farmland. High numbers in scrub and woodland raised the overall density on c1,700 ha in East Lothian in 1984 to 10 territories/km^2 (da Prato 1985).

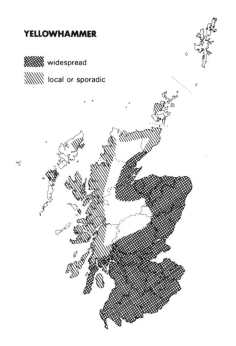

YELLOWHAMMER

▓ widespread

▨ local or sporadic

There has been a slight decline on the periphery of the breeding range in the last 10–15 years. In Orkney, where the Yellowhammer was 'not uncommon' in the 1940s (B&R), there was no record of breeding in 1976–82, and only two reports earlier in the 1970s. Breeding was confirmed in the Outer Hebrides on two occasions in 1970–71, on North Uist and at Stornoway, but has not occurred again; B&R knew of no proved breeding on these islands. On Skye Yellowhammers are widespread but not common, and only one pair was found on Raasay in 1982. Breeding occurs fairly regularly on Eigg, Islay and Jura, and at least sporadically on Mull, but has not been recorded on any of the other Inner Hebridean islands (Reed *et al* 1983). Only a few pairs

now nest on Arran, where the species was formerly much more abundant. There is little evidence of any overall change in status on the mainland although there have probably been local decreases in arable areas. Numbers have been increasing steadily in Ayr (Hogg 1983) and according to BTO data (mostly English) have been doing so in Britain as a whole.

In autumn Yellowhammers flock, sometimes joining up with Chaffinches and Greenfinches, and groups of 50–150 are often seen around farmyards and grain stores. Early autumn ploughing has reduced the importance of stubble fields for winter feeding. Passage through the Northern Isles occurs most years, in both spring and autumn, but the numbers involved are very small. Recent Fair Isle daily peaks have not exceeded 15 in autumn and ten in spring; numbers on the Isle of May are even smaller. A few birds occasionally winter in Shetland, Orkney and the Outer Hebrides. These immigrants probably come from Scandinavia; birds from the northern edge of the European breeding range, which extends from beyond the Arctic Circle to the northern shore of the Mediterranean, move south in winter. The few ringing recoveries suggest that native birds are largely sedentary, making only short seasonal movements.

Cirl Bunting *Emberiza cirlus*   [1]

Vagrant (Europe, including England) – recorded in six years, 1920–83.

Although breeding regularly in well-sheltered farmland in southern and southwestern England, the Cirl Bunting seldom occurs in Scotland, and has been reported only twice since 1950. The earliest record is of a pair in Argyll in June 1920 and there were reports from Ayr and Angus in 1928, and the Isle of May in 1935 (two about three weeks apart) and 1947 (three together). The most recent records are of a male at the Mull of Galloway on 17 August 1969 and one caught on the Isle of May on 11 June 1976. The earlier east coast records were in autumn and those from the west in spring and summer. This species has decreased during the last 100 years in England but has shown no significant change elsewhere in its breeding range, which extends from northwest Africa and Mediterranean Europe north to northern France and east into Asia Minor. It is normally sedentary and the only clue to the origin of Scottish vagrants comes from the 1976 Isle of May bird, which had been ringed the previous summer at Beachy Head, Sussex.

Ortolan Bunting *Emberiza hortulana*

Scarce and decreasing passage visitor, mainly in spring.

B&R described this species as 'a not uncommon passage migrant' and considered that it was 'no doubt frequently overlooked'. Despite the subsequent big increase in numbers of observers the Ortolan Bunting is still so infrequently recorded as to suggest that it occurs here more as a vagrant than as a true passage migrant. This is all the more surprising since it is widely distributed on the Continent, breeding in fairly dry open country or on forest edge from Fenno-Scandia south to Iberia. It winters in Africa south of the Sahara and in southern Arabia.

Ortolan Buntings occur much more regularly in spring than in autumn; of the 300 or so reported in 1968–83, over 200 were on spring passage. Annual totals vary greatly, ranging from under ten to over 100. 1969 was a notable year, with passage starting on 2 May, a few days earlier than usual, and an exceptional influx of 32 on Fair Isle the following day; totals of 60+ and 50 had been recorded in Fair Isle and Shetland by early June. Numbers have since been lower; in 1982 only eight were reported and there were no records at all in 1983. Most spring records are for May, with a few in April and early June. Autumn passage is more drawn-out, with extreme dates 22 August and 20 October, but most occurrences are in September and the first week of October. Autumn totals rarely exceed ten, and occasionally there are no reports at all at that season.

Most records of the Ortolan Bunting come from Fair Isle and Shetland, with smaller numbers from Orkney and the Isle of May. Elsewhere it occurs – or is recorded – only rarely. Between 1968 and 1983 there were mainland records from Aberdeen, Kincardine, Caithness and Sutherland, while B&R quote earlier reports from St Kilda, Easter Ross and the Bass Rock.

Cretzschmar's Bunting *Emberiza caesia*

Vagrant (E Mediterranean) – two records, 1967 and 1979.

The only British records of this species are from Fair Isle, on 10–20 June 1967 and 9–10 June 1979 (BB 62:144–148; 74:532–3). Both birds were males. Cretzschmar's Bunting is similar in appearance to an Ortolan Bunting and careful examination is needed to confirm its identification. This is a species of very limited distribution, breeding in dry country around the eastern Mediterranean and win-

tering in northeast Africa and southern Arabia; it is consequently rather surprising that it should occur here naturally. The first Fair Isle bird was closely examined in the hand for signs of feather damage which might suggest that it had escaped from captivity, but none were found.

Yellow-browed Bunting *Emberiza chrysophrys*

Vagrant (NE Asia/China) – one record, 1980.

The first accepted British record of this species was of a male on Fair Isle from 12 to 23 October 1980 (BB 76:217–225). There had been one earlier report, from Norfolk in 1975 (BB 69:358), but this was found unacceptable by the BOU Records Committee.

Rustic Bunting *Emberiza rustica*

Vagrant (NE Europe/Asia) – almost annual, in very small numbers.

In 1968–83 about 66 Rustic Buntings were reported – 40 in spring and 25 in autumn. 1968 and 1977 were the only completely blank years and the annual total exceeded ten only once, in 1980. Most reports come from the Northern Isles, but since 1968 there have also been records from the Isle of May, St Kilda, Islay and St Abbs (including one on the unusually early date of 7 April) and there is a 1965 record from South Uist. B&R give mainland records for Aberdeen and Sutherland, in 1905 (the 'end of March' is still the earliest British arrival date) and 1906 respectively. There is no regularity in the pattern of occurrences. In some years all reports are in spring, mostly in May but a few in early June; in others the majority are in autumn, generally between 20 September and 15 October, but with stragglers to 8 November.

Although breeding as far west as Sweden, the Rustic Bunting winters far to the east, in China and Japan. One ringed on Fair Isle in June 1963 did its best to move on in the right direction, travelling nearly 3,000 km southeast to Chios, in the Aegean Sea, by mid October that year. Although much whiter on throat and underparts, this species could be mistaken for a Reed Bunting on casual inspection.

B

Little Bunting *Emberiza pusilla*

Vagrant (N Fenno-Scandia/Siberia) – annual in very small numbers, mainly in autumn.

The first dated record of the Little Bunting given by B&R is for Pentland Skerries in 1903; of Fair Isle they say the species 'is not uncommon in the autumn', and conclude that it is a passage migrant, probably much overlooked. Like a small Reed Bunting in appearance, it could easily be missed in a mixed flock of finches or buntings. From present knowledge of its breeding distribution and migrations, it seems likely that the small numbers appearing in Britain should nevertheless be regarded as vagrants from the western edge of the breeding range rather than regular passage migrants. The Little Bunting breeds in forest/shrub tundra from Lapland eastwards through Siberia and Russia, and migrates southeastwards to winter in China and southeast Asia.

In 1968–84 this species was recorded annually in autumn, in totals ranging from one to 22, but only six times in spring (max. total only three). The spring records are for April/May, and several refer to singing males. Most of the 75 + autumn reports in this period were between mid September and the end of the third week in October, with three in November and two in December. One of the November records was a bird trapped at a reedbed roost in Dingwall in 1975 and there was a January report from the same site in 1977. These records, together with a sighting at Caerlaverock in February 1981 and several recent English records in December/April, suggest that Little Buntings may occur in winter more often than is realised. Although the great majority of reports are from Fair Isle and Shetland, Little Buntings have also been recorded since 1968 in Orkney, Caithness, Sutherland, Easter Ross, Inverness, Angus, Fife, the Isle of May, East Lothian, Berwick, Ayr and Dumfries – and at sea in the Forties oilfield. There are older records from St Kilda and North Uist.

[Chestnut Bunting *Emberiza rutila*]

Escape/vagrant (Category D)

There has been only one British record of this species, an adult male on Foula on 9–14 July 1974 (BB 70:444). The Chestnut Bunting, an East Asian species, is an inexpensive cagebird and it is probable that the Foula bird was an escape rather than a genuine vagrant.

Yellow-breasted Bunting *Emberiza aureola*

Vagrant (NE Europe/N Asia) – annual, in very small numbers and mostly in the Northern Isles.

B&R give the earliest record of the Yellow-breasted Bunting as 1907, on Fair Isle, and most subsequent reports are from there or from Shetland; the species has occurred annually on Fair Isle since 1971. It has also been reported from Orkney, St Kilda, Fife, the Isle of May, Tiree and Wester Ross, where a singing male was found in June 1982 (BB 76:526). Apart from the last mentioned, and one on Fair Isle in July 1980, all 1968–84 records were between 26 August and 4 October; most related to females or immatures, which resemble small, pale and little-streaked Yellowhammers. Annual totals in 1971–84 did not exceed nine and were usually less than five. The Yellow-breasted Bunting breeds in damp, open shrubland from Finland eastwards, and migrates in a southeasterly direction to winter in India and southeast Asia.

Reed Bunting *Emberiza schoeniclus*

Resident, widespread on the mainland and has recently expanded its range in the Hebrides and Northern Isles. Passage and winter visitor, probably in only small numbers.

Although most often associated with wetlands, the Reed Bunting now also nests in drier habitats such as farmland. Its gradual spread into drier situations was first noted in England in the 1930s (Bell 1969); in many areas it now occupies similar habitats to the Yellowhammer and Corn Bunting and occurs quite regularly in young conifer plantations – at up to 300 m asl in Glen Trool. B&R considered that the population was increasing during the early part of the century, and it was probably during a period of high numbers that the expansion into new habitats took place (Atlas). The recent extension in range has been comparatively slight, as Reed Buntings were already widespread by the middle of the century. Breeding now occurs at least sporadically on Tiree, Eigg, Muck, Rhum and Canna, for which B&R had few or no records. Although little information on numbers is available it seems likely that the population on the Outer Hebrides and Orkney has been increasing slowly but steadily. In Shetland, where nesting was first recorded in 1948, and there were no records north of Mainland in 1968–72, the Reed Bunting now breeds regularly

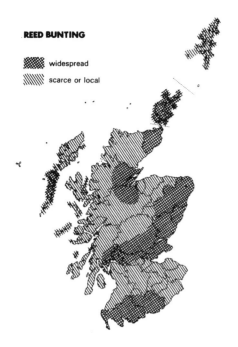

REED BUNTING

▨ widespread

⫽ scarce or local

on Unst (at least five pairs in 1981) and has also bred on Yell and Fetlar. Many of those breeding in the north appear to move away in winter, but there is no evidence that native birds emigrate, most ringing recoveries being within 100 km of the nesting area.

Although widespread, the Reed Bunting is an abundant breeding bird only locally, in areas of particularly suitable habitat. In East Lothian in 1984 da Prato (1985) found fewer than 0.5 territories/km² on 1,600 ha of farmland, but up to 13/km² in areas with extensive scrub. More than 65 pairs have been counted at Loch Ken, 31 pairs at Lochwinnoch and 28 at Aberlady Bay, the latter mainly in rank meadow-sweet and tall grass. Numbers fluctuate considerably and are affected by severe winters; in Britain as a whole the CBC index dropped by about 50% between 1968–72 and the early 1980s.

In autumn and winter large roosts are established in some areas, often in reedbeds: the highest counts reported in recent years have been 500 at Guardbridge and 300 at Glencaple, but flocks of 50–100 are more usual. It is not known whether these winter flocks consist solely of native birds or include some immigrants. Passage occurs annually in the Northern Isles, and less regularly on the Isle of May, in March/June and October/November. Peak daily counts since 1970 have been 250 on Fair Isle and 70 on the Isle of May in spring (May); autumn numbers are generally lower. Many of these passage birds presumably come from Scandinavia, where the Reed Bunting nests well beyond the Arctic Circle; those breeding in the northern and eastern part of the extensive Eurasian range are migratory. Migrants ringed on Fair Isle have been recovered in winter in France, so at least some of those reaching Scotland move on to winter further south.

Pallas's Reed Bunting *Emberiza pallasi*

Vagrant (Siberia) – two records, 1976 and 1981.

The only British records of this Twite-sized relative of the Reed Bunting are from Fair Isle. The first, an adult female, was there from 29 September to 11 October 1976 (BB 73:402–408; see also 73:400–401), and the second, a first-year bird, on 17–18 September 1981. Both birds were trapped and examined in the hand; being rather nondescript in appearance and skulking in habit they might easily have escaped detection anywhere other than Fair Isle. The 1981 bird was considered likely to belong to the northern race *E.p. polaris*.

[Red-headed Bunting *Emberiza bruniceps*]

Escape/vagrant (Category D)

The Red-headed Bunting is widely kept as a cagebird; although the possibility of true vagrancy cannot be ruled out, most – if not all – occurrences in Britain are currently regarded as escapes from captivity. In the wild the Red-headed Bunting breeds from southeast Russia eastwards into China and south to Iran, and winters in northern India. The first British record was a male shot on North Ronaldsay in June 1931, and most subsequent reports have also come from the northern half of the country. There are records from Shetland, Fair Isle, Orkney, Outer Hebrides (St Kilda, North and South Uists, Shiants), Inner Hebrides (Rhum – in several years, Mull, Tiree), Arran, Ailsa Craig, the Isle of May, and most mainland counties from Perth and Argyll northwards. South of the Tay/Clyde the only mainland records are from East Lothian, Kinross, Roxburgh, Ayr and Kirkcudbright. Red-headed Buntings have been recorded almost annually since 1968; most reports are in May/September but there is one March record (Caithness 1974) and one in October (Rhum 1970). The maximum reported in any one year has been 12, and adult males (which are much more easily identified than the duller females) are in the majority. A review of Red-headed Buntings on Fair Isle in 1950–67, by Roy Dennis, is given in BB 61:41–43.

Black-headed Bunting *Emberiza melanocephala*

Vagrant/escape (SE Europe) – less than annual, in very small numbers.

Like many of the attractively coloured buntings, the Black-headed Bunting is quite widely kept in captivity and many of those occurring in Britain are thought to be escapes. In the wild it breeds in scrub and open forest, from Italy eastwards through Asia Minor and Iran, and winters in India. The first record given by B&R, in November 1886 in Fife, is much later in the year than any of the subsequent records. Four more Black-headed Buntings were reported in the first half of this century, one on the Isle of May and the rest on Fair Isle, where there was another in 1951. The species occurred in six years during the 1960s and in nine during the 1970s. There were no reports in 1980–81; an adult female on Fair Isle on 12 September 1982 and a male there on 4 June 1984 were accepted as genuine vagrants. Nearly two-thirds of the records are in spring, from 12 May to July; apart from one October record, the remainder are in August/ September. The most recorded in one year has been four, in 1973. All recent reports have been from Shetland and Fair Isle, but there are earlier records from Orkney, Harris, Islay and the island of Seil in Argyll.

Corn Bunting *Miliaria calandra*

Very local and declining resident; numbers are highest in arable areas near east and northeast coasts, in the southwest and on the southern Outer Hebrides. Flocking occurs in winter, especially in severe weather.

The breeding population has declined markedly since 1950 and the Corn Bunting is now absent from many of its former haunts, and scarce and local in most of the areas still occupied. Because information on numbers and distribution was scanty, the early part of the decline was poorly documented, but there is ample evidence of the scale of decrease since 1970. In the early 1950s Corn Buntings nested on at least ten Shetland islands (Venables & Venables 1955), by 1977 only two or three pairs were present (Berry & Johnston 1980), and in 1982 only a single bird was recorded. On Fair Isle breeding had ceased by the late 1930s, but a few birds were recorded in most years until recently; one in October 1982 was the first since 1979. In Orkney the Corn Bunting was still fairly common in the 1930s but by 1970 was scarce except on Sanday (Balfour 1972); in 1982 about ten singing males were recorded, one each on Mainland and Stronsay and the rest on Sanday. In Caithness and Sutherland there has been a rapid decrease since 1968–72 and the species is probably now extinct; the last report of a juvenile in Caithness was in 1978, and in Sutherland none have been seen since 1972 in the Dornoch area, which held c20 pairs in the 1960s. From the Dornoch Firth to Banff a few still breed on the coastal strip, but numbers are down since the early 1970s.

From Aberdeen to East Lothian the Corn Bunting is des-

cribed by local recorders as declining but still 'common' or 'fairly common' near the coast, though several traditional sites in East Lothian have recently been deserted (Brown *et al* 1984). Inland breeding has decreased slightly since 1968–72 in Aberdeen, where the Buchan plain is the main stronghold; in 1981–84 breeding was confirmed in 30 10 km squares and noted as probable or possible in a further 19 squares (NE Atlas). No information on numbers is available for Fife, but a census in the Lothians in 1982 produced records of 37 singing males, with birds present in only ten 10 km squares, as against 25 occupied squares in 1968–72; average density over all areas of suitable habitat was estimated as one singing male per 11.9 sq km, but was higher locally, eg one per 2.7 sq km near Tranent (Brown *et al* 1984). In Berwick four or five territories were occupied near Eyemouth in 1980 but only one by 1983, while the last record for Roxburgh was in 1980. Breeding had ceased in Selkirk and Peebles by 1968–72. Small numbers currently breed in coastal Dumfries, in Kirkcudbright around Castle Douglas, and in south and southeast Wigtown and the Mull of Galloway. The estimated 24 singing males located in Ayrshire in 1978 were mostly in the centre and west of the county; around West Kilbride, where there were five occupied territories in 1966, none have been seen since 1975, nor around Crosshill since 1979 (Hogg 1983). In Renfrew there has been a big decrease since 1976, from an estimated 50 singing males to just three in 1982, while in Lanark, Dunbarton and Stirling a very few pairs remain. There are now no Corn Buntings in Clackmannan during the breeding season, and only one singing male was reported in Perth in 1981–83. About six regular sites are known in Kinross, but a decrease is taking place there too. There are no recent records from the west mainland coast north of the Clyde, nor from Arran, and few Corn Buntings have been seen on Bute or the Cumbraes since the late 1970s (Gibson *et al* 1980).

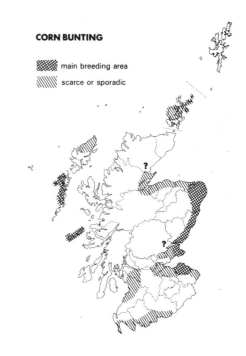

CORN BUNTING

main breeding area

scarce or sporadic

Tiree and Coll are the only Inner Hebridean islands on which the species still nests; it last bred in Skye in 1970, by which time it had already vanished from Mull, Islay and the Small Isles (Cadbury, in Reed *et al* 1983). In the Outer Hebrides the Lewis population was down to four or five singing males – in the Back, Ness and Bragar areas – by 1980, and none now nest in Harris. The very local population in the southern islands was estimated in 1983 as c50+ occupied territories (C. Spray).

In Scotland the Corn Bunting reaches the northern limit of its breeding range, which extends from southern Scandinavia and the eastern Baltic to north Africa. In parts of its range it occurs in steppe country at over 2,000 m asl but its British distribution is very closely associated with that of lowland farming, and Sharrock (Atlas) has suggested that its presence in much of central and western Europe is a direct result of man's influence in creating the agricultural landscape on which it is now dependent. It requires cereal crops or long grass for nesting in and prominent song-posts; overhead or fence wires are apparently preferred, but stone dykes, bushes and tall herbs are also used. Agricultural changes, including earlier harvesting of grass and more widespread use of sprays on growing cereals, have probably been partly responsible for the population decline, which started prior to 1950. Goodbody (1955) noted that nestlings appeared to be fed exclusively on small green caterpillars gathered off young corn; today few corn crops are allowed to carry high insect numbers. It is possible that climatic factors are also involved (Parslow 1973).

Although B&R regarded the Corn Bunting as a partial migrant and a summer visitor to the breeding areas, there is no evidence – apart from occasional occurrences on Fair Isle and some other islands – that birds travel any distance. Presumably they simply move into districts which offer the best feeding, with small breeding groups from different areas congregating in flocks during the winter. Flock size has decreased in most areas over the last 15 years. In the late 1960s there were reports of up to 300 at a roost in Kirkcudbright, 200 in Wigtown, and 100+ in Dumfries, but the only recent count of more than 50 in those counties is of 76 near the Mull of Galloway in December 1982. Flocks of 100–250 still occur fairly regularly in Aberdeen and Fife and occasionally around the Moray Basin and in Berwick, but since the mid 1970s no count in the Lothians or Angus has exceeded 100.

Studies of the Corn Bunting's habitat and breeding habits have been made in Aberdeen (Goodbody 1955) and Sutherland (Macdonald 1965). A strong preference for well-cultivated arable land was apparent in both these areas, while farmland intermixed with moorland or woodland was avoided. Polygamy was recorded in several cases, each involving only two females per cock. In southern England males have been known to mate with up to seven females and it has been suggested that this difference is related to the availability of food for chick-rearing.

Rose-breasted Grosbeak *Pheucticus ludovicianus*

Vagrant (N America) – one record, 1983.

This species was first recorded in Scotland on 7 October 1983, when a first-winter male was found on South Uist, Outer Hebrides; it died in care the following day (BB 77:560). There were also occurrences in Ireland (two), Wales (one) and Scilly (one) the same month. The Rose-breasted Grosbeak, which breeds in central North America and winters in Mexico and northwestern South America, was first recorded in Ireland in 1957.

[Black-headed Grosbeak *Pheucticus melanocephalus*]

A male of this American species caught on Fair Isle on 24 August 1969 proved to be an escaped cagebird (SB 6:122).

[Blue Grosbeak *Guiraca caerulea*]

Escape/vagrant (Category D)

Like several other colourful North American seed-eaters, the Blue Grosbeak is sometimes kept as a cagebird and British records are considered most likely to refer to escapes. There have been only three occurrences, all males: the first on Out Skerries, Shetland, from mid to 26 August 1970 and the others at Kiltarlity, Inverness, on 10–11 March 1972, and Ettrick, Selkirk on 22 May 1977.

[Indigo Bunting *Passerina cyanea*]

Escape/vagrant (Category D)

The sole Scottish record of this North American sparrow-sized bunting, is of a male on Fair Isle from 3 to 7 August 1964. The only other British record to date is of one in Essex in 1973. The very beautiful blue breeding plumage of the male explains why this is a popular cagebird.

[Lazuli Bunting *Passerina amoena*]

A bird of this American species shot in Orkney in June 1964 was regarded as an escape (BB 57:340), as have been the six or so subsequently reported.

[Varied Bunting *Passerina versicolor*]

A male trapped on Fair Isle in October 1974 was undoubtedly an escaped cagebird (SB 8:466).

[Painted Bunting *Passerina ciris*]

Escape/vagrant (Category D)

The four records of this very colourful American species are all from Shetland. The first two were males (on Mainland in May 1972 and Yell in July 1978), the third a female (Fair Isle in June/July 1979) and the most recent one on Noss on 8 June 1981.

Bobolink *Dolichonyx oryzivorus*

Vagrant (N America) – one record, 1975.

A bunting-like bird of northern meadowlands, the Bobolink migrates through the southeastern USA to winter in South America; all the British occurrences have been during the autumn passage period. The only Scottish record is from Out Skerries on 18 September 1975 (BB 70:222). Since the first Bobolink was recorded in Scilly in 1962 there have been several more occurrences there and two in Ireland.

Northern (Baltimore) Oriole *Icterus galbula*

Vagrant (N America) – one record, 1974.

Nearly all the thirteen British and Irish records of the Northern Oriole have been in September/October, when the birds are moving south from the woodlands of USA and southern Canada to their wintering quarters in Central America. The only accepted record for Scotland is of an immature on Fair Isle on 19–20 September 1974 (BB 68:330), but B&R include this species in an Appendix listing birds considered to be escapes (no details are given). There is a good description of a female in Sharrock & Grant (1982); the orange and black male is very distinctive.

Appendices

APPENDIX 1: List of Scottish sites referred to in the text (grid squares are shown on the map on p. 67, old counties and regions on the map on p. 65). See maps on pages 16, 22 and 36 for approx. locations of principal mountain ranges, islands, estuaries and rivers. Note: many of the grid references are taken from The Ordnance Survey Atlas of Great Britain (1982) and are approximate only.

Site	Grid ref.	Old county	Region
Aberlady Bay	NT 4581	E Lothian	Lothian
Abernethy Forest	NH 9918	Inverness	Highland
Aboyne	NO 5298	Aberdeen	Grampian
Achnahaird	NC 0213	Ross & Cromarty	Highland
Achnasheen	NH 1658	Ross & Cromarty	Highland
Ailsa Craig	NX 0199	Ayr	Strathclyde
Aird's Point	NX 9966	Dumfries	Dumfries & Galloway
Alloa	NS 8893	Clackmannan	Central
Almondbank	NO 0626	Perth	Tayside
Almond Estuary	NT 1877	West/Midlothian	Lothian
Altnaharra	NC 5635	Sutherland	Highland
Am Balg	NC 1866	Sutherland	Highland
Angus glens	NO 26–47	Angus	Tayside
Annandale	NY 1292	Dumfries	Dumfries & Galloway
Annan Waterfoot	NY 1864	Dumfries	Dumfries & Galloway
Applecross	NG 7144	Ross & Cromarty	Highland
Ardersier	NH 7854	Inverness	Highland
Ardfin, Jura	NR 4763	Argyll	Strathclyde
Ardmore	NS 3178	Dunbarton	Strathclyde
Ardnamurchan	NM 5766	Argyll	Strathclyde
Ardnave Loch, Islay	NR 2873	Argyll	Strathclyde
Ardrossan	NS 2342	Ayr	Strathclyde
Ardwell Bay	NX 0743	Wigtown	Dumfries & Galloway
Auskerry	HY 6716	Orkney	Orkney
Avoch	NH 6955	Ross & Cromarty	Highland
Ayton	NT 9260	Berwick	Borders
Badachro	NG 7873	Ross & Cromarty	Highland
Badanloch	NC 7933	Sutherland	Highland
Badbea	ND 0819	Caithness	Highland
Badenoch	NN 7091	Inverness	Highland
Balallan, Lewis	NB 2920	Ross & Cromarty	Western Isles
Balgavies Loch	NO 5351	Angus	Tayside
Balgray Reservoir	NS 5157	Renfrew	Strathclyde
Ballantrae	NX 0882	Ayr	Strathclyde
Ballinluig	NN 9852	Perth	Tayside
Balranald, North Uist	NF 7169	Inverness	Western Isles
Banchory	NO 6995	Kincardine	Grampian
Barbaraville	NH 7472	Ross & Cromarty	Highland
Barnsness	NT 7277	E. Lothian	Lothian
Barra	NF 6801	Inverness	Western Isles
Barr Loch	NS 3457	Renfrew	Strathclyde

Site	Grid ref.	Old county	Region
Barry	NO 5533	Angus	Tayside
Bass Rock	NT 6087	E. Lothian	Lothian
Bearasay, Lewis	NB 1242	Ross & Cromarty	Western Isles
Beauly Firth	NH 6147	Invern./Ross & Crom.	Highland
Benbecula	NF 8251	Inverness	Western Isles
Ben Lawers	NN 6341	Perth	Tayside
Bennachie	NJ 6522	Aberdeen	Grampian
Ben Nevis	NN 1671	Inverness	Highland
Ben Rinnes	NJ 2435	Banff	Grampian
Ben Wyvis	NH 4668	Ross & Cromarty	Highland
Berneray	NL 5680	Inverness	Western Isles
Berriedale	ND 1223	Caithness	Highland
Berstane Wood	HY 4610	Orkney	Orkney
Binscarth	HY 3414	Orkney	Orkney
Birsay Moor	HY 3719	Orkney	Orkney
Blackdog	NJ 9514	Aberdeen	Grampian
Black Isle	NH 6557	Ross & Cromarty	Highland
Black Loch (Renfrew)	NM 9231	Renfrew	Strathclyde
Blackshaw Bank	NY 0563	Dumfries	Dumfries & Galloway
Blairgowrie Lochs	NO 04–24	Perth	Tayside
Bluemull Sound	HP 5503	Shetland	Shetland
Boardhouse Loch	HY 2726	Orkney	Orkney
Boat of Garten	NH 9419	Inverness	Highland
Bogside Flats	NS 8353	Ayr	Strathclyde
Bo'ness	NS 9981	W. Lothian	Lothian
Borgie Forest	NC 6655	Sutherland	Highland
Bornish, S. Uist	NF 7329	Inverness	Western Isles
Borthwick	NT 3659	Midlothian	Lothian
Borve, Harris	NG 0294	Inverness	Western Isles
Braco	NN 8309	Perth	Tayside
Braemar	NO 1591	Aberdeen	Grampian
Braes, Skye	NG 5234	Inverness	Highland
Bressay	HU 5040	Shetland	Shetland
Broad Bay, Lewis	NB 5035	Ross & Cromarty	Western Isles
Brora	NC 9003	Sutherland	Highland
Buchan – Cliffs/Ness	NK 1342	Aberdeen	Grampian
Buddon	NO 5530	Angus	Tayside
Burghead Bay	NJ 0767	Moray	Grampian
Burntisland Bay	NT 2385	Fife	Fife
Burra Isle	HU 3734	Shetland	Shetland
Burrastow	HU 2247	Shetland	Shetland
Bute Lochs (Ascog/Quien/Fad)	NS 0862	Bute	Strathclyde
Butt of Lewis	NB 5166	Ross & Cromarty	Western Isles
Caerlaverock	NY 0565	Dumfries	Dumfries & Galloway
Cairnbulg	NK 0365	Aberdeen	Grampian
Cairn Gorm	NJ 0004	Inverness/Aberdeen	Highland/Grampian
Cairnryan	NX 0668	Wigtown	Dumfries & Galloway
Cairnsmore of Fleet	NX 5066	Kirkcudbright	Dumfries & Galloway
Cairnwell	NO 1377	Perth/Aberdeen	Tayside/Grampian
Calgary, Mull	NM 3751	Argyll	Strathclyde
Cambus	NS 8593	Clackmannan	Central
Cameron Reservoir	NO 4711	Fife	Fife
Campsie Fells	NS 6082	Stirling	Central/Strathclyde
Canna	NG 2405	Inverness	Highland
Cape Wrath	NC 2574	Sutherland	Highland
Cardross	NS 3477	Dunbarton	Strathclyde
Carlingwark Loch	NX 7661	Kirkcudbright	Dumfries & Galloway
Carnoustie	NO 5634	Angus	Tayside
Carradale	NR 8138	Argyll	Strathclyde
Carr Craig	NT 1983	Fife	Fife
Carron Valley	NS 6985	Stirling	Central
Carsebreck	NN 8699	Perth	Tayside
Carse of Gowrie	NO 2726	Perth	Tayside
Carsethorn	NX 9959	Kirkcudbright	Dumfries & Galloway
Carsluith	NX 4854	Kirkcudbright	Dumfries & Galloway
Carstairs	NS 9345	Lanark	Strathclyde
Castle Douglas	NX 7662	Kirkcudbright	Dumfries & Galloway

Site	Grid ref.	Old county	Region
Castle Forbes Loch	NJ 6219	Aberdeen	Grampian
Castle Loch, Lochmaben	NY 0881	Dumfries	Dumfries & Galloway
Castle Loch, Mochrum	NX 2853	Wigtown	Dumfries & Galloway
Castle Semple Loch	NS 3658	Renfrew	Strathclyde
Catfirth	HU 4354	Shetland	Shetland
Cheviots	NT 60–92	Roxburgh	Borders
Clo Mor	NC 3073	Sutherland	Highland
Cockburnspath	NT 7770	Berwick	Borders
Cockenzie	NT 4075	E. Lothian	Lothian
Coldingham	NT 9065	Berwick	Borders
Coldstream	NT 8439	Berwick	Borders
Colgrave Sound	HU 5789	Shetland	Shetland
Coll	NM 1957	Argyll	Strathclyde
Collieston	NK 0328	Aberdeen	Grampian
Colonsay	NR 3794	Argyll	Strathclyde
Copinsay	HY 6101	Orkney	Orkney
Corby Loch	NJ 9214	Aberdeen	Grampian
Correen Hills	NJ 5222	Aberdeen	Grampian
Corsemaul	NJ 4040	Aberdeen	Grampian
Corsewall Point	NW 9872	Wigtown	Dumfries & Galloway
Cotehill Loch	NK 0229	Aberdeen	Grampian
Coupar Angus	NO 2139	Perth	Tayside
Cousland	NT 3768	Midlothian	Lothian
Covesea	NJ 1870	Moray	Grampian
Cowal Peninsula	NS 0080	Argyll	Strathclyde
Craigleith	NT 5586	E. Lothian	Lothian
Crail	NO 6107	Fife	Fife
Crammag Head	NX 0834	Wigtown	Dumfries & Galloway
Cree Valley	NX 38–46	Kirkcudbright	Dumfries & Galloway
Crianlarich	NN 3825	Perth	Central
Crieff	NN 8621	Perth	Tayside
Cromarty Firth	NH 6667	Ross & Cromarty	Highland
Crosshill (Ayr)	NS 3206	Ayr	Strathclyde
Culbin Bar / Sands	NH 9763	Moray	Grampian
Culross	NS 9885	Fife	Fife
Dales Voe	HU 4545	Shetland	Shetland
Dalgety Bay	NT 1783	Fife	Fife
Dalmally	NN 1527	Argyll	Strathclyde
Dalmeny	NT 1477	W. Lothian	Lothian
Dalmore	NH 6668	Ross & Cromarty	Highland
Dalry, Castle Douglas	NX 6281	Kirkcudbright	Dumfries & Galloway
Dalswinton	NX 9385	Dumfries	Dumfries & Galloway
Deerness	HY 5606	Orkney	Orkney
Deeside	NO 19–79	Aberdeen	Grampian
Denny	NS 8182	Stirling	Central
Dighty Mouth	NO 4831	Angus	Tayside
Dingwall	NH 5458	Ross & Cromarty	Highland
Doonfoot	NS 3319	Ayr	Strathclyde
Dores	NH 5934	Moray	Grampian
Dornoch Firth	NH 8488	Sutherland	Highland
Doune	NN 7201	Perth	Tayside
Drummond Pond	NN 8418	Perth	Tayside
Drumnadrochit	NH 5029	Inverness	Highland
Drumochter	NN 6376	Perth / Inverness	Tayside / Highland
Duddingston Loch	NT 2972	Midlothian	Lothian
Dumbarton	NS 4075	Dunbarton	Strathclyde
Dun (St Kilda)	NF 1097	Inverness	Western Isles
Dunbar	NT 6878	E. Lothian	Lothian
Dunblane	NN 7801	Perth	Tayside
Duncansby Head	ND 4073	Caithness	Highland
Duncow	NX 9683	Dumfries	Dumfries & Galloway
Dundrennan	NX 7447	Kirkcudbright	Dumfries & Galloway
Dunnet Bay / Head	ND 2076	Caithness	Highland
Dunragit	NX 1557	Wigtown	Dumfries & Galloway
Dunrossness	HU 3815	Shetland	Shetland
Duns	NT 7853	Berwick	Borders
Dun's Dish	NO 6461	Angus	Tayside

Site	Grid ref.	Old county	Region
Dupplin Loch	NO 0320	Perth	Tayside
Durness	NC 4067	Sutherland	Highland
Eaglesham	NS 5751	Renfrew	Strathclyde
Earlshall	NO 4621	Fife	Fife
Earnmouth	NO 1918	Perth	Tayside
East Fenton	NT 5281	E. Lothian	Lothian
Eastpark	NY 0565	Dumfries	Dumfries & Galloway
Echnaloch Bay	ND 4797	Orkney	Orkney
Eday	HY 5634	Orkney	Orkney
Eddleston	NT 2447	Peebles	Borders
Eden Estuary	NO 4818	Fife	Fife
Edrom	NT 8255	Berwick	Borders
Eigg	NM 4786	Inverness	Highland
Elie Bay	NO 4900	Fife	Fife
Ellon	NJ 9530	Aberdeen	Grampian
Endrick Mouth	NS 4489	Dunbarton / Stirling	Strathclyde / Central
Errol	NO 2522	Perth	Tayside
Erskine	NS 4670	Renfrew	Strathclyde
Eskdalemuir	NY 2597	Dumfries	Dumfries & Galloway
Ettrick	NT 2714	Selkirk	Borders
Eyebroughty	NT 4986	E. Lothian	Lothian
Eyemouth	NT 9464	Berwick	Borders
Eynhallow	HY 3529	Orkney	Orkney
Fair Isle	HZ 2172	Shetland	Shetland
Fala Flow	NT 4258	Midlothian	Lothian
Faraid Head	NC 3971	Sutherland	Highland
Fetlar	HU 6291	Shetland	Shetland
Fidra	NT 5186	E. Lothian	Lothian
Fife Ness	NO 6309	Fife	Fife
Findhorn Bay / Bar	NJ 0462	Moray	Grampian
Finstown	HY 3513	Orkney	Orkney
Flanders Moss	NS 6398	Perth	Central
Flannan Isles	NA 7146	Ross & Cromarty	Western Isles
Flotta	ND 3593	Orkney	Orkney
Forest of Ae	NX 9991	Dumfries	Dumfries & Galloway
Forest of Birse	NO 5291	Aberdeen	Grampian
Forfar Loch	NO 4550	Angus	Tayside
Forth Estuary	NT 0085	Fife/Stirling/W. Lothian	Fife/Lothian/Central
Forth Islands	NT 28–68	Mid & E. Lothian	Lothian
Foula	HT 9638	Shetland	Shetland
Fowlsheugh	NO 8880	Kincardine	Grampian
Fraserburgh	NJ 9966	Aberdeen	Grampian
Gadloch	NS 6571	Lanark	Strathclyde
Gairloch	NG 8076	Ross & Cromarty	Highland
Gairsay Sound	HY 4424	Orkney	Orkney
Gareloch	NS 2486	Dunbarton	Strathclyde
Gargunnock Hills	NS 6891	Stirling	Central
Garry Valley (Perth)	NN 7070	Perth	Tayside
Gartfairn	NS 4390	Dunbarton	Central
Gartmorn Dam	NS 9394	Clackmannan	Central
Garvellachs	NM 6511	Argyll	Strathclyde
Garynahine, Lewis	NB 2331	Ross & Cromarty	Western Isles
Gelston	NX 7758	Kirkcudbright	Dumfries & Galloway
Gifford	NT 5368	E. Lothian	Lothian
Gigha	NR 6449	Argyll	Strathclyde
Girdleness	NJ 9705	Aberdeen / Kincardine	Grampian
Girvan Valley	NS 3000	Ayr	Strathclyde
Gladhouse Reservoir	NT 2953	Midlothian	Lothian
Glen Affric	NH 1922	Inverness	Highland
Glen App	NX 0774	Ayr	Strathclyde
Glenbuck Loch (Ayr)	NS 7429	Ayr	Strathclyde
Glencaple	NX 9968	Dumfries	Dumfries & Galloway
Glen Clova	NO 3570	Angus	Tayside
Glencoe	NN 1557	Inverness	Highland
Glen Devon Res.	NN 9204	Perth	Tayside
Glen Dye	NO 6384	Kincardine	Grampian
Glen Falloch	NN 3622	Perth	Central

Site	Grid ref.	Old county	Region
Glen Garry	NH 1300	Inverness	Highland
Glenkens	NX 5887	Kirkcudbright	Dumfries & Galloway
Glen Lyon	NN 5646	Perth	Tayside
Glenmore	NH 9709	Inverness	Highland
Glen Moriston	NH 2411	Inverness	Highland
Glen Nant	NN 0128	Argyll	Strathclyde
Glen Quoich	NH 0206	Aberdeen	Grampian
Glen Tanar	NO 4594	Aberdeen	Grampian
Glen Tilt	NN 9172	Perth	Tayside
Glen Trool	NX 4080	Kirkcudbright	Dumfries & Galloway
Golspie	NH 8399	Sutherland	Highland
Gosford Bay	NT 4478	E. Lothian	Lothian
Gourdon	NO 8270	Kincardine	Grampian
Grangemouth	NS 9281	Stirling	Central
Grantown-on-Spey	NJ 0327	Inverness	Highland
Great Bernera, Lewis	NB 1635	Ross & Cromarty	Western Isles
Great Cumbrae	NS 1656	Bute	Strathclyde
Great Glen	NN 12–NH 53	Inverness	Highland
Green Holms	HY 5227	Orkney	Orkney
Greenlaw	NT 7145	Berwick	Borders
Grogarry, S. Uist	NF 7739	Inverness	Western Isles
Grutness	HU 4009	Shetland	Shetland
Gullane Bay	NT 4783	E. Lothian	Lothian
Haddo Lochs	NJ 8634	Aberdeen	Grampian
Halligarth, Unst	HP 6209	Shetland	Shetland
Hallival, Rhum	NM 3996	Inverness	Highland
Hamilton	NS 7225	Lanark	Strathclyde
Handa	NC 1348	Sutherland	Highland
Harperrig	NT 0961	Midlothian	Lothian
Hascosay	HU 5592	Shetland	Shetland
Hawick	NT 5014	Roxburgh	Borders
Hermaness, Unst	HP 6018	Shetland	Shetland
Hightae Loch	NY 0979	Dumfries	Dumfries & Galloway
Hillwell	HU 3714	Shetland	Shetland
Hirsel	NT 8240	Berwick	Borders
Hirta (St Kilda)	NF 0999	Inverness	Western Isles
Holm of Huip	HY 6331	Orkney	Orkney
Holm of Papa Westray	HY 5253	Orkney	Orkney
Hopeman	NJ 1469	Moray	Grampian
Horse Island	NS 2142	Ayr	Strathclyde
Horse of Burravoe, Yell	HU 5381	Shetland	Shetland
Hoselaw Loch	NT 8031	Roxburgh	Borders
Hound Point	NT 1579	W. Lothian	Lothian
Howmore Estuary, S. Uist	NF 7536	Inverness	Western Isles
Hoy	ND 2596	Orkney	Orkney
Hule Moss	NT 7149	Berwick	Borders
Hunterston	NS 1952	Ayr	Strathclyde
Inchcailloch	NS 4190	Stirling	Central
Inchgarvie	NT 1379	W. Lothian	Lothian
Inchkeith	NT 2982	Fife	Fife
Inchmarnock	NS 0159	Bute	Strathclyde
Inchmickery	NT 2080	Midlothian	Lothian
Inchmoan	NS 3791	Dunbarton	Strathclyde
Insh Marshes	NH 8002	Inverness	Highland
Inverewe	NG 8682	Ross & Cromarty	Highland
Invergordon	NH 7168	Ross & Cromarty	Highland
Invergowrie Bay	NO 3430	Perth	Tayside
Iona	NM 2723	Argyll	Strathclyde
Irvine Flats	NS 3038	Ayr	Strathclyde
Isay, Skye	NG 2157	Inverness	Highland
Isle of May	NT 6599	Fife	Fife
Islesteps	NX 9772	Dumfries	Dumfries & Galloway
Jura	NR 5683	Argyll	Strathclyde
Kennet Pans	NS 9188	Clackmann. / Stirling	Central
Kergord	HU 3954	Shetland	Shetland
Kilconquhar Loch	NO 4801	Fife	Fife
Kilcreggan	NS 2380	Dunbarton	Strathclyde

Site	Grid ref.	Old county	Region
Kilmacolm	NS 3569	Renfrew	Strathclyde
Kinclaven	NO 1538	Perth	Tayside
Kincraig	NH 8305	Inverness	Highland
Kinghorn	NT 2686	Fife	Fife
Kingoodie	NO 3329	Perth	Tayside
Kingussie	NH 7500	Inverness	Highland
Kinlochbervie	NC 2156	Sutherland	Highland
Kinloch Rannoch	NN 6658	Perth	Tayside
Kinloss	NJ 0661	Moray	Grampian
Kinmount	NY 1368	Dumfries	Dumfries & Galloway
Kinnaber	NO 7261	Angus	Tayside
Kinneil	NS 9881	W. Lothian	Lothian
Kintyre	NR 7540	Argyll	Strathclyde
Kirkcaldy	NT 2791	Fife	Fife
Kirkconnell Flow	NX 9769	Kirkcudbright	Dumfries & Galloway
Kirkconnell Merse	NX 9868	Kirkcudbright	Dumfries & Galloway
Kirkcudbright Bay	NX 6747	Kirkcudbright	Dumfries & Galloway
Kirkwall	HY 4410	Orkney	Orkney
Kirroughtree	NX 4473	Kirkcudbright	Dumfries & Galloway
Kyleakin	NG 7526	Inverness	Highland
Kyle of Sutherland	NH 5795	Sutherland	Highland
Lady Isle	NS 2729	Ayr	Strathclyde
Lairg	NC 5806	Sutherland	Highland
Lairig Ghru	NH 9603	Inverness / Aberdeen	Highland / Grampian
Lake of Menteith	NS 5699	Perth	Central
Lammermuir Hills	NT 5863	E. Lothian / Berwick	Lothian / Borders
Lanark	NS 8843	Lanark	Strathclyde
Langholm	NY 3684	Dumfries	Dumfries & Galloway
Largo Bay	NO 4201	Fife	Fife
Lauriston	NX 6864	Kirkcudbright	Dumfries & Galloway
Leadhills	NS 8814	Lanark	Strathclyde
Leith Docks	NT 2676	Midlothian	Lothian
Lenzie	NS 6571	Dunbarton	Strathclyde
Lerwick	HU 4741	Shetland	Shetland
Leswalt	NX 0263	Wigtown	Dumfries & Galloway
Leuchars	NO 4521	Fife	Fife
Leven	NO 3700	Fife	Fife
Linlithgow Loch	NS 9977	W. Lothian	Lothian
Linn of Dee	NO 0589	Aberdeen	Grampian
Little Cumbrae	NS 1451	Bute	Strathclyde
Littleferry	NH 8095	Sutherland	Highland
Little Gruinard Bay	NG 9486	Ross & Cromarty	Highland
Little Linga	HY 6030	Orkney	Orkney
Lochaber	NN 1492	Inverness	Highland
Loch Achray	NN 5106	Perth	Central
Locharbriggs	NX 9980	Dumfries	Dumfries & Galloway
Loch Arkaig	NN 0891	Inverness	Highland
Lochar Moss	NY 0371	Dumfries	Dumfries & Galloway
Loch Asta	HU 4141	Shetland	Shetland
Loch Bee, S. Uist	NF 7743	Inverness	Western Isles
Loch Brora	NC 8508	Sutherland	Highland
Loch Brow	HU 3815	Shetland	Shetland
Loch Butterstone	NO 0646	Perth	Tayside
Loch Carron	NG 8735	Ross & Cromarty	Highland
Loch Connell	NX 0168	Wigtown	Dumfries & Galloway
Loch Davan	NJ 4400	Aberdeen	Grampian
Loch Doon	NX 4997	Ayr	Strathclyde
Loch Druidibeg, S. Uist	NF 7937	Inverness	Western Isles
Loch Earn	NN 6423	Perth	Tayside
Loch Eye	NH 8379	Ross & Cromarty	Highland
Loch Faskally	NN 9258	Perth	Tayside
Loch Fitty	NT 1291	Fife	Fife
Loch Fleet	NH 7896	Sutherland	Highland
Loch Garten	NH 9718	Inverness	Highland
Loch Helly	NT 2092	Fife	Fife
Lochgilphead	NR 8687	Argyll	Strathclyde
Loch Gruinart, Islay	NR 2870	Argyll	Strathclyde

Site	Grid ref.	Old county	Region
Loch Hallan, S. Uist	NF 7422	Inverness	Western Isles
Loch of Harray	HY 2915	Orkney	Orkney
Lochs Heilen & St Johns	ND 2070	Caithness	Highland
Loch Hillwell	HU 3714	Shetland	Shetland
Loch Indaal, Islay	NR 2758	Argyll	Strathclyde
Loch Insh	NH 8304	Inverness	Highland
Loch Ken	NX 6474	Kirkcudbright	Dumfries & Galloway
Loch Kinord	NO 4499	Aberdeen	Grampian
Loch of Kinnordy	NO 3654	Angus	Tayside
Loch Laggan	NN 4886	Inverness	Highland
Loch Leven	NO 1401	Kinross	Tayside
Loch of Lindores	NO 2616	Fife	Fife
Loch Linnhe	NM 9354	Argyll / Inverness	Strathclyde / Highland
Loch of Lintrathen	NO 2754	Angus	Tayside
Loch Lomond	NS 3598	Dunbarton / Stirling	Strathclyde / Central
Loch of the Lowes (Perth)	NO 0544	Perth	Tayside
Loch Maddy, N. Uist	NF 9368	Inverness	Western Isles
Loch Mallachie	NH 9617	Inverness	Highland
Loch Maree	NG 9570	Ross & Cromarty	Highland
Loch Morar	NM 7790	Inverness	Highland
Loch Muick	NO 2882	Aberdeen	Grampian
Lochnagar	NO 2485	Aberdeen	Grampian
Loch Ore	NT 1695	Fife	Fife
Loch Rangag	ND 1741	Caithness	Highland
Loch Rannoch	NN 5857	Perth	Tayside
Loch Riddon	NS 0076	Argyll	Strathclyde
Loch Ryan	NX 0565	Wigtown	Dumfries & Galloway
Loch Scarmclate	ND 1859	Caithness	Highland
Loch Shin	NC 4816	Sutherland	Highland
Loch of Skene	NJ 7807	Aberdeen	Grampian
Loch Sligachan, Skye	NG 5132	Inverness	Highland
Loch Soulseat	NX 1058	Wigtown	Dumfries & Galloway
Loch Spiggie	HU 3716	Shetland	Shetland
Loch Spynie	NJ 2366	Moray	Grampian
Loch of Stenness	HY 2813	Orkney	Orkney
Loch Stiapavat, Lewis	NB 5264	Ross & Cromarty	Western Isles
Loch of Strathbeg	NK 0758	Aberdeen	Grampian
Loch Tarbert	NR 5481	Argyll	Strathclyde
Loch Tay	NN 6838	Perth	Tayside
Loch Tingwall	HU 4142	Shetland	Shetland
Loch an Tomain, N. Uist	NF 9261	Inverness	Western Isles
Loch Tummel	NN 8259	Perth	Tayside
Loch Ussie	NH 5057	Ross & Cromarty	Highland
Loch Watten	ND 2256	Caithness	Highland
Lochwinnoch	NS 3558	Renfrew	Strathclyde
Lomond Hills	NO 2106	Fife / Kinross	Fife / Tayside
Longannet	NS 9485	Fife	Fife
Longman Bay	NH 6846	Inverness	Highland
Longniddry	NT 4476	E. Lothian	Lothian
Lossiemouth	NJ 2370	Moray	Grampian
Lowther Hills	NS 8910	Lanark / Dumfries	Strathcl. / Dumf. & Gall.
Luce Bay	NX 2244	Wigtown	Dumfries & Galloway
Luggiebank	NS 7672	Lanark / Dunbarton	Strathclyde
Lunan Bay	NO 7051	Angus	Tayside
Luskentyre, Harris	NG 0699	Inverness	Western Isles
Lybster	ND 2535	Caithness	Highland
Lyness, Hoy	ND 3094	Orkney	Orkney
Machrihanish	NR 6220	Argyll	Strathclyde
Mar Forest	NO 0292	Aberdeen	Grampian
Marwick Head	HY 2225	Orkney	Orkney
Mellerstain	NT 6439	Berwick	Borders
Melvich	NC 8864	Sutherland	Highland
Mersehead Sands	NX 9153	Dumfries	Dumfries & Galloway
Methil Docks	NT 3699	Fife	Fife
Mey	ND 2872	Caithness	Highland
Mingulay	NL 5683	Inverness	Western Isles
Mochrum Loch	NX 3053	Wigtown	Dumfries & Galloway

Site	Grid ref.	Old county	Region
Monach Isles	NF 6262	Inverness	Western Isles
Monadhliaths (hills)	NH 6710	Inverness	Highland
Montrose Basin	NO 6957	Angus	Tayside
Moorfoots	NT 3452	Midlothian / Peebles	Lothian / Borders
Moray Basin	NO 0080	Suth/R&C/Nairn/Moray/Inv.	Grampian/Highland
Moray Firth	NH 9467	Moray / Ross & Crom.	Grampian / Highland
Morrich More	NJ 8384	Ross & Cromarty	Highland
Morton Lochs	NO 4627	Fife	Fife
Mossbank	HU 4575	Shetland	Shetland
Moss of Cree	NX 4360	Kirkcudbright	Dumfries & Galloway
Mount Battock	NO 5484	Abdn / Kinc / Angus	Grampian / Tayside
Mousa	HU 4624	Shetland	Shetland
Muchalls	NO 9091	Kincardine	Grampian
Muck	NM 4179	Inverness	Highland
Muckle Skerry	ND 4678	Orkney	Orkney
Mull of Galloway	NX 1530	Wigtown	Dumfries & Galloway
Munlochy Bay	NH 6752	Ross & Cromarty	Highland
Murcar	NJ 9412	Aberdeen	Grampian
Musselburgh	NT 3472	Midlothian	Lothian
Nairn Bar	NH 9260	Nairn	Highland
Neidpath	NT 2340	Peebles	Borders
Nethy Bridge	NJ 0020	Inverness	Highland
Newburgh (Abdn)	NJ 9925	Aberdeen	Grampian
New Cumnock	NS 6113	Ayr	Strathclyde
New Luce	NX 1764	Wigtown	Dumfries & Galloway
Newton, N. Uist	NF 8977	Inverness	Western Isles
Nigg Bay	NH 7771	Ross & Cromarty	Highland
Nithsdale	NX 8991	Dumfries	Dumfries & Galloway
North Berwick	NT 5485	E. Lothian	Lothian
North Esk (Angus)	NO 6060	Angus / Kincardine	Tayside / Grampian
North Kessock	NH 6548	Ross & Cromarty	Highland
North Roe	HU 3487	Shetland	Shetland
North Rona	HW 8233	Ross & Cromarty	Western Isles
North Ronaldsay	HY 7654	Orkney	Orkney
North Solway	NY 1060	Dumfries / Kirkcud.	Dumfries & Galloway
North Uist	NF 8370	Inverness	Western Isles
Noss	HU 5539	Shetland	Shetland
Noss Head	ND 3854	Caithness	Highland
Noup of Westray	HY 3950	Orkney	Orkney
Ochils (hills)	NN 90–NO 10	Perth / Kinross / Clack.	Tayside / Central
Oronsay	NR 3588	Argyll	Strathclyde
Out Skerries	HU 6771	Shetland	Shetland
Paisley Glen	NS 4860	Renfrew	Strathclyde
Papa Stour	HU 1660	Shetland	Shetland
Papa Westray	HY 4952	Orkney	Orkney
Pentland Firth / Skerries	ND 4777	Orkney	Orkney
Pentland Hills	NT 1459	Midlothian	Lothian
Pladda, Arran	NS 0219	Bute	Strathclyde
Poolewe	NG 8580	Ross & Cromarty	Highland
Portmore Loch	NT 2650	Peebles	Borders
Portobello	NT 3073	Midlothian	Lothian
Port o' Warren	NX 8853	Kirkcudbright	Dumfries & Galloway
Port Seton	NT 4076	E. Lothian	Lothian
Possil Marsh	NS 5870	Lanark	Strathclyde
Powfoot	NY 1465	Dumfries	Dumfries & Galloway
Priest Island	NB 9202	Ross & Cromarty	Highland
Priestside	NY 1164	Dumfries	Dumfries & Galloway
Queensferry	NT 1278	W. Lothian	Lothian
Quendale	HU 3713	Shetland	Shetland
Raasay	NG 5436	Inverness	Highland
Rannoch Moor	NN 3852	Perth / Argyll	Tayside / Highland
Rattray Head	NK 1057	Aberdeen	Grampian
Rescobie Loch	NO 5252	Angus	Tayside
Rhinns of Galloway	NX 05–36	Wigtown	Dumfries & Galloway
Rhinns of Islay (lighthouse)	NR 1651	Argyll	Strathclyde
Rhunahaorine	NR 7084	Argyll	Strathclyde
River Kelvin	NS 6775	Stirl / Dunbart / Lanark	Strathclyde

Site	Grid ref.	Old county	Region
River Teviot	NT 6424	Roxburgh	Borders
River White Cart	NS 5060	Renfrew	Strathclyde
Rosehearty	NJ 9367	Aberdeen	Grampian
Rosyth	NT 1183	Fife	Fife
Rousay	HY 4030	Orkney	Orkney
Rowbank Reservoir	NS 3957	Renfrew	Strathclyde
Rumster Forest	ND 2038	Caithness	Highland
Ruthwell	NY 1067	Dumfries	Dumfries & Galloway
St Abbs / St Abb's Head	NT 9169	Berwick	Borders
St Andrews Bay	NO 5318	Fife	Fife
St Cyrus	NO 7464	Kincardine	Grampian
St Fergus	NK 0951	Aberdeen	Grampian
St Kilda (Hirta, Soay, Dun.)	NF 0999	Inverness	Western Isles
St Serf's Island	NO 1500	Kinross	Tayside
Salisbury Crags	NT 2773	Midlothian	Lothian
Sanda	NR 7204	Argyll	Strathclyde
Sanday	HY 6841	Orkney	Orkney
Sands of Forvie	NK 0126	Aberdeen	Grampian
Sandside (Caith)	NC 9666	Caithness	Highland
Sandside Bay, Deerness	HY 5906	Orkney	Orkney
Sandwater	HU 4154	Shetland	Shetland
Saxa Vord, Unst	HP 6316	Shetland	Shetland
Scalloway	HU 4039	Shetland	Shetland
Scapa Flow	ND 3899	Orkney	Orkney
Scar (Scare) Rocks	NX 2634	Wigtown	Dumfries & Galloway
Scatness	HU 3809	Shetland	Shetland
Scone	NO 1126	Perth	Tayside
Scourie	NC 1544	Sutherland	Highland
Seafield	NT 3074	Midlothian	Lothian
Seamill	NS 2047	Ayr	Strathclyde
Shapinsay	HY 5018	Orkney	Orkney
Shewalton	NS 3336	Ayr	Strathclyde
Shiant Islands	NG 4298	Ross & Cromarty	Western Isles
Shin Forest	NC 5601	Sutherland	Highland
Sinclair's Bay	ND 3656	Caithness	Highland
Skinflats	NS 9083	Stirling	Central
Slains Lochs	NK 0230	Aberdeen	Grampian
Small Isles	NR 5468	Inverness	Highland
Smoo Cave	NC 4267	Sutherland	Highland
Soay (St Kilda)	NA 0601	Inverness	Western Isles
Sound of Barra	NF 7510	Inverness	Western Isles
Sound of Gigha	NR 6747	Argyll	Strathclyde
Sound of Jura	NR 6480	Argyll	Strathclyde
Sound of Taransay (Harris)	NB 0500	Inverness	Western Isles
Southerness Point	NX 9754	Kirkcudbright	Dumfries & Galloway
South Ronaldsay	ND 4590	Orkney	Orkney
South Uist	NF 7829	Inverness	Western Isles
South Walls	ND 3189	Orkney	Orkney
Spey Bay / Speymouth	NJ 3866	Moray	Grampian
Spinningdale	NH 6789	Sutherland	Highland
Stanley Dam	NS 4661	Renfrew	Strathclyde
Stenhouse Reservoir	NT 2187	Fife	Fife
Stonehaven	NO 8685	Kincardine	Grampian
Stormont Loch	NO 1843	Perth	Tayside
Stornoway Woods (Lewis)	NB 4233	Ross & Cromarty	Western Isles
Straloch	NJ 8621	Aberdeen	Grampian
Stranraer Lochs	NX 1060	Wigtown	Dumfries & Galloway
Strath Ardle	NO 1055	Perth	Tayside
Strathbraan	NN 9739	Perth	Tayside
Strathearn	NN 8818	Perth	Tayside
Strathfarrar	NH 2538	Inverness	Highland
Strathspey	NJ 1536	Inverness	Highland
Stroma	ND 3577	Caithness	Highland
Strome Islands	NG 8030	Ross & Cromarty	Highland
Stromness	HY 2509	Orkney	Orkney
Stronsay	HY 6525	Orkney	Orkney
Sula Sgeir	HW 6230	Ross & Cromarty	Highland

Site	*Grid ref.*	*Old county*	*Region*
Sule Skerry	HX 6224	Orkney	Orkney
Sullom Voe	HU 3773	Shetland	Shetland
Sumburgh	HU 4009	Shetland	Shetland
Summer Isles	NB 9706	Ross & Cromarty	Highland
Summerston	NS 5772	Lanark	Strathclyde
Switha Holm	ND 3792	Orkney	Orkney
Swona	ND 3884	Orkney	Orkney
Tain Bay	NH 7883	Ross & Cromarty	Highland
Tantallon	NT 5985	E. Lothian	Lothian
Tarbat Ness	NH 9487	Ross & Cromarty	Highland
Tarradale	NH 5548	Ross & Cromarty	Highland
Tayinloan	NR 6945	Argyll	Strathclyde
Taymouth Castle	NN 7846	Perth	Tayside
The Lamb	NT 5386	E. Lothian	Lothian
Tentsmuir / Forest	NO 4825	Fife	Fife
Thornhill (Dumf.)	NX 8795	Dumfries	Dumfries & Galloway
Threave	NX 7362	Kirkcudbright	Dumfries & Galloway
Thurso	ND 1168	Caithness	Highland
Tiree	NM 0045	Iverness	Highland
Tomintoul	NJ 1618	Banff	Grampian
Tranent	NT 4072	E. Lothian	Lothian
Treshnish Isles	NM 2741	Argyll	Strathclyde
Tresta	HU 3651	Shetland	Shetland
Troon	NS 3230	Ayr	Strathclyde
Trossachs	NN 4907	Perth	Central
Troup Head	NJ 8267	Banff	Grampian
Tullibody Island	NS 8692	Clackmannan	Central
Turnberry	NS 2005	Ayr	Strathclyde
Turriff	NJ 7249	Aberdeen	Grampian
Tweedsmuir Hills	NT 1020	Peebles	Borders
Tyninghame Estuary	NT 6179	E. Lothian	Lothian
Udale Bay	NH 7166	Ross & Cromarty	Highland
Unst	HP 6009	Shetland	Shetland
Vaila	HU 2346	Shetland	Shetland
Vane Crags / Farm	NT 1699	Kinross	Tayside
Virkie	HU 3911	Shetland	Shetland
Voe	HU 4063	Shetland	Shetland
Vorran (Verran) Island, S. Uist	NF 7335	Inverness	Highland
Walton Dam	NS 6686	Stirling	Central
West Freugh	NX 1154	Wigtown	Dumfries & Galloway
Westhaven	NO 5734	Angus	Tayside
Westray	HY 4546	Orkney	Orkney
West Water Reservoir	NT 1152	Peebles	Borders
Whalsay	HU 5663	Shetland	Shetland
Whiteness head	NH 8058	Nairn	Highland
Whiteness Voe	HU 3943	Shetland	Shetland
Wigtown Merse	NX 4454	Wigtown	Dumfries & Galloway
Woodhall	NX 6867	Kirkcudbright	Dumfries & Galloway
Wood of Cree	NX 3870	Kirkcudbright	Dumfries & Galloway
Yarrow	NT 3527	Selkirk	Borders
Yell / Yell Sound	HU 4087	Shetland	Shetland
Yetholm Loch	NT 8028	Roxburgh	Borders
Ythan Estuary	NJ 0027	Aberdeen	Grampian

APPENDIX 2

(a) List of local bird reports, with date of first year covered. (Most are available for reference in the SOC's Waterston Library).

Angus and South Kincardine Bird Report (in Angus Wildlife Review) – 1974
Argyll Bird Report – 1980
Arran Bird Report & Checklist – 1978
Ayrshire Bird Report – 1976
Birds of the Outer Hebrides (in Hebridean Naturalist) – 1981
Borders Bird Report – 1979
Caerlaverock Bird Report – 1979
Caithness Bird Report – 1975 (local circulation only until 1984)
Clyde Area Bird Report (initially covering Dunbarton, W Stirling, Renfrew, Lanark, Ayr, Arran & Bute; now only the first four) – 1973
Fair Isle Bird Observatory Report – 1948
Fife & Kinross Bird Report – 1980
Forth Area Bird Report (covering Clacks, Stirling and SW Perth – in Forth Naturalist & Historian) – 1980

Highland Bird Report (Ross & Inverness) – 1983
Isle of May Bird Observatory & Field Station Report – 1935 (1935–38, 1946–50 and 1953–56 in Scot Nat; 1951–52 in Fair Isle BO Report; 1957 in Edin Bird Bull; since 1958 in Scot Birds. No reports 1939–45)
Loch Lomond Bird Report – 1972
Lothian Bird Report – 1979
North-East Scotland Bird Report – 1974
Orkney Bird Report – 1974
Shetland Bird Report – 1969
Stirling & Clackmannan Bird Report (in Forth Naturalist & Historian) – 1974
Perthshire Bird Report – 1974
West Dunbartonshire Bird Report (in West Dunbartonshire Naturalist Report) – 1973

(b) Local ringing groups active in 1985; most produce annual reports. (Contact details can be obtained from the SOC or BTO).

Clyde Ringing Group
Grampian Ringing Group
Highland Ringing Group
Lothian Ringing Group (formerly Edinburgh Ringing Group)

North Solway Ringing Group
Orkney Ringing Group
Shetland Ringing Group
Tay Ringing Group

APPENDIX 3: *Island checklists*

Those listed are the most recently published; most give references to earlier accounts. The SOC's Waterston Library also holds a number of unpublished reports; the most comprehensive of these are included below in square brackets.

SHETLAND
GENERAL: *A Guide to Shetland Birds* R. J. Tulloch & F. Hunter (1979) The Shetland Times.
   Table 29. Birds, in *The Natural History of Shetland*, R. J. Berry & J. L. Johnston (1980) Collins.
   A bibliographically annotated checklist of the birds of Shetland. N. Hamilton (1981) ITE.
FAIR ISLE: *Fair Isle and its Birds* K. Williamson (1965) Oliver & Boyd.
   The Birds of Fair Isle J. F. Holloway & R. H. F. Thorne (1981) The Shetland Times.
FOULA: *The Birds of Foula* R. W. Furness (1983) The Brathay Hall Trust.
UNST: The moorland birds of Unst, Shetland. K. Williamson (1951) *Scot. Nat.* 63: 37–44.
   Some breeding birds of Unst. G. Bundy (1978) *Scot. Birds* 10: 38–50.
YELL: Some notes on the birds of Yell. C. Inkster (1951) *Bird Notes* 24: 227–231, 270–273.
   Breeding seabirds on the Yell Sound Islands, Shetland. J. A. Fowler *et al.* (1984) *Scot. Birds* 13: 77–83.
OUT SKERRIES: Birds on Out Skerries, Shetland, 1966. R. J. Tulloch (1967) *Scot. Birds* 4: 467–480.
WHALSAY: Breeding birds of Whalsay, Shetland. J. H. Simpson (1968) *Scot. Birds* 5: 170–171.

ORKNEY
GENERAL: *The Birds of Orkney*. C. J. Booth *et al.* (1984) The Orkney Press.
SULE SKERRY: An Ornithological Survey of Sule Skerry, 1975. A. C. Blackburn & D. Budworth (1976) *Seabird Report* 5: 27–33.
   A visit to Stack Skerry and Sule Skerry. D. M. Stark (1967) *Scot. Birds* 4: 548–553.

OUTER HEBRIDES
GENERAL: *Birds of the Outer Hebrides* P. Cunningham (1983) The Melven Press.
   Birds of the Outer Hebrides, in *The Natural Environment of the Outer Hebrides* Ed. J. Morton Boyd (1979) *Proc. R. Soc. Edinb.* 77B: 407–475.
ROCKALL: A midsummer visit to Rockall. R. M. Lockley & S. Marchant (1951) *Brit. Birds* 44: 373–383.
ST KILDA: *Birds of St Kilda*. M. P. Harris & S. Murray (1978) ITE.
   The birds of Boreray, St Kilda. N. Duncan *et al.* (1982) *Seabird Report* 6: 18–25.
NORTH RONA & SULA SGEIR: The birds of North Rona in 1958 with notes on Sula Sgeir. T. B. Bagenal & D. E. Baird (1959) *Bird Study* 6: 153–174.
   The breeding birds of North Rona. M. J. H. Robson (1968) *Scot. Birds* 5: 126–155.
   Report on BOU supported expedition to North Rona and

Sula Sgeir, 1972. P. G. H. Evans (1973) *Ibis* 115: 476–478.

FLANNAN IS.: A description of the Flannan Isles and their birds. A. Anderson *et al.* (1961) *Bird Study* 8: 71–88.

MONACH IS.: The birds of the Monach Isles, Outer Hebrides. I. R. Hepburn *et al* (1977) *Bird Study* 24: 25–43.

SKERRYVORE: Some notes from Skerryvore lighthouse. W. A. Cameron (1962) *Scot. Birds* 2: 74–76.

BERNERAY, ETC (South of Barra): Notes on the birds of Berneray, Mingulay and Pabbay. A. W. Diamond *et al.* (1965) *Scot. Birds* 3: 397–404.

BEARASAY (L. Roag, Lewis): Notes on the birds of Bearasay, Lewis. M. Robson & P. Wills (1963) *Scot. Birds* 2: 410–414.

GASKER (off N Harris): Notes on the islet of Gasker. R. Atkinson & B. Roberts (1952) *Scot. Nat.* 64: 129–137.

SHIANT IS.: Birds of the Shiant Islands, Outer Hebrides. M. de L. Brooke (1973) *Bird Study* 20: 197–206.

INNER HEBRIDES

GENERAL: The birds of the Inner Hebrides, in *The Natural Environment of the Inner Hebrides*. Ed. J. Morton Boyd. (1983) *Proc. R. Soc. Edinb.* 83B: 449–504.

Birds of the Small Isles – Muck, Eigg, Rhum and Canna. P. R. Evans & W. U. Flower (1967) *Scot. Birds* 4: 404–445.

A Guide to the Birds of Mid-Argyll, Kintyre and Cowal. C. McLaren & G. Murray (1985) Droineach Press.

ISLAY: *Birds in Islay.* C. G. Booth (1981) Argyll Reproductions Ltd.

COLONSAY: [The birds of Colonsay & Oronsay, 1970–83. J. & P. M. Clarke (1983)]

COLL: Breeding birds of Coll 1969–70. J. G. Blatchford (1971) *Scot. Birds* 6: 271–274.

TIREE: Breeding birds of Tiree, 1969. B. Gillam & G. R. Jacobs (1971) *Scot. Birds* 6: 267–271.

RHUM: *Checklist of birds of Rhum.* J. A. Love (1984) NCC.

CANNA: The birds of Canna. R. Carrick & G. Waterston (1939) *Scot. Nat.* 1939: 5–22.

[The birds of Canna. R. L. Swann (1983)]

EIGG: [Notes on the birds of the Island of Eigg. W. J. Edwards (1977)]

SKYE: Notes on the birds of Skye. I. M. Murray (1954) *Glasgow Bird Bull.* 3: 6–13.

Flora and fauna, Chapter 3 in *Skye.* F. C. Sillar & R. Meyler (1973) David & Charles.

[Raasay bird report for summer 1982. ed. R. H. Hogg (1983)]

MULL & IONA: *The birds of Iona and Mull.* H. D. Graham (1890) David Douglas.

JURA: *Birds in Jura.* C. G. Booth (1976) MacKenzie Sproat.

CLYDE ISLANDS

GENERAL: *A Regional Checklist of Clyde Birds.* J. A. Gibson (1981) Scottish Wildlife Trust (Glasgow).

ARRAN: *The Birds of Arran* (Annual Report and Checklist) Ed. M. H. Dunn (1984) Isle of Arran Natural History Society.

The breeding birds of Pladda. J. A. Gibson (1954) *Glasgow Bird Bull.* 3: 55–56.

BUTE & CUMBRAES: Notes on the birds of the island of Bute 1927–1980. J. A. Gibson *et al* (1980) *Trans. Buteshire Nat. Hist. Soc.* 21: 69–92.

Notes on the birds of Great Cumbrae. J. A. Gibson (1957) *Glasgow Bird Bull.* 6: 5–6.

The breeding birds of the Little Cumbrae. J. A. Gibson (1969) *Trans. Buteshire Nat. Hist. Soc.* 17: 109–114.

The breeding birds of Inchmarnock. J. A. Gibson & C. J. Gordon (1953) *Glasgow Bird Bull.* 2: 54–57.

SMALL ISLANDS: The breeding birds of the small Clyde islands. J. A. Gibson (1958) *Glasgow Bird Bull.* 7: 99–116.

Recent notes on the breeding birds of some small Buteshire islands. J. A. Gibson (1969) *Trans. Buteshire Nat. Hist. Soc.* 17: 119–123.

SANDA: An ornithological survey of the Sanda Islands, Kintyre in 1980. E. J. Maguire (1981) NCC.

AILSA CRAIG: The breeding distribution, population and history of the birds of Ailsa Craig. J. A. Gibson (1951, 1952) *Scot. Nat.* 1951: 73–100, 159–177; 1952: 61.

OTHER WEST AND NORTH COAST ISLANDS

SOLWAY: Birds of the Scar Rocks – the Wigtownshire gannetry. J. G. Young (1968) *Scot. Birds* 5: 204–208.

ARGYLL: Natural history notes from the Isle of Gigha. L. J. Rintoul & E. V. Baxter (1950) *Scot. Nat.* 62: 93–97.

Birds of the Garvelloch Isles. M. H. Dunn *et al.* (1954) *Glasgow Nat.* 1953: 142–145.

SUTHERLAND: Handa Bird Reserve 1962. G. Waterston (1963) *Scot. Birds* 2: 360–363.

Notes on Eilean Bulgach. J. Fisher & K. Piercy (1950) *Scot. Nat.* 62: 26–30.

Notes from Island Roan. I. R. Downhill (1963, 1965) *Scot. Birds* 2: 351–357; 3: 404–405.

FORTH ISLANDS

ISLE OF MAY: *The Isle of May.* W. J. Eggeling (2nd ed. 1985, includes *Scot. Birds* Special Supplement 1974) Lorien Press.

BASS ROCK: Notes from the Bass Rock. J. B. Nelson (1962, 1963) *Scot. Birds* 2: 76–78, 357–360.

Forth Island bird counts in 1984. R. W. J. Smith (1985) *Edin. Nat. Hist. Soc. J.* 1984: 31–32.

[Inchkeith: Breeding birds at Inchkeith 1957–1960. W. A. Cameron (1960)]

APPENDIX 4: *Wetlands provisionally listed by IWRB as being of international importance for waterfowl (Based on Scott 1980)*

1 North Fetlar, Shetland – breeding seabirds and waders
2 Papa Stour, Shetland – breeding terns
3 Foula, Shetland – breeding seabirds
4 West Westray and North Hill, Papa Westray, Orkney – breeding seabirds
5 Lochs of Harray, Stenness & Skaill, Orkney – wintering wildfowl
6 Pentland Skerries, Orkney – breeding terns
7 Shiant Is, Western Isles – wintering Barnacle Geese
8 Islands in Sound of Harris, Western Isles – wintering Barnacle Geese
9 Monach Is, Western Isles – wintering Barnacle Geese
10 Lochs Hallan & Kilpheder, S Uist, Western Isles – wintering White-fronted Geese
11 Islands in Sound of Barra, Western Isles – wintering Barnacle Geese
12 Rabbit Islands, Highland – wintering Barnacle Geese
13 Caithness Lochs, including Calder, Brubster, Heilen & Meadie, Highland – wintering wildfowl
14 Lower Dornoch Firth including Morrich More, Highland – passage and wintering wildfowl
15 Loch Eye, Highland – wintering wildfowl
16 Cromarty Firth, Highland – passage and wintering wildfowl and waders
17 Black Isle, Highland – feeding area for Greylag Geese
18 Beauly Firth – wintering wildfowl
19 Moray Firth including Findhorn Bay, Highland & Grampian – passage and wintering wildfowl and waders
20 Loch Insh and Fens, Highland – passage and wintering Whooper Swans
21 Isay, Skye, Highland – wintering Barnacle Geese
22 Loch Sheil and Kentra Moss, Highland – wintering White-fronted Geese
23 Tiree and Coll, Strathclyde – wintering White-fronted and Barnacle Geese
24 Treshnish Islands, Strathclyde – wintering Barnacle Geese
25 Islay, especially Lochs Gruinart & Indaal, Strathclyde – wintering White-fronted and Barnacle Geese
26 Dumbuck Foreshore – Pillar Bank, Inner Clyde, Strathclyde – passage and wintering wildfowl and waders
27 Lochlyoch & Cleuch Reservoirs, Strathclyde – roost for Pink-footed Geese
28 Haughs of Clyde: Carnwath and Libberton, Strathclyde – feeding area for Pink-footed Geese
29 Rhunahaorine, Strathclyde – wintering White-fronted Geese
30 Machrihanish & Tangy Loch, Strathclyde – wintering White-fronted Geese

31 Loch Spynie, Grampian – wintering Greylag Geese
32 Dee Valley & Loch Skene, Grampian – wintering Greylag Geese
33 Loch of Strathbeg, Grampian – passage and wintering wildfowl
34 Ythan Estuary & Sands of Forvie, Grampian – breeding terns, passage and wintering wildfowl
35 Montrose Basin, Tayside – passage and wintering waders
36 Forfar Lochs: Rescobie & Balgavies, Tayside – wintering Greylag & Pink-footed Geese
37 Loch of Kinnordy, Tayside – wintering Greylag Geese
38 Loch of Lintrathen, Tayside – wintering Greylag Geese
39 Tay-Isla Valley, including Monk Myre, Stormont Loch, Marlee Loch, Meikleour and Loch of Clunie, Tayside – wintering Greylag Geese
40 Drummond Pond & Benny Beg Pond, Tayside – wintering Greylag Geese
41 Carsebreck & Rhynd Lochs, Tayside – wintering Pink-footed and Greylag Geese
42 Dupplin Lochs, Tayside – passage and wintering wildfowl
43 Loch Leven, Tayside – passage and wintering wildfowl
44 Tay Estuary, Tayside & Fife – passage and wintering wildfowl and waders
45 Tentsmuir Point & Abertay Sands, Fife – wintering wildfowl
46 Eden Estuary, Fife – passage and wintering wildfowl and waders
47 Cameron Reservoir, Fife – roost for Pink-footed Geese
48 Loch Mahaick, Central – roost for Pink-footed Geese
49 Loch Lomond at Endrick Mouth, Central – wintering wildfowl
50 Flanders Moss, Central – wintering Pink-footed and Greylag Geese
51 Firth of Forth, Forth Islands & Aberlady Bay, Central, Lothian & Fife – breeding seabirds, passage and wintering wildfowl and waders
52 Duddingston Loch, Lothian – day roost for wintering Pochard (but note recent decrease)
53 Gladhouse Reservoir, Lothian – roost for Pink-footed Geese
54 Westwater & Baddingsgill Reservoirs, Borders – roost for Pink-footed Geese
55 Fala Flow, Lothian – roost for Pink-footed Geese
56 Greenlaw Moor & Hule Moss, Borders – roost for Pink-footed Geese
57 Hoselaw Loch, Borders – wintering Greylag Geese
58 White Loch (Lochinch) & area, Dumfries & Galloway – wintering geese
59 Loch Ken & Dee Marshes, Dumfries & Galloway – wintering wildfowl
60 Upper Solway Flats & Marshes including Blackshaw Bank, Nith Merses, & Caerlaverock, Dumfries & Galloway – passage and wintering wildfowl and waders.

APPENDIX 5: *Species new to Scotland reported from January to mid-October 1985. Note that all these records are subject to acceptance by BBRC.*

Eleanor's Falcon *Falco eleanorae* (Mediterranean & NW Africa)
One on South Uist on 14 May 1985. Two previous British records, both in England, in 1977 and 1981.

Forster's Tern *Sterna forsteri* (N America)
One at Musselburgh, Lothian, first seen 6 October 1985. First recorded in Britain in 1980; five more British and Irish reports by the end of 1984.

Little Swift *Apus affinis* (Africa & Middle East)
One at St Andrews, Fife, on 29 May 1985. First recorded in Ireland in 1967; four more British and Irish records by the end of 1984.

Cedar Waxwing *Bombycilla cedrorum* (America)
One on Noss, Shetland, on 27 June 1985. The first record for Britain and Europe.

Brown Shrike *Lanius cristatatus* (Far East)
An adult male at Sumburgh, Shetland, on 30 September 1985. The first record for Britain and Europe.

Daurian Starling *Sturnus sturninus* (Asia)
One on Fair Isle on 7 May 1985. The first record for Britain and Europe.

Red-eyed Vireo *Vireo olivaceous* (N America)
One at Wick, Caithness, on 13 October 1985. There have been c22 previous British and Irish records.

Chestnut-sided Warbler *Dendroica pensylvanica* (N America)
One on Fetlar, Shetland, on 20 September 1985. The first record for Britain and Europe.

Blackpoll Warbler *Dendroica striata* (N America)
One on Whalsay, Shetland, on 30 September 1985. There have been c21 previous British and Irish records.

Bibliography

GENERAL

Anon. 1982. *Economic Report on Scottish Agriculture.* DAFS.

Anon. 1983. *Scotland: census of woodlands and trees, 1979–82.* Forestry Commission.

Anon. 1984. *Nature Conservation in Great Britain.* NCC.

Anon. 1984. *Scotland: land use and physical features.* Factsheet 19, Scottish Information Office.

Barber, D. (ed.) 1970. *Farming and wildlife: a study in compromise.* RSPB.

Bunce, R. G. H. & Last, F. T. 1981. How to characterize the habitats of Scotland, in *Edinburgh Centre for Rural Economy Annual Report 1980–81.*

Curry-Lindahl, K., Watson, A. & Watson, A. D. 1982. *The Future of the Cairngorms.* Aberdeen, The North East Mountain Trust.

Darling, F. F. & Boyd, J. M. 1969. *The Highlands and Islands.* Collins.

Elton, C. S. 1966. *The Pattern of Animal Communities.* Methuen.

Goodier, R. & Bunce, R. G. H. 1977. The native pinewoods of Scotland: the current state of the resource, pp 78–87 in *Native Pinewoods of Scotland* (ed. Bunce), ITE.

Langdale-Brown, I., Jennings, G., Crawford, C. L., Jolly, G. M. & Muscott, J. 1980. Lowland agricultural habitats (Scotland): air photo analysis of change. Unpub. report to NCC.

Mellanby, K. 1967. *Pesticides and Pollution.* Collins.

Mellanby, K. 1981. *Farming and Wildlife.* Collins.

Munro, R. W. 1973. *Johnston's Gazetteer of Scotland.* Johnston & Bacon.

Nethersole-Thompson, D. & Watson, A. 1981. *The Cairngorms.* The Melven Press.

Nicholson, M. 1970. *The Environmental Revolution.* Hodder & Stoughton.

Pollard, E., Hooper, M. D. & Moore, N. W. 1974. *Hedges.* Collins.

Ratcliffe, D. A. 1977. *A Nature Conservation Review.* Cambridge University Press.

Ritchie, W. & Mather, A. S. 1984. *The Beaches of Scotland.* Countryside Commission for Scotland.

Watson, A., Bayfield, N. & Moyes, S. M. 1970. Research on human pressures on Scottish mountain tundra, soils and animals, in *Proc. Conf. on Productivity and Conservation in Northern Circumpolar Lands,* pp 256–266. IUCN.

Whittow, J. B. 1977. *Geology and Scenery in Scotland.* Penguin Books.

Wildlife and Countryside Act 1981. HMSO.

Wise, M. & Butlin, R. (comp.). *The Ordnance Survey Atlas of Great Britain.* OS / Country Life.

ORNITHOLOGICAL

Ainslie, J. A. & Atkinson, R. 1937. On the breeding habits of Leach's Fork-tailed Petrel. *Brit. Birds* 30:234–248.

Allen, R. H. 1974. The Mersey Ducks since 1950. *Cheshire Bird Report* 1973:31–33.

Allison, A., Newton, I. & Campbell, C. 1974. *Loch Leven National Nature Reserve.* WAGBI.

Anderson, A. 1965. Moorhens at Newburgh. *Scot. Birds* 3:230–233.

Anderson, A. 1975. A method of sexing Moorhens. *Wildfowl* 26:77–82.

Anderson, A. 1982. The establishment and growth of a new Fulmar colony on sand dunes. *Bird Study* 29:189–194.

Angus, S. (ed.) 1983. *Sutherland Birds.* The Northern Times Ltd.

Arnold, G. W. 1983. The influence of ditch and hedgerow structure, length of hedgerows, and area of woodland and garden on bird numbers on farmland. *J. appl. Ecol.* 20:731–750.

Atkinson, N. 1981. Way Down Upon the Swanee River. *Scottish Wildlife* 17:21.

Atkinson, N. K. 1982. The Little Tern at St Cyrus National Nature Reserve. *Seabird Report* 6:86–92.

Atkinson, N. K., Davies, M. & Prater, A. J. 1978. The winter distribution of Purple Sandpipers in Britain. *Bird Study* 25:223–228.

Atkinson, N. K., Summers, R. & Greenwood, J. J. D. 1981. Population, movements and biometrics of the Purple Sandpiper *Calidris maritima* in eastern Scotland. *Ornis Scand.* 12:18–27.

Atkinson-Willes, G. L. 1963. *Wildfowl in Great Britain.* HMSO.

Badenoch, C. O. 1980. A report on the wintering and one mortality incident among Whooper Swans (*Cygnus cygnus*) on the River Teviot, Roxburgh. *Hist. Berwickshire Nat. Club* 41:221–226.

Bagenal, T. B. & Baird, D. E. 1959. The birds of North Rona and notes on Sula Sgeir. *Bird Study* 6:153–174.

Baillie, S. R. & Mead, C. J. 1982. The effect of severe oil pollution during the winter of 1980–81 on British and Irish auks. *Ringing & Migration* 4:33–44.

Baillie, S. R. & Milne, H. 1982. The influence of female age on breeding in the Eider. *Bird Study* 29:55–66.

Baillie, S. R. & Swann, R. L. 1980. The extent of post-juvenile moult in the Blackbird. *Ringing & Migration* 3:21–26.

Bainbridge, I. P. & Minton, C. D. T. 1978. The migration and mortality of the Curlew in Britain and Ireland. *Bird Study* 25:39–50.

Balfour, E. 1968. Fieldfares breeding in Orkney. *Scot. Birds* 5:31–32.

Balfour, E. 1972. *Orkney Birds: status and guide*. Charles Senior.

Balfour, E., Anderson, A. & Dunnet, G. M. 1967. Orkney Cormorants – their breeding distribution and dispersal. *Scot. Birds* 4:481–493.

Balfour, E. & Cadbury, J. 1974. A population study of the Hen Harrier, *Circus cyaneus*, in Orkney, pp 122–128 in *The natural environment of Orkney* (ed. Goodier), NCC.

Balfour, E. & Cadbury, J. 1979. Polygyny, spacing and sex ratio among Hen Harriers, *Circus cyaneus* (L) in Orkney, Scotland. *Ornis Scand.* 10:133–141.

Barnes, J. A. G. 1961. The winter status of the Lesser Black-backed Gull 1959–60. *Bird Study* 8:127–147.

Barrett, J. 1983. Moray Firth Seaducks: winters 1981–82 and 1982–83. Report to Britoil of surveys carried out by RSPB.

Barrett, J. & Barrett, C. F. 1985. Divers in the Moray Firth, Scotland. *Scot. Birds* 13:149–154.

Batten, L. A. & Marchant, J. H. 1976. Bird population changes for the years 1973–74. *Bird Study* 23:11–20.

Baxter, E. V. & Rintoul, L. J. 1953. *The Birds of Scotland*. Oliver & Boyd.

Bell, B. D. 1969. Some thoughts on the apparent ecological expansion of the Reed Bunting. *Brit. Birds* 62:209–218.

Bell, M. V. 1981. Wintering wildfowl at the Loch of Strathbeg. *North-East Scot. Bird Report* 1980:37–42.

Benton, C., Khan, F., Monaghan, P., Richardson, W. N. & Shedden, C. B. 1983. The contamination of a major water supply by gulls. *Water Res.* 7:789–798.

Berry, R. J. & Davis, P. E. 1970. Polymorphism and behaviour in the Arctic Skua (*Stercorarius parasiticus* (L)). *Proc. R. Soc. B.*, 175:255–267.

Berry, J. & Johnston, J. L. 1980. *The Natural History of Shetland*. Collins.

Bibby, C. J. & Green, R. E. 1981. Autumn migration strategies of Reed and Sedge Warblers. *Ornis Scand.* 12:1–12.

Bignal, E. M. 1980. Observations on the Shelduck population of the Loch Lomond NNR. Unpub. report to NCC.

Birkhead, M. E. 1981. The social behaviour of the Dunnock. *Ibis* 123:75–84.

Birkhead, T. R. 1974. Movements and mortality rates of British Guillemots. *Bird Study*: 241–253

Birkhead, T. R. 1984. Distribution of the bridled form of the Common Guillemot *Uria aalge* in the North Atlantic. *J. Zool. Lond.* 202:165–176.

Black, J. M. & Rees, E. C. 1984. The structure and behaviour of the Whooper Swan population wintering at Caerlaverock, Dumfries & Galloway, Scotland. *Wildfowl* 35:21–34.

Blake, B. F., Tasker, M. L., Hope Jones, P., Dixton, T. J., Mitchell, R. & Langslow, D. R. 1984. *Seabird distribution in the North Sea*. NCC.

Blurton Jones, N. G. 1956. Census of breeding Canada Geese. *Bird Study* 3:135–170.

Boddington, D. 1960. Unusual mortality of young Puffins on St Kilda, 1959. *Scot. Birds* 1:218–220.

Booth, C. G. 1981. *Birds in Islay*. Argyll Reproductions Ltd.

Booth, C. J. 1979a. A study of Ravens in Orkney. *Scot. Birds* 10:261–267.

Booth, C. J. 1979b. Golden Plover survey 1977–8. *Orkney Bird Report* 1978–79.

Booth, C. J. 1982. Fledging success of some Red-throated Divers in Orkney. *Scot. Birds* 12:33–38.

Booth, C., Cuthbert, M. & Reynolds, P. 1984. *The Birds of Orkney*. The Orkney Press.

Bourne, W. R. P. 1968. Arctic auks on the Scottish coast. *Scot. Birds* 5:104–107.

Bourne, W. R. P. 1980. The midnight descent, dawn ascent and re-orientation of landbirds migrating across the North Sea in autumn. *Ibis* 122:536–540.

Bourne, W. R. P. 1982. The manner in which wind drift leads to seabird movements along the east coast of Scotland. *Ibis* 124:81–88.

Bourne, W. R. P. & Harris, M. P. 1979. Birds of the Hebrides:seabirds. *Proc. R. Soc. Edinb.* 77B:445–475.

Bourne, W. R. P. & Patterson, I. J. 1962. The spring departure of Common Gulls *Larus canus* from Scotland. *Scot. Birds* 1:3–17.

Bourne, W. R. P. & Smith, A. J. M. 1974. Threats to Scottish Sandwich Terns. *Biol. Conserv.* 6:222–224.

Bowes, A., Lack, P. C. & Fletcher, M. R. 1984. Wintering gulls in Britain, January 1983. *Bird Study* 31:161–170.

Boyd, H. 1955. The role of tradition in determining the winter distribution of Pinkfeet in Britain. *Wildfowl Trust Ann. Report* 7:107–122.

Boyd, H. 1968. Barnacle Geese in the west of Scotland, 1957–67. *Wildfowl* 19:96–107.

Boyd, H. & Eltringham, S. K. 1962. The Whooper Swan in Great Britain. *Bird Study* 9:217–241.

Boyd, H. & Ogilvie, M. A. 1972. Icelandic Greylag Geese wintering in Britain in 1960–71. *Wildfowl* 23:64–82.

Boyd, J. M. 1956. Fluctuations of Common Snipe, Jack Snipe and Golden Plover in Tiree, Argyllshire. *Bird Study* 3:105–118.

Branson, N. J. B. A. & Minton, C. D. T. 1976. Moult, measurements and migrations of the Grey Plover. *Bird Study* 23:257–266.

Branson, N. J. B. A., Ponting, E. D. & Minton, C. D. T. 1978. Turnstone migrations in Britain and Europe. *Bird Study* 25:181–187.

Brazil, M. 1981. The behavioural ecology of *Cygnus cygnus cygnus* in central Scotland. *Proc. 2nd Int. Swan Symp., Sapporo 1980*: 273–291.

Brazil, M. 1983. Preliminary results from a study of Whooper Swan movements using neck collars. *J. Coll. of Dairying* (Japan) 10:79–90.

Broad, R. A. 1974. Contamination of birds with Fulmar oil. *Brit. Birds* 67:297–301.

Broad, R. & Ewins, P. 1984. Fair Isle Tysties – a progress report. *Fair Isle Bird Obs. Report* 1983:19–22.

Brodie, J. 1974. Evening assembly of Starlings at a winter roost. *Scot. Birds* 8:63–71.

Brown, A. W. & Brown, L. M. 1984a. The status of the Mute Swan in the Lothians. *Scot. Birds* 13:8–15.

Brown, A. W. & Brown, L. M. 1984b. Mink in a Black-headed Gull colony in the Pentlands. *Scot. Birds* 13:89.

Brown, A. W. & Brown, L. M. 1985. The Scottish Mute Swan Census 1983. *Scot. Birds* 13:140–148.

Brown, A. W., Leven, M. R. & da Prato, S. R. D. 1984. The status of the Corn Bunting in the Lothians. *Scot. Birds* 13:107–111.

Brown, L. H. & Watson, A. 1964. The Golden Eagle in relation to its food supply. *Ibis* 106:78–100.

Brown, P. & Waterston, G. 1962. *The return of the Osprey*. Collins.

Brown, R. L. 1934. Breeding habits and numbers of Kingfishers in Renfrewshire. *Brit. Birds* 27:256–258.

Bryant, D. M. 1978a. Moulting Shelducks on the Forth Estuary. *Bird Study* 25:103–108.

Bryant, D. M. 1978b. Environmental influences on growth and survival of nestling House Martins. *Ibis* 120:271–283.

Bryant, D. M. 1979. Reproductive costs in the House Martin. *J. Anim. Ecol.* 48:655–675.

Bryant, D. M. 1980. A report on the bird fauna of the Forth Estuary mudflats: January-February 1980. Unpub. report of Stirling University.

Bryant, D. M. & Leng, J. 1975. Feeding distribution and behaviour of Shelduck in relation to food supply. *Wildfowl* 26:20–30.

Bryant, D. M. & McLusky, D. S. 1977. Invertebrate and bird fauna of the Forth Estuary mudflats. Report of Stirling University & NCC.

Buckland, S. T. & Knox, A. G. 1980. Brambling breeding in Scotland. *Brit. Birds* 73:360–361.

Bullock, I. D. & Gomersall, C. H. 1981. The breeding populations of terns in Orkney and Shetland in 1980. *Bird Study* 28:187–200.

Bundy, G. 1976. Breeding biology of the Red-throated Diver. *Bird Study* 23:249–256.

Bundy, G. 1978a. Some breeding birds of Unst. *Scot. Birds* 10:38–50.

Bundy, G. 1978b. Breeding Red-throated Divers in Shetland. *Brit. Birds* 71:199–208.

Bundy, G. 1979. Breeding and feeding observations on Black-throated Divers. *Bird Study* 26:33–36.

Bunn, D. S., Warburton, A. B. & Wilson, R. D. S. 1982. *The Barn Owl.* Poyser.

Burn, D. M. & Mather, J. R. 1974. The White-billed Diver in Britain. *Brit. Birds* 67:257–296.

Burton, H., Evans, T. L. & Weir, D. N. 1970. Wrynecks breeding in Scotland. *Scot. Birds* 6:154–156.

Buxton, E. J. M. 1962. The inland breeding of the Oystercatcher in Great Britain 1958–59. *Bird Study* 8:194–209.

Buxton, N. E. 1982a. Wintering coastal waders of Lewis and Harris. *Scot. Birds* 12:38–43.

Buxton, N. E. 1982b. Wintering waders on the Atlantic shores of the Uists and Benbecula. *Scot. Birds* 12:106–113.

Buxton, N. E. 1983. Unnatural mortality of Red-throated Divers. *Scot. Birds* 12:227–228.

Buxton, N. E. 1985. The current status and distribution of terns in the Outer Hebrides. *Scot. Birds* 13:172–178.

Buxton, N. E. & Young, C. M. 1981. The food of the Shelduck in north-east Scotland. *Bird Study* 28:41–48.

Cadbury, C. J. 1980a. *Silent Death: the destruction of birds and mammals through the deliberate misuse of poisons in Britain.* RSPB.

Cadbury, C. J. 1980b. The status and habitats of the Corncrake in Britain 1978–79. *Bird Study* 27:203–218.

Cadbury, C. J. 1981. Nightjar census methods. *Bird Study* 28:1–4.

Campbell, B. 1954. The breeding distribution and habitats of the Pied Flycatcher in Britain. *Bird Study* 1:81–100, 2:24–31, 2:179–191.

Campbell, B. 1965. The British breeding distribution of the Pied Flycatcher 1953–62. *Bird Study* 12:305–318.

Campbell, B. 1974. *The Crested Tit.* Forestry Commission.

Campbell, L. H. 1977. Local variations in the proportion of adult males in flocks of Goldeneye wintering in the Firth of Forth. *Wildfowl* 28: 77–80.

Campbell, L. H. 1978. *Report of the Forth Ornithological Working Party.* NCC.

Campbell, L. H. 1978a. Diurnal and tidal behaviour patters of Eiders wintering at Leith. *Wildfowl* 29:147–152.

Campbell, L. H. 1978b. Patterns of distribution and behaviour of flocks of seaducks wintering at Leith and Musselburgh, Scotland. *Biol. Conserv.* 14:111–124.

Campbell, L. H. 1979. Forth Sea Ducks 1978–79. Report to NCC.

Campbell, L. 1980. The impact of an oilspill in the Firth of Forth on Great Crested Grebes. *Scot. Birds* 11:43–48.

Campbell, L. H. 1984. The impact of changes in sewage treatment on seaducks wintering in the Firth of Forth, Scotland. *Biol. Conserv.* 28:173–180.

Campbell, L. H. & Milne, H. 1983. Moulting Eiders in eastern Scotland. *Wildfowl* 34:105–107.

Castle, M. E. 1977. Rookeries in Scotland – 1975. *Scot. Birds* 9:327–334.

Catley, G. P. & Hursthouse, D. The Parrot Crossbill in Britain, with special reference to the influx of 1982–83. *Brit. Birds* 78:482–505.

Chandler, R. J. 1981. Influxes into Britain and Ireland of Red-necked Grebes and other waterbirds during winter 1978/79. *Brit. Birds* 74:55–81.

Clancey, P. A. 1935. On the habits of the Kingfisher. *Brit. Birds* 28:295–301.

Clark, F. & McNeil, D. A. C. 1980. Cliff-nesting colonies of House Martins in Great Britain. *Ibis* 122:27–42.

Clark, N. A., Turner, B. S. & Young, J. F. 1982. Spring passage of Sanderlings *Calidris alba* on the Solway Firth. *Wader Study Group Bull.* 36:10–11.

Cook, A. 1975. Changes in the Carrion / Hooded Crow hybrid zone and the possible importance of climate. *Bird Study* 22:165–168.

Cook, M. J. H. 1982. Breeding status of the Crested Tit. *Scot. Birds* 12:97–106.

Cooke, A. S., Bell, A. A. & Haas, M. B. 1982. *Predatory birds, pesticides and pollution.* ITE.

Corkhill, P. 1980. Golden Eagles on Rhum. *Scot. Birds* 11:33–43.

Coulson, J. C. 1963. The status of the Kittiwake in the British Isles. *Bird Study* 10:147–179.

Coulson, J. C. 1983. The changing status of the Kittiwake *Rissa tridactyla* in the British Isles, 1969–1979. *Bird Study* 30:9–16.

Coulson, J. C. & Brazendale, M. G. 1968. Movements of Cormorants ringed in the British Isles and evidence of colony specific dispersal. *Bird Study* 61:1–21.

Coulson, J. C., Duncan, N. & Thomas, C. 1982. Changes in the breeding biology of the Herring Gull (*Larus argentatus*) induced by reduction in the size and density of the colony. *J. Anim. Ecol.* 51:739–756.

Coulson, J. C., Monaghan, P., Butterfield, J. E. L., Duncan, N., Ensor, K., Shedden, C. & Thomas, C. 1984. Scandinavian Herring Gulls wintering in Britain. *Ornis Scand.* 15:79–88.

Cowley, E. 1979. Sand Martin population trends in Britain, 1965–1978. *Bird Study* 26:113–116.

Cowper, C. N. L. 1973. Breeding distribution of Grey Wagtails, Dippers and Common Sandpipers on the Midlothian Esk. *Scot. Birds* 7:302–306.

Cramp, S., Bourne, W. R. P. & Saunders, D. 1974. *The Seabirds of Britain and Ireland.* Collins.

Cramp, S. & Simmons, K. E. L. 1977 *et seq. The Birds of the Western Palearctic.* Vols. 1–3. Oxford University Press.

Crummy, J. A. 1981. Mute Swans in Lewis. *Hebrid. Nat.* 5:66.

Cunningham, P. 1983. *Birds of the Outer Hebrides.* The Melven Press.

Currie, A. 1981. The vegetation of gulleries on moorland in Lewis, Outer Hebrides. *Hebrid. Nat.* 5:41–49.

da Prato, S. R. D. 1980. How many Lesser Whitethroats breed in the Lothians? *Scot. Birds* 11:108–112.

da Prato, S. R. D. 1981. Warblers in mixed passerine flocks in summer. *Brit. Birds* 74:513–515.

da Prato, S. R. D. 1985. The breeding birds of agricultural land in southeast Scotland. *Scot. Birds* 13:203–216

da Prato, E. S. & da Prato, S. R. D. 1979. (Letter) Wader counting on the rocky shores of East Lothian. *Scot. Birds* 10:184–186.

da Prato, S. R. D. & da Prato, E. S. 1980. The seabirds of Berwickshire. *Scot. Birds* 11:13–20.

da Prato, S. R. D. & da Prato, E. S. 1983. Movements of Whitethroats *Sylvia communis* ringed in the British Isles. *Ringing & Migration* 4:193–210.

da Prato, S. R. D. & da Prato, E. S. 1984. Censusing Robins in winter: a test using colour-ringed birds. *Ornis Scand.* 15:248–252.

da Prato, S. R. D., da Prato, E. S. & Chittenden, D. E. 1980. Redwing migration through the British Isles. *Ringing & Migration* 3:9–20.

da Prato, S. R. D., Dickson, J. M. & Symonds, F. L. 1981. Sandwich Terns in the Firth of Forth in winter. *Scot. Birds* 11:226–227.

da Prato, S. & Langslow, D. R. 1976. Breeding Dippers and their food supply. *Edinb. Ringing Gp. Report* 4:33–36.

Davenport, D. L. 1979. Spring passage of skuas at Balranald, North Uist. *Scot. Birds* 10:216–221.

Davenport, D. L. 1982. Influxes into Britain of Hen Harriers, Long-eared Owls and Short-eared Owls in winter 1978/79. *Brit. Birds* 75:309–316.

Davidson, N. C. 1982. Changes in the body condition of Redshanks during mild winters: an inability to regulate reserves? *Ringing and Migration* 4:51–62.

Davidson, N. C. & Evans, P. R. 1982. Mortality of Redshanks and Oystercatchers from starvation during severe weather. *Bird Study* 29:183–188.

Dean, A. R. 1984. Origins and distribution of British Glaucous Gulls. *Brit. Birds*. 77:165–166.

Dennis, R. 1964. Capture of moulting Canada Geese on the Beauly Firth. *Wildfowl Trust Ann. Report* 15:71–74.

Dennis, R. H. 1966. Great and Cory's Shearwaters at Fair Isle. *Scot. Birds* 4:218–222.

Dennis, R. H. 1973. Possible interbreeding of Slavonian Grebe and Black-necked Grebe in Scotland. *Scot. Birds* 7:307–308.

Dennis, R. H. 1983. Population studies and conservation of Ospreys in Scotland. pp 207–214 in *Biology and Management of Bald Eagles and Ospreys*. Montreal.

Dennis, R. H. 1983a. Purple Sandpipers breeding in Scotland. *Brit. Birds* 76:563–566.

Dennis, R. H. 1984. *Birds of Badenoch and Strathspey*. Roy Dennis Enterprises.

Dennis, R. H. & Dow, H. 1984. The establishment of a population of Goldeneye *Bucephala clangula* breeding in Scotland. *Bird Study* 31:217–222.

Dennis, R. H., Ellis, P. M., Broad, R. A. & Langslow, D. R. 1984. The status of the Golden Eagle in Britain. *Brit. Birds* 77:592–607.

Densley, M. 1977. Ross's Gulls in Britain. *Scot. Birds* 9:334–342.

Dick, D. 1985. Raptor Study Group Reports. *Scot. Birds* 13: 162–166.

Dick, S. J. A., Pienkowski, M. W., Waltner, M. & Minton, C. T. D. 1976. Distribution and geographical origins of Knot *Calidris canutus* wintering in Europe and Africa. *Ardea* 64:22–47.

Dickson, R. C. 1975. Wintering grebes in Loch Ryan. *Scot. Birds* 8:379–381.

Dobson, R. H. 1985. Manx Shearwaters breeding on the Isle of Muck. *Glasgow Nat.* 20:491

Dougall, T. W. & North, P. M. 1983. Problems of censusing Long-tailed Tits (*Aegithalos caudatus*) by the mapping method. *The Ring* 114–115:88–97.

Duncan, N. 1981. The Lesser Black-backed gull on the Isle of May. *Scot. Birds* 11:180–188.

Duncan, W. N. M. & Monaghan, P. 1977. Infidelity to the natal colony by breeding Herring Gulls. *Ringing & Migration* 1:166–172.

Dunnet, G. M. 1982. Oil pollution and seabird populations. *Phil. Trans. R. Soc. Lond. B* 297:413–427.

Dunnet, G. M. & Ollason, J. C. 1978. The estimation of survival rate in the Fulmar *Fulmarus glacialis*. *J. Anim. Ecol.* 47:507–520.

Dunnet, G. M. & Ollason, J. C. 1982. The feeding dispersal of Fulmars *Fulmarus glacialis* in the breeding season. *Ibis* 124:359–361.

Dunnet, G. M., Ollason, J. C. & Anderson A. 1979. A 28-year study of breeding Fulmars (*Fulmarus glacialis*) in Orkney. *Ibis* 121:293–300.

Dunnet, G. M. & Patterson, I. J. 1968. The Rook problem in north-east Scotland, pp 119–139 in *The Problems of Birds as Pests*. Academic Press.

Dunthorn, A. A. 1971. The predation of cultivated mussels by Eiders. *Bird Study* 18:107–112.

Durman, R. (ed.) 1976. *Bird Observatories in Britain and Ireland*. Berkhamsted, Poyser.

Durman, R. F. 1977. Ring Ousels in the Pentlands. *Edinb. Ringing Gp. Rep.* 5:24–27.

Eggeling, W. J. 1960. *The Isle of May*. Oliver & Boyd.

Eggeling, W. J. 1974. The birds of the Isle of May. *Scot. Birds* 8:93–148.

Elkins, N. 1965. The effects of weather on the Long-tailed Duck in Lewis. *Bird Study* 12:132–134.

Elkins, N. 1983. *Weather and Bird Behaviour*. Poyser.

Elkins, N. & Williams, M. R. 1972. Aspects of seabird movement off northeast Scotland. *Scot. Birds* 7:66–75.

Elkins, N. & Williams, M. R. 1974. Shag movements in Northeast Scotland. *Bird Study* 21:149–151.

Ellis, P. 1983. The phenomenal migrant fall of October 1982. *Scot. Birds* 12:246–251.

Evans, P. G. H. (ed.) 1980. *Auk Censusing Manual*. Seabird Group.

Evans, P. R. 1968. Autumn movements and orientation of waders in north-east England and southern Scotland, studied by radar. *Bird Study* 15:53–64.

Evans, P. R. 1984. The British Isles, pp 261–275 in *Coastal Waders and Wildfowl in Winter* (ed. Evans, Goss-Custard & Hale). Cambridge University Press.

Evans, P. R. & Flower, W. U. 1967. The birds of the Small Isles. *Scot. Birds* 4:404–445.

Evans, P. R. & Pienkowski, M. W. 1982. Behaviour of Shelduck *Tadorna tadorna* (*L*) in a winter flock: does regulation occur? *J. Anim. Ecol.* 51:241–262.

Everett, M. J. 1967. Waxwings in Scotland 1965/6 and 1966/7. *Scot. Birds* 4:534–548.

Everett, M. J. 1971. Breeding status of Red-necked Phalaropes in Britain and Ireland. *Brit. Birds* 64:293–302.

Everett, M. J. 1971a. The Golden Eagle survey in Scotland in 1964–68. *Brit. Birds* 64:49–56.

Everett, M. J. 1982. Breeding Great and Arctic Skuas in Scotland in 1974–75. *Seabird Report* 6:50–58.

Ewins, P. J. 1985. Growth, diet and mortality of Arctic Tern *Sterna paradisaea* chicks in Shetland. *Seabird* 8:59–68.

Ewins, P. J. 1985a. Colony attendance and censusing of Black Guillemots (*Cepphus grylle*) in Shetland. *Bird Study* 32:176–185.

Ewins, P. J. & Perrins, C. M. 1982. Tystie studies in Shetland, 1982. Interim report to the Shetland Oil Terminal Environmental Advisory Group.

Ewins, P. J. & Tasker, M. L. 1985. The breeding distribution of Black Guillemots (*Cepphus grylle*) in Orkney and Shetland, 1982–84. *Bird Study* 32:186–193.

Fair Isle Bird Observatory Reports 1962–83.

Feare, C. J. 1978. The ecology of damage by Rooks. *Ann. Appl. Biol.* 88:329–334.

Feare, C. 1984. *The Starling*. Oxford University Press.

Ferns, P. N. 1980a. The spring migration of Ringed Plovers through Britain in 1979. *Wader Study Group Bull.* 29:10–13.

Ferns, P. N. 1980b. The spring migration of Sanderlings *Calidris alba* through Britain in 1979. *Wader Study Group Bull.* 30:22–25.

Ferns, P. N. 1981a. The spring migration of Turnstones through Britain in 1979. *Wader Study Group Bull.* 31:36–40.

Ferns, P. N. 1981b. The spring migration of Dunlins through Britain in 1979. *Wader Study Group Bull.* 32:14–19.

Fisher, J. 1966. The Fulmar population of Britain and Ireland, 1959. *Bird Study* 13:5–76.

Flegg, J. J. M. 1973. A study of Treecreepers. *Bird Study* 20:287–302.

Flegg, J. J. M. & Cox, J. G. 1972. Movement of Black-headed Gulls from colonies in England and Wales. *Bird Study* 29:228–240.

Flegg, J. J. M. & Glue, D. E. 1973. A Water Rail study. *Bird Study* 20:69–79.

Flegg, J. J. M. & Glue, D. E. 1975. The nesting of the Ring Ousel. *Bird Study* 22:1–8.

Fowler, J. A. 1979. Manx Shearwaters breeding on Yell. *Shetland Bird Report* 1979: 47–48.

Fowler, J. A. 1982. Leach's Petrel present on Ramna Stacks, Shetland. *Seabird Report* 6:93.

Fowler, J. A., Okill, J. D. & Marshall, B. 1982. A retrap analysis of Storm Petrels tape-lured in Shetland. *Ringing & Migration* 4:1–7.

Fowler, J. A. & Swinfen, R. 1984. Scottish Storm Petrels in Iceland. *Scot. Birds* 13:52.

Fox, A. D., Madsen, J. & Stroud, D. A. 1983. A review of the summer ecology of the Greenland White-fronted Goose *Anser albifrons flavirostris*. *Dansk. Orn. Foren. Tidsskr.* 77:43–55.

Fox, A. D. & Stroud, D. A. (in press). A preliminary inventory of Greenland White-fronted Goose wintering sites in Britain. Aberystwyth, Greenland White-fronted Goose Study.

Fuller, R. J. 1981. The breeding habitats of waders on North Uist machair. *Scot. Birds* 11:142–152.

Fuller, R. J. 1982. *Bird Habitats in Britain*. Poyser.

Fuller, R. J. & Lloyd, D. 1981. The distribution and habitats of wintering Golden Plovers in Britain, 1977–1978. *Bird Study* 28:169–186.

Fuller, R. J., Marchant, J. H. & Morgan, R. A. 1985. How representative of agricultural practice in Britain are Common Birds Census farmland plots? *Bird Study* 32:63–77.

Fuller, R. J., Wilson, J. R. & Coxon, P. 1979. Birds of the Outer Hebrides: the waders. *Proc. R. Soc. Edinb.* 77B:419–430.

Furness, R. W. 1978. Movements and mortality rates of Great Skuas ringed in Shetland. *Bird Study* 25:229–238.

Furness, R. W. 1981a. Colonization of Foula by Gannets. *Scot. Birds* 11:211–213.

Furness, R. W. 1981b. Seabird populations on Foula. *Scot. Birds* 11:237–253.

Furness, R. W. 1982. Methods used to census skua colonies. *Seabird Report* 6:44–47.

Furness, R. W. 1983. *Foula, Shetland: The Birds of Foula*. The Brathay Hall Trust.

Furness, R. W. & Baillie, S. R. 1981. Age ratios, wing length and moult as indicators of the population structure of Redshank wintering on British estuaries. *Ringing & Migration* 3:123–132.

Furness, R. W. & Galbraith, H. 1980. Numbers, passage and local movements of Redshanks *Tringa totanus* on the Clyde Estuary as shown by dye-marking. *Wader Study Group Bull.* 29:19–22.

Furness, R. W., Monaghan, P. & Shedden, C. 1981. Exploitation of a new food source by the Great Skua in Shetland. *Bird Study* 28:49–52.

Galbraith, H. 1977. The post-nuptial moult of a migratory population of Pied Wagtails. *Ringing & Migration* 1:184–186.

Galbraith, H. 1981. Fluctuations in breeding Shags on the Isle of May. *Scot. Birds* 11:193–194.

Galbraith, H. 1983. The diet and feeding ecology of breeding Kittiwakes *Rissa tridactyla*. *Bird Study* 30:109–120.

Galbraith, H. & Furness, R. W. 1983. Breeding waders on agricultural land. *Scot. Birds* 12:148–153.

Galbraith, H., Furness, R. W. & Fuller, R. J. 1984. Habitats and distribution of waders breeding on Scottish agricultural land. *Scot. Birds* 13:98–107.

Galbraith, H., Russell, S. & Furness, R. W. 1981. Movements and mortality of Isle of May Shags as shown by ringing recoveries. *Ringing & Migration* 3:181–189.

Galbraith, H. & Tyler, S. J. 1982. Movements and mortality of the Dipper as shown by ringing recoveries. *Ringing & Migration* 4:9–14.

Gardarsson, A. & Skarphedinsson, K. H. 1984. A census of the Icelandic Whooper Swan population. *Wildfowl* 35:37–47.

Garden, E. A. 1958. The national census of Heronries in Scotland 1954, with a summary of the 1928/29 census. *Bird Study* 5:90–109.

Gibson, I. 1978. Numbers of birds on the Clyde Estuary, in *Nature Conservation Interests in the Clyde Estuary*. NCC.

Gibson, J. A. 1951. The Breeding distribution, population and history of the birds of Ailsa Craig. *Scot. Nat.* 63:73–100.

Gibson, J. A. 1979. The breeding birds of the Clyde Area: supplementary notes. *Western Nat.* 8:27–45.

Gibson, J. A. 1981. A regional checklist of Clyde birds. Scot. Wildlife Trust (Glasgow).

Gibson, J. A., Hopkins, I. & Stephen, A. 1980. Notes on the birds of the Island of Bute 1927–80. *Trans. Buteshire Nat. Hist. Soc.* 21:69–92.

Gochfeld, M. 1983. The Roseate Tern: world distribution and status of a threatened species. *Biol. Conserv.* 25:103–125.

Gomersall, C. H., Morton, J. S. & Wynde, R. M. 1984. Status of breeding Red-throated Divers in Shetland, 1983. *Bird Study* 31:223–229.

Goodbody, I. M. 1955. Field notes on the Corn Bunting (*Emberiza calandra*): habitat and distribution in Aberdeenshire. *Scot. Nat.* 1955:90–97.

Grant, J. P. 1984. (Letter) Crested Tits on Deeside. *Scot. Birds* 13:54–55.

Green, G. H. (ed.) 1984. *A survey of waders breeding on the west coast of the Uists and Benbecula (Outer Hebrides), 1983*. Wader Study Group & NCC.

Green, P. 1985. Some results from the use of a long life radio transmitter package on corvids. *Ringing & Migration* 6:45–51

Green, R. 1976. Breeding behaviour of Ospreys *Pandion haliaetus* in Scotland. *Ibis* 118:475–490.

Greenwood, J. J. D., Donally, R. J., Feare, C. J., Gordon, N. J. & Waterston, G. 1971. A massive wreck of oiled birds: northeast Britain, winter 1970. *Scot. Birds* 6:235–255.

Greenwood, J. J. D. & Hubbard, S. F. 1979. Breeding of Blue Tits in relation to food supply. *Scot. Birds* 10:268–271.

Gribble, F. C. 1983. Nightjars in Britain and Ireland in 1981. *Bird Study* 30:165–176.

Grierson, J. 1961. Little Gulls in Angus and Fife. *Scot. Birds* 1:362–367.

Grierson, J. 1962. A check-list of the birds of Tentsmuir, Fife. *Scot. Birds* 2:113–164.

Halliday, J. B., Curtis, D. J., Thompson, D. B. A., Bignal, E. M. & Smyth, J. C. 1982. The abundance and feeding distribution of Clyde Estuary shorebirds. *Scot. Birds* 12:65–72.

Halliday, K. C. R. 1970. Notes on a Great Grey Shrike wintering in Lanarkshire. *Scot. Birds* 6:22–23.

Hamadani, H. M. 1978. Population dispersion and reproduction of the Jackdaw in Northeast Scotland. PhD thesis, University of Aberdeen.

Hamilton, F. D. 1962. Census of Black-headed Gull colonies in Scotland, 1958. *Bird Study* 9:72–80.

Hamilton, G. A. & Stanley, P. I. 1975. Further cases of poisoning of wild geese by an organophosphorus winter wheat seed treatment. *Wildfowl* 26:49–54.

Hardey, J. & Rae, R. 1979. Nest sites used by Grampian Dippers. *Grampian Ring. Gp. Report* 2:22–23.

Hardey, J., Rae, R. & Rae, S. 1978. Breeding success of Dippers in the Grampian Region. *Grampian Ring. Gp. Report* 1:23–25.

Hardy, A. R. 1977. Hunting ranges and feeding ecology of owls in farmland. PhD thesis, University of Aberdeen.

Hardy, A. R. & Minton, S. D. T. 1980. Dunlin migration in Britain and Ireland. *Bird Study* 27:81–92.

Harris, J. 1960. Some observations on the Capercaillie. *Scot. Birds* 1:283–286.

Harris, M. P. 1976. The seabirds of Shetland in 1974. *Scot. Birds* 9:37–68.

Harris, M. P. 1977. Puffins on the Isle of May. *Scot. Birds* 9:285–290.

Harris, M. P. 1982. The breeding seasons of British Puffins. *Scot. Birds* 12:11–17.

Harris, M. P. 1984a. Movements and mortality patterns of North Atlantic Puffins as shown by ringing. *Bird Study* 31:131–140.

Harris, M. P. 1984b. *The Puffin*. Poyser.

Harris, M. P. & Galbraith, H. 1983. Seabird populations of the Isle of May. *Scot. Birds* 12:174–180.

Harris, M. P. & Murray, S. 1978. *Birds of St Kilda*. ITE.

Harris, M. P. & Murray, S. 1981. Monitoring of Puffin numbers at Scottish colonies. *Bird Study* 28:15–20.

Harris, M. P. & Osborn, D. 1982. Effect of polychlorinated biphenyl on the survival and breeding of Puffins. *J. appl. Ecol.* 18:471–479.

Harris, M. P. & Wanless, S. 1984. The effect of the wreck of seabirds in February 1983 on auk populations on the Isle of May (Fife). *Bird Study* 31:103–110.

Harris, M. P., Wanless, S. & Rothery, P. 1983. Assessing changes in the numbers of Guillemots *Uria aalge*. *Bird Study* 30:57–66.

Harrison, C. 1982. *An Atlas of the Birds of the Western Palaearctic.* Collins.

Harrison, J. G. (ed.) 1974. *Caerlaverock: conservation and wildfowling in action.* WAGBI.

Headlam, C. G. 1971. Whimbrels breeding at sea-level in northern Highlands. *Scot. Birds* 6:279–280.

Henderson, A. C. B. 1983. Numbers of Corncrakes and habitat use in the Uists, Outer Hebrides, 1983. Unpub. report to RSPB.

Henty, C. 1975. The birds of Strathbraan 1905–74: a salute to Charles Macintosh. *Scot. Birds* 8:344–355.

Heppleston, P. B. 1971. The feeding ecology of Oystercatchers (*Haematopus ostralegus*) in winter in northern Scotland. *J. Anim. Ecol.* 40:651–672.

Heppleston, P. B. 1972. The comparative breeding ecology of Oystercatchers (*Haematopus ostralegus* L.) in inland and coastal habitats. *J. Anim. Ecol.* 41:23–51.

Heppleston, P. B. 1981. The Curlew in Orkney. *Orkney Bird Report* 1980:35–38.

Herfst, M. & Richardson, M. G. 1982. Whimbrel and wader survey in parts of Unst, Shetland. Unpub. report to NCC.

Heubeck, M. & Richardson, M. G. 1980. Bird mortality following the *Esso Bernicia* oil spill, Shetland, December 1978. *Scot. Birds* 11:97–107.

Hewson, R. 1967. The Rock Dove in Scotland in 1965. *Scot. Birds* 4:359–371.

Hewson, R. 1970. Wintering home range and feeding habits of a Great Grey Shrike in Morayshire. *Scot. Birds* 6:18–22.

Hewson, R. & Leitch, A. F. 1982. The spacing and density of Hooded Crow nests in Argyll (Strathclyde). *Bird Study* 29:235–238.

Hickling, R. (ed.) 1983. *Enjoying Ornithology.* Poyser.

Hirons, G. J. M. 1976. A population study of the Tawny Owl (*Strix aluco*) and its main prey in woodland. D.Phil. thesis, Oxford University.

Hirons, G. J. M. 1981. Breeding behaviour of *Scolopax rusticola*. Unpub. report to NERC.

Hockey, P. 1983. Status and sex ratio of Pochard wintering at Edinburgh. *Scot. Birds* 12:143–148.

Hogg, A. 1983. *Birds of Ayrshire.* Glasgow University.

Hope Jones, P. 1975. The migration of Redstarts through and from Britain. *Ringing and Migration* 1:12–17.

Hope Jones, P. 1977. Counting cliff breeding seabirds at sample sites. Unpub. NCC report.

Hope Jones, P. 1979. Roosting behaviour of Long-tailed Ducks in relation to possible oil pollution. *Wildfowl* 30:155–158.

Hope Jones, P. 1980. Beached birds at selected Orkney beaches 1976–8. *Scot. Birds* 11:1–12.

Hope Jones, P. & Kinnear, P. K. 1979. Moulting Eiders in Orkney and Shetland. *Wildfowl* 30:109–113.

Hopkins, P. G. & Coxon, P. 1979. Birds of the Outer Hebrides: waterfowl. *Proc. R. Soc. Edinb.* 77B:431–444.

Houston, D. 1977. The effect of Hooded Crows on hill sheep farming in Argyll. *J. appl. Ecol.* 14:1–15, 17–29.

Howard, R. & Moore, A. 1984. *A Complete Checklist of the Birds of the World.* Macmillan.

Hudson, R. 1972. Collared Doves in Britain and Ireland during 1965–70. *Brit. Birds* 65:139–155.

Hudson, R. 1976. Ruddy Ducks in Britain. *Brit. Birds* 69:132–143.

Hudson, R. & Mead, C. J. 1984. Origins and ages of auks wrecked in eastern Britain in February-March 1983. *Bird Study* 31:89–94.

Hughes, S. W. M., Bacon, P. & Flegg, J. J. M. 1979. The 1975 census of the Great Crested Grebe in Britain. *Bird Study* 26:213–226.

Hutchison, C. D. & Neath, B. 1978. Little Gulls in Britain and Ireland. *Brit. Birds* 71:563–582.

International Wildfowl Research Bureau. 1980. *Conference on the Conservation of Wetlands of International Importance Especially as Waterfowl Habitat.* IWRB.

Isle of May Bird Observatory and Field Station Reports, 1958–84, in *Scot. Birds.*

Jackson, E. 1966. The birds of Foula. *Scot. Birds* 4:1–60.

Jenkins, D., French, D. D. & Conroy, J. W. H. 1984. Song birds in some semi-natural pine woods in Deeside, Aberdeenshire, in 1980–83. *ITE Annual Report* 1983:75–78.

Jenkins, D., Murray, M. G. & Hall, P. 1975. Structure and regulation of a Shelduck (*Tadorna tadorna* (L)) population. *J. Anim. Ecol.* 44:201–231.

Jones, G. 1985. Parent:offspring resource allocation strategies in birds; studies on swallows (*Hirundinidae*). Ph.D. thesis, University of Stirling.

Jones, P. H. & Tasker, M. L. 1982. *Seabird movements at coastal sites around Great Britain and Ireland 1978–80.* Report of NCC & Seabird Group.

Kerbes, R. H., Ogilvie, M. A. & Boyd, H. 1971. Pink-footed Geese of Iceland and Greenland: a population review based on an aerial survey of pjorsarver in June, 1970. *Wildfowl* 22:5–17.

Keymer, R. 1980. The breeding birds of Red Moss Nature Reserve. *Edinb. Ring. Group Report* 8:10–17.

Kinnear, P. K. 1978. The status of Red-necked and Slavonian Grebes in Shetland. *Shetland Bird Report* 1978:58–62.

Knight, R. C. & Haddon, P. C. 1983. *A guide to Little Tern conservation.* RSPB.

Knox, A. 1977. The status of the Red-breasted Merganser in North-East Scotland. *North-East Scot. Bird Report* 1977:35.

Knox, A. 1983. The Crested Tit on Deeside. *Scot. Birds* 12:255–258.

Lang, J. T. 1982. The European Community Directive on Bird Conservation. *Biol. Conserv.* 22:11–25.

Langham, N. P. E. 1971. Seasonal movements of British Terns in the Atlantic Ocean. *Bird Study* 18:155–175.

Langslow, D. R. 1977. Weight increases and behaviour of Wrynecks on the Isle of May. *Scot. Birds* 9:262–267.

Langslow, D. R. 1978. Recent increases of Blackcaps at bird observatories. *Brit. Birds* 71: 345–354.

Langslow, D. R. 1979. Movements of Blackcaps ringed in Britain and Ireland. *Bird Study* 26:239–252.

Lea, D. 1980. Seafowl in Scapa Flow, Orkney 1974–1978. Unpub. report to NCC.

Lea, D. & Bourne, W. R. P. 1975. The birds of Orkney, in *The Natural Environment of Orkney* (ed. Goodier), NCC.

Leach, I. H. 1981. Wintering Blackcaps in Britain and Ireland. *Bird Study* 28:5–14.

Little, B. & Furness, R. W. 1985. Long-distance moult migration by British Goosanders *Mergus merganser. Ringing & Migration* 6:77–82.

Lloyd, C. 1974. Movements and survival of British Razorbills. *Bird Study* 21:102–116.

Lloyd, C. 1975. Timing and frequency of census counts of cliff-nesting auks. *Brit. Birds* 68:507–513.

Lloyd, C. S. 1984. A method of assessing the relative importance of seabird breeding colonies. *Biol. Conserv.* 28:155–172.

Lloyd, C. S., Bibby, C. J. & Everett, M. J. 1975. Breeding terns in Britain and Ireland in 1969–74. *Brit. Birds* 68:221–237.

Lockie, J. D. 1955. The breeding habits of Short-eared Owls after a vole plague. *Bird Study* 2:53–69.

Lockie, J. D. 1955a. The breeding and feeding of Jackdaws and Rooks with notes on Carrion Crows and other Corvidae. *Ibis* 97:341–369.

Lockie, J. D. 1964. The breeding density of the Golden Eagle and Fox in relation to food supply in Wester Ross, Scotland. *Scot. Nat.* 71:67–77.

Lockie, J. D., Ratcliffe, D. A. & Balharry, R. 1969. Breeding success and organo-chlorine residues in Golden Eagles in West Scotland. *J. appl. Ecol.* 6:381–389.

Lok, C. M. & Vink, J. A. J. 1979. Lodnaver-Kjalkaver (Central Iceland), a hitherto unrecognized important breeding area of the Pink-footed Goose, *Anser brachyrhynchus*. *Le Gerfault* 69:447–459.

Love, J. A. 1978. Leach's and Storm Petrels on North Rona 1971–1974. *Ringing & Migration* 2:15–19.

Love, J. A. 1981. An island population of Blue Tits. *Bird Study* 28:63–64.

Love, J. A. 1983. *The Sea Eagle*. Cambridge University Press.

Love, J. A. & Summers, R. W. 1973. Breeding biology of Magpies in Aberdeenshire. *Scot. Birds* 7:399–403.

Lynch, B. M. 1984. Mass and population structure of emigrating Swallows (*Hirundo rustica*) in roosts in Tayside (1972–1982). *Tay Ring. Group Report* 1982–83:28–44.

Lyster, I. H. J. 1971. Waxwings in Scotland 1970/71. *Scot. Birds* 6:420–438.

Macdonald, D. 1965. Notes on the Corn Bunting in Sutherland. *Scot. Birds* 3:235–246.

Macdonald, J. W. & Standring, K. T. 1978. An outbreak of botulism in gulls on the Firth of Forth, Scotland. *Biol. Conserv.* 14:149–155.

Macdonald, M. A. 1977. An analysis of the recoveries of British-ringed Fulmars. *Bird Study* 24:208–214.

Mackenzie, J. M. D. 1952. Fluctuations in the numbers of British tetraonids. *J. Anim. Ecol.* 21:128–153.

McKilligan, N. G. 1980. The winter exodus of the Rook from a Scottish Highland valley. *Bird Study* 27:93–100.

Macmillan, A. T. 1964. The Waxwing invasion of October-November 1963. *Scot. Birds* 3:180–194.

Macmillan, A. T. 1965. The Collared Dove in Scotland. *Scot. Birds* 3:292–301.

Macmillan, A. T. 1975. Scottish records of the White-billed Diver. *Scot. Birds* 8:377–379.

McMillan, R. L. 1979. An investigation into the commercial harvesting of the Tay reedbeds. MA(Hons) dissertation, University of Dundee.

Magee, J. D. 1965. The breeding distribution of the Stonechat in Britain and the causes of its decline. *Bird Study* 12:83–89.

Maguire, E. 1978. Breeding of Storm Petrel and Manx Shearwater in Kintyre, Argyll. *Western Nat.* 7:63–66.

Mainwood, A. R. 1976. The movements of Storm Petrels as shown by ringing. *Ringing & Migration* 1:98–104.

Mainwood, A. R. 1979. Ringed Plover. *Highland Ring. Group Report* 3:22.

Marchant, J. H. (ed.) 1981. Birds of Estuaries Enquiry 1976–77 to 1978–79. BTO.

Marchant, J. H. 1983. Bird population changes for the years 1981–82. *Bird Study* 30:127–133.

Marchant, J. H. & Hyde, P. A. 1980a. Population changes for waterways birds 1978–79. *Bird Study* 27:179–182.

Marchant, J. H. & Hyde, P. A. 1980b. Aspects of the distribution of riparian birds on waterways in Britain and Ireland. *Bird Study* 27:183–202.

Marquiss, M. 1983. Eggshell thinning and the breakage and non-hatch of eggs at Grey Heron colonies. *ITE Annual Report* 1982:70–74.

Marquiss, M. & Newton, I. 1982. A radio-tracking study of the ranging behaviour and dispersion of European Sparrowhawks *Accipiter nisus*. *J. Anim. Ecol.* 51:111–133.

Marquiss, M. & Newton, I. 1982a. The Goshawk in Britain. *Brit. Birds* 75;243–260.

Marquiss, M., Newton, I. & Ratcliffe, D. A. 1978. The decline of the Raven, *Corvus corax*, in relation to afforestation in southern Scotland and northern England. *J. appl. Ecol.* 15:129–144.

Marquiss, M., Nicoll, M. & Brockie, K. 1983. Scottish Herons and the 1981/82 cold winter. *BTO News* 125:4–5.

Massie, J. 1984. Tawny Owl study. *Grampian Ring. Gp.Rep.* 4:41–51.

Mead, C. J. 1973. Movements of British raptors. *Bird Study* 20:259–286.

Mead, C. J. 1974. The results of ringing auks in Britain and Ireland. *Bird Study* 21:45–86.

Mead, C. J. 1979. Colony fidelity and interchange in the Sand Martin. *Bird Study* 26:99–106.

Mead, C. J. & Harrison, J. D. 1979a. Sand Martin movements within Britain and Ireland. *Bird Study* 26:73–87.

Mead, C. J. & Harrison, J. D. 1979b. Overseas movements of British and Irish Sand Martins. *Bird Study* 26:87–98.

Mead, C. J., North, P. M. & Watmough, B. R. 1979. The mortality of British Grey Herons. *Bird Study* 26:13–22.

Mearns, R. 1983. The diet of the Peregrine *Falco peregrinus* in south Scotland during the breeding season. *Bird Study* 30:81–90.

Mearns, R. 1983a. The status of the Raven in southern Scotland and Northumbria. *Scot. Birds* 12:211–218.

Mearns, R. 1984. Winter sightings of Peregrines at Caerlaverock. *Scot. Birds* 13:73–77.

Mearns, R. & Newton, I. 1984. Turnover and dispersal in a Peregrine *Falco peregrinus* population. *Ibis* 126:347–355.

Meek, E. R. & Little, B. 1977a. The spread of the Goosander in Britain and Ireland. *Brit. Birds* 70:229–237.

Meek, E. R. & Little, B. 1977b. Ringing studies of Goosanders in Northumberland. *Brit. Birds* 70:273–283.

Meek, E. R., Booth, C. J., Reynolds, P. & Ribbands, B. 1985. Breeding skuas in Orkney. *Seabird* 8:21–33.

Merrie, T. D. H. 1978. Relationship between spatial distribution of breeding divers and the availability of fishing waters. *Bird Study* 25:119–122.

Meyer, J. 1981. Easy pickings. *Birds* 8:51–53.

Mills, D. 1962. The Goosander and Red-breasted Merganser in Scotland. *Wildfowl Trust Ann. Report* 13:79–92.

Mills, D. 1965. The distribution and food of the Cormorant in Scottish inland waters. *Freshwater Salmon Fisheries Res.* No. 5.

Mills, D. 1969. The food of the Cormorant at two breeding colonies on the east coast of Scotland. *Scot. Birds* 5:268–276.

Mills, D. 1969a. The food of the Shag in Loch Ewe, Ross-shire. *Scot. Birds* 5:264–268.

Milne, H. 1974. Breeding numbers and reproductive rate of Eiders at the Sands of Forvie National Nature Reserve, Scotland. *Ibis* 116:194–210.

Milne, H. 1977, 1980, 1981. Air survey of moulting seaducks off east coast Scotland. Unpub. reports to NCC.

Milne, H. & Campbell, L. H. 1973. Wintering seaducks off the east coast of Scotland. *Bird Study* 20:153–172.

Milsom, T. P. & Watson, A. 1984. Numbers and spacing of summering Snow Buntings and snow cover in the Cairngorms. *Scot. Birds* 13:19–23.

Mitchell, J. 1977. Observations on the Common Scoter on Loch Lomond. *Loch Lomond Bird Report* 5:8–13.

Mitchell, J. 1980. The Black-headed Gull – its rise and decline as a breeding species on Loch Lomondside and neighbouring areas. *Loch Lomond Bird Report* 8:15–19.

Mitchell, J. 1981. The decline of the Raven as a breeding species in central Scotland. *Forth Nat. & Hist.* 6:35–42.

Mitchell, J. 1984. Loch Lomondside Buzzard Survey 1983. *Loch Lomond Bird Report* 12:11–15.

Mitchell, J. 1985. The Ruby-crowned Kinglet at Loch Lomond. *Scot. Nat.* 1983:81–90

Moore, N. W. 1957. The past and present status of the Buzzard in the British Isles. *Brit. Birds* 50:173–197.

Moreau, R. E. 1972. *The Palearctic-African Bird Migration Systems.* Academic Press.

Morgan, R. & Glue, D. 1977. Breeding, mortality and movements of Kingfishers. *Bird Study* 24:15–24.

Moser, M. 1984. *Solway winter shorebird survey 1982–84.* NCC & RSPB.

Moss, D. 1978a. Diversity of woodland song-bird populations. *J. Anim. Ecol.* 47:521–527.

Moss, D. 1978b. Song-bird populations in forestry plantations. *Quart. J. Forestry* 72:5–14.

Moss, D. 1979. Even-aged plantations as a habitat for birds, pp 413–427 in *The Ecology of Even Aged Plantations* (ed. Ford *et al*). Edinburgh.

Moss, D., Taylor, P. N. & Easterbee, N. 1979. The effect on song-bird populations of upland afforestation with spruce. *Forestry* 52:129–147.

Moss, R. 1980. Why are Capercaillie cocks so big? *Brit. Birds* 73:440–447.

Moss, R., Weir, D. & Jones, A. 1979. Capercaillie management in Scotland, pp 140–155 in *Woodland Grouse 1978* (ed. Lovel). World Pheasant Association.

Mudge, G. P. 1979. The cliff-breeding birds of east Caithness in 1977. *Scot. Birds* 10:247–261.

Mudge, G. P. & Allen, D. S. 1980. Wintering seaducks in the Moray and Dornoch Firths, Scotland. *Wildfowl* 31:123–130.

Munro, J. H. B. 1971. Scottish winter Rook-roost survey – southern Scotland. *Scot. Birds* 6:438–443.

Munro, J. H. B. 1975. Scottish winter Rook-roost survey – central and northern Scotland. *Scot. Birds* 8:309–314.

Murray, S. 1981. A count of Gannets on Boreray, St Kilda. *Scot. Birds* 11:205–211.

Murray, S. & Wanless, S. 1983. The Ailsa Craig gannetry in 1982. *Scot. Birds* 12:225–226.

Murton, R. K. 1965. *The Woodpigeon.* Collins.

Nelson, B. 1978. *The Gannet.* Poyser.

Nethersole-Thompson, D. 1951. *The Greenshank.* Collins.

Nethersole-Thompson, D. 1966. *The Snow Bunting.* Oliver & Boyd.

Nethersole-Thompson, D. 1973. *The Dotterel.* Collins.

Nethersole-Thompson, D. 1975. *Pine Crossbills.* Poyser.

Nethersole-Thompson, D. 1976. Recent distribution, ecology and breeding of Snow Buntings in Scotland. *Scot. Birds* 9:147–162.

Nethersole-Thompson, D. & Nethersole-Thompson, M. 1979. *Greenshanks.* Poyser.

Newton, I. 1972. *Finches.* Collins.

Newton, I. 1972a. Birds of prey in Scotland; some conservation problems. *Scot. Birds* 7:5–23.

Newton, I. 1982. Birds and Forestry, in *Forestry and Conservation* (ed. Harris). Royal Forestry Society.

Newton, I. 1983. Sparrowhawk – pesticide recovery. *Birds* 9:27.

Newton, I. 1984. Raptors in Britain – a review of the last 150 years. *BTO News* 131:6–7.

Newton, I. & Campbell, C. R. G. 1970. Goose studies at Loch Leven in 1967/68. *Scot. Birds* 6:5–18.

Newton, I. & Campbell, C. R. G. 1973. Feeding of geese on farmland in east-central Scotland. *J. appl. Ecol.* 10:781–801.

Newton, I. & Campbell, C. R. G. 1975. Breeding of ducks at Loch Leven, Kinross. *Wildfowl* 26:83–103.

Newton, I. & Haas, M. B. 1984. The return of the Sparrowhawk. *Brit. Birds* 77:47–70.

Newton, I., Meek, E. R., & Little, B. Breeding season foods of Merlins *Falco columbarius* in Northumbria. *Bird Study* 31:49–56.

Newton, I. & Marquiss, M. 1982. Food, predation and breeding season in Sparrowhawks (*Accipiter nisus*). *J. Zool. Lond.* 197:221–240.

Newton, I. & Marquiss, M. 1983. Dispersal of Sparrowhawks between birthplace and breeding place. *J. Anim. Ecol.* 52:463–479.

Newton, I., Marquiss, M., Weir, D. N. & Moss, D. 1977. Spacing of Sparrowhawk nesting territories. *J. Anim. Ecol.* 46:425–441.

Newton, I. & Moss, D. 1977. Breeding birds of Scottish pinewoods, pp. 26–34 in *Native Pinewoods of Scotland* (ed. Bunce & Jeffers). ITE.

Newton, I., Thom, V. M. & Brotherston, W. 1973. Behaviour and distribution of wild geese in south-east Scotland. *Wildfowl* 24:111–121.

Nicholl, M. 1980. Grey Wagtail breeding biology. *Tay Ring. Group Report* 1978–79:40–44.

Nicoll, M. & Summers, R. W. 1980. A study of Redshanks in Eastern Scotland. *Tay Ring. Group Report* 1978–79:18–24.

Norman, R. K. & Saunders, D. R. 1969. Status of Little Terns in Great Britain and Ireland in 1967. *Brit. Birds* 62:4–13.

Norris, C. A. 1947. Report on the distribution and status of the Corn Crake. *Brit. Birds* 40:226–244.

North, P. M. 1979. Relating Grey Heron survival rates to winter weather conditions. *Bird Study* 26:23–28.

North Sea Bird Club Bulletins. Aberdeen.

O'Connor, R. J. & Mead, C. J. 1984. The Stock Dove in Britain, 1930–80. *Brit. Birds* 77:181–201.

O'Donald, P. 1983. *The Arctic Skua.* Cambridge University Press.

Ogilvie, M. A. 1962. Movements of Shoveler ringed in Britain. *Wildfowl Trust Ann. Report* 13:65–69.

Ogilvie, M. A. 1969. Bewick's Swans in Britain and Ireland. 1959–69. *Brit. Birds* 62:505–522.

Ogilvie, M. A. 1977. The numbers of Canada Geese in Britain, 1976. *Wildfowl* 28:27–34.

Ogilvie, M. A. 1978. *Wild Geese.* Poyser.

Ogilvie, M. A. 1981. The Mute Swan in Britain, 1978. *Bird Study* 28:87–106.

Ogilvie, M. A. 1983. Wildfowl of Islay. *Proc. R. Soc. Edinb.* 84B:473–489.

Ogilvie, M. A. 1983a. The numbers of Greenland Barnacle Geese in Britain and Ireland. *Wildfowl* 34:77–88.

Ogilvie, M. A. & Atkinson-Willes, G. W. 1983. Wildfowl of the Inner Hebrides. *Proc. R. Soc. Edinb.* 84B:491–504.

Ogilvie, M. A. & Boyd, H. 1975. Greenland Barnacle Geese in the British Isles. *Wildfowl* 26:139–147.

Ogilvie, M. A. & Boyd, H. 1976. The numbers of Pink-footed and Greylag Geese wintering in Britain: observations 1969–75 and predictions 1976–80. *Wildfowl* 27:63–75.

Okill, J. D. & Ewins, P. J. 1977. The food of Long-eared Owls in winter. *Shetland Bird Report* 1977:48–50.

Oliver, D. W. 1980. Lapwings in Tayside. *Tay Ring. Group Report* 1978–9:53–57.

Ollason, J. C. & Dunnet, G. M. 1978. Age, experience and other factors affecting the breeding success of the Fulmar, *Fulmarus glacialis*, in Orkney. *J. Anim. Ecol.* 47:961–976.

Osborne, P. 1984. Bird numbers and habitat characteristics in farmland hedgerows. *J. appl. Ecol.* 21:63–82.

Owen, M. 1982. Population dynamics of Svalbard Barnacle Geese. *Aquila* 89:229–247.

Owen, M., Atkinson-Willes, G. L. & Salmon, D. 1986. *Wildfowl in Great Britain* (Second Edition). Cambridge University Press.

Owen, M. & Campbell, C. R. G. 1974. Recent studies on Barnacle Geese at Caerlaverock. *Scot. Birds* 8:181–93.

Owen, M. & Kerbes, R. H. 1971. The autumn food of Barnacle Geese at Caerlaverock National Nature Reserve. *Wildfowl* 22:114–119.

Owen, M. & Williams, G. 1976. Winter distribution and habitat requirements of Wigeon in Britain. *Wildfowl* 27:83–90.

Parr, R. 1980. Population study of Golden Plover *Pluvialis apricaria*, using marked birds. *Ornis Scand.* 11:179–189.

Parslow, J. 1973. *Breeding Birds of Britain and Ireland.* Poyser.

Parsons, J. & Duncan, N. 1978. Recoveries and dispersal of Herring

Gulls from the Isle of May. *J. Anim. Ecol.* 47:993–1005.

Patterson, I. J. 1982. *The Shelduck: a study in behavioural ecology.* Cambridge University Press.

Patterson, I. J., Makepeace, M. & Williams, G. 1983. Limitation of local population size in the Shelduck. *Ardea* 71:105–116.

Patton, D. L. H. & Frame, J. 1981. The effect of grazing in winter by wild geese on improved grassland in west Scotland. *J. appl. Ecol.* 18:311–325.

Payne, A. & Watson, A. 1983. Work on Golden Eagle and Peregrine in north-east Scotland in 1982. *Scot. Birds* 12:159–162.

Payne, A. & Watson, A. 1984. Work on Golden Eagle and Peregrine in north-east Scotland in 1983. *Scot. Birds* 13:24–26.

Pemberton, J. E. (ed.) 1984. *The Birdwatcher's Yearbook 1984.* Buckingham Press.

Pennie, I. D. 1950, 1951. The history and distribution of the Capercaillie in Scotland. *Scot. Nat.* 62:65–87, 157–178; 63:4–17, 135.

Pennie, I. D. 1962. A century of bird-watching in Sutherland. *Scot. Birds* 2:167–192.

Perrins, C. 1979. *British Tits.* Collins.

Peterson, R. T. 1959. *A Field Guide to the Birds.* Houghton Mifflin.

Petty, S. J. 1983. A study of Tawny Owls *Strix aluco* in an upland spruce forest. *Ibis* 125:592.

Phillips. J. H. 1963. The distribution of the Sooty Shearwater around the British Isles. *Brit. Birds* 56:197–203.

Phillips, J. S. 1970. Inter-specific competition in Stonechat and Whinchat. *Bird Study* 17:320–324.

Phillips, J. S. 1973. Stonechats in young forestry plantations. *Bird Study* 20:82–84.

Phillips, J. S. 1976. Survival and local movement in young Stonechats. *Bird Study* 23:57–58.

Pickup, C. 1982. A survey of Greylag Geese (*Anser anser*) in the Uists. Unpub. report to NCC.

Picozzi, N. 1975. A study of the Carrion/Hooded Crow in north-east Scotland. *Brit. Birds* 68:409–419.

Picozzi, N. 1978. Dispersion, breeding and prey of the Hen Harrier *Circus cyaneus* in Glen Dye, Kincardineshire. *Ibis* 120:498–508.

Picozzi, N. 1981. The importance of polygyny to Hen Harriers. *ITE Annual Report 1981*:49–51.

Picozzi, N. 1982. Change of Crow hybrid zone in Kincardineshire. *Scot. Birds* 12:23–24.

Picozzi, N. 1983. Growth and sex of nestling Merlins in Orkney. *Ibis* 125:377–382.

Picozzi, N. 1984a. Breeding biology of polygynous Hen Harriers in Orkney. *Ornis Scand.* 15:1–10.

Picozzi, N. 1984b. Sex ratio, survival and territorial behaviour in polygynous Hen Harriers in Orkney. *Ibis* 126:356–365.

Picozzi, N. & Cuthbert, M. F. 1982. Observations and food of Hen Harriers at a winter roost in Orkney. *Scot. Birds* 12:73–80.

Picozzi, N. & Hewson, R. 1970. Kestrels, Short-eared Owls and Field Voles in Eskdalemuir in 1970. *Scot. Birds* 6:185–190.

Picozzi, N. & Watson, J. 1985. Breeding by an Orkney Hen Harrier on the Scottish mainland. *Scot. Birds* 13:187.

Picozzi, N. & Weir, D. 1974. Breeding biology of the Buzzard in Speyside. *Brit. Birds* 67:199–210.

Picozzi, N. & Weir, D. 1976. Dispersal and causes of death of Buzzards. *Brit. Birds* 69:193–201.

Pienkowski, M. W. & Clark, H. 1979. Preliminary results of winter dye-marking in the Firth of Forth, Scotland. *Wader Study Gp. Bull.* 27:16–18.

Pienkowski, M. W. & Dick, W. J. A. 1975. The migration and wintering of Dunlin *Calidris alpina* in north-west Africa. *Ornis Scand.* 6:151–167.

Pienkowski, M. W. & Evans, P. R. 1979. The origins of Shelducks moulting on the Forth. *Bird Study* 26:195–196.

Pienkowski, M. W. & Evans, P. R. 1982. Breeding behaviour, productivity and survival of colonial and non-colonial Shelducks *Tadorna tadorna* (L). *Ornis Scand.* 13:101–116.

Pienkowski, M. W. & Pienkowski, A. 1983. WSG Project on the movement of wader populations in western Europe: eighth progress report. *Wader Study Gp. Bull.* 38:13–22.

Potts, G. R. 1970. Recent changes in the farmland fauna with special reference to the decline of the Grey Partridge. *Bird Study* 17:145–166

Potts, G. R. 1980. The effects of modern agriculture, nest predation and game management on the population ecology of partridges (*Perdix perdix* and *Alectoris rufa*). *Adv. Ecol. Res.* 11:2–79.

Potts, G. R. 1983. The Grey Partridge situation. *Game Conservancy Ann. Review* 14:24–28.

Pounder, B. 1971. Wintering Eiders in the Tay estuary. *Scot. Birds* 6:407–419.

Pounder, B. 1974. Breeding and moulting Eiders in the Tay region. *Scot. Birds* 8:159–176.

Pounder, B. 1976. Wintering flocks of Goldeneye at sewage outfall sites in the Tay estuary. *Bird Study* 23:121–131.

Pounder, B. 1976. Waterfowl at effluent discharges in Scottish coastal waters. *Scot. Birds* 9:5–36.

Prater, A. J. 1975. The wintering population of the Black-tailed Godwit. *Bird Study* 22:169–176.

Prater, A. J. 1976. The breeding population of the Ringed Plover in Britain. *Bird Study* 23:155–161.

Prater, A. J. 1981. *Estuary Birds of Britain and Ireland.* Poyser.

Prater, A. J. & Davies, M. 1978. Wintering Sanderlings in Britain. *Bird Study* 25:33–38.

Prestt, I. 1965. An enquiry into the recent breeding status of some smaller birds of prey and crows in Britain. *Bird Study* 12:196–221.

Prestt, I. & Mills, D. H. 1966. A census of the Great Crested Grebe in Britain, in 1965. *Bird Study* 13:163–203.

Pym, A. 1982. Identification of Lesser Golden Plover and status in Britain and Ireland. *Brit. Birds* 75:112–124.

Pyman, G. A. 1959. The status of Red-crested Pochard in the British Isles. *Brit. Birds* 52:42–56.

Radford, M. C. 1960. Common Gull movements shown by ringing returns. *Bird Study* 7:81–93.

Ratcliffe, D. A. 1962. Breeding densities in the Peregrine *Falco peregrinus* and Raven *Corvus corax. Ibis* 104:13–39.

Ratcliffe, D. A. 1976. Observations on the breeding of the Golden Plover in Great Britain. *Bird Study* 23:63–116.

Ratcliffe, D. A. 1977a. Uplands and birds – an outline. *Bird Study* 24:140–158.

Ratcliffe, D. A. 1980. *The Peregrine Falcon.* Poyser.

Ratcliffe, D. A. 1984. The Peregrine breeding population of the United Kingdom in 1981. *Bird Study* 31:1–18.

Redfern, C. P. F. 1982. Lapwing nest sites and chick mobility in relation to habitat. *Bird Study* 29:201–208.

Reed, T. M., Currie, A. & Love, J. A. 1983. The birds of the Inner Hebrides. *Proc. R. Soc. Edinb.* 84B:449–472.

Reed, T. M., Langslow, D. R. & Symonds, F. L. 1983a. Breeding waders of the Caithness flows. *Scot. Birds* 12:180–186.

Reed, T. M., Langslow, D. R. & Symonds, F. L. 1983b. Arctic Skuas in Caithness, 1979 and 1980. *Bird study* 30:24–26.

Reynolds, C. M. 1979. The heronries census; 1972–1977 population changes and a review. *Bird Study* 26:7–12.

Reynolds, P. 1982. Wintering Whooper Swans. *Orkney Bird Report* 1981:49–54.

Richardson, M. G. 1985. Status and distribution of the Kittiwake in Shetland, 1981. *Bird Study* 32:11–18.

Richardson, M. G., Dunnet, G. M. & Kinnear, P. K. 1981. Monitoring seabirds in Shetland. *Proc. R. Soc. Edinb.* 80B:157–179.

Richardson, M. G., Heubeck, M., Lea, D. & Reynolds, P. 1982. Oil pollution, seabirds and operational consequences, around the Northern Isles of Scotland. *Environmental Conserv.* 9:315–321.

Riddiford, N. 1983. Recent declines of Grasshopper Warblers *Locustella naevia* at British bird observatories. *Bird Study* 30:143–148

Riddiford, N. & Findley, P. 1981. *Seasonal Movements of Summer Migrants.* BTO Guide No. 18.

Riddle, G. S. 1979. The Kestrel in Ayrshire 1970–78. *Scot. Birds* 10:201–216.

Robertson, I. S. 1982. The origin of migrant Merlins on Fair Isle. *Brit. Birds* 75:108–111.

Rogers, M. J. 1982. Ruddy Shelducks in Britain in 1965–79. *Brit. Birds* 74:446–455.

RSPB. 1979. *Marine Oil Pollution and Birds.* RSPB.

RSPB. 1984. *Hill Farming and Birds: a survival plan.* RSPB.

Ruttledge, R. F. & Ogilvie, M. A. 1979. The past and current status of the Greenland White-fronted Goose in Ireland and Britain. *Irish Birds* 1:293–363.

Sage, B. L. & Vernon, J. D. R. 1978. The 1975 national survey of rookeries. *Bird Study* 25:64–86.

Salmon, D. (ed.) 1981 *et seq. Wildfowl and Wader Counts* (WWC) 1979–80 to 1982–83. The Wildfowl Trust.

Salmon, D. & Moser, M. E. (eds.) 1984. *Wildfowl and Wader Counts* 1983–84. The Wildfowl Trust.

Sandeman, G. L. 1963. Roseate and Sandwich Tern colonies in the Forth and neighbouring areas. *Scot. Birds* 2:286–293.

Sandeman, P. W. 1982. Inland colonies of Lesser Black-backed Gulls. *Scot. Birds* 12:119–120.

Scott, D. A. 1980. A preliminary inventory of wetlands of importance for waterfowl in West Europe and Northwest Africa. *IWRB Special Publication No. 2*

Sharrock, J. T. R. 1972. Habitat of Redwings in Scotland. *Scot. Birds* 7:208–209.

Sharrock, J. T. R. (comp.) 1976. *The Atlas of Breeding Birds in Britain and Ireland.* (BTO/IWC) Berkhamsted, Poyser.

Sharrock, J. T. R. & Grant, P. J. 1982. *Birds New to Britain and Ireland.* Calton, Poyser.

Sharrock, J. T. R. & Sharrock, E. M. 1976. *Rare Birds in Britain and Ireland.* Berkhamsted, Poyser.

Shaw, G. 1976. The breeding bird community of the hillside oak-woods of Glen Falloch. *Western Nat.* 5:41–51.

Shaw, G. 1978. The breeding biology of the Dipper. *Bird Study* 25:149–160.

Shaw, G. 1979. Functions of Dipper roosts. *Bird Study* 26:171–178.

Simms, E. 1971. *Woodland Birds.* Collins.

Smith, A. J. M. 1975. Studies of breeding Sandwich Terns. *Brit. Birds* 68:142–156.

Smith, K. W. 1983. The status and distribution of waders breeding on wet lowland grasslands in England and Wales. *Bird Study* 30:177–192.

Smith, R. W. J. 1969. Scottish Cormorant colonies. *Scot. Birds* 5:363–378.

Smith, R. W. J. 1974. SOC Great Crested Grebe enquiry 1973. *Scot. Birds* 8:151–159.

Snow, D. W. (ed.) *The Status of Birds in Britain and Ireland* (BOU). Blackwell.

Spencer, R. 1975. Changes in the distribution of recoveries of ringed Blackbirds. *Bird Study* 22:177–190.

Spray, C. J. 1981. An isolated population of *Cygnus olor* in Scotland. *Proc. 2nd Int. Swan Symp., Sapporo 1980:*191–208.

Spray, C. 1982. Movements of Mute Swans from Scotland to Ireland. *Irish Birds* 2:82–84.

Spray, C. 1983. East of Scotland Mute Swan study. *Scot. Birds* 12:197–198.

Standring, K. 1978. Winter birds of the Clyde Estuary: a comparison with other estuaries, in *Nature Conservation Interests in the Clyde Estuary.* NCC.

Stanley, P. I. & Minton, C. D. T. 1972. The unprecedented west-ward migration of Curlew Sandpipers in autumn 1969. *Brit. Birds* 65:365–380.

Steventon, D. J. 1982. Shiants Razorbills: movements, first year survival and age of first return. *Seabird Report* 6:105–109.

Stewart, A. G. 1970. The seabird wreck – autumn 1969. *Scot. Birds* 6:142–149.

Stowe, T. J. 1982. Recent population trends in cliff-breeding sea-birds in Britain and Ireland. *Ibis* 124:502–510.

Stowe, T. J. & Harris, M. P. 1984. Status of Guillemots and Razor-bills in Britain and Ireland. *Seabird* 7:5–18.

Stroud, D. A. 1984. Status of Greenland White-fronted Geese in Britain, 1982/83. *Bird Study* 31:111–116.

Stroud, D. A. 1985. Interim report of the British Greenland White-fronted Goose census autumn 1984. Aberystwyth, Greenland White-fronted Goose Study.

Summers, R. W. 1975. Sanderlings in the Firth of Tay. *Tay Ring. Group Report* 1974:12–14.

Summers, R. W., Atkinson, N. K. & Nicoll, M. 1975. Wintering wader populations on the rocky shores of eastern Scotland. *Scot. Birds* 8:299–308.

Summers, R. W., Atkinson, N. K. & Nicoll, M. 1975a. Aspects of Turnstone ecology in Scotland. *Tay Ringing Group Report* 1974:3–10.

Summers, R. W. & Buxton, N. E. 1979. Autumn waders in the Outer Hebrides. *Western Nat.* 8:75–82.

Summers, R. W. & Buxton, N. E. 1983. Winter wader populations on the open shores of northern Scotland. *Scot. Birds* 12:206–211.

Summers, R. W., Corse, C. V., Meek, E. R., Moore, P. & Nicoll, M. 1984. The value of single counts of waders on rocky shores. *Wader Study Gp. Bull.* 41:7–9

Swann, R. L. 1975. Communal roosting of Robins in Aberdeen-shire. *Bird Study* 22:93–98.

Swann, R. L. 1975a. Seasonal variations in suburban Blackbird roosts in Aberdeen. *Ringing & Migration* 1:37–42.

Swann, R. L. 1981. Bar-tailed Godwits on the Moray Firth. *Highland Ring. Group Report* 4:3–4.

Swann, R. I. 1983a. Redwings in a Highland glen. *Scot. Birds* 12:260–261.

Swann, R. L. 1983b. *The Birds of Canna.* Duplicated checklist.

Swann, R. L. 1985. Highland Oystercatchers. *Ringing & Migration* 6:55–59.

Swann, R. L. & Ramsay, A. D. K. 1976. Scottish Shearwaters. *Seabird Report* 5:38–41.

Swann, R. L. & Ramsay, A. D. K. 1979. An analysis of Shag recover-ies from North West Scotland. *Ringing & Migration* 2:137–143.

Swann, R. L. & Ramsay, A. D. K. 1983. Movements from and age of return to an expanding Scottish Guillemot colony. *Bird Study* 30:207–214.

Swann, R. L. & Ramsay, A. D. K. 1984. Long-term seabrd monitor-ing on the Isle of Canna. *Scot. Birds* 13:40–47.

Symonds, F. L. & Langslow, D. R. 1984. Geographical origins and movements of shorebirds using the Firth of Forth. *Ringing & Migration* 5:145–152.

Symonds, F. L., Langslow, D. R. & Pienkowski, M. W. 1984. Move-ments of wintering shorebirds within the Firth of Forth: species differences in usage of an intertidal complex. *Biol. Conserv.* 28:187–215.

Tasker, M. L. 1984. The breeding population of Tysties. *Orkney Bird Report* 1983:50–55.

Tasker, M. L. & Reynolds, P. 1983. A survey of Tystie (Black Guillemot) *Cepphus grylle* distribution in Orkney, April 1983. Unpub. report to NCC.

Tasker, M. L. & Webb, A. 1984. A survey of Tystie (Black Guille-mot) *Cepphus grylle* distribution in Orkney, April 1984. Unpub. report to NCC.

Tay & Orkney Ringing Groups. 1984. *The Shore-birds of the Orkney Islands.* Tay Ring. Group.

Taylor, I. R. 1978. Wader migration in the Upper Forth Estuary. *Wader Study Gp. Bull.* 22:11–16.

Taylor, K. 1983. Buzzard – regaining lost ground. *Birds* 9:27.

Taylor, K. & Marchant, J. H. 1983. Population changes for water-ways birds 1981–82. *Bird Study* 30:121–126

Taylor, K. & Reid, J. B. 1981 Earlier colony attendance by Guillemots and Razorbills. *Sot. Birds* 11:173–180.

Thom, V. M. 1966. Perthshire Heronries. *Trans. Perth. Soc. Nat. Sci.* 11:28–29.

Thom, V. M. 1969. Wintering duck in Scotland 1962–68. *Scot. Birds* 5:417–466.

Thom, V. M. & Cameron, E. D. 1981. The spread of the Great Crested Grebe (*Podiceps cristatus*) as a breeding species in Perthshire. *Trans. Perth. Soc. Nat. Sci.* 13:27–29.

Thomas, G. J. 1982. Breeding terns in Britain and Ireland, 1975–79. *Seabird Report* 1977–81:59–69.

Thomson, A. L. 1975. Dispersal of first-year Gannets from the Bass Rock. *Scot. Birds* 8:295–298.

Thompson, D. B. A. 1981. Feeding behaviour of wintering Shelduck on the Clyde Estuary. *Wildfowl* 32:88–98.

Thompson, D. B. A 1982. The abundance and distribution of intertidal invertebrates, and an estimation of their selection by Shelduck. *Wildfowl* 33:151–158.

Thompson, D. B. A. & Thompson, P. S. 1980. Breeding Manx Shearwaters on Rhum: an updated population assessment in selected areas. *Hebridean Nat.* 4:54–65.

Thompson, P. S. 1983. Dotterel numbers and breeding in the Central Grampians. *Scot. Birds* 12:190–191.

Thorpe, R. I. 1981. Spring passage of skuas at Handa. *Scot. Birds* 11:225.

Tulloch, R. J. 1968. Snowy Owls breeding in Shetland in 1967. *Brit. Birds* 61:119–132.

Tulloch, R. J. 1969. Snowy Owls breeding in Shetland. *Scot. Birds* 5:244–257.

Tulloch, R. J. 1970. Notes on the feeding behaviour of a captive Great Grey Shrike. *Scot. Birds* 6:24–25.

Tulloch, R. J. & Hunter, F. *A Guide to Shetland Birds.* The Shetland Times.

Turner, A. K. & Bryant, D. M. 1979. Growth of nestling Sand Martins. *Bird Study* 26:117–122.

Tyler, S. J. 1979. Mortality and movements of Grey Wagtails. *Ringing & Migration* 2:122–131.

Underwood, L. A. & Stowe, T. J. 1984. Massive wreck of seabirds in eastern Britain, 1983. *Bird Study* 31:79–88.

Venables, L. S. V. & Venables, U. M. 1955. *Birds and Mammals of Shetland.* Oliver & Boyd.

Vernon, J. D. R. 1969. Spring migration of the Common Gull in Britain and Ireland. *Bird Study* 16:101–107.

Verrall, K. & Bourne, W. R. P. 1982. Seabird movements around western Islay. *Scot. Birds* 12:3–11.

Village, A. 1981. The diet and breeding of Long-eared Owls in relation to vole numbers. *Bird Study* 28:215–224.

Village, A. 1982a. The diet of the Kestrel in relation to vole abundance. *Bird Study* 29:129–138.

Village, A. 1982b. The home range and density of Kestrels in relation to vole abundance. *J. Anim. Ecol.* 51:413–428.

Village, A. 1983. The role of nest-site availability and territorial behaviour in limiting the breeding density of Kestrels. *J. Anim. Ecol.* 52:635–645.

Village, A. 1984. Problems of estimating Kestrel breeding density. *Bird Study* 31:121–125.

Vinicombe, K. E. 1985. Ring-billed Gulls in Britain and Ireland. *Brit. Birds* 78:327–337.

Voous, K. H. 1960. *Atlas of European Birds.* Nelson.

Walker, A. F. G. 1970. The moult migration of Yorkshire Canada Geese. *Wildfowl* 21:99–104.

Wallace, D. I. M. 1981. Baikal Teal on Fair Isle. *Brit. Birds* 74:321–326.

Wanless, S. 1983. Seasonal variation in the numbers and condition of Gannets *Sula bassana* dying on Ailsa Craig, Scotland. *Bird Study* 30:102–108.

Wanless, S., French, D. D., Harris, M. P. & Langslow, D. R. 1982. Detection of annual changes in the numbers of cliff-nesting seabirds in Orkney 1976–80. *J. Anim. Ecol.* 51:785–795.

Wanless, S. & Harris, M. P. 1984. Effect of date on counts of nests of Herring and Lesser Black-backed Gulls. *Ornis Scand.* 15:89–94.

Wanless, W. & Wood, V. E. 1982. St Kilda Gannets in 1980. *Scot. Birds* 12:120–121.

Warnes, J. M. 1983. The status of the Chough in Scotland. *Scot. Birds* 12:238–246.

Waterston, G. 1971. *Ospreys in Speyside.* RSPB.

Watson, A. 1965a. Research on Scottish Ptarmigan. *Scot. Birds* 3:331–349.

Watson, A. 1965b. The food of Ptarmigan (*Lagopus mutus*) in Scotland. *Scot. Nat.* 71:60–66.

Watson, A. 1965c. A population study of Ptarmigan (*Lagopus mutus*) in Scotland. *J. Anim. Ecol.* 3:135–172

Watson, A. 1967. The Hatton Castle rookery and roost in Aberdeenshire. *Bird Study* 14:116–119.

Watson, A. 1972. The behaviour of the Ptarmigan. *Brit. Birds* 65:6–26, 93–117.

Watson, A. 1979. Bird and mammal numbers in relation to human impact at ski lifts on Scottish hills. *J. appl. Ecol.* 16:753–764.

Watson, A. 1981. Effects of human impact on Ptarmigan and Red Grouse near ski lifts in Scotland. *ITE Annual Report* 1981:51.

Watson, A. 1982. Work on Golden Eagle and Peregrine in NE Scotland. *Scot. Birds* 12:54–56.

Watson, A. 1983. Grouse in a downward spiral. *The Field*, Dec. 1983:1253.

Watson, A. & Morgan, N. C. 1964. Residues of organo-chlorine insecticides in a Golden Eagle. *Brit. Birds* 57:314–344.

Watson, A. & Moss, R. 1980. Advances in our understanding of the population dynamics of Red Grouse from a recent fluctuation in numbers. *Ardea* 68:103–111, 113–119.

Watson, D. 1977. *The Hen Harrier.* Poyser.

Watson, D. In press. Bean Geese in south-west Scotland. *Scot. Birds.*

Watson, J. 1979. Food of Merlins nesting in young conifer forest. *Bird Study* 26:253–258.

Waugh, D. R. 1979. The diet of Sand Martins during the breeding season. *Bird Study* 26:123–128.

Weir, D. N. 1978 (Letter) Effects of poisoning on Raven and raptor populations. *Scot. Birds* 10:31.

Weir, D. & Picozzi, N. 1983. Dispersion of Buzzards in Speyside. *Brit. Birds* 76:66–78.

Williamson, K. 1965. *Fair Isle and its Birds.* Oliver & Boyd.

Williamson, K. 1969. Bird community in woodland habitats in Wester Ross, Scotland. *Quart. J. For.* 63:305–328.

Williamson, K. 1973. Habitat of Redwings in Wester Ross. *Scot. Birds* 7:268–269.

Williamson, K. 1974. Breeding birds in the deciduous woodlands of Mid-Argyll, Scotland. *Bird Study* 21:29–44.

Williamson, K. 1975. Bird colonisation of new plantations on the moorland of Rhum, Inner Hebrides. *Quart. J. For.* 69:157–168.

Williamson, K. 1975a. Birds and climatic change. *Bird Study* 22:143–164.

Williamson, K. 1976. Bird life in the Wood of Cree, Galloway. *Quart. J. For.* 70:206–215.

Wilson, J. R. 1978. Agricultural influences on waders nesting on the South Uist machair. *Bird Study* 25:198–206.

Wilson, J. R. 1981. The migration of High Arctic shorebirds through Iceland. *Bird Study* 28:21–32.

Wilson, J. R., Czaikowski, M. A. & Pienkowski, M. W. 1980. The migration through Europe and wintering in West Africa of Curlew Sandpiper. *Wildfowl* 31:107–122.

Winstanley, D., Spencer, R. & Williamson, K. 1974. Where have all the Whitethroats gone? *Bird Study* 21:1–14.

Woolhouse, M. E. J. 1983. The theory and practice of the species-area effect, applied to the breeding birds of British woods. *Biol. Conserv.* 27:315–332.

Wormell, P. 1976. The Manx Shearwaters of Rhum. *Scot. Birds* 9:103–118.

Wright, M. 1980. Caerlaverock Bird Report No. 2. Report to NCC.

Wynne-Edwards, V. C. 1953. Leach's Petrels stranded in Scotland in October–November 1952. *Scot. Nat.* 65:167–189.

Yapp, W. B. 1962. *Birds and Woods.* Oxford University Press.

Yapp, W. B. 1974. Birds of the northwest Highland birchwoods. *Scot. Birds* 8:16–31.

Young, J. G. 1968. Birds of the Scar rocks. *Scot. Birds* 5:204–208.

Young, J. G. 1972. Distribution, status and movements of feral Greylag Geese in southwest Scotland. *Scot. Birds* 7:170–182.

Young, J. G. 1984. Large, temporary roost of Jays in Midlothian. *Scot. Birds* 13:88.

Zonfrillo, B. 1982. Response of Storm Petrels to calls of other species. *Scot. Birds* 12:85–86.

Zonfrillo, B. 1983. Isle of May Storm Petrel Movements. *The Seabird Gp. Newsletter* No. 38.

Zwickel, F. C. 1966. Winter food habits of Capercaillie in north-east Scotland. *Brit. Birds* 59:325–336.

General Index

Species Index

Page reference of species accounts in bold numerals